Solar System Data

Planet	Distance From Sun (m)	Mass (kg)	Mean Radius (m)	Orbital Period (sec)
Mercury	5.79×10^{10}	3.18×10^{23}	2.43×10^{6}	7.60×10^{6}
Venus	1.08×10^{11}	4.88×10^{24}	6.06×10^{6}	1.94×10^{7}
Earth	1.51×10^{11}	5.98×10^{24}	6.38×10^{6}	3.16×10^{7}
Mars	2.28×10^{11}	6.42×10^{23}	3.37×10^{6}	5.94×10^{7}
Jupiter	7.78×10^{11}	1.90×10^{27}	6.99×10^{7}	3.74×10^{8}
Saturn	1.43×10^{12}	5.68×10^{26}	5.85×10^{7}	9.35×10^{8}
Uranus	2.87×10^{12}	8.68×10^{25}	2.33×10^{7}	2.64×10^{9}
Neptune	4.50×10^{12}	1.03×10^{26}	2.21×10^{7}	5.22×10^{9}
Pluto	5.91×10^{12}	1.3×10^{22}	1.2×10^{6}	7.8×10^{9}

Moon: Mass = 7.36×10^{22} kg Mean Radius = 1.74×10^{6} m Distance From Earth: 3.84×10^{8} m

Sun: Mass = 1.99×10^{30} kg Mean Radius = 6.96×10^{8} m

The Greek Alphabet

Alpha	Α	α	Iota	Ι	ι	Rho	Ρ	ρ
Beta	Β	β	Kappa	Κ	κ	Sigma	Σ	σ
Gamma	Γ	γ	Lambda	Λ	λ	Tau	Τ	τ
Delta	Δ	δ	Mu	Μ	μ	Upsilon	Υ	υ
Epsilon	Ε	ε	Nu	Ν	ν	Phi	Φ	ϕ
Zeta	Ζ	ζ	Xi	Ξ	ξ	Chi	Χ	χ
Eta	Η	η	Omicron	Ο	ο	Psi	Ψ	ψ
Theta	Θ	θ	Pi	Π	π	Omega	Ω	ω

Exploring Creation

with

Physics
2nd Edition

by Dr. Jay L. Wile

This work is dedicated to my Parents, Howard and Joan Wile.

None of my books would have been possible without your love and encouragement.

Exploring Creation With Physics, 2nd Edition

Published by
Apologia Educational Ministries, Inc.
1106 Meridian Plaza, Suite 220
Anderson, IN 46016
www.apologia.com

Manufactured in the United States of America
Eleventh Printing, January 2017

ISBN: 978-1-932012-42-2

Manufactured in the United States by LSC Communications

Cover photos: Equations *© 1999 Photodisc, Inc.* Earth, moon, and space shuttle *courtesy of NASA*

Cover design by Kim Williams

Any photo or illustration not credited in the text was done by the author.

INSTRUCTIONAL SUPPORT

Did you know that in addition to publishing award-winning curriculum Apologia also offers instructional support? We believe in helping students achieve their full potential, whatever their learning style. When you choose an Apologia curriculum, you are not just selecting a textbook. Every course has been designed with the student's needs in mind.

INDEPENDENT LEARNERS

Apologia textbooks and notebooks are written to the student in a conversational tone so that young people can easily navigate through the curriculum on their own. Apologia curriculum helps students methodically learn, self-check, and master difficult concepts before moving on.

AUDITORY LEARNERS

Sometimes students learn best when they can see and hear what they're studying. **Apologia Audio Books** are the complete text of the course read aloud. Students can follow along with the audio while reading or continue learning when they're away from home by listening in the car.

VISUAL LEARNERS

Sometimes subject matter is easier to comprehend when the topic is animated and presented by a knowledgeable instructor. **Apologia Video Instructional DVDs** enhance the student's education with more than 20 hours of instruction, including on-location video footage, PowerPoint lectures, animated diagrams of difficult concepts, and video presentations of all experiments.

SOCIAL LEARNERS

Some students learn best when they are able to interact with others in an online setting and ask questions of a live instructor. With **Apologia Online Academy**, students can interact in real time with both their classmates and a professional instructor in a structured virtual classroom. Also, we offer recordings of all our live classes on the Apologia Online Academy Video-On-Demand Channel.

At Apologia, we believe in homeschooling. We are here not only to support your endeavors, but also to help you and your student thrive! Find out more at apologia.com.

STUDENT NOTES

Exploring Creation With Physics, 2nd Edition

You are about to embark upon an amazing journey! In this course, you will be introduced to the fascinating subject of physics. You will learn the basics about the universe around you and how it works. Although the course will be hard work, you will learn some truly amazing things. Hopefully, these things will help you develop an even deeper appreciation for the wonderful creation that God has given us! I hope that you enjoy taking this course as much as I have enjoyed writing it.

Pedagogy of the Text

This text contains 16 modules. Each module should take you about 2 weeks to complete, as long as you devote 45 minutes to an hour of every school day to studying physics. At this pace, you will complete the course in 32 weeks. Since most people have school years which are longer than 32 weeks, there is some built-in "flex time." You should not rush through a module just to make sure that you complete it in 2 weeks. Set that as a goal, but be flexible. Some of the modules might come harder to you than others. On those modules, take more time on the subject matter.

To help you guide your study, there are several student exercises which you should complete:

- The "On Your Own" problems should be solved as you read the text. The act of working out these problems will cement in your mind the concepts you are trying to learn. Complete solutions to these problems appear at the end of each module. Once you have solved an "On Your Own" problem, turn to the end of the module and check your work. If you did not get the correct answer, study the solution to learn why.

- The review questions are conceptual in nature and should be answered after you have completed the entire module. They will help you recall the important concepts from the reading.

- The practice problems should also be solved after the module has been completed, allowing you to review the important quantitative skills from the module.

Your teacher/parent has the solutions to the review questions and practice problems.

Any information that you must memorize is centered in the text and put in boldface type. In addition, all definitions presented in the text need to be memorized. Words that appear in boldface type (centered or not) in the text are important terms that you should know. Finally, if any student exercise requires the use of a formula or skill, you must have that memorized for the test. Often, physical constants or other well-known physical data will be required in order to solve problems. Generally, such constants and data are given to you on the test. However, you will be required to memorize a few of them. If you are required to memorize a physical constant or any other piece of information, it will be centered and written in boldface type in the module. If any physical constants or data are given in the practice problems and review questions, they will also be given on the tests.

Learning Aids

Extra material is available to aid you in your studies. For example, Apologia Educational Ministries, Inc. has produced a multimedia companion CD that accompanies this course. It contains videos of experiments that you would probably not be able to perform yourself. These experiments demonstrate concepts that are discussed in the course. In addition, it contains animated white board solutions of many of the example problems in the course, along with an audio explanation that is different from the explanation given in this book. Thus, if you are having trouble understanding how I worked a certain example problem, you might find more explanation on the multimedia CD. The following graphic in the book:

indicates that there is a video or animation on the CD that further explains a concept or example problem.

The CD also contains audio pronunciations of the technical words used in this book. Even though the book gives pronunciation guides for most of the technical words used, nothing beats actually hearing someone say the word! As you read through the book, you will see words that have pronunciation guides in parentheses. If you would like to hear one of those words pronounced for you, you will find it on the multimedia companion CD.

In addition to the multimedia companion CD, there is a special website for this course that you can visit. The website contains links to web-based materials related to the course. These links are arranged by module, so if you are having trouble with a particular subject in the course, you can go to the website and look at the links for that module. Most likely, you will find help there. In addition, there are answers to many of the frequently-asked questions regarding the material. For example, many people ask us for examples of how to properly record experiments in your laboratory notebook. Those examples can be found at the website. Finally, if you are enjoying a particular module in the course and would like to learn more about it, there are links which will lead you to advanced material related to that module.

To visit the website, go to the following address:

http://www.apologia.com/bookextras

When you get to the address, you will be asked for a password. Type the following into the password box:

Physicsisphun

Be sure that you do not put spaces between any of the letters and that the first letter is capitalized. When you click on the button labeled "Submit," you will be sent to the course website. You must use Internet Explorer 5.1 or higher to view this website.

There are also several items at the end of the book that you will find useful in your studies. There is a glossary that defines many of the terms used in the course and an index that will tell you where topics can be found in the course. In addition, there are three appendices in the course. Appendix A lists formulas and laws that are found throughout the reading. Appendix B contains extra problems for each module of the course. If you are having difficulty with a certain type of problem, you can get more practice by solving the related problems in Appendix B. Your parent/teacher has the worked-out solutions for those problems. Appendix C contains a complete list of all of the supplies you need to perform the experiments in this course.

Experiments

The experiments in this course are designed to be done as you are reading the text. I recommend that you keep a notebook of these experiments. This notebook serves two purposes. First, as you write about the experiment in the notebook, you will be forced to think through all of the concepts that were explored in the experiment. This will help you cement them into your mind. Second, certain colleges might actually ask for some evidence that you did, indeed, have a laboratory component to your physics course. The notebook will not only provide such evidence but will also show the college administrator the quality of your physics instruction. I recommend that you perform the experiments in the following way:

- When you get to an experiment, read through it in its entirety. This will allow you to gain a quick understanding of what you are to do.

- Once you have read the experiment, start a new page in your laboratory notebook. The first page should be used to write down all of the data taken during the experiment. What do I mean by "data"? Any observations or measurements you make during the experiment are considered data. Thus, if you see a ball roll down a ramp with increasing speed, that is part of the experiment's data and should be written down. If you measure the time it takes for something to happen during the experiment, that is part of the experiment's data and should be written down. In addition, any calculation that you are asked to do as a part of the experiment should be done on this page.

- When you have finished the experiment and any necessary calculations, write a brief report in your notebook, right after the page where the data and calculations were written. The report should be a brief discussion of what was done and what was learned. You should not write a step-by-step procedure. Instead, write a brief summary that will allow someone who has never read the text to understand what you did and what you learned.

PLEASE OBSERVE COMMON SENSE SAFETY PRECAUTIONS! The experiments in this course are no more dangerous than most normal, household activity. Remember, however, that the vast majority of accidents do happen in the home. Chemicals used in the experiments should never be ingested; hot liquids and flames should be regarded with care; and all physics experiments should be performed while wearing eye protection such as safety glasses or goggles.

Although the items used in the experiments are common substances that you either have around the home or can purchase at a hardware store, it is important to be prepared for each module. You will probably have to purchase at least some of the items you need for the experiments, and you might have trouble finding one or more of them. Thus, you should look at Appendix C regularly to see what items you will need to perform experiments in the upcoming modules. That way, you can begin shopping for them before you actually need them!

Exploring Creation With Physics

Table of Contents

Introductory Remarks

In this course, you will study the science of physics, which is often referred to as the "fundamental science." Why is it called that? Well, as Ernest Rutherford (pictured below) once said, "All science is either physics or stamp collecting" (J. B. Birks, *Rutherford at Manchester* [New York: W. A. Benjamin, 1962], 108). What he meant was quite simple. In principle, all fields of science can be reduced to physics. Since physics attempts to understand in detail how everything in the universe interacts with everything else, any phenomenon in nature is controlled by the laws of physics.

Image in the public domain

Ernest Rutherford

Ernest Rutherford (1871-1937) was born in New Zealand and was educated at both the University of New Zealand and Cambridge University. He determined that there are three types of naturally-occurring radiation, and he named them "alpha," "beta," and "gamma." We now know that alpha particles are helium nuclei, beta particles are electrons, and gamma rays are high-energy photons (light). Rutherford is probably most famous for his experiments on the structure of the nucleus. By bombarding a gold foil with alpha particles and watching how the alpha particles were deflected by the foil, he concluded that the atom is composed of a dense, positively-charged nucleus around which electrons orbit. He was also the first to produce an artificial nuclear reaction. Rutherford was awarded the Nobel Prize in Chemistry in 1908.

If Rutherford's statement is true, why do we have other fields of science? Why doesn't everyone just study physics? Well, this is what the "stamp collecting" part of Rutherford's quote means. Even though the laws of physics apply to all fields of science, there are many, many aspects of nature that are simply too complex to explain in terms of physics. For example, even the simplest life form in the universe is incredibly complicated. A single-celled creature such as an amoeba has hundreds of thousands of processes that work together to keep it alive. It is simply too complicated to explain each of these processes and how they interact in detail. As a result, the science of biology simply collects all of the facts related to how an amoeba functions, much like a stamp collector collects stamps.

In other words, although the underlying principles which control all of the processes that occur in an amoeba obey the laws of physics, the specifics of how they function and interact are far too complex to understand in detail. As a result, the science of biology collects the facts that we know about an amoeba and tries to draw conclusions from those facts. If, at some point in the future, humankind has the ability to explain such complex systems in terms of physics, the science of biology may not be necessary, because physics may be able to explain everything regarding living systems. Thus, physics is called the fundamental science because it forms the basis of all other fields of science.

Of course, if you are going to attempt to study and understand the details of how things interact in nature, you will have to do a lot of observation and experimentation. One of the most important aspects of observation and experimentation is measurement; thus, you will be making and using a lot of measurements in this course. As a result, you need to become very comfortable with the process of making measurements and the language that revolves around those measurements. You must also be comfortable with using those measurements in mathematical equations and making sure that you report the results of the equations with a precision that reflects the precision of the original measurements.

If you have already taken a good chemistry course, you have covered what you need to know about measurements and how to use them in mathematical equations. If your previous chemistry course was *Exploring Creation with Chemistry*, you covered all of these topics in that course's first module. However, if you did not take that course, or if you think you might have forgotten some of the material, I will quickly summarize the skills that you need to know. If the summary contains anything that you do not understand, you can visit the course website mentioned in the "Student Notes" portion of the text. When you log into that website, you will see a link that takes you to an electronic version of the first module of *Exploring Creation with Chemistry*. That module gives full explanations for each of the skills that I will discuss in the sections that follow.

The Metric System

In this course, you will use the English system of units occasionally, but you will primarily use the metric system of units. Thus, you must be familiar with the metric units for mass, distance, and time, as well as the prefixes which are used to modify the size of the units.

In 1960, an international committee established the standard units for the measurement of fundamental quantities in science. This standard is called the **System Internationale (SI)** set of units. In this course, it will be most helpful to use SI units. The SI unit for mass is the *kilogram*; the SI unit for distance is the *meter*; and the SI unit for time is the *second*. Later in the course, a few more SI units will be introduced.

The Factor-Label Method

Often, you will come across measurements in the English system that must be converted into the metric system, or you will run into measurements that are in the metric system but are not SI units. Thus, you need to be very familiar with converting from one unit to another. The best way to convert between units is the **factor-label method**, and you must understand this method to take this course.

A quick example of the factor-label method will help illustrate what you need to know. Suppose you need to convert the mass of an object from 4,523 centigrams into the SI unit for mass, which is the kilogram. Here's how you would do it using the factor-label method:

$$\frac{4{,}523\ \cancel{\text{cg}}}{1} \times \frac{0.01\ \cancel{\text{g}}}{1\ \cancel{\text{cg}}} \times \frac{1\ \text{kg}}{1{,}000\ \cancel{\text{g}}} = 0.04523\ \text{kg}$$

If you do not understand how I set that up, why the units cancel the way I have canceled them, or how to get the answer, you need to review the factor-label method.

Using Units in Mathematical Equations

Physics and math are intimately linked. As you progress through this course, you will be using mathematical equations to analyze a host of physical situations. As a result, you need to be completely comfortable using units in mathematical equations. When you add or subtract measurements, you cannot add them unless the units are the same. Thus, an equation like 1.2 m + 3.4 kg is meaningless. There is no way you can add those two measurements.

However, you can multiply or divide measurements whether or not the units are the same. If you have a box with a length of 0.50 m, a height of 0.25 m, and a length of 0.45 m, you can multiply the length, width, and height together to calculate that the box has a volume of 0.056 m^3. If that box has a mass of 5.1 kg, you can divide the mass by the volume to find out that the density of the box is 91 kg/m^3. If you do not understand why the unit for the volume is m^3 or why the unit for the density is kg/m^3, you need to review the use of units in mathematical equations.

Making Measurements

In this course, you will be making some measurements of your own. Thus, you need to know how to read measuring instruments and how to report your measurements with the proper precision. A metric ruler, for example, is usually marked off in increments of 0.1 cm, or 1 mm. However, because you can estimate in between those marks, you can report your answer to a precision of 0.01 cm. Consider, for example, the situation below:

Illustration by Megan Whitaker

The blue ribbon in the figure above is 3.45 cm long. If you do not understand how I got that measurement or why the ribbon starts on the 1 cm mark rather than at the beginning of the ruler, you need to review the process of making measurements.

Accuracy, Precision, and Significant Figures

There is a big difference between the **accuracy** of a measurement and the **precision** of a measurement. You need to understand the difference. You also need to understand how to use **significant figures** to determine the precision of a measurement as well as to determine where to round off your answers when you are working problems. So that you can easily refer back to them, I will summarize the rules of significant figures below.

In order to determine whether or not a figure is significant, you simply follow this rule:

A digit within a number is considered to be a significant figure if:

I. It is non-zero OR
II. It is a zero that is between two significant figures OR
III. It is a zero at the end of the number *and* to the right of the decimal point

When using measurements in mathematical equations, you must follow these rules:

Adding and Subtracting with Significant Figures: **When adding and subtracting measurements, round your answer so that it has the same precision as the *least precise* measurement in the equation.**

Multiplying and Dividing with Significant Figures: **When multiplying and dividing measurements, round the answer so that it has the *same number of significant figures as the measurement with the fewest significant figures.***

To quickly review how these rules work, consider the following subtraction problem:

$$546.2075 \text{ kg} - 87.61 \text{ kg}$$

The answer to this problem is 458.60 kg. The first number has its last significant figure in the ten thousandths place, while the second has its last significant figure in the hundredths place. Since the second number has the lowest precision, the answer must have the same precision, so the answer must have its last significant figure in the hundredths place. Compare that to the following division problem:

$$\text{Speed} = 3.012 \text{ miles} \div 0.430 \text{ hours}$$

The answer is 7.00 miles/hour. The first number has four significant figures, while the second number has three. Thus, the answer must have three significant figures. If any of this discussion is confusing, please review the concept of significant figures.

Scientific Notation

Since reporting the precision of a measurement is so important, we need to be able to develop a notation system that allows us to do this no matter what number is involved. Suppose you work out an equation, and the answer turns out to be 100 g. However, suppose you need to report that measurement to three significant figures. The number "100" has only one significant figure. So how can you report it to three significant figures? For that, you use scientific notation. If you report 100 g as 1.00×10^2 g, the two zeros are now significant because of the decimal place, so the answer now has three significant figures. If you need to report "100" with two significant figures, you could once again use scientific notation, but this time, you would have only one zero after the decimal: 1.0×10^2 g. You must be very comfortable using scientific notation and determining the significant figures in a number that is expressed in scientific notation.

Mathematical Preparation

In addition to the concepts discussed above, there are certain mathematical skills I am going to assume that you know. You should be very comfortable with algebra, and you need to know the three basic trigonometric functions (sine, cosine, and tangent) and how they are defined on a right triangle. You also need to be familiar with the inverses of those functions ($\sin^{-1}$, $\cos^{-1}$, and $\tan^{-1}$). Please do not go any further in this course until you are comfortable with everything I have mentioned so far. Once again, there is a good review of these concepts (not including the algebra and trigonometry mentioned in this section) posted on the course website.

MODULE #1: Motion In One Dimension

Introduction

As I said in my introductory remarks, the science of physics attempts to explain everything that is observed in nature. Now of course, this is a monumentally impossible task, but physicists nevertheless do the best job that they possibly can. Over the last three thousand years, remarkable advances have been made in explaining the nature of the world around us, and in this physics course, we will learn about many of those advances. This module will concentrate on describing *motion*.

If you look around, you will see many things in motion. Trees, plants, and sometimes bits of garbage blow around in the wind. Cars, planes, animals, insects, and people move about from place to place. You should have learned in chemistry that even objects which appear stationary are, in fact, filled with motion because their component molecules or atoms are moving. In short, the world around us is alive with motion.

In fact, Thomas Aquinas (uh kwy' nus) listed the presence of motion as one of his five arguments for the existence of God. He said that based on our experience, we have found that motion cannot occur without a mover. In other words, in order for something to move, there must be something else that moves it. When a rolling ball collides with a toy car, the car will move because the ball gave it motion. But, of course, the ball would not have been rolling to begin with if it had not been pushed or thrown. Thus, Aquinas says that our practical experience indicates that any observable motion should be traceable back to the original mover. When the universe began, then, something had to be there to start all of the motion that we see today. Aquinas says that God is this "original mover."

While philosophers and scientists can mount several objections to Thomas Aquinas's argument, it nevertheless demonstrates how important motion is in the universe. Thus, it is important for us to be able to study and understand motion. In this module, we will attempt to understand the most basic type of motion: motion in one dimension. Remember from geometry what "one dimension" means. If an object moves in one dimension, it moves from one point to another in a straight line. In this module, therefore, we will attempt to understand the motion of objects when they are constrained to travel straight from one point to another.

Image in the public domain

FIGURE 1.1
Thomas Aquinas

Thomas Aquinas (1225-1274) was an Italian philosopher and Roman Catholic theologian. He was a prolific writer, being credited with about eighty important works. In his work entitled *Summa Theologica*, he cites five arguments for the existence of God. The first one is summarized as follows:

"It is certain, and evident to our senses, that in the world some things are in motion. Now whatever is in motion is put in motion by another...If that by which it is put in motion be itself put in motion, then this also must needs be put in motion by another, and that by another again. But this cannot go on to infinity...Therefore it is necessary to arrive at a first mover, put in motion by no other; and this everyone understands to be God." (*Summa Theologica*, Second and Revised Edition, 1920; retrieved from http://www.newadvent.org/summa/100203.htm on 11/14/2003)

Distance and Displacement

When studying the motion of an object, there are a few very fundamental questions you can ask: Where is the object? How fast is it moving? How is the object's motion changing? In physics terminology, we say that the answers to these questions are the object's **position**, **velocity**, and **acceleration**. You might also ask how the object's position has changed. Physicists call that **displacement**.

Displacement - The change in an object's position

I will discuss velocity and acceleration in upcoming sections of this module. For right now, I want to concentrate on displacement.

Suppose you are sitting on the sofa reading a book (maybe even this one), and you suddenly decide that you want to go to the refrigerator for a drink. You get up, and you move to the refrigerator, which is 10 meters away from the sofa. You get your drink and then walk 10 meters back to the sofa. How much distance did you travel in your quest for liquid refreshment? Well, you walked 10 meters there and 10 meters back, so you walked a total of 20 meters. After everything was finished, what was your total displacement? *It was zero meters*!! You see, before everything began, you were at the sofa. Since you started there, we can define it as your initial position. You moved to the refrigerator, at which point you were 10 meters displaced from the sofa. However, when you turned around and came back, you ended up at exactly the same point from which you started. In the end, then, you were 0 meters from your starting position; thus, your displacement was 0 meters.

You see, then, that the concept of displacement includes information about direction, whereas the concept of distance does not. In the situation we just imagined, you walked a *distance* of 20 meters, but your *displacement* was 0 because you walked 10 meters in one direction and then another 10 meters in precisely the opposite direction. Since the displacement in one direction canceled the displacement in the opposite direction, your total displacement was 0. When a physical quantity carries information concerning direction we call it a **vector** (vek' ter) **quantity**. When the physical quantity does not carry information concerning direction, we call it a **scalar** (skay' ler) **quantity**.

Vector quantity - A physical measurement that contains directional information

Scalar quantity - A physical measurement that does not contain directional information

Thus, distance is a scalar quantity, and displacement is a vector quantity.

When dealing with displacement, we must find some mathematical way to denote the direction that is inherent in the measurement. The way we will do this is to label displacement in one direction positive and displacement in the opposite direction negative. That way, when you add displacements together, motion in one direction will cancel motion in the opposite direction. Thus, we could say that in the situation discussed above, your displacement was +10 meters when you moved from the sofa to the refrigerator and -10 meters when you moved the opposite direction from the refrigerator to the sofa. Your total displacement, then, was +10 meters plus -10 meters, which is 0.

What's really nice about this mathematical way of noting direction is that it doesn't really matter which direction you label as positive or which you label as negative. We could just as easily have said that your displacement when you arrived at the refrigerator was -10 meters. That would mean that your displacement when you moved from the refrigerator to the couch was +10 meters. The total displacement would still be 0. Thus, it doesn't matter which direction you label as positive, as long as you keep it consistent. To make sure that you understand what I mean, consider Figure 1.2:

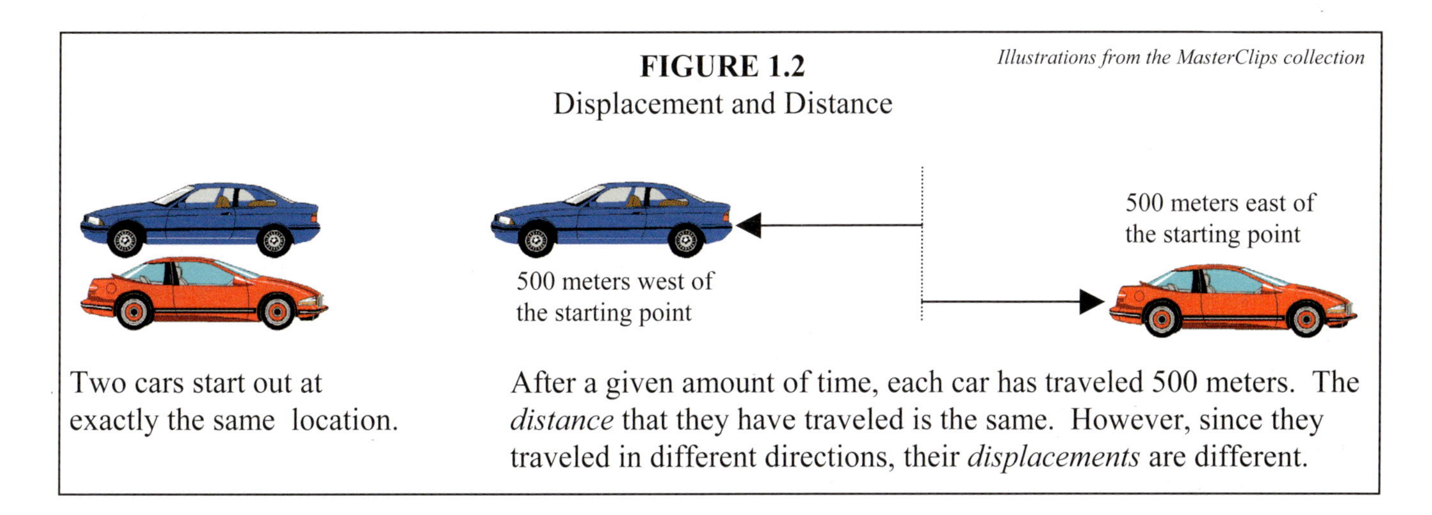

FIGURE 1.2
Displacement and Distance

In the figure, two cars start at the same location, but the blue car is heading west, while the red car is heading east. After a given amount of time, each car has traveled a total distance of 500 meters. Although both cars have traveled the same distance, their displacements are not the same, because they traveled in opposite directions. Suppose we define east as the positive direction. In that case, the red car would have a displacement of +500 meters, and the blue car would have a displacement of -500 meters. Alternatively, we could define west as the positive direction. If we did that, the blue car would have a displacement of +500 meters, and the red car would have a displacement of -500 meters.

Does it matter which car has a negative displacement and which has a positive displacement? *No, it really doesn't!* Which direction you define as positive is not important. The only thing that is important is that you *remember the definition and use it consistently throughout your analysis*. If you define east as positive, that's fine, but just remember that in the end, any displacement which ends up positive means the displacement is to the east, and any displacement that ends up negative means the displacement is to the west. Alternatively, if you define west as positive, just remember that any displacement which turns out to be positive means the displacement is to the west, and any displacement that ends up negative means the displacement is to the east.

This can get a little confusing if you are not completely comfortable with the idea of using positive and negative signs to denote direction, so I want to show you how to keep all of this straight. Study the following example to see that the final answer is really independent of which direction you define as negative, as long as you are consistent in your definition. After you have studied the example, solve the "On Your Own" problem that follows it to make sure you understand this important concept.

EXAMPLE 1.1

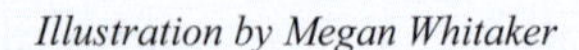

Illustration by Megan Whitaker

A child is 5.0 meters away from a wall and rolls a ball towards it. The ball hits the wall and bounces back, rolling 3.3 meters before coming to a halt. What is the total distance covered by the ball? What is the ball's displacement ?

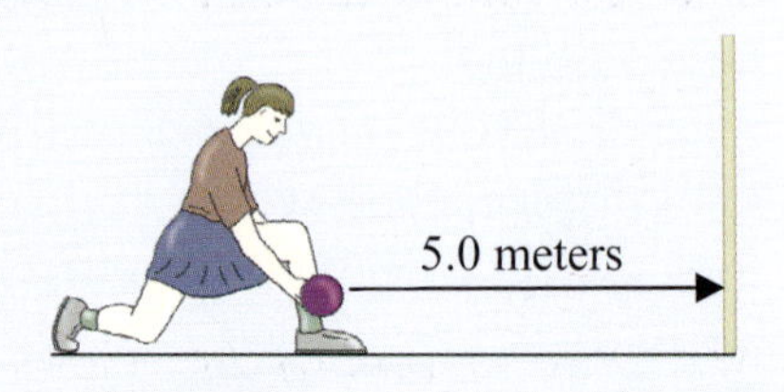

The total distance is easy to calculate. The ball rolled 5.0 meters to reach the wall and 3.3 meters in the other direction after bouncing back. The total distance is calculated as follows:

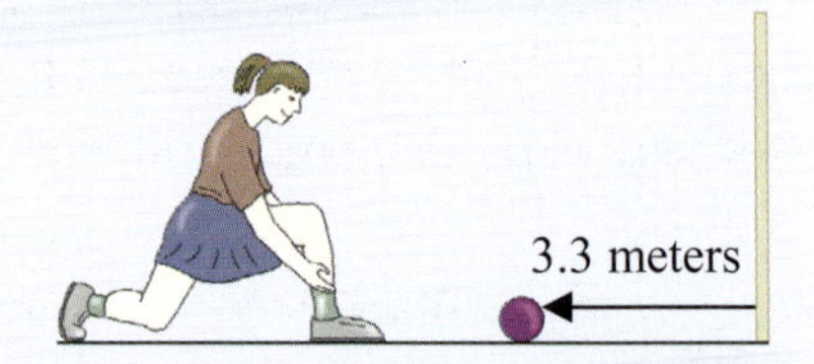

Total distance = 5.0 meters + 3.3 meters = 8.3 meters

Calculating the displacement is a bit more difficult, however. To do this, we must first define directions. I will say that motion from the child to the wall represents positive displacement while motion from the wall to the child is negative displacement. Thus, the ball first had a displacement of +5.0 meters and then a displacement of -3.3 meters. The total displacement, then, is:

Total displacement = 5.0 meters + -3.3 meters = 1.7 meters

The displacement is positive, so the ball is <u>1.7 meters away from the child, in the direction of the wall</u>.

Alternatively, I could have said that motion from the child towards the wall represented *negative* displacement. In that case, the ball would have had a -5.0 meters displacement followed by a +3.3 meters displacement. This would indicate a total displacement of -1.7 meters. You might think that this is a different answer than the one I got previously, because this one is negative. Remember, however, what negative displacement means in this case. It means displacement *from the child towards the wall.* Thus, my answer is still <u>1.7 meters away from the child, in the direction of the wall</u>. As long as you stay consistent, then, your answer will be the same regardless of which direction you say is positive and which is negative. The trick is to give your answer in relation to the initial position, not with just a positive or negative sign.

ON YOUR OWN

1.1 An ant starts at his anthill and walks 15.2 cm to a crust of bread. He takes the bread, turns around, and walks back towards his anthill. He stops after he has traveled 3.8 cm and eats part of the crust of bread. What is the total distance he has traveled up to that point? What is the total displacement?

<u>Speed and Velocity</u>

Now that you have some idea of what displacement is, you can begin to learn about velocity.

<u>Velocity</u> - The time rate of change of an object's position

This definition may sound a bit strange, but it is really easy to understand. Velocity simply tells us how quickly an object's position is changing. That's what "time rate of change" means. In order to determine this, all you need to do is take the change in position and divide it by the time it took to make that change. Mathematically, we could say:

$$\mathbf{v} = \frac{\Delta \mathbf{x}}{\Delta t} \quad (1.1)$$

where "**v**" represents the velocity, "**x**" represents the position, and "t" represents time. The symbol "Δ" represents the capital Greek letter "delta" and means "change in." Thus, "Δ**x**" means the change in position, while "Δt" means the change in time. Now remember, the change in an object's position is defined as its displacement, so "Δ**x**" also means "displacement."

There are two very important things you need to learn about Equation (1.1). First, since we calculate velocity by taking displacement (usually measured in meters) and dividing by time (usually measured in seconds), the SI unit for velocity is meters/second ("meters per second"). Thus, if I travel for 30.0 seconds and my total displacement during that time is 60.0 meters to the west, my velocity is 60.0 meters ÷ 30.0 seconds, or 2.00 meters/second (abbreviated as m/sec) to the west.

The second thing you need to learn about this equation is that velocity and displacement are both vector quantities. You have already learned that about displacement, and since you use displacement to calculate velocity, it only makes sense that velocity is also a vector quantity. Whenever you use velocity, then, you must be sure to keep track of direction. Mathematically, we will do it the same way we did with displacement. Motion in one direction will be noted as positive velocity, while motion in the opposite direction will be written as negative velocity.

What about time in Equation (1.1)? Is it a vector or a scalar quantity? Well, if you think about it, time only goes one way. As far as we can tell, time cannot go in reverse. Thus, since time does not have a direction attached to it, it is considered a scalar quantity. This is why I have written "**v**" and "**x**" in boldfaced type but kept "t" in normal type. The boldfaced type indicates that "**v**" and "**x**" are vector quantities. Since t is not in boldfaced type, you can assume it is not a vector. This kind of notation will exist throughout the rest of the course. When I write a variable in boldfaced type, it will mean that the variable is a vector quantity. If the variable is not in boldfaced type, it will be considered a scalar quantity.

Now it is very important that you do not confuse the concept of velocity with the concept of speed. Just as distance and displacement are different quantities, velocity and speed are also different quantities.

Speed - The time rate of change of the distance traveled by an object

In other words, to determine the speed of an object, you take the total distance traveled and divide by the time it took to travel that distance. Mathematically, we could say:

$$\text{speed} = \frac{\Delta d}{\Delta t} \quad (1.2)$$

where "d" represents distance, and "t" represents time. Notice that none of the variables in this equation are written in boldfaced type. This indicates that there are no vectors in Equation (1.2), and that is the main difference between velocity and speed. While velocity is a *vector quantity*, speed is a *scalar quantity*. Thus, although Equations (1.1) and (1.2) look very similar, speed and velocity are quite different, because one is a vector and one is not. Let's study a couple of examples to make sure you understand these distinctions.

EXAMPLE 1.2

You hop on your bicycle and pedal 151.1 meters to the end of your street in 25.2 seconds. You then turn around and pedal back to where you started. If the return trip takes 27.1 seconds, what was your speed and what was your velocity over the course of the entire bike ride?

We will solve for speed first, because that's a little easier. According to Equation (1.2), we can figure out speed by taking the distance traveled (Δd) and dividing by the time it took to travel that distance (Δt). If the street is 151.1 meters long and you traveled to the end and back, you traveled a total distance of 151.1 m + 151.1 m, or 302.2 m. The total time it took to travel that distance was 25.2 seconds + 27.1 seconds or 52.3 seconds. Thus, according to Equation (1.2):

$$\text{speed} = \frac{\Delta d}{\Delta t} = \frac{302.2 \text{ m}}{52.3 \text{ sec}} = 5.78 \frac{\text{m}}{\text{sec}}$$

Now remember, we must take significant figures into account when doing calculations. In this case, we are dividing, so we count significant figures. Since 302.2 has four significant figures, and 52.3 has three significant figures, we must report our answer to three significant figures. That's why the speed over the course of the entire trip was <u>5.78 m/sec</u>.

Calculating velocity, however, is quite another matter. Velocity is determined by taking the displacement and dividing by the time it took to achieve that displacement. By the time that the bike ride was over, your displacement was zero, because you ended up back where you started. Thus, Equation (1.1) becomes:

$$\mathbf{v} = \frac{\Delta \mathbf{x}}{\Delta t} = \frac{0 \text{ m}}{52.3 \text{ sec}} = 0 \frac{\text{m}}{\text{sec}}$$

In the end, then, while your total speed was considerable (5.78 m/sec), your velocity was zero! It might sound strange that you could ride a bike with zero velocity, but once again, remember that velocity is a vector quantity. When your velocity is zero, it means simply that your total displacement was zero. Thus, even though you pedaled a lot, you ended up going nowhere by the end of your ride, so your displacement and velocity were both zero!

A sprinter runs the 200-meter (2.00×10^2 m) dash in 24.00 seconds. He then turns around and walks 15 meters back towards the starting line in order to talk to his coach. Because he is so tired, it takes him 25 seconds to walk that 15 meters. What was the sprinter's velocity during the 200-meter dash? What was his velocity when he walked back to talk to the coach? What was his velocity for the entire trip?

In this case, we are asked to calculate velocity, so we will only be using Equation (1.1). Once again, we are dealing with vector quantities here, so we must define direction. I will call motion from the starting line to the finish line positive motion. This makes motion from the finish line to the starting line negative motion. The first part of the question asks us to calculate the sprinter's velocity during the 200-meter dash. During that time, the sprinter was moving from the starting line to the finish line. Thus, his displacement was 2.00 x 10^2 meters. It took him 24.00 seconds to make the run, so Equation (1.1) becomes:

$$\mathbf{v} = \frac{\Delta\mathbf{x}}{\Delta t} = \frac{2.00 \times 10^2 \text{ m}}{24.00 \text{ sec}} = 8.33 \frac{\text{m}}{\text{sec}}$$

Three significant figures in 2.00 x 10^2 m limit our answer to three significant figures. We could therefore say that his velocity was <u>8.33 m/sec in the direction of the finish line</u>.

The second part of the question asks us to calculate his velocity as he is walking back to speak with his coach. During that time, he walked towards the starting line, so his displacement was negative:

$$\mathbf{v} = \frac{\Delta\mathbf{x}}{\Delta t} = \frac{-15 \text{ m}}{25 \text{ sec}} = -0.60 \frac{\text{m}}{\text{sec}}$$

Note that both -15 m and 25 seconds have two significant figures, so our answer can have only two significant figures. Thus, we must say that his velocity was <u>0.60 m/sec towards the starting line</u>.

Finally, the problem asks us to determine his velocity over the entire trip. You might think that we could simply average the two velocities that we already calculated, but that would not be correct. The only way we can properly calculate the velocity is to determine the displacement and then divide by the time that elapsed. When the sprinter finished the race, his displacement was 2.00 x 10^2 meters. However, when he walked back to talk to his coach, his displacement changed by -15 meters. Thus, his total displacement was 2.00 x 10^2 m + -15 m = 185 m in the direction of the finish line. Now before we use this in Equation (1.1), I want to make sure that you understand how the significant figures work. Since we are adding these numbers, we use the rule of addition and subtraction, which says that you report your answer with the same precision as the least precise number in the problem. The initial displacement (2.00 x 10^2 m) has its last significant figure in the ones place. Don't let the scientific notation fool you: 0.01 x 10^2 = 1. Thus, the second zero after the decimal is in the ones place. The second displacement (-15 m) has its last significant figure in the ones place. Thus, the precision of each measurement is to the ones place, so our answer must be reported to the ones place. That's why it is 185 m. If this does not make sense, you might want to review the section on significant figures in the module that is posted on the course website.

Now that we have the significant figures out of the way, let's finish the problem. The total time it took to achieve a displacement of 185 m was 24.00 seconds + 25 seconds = 49 seconds. Once again, note that because 25 sec is precise only to the ones place, our answer must be reported to the ones place. Using the displacement and change in time that we just calculated, Equation (1.1), becomes:

$$\mathbf{v} = \frac{\Delta\mathbf{x}}{\Delta t} = \frac{185 \text{ m}}{49 \text{ sec}} = 3.8 \frac{\text{m}}{\text{sec}}$$

Since the velocity is positive, we know that even though he walked back a little, his overall velocity was still <u>3.8 m/sec in the direction of the finish line</u>.

Make sure you understand these concepts by solving the following "On Your Own" problem.

ON YOUR OWN

1.2 A mail carrier drives down a street delivering mail. She travels 3.00×10^2 meters down the street in 332 seconds. She then turns around and heads back up the street, but because of the way the mailboxes are placed, she only needs to travel 208 meters in that direction, and that trip takes her only 2.30×10^2 seconds. What was her velocity as she traveled down the street? What was it as she traveled up the street? What was her velocity for the entire trip?

Now of course, Equation (1.1) has more applications than the ones you have seen so far. Study the next example and solve the "On Your Own" problem that follows in order to see how other types of problems can be solved using this equation.

EXAMPLE 1.3

A jogger runs down a long, straight country road at 2.3 m/sec. If she jogs in that direction for 15.3 minutes, how far does she run?

Part of the trick to solving physics problems is learning how to read the question so that you see what you are trying to solve for. In this example, a couple of words should jump out at you. First, you are given a speed, but you are also given direction because the words "straight" and "down" are used. Thus, the 2.3 m/sec is a velocity, because direction is included. The problem also gives you time, but it is not in units that are consistent with the velocity. The velocity is given in m/sec, but the time is given in minutes. To be able to use both of these pieces of information in any solution, the units must be consistent. We therefore must convert one of these quantities into different units. Since m/sec is the standard, I won't convert it. Instead, I will convert 15.3 minutes into seconds:

$$\frac{15.3\ \cancel{\text{min}}}{1} \times \frac{60\ \text{sec}}{1\ \cancel{\text{min}}} = 918\ \text{sec}$$

Remember that the "60" and the "1" in the conversion relationship are exact. There are exactly 60 seconds in a minute. Thus, both "60" and "1" really have an infinite number of significant figures (60.000... and 1.000...), even though they are not listed. Thus, the number of significant figures in the original measurement limits the number of significant figures in the answer.

Now that we have our units straight, we can continue. The problem asks us to determine how far the jogger will go. Well, "how far" is another way of saying "how much displacement." After all, if she runs, say, 100 meters, her displacement from the place that she started will be 100 meters. Thus, we are given velocity and time and asked to determine displacement. Equation (1.1) relates these three quantities. We will therefore use Equation (1.1), substituting the values that we already know:

$$\mathbf{v} = \frac{\Delta \mathbf{x}}{\Delta t}$$

$$2.3\ \frac{\text{m}}{\text{sec}} = \frac{\Delta\mathbf{x}}{918\ \text{sec}}$$

Now we can use algebra to rearrange this equation and solve for the displacement ($\Delta\mathbf{x}$):

$$2.3\ \frac{\text{m}}{\cancel{\text{sec}}} \times 918\ \cancel{\text{sec}} = \Delta\mathbf{x}$$

$$2.1 \times 10^3\ \text{m} = \Delta\mathbf{x}$$

So the jogger's displacement is 2.1 x 10^3 m down the road. Thus, the jogger ran 2.1 x 10^3 m. Please note that I did not need to use scientific notation in the answer. I could have said 2,100 m. Either way, the displacement is the same, and the number of significant figures is the same. So either answer is correct.

Do you see how we solved this problem? We read it carefully, and we picked out words that told us what quantities we had and what quantities we needed to determine. We then found an equation that related these quantities and used algebra to solve the equation. This is the way you solve physics problems. Try it yourself with the following "On Your Own" problem.

ON YOUR OWN

1.3 A boat travels straight down a river at a speed of 15 m/sec. If the boat travels a distance of 34.1 km, how long was the boat ride?

Average and Instantaneous Velocity

In "On Your Own" problem 1.2 and in the example preceding it, we got answers that you might think are a bit strange. In the example, for instance, the sprinter's velocity over the entire trip was 3.8 m/sec in the direction of the finish line. You might find it odd that despite the fact that the sprinter traveled in both directions, his overall velocity was in the direction of the finish line. If you find it strange, don't worry. That's because we haven't discussed the difference between instantaneous and average velocity. We'll do that now.

Instantaneous velocity - The velocity of an object at one moment in time

Average velocity - The velocity of an object over an extended period of time

These two concepts of velocity are quite different. To see how different they are, perform the following experiment.

EXPERIMENT 1.1
Measuring Average Velocity

Note: A sample set of calculations is available in the solutions and tests guide. It is with the solutions to the practice problems.

Supplies:

- Safety goggles
- A stopwatch (A watch with a second hand will do.)
- A pile of books between 6 and 9 centimeters thick
- A wooden board, about 1 meter long (Any long, flat surface that you can prop up on one end will do. It needs to be as smooth as possible.)
- A pencil (Anything that you can use to mark the board will do.)
- A ball that will easily roll down the board

1. Choose the smoothest side of the board and clear it of any debris.
2. Make a mark on the board in the center. Make sure the mark is easy to see.
3. Prop the board up on one end with the books, so that the board forms an incline as shown below. In a moment, you will be rolling the ball down the incline. Your experiment should look something like this:

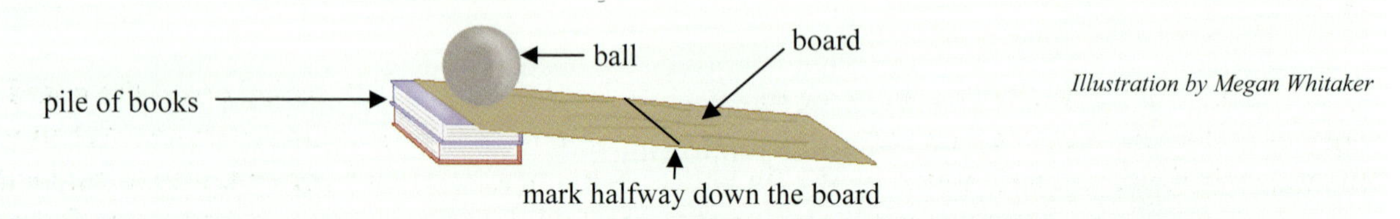

Illustration by Megan Whitaker

4. Measure the distance from the top of the board to the mark halfway down the board. Make sure you record the distance to the proper precision. Since you can estimate between the lines, most metric rulers can be read to 0.01 cm. Call this distance "d_1."
5. Measure the distance from the mark to the end of the board as well, once again writing your answer with the proper precision. Call it "d_2." If you really made the mark in the center of the board, d_1 and d_2 should be the same. If not, don't worry about it. They do not have to be equal.
6. Once you have set up your experiment and made both distance measurements, hold the ball on the very top of the board and be ready to release it. At the exact moment that you release the ball, start the stopwatch. Stop the watch when the ball hits the mark.
7. Write down the time you measured. Be as precise as the stopwatch allows.
8. Repeat this measurement four more times. After you have a total of five measurements for the time, average them and write down your answer. Why did I have you measure the same thing five times and average the result? Well, there are many errors which can occur when you make these kinds of measurements. Most likely, you did not start the stopwatch at *exactly* the time that you released the ball. You probably started it a bit before or a bit after. In the same way, you probably did not stop it at *exactly* the time that the ball reached the mark. You probably stopped it shortly before or shortly after. These types of errors (called "random errors") make a single measurement inaccurate. However, if you make several such measurements and average the results, the random errors in the individual measurements will (to some extent) cancel out, making the average a better estimate of the true value. The more measurements you make, the better this works.
9. Once you have that average, divide it into the distance from the top of the board to the first mark (d_1). Let's say that motion down the board is positive. That way, the distance you measured is also the ball's displacement. Thus, the calculation you just made took displacement and divided it by time, which gives you the velocity of the ball as it traveled from the top of the board to the first mark. Call this velocity v_1.

10. Hold the ball at the top of the board again, and be ready to release it. This time, however, *do not start the stopwatch until the ball hits the first mark.* Stop the watch when the ball hits the end of the board. Do this measurement a total of five times as well, and once again, average the results.
11. Take the average and divide it into d_2. This will give you the velocity of the ball as it traveled down the second half of the board. Call it v_2.
12. Finally, do the same thing again, this time starting the watch the instant that you release the ball and stopping the watch once the ball hits the end of the board.
13. Average the five results and divide that average into the total length of the board ($d_1 + d_2$). This is the velocity of the ball over the entire trip. Call it v_3.
14. Clean up your mess, but save the supplies, because you will use them again in Experiment 1.2.

Now let's figure out what this experiment demonstrates. Compare your three velocities. If you did the experiment correctly, v_1 should be less than v_2. In addition, v_3 should be between v_1 and v_2. Why? Well, the ball was speeding up the whole time it traveled down the board. Thus, v_1 is the lowest because the ball had not sped up all of the way by the time it hit the first mark. V_2 was larger than v_1 because the ball had more time to speed up traveling down the second half of the board. The total velocity (v_3) was the *average* of the two velocities you measured. That's why it falls in between them.

So, which velocity (v_1, v_2, or v_3) is the velocity of the ball as it traveled down the board? The answer is that *all three* are. However, they are *measured over different time intervals*. When we take the total displacement and divide by the time it takes to make that displacement, we are calculating the *average velocity*. V_1 is the average velocity of the ball while it traveled down the first half of the board; v_2 is the average velocity of the ball as it traveled down the second half of the board; and v_3 is the average velocity as the ball traveled down the entire board.

Now suppose we divided the board into five sections instead of just two, and suppose we measured the velocity of the ball as it traveled through each of the five sections, determining v_1 - v_5, as well as the average across the entire board, v_6. You can probably predict the results: v_1 would be the smallest, v_2 - v_4 would each be progressively larger, and v_5 would be the largest. In addition, v_6 would be in between v_1 and v_5.

Now suppose that we were able to divide the board into an *infinite* number of *extremely tiny* sections, and suppose further that we could measure the velocity as the ball traveled through each of these infinitesimally small sections. What would we have then? Well, we would have an incredibly bored and frustrated student, but we would also have an infinite number of velocities, each of which would be slightly greater than the one before. At that point, however, we would no longer have *average* velocities. We would have *instantaneous* velocities.

That's the difference between instantaneous and average velocity. Instantaneous velocity represents the velocity measured over an infinitesimally small time interval. Of course, it is impossible for us to measure instantaneous velocity, but the smaller the time interval, the closer the average velocity is to the instantaneous velocity. Consider the results of your experiment again. V_3 was the average velocity as the ball traveled down the entire board. V_1 was the average velocity as the ball traveled down the first half of the board, and v_2 was the average velocity as the ball traveled down the second half of the board. Since v_1 and v_2 were measured over shorter time intervals than v_3, v_1 and v_2 are *closer* to instantaneous velocities than is v_3. If we divided up the board into five sections instead of two, the velocities v_1 - v_5 would each be closer to instantaneous velocities than were the v_1 and v_2 you measured in your experiment.

This is why you can get strange answers like the one we got in Example 1.2. The last velocity that we calculated in the example was the average velocity of the sprinter. This, in effect, averaged the positive and negative velocities that we calculated in the first part of the example. Since the sprinter ran faster and longer in the positive direction, the average velocity turned out to be positive, even though the sprinter traveled in both directions. Thus, average velocity is calculated over a long time span, while instantaneous velocity is calculated over an infinitely short time span. Since it is impossible to measure displacement and time over an infinitely short time span, we can never really measure instantaneous velocity. Thus, all of the velocity measurements that we can make are really *average* velocities. However, the smaller the time span that we use to measure average velocity, the closer the average velocity is to the instantaneous velocity.

Although it is not possible to *measure* instantaneous velocity, we can estimate it rather easily by reading a graph. Consider, for example, the graph in Figure 1.3:

FIGURE 1.3
A Position-Versus-Time Graph

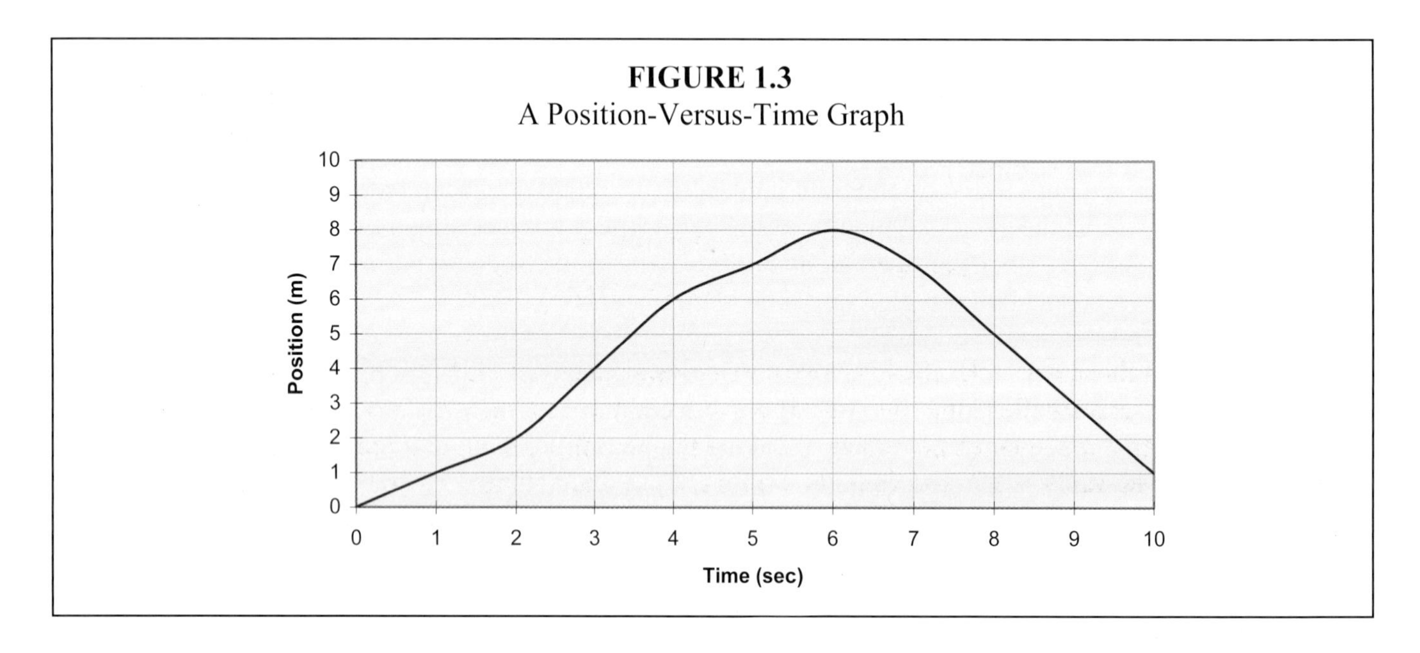

In this graph, the position of an object is plotted on the y-axis while time is plotted on the x-axis. Thus, the curve represents an object's position at various time intervals. If you look at the graph, you will see that the object starts at zero and then moves in a positive direction to a maximum position of 8.0 meters from its starting point. It reaches that maximum position in 6.0 seconds. At that point, the object's position begins to decrease. The only way that can happen is if it begins to move back towards the starting point. In other words, it begins to move in the *negative* direction. Thus, after it reached a position of 8.0 meters from its starting point, the object must have turned around and moved in the opposite direction.

Now, despite the fact that the velocity is not plotted in this graph, it can be determined from the graph. In fact, you can actually get a good feel for the meaning of instantaneous velocity by looking at this graph. I'm getting ahead of myself, however. How can you determine velocity from such a graph? Well, according to Equation (1.1), you can calculate velocity by taking the change in position (the displacement) and dividing by the change in time. On this graph, position is plotted on the y-axis and time is plotted on the x-axis. Thus, to get velocity, we need to take the change in the y coordinate and divide it by the change in the x coordinate. What's another name for the quantity you get when you take the change in y and divide by the change in x? It's the *slope!* Thus, we come to a very important fact:

The slope of a position-versus-time curve is the velocity.

What does this mean? Well, we can calculate the slope of the curve in Figure 1.3, and that represents the velocity of the object. Thus, suppose we looked at the object's position at a time of 1.0 seconds. According to the graph, the position is 1.0 m from the starting point. At 6.0 seconds, however, the position is 8.0 meters from the starting point. Thus, the slope of the curve during that time interval is:

$$\text{slope} = \frac{\text{rise}}{\text{run}} = \frac{8.0\text{ m} - 1.0\text{ m}}{6.0\text{ sec} - 1.0\text{ sec}} = 1.4\frac{\text{m}}{\text{sec}}$$

Now remember, since the slope of a position-versus-time curve is the velocity, the average velocity of the object during the time interval of 1.0 second to 6.0 seconds was 1.4 m/sec.

On the other hand, suppose we examined the time interval between 0.0 and 1.0 seconds. At zero seconds, the position was 0.0, while at 1.0 seconds, it was 1.0 m. The slope of the curve over that time frame (which is the velocity), then, is:

$$\text{slope} = \mathbf{v} = \frac{\text{rise}}{\text{run}} = \frac{1.0\text{ m} - 0.0\text{ m}}{1.0\text{ sec} - 0.0\text{ sec}} = 1.0\frac{\text{m}}{\text{sec}}$$

Which of those two velocities is closest to an instantaneous velocity? The second one, because it is calculated over a smaller time interval. If we reduced the time interval even more, we would get even closer to a true instantaneous velocity. Taking this reasoning to an extreme, when the time interval is infinitesimally small, the velocity would truly be instantaneous. Thus, if we were to look at a position-versus-time curve at a single point in time, we could estimate the instantaneous velocity by estimating the slope of the curve at that point.

Now if all of this seems a bit confusing, don't worry about it. We'll get lots of practice at examining such graphs, so you'll become a veritable expert at this stuff. Let's look at Figure 1.3 again and look at another time interval. Specifically, let's look at the time interval between 5.9 and 6.1 seconds. What's the velocity during that time interval? Well, according to the graph, the object's position seems to stay steady at 8.0 m during that time interval. The velocity, then, is:

$$\text{slope} = \mathbf{v} = \frac{\text{rise}}{\text{run}} = \frac{8.0\text{ m} - 8.0\text{ m}}{6.1\text{ sec} - 5.9\text{ sec}} = 0.0\frac{\text{m}}{\text{sec}}$$

This velocity is very close to instantaneous velocity, because the time interval is very short.

In order to make you truly sick of all of this, let's look at one more time interval. What is the average velocity during the time interval of 7.0 seconds to 10.0 seconds? According to the graph, the object's position falls from 7.0 m to 1.0 m over that time interval. The velocity, then, is:

$$\text{slope} = \mathbf{v} = \frac{\text{rise}}{\text{run}} = \frac{1.0\text{ m} - 7.0\text{ m}}{10.0\text{ sec} - 7.0\text{ sec}} = -2.0\frac{\text{m}}{\text{sec}}$$

What does the negative sign mean? It means that the object is *moving in the opposite direction* during this time interval compared to the others we have examined so far. So you see, we can learn a lot about the velocity of an object by looking at a position-versus-time graph.

Before we move on, I need to make a quick point about significant figures. Note that in all of the calculations I have done so far, I have reported both the position and the time with a precision that goes out to the tenths place. Why did I do that? Well, when you learned about making measurements, you should have been told that when you read numbers from a scale (or a graph), you should be able to estimate in between the markings of the scale (or graph). This gives you one more decimal place than what is marked off. Notice that the graph in Figure 1.3 is marked off in meters and seconds. By estimating between the marks, we can determine position to tenths of a meter and time to tenths of a second. Thus, that is the precision that I must use when I read the position and time from the graph.

Now, let's move on to thinking about the *instantaneous* velocity of the object. Once again, look at the graph in Figure 1.3. You should remember from algebra that the steeper a curve rises or falls, the larger its slope is. If the curve rises, its slope is positive, and if the curve falls, its slope is negative. Finally, when the curve is flat, its slope is zero. If we remember these facts, we can answer some pretty fundamental questions about the motion of an object when examining a position-versus-time graph.

For example, when does the object reach its maximum speed? Well, if we look at the figure, the graph seems to be steepest between 3.0 and 3.8 seconds. During that time interval, the speed is at its maximum value. Now remember, it doesn't matter whether the curve is rising or falling when trying to determine maximum speed. If the curve happens to be steepest as it is falling, that would be the maximum speed. Since the negative sign simply tells us direction, we don't consider it when determining the speed, because speed is not a vector quantity.

Where is the object's speed at its minimum? Once again, we don't worry about whether the velocity is positive or negative, because the sign just tells us direction. Thus, the velocity is lowest where the curve is the least steep. That obviously occurs from 5.9 to 6.1 seconds, where the curve is flat.

We can also compare instantaneous velocities using the graph in Figure 1.3. For example, which is larger, the instantaneous velocity at 4.5 seconds or the instantaneous velocity at 8.5 seconds? Once again ignoring the positive and negative signs because they simply tell us direction, the curve is obviously steeper at 8.5 seconds than it is at 4.5 seconds. Therefore, the instantaneous velocity of the object is greater at 8.5 seconds than it is at 4.5 seconds.

Sometimes, we can actually give a value for the instantaneous velocity by simply looking at the graph. For example, what is the instantaneous velocity at 6.0 seconds? At that time, the curve is flat. Whenever a curve is flat, its slope is zero. Thus, the instantaneous velocity of the object at 6.0 seconds is zero. Also, consider the time interval between 7.0 seconds and 10.0 seconds. During that time, the curve looks like a straight line. Well, in algebra you should have learned that the slope of a straight line is the same no matter where you are on the line. Since we already calculated that the slope of the curve during this time interval is -2.0 m/sec, we can say that the slope of the curve at any point from 7.0 seconds to 10.0 seconds is -2.0 m/sec. Thus, the instantaneous velocity at, say, 8.2 seconds is also -2.0 m/sec.

Since we can learn so much about the velocity of an object from these curves, we need to study them in detail. Study the next example and then do the "On Your Own" problems that follow in order to make sure you can interpret graphs like these.

EXAMPLE 1.4

Consider an object in motion whose position-versus-time graph is as follows:

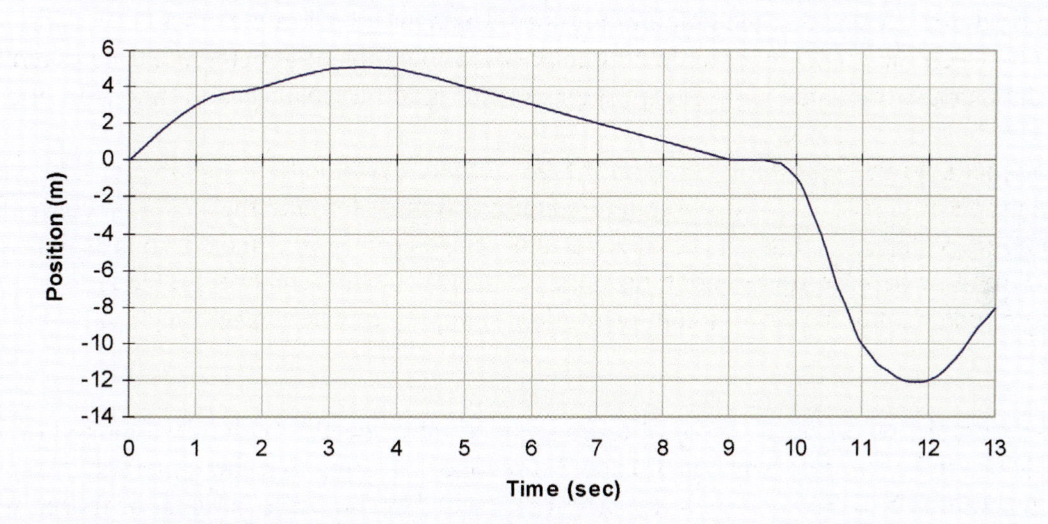

During what time interval is the object's speed the greatest?

To answer this question, we simply look for the steepest part of the graph. The curve is clearly steepest from 10.0 to 11.0 seconds, so that's the time interval in which the object is moving with the greatest speed. Now remember, the fact that the slope is negative during that time interval simply tells us the direction in which the object is moving. We don't consider the *sign* of the slope in determining the *speed* of the object. Thus, even though the *velocity* is negative, the object's *speed* is greatest from 10.0 to 11.0 seconds.

How many times did the object change directions?

The slope of the curve starts out positive (because the curve is rising), so the object begins by moving in a positive direction. At 4.0 seconds, however, the slope becomes negative (because the curve is falling). This means that the object starts to move in the opposite direction at that time, because the velocity changed direction. That is the first time the object changes direction. At 11.8 seconds, the slope changes from negative back to positive, indicating another direction change. Thus, the object changed direction twice.

What is the instantaneous velocity of the object at 11.8 seconds?

At 11.8 seconds, the curve is flat. Thus, the velocity is 0.0 m/sec.

What is the instantaneous velocity of the object at 6.0 seconds?

During the interval of 5.0 to 9.0 seconds, the curve looks like a straight line. Thus, the slope at any point along that part of the curve is the same. We can therefore calculate the average velocity from 5.0 to 9.0 seconds and, since the velocity stays the same throughout that time interval, that will also be the instantaneous velocity at any time during that interval, including 6.0 seconds.

To read the numbers from the graph, we need to realize that the graph is marked off in meters and seconds, so I can read both the position and the time to the tenths place. The position at 5.0 seconds is 4.0 meters according to the graph, and the position at 9.0 seconds is 0.0 meters. The average velocity, then, is:

$$\text{slope} = \mathbf{v} = \frac{\text{rise}}{\text{run}} = \frac{0.0\text{ m} - 4.0\text{ m}}{9.0\text{ sec} - 5.0\text{ sec}} = -1.0\,\frac{\text{m}}{\text{sec}}$$

This means that the instantaneous velocity at 6.0 seconds is also <u>-1.0 m/sec</u>.

ON YOUR OWN

Consider the following position-versus-time curve:

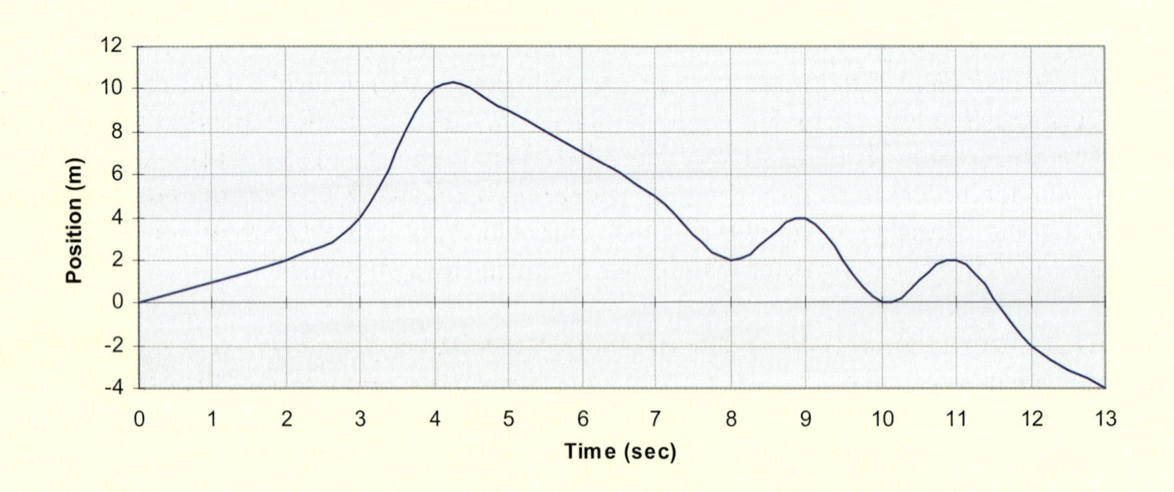

1.4 Is the object moving faster at 3.5 seconds or at 8.5 seconds?

1.5 How many times does the object change directions?

1.6 What is the instantaneous velocity at 1.0 seconds?

Before we leave this section, I must point something out. Often when students are studying introductory physics, they think that the problems they work out are useless exercises. Nothing could be further from the truth! Nearly every aspect of physics has practical applications. Students, however, are not knowledgeable enough to realize what they are. For example, students often complain that the position-versus-time graphs we just learned are a waste of time because there are no practical uses for them. How wrong these students are!

Race-car drivers spend hours of time on the track, trying to determine the best way to negotiate the curves and straightaways to get the best time possible. It turns out that while they are on the track, computers keep measuring the car's position and time. At the end of the run, the driver and his team study the position-versus-time graph very carefully. You see, by looking at the slope of the curve, the driver can easily see where the car slowed and where it sped up. If the car was slowing down in the wrong place, studying the position-versus-time curve will show that, and the driver can adjust his strategy accordingly. So, now that you understand these position-versus-time curves, you could help a race car driver develop strategies for his next race!

Velocity Is Relative

Now if all of this velocity talk hasn't been confusing enough, there is one more concept that we must cover. One of the most important things to realize about velocity is that it is *relative*. What does that mean? The best way to illustrate how velocity is relative is by considering an example. Let's suppose you've just finished visiting your grandmother's house, and you get in the family car to drive away. You are riding in the passenger's seat next to your father, who is driving. Your grandmother, sorry to see you go, has come out of the house and is standing in front of the car waving good-bye. As the car backs out of the driveway, you are looking at your grandmother, waving good-bye as well. Now, answer this one simple question: Are you moving?

Your first instinct is probably to say, "Well yes, of course I'm moving, because the car is backing out of the driveway!" Wait a minute, though. Aren't you actually sitting still? If your father looks at you, does he think that you're moving? Probably not. After all, as far as he can see, you are sitting still right next to him. You don't seem to move at all. From your grandmother's point of view, however, you are moving. You are moving away from her. That's the point. As far as your father is concerned, you don't seem to be moving at all. From your grandmother's point of view, however, you are, indeed moving. Thus, your father thinks that your velocity is zero, while your grandmother sees that you have a velocity that is greater than zero and directed away from her.

This is what we mean when we say that velocity is relative. It depends on who is observing that velocity. Since your father is in the car with you, you are both moving with the exact same velocity. As a result, your position relative to him never changes. When position doesn't change, velocity is zero. Thus, your father thinks that your velocity is zero. On the other hand, your position relative to your grandmother is changing. As a result, she sees a velocity greater than zero, directed away from her. Thus, velocity can only be determined relative to an observer.

"Now wait a minute," you might be saying, "don't I *really* know that my father and I are moving? After all, we are in the car." The answer to that is definitely not. It is really impossible for us to say what is moving and what is sitting still. For example, consider your grandmother's house. Is it moving? You would probably say that it is not. However, suppose I were on the moon observing her house through a powerful telescope. In order to continue to observe her house, I would have to constantly change the direction in which my telescope is pointing. Why? Because relative to the moon, her house *is* moving. Thus, motion truly is relative. There is no way for us to point at something and say that it is moving. Relative to us (or some other observer) it might be moving, but relative to another observer, it might very well be sitting still. Thus, the best we can do is say that relative to *us*, the object is moving.

What an observer actually sees, then, is the difference between his velocity and the velocity of what he is observing. Let's go back to the situation we were just discussing. From your grandmother's point of view, her velocity was zero. You, on the other hand, were moving away from her in the car. The velocity she saw was the difference between her velocity (0) and your velocity. Thus, she observed you moving. Your father, however, was moving with the car and had exactly the same velocity (relative to your grandmother) that you did. The difference between his velocity and your velocity, then, was zero, and that's why from your father's viewpoint, you were not moving. See if you understand this concept by studying the following example and performing the "On Your Own" problem afterwards.

Illustrations from the MasterClips collection

EXAMPLE 1.5

A car and a truck are approaching each other on a 2-lane road (see diagram below). The speedometer in the car reads 56 mph, and the speedometer in the truck reads 45 mph. If you were standing at the side of the road watching this situation, what velocity would you observe for the car? What velocity would you observe for the truck? What velocity does the driver of the car observe for the truck? What velocity does the driver of the truck observe for the car?

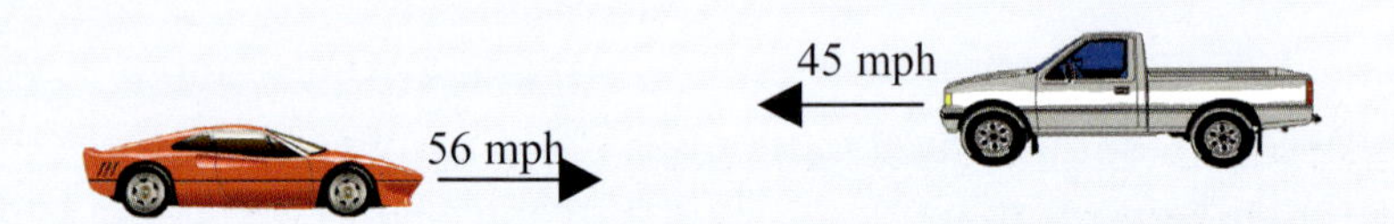

What would you see? You are standing still, so as far as you are concerned, your velocity is zero. If we define motion to your right as positive, you see the car moving at 56 mph - 0 mph = 56 mph. Thus, <u>according to you, the car is moving to your right at 56 mph</u>. Since the truck is moving to your left, you see its velocity as -45 mph - 0 mph = -45 mph. As a result, <u>you see the truck traveling at 45 mph to your left</u>. The driver in the car, however, is already moving. As he looks at the truck, he has no idea what its speedometer reads. What he does see, however, is that the truck is approaching very quickly. As all observers do, he sees the difference between his velocity and the truck's velocity. Thus, the velocity he observes is -45 mph - 56 mph = -101 mph. According to our definition of positive and negative, this means that <u>the driver of the car observes the truck moving to your left at 101 mph</u>. Finally, the truck driver also observes the difference between the car's velocity and his velocity. Thus, the truck driver observes a velocity of 56 mph - (-45 mph) = 101 mph. The positive sign means that the motion is to your right. Thus, <u>the truck driver sees the car moving to your right at 101 mph</u>.

In the end, then, the velocity of the car and truck depend on the observer. As the stationary person in the example, you observed one set of velocities, while the drivers observed another. That's what I mean when I say that velocity is relative. It depends on the observer. As a point of mathematical clarification, when you are calculating the difference in velocities, always take the velocity of what is being observed minus the velocity of the observer. That way, your signs will always work out to the proper directions.

ON YOUR OWN

1.7 A boat is traveling up a river against the current. A boy on a raft is floating down the river with the current. They are both being observed by a fisherman sitting on the shore. The fisherman observes the boat traveling 15 m/sec up the river. He also notices that the boy and his raft have a velocity of 3 m/sec down the river. What is the velocity of the raft as observed by someone on the boat? What is the velocity of the boat as observed by the boy on the raft?

Acceleration

We now come to the last concept we need to cover in this module: **acceleration**.

Acceleration - The time rate of change of an object's velocity

Does this definition sound similar to the one for velocity? It should. Just as velocity measures how an object's position varies with time, acceleration measures how an object's velocity changes with time. The mathematical definition of acceleration is as follows:

$$\mathbf{a} = \frac{\Delta \mathbf{v}}{\Delta t} \qquad (1.3)$$

where "**a**" is the acceleration, "**v**" is the velocity, and "t" is time. Once again, since **a** and **v** are in boldfaced type, they are vector quantities.

What units are attached to acceleration? Well, we already know that velocity has the SI unit of m/sec. In order to get acceleration, you take the change in velocity (which still has units of m/sec) and divide by time (which has the SI unit of seconds). What happens when you take m/sec and divide by sec? You get m/sec^2 (meters per second squared). This is the SI unit for acceleration.

Since Equation (1.3) tells us that acceleration is a vector, we need to be sure we understand all of the implications of this fact. When you hear the term "acceleration" in everyday language, it means "speed up." For example, when a driver increases the velocity of a car, we say that the car accelerated. In physics, though, acceleration does not have to mean "speed up." It can also mean "slow down." After all, acceleration just tells us how the velocity of an object is changing. If the velocity is decreasing, then it is changing, and thus there is acceleration.

When we see acceleration, then, how will we know whether it is causing an increase in velocity (speeding the object up) or a decrease in velocity (slowing the object down)? Actually, it is quite simple. If the acceleration and velocity have opposite signs, the object is slowing down. If they have identical signs, the object is speeding up. Thus, if an object has a velocity of -3.2 m/sec and an acceleration of 0.1 m/sec^2, the object is slowing down. Alternatively, a velocity of 13.2 m/sec and an acceleration of 2.2 m/sec^2 mean that the object is speeding up. That's the vector nature of acceleration. If acceleration and velocity have the same direction, the acceleration is increasing the velocity. Alternatively, if the acceleration and velocity are pointed in opposite directions, the acceleration is decreasing the velocity. Perform the following experiment to help you understand what acceleration is all about.

EXPERIMENT 1.2
Measuring an Object's Acceleration

Note: A sample set of calculations is available in the solutions and tests guide. It is with the solutions to the practice problems.

Supplies:

- Safety goggles
- A stopwatch (A watch with a second hand will do.)
- A pile of books between 18 and 27 centimeters thick
- A wooden board, about 1 meter long (Any long, flat surface that you can prop up on one end will do. It needs to be as smooth as possible.)
- A pencil (Anything that you can use to mark the board will do.)
- A ball that will easily roll down the board
- A few extra books
- Masking tape or electrical tape
- An uncarpeted floor

1. Construct the same experimental setup that you had for Experiment 1.1. This time, however, use the tape to make a mark on the floor exactly 1.00 meter from the end of the board.
2. Hold the ball at the top of the board and release it. Do not start the stopwatch until the instant that the ball rolls off of the board and onto the floor. Stop the watch when the ball reaches the tape. In this way, you have measured the time it takes for the ball to roll one meter once it has left the end of the board.
3. Just as you did in Experiment 1.1, make this measurement five times and average the result.
4. Take that average and divide it into 1.00 m. This measures the average velocity of the ball once it rolls off of the board.
5. If you think about it, the ball rolls down the board because of gravity. We'll discuss that subject several times throughout this course, so I don't want to talk about gravity itself in depth at this time. Nevertheless, you should be aware that the reason the ball rolls down the board is that gravity is pulling it down. Since gravity is pulling down on the ball, the ball accelerates. It starts with a velocity of zero (because you held it still to begin with), and it rolls off of the board with a large velocity. Since velocity changed, by definition, there must have been acceleration. Gravity supplies that acceleration. Once the ball leaves the board, however, gravity can no longer accelerate it. Therefore, the ball rolls across the floor with a relatively constant velocity. Now, of course, the ball eventually slows down and stops because it either runs into something or because of *friction*, which we will explore in a later module. For the first meter after it rolls off the board, however, it is a reasonably good assumption that the ball rolls with a constant velocity, as long as the floor that you set the experiment on is not carpeted. Thus, the velocity that you measured is approximately the same as the velocity the ball had when it rolled off the end of the board.
6. Hold the ball at the top of the board again and release it. This time, start the watch as soon as you release the ball and stop it when the ball reaches the end of the board. Once again, make this measurement five times and average the result. Do not calculate any velocities. You are only measuring time in this portion of the experiment.
7. What does this measurement represent? Well, it represents the time it takes for the ball to roll down the board. What's so important about that? Think about it. The ball started (at the top of the board) with a velocity of zero and ended (at the bottom of the board) with the velocity that you measured in the first part of this experiment. Thus, it must have accelerated. When did that acceleration take place? When the object was on the board. Remember, the velocity of the ball

stayed constant once it rolled off of the board. This means that all of its acceleration took place while it was on the board. Therefore, we know the beginning velocity (0), and the ending velocity (the velocity that you measured in the first part of this experiment). If we subtract the former from the latter, we will get $\Delta \mathbf{v}$, the change in velocity while the ball was on the board. The time that you just measured is the time interval over which the ball stayed on the board, or Δt. Take your value for $\Delta \mathbf{v}$ and divide it by Δt, and you get the acceleration that the ball experienced!

8. Add 6-9 more centimeters of books to the book pile so that the board tilts more steeply. Repeat the entire experiment, so that you get a new value for acceleration.
9. Add another 6-9 cm worth of books to the pile and repeat the experiment one more time to get yet another value for the ball's acceleration.
10. Clean up your mess.

Now that you have completed the experiment, compare the three accelerations that you measured. The first one should be the smallest, the second one should be larger, and the third one should be the largest. That should not surprise you. As you increase the tilt of the board, gravity can pull the ball along the surface of the board more effectively. As a result, the ball's acceleration increases. This makes the ball travel along the board more quickly so that it has a greater velocity when it reaches the end of the board. That's what you saw in the experiment.

That conclusion was not the major goal of the experiment, however. The major goal was to show you how to measure an object's acceleration. You measured its initial velocity (0), its final velocity (the velocity at the end of the board), and the time it took for that change in velocity to occur. By taking the change in velocity and dividing by the time over which the change occurred, you got the acceleration. The fact that your measurement increased as the tilt of the board increased was simply an indication that you did, indeed, measure the ball's acceleration.

So we see that acceleration is the agent by which velocity change occurs. Study the following examples and solve the "On Your Own" problems that appear afterward so that you are sure to have a firm grasp of the concept of acceleration.

EXAMPLE 1.6

A car is moving with a velocity of 25 m/sec to the east. The driver suddenly sees a deer in the middle of the road and slams on the brakes. The car comes to a halt in 2.1 seconds. What was the car's acceleration?

This problem is a straightforward application of Equation (1.3). The problem says that the car starts with a velocity of 25 m/sec east and ends up stopping (v = 0). Thus, we can subtract the initial velocity from the final velocity to get $\Delta \mathbf{v}$:

$$\Delta \mathbf{v} = \mathbf{v}_{\text{final}} - \mathbf{v}_{\text{initial}} = 0 \text{ m/sec} - 25 \text{ m/sec} = -25 \text{ m/sec}$$

The problem also gives us time, so to calculate the acceleration, all we have to do is plug these numbers into Equation (1.3):

$$\mathbf{a} = \frac{\Delta \mathbf{v}}{\Delta t}$$

$$\mathbf{a} = \frac{-25\ \frac{\text{m}}{\text{sec}}}{2.1\ \text{sec}} = -12\ \frac{\text{m}}{\text{sec}^2}$$

What does the negative mean? Well, since we made the initial velocity positive, that defined motion to the east as positive. The fact that the acceleration is negative means that the acceleration is pointed in the *opposite* direction. Thus, the car's acceleration was $\underline{12\ \text{m/sec}^2 \text{ to the west}}$. Since the velocity and acceleration are pointed in different directions, the car was slowing down. Of course, you already know that the car was slowing down, as the driver was trying to stop. However, this problem illustrates what I have already discussed: when the acceleration and velocity are pointed in opposite directions, the speed will decrease.

In the next module, we will learn that when objects are dropped, they fall straight down with an acceleration of 9.8 m/sec^2. If a ball is dropped with no initial velocity, how long would it take to accelerate to a downward velocity of 11.0 m/sec?

This problem tells us acceleration and the change in velocity and asks us to calculate the time over which the change occurred. The velocity starts at 0 m/sec and ends at 11.0 m/sec. Thus, we can calculate $\Delta\mathbf{v}$:

$$\Delta \mathbf{v} = \mathbf{v}_{\text{final}} - \mathbf{v}_{\text{initial}} = 11.0\ \text{m/sec} - 0\ \text{m/sec} = 11.0\ \text{m/sec}$$

Before I go on, I want to make a quick point about significant figures. This equation might pose a dilemma for you when trying to determine how many significant figures $\Delta\mathbf{v}$ should have. After all, how many significant figures does 0 m/sec have? Well, when you read a statement like "no initial velocity" or "it comes to a halt," you have to assume that the object is not moving at all. Thus, you must assume that its velocity is *exactly* 0.00000000… m/sec. As a result, it is infinitely precise and has an infinite number of significant figures. Thus, the precision with which we report our answer depends only on the *other* numbers in the problem, not the zero. That's why I reported my answer to the tenths place, because the other number in the problem has its last significant figure in the tenths place. Now please understand that the ball probably doesn't have *exactly* zero velocity. The person dropping the ball probably cannot hold her hand perfectly still, for example. Thus, the ball probably has some small initial velocity. However, compared to our other measurements, the size of that velocity is most likely insignificant, so it is safe to assume that a velocity of zero is exact, at least as far as we are concerned.

Now that we have acceleration and $\Delta\mathbf{v}$, we can use Equation (1.3) to solve for time:

$$\mathbf{a} = \frac{\Delta \mathbf{v}}{\Delta t}$$

$$9.8\ \frac{\text{m}}{\text{sec}^2} = \frac{11.0\ \frac{\text{m}}{\text{sec}}}{\Delta t}$$

$$\Delta t = \frac{11.0 \frac{\text{m}}{\text{sec}}}{9.8 \frac{\text{m}}{\text{sec}^2}} = 1.1 \text{ sec}$$

Notice how the units work out here. The meters cancel, and the seconds in the velocity unit cancels the square in "sec^2" of the acceleration unit, leaving the unit as seconds. That's good, since we are solving for time. Thus, it takes the ball 1.1 seconds to accelerate to a velocity of 11.0 m/sec downwards. Now 11.0 m/sec is about the same as 25 mph, so things that fall speed up quickly!

ON YOUR OWN

1.8 A sprinter starts from rest and, in 3.4 seconds, is traveling with a velocity of 16 m/sec east. What is the sprinter's acceleration?

1.9 A race car accelerates at -7.2 m/sec^2 when the brakes are applied. If it takes 3.1 seconds to stop the car when the brakes are applied, how fast was the car originally going?

1.10 In Experiment 1.2, we made an assumption that the velocity of the ball was constant while it was rolling from the end of the board to the tape. However, we know that this assumption is wrong to some extent, because we know that given enough time, the ball will eventually stop rolling. Describe a way that we could use the same experimental setup to evaluate the validity of this assumption.

Average and Instantaneous Acceleration

Since the equations for velocity and acceleration are similar, you might expect that acceleration, like velocity, can be defined as average or as instantaneous. Just like velocity, when the time interval is large, the acceleration is an average. When the time interval is infinitely short, however, the acceleration is instantaneous. Just like velocity, the only real way to determine instantaneous acceleration is by studying graphs.

What kinds of graphs will we study in this case, however? Well, since acceleration tells us how velocity changes with time, we should examine velocity-versus-time graphs. If we plot velocity on the y-axis and time on the x-axis, the slope of the curve will be the acceleration.

The slope of a velocity-versus-time curve is the acceleration.

Since the methods for studying velocity-versus-time curves are identical to the ones we used to analyze position-versus-time curves, I will not explain them all over again. Instead, study the next example and solve the "On Your Own" problems that follow to make sure you can analyze these graphs as well.

EXAMPLE 1.7

A race car's motion is given by the following graph:

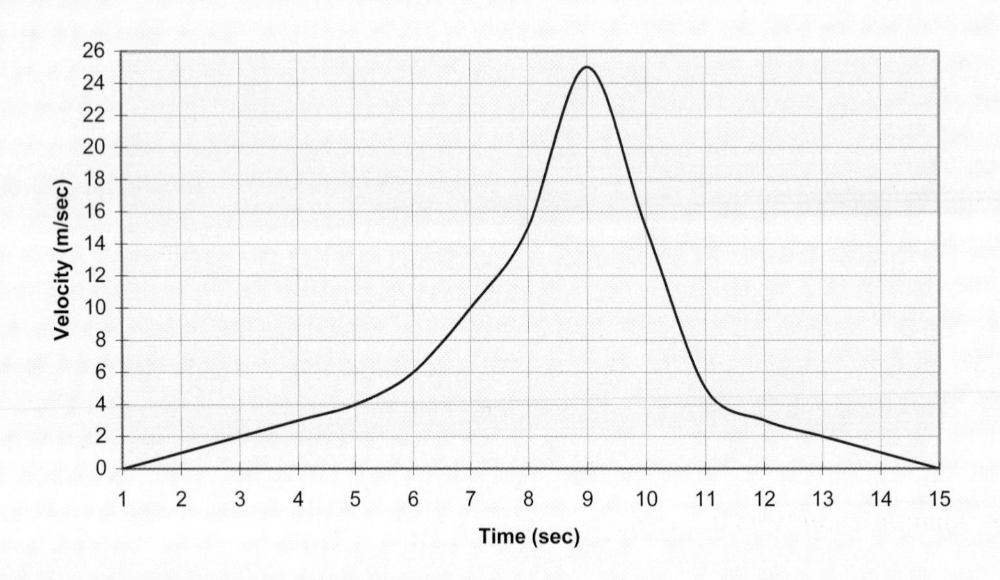

Over what time interval is the car speeding up?

The car speeds up when acceleration and velocity have the same sign. According to the graph, velocity is always positive. This means that in order to be speeding up, the acceleration must also be positive. Thus, the car is speeding up when the curve is rising. This occurs during the time interval of 1.0 to 9.0 seconds. The car is slowing down from 9.0 to 15.0 seconds.

When is the car's acceleration zero?

The slope of a curve is zero when the curve is flat. This happens briefly at 9.0 seconds.

What is the instantaneous acceleration of the car at 3.0 seconds?

The curve looks like a straight line from 1.0 to 4.0 seconds. Thus, the slope of the curve at any point during that time interval is the same as the average slope. At 1.0 second, the velocity is 0.0 m/sec. At 4.0 seconds, the velocity is 3.0. The average slope, then, is:

$$\text{slope} = \frac{\text{rise}}{\text{run}} = \frac{3.0\,\frac{\text{m}}{\text{sec}} - 0.0\,\frac{\text{m}}{\text{sec}}}{4.0\text{ sec} - 1.0\text{ sec}} = 1.0\,\frac{\text{m}}{\text{sec}^2}$$

This slope is the same throughout that entire time interval, so at 3.0 seconds, the acceleration is 1.0 m/sec^2.

ON YOUR OWN

Consider an object whose motion is described by the following graph:

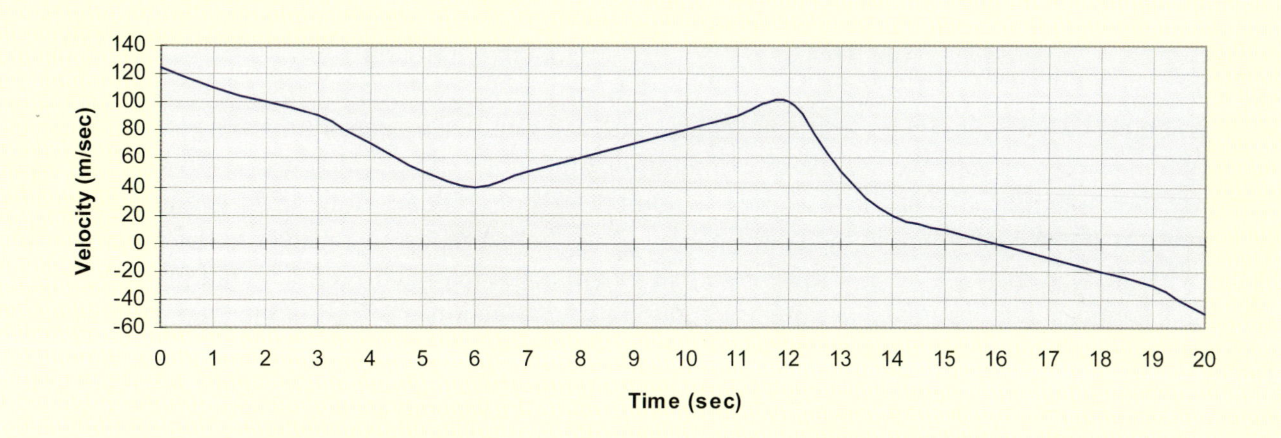

1.11 During what time intervals is the object's speed increasing?

1.12 When is the object's acceleration zero?

Before we finish this module, I need to make two points. First, there is one special property of a velocity-versus-time curve. The area under such a curve represents the object's displacement. Thus, if I could take the velocity-versus-time curve above and somehow calculate how much area exists under the line, I would be able to determine the final displacement of the object. Now, of course, you have no way of doing this, so you don't have to worry. I won't ask you any questions about this. It turns out, however, that the mathematical field of calculus is devoted to two things: calculating the slope of curves and *the area under curves*. Thus, when you learn calculus, you will learn another way to analyze these graphs.

The last point I need to make is rather important. If you solved "On Your Own" problem 1.12 correctly, you found that there were two times that the object had zero acceleration: approximately 6.0 seconds and 11.8 seconds. What were the object's velocities at those two times? They were 40 m/sec and 102 m/sec, respectively. Note that *although the acceleration was zero at these times, the velocity was not*. This is an important point and cannot be overemphasized. It is very tempting to say that velocity is zero when acceleration is zero. Although that is indeed possible, it is *not necessarily true*.

The converse of this statement is just as true and just as important. In the "On Your Own" section above, what was the velocity of the object at 16 seconds? It was zero. Was the acceleration zero? No, it was negative. We see, then, that acceleration does not have to be zero when the velocity is zero. Acceleration is the *change* in velocity. Thus, it is very possible for one to be zero and the other to be non-zero.

If velocity is zero, acceleration does not have to be zero.
If acceleration is zero, velocity does not have to be zero.

Plant this fact in your head, or you will be really lost in the next module!

ANSWERS TO THE "ON YOUR OWN" PROBLEMS

1.1 The total distance is easy to calculate. The ant crawled 15.2 centimeters in one direction and 3.8 centimeters in the other. The total distance then, is simply:

$$\text{Total Distance} = 15.2 \text{ cm} + 3.8 \text{ cm} = 19.0 \text{ cm}$$

Now remember, we have to take significant figures into account when determining the answer. Since we are adding two numbers, we use the rule of addition and subtraction, which tells us to report our answer to the same precision as the least precise number in the problem. Both 15.2 cm and 3.8 cm have their last significant figure in the tenths place. Thus, I must report my answer to the tenths place. That's why the answer is 19.0 cm. Please note that 19 cm is not really correct. It is not precise enough. The measurements given are precise enough for us to report the digit in the tenths place, even if it happens to be zero. In the same way, 19.00 cm would also not be correct, as it is too precise for the measurements given.

Calculating the displacement is a bit more difficult. To do this, we must first define direction. I will say that motion from the anthill to the bread results in positive displacement while motion from the bread to the anthill results in negative displacement. Thus, the ant first had a displacement of +15.2 cm and then a displacement of -3.8 cm. The total displacement, then, is:

$$\text{Total Displacement} = 15.2 \text{ cm} + -3.8 \text{ cm} = 11.4 \text{ cm}$$

Once again, since both of the measured distances have their last significant figure in the tenths place, the answer must be reported to the tenths place. This is a positive displacement, which means that the ant is 11.4 cm away from the anthill, in the direction of the bread.

Note that saying 11.4 cm isn't good enough. With the opposite definition of positive and negative displacement, another person would have gotten -11.4 cm. Both answers would be correct, depending on the definition of direction. Thus, we must give the answer in relation to the points in the problem, so that the answer is independent of our definition of positive and negative direction.

1.2 In this problem, we are asked to calculate velocity, so we will be using Equation (1.1). Once again, we are dealing with vector quantities here, so we must define direction. I will call motion down the street positive motion and motion up the street negative. The first part of the question asks us to calculate the mail carrier's velocity while she travels down the street. Well, during that time, her displacement was 3.00×10^2 meters. It took her 332 seconds to travel down the street, so Equation (1.1) becomes:

$$\mathbf{v} = \frac{\Delta \mathbf{x}}{\Delta t} = \frac{3.00 \times 10^2 \text{ m}}{332 \text{ sec}} = 0.904 \frac{\text{m}}{\text{sec}}$$

We could therefore say that her velocity was 0.904 m/sec down the street. The second part of the question asks us to calculate her velocity as she is traveling up the street. During that time, her displacement was negative, so Equation (1.1) becomes:

$$\mathbf{v} = \frac{\Delta \mathbf{x}}{\Delta t} = \frac{-208 \text{ m}}{2.30 \times 10^2 \text{ sec}} = -0.904 \frac{\text{m}}{\text{sec}}$$

Thus, we could say that her velocity was <u>0.904 m/sec up the street</u>. Finally, the problem asks us to determine her velocity over the entire trip. Well, in order to determine velocity, we must first determine displacement. The mail carrier's total displacement was 3.00×10^2 m + -208 m = 92 m. The total time it took to achieve that displacement was 332 sec + 2.30×10^2 sec = 562 sec. Equation (1.1), then, becomes:

$$\mathbf{v} = \frac{\Delta \mathbf{x}}{\Delta t} = \frac{92\text{ m}}{562\text{ sec}} = 0.16\,\frac{\text{m}}{\text{sec}}$$

Since the velocity is positive, we know that even though the mail carrier traveled in both directions, her overall velocity was <u>0.16 m/sec down the street</u>.

1.3 The problem gives us a speed and a direction. This means that the 15 m/sec is actually a velocity. In addition, we are told how far the boat travels (34.1 km). If we consider the place the boat started as our point of reference, then this distance is actually the displacement during the boat ride ($\Delta \mathbf{x}$). The problem, however, is that the units do not match. Velocity is in m/sec while displacement is in km. We need to gets these units into agreement, so we need to convert km into m:

$$\frac{34.1\,\cancel{\text{km}}}{1} \times \frac{1{,}000\text{ m}}{1\,\cancel{\text{km}}} = 3.41 \times 10^4\text{ m}$$

Now we can substitute into Equation (1.1), use algebra to rearrange the equation, and solve for time:

$$\mathbf{v} = \frac{\Delta \mathbf{x}}{\Delta t}$$

$$15\,\frac{\text{m}}{\text{sec}} = \frac{3.41 \times 10^4\text{ m}}{\Delta t}$$

$$\Delta t = \frac{3.41 \times 10^4\,\cancel{\text{m}}}{15\,\frac{\cancel{\text{m}}}{\text{sec}}} = 2.3 \times 10^3\text{ sec}$$

Thus, the boat ride took <u>2.3×10^3</u> seconds, or 38 minutes.

1.4 The slope of the curve is steeper at 3.5 seconds than at 8.5 seconds, so <u>the object is moving faster at 3.5 seconds</u>.

1.5 The slope changes from positive to negative at 4.3 seconds. This represents one direction change. The slope changes from negative to positive at 8.0 seconds, representing the second direction change. It changes from positive to negative at 9.0 seconds and then again from negative back to positive at about 10.2 seconds. These represent the third and fourth direction changes. Finally, at 11.0 seconds, the slope changes from positive to negative. This is the fifth (and last) direction change. Thus, <u>the object changed directions five times</u>.

1.6 During the interval of 0.0 to 2.0 seconds, the curve looks like a straight line. Thus, the slope at any point along that part of the curve is the same. We can therefore calculate the average velocity from 0.0 to 2.0 seconds and, since the velocity stays the same throughout that entire time interval, it will also be the instantaneous velocity at any time during that interval, including 1.0 seconds. To read from the graph, we have to realize that it is marked off in seconds and two-meter intervals, so by estimating in between the marks, we can report our positions and times to the tenths place.

The position at 0.0 seconds is 0.0 meters according to the graph. At 2.0 seconds, the displacement is 2.0 meters. The average velocity, then, is:

$$\text{slope} = \mathbf{v} = \frac{\text{rise}}{\text{run}} = \frac{2.0\text{ m} - 0.0\text{ m}}{2.0\text{ sec} - 0.0\text{ sec}} = 1.0\,\frac{\text{m}}{\text{sec}}$$

This means that the instantaneous velocity at 1.0 second is also 1.0 m/sec.

1.7 We have the velocities relative to the fisherman, so we can use them to determine the velocities of the raft and the boat relative to each other. We will say that motion up the river is positive and motion down the river is negative. Thus, the boat is traveling at 15 m/sec, and the raft is traveling at -3 m/sec.

To determine the velocity of an object relative to another, we take the velocity of the thing being observed and subtract from it the velocity of the observer. Therefore, a person on the boat observes the raft moving at -3 m/sec - 15 m/sec = -18 m/sec. Since negative means motion down the river, the people on the boat observe the raft moving 18 m/sec down the river. The boy on the raft, however, observes the boat moving at a velocity of 15 m/sec - (-3 m/sec) = 18 m/sec. Since positive means motion up the river, the boy observes the boat moving 18 m/sec up the river. Please note that I could have defined motion up the river as negative and motion down the river as positive. If I did that, the answers would end up with different signs, but once I translated the signs into "up the river" and "down the river," the answers would end up the same.

1.8 This problem is a straightforward application of Equation (1.3). The problem says that the sprinter starts from rest ($\mathbf{v} = 0$) and sprints to a velocity of 16 m/sec. Thus, we can subtract the initial velocity from the final velocity to get $\Delta\mathbf{v}$:

$$\Delta\mathbf{v} = \mathbf{v}_{\text{final}} - \mathbf{v}_{\text{initial}} = 16\text{ m/sec} - 0\text{ m/sec} = 16\text{ m/sec}$$

The problem also gives us time, so to calculate the acceleration, all we have to do is plug these numbers into Equation (1.3):

$$\mathbf{a} = \frac{\Delta\mathbf{v}}{\Delta t}$$

$$\mathbf{a} = \frac{16\,\frac{\text{m}}{\text{sec}}}{3.4\text{ sec}} = 4.7\,\frac{\text{m}}{\text{sec}^2}$$

Since acceleration and velocity have the same signs, we know that they are pointed in the same direction. Therefore, the sprinter was speeding up, and the sprinter's acceleration was 4.7 m/sec^2 east.

1.9 This problem tells us acceleration, time, and final velocity and asks us to calculate the initial velocity. We can do this by calculating $\Delta\mathbf{v}$, using Equation (1.3):

$$\mathbf{a} = \frac{\Delta\mathbf{v}}{\Delta t}$$

$$-7.2\ \frac{\text{m}}{\text{sec}^2} = \frac{\Delta\mathbf{v}}{3.1\ \text{sec}}$$

$$\Delta\mathbf{v} = -7.2\ \frac{\text{m}}{\text{sec}^{\cancel{2}}} \times 3.1\ \cancel{\text{sec}} = -22\ \frac{\text{m}}{\text{sec}}$$

Now that we have $\Delta\mathbf{v}$, we can use the definition of $\Delta\mathbf{v}$ to solve for the initial velocity:

$$\Delta\mathbf{v} = \mathbf{v}_{\text{final}} - \mathbf{v}_{\text{initial}}$$

$$-22\ \text{m/sec} = 0\ \text{m/sec} - \mathbf{v}_{\text{initial}}$$

$$\mathbf{v}_{\text{initial}} = 22\ \text{m/sec}$$

Thus, the car was originally traveling at 22 m/sec. Notice that the velocity and acceleration have different signs. They should, since the car slowed down!

1.10 To test the assumption, put another piece of tape at 0.500 meters from the edge of the board. Then, measure the average velocity of the ball as it travels from the end of the board to the first piece of tape. Next, measure the average velocity of the ball as it travels from the first piece of tape to the second piece of tape. Compare the two velocities. If the assumption is good, the velocities should be roughly equal, indicating that the ball did not slow down significantly between the first half of its trip and the second half of its trip. However, if the second velocity is significantly lower than the first, then the ball *did* slow down considerably over the course of 1 meter, and the assumption was not valid.

1.11 Be very careful solving this one. Remember, the object will speed up whenever velocity and acceleration have the same signs. From the time interval of 6.0 seconds to 11.8 seconds, the velocity is positive and the acceleration (slope) is positive. Thus, the object speeds up in that interval. You might be tempted to say that this is the only interval in which the car speeds up, but you would be wrong. From 16.0 seconds to 20.0 seconds, the velocity and acceleration are both negative. Thus, the object is speeding up then as well. Therefore, there are two time intervals during which the object speeds up, 6.0 - 11.8 seconds and 16.0 - 20.0 seconds. (Your numbers may be slightly different from mine, since you are reading from a graph. That's fine.)

1.12 The acceleration is zero wherever the curve is flat. That happens at about 6.0 seconds and 11.8 seconds. (Your numbers may be slightly different from mine, since you are reading from a graph. That's fine.)

REVIEW QUESTIONS

1. What is the main difference between a scalar quantity and a vector quantity?

2. On a physics test, the first question asks the students to calculate the acceleration of an object under certain conditions. Two students answer this question with the same number, but the first student's answer is positive while the second student's answer is negative. The teacher says that they both got the problem 100% correct. How is this possible?

3. Which is a vector quantity: speed or velocity?

4. What is the main difference between instantaneous and average velocity?

5. What physical quantity is represented by the slope of a position-versus-time graph?

6. What do physicists mean when they say that velocity is "relative?"

7. You are reading through someone else's laboratory notebook, and you notice a number written down: 12.3 m/sec^2. Even though it is not labeled, you should immediately be able to tell what physical quantity the experimenter measured. What is it?

8. Another experiment in the same laboratory notebook says that an object has a 1.4 m/sec^2 acceleration when it has a -12.6 m/sec velocity. At that instant in time, is the object speeding up or slowing down?

9. What kinds of graphs do you study if you are interested in learning about acceleration?

10. An object's velocity is zero. Does this mean its acceleration is zero? Why or why not?

PRACTICE PROBLEMS

1. A delivery truck travels down a straight highway for 35.4 km to make a delivery. On the way back, the truck has engine trouble, and the driver is forced to stop and pull off the road after traveling only 13.2 km back towards its place of business. How much distance did the driver cover? What is his final displacement?

2. If the driver in the above problem took 21.1 minutes to reach the delivery point and broke down 7.5 minutes into the return trip, what was the average speed? What was the driver's average velocity?

3. A plane flies straight for 672.1 km and then turns around and heads back. The plane then lands at an airport that is only 321.9 km away from where the pilot turned around. If the plane's average velocity over the entire trip was 42 m/sec, how much time did the entire trip take?

4. An athlete runs 1600.0 meters down a straight road. Over the first 800.0 meters, the runner's average velocity is 6.50 m/sec. Over the remaining 800.0 meters, his average velocity is 4.30 m/sec. What is the runner's average velocity over the entire race? [Be careful on this one. Remember what Equation (1.1) tells you.]

Questions 5 and 6 refer to the figure below:

A car's motion is described by the following position-versus-time curve:

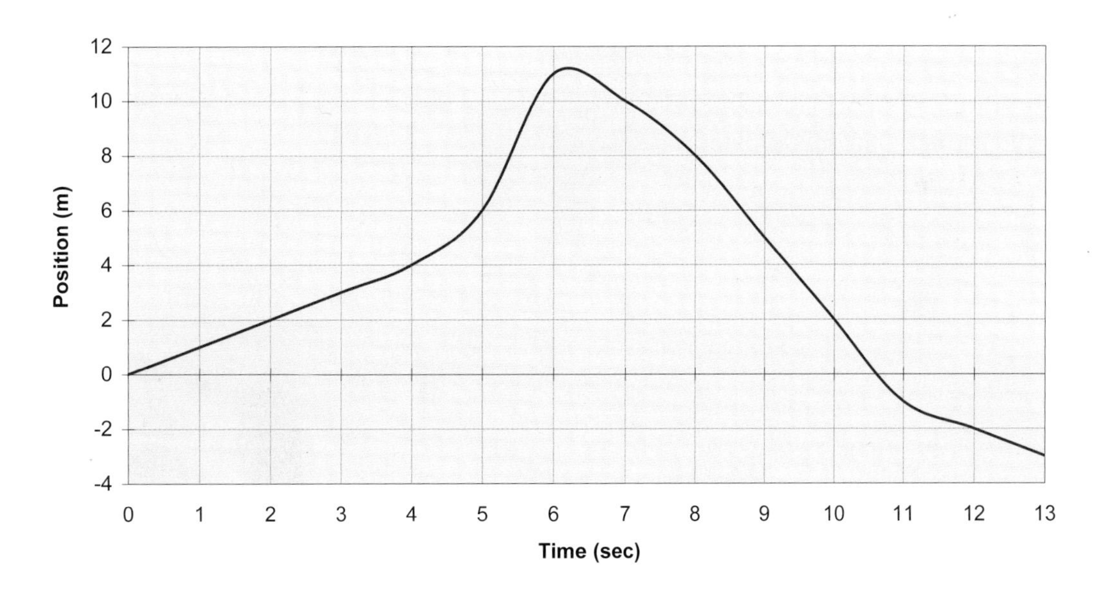

5. At approximately what time does the car change its direction?

6. Over what time interval is the car moving the fastest?

7. A train is traveling with an initial velocity of 20.1 m/sec. If the brakes can apply a maximum acceleration of -0.0500 m/sec^2, how long will it take the train to stop?

Questions 8 - 10 refer to the figure below:

A runner's motion is described by the following velocity-versus-time graph:

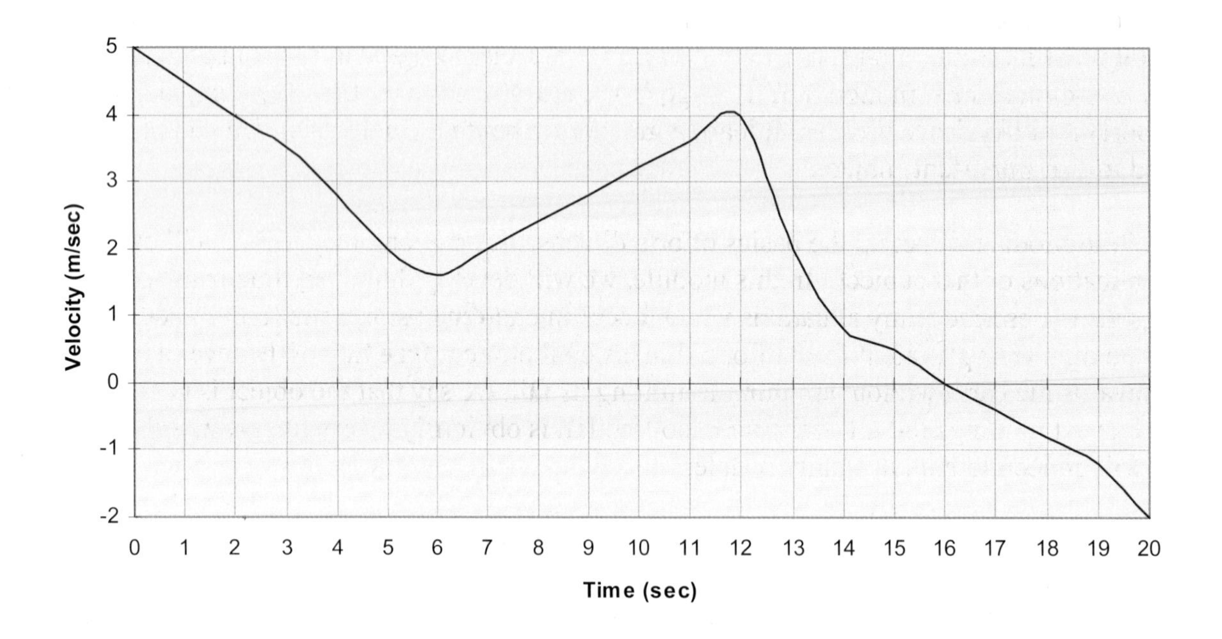

8. Over what time intervals is the runner slowing down?

9. What is the runner's acceleration at 6.0 seconds?

10. What is the runner's acceleration at 1.0 seconds?

MODULE #2: One-Dimensional Motion Equations and Free Fall

Introduction

Are you sick and tired of one-dimensional motion yet? I hope not, because we have another entire module of it ahead! You see, while Module #1 laid a good foundation for understanding the basics of one-dimensional motion, it did not go into enough depth or detail. One-dimensional motion is so important in physics that we must learn a great deal about it. As a result, this module is also dedicated to that important subject.

While Module #1 set up the basics of one-dimensional motion, this module will explore the many applications of that subject. In this module, we will develop three very important equations that will allow us to describe many situations which occur in one-dimensional motion. In addition, we will concentrate on a very prevalent form of one-dimensional motion: **free fall**. Whenever any object is falling towards the earth without anything inhibiting its fall, we say that the object is in free fall. If you think about it, things fall all of the time, so free fall is obviously a very important subject. That's why we will concentrate on it in this module.

Relating Velocity, Acceleration, and Time

In order to be able to solve any type of one-dimensional motion problem, we will have to learn three basic equations. The first of those three is based on Equation (1.3):

$$\mathbf{a} = \frac{\Delta \mathbf{v}}{\Delta t} \tag{1.3}$$

As we learned in the previous module, "**a**" is the acceleration, "$\Delta\mathbf{v}$" is the change in velocity, and "Δt" is the change in time. Now remember, $\Delta\mathbf{v}$ can be calculated by subtracting the initial velocity from the final velocity, and Δt can be calculated in a similar manner:

$$\Delta \mathbf{v} = \mathbf{v}_{final} - \mathbf{v}_{initial} \tag{2.1}$$

$$\Delta t = t_{final} - t_{initial} \tag{2.2}$$

Let's take Equations (2.1) and (2.2) and substitute them into Equation (1.3). When we do that, the equation looks a little different:

$$\mathbf{a} = \frac{\mathbf{v}_{final} - \mathbf{v}_{initial}}{t_{final} - t_{initial}} \tag{2.3}$$

Now we're going to make a simplifying assumption. Let's assume that whenever a situation begins, $t_{initial} = 0$. That way, the equation is a bit simpler:

$$\mathbf{a} = \frac{\mathbf{v}_{final} - \mathbf{v}_{initial}}{t_{final}} \tag{2.4}$$

Next, let's rearrange the equation using algebra:

$$\mathbf{v}_{\text{final}} = \mathbf{v}_{\text{initial}} + \mathbf{a}t_{\text{final}} \tag{2.5}$$

As a final adjustment to the equation, I am going to rename a couple of variables. Since most problems want to know things about the final state of a situation, I am going to rename $\mathbf{v}_{\text{final}}$ to just $\mathbf{v}$ and t_{final} to just t. They mean the exact same thing; I am just making them easier to write. Finally, I will also make $\mathbf{v}_{\text{initial}}$ easier to write by calling it $\mathbf{v}_o$ (usually called "v naught"). With these new symbols, the equation becomes:

$$\mathbf{v} = \mathbf{v}_o + \mathbf{a}t \tag{2.6}$$

Equation (2.6) is the first of our three new one-dimensional motion equations, and it is very important. That's why I have put it in a box. What does it mean? It means that I can figure out the final velocity (**v**) of an object by taking its initial velocity ($\mathbf{v}_o$) and adding the acceleration (**a**) times the time (t). You will use this equation over and over in this module and the ones to come, so you should commit it to memory now!

As a brief aside, everything that you've read from the beginning of this section to the end of the previous paragraph is called a "derivation." In the science of physics, we do this quite a bit. In general, a derivation starts with something that you already know [Equation (1.3) in this case] and uses mathematics and reasoning to come up with something brand new [Equation (2.6) in this case]. Although I presented the derivation of Equation (2.6) as a means of helping you understand where the equation comes from, there is no need for you to be able to reproduce such a derivation on your own. Thus, although you should understand the reasoning I just went through, the only thing you really need to remember is Equation (2.6) and how to use it.

This will be the general philosophy for this course when it comes to derivations. If I present a derivation, you should try to understand it, but you will not be expected to reproduce it or any of the equations in it. The only thing you will be expected to really remember is the final result and how to use it. So don't worry if a few of my derivations go over your head. Just be sure you can use the final equation that is produced. Now that we have that out of the way, let's see how we can use Equation (2.6).

EXAMPLE 2.1

A car slams on the brakes to avoid hitting an obstacle in the road. If the car has a maximum deceleration of 2.3 m/sec^2 and it takes 15 seconds for the car to stop, how fast was the car initially traveling?

This problem tells us that the deceleration of the car is 2.3 m/sec^2. The word "deceleration" just means that the car is slowing down. In the previous module, we learned that when a car is slowing down, the acceleration and velocity have opposite signs. Thus, if we define the direction that the car is moving as positive, this tells us that acceleration is -2.3 m/sec^2. The final velocity must be zero, because the car is stopped after everything is said and done. The problem also gives us time and asks us to calculate the initial velocity. Well, Equation (2.6) relates these three quantities, so we will use it:

$$\mathbf{v} = \mathbf{v}_o + \mathbf{at}$$

$$0 = \mathbf{v}_o + (-2.3\frac{m}{sec^2})\cdot(15\ sec)$$

Using algebra, we can rearrange the equation to solve for $\mathbf{v}_o$:

$$\mathbf{v}_o = (2.3\frac{m}{sec^{\cancel{2}}})\cdot(15\ \cancel{sec})$$

$$\mathbf{v}_o = 35\frac{m}{sec}$$

Therefore, the car must have been traveling with an initial velocity of 35 m/sec.

This example should have looked familiar to you. We did a problem almost identical to this one in the previous module, and we used Equation (1.3) to solve it. Well, Equation (2.6) is based on Equation (1.3), so you should not be surprised that we can solve the same types of problems with it. If that's the case, why bother with Equation (2.6)? Why not just use Equation (1.3)? The main reason is that Equation (2.6) goes well with the next two equations that we are going to learn. Also, as we will see in a while, Equation (1.3) is a definition and is therefore a reliable equation under any conditions. Equation (2.6), however, is only valid when the acceleration remains constant. We will discuss this in detail a bit later. For now, make sure you can use Equation (2.6) by solving the following "On Your Own" problem.

Illustrations from the MasterClips collection

ON YOUR OWN

2.1 A car traveling at 35 m/sec (about 78 mph) passes a police car which is moving at 23 m/sec (about 51 mph). The policeman does not have a radar gun, so he immediately starts to speed up in order to match the speeder's velocity. That way, he will be able to measure how fast the speeder is moving. If the police car can accelerate at a rate of 2.5 m/sec^2, how long will it take for the policeman to match the speeder's velocity?

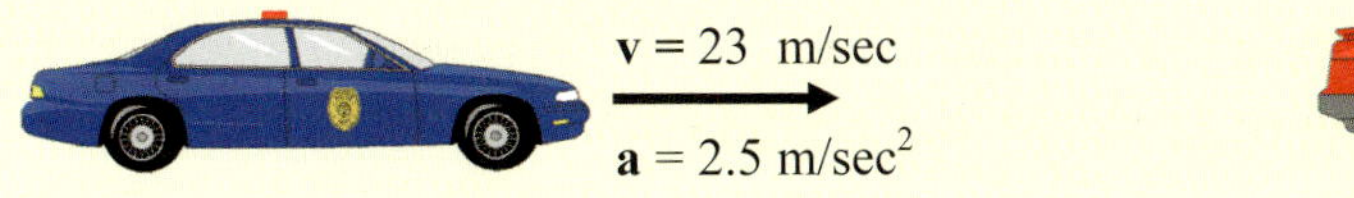

Relating Velocity, Acceleration, and Displacement

Although Equation (2.6) is very useful, we need to be able to relate velocity and acceleration to something other than time. Since displacement is one of the fundamental quantities that we can measure with regard to motion, it is only natural that we try to find some way of relating velocity and acceleration to displacement. In order to do this, we need to blend a few equations together. We start with our definition of average velocity, Equation (1.1):

$$\mathbf{v} = \frac{\Delta \mathbf{x}}{\Delta t} \tag{1.1}$$

Using the definition of Δt, we can expand this equation to read:

$$\mathbf{v} = \frac{\Delta \mathbf{x}}{t_{final} - t_{initial}} \tag{2.7}$$

Now, just as we did in the previous derivation, we will assume that our timing clock starts whenever the problem starts, which allows us to say that the initial time is zero. We will also say that since the final value of time is the only interesting value to us, we will simply call it "t." Equation (2.7) then becomes:

$$\mathbf{v} = \frac{\Delta \mathbf{x}}{t} \tag{2.8}$$

Now we need to bring in another equation. Remember that according to our discussion in Module #1, the velocity calculated with Equation (1.1) is the *average* velocity. Equation (1.1) can only give us the instantaneous velocity if Δt is infinitesimally small. Since we cannot really measure an infinitesimally small Δt, and since Equation (2.8) comes from Equation (1.1), Equation (2.8) also gives us the *average* velocity. Well, if you think about it, there is another way that we can calculate the average velocity. If the object we are studying starts off with an initial velocity of $\mathbf{v}_o$ and ends up with a final velocity of **v**, then, according to the definition of average, we could calculate the average velocity this way:

$$\mathbf{v}_{avg} = \frac{\mathbf{v} + \mathbf{v}_o}{2} \tag{2.9}$$

Since the velocity calculated with Equation (2.9) is the same as the velocity calculated with Equation (2.8), we can use the right side of Equation (2.8) and substitute it in for the left side of Equation (2.9):

$$\frac{\Delta \mathbf{x}}{t} = \frac{\mathbf{v} + \mathbf{v}_o}{2} \tag{2.10}$$

Now we will rearrange this equation to solve for t:

$$t = \frac{2 \cdot \Delta \mathbf{x}}{\mathbf{v} + \mathbf{v}_o} \tag{2.11}$$

Believe it or not, we're still not done yet. We have one more equation to use: the definition of acceleration:

$$\mathbf{a} = \frac{\Delta \mathbf{v}}{\Delta t} \tag{1.3}$$

Using the same assumptions that we did the last time we used this equation, we get:

$$\mathbf{a} = \frac{\mathbf{v} - \mathbf{v}_o}{t} \qquad (2.12)$$

Now we'll take this equation and solve for t:

$$t = \frac{\mathbf{v} - \mathbf{v}_o}{\mathbf{a}} \qquad (2.13)$$

Look what we have now. Equation (2.11) and Equation (2.13) both give different expressions for the same thing: time. Thus, we can take the right side of Equation (2.11) and substitute it for the left side of Equation (2.13):

$$\frac{2 \cdot \Delta\mathbf{x}}{\mathbf{v} + \mathbf{v}_o} = \frac{\mathbf{v} - \mathbf{v}_o}{\mathbf{a}} \qquad (2.14)$$

Using algebra to rearrange the equation gives us:

$$\mathbf{v}^2 = \mathbf{v}_o^2 + 2\mathbf{a} \cdot \Delta\mathbf{x} \qquad (2.15)$$

Just as I said after the last derivation, it's not all that important for you to have followed each and every step of what you just read. Indeed, as far as you could tell, there may have been no reason for some of the mathematical manipulations that I went through. That's fine. The real point here is that I took three equations [Equation (1.1), Equation (1.3), and Equation (2.9)] and blended them together to come up with a brand new equation that relates four variables which we had never related before: initial velocity, final velocity, acceleration, and displacement. In the end, then, Equation (2.15) is the only one that you really need to know and understand.

Well, now that we've derived this equation, what good is it? Study the following examples and you'll find out!

EXAMPLE 2.2

Van illustration from the MasterClips collection
Deer illustration copyright GifArt.com

A woman is driving down the road at 20.0 m/sec (about 45 mph) when she suddenly sees a deer in her headlights. The moment that she sees the deer, she slams on the brakes, which have a maximum deceleration capacity of 2.1 m/sec^2. How far will she travel before she stops?

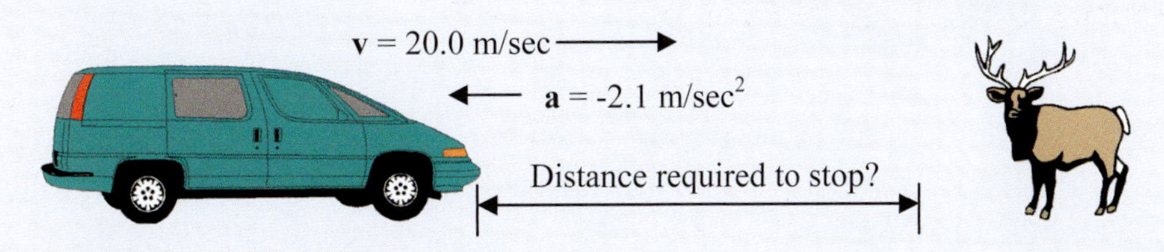

As I have said before, the trick to solving physics problems is to determine *everything* that the problem gives you, as well as what the problem wants you to figure out. Then, you need to find an equation (or a series of equations) that relates these things. In this problem, we are given three things: the van's initial velocity (20.0 m/sec), the van's acceleration (-2.1 m/sec^2: remember, deceleration means the acceleration opposes the velocity), and the fact that the van is going to stop (which means that the final velocity is zero).

Now that we know what we are given, what do we need to calculate? The problem asks us to determine how far the woman travels before she stops. In other words, we want to know the displacement from the point at which she sees the deer to the point at which the van stops. Thus, we are given initial velocity ($\mathbf{v_o}$), final velocity ($\mathbf{v}$), and acceleration ($\mathbf{a}$); and we are asked to calculate displacement ($\Delta\mathbf{x}$). These three things are related by Equation (2.15), so we can use it to solve the problem:

$$\mathbf{v}^2 = \mathbf{v}_o^2 + 2\mathbf{a}\cdot\Delta\mathbf{x}$$

$$0^2 = (20.0\ \frac{\text{m}}{\text{sec}})^2 + 2\cdot(-2.1\frac{\text{m}}{\text{sec}^2})\cdot(\Delta\mathbf{x})$$

We can now rearrange this equation using algebra to get the answer:

$$\Delta\mathbf{x} = \frac{-(20.0\ \frac{\text{m}}{\text{sec}})^2}{2(-2.1\frac{\text{m}}{\text{sec}^2})} = \frac{-4.00\times10^2\ \frac{\text{m}^{\cancel{2}}}{\cancel{\text{sec}^2}}}{-4.2\ \frac{\cancel{\text{m}}}{\cancel{\text{sec}^2}}} = 95\ \text{m}$$

It therefore takes the driver 95 m to stop the van. What we have just calculated is called the "stopping distance" for that van at that velocity under the conditions specified. This is a very important thing to be able to calculate, because a driver needs to know how far his or her van will travel before he or she can get it stopped.

Athletes often use a radar gun to help them determine their acceleration. From a dead stop, they run as fast as they can. After a certain distance, the radar gun measures their velocity. This allows the trainer to determine the acceleration of the athlete. A sprinter starts from rest and runs 10.0 yards. If the radar gun says that his final velocity was 26 ft/sec, what was the runner's acceleration?

Illustrations copyright GifArt.com

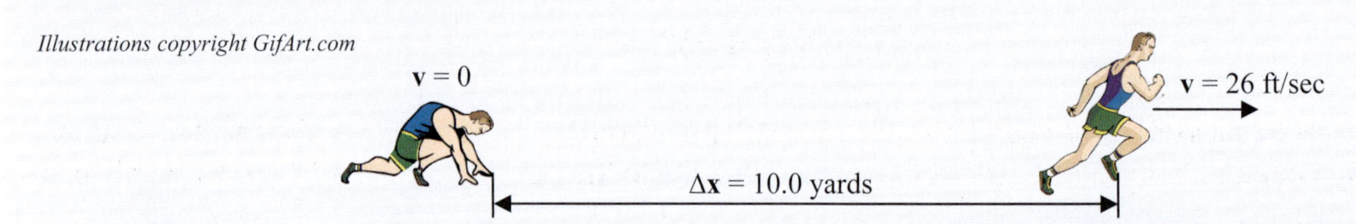

Once again, we need to know what we are given and what we are calculating. We are told that the runner is starting from rest. This means the initial velocity is zero. We are also told the final velocity (26 ft/sec) and the displacement (10.0 yards). We are asked to calculate the acceleration. Since Equation (2.15) relates these three quantities, we can use it to solve the problem. We first have to worry about units, however.

Now don't be alarmed that these aren't metric units. I will throw odd units at you every once in a while just to keep you on your toes. The unit ft/sec is an acceptable velocity unit, because it contains a displacement unit divided by a time unit. Yards are also acceptable units for displacement, but we can't use them in this problem. The displacement unit used in our velocity is feet. Thus, we either need to change feet to yards or yards to feet in order to keep our units consistent. I will choose the latter:

$$\frac{10.0\ \cancel{\text{yds}}}{1} \times \frac{3\ \text{ft}}{1\ \cancel{\text{yd}}} = 30.0\ \text{ft}$$

Now that our units are consistent, we can put our numbers into Equation (2.15):

$$\mathbf{v}^2 = \mathbf{v}_o^2 + 2\mathbf{a} \cdot \Delta\mathbf{x}$$

$$(26\ \frac{\text{ft}}{\text{sec}})^2 = 0^2 + 2 \cdot (\mathbf{a}) \cdot (30.0\ \text{ft})$$

Now we use algebra to rearrange the equation and come up with the answer :

$$\mathbf{a} = \frac{(26\ \frac{\text{ft}}{\text{sec}})^2}{2 \cdot (30.0\ \text{ft})} = \frac{676\ \frac{\text{ft}^{\cancel{2}}}{\text{sec}^2}}{60.0\ \cancel{\text{ft}}} = 11\ \frac{\text{ft}}{\text{sec}^2}$$

The athlete, then, can accelerate at 11 ft/sec^2. Once again, do not be concerned that the units are not metric. The unit ft/sec^2 is an acceptable unit for acceleration because it contains a displacement unit divided by the square of a time unit. Also, do not be concerned that I did not follow the rules of significant figures when I squared 26 ft/sec. I just wanted to illustrate the step of squaring the velocity so that you would see how the units worked out. I went ahead and kept all of the digits when I squared 26, because you should typically wait until the end of an equation to worry about significant figures. Thus, once I was completely done with the equation, I realized that the two significant figures in 26 ft/sec limited my answer to two significant figures. However, while plugging through the math to get to the final answer, I kept all of the digits.

Make sure that you can use Equation (2.15) by solving the following "On Your Own" problem.

ON YOUR OWN

2.2 A car has a maximum acceleration of 1.15×10^4 miles/hour^2. If it starts from rest and accelerates as quickly as possible for 5.00×10^2 yards, what will its velocity be? (There are 1.760×10^3 yards in a mile).

Relating Displacement, Velocity, Acceleration, and Time

We now come to the last equation we need in order to study one-dimensional motion. In order to derive this equation, however, I must go back to something I told you in Module #1. Back when we were studying velocity-versus-time graphs, I told you that the area under a velocity-versus-time curve is equal to the displacement caused by the motion. For example, look at the following velocity-versus time-graph:

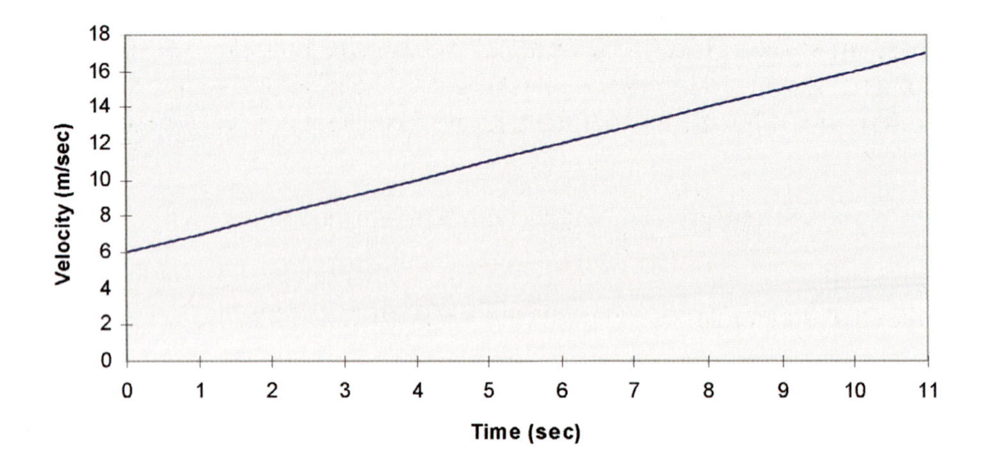

As I said in Module #1, the area under the line will be equal to the displacement caused by the motion depicted in the graph. Well, how in the world can we determine the area under that line? First, let's split the space under the line into two components, a rectangular component and a triangular one:

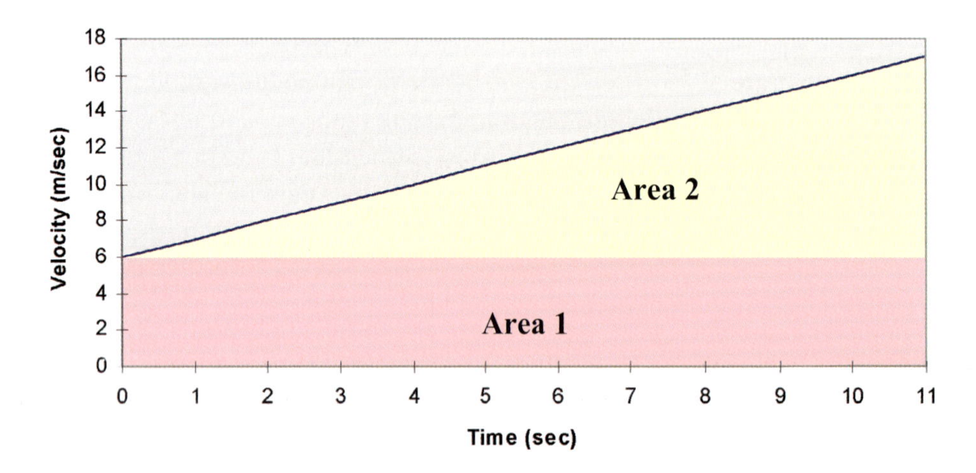

If we add the area in the pink rectangular region (Area 1) to the yellow triangular region (Area 2), we would have the total area under the line in the graph, which is equal to the displacement caused by the motion. Thus, if we can determine the area of the rectangle and add it to the area of the triangle, we can get the total displacement caused by the motion depicted in the graph.

Actually, it's not hard to calculate Area 1 and Area 2. After all, Area 1 is just a rectangle. The area of a rectangle is equal to the rectangle's length times its width. What's the length of the pink rectangle? According to the x-axis, it's 11.0 seconds. What's the width? According to the y-axis, it's 6.0 m/s. Thus, the area of the pink rectangle is:

$$\text{Area 1} = (11.0\ \cancel{\text{sec}})\cdot(6.0\ \frac{\text{m}}{\cancel{\text{sec}}}) = 66\ \text{m}$$

Notice how the units work out. When seconds are multiplied by meters per seconds, the seconds drop out, leaving meters, which is the unit for displacement.

Well, we're halfway home. Now all we have to do is determine the area of the yellow triangular region. Once again, this is not too hard, because the area of a triangle is one-half the length of the triangle's base times the height of the triangle. What's the length of this triangle's base? According to the x-axis, it's 11.0 sec. What's the height? The triangle starts when y = 6.0 m/sec and stops when y = 16.0 m/sec. This means that the triangle has a height of 10.0 m/sec. Thus, the area is:

$$\text{Area 2} = \frac{1}{2}\cdot(11.0\ \cancel{\text{sec}})\cdot(10.0\ \frac{\text{m}}{\cancel{\text{sec}}}) = 55.0\ \text{m}$$

In the end, then, the total area under the line is Area 1 + Area 2, or 66 m + 55.0 m = 121 m.

So we have learned from this exercise that for *this particular graph*, the total displacement is 121 meters. It would be nice, however, to be able to determine the displacement for *any* situation, not just the one depicted in the graph we studied. It turns out that we can do this. Let's look at the graph again, but this time, we'll be a little more general:

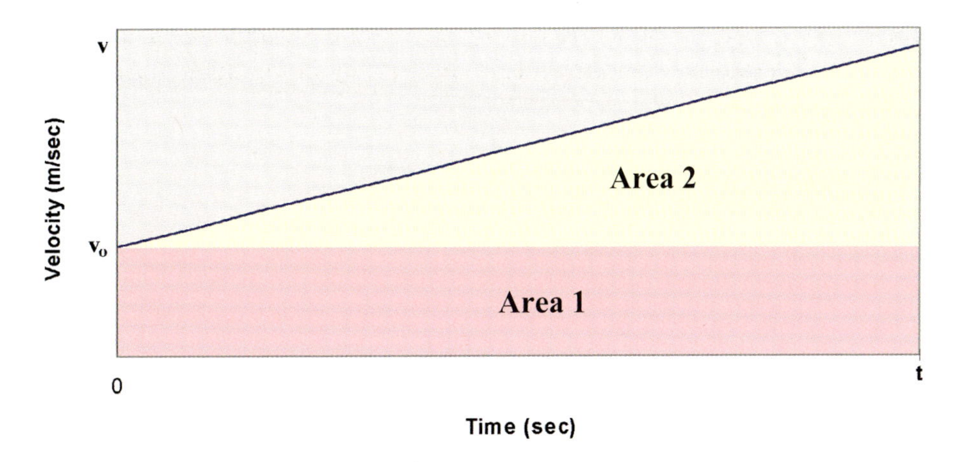

Notice how this graph can now be applied to many other situations. Rather than having the actual time span on the graph, we now just say that the motion occurred from time zero to a time of "t." In the same way, the velocity at time = 0 is just called the initial velocity ($\mathbf{v}_o$) and the velocity at the end of the motion is just called the final velocity (**v**). Now we can do the same thing we did before.

The area of the pink region is the length of the rectangle times its width. According to the x-axis, the motion occurred from x = 0 to x = t. Thus, the length of the rectangle is t. According to the y-axis, the rectangle starts at y = 0 and ends at y = v_o. Thus, the width is v_o. This makes the area:

$$\text{Area 1} = \mathbf{v}_o \cdot t \tag{2.16}$$

Now all we need to do is to determine the area of the triangle. According to the x - axis, the triangle starts at x = 0 and ends at x = t. This makes the length of the base equal to t. According to the y-axis, the triangle starts at y= $\mathbf{v}_o$ and ends at y = **v**. The height of the triangle, then, is **v** - $\mathbf{v}_o$, which we can also call Δ**v**. This makes the area:

$$\text{Area 2} = \frac{1}{2}\cdot\Delta\mathbf{v}\cdot t \tag{2.17}$$

Before we add these two areas together to get the total area under the line, I want to manipulate Equation (2.17) a bit. Notice that Equation (2.17) has $\Delta\mathbf{v}$ in it. We know another equation that contains $\Delta\mathbf{v}$:

$$\mathbf{a} = \frac{\Delta\mathbf{v}}{\Delta t} \tag{1.3}$$

Let's rearrange this equation to solve for $\Delta\mathbf{v}$:

$$\Delta\mathbf{v} = \mathbf{a} \cdot \Delta t \tag{1.3a}$$

Also, remember that since we defined the starting time as 0 and the ending time as t, we can replace Δt with t:

$$\Delta\mathbf{v} = \mathbf{a} \cdot t \tag{1.3b}$$

We will use this expression for $\Delta\mathbf{v}$ in Equation (2.17):

$$\text{Area } 2 = \frac{1}{2} \cdot \mathbf{a} \cdot t \cdot t$$

$$\text{Area } 2 = \frac{1}{2} \cdot \mathbf{a} \cdot t^2 \tag{2.18}$$

Now all we have to do is add these two areas together. This will determine the total area under the line, which will also be equal to the displacement caused by the motion:

$$\Delta\mathbf{x} = \mathbf{v}_0 t + \frac{1}{2}\mathbf{a}t^2 \tag{2.19}$$

Once again, if you were lost throughout the course of this derivation, don't let it bother you. Just realize that using the fact that the area under a velocity-versus-time curve is the displacement, we derived a general equation that relates the displacement to the initial velocity, the acceleration, and the time. This equation and its use are very important. Study the example and solve the "On Your Own" problem that follows to make sure you understand Equation (2.19) and how to use it.

EXAMPLE 2.3

A plane starts from rest and begins to accelerate down the runway at 2.3 m/sec^2. If it takes off in 28 seconds, how far did it travel down the runway?

This is a straightforward application of Equation (2.19). The problem gives us the initial velocity (0), the time (28 sec) and the acceleration (2.3 m/sec^2). It asks us to calculate the displacement down the runway. Equation (2.19) relates all of these quantities:

$$\mathbf{\Delta x} = \mathbf{v}_o t + \frac{1}{2}\mathbf{a}t^2$$

$$\mathbf{\Delta x} = (0) \cdot 28\,\text{sec} + \frac{1}{2} \cdot (2.3\,\frac{\text{m}}{\text{sec}^2}) \cdot (28\,\text{sec})^2 = \frac{1}{2} \cdot (2.3\,\frac{\text{m}}{\cancel{\text{sec}^2}}) \cdot (784\,\cancel{\text{sec}^2})$$

$$\mathbf{\Delta x} = \underline{9.0 \times 10^2\ \text{m}}$$

Notice that once again I did not worry about significant figures until I got to the end of the equation. Since 2.3 and 28 each have two significant figures, the answer can have only two. Notice also that the *only* way to report the answer to two significant figures is with scientific notation, as the decimal equivalent (900 m) has only one significant figure.

ON YOUR OWN

2.3 A race car starts at rest and travels 1,321 ft (a quarter of a mile) in 11 seconds. What was the car's acceleration?

Using Our Equations for One-Dimensional Motion

By now, you have learned five powerful equations that can go a long way in describing many types of one-dimensional motion. These five equations are summarized below:

$$\mathbf{v} = \frac{\mathbf{\Delta x}}{\Delta t} \quad (1.1)$$

$$\mathbf{a} = \frac{\mathbf{\Delta v}}{\Delta t} \quad (1.3)$$

$$\mathbf{v} = \mathbf{v}_o + \mathbf{a}t \quad (2.6)$$

$$\mathbf{v}^2 = \mathbf{v}_o^2 + 2\mathbf{a} \cdot \mathbf{\Delta x} \quad (2.15)$$

$$\mathbf{\Delta x} = \mathbf{v}_o t + \frac{1}{2}\mathbf{a}t^2 \quad (2.19)$$

If you can understand and use these equations, then you can analyze many kinds of one-dimensional motion. I say "many kinds" because these equations do have their limits.

Consider, for example, the last one, Equation (2.19). We derived this equation by using a velocity-versus-time graph. What was the shape of the curve in that graph? It was a straight line. When a graph has a straight line, what can we say about its slope? The slope is constant. We learned in Module #1 that the slope of a velocity-versus-time curve is the acceleration. What does this tell us?

Since Equation (2.19) was derived from a graph which described motion with constant acceleration, the equation is only good in situations where there is constant acceleration. In other words, if a car is continually speeding up and slowing down, then the acceleration is not constant, and Equation (2.19) does us no good. We can only use it under conditions of constant acceleration.

It turns out that this is true for all three of the equations we derived in this module [Equations (2.6), (2.15), and (2.19)]. All three of these equations are valid only under conditions of constant acceleration. This is an incredibly important fact and one that must be kept in mind:

Equations (2.6), (2.15), and (2.19) are valid only when the acceleration is constant.

Of course, all problems that we will do in this chapter assume a constant acceleration. It is just good for you to be aware of the limitations of any equation that you use.

You've already had some practice using these powerful equations, but the practice was a bit easy. After all, in each "On Your Own" problem you solved, you knew what equation to use. It was obviously going to be the equation you had just been studying. The real trick to being able to use these equations, however, is being able to *determine* which equation to use. Therefore, I want you to study the next two example problems and then solve the "On Your Own" problems that follow. In each problem, you will have to determine which of the equations to use, making the solution just a little more difficult.

Illustrations copyright GifArt.com

EXAMPLE 2.4

A car is traveling down the road at 27 m/sec, when the driver suddenly sees a cat in the road ahead. The driver immediately hits the brakes, which apply a deceleration of 7.5 m/sec². If the cat is 50.0 m away when the driver hits the brakes and is too scared to move, will the cat get hit by the car?

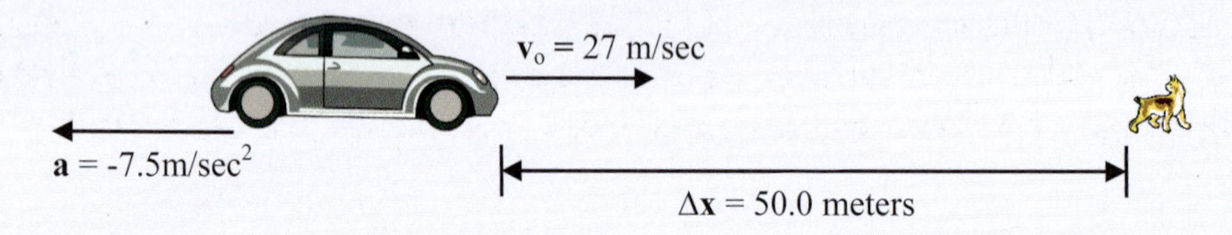

As I have said before, the trick to solving physics problems is to recognize what is given and what you need to determine. After you have done that, you can usually find an equation that will allow you to solve for the answer. What do we need to determine here? We need to know if the car can stop within 50.0 meters. If it can, the cat is safe. If not, the cat has some real problems. What do we solve for to determine this? Well, we need to know the car's displacement from the time the driver hits the brakes to the time the car stops. That's **Δx**.

What are we given? We are told the initial velocity (27 m/sec) and the acceleration (-7.5 m/sec^2). The acceleration is negative because the car is decelerating, which means the sign of the acceleration is opposite that of the velocity. Even though it is not explicitly stated, we also know that the final velocity is 0, because the car has to stop to avoid hitting the poor little kitty.

Now that we know what we are solving for (displacement) and what we have (initial velocity, final velocity, and acceleration), we need to determine what equation to use. We need an equation that relates these quantities and *only* these quantities. Why? Well, we already have one unknown, and we can't use an equation to solve for any more than one unknown. Which equation relates these and only these quantities? It's Equation (2.15):

$$\mathbf{v}^2 = \mathbf{v}_o^2 + 2\mathbf{a} \cdot \Delta\mathbf{x}$$

The last thing left to do is check the units. Are they all consistent? Yes, because all time units are in seconds and all displacement units are in meters. Plugging the values that we have into our equation gives us:

$$0^2 = (27\frac{\text{m}}{\text{sec}})^2 + 2(-7.5\frac{\text{m}}{\text{sec}^2}) \cdot \Delta\mathbf{x}$$

$$\Delta\mathbf{x} = \frac{-729\frac{\text{m}^{\not{2}}}{\not{\text{sec}}^{\not{2}}}}{-15\frac{\not{\text{m}}}{\not{\text{sec}}^{\not{2}}}} = 49\text{ m}$$

This means that it takes 49 meters for the car to stop. Since the cat was 50.0 meters away when the driver hit the brakes, the car just barely misses the cat. You really didn't think I would let the car hit the cat, did you?

If a car starts from rest and accelerates at a rate of 1.2 m/sec^2, how long will it take for the car to travel 0.18 km?

In order to solve this problem, we need to know what we are given and what we need to determine. The problem asks how long it will take for the car to travel 0.18 km. Obviously, then, we are solving for time. In order to determine this, we are given the initial velocity (0), the acceleration (1.2 m/sec^2), and the displacement (0.18 km). In order to get an answer, then, we need to find an equation that relates these quantities and only these quantities. Equation (2.19) fits the bill:

$$\Delta\mathbf{x} = \mathbf{v}_o t + \frac{1}{2}\mathbf{a}t^2$$

Before we plug our values into the equation, however, we need to determine whether or not all of the units agree. They do not. While the acceleration has the meters unit in it, the displacement is given in kilometers. Thus, we need to change kilometers to meters or meters to kilometers in order to make everything consistent. Since meters is the SI displacement unit, we will convert km to m.

$$\frac{0.18\ \not{\text{km}}}{1} \times \frac{1{,}000\text{ m}}{1\ \not{\text{km}}} = 180\text{ m}$$

Remember, since the problem does not ask for any specific units in the answer, I could just as easily have converted meters per second into kilometers per second. Either conversion would have worked, as long as the units ended up the same.

Now that all of our units are consistent, we can plug our numbers into Equation (2.19) and solve for time:

$$180 \text{ m} = (0) \cdot \text{t} + \frac{1}{2} \cdot (1.2 \frac{\text{m}}{\text{sec}^2}) \cdot \text{t}^2$$

$$\text{t}^2 = \frac{2 \cdot 180 \cancel{\text{m}}}{1.2 \frac{\cancel{\text{m}}}{\text{sec}^2}}$$

$$\text{t}^2 = 3.0 \times 10^2 \text{ sec}^2$$

$$\text{t} = 17 \text{ sec}$$

This tells us that the car takes 17 seconds to travel 0.18 km.

Illustration by Megan Whitaker

ON YOUR OWN

2.4 If a car can accelerate from rest to a velocity of 60.0 miles per hour in 10.0 seconds, what is its acceleration?

2.5 In order to take off, a certain plane needs to start from rest and achieve a velocity of 1.50×10^2 miles per hour before it reaches the end of the runway. If the plane's acceleration is 2.00×10^4 miles/hr^2, what is the minimum length needed for a runway?

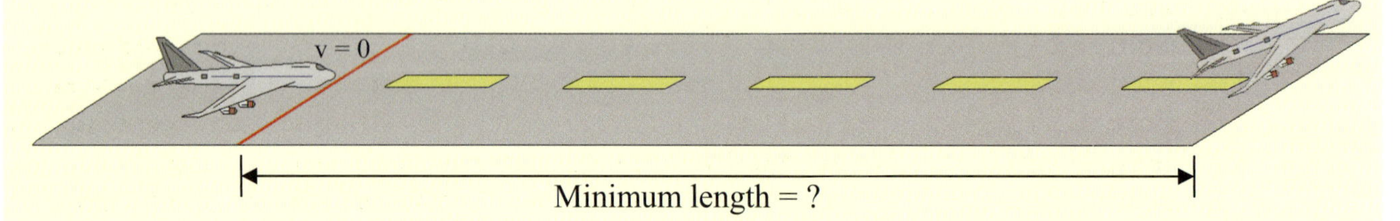

2.6 A biker can maintain a constant acceleration of 0.030 m/sec^2. If the biker starts from rest, how far can she travel if she keeps that acceleration up for 5.0 minutes?

Free Fall

You might be wondering just how practical all of this stuff really is. Remember, to apply the things that we have learned to any situation, two conditions must exist. First, all motion must occur in a straight line. After all, we are talking about one-dimensional motion here. Things can only travel in one straight direction or in precisely the opposite direction. No curves, no bends in the road, nothing. All motion has to be perfectly straight. Second, the acceleration of any body that we study must be constant. If the acceleration is not constant, then the last three equations that we derived [Equations (2.6), (2.15), and (2.19)] are not applicable.

How often does this really happen? How often do things actually travel in straight lines with constant acceleration? Well, there is one situation in which both of these conditions always apply: **free fall**.

Free fall - The motion of an object when it is falling solely under the influence of gravity

If you hold a ball high in the air and then let it go, from the time it leaves your hand to the time it hits the floor, we say that it is experiencing free fall.

In principle, we all know why things fall when they are dropped. The earth tends to attract other bodies to itself. We call this attraction "gravity." We will be learning a great deal about gravity in an upcoming module, so for right now, I just want to concentrate on the aspect of gravity that is relevant to our discussion of one-dimensional motion. When an object is falling near the surface of the earth, it is accelerating straight down at a constant acceleration of 9.8 m/sec^2, which is also equal to 32 $feet/sec^2$. This is an important fact:

Objects falling near the surface of the earth experience a constant acceleration of 9.8 m/sec^2 (32 $feet/sec^2$) straight down.

The value of 9.8 m/sec^2 or 32 $feet/sec^2$ is called the "acceleration due to gravity" and it is often given the abbreviation "g." Thus, you will often see physics books say "$g = 9.8\ m/sec^2$." To understand how this works, study Figure 2.1.

Digital Composite Photo by Kathleen J. Wile

FIGURE 2.1
A Tennis Ball in Free Fall

In this composite picture, a tennis ball is being dropped by a man standing on a ladder. The image of the tennis ball is shown at several equal time intervals during the fall. Notice that in each time interval, the ball travels farther than it did in the previous time interval. Why is that? Well, remember that while the ball is in free fall, it experiences a *constant* acceleration. Think about what that means. The tennis ball starts out with zero initial velocity. During the first time interval, the acceleration due to gravity increases the velocity. Thus, the ball begins to move. However, it cannot move very far, as its average velocity during that time interval is fairly small. During the second time interval, the tennis ball already has some velocity, and the acceleration due to gravity adds even more. As a result, the average velocity of the ball is much higher during the second time interval. During the third time interval, the ball starts out with even more velocity, and the acceleration adds still more to the velocity. This continues until the ball hits the floor. If the *velocity* were constant during free fall, the ball would travel the same distance in each time interval. However, the *acceleration* is constant in free fall, so the ball travels farther during each successive time interval.

Do you see how this relates to our one-dimensional motion discussion? If the acceleration is straight down, the object will most likely move in a straight line. Since the acceleration has a fixed value, it is constant. Thus, both conditions necessary for our one-dimensional motion equations are satisfied. You might also want to know what "near the surface of the earth" means. Well, as we will learn later, the acceleration due to gravity changes as the distance from the earth changes. However, as long as the object in question is near the surface of the earth, the acceleration due to gravity is a constant 9.8 m/sec^2. When a physicist says "near the surface of the earth," she generally means the object cannot be more than a few kilometers above the ground. At an altitude of roughly 40 km, for example, the acceleration due to gravity is only about 1% lower than the acceleration due to gravity at sea level. Since most of us do not often find ourselves higher than a few kilometers above sea level, it is safe to assume that for our everyday experiences, the acceleration due to gravity is 9.8 m/sec^2.

Now, before we actually see how we can use our one-dimensional motion equations to analyze free fall, I need to point something out. Notice that I did not mention *anything* about the nature of the object when I indicated the value for the acceleration due to gravity. This is because the acceleration due to gravity is *completely independent* of the nature of the object that is falling, as long as the object has mass. This is another important fact:

The acceleration due to gravity is independent of the nature of the object experiencing free fall, as long as the object has mass.

In other words, all objects, regardless of their shape, size, or composition, experience the same acceleration due to gravity: 9.8 m/sec^2.

(The multimedia CD has a video demonstration of this fact.)

This fact goes against your everyday experience, so it requires a bit more discussion. Suppose you hold a rock in one hand and a feather in the other. Suppose further that you let them go at precisely the same time. Which one will hit the ground first? From our everyday experience, we say that the rock will fall very quickly, while the feather will float down slowly. Obviously, then, the rock will hit the ground first.

Let's take this situation and examine it from a physics point of view. If I hold the rock and the feather steady and then release them at the same instant, what are their initial velocities? Since they were both held still before being released, their initial velocities are both zero. Thus, both the feather and the rock have exactly the same initial velocities. Once you release them, however, the rock starts traveling much faster than the feather. In other words, over the same time interval, the rock's velocity becomes much greater than that of the feather. We have already learned that the only thing that can change velocity is acceleration. Thus, the rock must have a faster acceleration due to gravity than the feather, right? Surprisingly, the answer to this question is "NO!" To find out why, perform the following experiment.

EXPERIMENT 2.1

The Acceleration Due to Gravity Is the Same for All Objects.

Supplies:

- Safety goggles
- A large (at least 21 cm by 27 cm), heavy book
- A small (about 10 cm by 10 cm) piece of paper

1. Hold the book in one hand and the paper in the other at arm's length, and make sure that they are at exactly the same height. Make sure that there are no obstructions beneath the two objects, so that they can fall to the floor without running into something.
2. Release them both at precisely the same instant. What happens?
3. Pick up the paper and book.
4. This time, place the piece of paper on top of the book and hold the book out at arm's length with both hands.
5. Release the book. What happens this time?
6. Put the paper and book away.

What happened in the experiment? Well, when you dropped the book and paper in the first part of the experiment, the book probably hit the ground much earlier than the paper. This might lead you to conclude that the paper experiences a smaller acceleration due to gravity than does the book. However, in the second part of the experiment, the piece of paper stayed on top of the book, falling just as fast as the book fell! Why did this happen? The paper was not stuck to the book; therefore, it did not *have* to stay on the book. If it were really experiencing a smaller acceleration due to gravity than the book, the book should have started traveling faster than the piece of paper, eventually pulling away from it. Instead, they both fell at exactly the same velocity. Why?

The explanation has to do with something we call **air resistance**. You see, when an object falls through the air, there are several gaseous molecules (like nitrogen and oxygen) and atoms (like argon) that are in the object's way. In order to fall, the object must shove the gaseous molecules and atoms out of its way. Well, the molecules and atoms resist this movement, and thus the object must force its way through them. A heavy object is much better at doing this than a light object. Therefore, heavy objects fall faster than light objects not because their acceleration due to gravity is larger, but because they are not as strongly affected by air resistance as light objects are.

The fact that light objects are affected by air resistance more than heavy objects is illustrated by the first part of the experiment. When you held the book and paper in each hand and dropped them, they were both subject to air resistance. Since the paper was much more affected by air resistance than the book, it fell more slowly, because it had a harder time shoving through the molecules and atoms in the air. When you placed the paper on top of the book in the second part of the experiment, however, the book shoved the molecules and atoms in the air out of the way. The paper, therefore, did not have to. As a result, it was not subject to air resistance. Under those circumstances, then, the paper and the book fell with the same velocity, demonstrating the fact that gravity accelerates all objects equally.

What we learn from the experiment, then, is that when we neglect air resistance, all objects falling near the surface of the earth accelerate equally at 9.8 m/sec^2 straight down. We will discuss air resistance in a bit, and you will explore factors other than weight which affect how strongly air resistance acts on an object. For now, however, I want to concentrate on free fall. Therefore, we will always neglect air resistance when solving problems that involve objects falling near the surface of the earth. For most of the objects we drop (balls, rocks, coins, etc.), this is a reasonable thing to do, because air resistance does not affect these objects very much.

In fact, any object that is affected greatly by air resistance cannot experience free fall when dropped near the surface of the earth. After all, remember the definition of free fall. It tells us that in order to really experience free fall, the object must be falling solely under the influence of gravity. Air resistance produces an influence other than that of gravity; thus, objects significantly affected by air resistance do not experience free fall when dropped near the surface of the earth. Since we are neglecting air resistance, then, when an object falls near the surface of the earth, we will consider it in free fall. Once again, this is only an assumption, because all objects are influenced to some degree by air resistance. However, if this effect is small, then free fall is a good assumption.

In the end, then, we will assume that objects falling near the surface of the earth are not greatly affected by air resistance. This means that the objects are in free fall. If that is the case, then all of our one-dimensional motion equations apply to the situation. It also means that we know the acceleration: 9.8 m/sec^2 (or 32 $feet/sec^2$) straight down.

Okay, now that we know the assumptions we will use while analyzing free fall, let's see how we can apply the equations we have already learned in order to find out something useful. For example, have you ever wondered how fast you can react to something that happens? Suppose you are driving a car and suddenly see an obstruction in the road ahead. How long will it take you to recognize the obstruction and move your foot from the accelerator to the brake? The time it takes you to do this is called your **reaction time**. As you might imagine, reaction time is different for different people. Some people react very quickly, and they make good race car drivers, airplane pilots, and tennis players. Other people just can't react quickly, and such fast-paced professions are not for them. Using the concept of free fall and our equations of one-dimensional motion, you can perform the following experiment to measure your own reaction time.

EXPERIMENT 2.2

Determining a Person's Reaction Time

Note: A sample set of calculations is available in the solutions and tests guide. It is with the solutions to the practice problems.

Supplies:

- Safety goggles
- A ruler, preferably metric
- Another person to help you

1. Get your helper to hold on to the ruler at the very top. Have him hold the ruler vertically, with his arm held straight out. You should stand facing your helper, slightly less than an arm's length away from the ruler.
2. Take your right hand (or your left one if you are left-handed) and hold your thumb and forefinger about 3 cm apart, as if you are ready to pinch someone.
3. Hold your hand so that the middle of the ruler is directly between your thumb and forefinger and so that one side of the ruler faces your thumb while the other side faces your forefinger (see drawing below). Line your thumb and forefinger up with one of the marks on the ruler, and remember what mark it is.

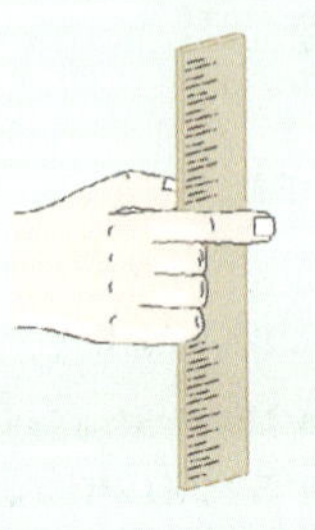

Illustration by Megan Whitaker

4. Once you are all set, tell your helper to release the ruler without warning you. The ruler will begin to fall. As soon as you notice it falling, grab the ruler by pinching your thumb and forefinger together.

5. Now look at the mark on the ruler where your thumb and forefinger ended up grasping it. Determine the distance in centimeters between this mark and the mark with which you lined up your thumb and forefinger. If you don't have a metric ruler, determine the distance in inches and then convert to centimeters by multiplying by 2.54.
6. Repeat this experiment five times and average the results. In the end, you will have the average displacement through which the ruler fell before you grasped it. We can use this displacement to determine your reaction time.
7. Do the entire experiment again, this time having your helper grasp the ruler while you drop it.
8. Put everything away.

Now that you have the average displacement of the ruler as it fell, we can determine the time it took for you to notice that the ruler was falling and grasp it with your thumb and forefinger. What was the initial velocity of the ruler? Well, since your helper was holding the ruler still before he dropped it, the ruler had an initial velocity of zero. Once your helper released it, however, the ruler was in free fall, so its acceleration was 9.8 m/sec^2. You also determined the ruler's displacement before you were able to recognize that it was falling and grasp it. Using this information, you can determine the time that took. How? Well, we have the initial velocity, the acceleration, and the displacement. We want to determine the time. Equation (2.19) relates these quantities:

$$\Delta \mathbf{x} = \mathbf{v}_0 t + \frac{1}{2}\mathbf{a}t^2$$

Before we do this, we need to make sure the units are consistent. You measured the distance that the ruler fell in centimeters, but the acceleration is in m/sec^2. These units aren't consistent, so we must convert one of them. Well, an acceleration of 9.8 m/sec^2 is the same as 980 cm/sec^2. We can use that value of the acceleration, because it is consistent with the distance you measured. Now we will put in the quantities that we know and rearrange the equation to solve for time. Since I don't know what displacement you ended up measuring, I will call it "d cm":

$$d\text{ cm} = (0)\cdot t + \frac{1}{2}\cdot(980\,\frac{\text{cm}}{\text{sec}^2})\cdot t^2$$

$$t^2 = \frac{2\cdot(d\text{ cm})}{980\,\frac{\text{cm}}{\text{sec}^2}}$$

$$t = \sqrt{\frac{2\cdot(d\,\cancel{\text{cm}})}{980\,\frac{\cancel{\text{cm}}}{\text{sec}^2}}}$$

So, if you take the distance you measured in cm, multiply by 2, divide by 980, and then take the square root of the answer, you will get your reaction time in seconds. Do this calculation so that you determine your reaction time and your helper's reaction time. Who reacted faster?

Do you see how to analyze free fall? Really, it's no different than what we have been doing, except that in the case of free fall, we always know the acceleration: 9.8 m/sec^2 or 32 ft/sec^2. So, we can analyze free fall situations the same way we have before, but the problems do not have to give us the acceleration, because we already know it. Study the next example and solve the "On Your Own" problems that follow to be sure you understand this.

EXAMPLE 2.5

A person drops a penny off a tall (0.150 km) building. If the person holds the penny still before dropping it, what will be the velocity of the penny at the instant that it hits the ground?

When the person holds the penny still and then drops it, its initial velocity is zero. Also, since the penny is in free fall, its acceleration is -9.8 m/sec^2. Since I have just said that the acceleration is negative, this means that I have defined down as the negative direction. We also know that, at the instant the penny hits the ground, it has traveled -0.150 km in free fall. This also has a negative value, since the penny travels 0.150 km down. So, we have initial velocity, acceleration, and displacement, and we want to determine the final velocity. Well, equation (2.15) relates these and only these quantities, so we can use it:

$$\mathbf{v}^2 = \mathbf{v}_o^2 + 2\mathbf{a} \cdot \Delta\mathbf{x}$$

Before we use the equation, however, we need to make sure that our units are consistent. Right now they are not, because the displacement is in km while the acceleration is in m/sec^2. I will convert km to m in order to make things consistent:

$$\frac{-0.150 \cancel{\text{km}}}{1} \times \frac{1{,}000 \text{ m}}{1 \cancel{\text{km}}} = -1.50 \times 10^2 \text{ m}$$

Now we can plug our numbers into the equation and solve for final velocity:

$$\mathbf{v}^2 = (0)^2 + 2 \cdot (-9.8 \frac{\text{m}}{\text{sec}^2}) \cdot (-1.50 \times 10^2 \text{ m})$$

$$\mathbf{v}^2 = 2{,}940 \frac{\text{m}^2}{\text{sec}^2}$$

$$\mathbf{v} = 54 \frac{\text{m}}{\text{sec}}$$

The penny, then, is traveling with a velocity of <u>54 m/sec</u> downwards at the instant it hits the ground. This velocity is equivalent to about 120 mph! That's why dropping objects from high places can be dangerous!

ON YOUR OWN

2.7 How long will it take a rock to hit the ground if it is dropped from the Leaning Tower of Pisa (height = 54.6 m)?

2.8 In order to determine the height of a bridge, a physics student drops a rock off the side of the bridge and times how long it takes for the rock to hit the water. If it takes 1.4 seconds for the rock to hit the water, what is the height of the bridge in feet?

A More Detailed Look at Free Fall

We've talked a lot about dropping objects, but that's not the only way to initiate free fall. When you throw a ball up in the air, the ball is also in free fall. That might sound a little strange to you, since the ball isn't actually falling right away. Nevertheless, if you think about it, even though the ball is initially going up, gravity is *trying* to make it fall. We also know that eventually, gravity will win, and the ball will start falling. Throughout the whole process, the ball's acceleration is a constant 9.8 m/sec^2 downward. Thus, even though it starts traveling upwards, it is still in free fall.

Let's study the motion of that ball a little more closely. Suppose you throw the ball up in the air with an initial velocity of 3.5 m/sec. What happens to its velocity as time goes on? Well, at first, the acceleration that the ball is experiencing opposes the velocity. After all, the ball was thrown up, but gravity always accelerates things down. So, if the initial velocity is positive 3.5 m/sec, the acceleration is -9.8 m/sec^2. Since acceleration and velocity have opposite signs, the ball slows down.

At some point however, the acceleration will succeed in slowing the ball down until the ball actually stops in mid-air. When it stops, it has reached the highest point of its journey. Now what happens? Does the ball just float up there with zero velocity? No, of course not. Even though the velocity of the ball is zero, its acceleration is not. The acceleration is still -9.8 m/sec^2. Thus, the ball stops only for an instant, and then it begins to fall. As it falls, its velocity is negative, so now velocity and acceleration both have the same direction. Under those circumstances the ball speeds up. So, when the ball starts falling, it also starts gaining speed.

Let's suppose you reach out and catch the ball at exactly the same height that you threw it. What will its velocity be? You might think that you would have to pull out your calculator and use one of our one-dimensional motion equations, but you really don't. Think about it. The ball travels upward for some distance, until the acceleration gets rid of all of the ball's positive velocity. As it falls back down, the acceleration starts giving it more and more velocity, only this time in the negative direction. If it travels downward exactly the same distance that it traveled upward, doesn't it make sense that the acceleration should be able to give back all of the speed that it took away to begin with? Well, that's exactly what happens.

When you catch the ball at the same height that you threw it, the ball has the same speed that it originally had when you threw it. Of course, since it is now traveling in the opposite direction, the sign of its velocity is reversed. So, if you throw the ball up at 3.5 m/sec, when the ball reaches the height at which it was thrown, its velocity will be -3.5 m/sec. This fact, combined with the fact that the ball's velocity is zero at its maximum height, is very important and needs to be remembered.

When an object is thrown upward in the presence of gravity, the object will reach its maximum height when the object's velocity equals zero. In addition, when it returns to the height from which it was thrown, its velocity will be equal to and opposite of its initial velocity.

Think about a practical application of this situation. Consider someone firing a gun up into the air, perhaps in celebration. You might think that since the gun was fired into the air, the bullet wouldn't hurt anyone. That's not necessarily true. Remember, the bullet has to fall back to the ground at some point. Suppose someone happens to be unlucky enough to be standing where the bullet falls back to earth. How fast will the bullet be traveling when it hits that person? Neglecting air resistance, it will be traveling as fast when it hits that person as it was when it left the gun. Thus, the speed of the impact for that person would be the same as if the person were shot from point-blank range! Thus, even firing a gun up in the air can be dangerous!

So you see that free fall problems can be rather complex. When objects are thrown up in the air, their initial velocity and their acceleration oppose each other. This leads to very complex motion. When solving problems like these, then, you need to make sure you keep track of your positive and negative direction definitions. Make sure you understand this by studying the following example.

Illustration from the MasterClips collection

EXAMPLE 2.6

A bullet is shot straight up in the air with an initial velocity of 335 m/sec. What is the maximum height that the bullet reaches before it starts to fall back to the ground?

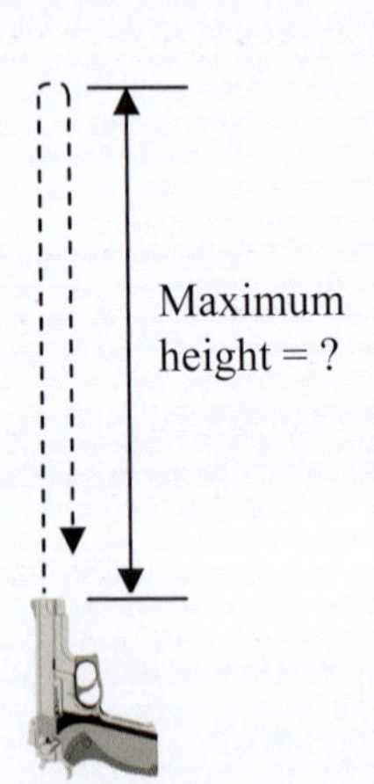

To solve this problem, we need to remember one of the facts we just learned. What do we know about the maximum height of any object that is initially moving upward? We know that at that maximum height, the velocity is zero. Thus, in this problem, we know the acceleration (because the object is in free fall), and we know the final velocity (because the final point of the problem is when the bullet reaches its maximum height). At that point, we know that velocity is zero. We are also given the initial velocity. We are then asked to determine the height, which is also the displacement. Thus, we need to use Equation (2.15):

$$\mathbf{v}^2 = \mathbf{v}_o^2 + 2\mathbf{a} \cdot \Delta\mathbf{x}$$

Before we plug our numbers into this equation, we need to do two things. First, we need to make sure that our units are consistent. They are. Also, we need to keep track of direction. If we say that the initial velocity is 335 m/sec, then we are, in effect, defining upward motion as positive, because the initial velocity is up. As a result, then, the acceleration is -9.8 m/sec^2 because acceleration due to gravity is always pointed down. Now we can plug into our equation:

$$0^2 = (335 \frac{\text{m}}{\text{sec}})^2 + 2 \cdot (-9.8 \frac{\text{m}}{\text{sec}^2}) \cdot \Delta\mathbf{x}$$

$$\Delta\mathbf{x} = \frac{112{,}225 \frac{\text{m}^{\cancel{2}}}{\cancel{\text{sec}^2}}}{19.6 \frac{\cancel{\text{m}}}{\cancel{\text{sec}^2}}} = 5.7 \times 10^3 \text{ m}$$

We know, then, that the bullet reaches a height of <u>5.7 x 10^3 m</u> before it starts to fall back to the ground.

A physics student stands on a bridge and throws a rock up in the air with an initial velocity of 2.3 m/sec. The student throws the rock from the side of the bridge, so it is free to fall into the water when it returns. What is the displacement of the rock after 0.60 seconds?

This problem gives us the initial velocity (2.3 m/sec) and the time (0.60 sec). Since the rock is in free fall, we also know its acceleration. Let's be careful, though. If the initial velocity is positive, we have defined up as the positive direction. This means that the acceleration is -9.8 m/sec^2. So, we have initial velocity, acceleration, and time. We want to determine displacement. Equation (2.19) is the tool to use :

$$\Delta\mathbf{x} = \mathbf{v}_0 t + \frac{1}{2}\mathbf{a}t^2$$

$$\Delta\mathbf{x} = (2.3\ \frac{\text{m}}{\cancel{\text{sec}}})\cdot(0.60\ \cancel{\text{sec}}) + \frac{1}{2}\cdot(-9.8\ \frac{\text{m}}{\cancel{\text{sec}^2}})\cdot(0.60\ \cancel{\text{sec}})^2$$

$$\Delta\mathbf{x} = 1.4\ \text{m} - 1.8\ \text{m} = -0.4\ \text{m}$$

Now according to our direction definition, up is positive and down is negative. Thus, a negative displacement means down. Since we always define displacement relative to the initial position, then -0.4 m means <u>0.4 meters below the point where the rock was thrown</u>.

Before we leave this problem, look at how the significant figures work here. Normally, you wait until the end of an equation to worry about significant figures. However, I cannot do that here, because I have to apply different rules in the middle of the equation. When I multiply 2.3 and 0.60, for example, I must count significant figures. However, when I *add* the result to the second term in the equation, I must take precision into account. Thus, I need to do the multiplication, then round according to the number of significant figures. That's how I end up with 1.4 m - 1.8 m. At that point, however, I must look at precision, and since each number goes out to the tenths place, my answer can only go out to the tenths place. That's why the final answer is 0.4 m.

Notice that once again, there is really nothing new here. As long as we remember the two facts that are in boldface type at the beginning of this section, and we keep careful track of our definitions of positive and negative motion, these free fall problems are just like the other free fall problems, which are also just like our one-dimensional motion problems. See if you can do all of this by solving the following "On Your Own" problems.

Illustration by Megan Whitaker

ON YOUR OWN

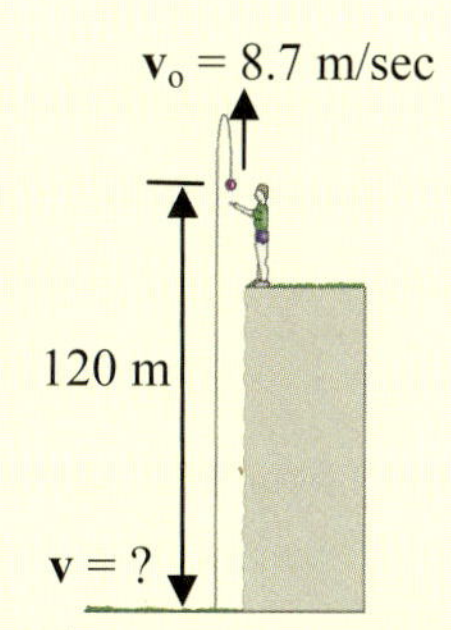

2.9 A student throws a rock at the edge of a cliff, the bottom of which is 120 meters below the student's hand (see illustration on the right). If the rock is thrown upward with a velocity of 8.7 m/sec, what will the rock's velocity be when it reaches the bottom of the cliff?

2.10 A ball is thrown up in the air and reaches a maximum height of 4.5 meters. What was its initial velocity?

Terminal Velocity

Before we leave this module, I want you to learn about a very interesting phenomenon known as **terminal velocity**. We've already discussed air resistance to some extent, but one thing I did not mention was that as an object's velocity increases, the air resistance it experiences also increases. The details behind this fact are a bit too complicated to explain here, but the fact itself leads to a very interesting effect.

Think about an object falling. As it falls, its velocity increases. As its velocity increases, the air resistance it experiences increases as well. In Experiment 2.1, we learned that increased air resistance decreases the acceleration that the object experiences. So, as the air resistance increases, the acceleration decreases. Well, the faster the object falls, the greater the air resistance. The air resistance eventually increases so much that the acceleration actually falls to zero.

What happens when the acceleration falls to zero? At that point, the velocity cannot increase anymore. Thus, no matter how much farther the object falls, its velocity will not increase, because the air resistance completely cancels out the acceleration due to gravity. The velocity at which this happens is called "terminal velocity."

Terminal velocity - The velocity a falling object has when, due to air resistance, its acceleration is reduced to zero. This is the maximum velocity a falling object subject to air resistance can achieve.

You learned in Experiment 2.1 that the weight of an object influences the extent to which air resistance affects it. Well, the more that air resistance affects an object, the *smaller* its terminal velocity will be. Thus, the weight of an object affects its terminal velocity. There are other factors that affect the air resistance and hence the terminal velocity experienced by an object. Perform the following experiment to learn about some of these factors.

EXPERIMENT 2.3
Factors That Affect Air Resistance

Supplies:

- Safety goggles
- A large (at least 21 cm by 27 cm), heavy book
- Four small (about 10 cm by 10 cm) pieces of paper, all the same size

1. Take the first piece of paper and wad it into a small ball.
2. Fold the second piece of paper once horizontally and once vertically, to make a piece of paper that is one-fourth of its original size. Don't do anything to the third and fourth pieces of paper.
3. Place the first, second, and third pieces of paper on the top of the book.
4. Like you did in Experiment 2.1, hold the book out at arm's length and drop it. Note how the pieces of paper fall.
5. Hold the first piece of paper (the wadded up ball) in one hand and the third piece of paper (one of the two that you have not modified) in the other hand.
6. Hold your hands at the same height, and hold the third piece of paper horizontal, so it is parallel to the floor.

7. Release both pieces of paper simultaneously. Note which one falls fastest.
8. Repeat steps 5-7, this time holding the second piece of paper (the folded one) in one hand and the third piece of paper in the other hand.
9. Repeat steps 5-7 again, this time holding the first piece of paper in one hand and the second piece of paper in the other hand.
10. Repeat steps 5-7 one more time. This time, however, hold the third piece of paper in one hand and the fourth piece of paper (the other one you did not modify) in the other hand. Hold the third piece of paper horizontally so that it is parallel to the floor and the fourth piece of paper vertically, so that it is perpendicular to the floor. Make sure that the bottom of the fourth piece of paper is at the same height as the third piece of paper.
11. Clean up your mess.

What happened in the experiment? Well, in step 4, you should have seen that the pieces of paper all fell at the same rate. This is nothing new, since the book significantly reduced air resistance for the pieces of paper, so they all experienced the same acceleration, 9.8 m/sec^2 downward. However, when you got rid of the book, the pieces of paper did not fall at the same rate, did they? The wadded-up piece of paper (the first piece) should have fallen the fastest. The folded piece of paper (the second piece) should have fallen faster than the flat piece of paper held horizontally (the third piece), but not as fast as the wadded-up piece of paper. Finally, the flat piece of paper held vertically should have fallen faster than the flat piece of paper held horizontally.

What can we make of these results? First, we can say that the *shape* of an object influences the extent to which air resistance affects it. After all, the pieces of paper all had the same weight. However, the more compact the shape, the faster it fell. Thus, the more compact the shape, the *lower* the effect of air resistance. The wadded up piece of paper, then, was not as strongly affected by air resistance as was the folded piece of paper. In turn, the folded piece of paper was not as strongly affected by air resistance as was the flat piece of paper held horizontally. What does that tell us about the terminal velocities of these pieces of paper? The terminal velocity of the wadded up piece of paper would be the highest, the terminal velocity of the folded piece of paper would be lower, and the terminal velocity of the flat piece of paper held horizontally would be lower still.

What about the last trial in the experiment? In that trial, both pieces of paper were the same shape, but the orientation at which they fell was different. The piece of paper that you held horizontally (parallel to the floor) fell much more slowly than the piece of paper you held vertically (perpendicular to the floor). Thus, the orientation of an object affects air resistance as well. The more surface area that the object exposes to the atoms and molecules that it must shove out of the way, the more the air resistance. This means that the terminal velocity of an object gets lower as the amount of surface area it exposes to the atoms and molecules that it must shove out of the way increases. In the end, then, we know at least three things that affect the terminal velocity of an object: its weight, shape, and orientation as it falls.

Because of the blend of their weight and shape, most insects that live on the ground have such a large air resistance that their terminal velocity is rather small. These insects can fall any distance and, because their terminal velocity is so small, they will suffer no damage when they hit the ground. Certain breeds of mice and other small rodents are the same. Their natural air resistance is so high that they reach terminal velocity very quickly when they start falling. As a result, they can fall any distance without being hurt. Needless to say, humans and most mammals do not have such large air resistance. So, although if humans and mammals fall long enough they will eventually reach a terminal velocity, that velocity will be deadly when they hit the ground!

Of course, you can artificially decrease the terminal velocity of a person (or anything else) by equipping it with something that increases its air resistance. Consider, for example, Figure 2.2.

Photo courtesy of NASA

FIGURE 2.2

The X-38 Experimental Crew Return Vehicle

The X-38 is an experimental ship designed to return the crew of the international space station back to earth. It would use a short burst from its engines to break orbit, but after that, it would simply fall back to earth. The aerodynamic shape of the craft allows for excellent control of the ship as it falls, and the parachute shown here (a 7,500 ft^2 parafoil) increases the surface area so that air resistance is quite high. This makes the terminal velocity low. As a result, even though the X-38 would fall *from space*, its low terminal velocity makes it a safe means for returning the crew of the international space station to earth. Although the X-38 project has been canceled by NASA, in the test flight shown here, it was dropped from a B-52 bomber at a height of 45,000 feet. Due to the high air resistance imposed by the parachute, however, it landed with a speed of less than 40 miles per hour!

A parachute will artificially decrease the terminal velocity of any object, because the parachute dramatically increases the surface area exposed to the atoms and molecules that must be shoved out of the way as the object falls. This increases air resistance considerably, which in turn, decreases terminal velocity.

I hope that you have learned a great deal about free fall in this module. We will be visiting the subject again, when we discuss gravity in a more detailed way. For right now, however, you have the basics. So complete the review questions, practice problems, and test so that you can cement what you have learned into your mind.

ANSWERS TO THE "ON YOUR OWN" PROBLEMS

2.1 This problem gives us the final velocity that the police car needs (35 m/sec). It also tells us the acceleration (2.5 m/sec^2) and the initial velocity (23 m/sec). Since the policeman is speeding up, the acceleration and velocity have the same sign. Thus, to make things easy, we will define the direction that the cars are moving as positive motion. The problem asks us to solve for time. Equation (2.6) relates all of those things, so that's what we'll use:

$$\mathbf{v} = \mathbf{v}_o + \mathbf{a}t$$

$$35\frac{\text{m}}{\text{sec}} = 23\frac{\text{m}}{\text{sec}} + (2.5\frac{\text{m}}{\text{sec}^2})t$$

Now we can use algebra to rearrange the equation and solve for time:

$$t = \frac{35\frac{\text{m}}{\text{sec}} - 23\frac{\text{m}}{\text{sec}}}{2.5\frac{\text{m}}{\text{sec}^2}} = \frac{12\frac{\cancel{\text{m}}}{\cancel{\text{sec}}}}{2.5\frac{\cancel{\text{m}}}{\text{sec}^{\cancel{2}}}} = 4.8 \text{ sec}$$

We've determined, therefore, that the policeman needs <u>4.8 sec</u> to match the velocity of the speeder. Notice what I had to do in order to keep track of significant figures. Since I am mixing addition/subtraction with multiplication/division, I could not wait until the end of the equation to deal with significant figures. I had to do the subtraction first and use precision to determine how to report my answer. Since both numbers have their last significant figure in the ones place, the answer goes out to the ones place. After that, I had to divide, so I counted significant figures to find that I had to report my answer to two significant figures.

2.2 In this problem, we are given the initial velocity (0), the displacement (5.00 x 10^2 yards), and the acceleration (1.15 x 10^4 miles/hour^2). We are asked to calculate the final velocity. These physical quantities are related by Equation (2.15), so that's what we'll use. Don't worry that the units are not metric. 1.15 x 10^4 miles/hour^2 is a legitimate acceleration unit (displacement divided by time squared) and yards are a legitimate displacement unit. The problem, however, is that the displacement units (yards and miles) do not agree. Fortunately, however, the problem gives us a conversion relationship, so I will convert yards into miles:

$$\frac{5.00\times10^2 \cancel{\text{yds}}}{1} \times \frac{1 \text{ mile}}{1.760\times10^3 \cancel{\text{yds}}} = 0.284 \text{ miles}$$

Now we can use Equation (2.15):

$$\mathbf{v}^2 = \mathbf{v}_o^2 + 2\mathbf{a}\cdot\Delta\mathbf{x}$$

$$\mathbf{v}^2 = 0^2 + 2\cdot(1.15\times10^4\frac{\text{miles}}{\text{hr}^2})\cdot(0.284 \text{ miles}) = 6.53\times10^3\frac{\text{miles}^2}{\text{hr}^2}$$

$$v = \sqrt{6.53 \times 10^3 \frac{\text{miles}^2}{\text{hr}^2}} = 80.8 \frac{\text{miles}}{\text{hr}}$$

With that kind of acceleration, the car can go from 0 to 80.8 mph in just 500 yards.

2.3 This problem gives us the initial velocity (0), the time (11 sec) and the displacement (1,321 ft). It asks us to calculate the acceleration. All of these quantities are related by Equation (2.19):

$$\Delta\mathbf{x} = \mathbf{v}_0 t + \frac{1}{2}\mathbf{a}t^2$$

$$1{,}321 \text{ ft} = (0) \cdot 11 \text{ sec} + \frac{1}{2} \cdot (\mathbf{a}) \cdot (11 \text{ sec})^2$$

$$1{,}321 \text{ ft} = \frac{1}{2} \cdot (\mathbf{a}) \cdot (121 \text{ sec}^2)$$

$$\mathbf{a} = \frac{2 \cdot 1{,}321 \text{ ft}}{121 \text{ sec}^2} = 22 \frac{\text{ft}}{\text{sec}^2}$$

The race car, then, had an acceleration of 22 ft/sec^2. The problem did not ask for the acceleration in any specific units, so there is no need to convert. These units are acceptable acceleration units because they contain a displacement unit divided by a time unit squared. Notice that once again, I did not worry about significant figures when I squared 11 sec. This is because since the first term of the equation drops out, the equation deals only with multiplication and division, so I waited until the end of the equation to determine the significant figures. Since the original numbers in the equation (1,321 and 11) had four and two significant figures, respectively, the answer can have only two significant figures.

2.4 In analyzing this problem, we are asked to solve for the car's acceleration. To do this, we are given some information. Since the car starts at rest, its initial velocity is 0. In addition, we know that the final velocity is 60.0 miles per hour. Finally, we are told that the change in velocity takes place over a period of 10.0 seconds. Thus, we are given initial velocity, final velocity, and time. With that information, we are supposed to calculate the acceleration. We therefore need an equation that relates only those quantities. Equation (2.6) fits the bill:

$$\mathbf{v} = \mathbf{v}_0 + \mathbf{a}t$$

Before we use this equation, however, we need to see if our units are consistent. They are not. Time is given in seconds while velocity is in miles per hour. In order to fix this, I will convert seconds into hours. Since there are no directions on what units the final answer should be expressed in, I am free to choose any unit scheme I want. Since car velocities are usually given in miles per hour, I choose to keep the hour unit and get rid of seconds:

$$\frac{10.0\ \cancel{\text{sec}}}{1} \times \frac{1\ \text{hour}}{3{,}600\ \cancel{\text{sec}}} = 0.00278\ \text{hours}$$

Please note that the conversion relationship we used here (1 hour = 3,600 sec) is exact. Thus, it has an infinite number of significant figures. That's why I kept three significant figures in my answer.

Now that our units are consistent, we can use Equation (2.6) :

$$60.0\ \frac{\text{miles}}{\text{hour}} = 0 + \mathbf{a} \cdot (0.00278\ \text{hours})$$

$$\mathbf{a} = \frac{60.0\ \frac{\text{miles}}{\text{hour}}}{0.00278\ \text{hours}} = 2.16 \times 10^4\ \frac{\text{miles}}{\text{hour}^2}$$

This car, then, has an acceleration of $\underline{2.16 \times 10^4\ \text{miles/hour}^2}$.

2.5 In this problem, we are asked to determine the minimum length of a runway. Well, if the plane travels down the runway, then the plane's displacement will be the minimum length that the runway can have. We are therefore solving for displacement. In order to do this, we were given the initial velocity (0), the final velocity (1.50×10^2 miles per hour), and the plane's acceleration (2.00×10^4 miles/hour^2). To solve the problem, we must find an equation that relates these and only these quantities. Equation (2.15) will work:

$$\mathbf{v}^2 = \mathbf{v}_o^2 + 2\mathbf{a} \cdot \Delta\mathbf{x}$$

Before we use it, however, we need to make sure that our units are consistent. They are. All displacement units are in miles and all time units are in hours. Thus, we can just plug things into our equation:

$$(1.50 \times 10^2\ \frac{\text{miles}}{\text{hour}})^2 = (0)^2 + 2 \cdot (2.00 \times 10^4\ \frac{\text{miles}}{\text{hour}^2}) \cdot \Delta\mathbf{x}$$

$$\Delta\mathbf{x} = \frac{2.25 \times 10^4\ \frac{\text{miles}^{\cancel{2}}}{\cancel{\text{hour}^2}}}{4.00 \times 10^4\ \frac{\cancel{\text{miles}}}{\cancel{\text{hour}^2}}} = 0.563\ \text{miles}$$

The runway, then, must be at least $\underline{0.563\ \text{miles}}$ long.

2.6 In this problem, we are asked to solve for how far the biker can travel. This means we are solving for displacement. We are given the biker's initial velocity (0), the acceleration (0.030 m/sec^2), and the time (5.0 min). To solve the problem, we need an equation that relates these and only these quantities. Equation (2.19) fits the bill:

$$\Delta\mathbf{x} = \mathbf{v}_o t + \frac{1}{2}\mathbf{a}t^2$$

Before we use the equation, however, we need to determine whether or not our units agree. They do not. Time is given in minutes, while the acceleration is given in m/sec^2. To make the units agree, I will convert minutes to seconds:

$$\frac{5.0\ \cancel{\text{min}}}{1} \times \frac{60\ \text{sec}}{1\ \cancel{\text{min}}} = 3.0 \times 10^2\ \text{sec}$$

Please note that the conversion relationship (1 min = 60 sec) is exact (1.000…. min = 60.000…. sec). Thus, it really has an infinite number of significant figures, making the 5.0 min the number with the fewest significant figures. That's why I kept two significant figures in my answer.

Now that we have our data in consistent units, we can use the equation:

$$\Delta\mathbf{x} = (0) \cdot (3.0 \times 10^2\ \text{sec}) + \frac{1}{2} \cdot (0.030\ \frac{\text{m}}{\text{sec}^2}) \cdot (3.0 \times 10^2\ \text{sec})^2$$

$$\Delta\mathbf{x} = 0.015\ \frac{\text{m}}{\cancel{\text{sec}^2}} \cdot 9.0 \times 10^4\ \cancel{\text{sec}^2} = 1.4 \times 10^3\ \text{m}$$

The biker, then, can travel <u>1.4 x 10^3 meters</u> in 5.0 minutes with that acceleration.

2.7 We know that the object is in free fall, because it is falling near the surface of the earth, and we always neglect air resistance. Thus, we know its acceleration to be -9.8 m/sec^2. A negative acceleration means we have defined downwards motion as negative. We also know the displacement is -54.6 meters. The displacement is also negative, because the object will travel 54.6 meters *down.* In addition, we know that if it is dropped, the initial velocity is zero. So we know displacement, initial velocity, and acceleration, and we want to determine the time. Equation (2.19) relates all of these quantities and, since all of our units are consistent, we can simply plug in our numbers and solve for time:

$$\Delta\mathbf{x} = \mathbf{v}_0 t + \frac{1}{2}\mathbf{a}t^2$$

$$-54.6\ \text{m} = (0) \cdot t + \frac{1}{2} \cdot (-9.8\ \frac{\text{m}}{\text{sec}^2}) \cdot (t)^2$$

$$t^2 = \frac{2 \cdot (-54.6\ \cancel{\text{m}})}{-9.8\ \frac{\cancel{\text{m}}}{\text{sec}^2}}$$

$$t = 3.3\ \text{sec}$$

So it takes <u>3.3 seconds</u> for an object to fall from the Leaning Tower of Pisa.

2.8 Since the rock is being dropped near the surface of the earth and since we ignore air resistance, we know the rock is in free fall. Its initial velocity is zero and its acceleration is -9.8 m/sec^2. However, since the problem specifically asks for the answer in feet, we need to use the other value for the acceleration due to gravity, -32 feet/sec^2. Since I have defined the acceleration as negative, this means that downward motion is negative. The problem also gives us the time it takes for the rock to fall, so Equation (2.19) is the one to use:

$$\Delta \mathbf{x} = \mathbf{v}_o t + \frac{1}{2}\mathbf{a}t^2$$

$$\Delta \mathbf{x} = (0)\cdot(1.4 \text{ sec}) + \frac{1}{2}\cdot(-32 \frac{\text{ft}}{\text{sec}^2})\cdot(1.4 \text{ sec})^2$$

$$\Delta \mathbf{x} = \frac{1}{2}\cdot(-32 \frac{\text{ft}}{\cancel{\text{sec}^2}})\cdot(1.96 \cancel{\text{sec}^2})$$

$$\Delta \mathbf{x} = -31 \text{ ft}$$

A negative displacement means that the rock traveled downward. Since the rock traveled down, the bridge is <u>31 feet</u> above the water. Notice that, once again, I did not worry about significant figures when I squared the 1.4 sec. That's because once the first term of the equation drops out, there are only multiplication and division in this equation, so I waited until the very end of the equation to deal with significant figures. Since both 1.4 and 32 each have two significant figures, the answer can have only two significant figures.

2.9 This problem gives us an initial velocity (8.7 m/sec). Now, if we keep the initial velocity as positive, this means that downward motion is negative. Thus, the acceleration is -9.8 m/sec^2. What do we do with the 120 meters? The problem asks us to determine the velocity when the rock reaches the bottom of the cliff. What will its displacement be at that point? Well, when the rock reaches the bottom of the cliff, it will be 120 meters below where it started. This means its displacement will be -120 m. So, we have initial velocity, acceleration, and displacement, and we want to determine final velocity. This means we will use Equation (2.15):

$$\mathbf{v}^2 = \mathbf{v}_o^2 + 2\mathbf{a}\cdot\Delta\mathbf{x}$$

$$\mathbf{v}^2 = (8.7 \frac{\text{m}}{\text{sec}})^2 + 2\cdot(-9.8 \frac{\text{m}}{\text{sec}^2})\cdot(-120 \text{ m})$$

$$\mathbf{v}^2 = 76 \frac{\text{m}^2}{\text{sec}^2} + 2{,}400 \frac{\text{m}^2}{\text{sec}^2} = 2{,}500 \frac{\text{m}^2}{\text{sec}^2}$$

$$\mathbf{v} = -5.0\times10^1 \frac{\text{m}}{\text{sec}}$$

Now why do I say that the velocity is negative? Remember from math that when you take a square root, you get both a positive and negative answer. Since we know that the rock is traveling down, we must choose the negative answer, because we defined negative as down. In the end, then, the rock is traveling down at 5.0×10^1 m/sec.

Notice how the significant figures work here. Since we are both adding and multiplying, we need to first do the multiplication (squaring 8.7 and multiplying the second term out). Then, we need to deal with significant figures before we add, since addition uses different rules. Thus, even though 8.7^2 is 75.69, I had to round it to 76 because 8.7 has only two significant figures. In the same way, the second term works out to 2,352, but it has to round to 2,400 because 9.8 and 120 have only two significant figures. Normally, I would not do this in the middle of an equation, but I must do it here, because once the multiplication is done, I have to add, and addition goes by different rules. When I add 76 and 2,400, I must go by precision, and 2,400 has its last significant figure in the hundreds place, so the answer must go out to the hundreds place. Thus, the answer is 2,500. When I took the square root, I went back to counting significant figures, since a square root involves multiplication. Thus, since 2,500 has only two significant figures, the answer can have only two significant figures. The answer must be expressed in scientific notation, since 50 has only one significant figure.

2.10 In this problem, we are given the maximum height (4.5 m) and asked to determine the initial velocity. Now, we know that at the maximum height, the ball's velocity is zero. We also know the acceleration due to gravity. Since the height is positive, the upward direction is positive. This means that the acceleration is -9.8 m/sec^2. So, we know final velocity, displacement, and acceleration, and we need to determine the initial velocity. This means that Equation (2.15) is once again our tool:

$$\mathbf{v}^2 = \mathbf{v}_o^2 + 2\mathbf{a} \cdot \Delta\mathbf{x}$$

$$(0)^2 = \mathbf{v}_o^2 + 2 \cdot (-9.8 \frac{\text{m}}{\text{sec}^2}) \cdot (4.5 \text{ m})$$

$$\mathbf{v}_o = 9.4 \frac{\text{m}}{\text{sec}}$$

Since we took a square root to get the answer, we have both a positive and a negative sign. Since the ball is thrown upwards, we choose the positive value. So the ball's initial velocity was 9.4 m/sec.

REVIEW QUESTIONS

1. What two conditions must be met before you can use the three equations we derived in this module?

2. What is air resistance? How does air resistance affect the acceleration of a falling object?

3. What is the definition of free fall?

4. A very picky physicist says that no object can really experience free fall when falling near the surface of the earth. Explain why the physicist is technically correct.

5. Even though the physicist in question 4 is technically correct, explain why we can still assume that most objects falling near the surface of the earth are, indeed, in free fall.

6. If an object is thrown up in the air, where is its velocity zero? When its velocity is zero, what is its acceleration?

7. On the moon, gravity is significantly weaker than on earth. Would you expect the acceleration due to gravity on the moon to be larger, smaller, or the same as compared to the acceleration due to gravity on earth?

8. A feather and a penny are put in a long glass tube and all of the air is sucked out of the tube. While inside the tube, will the feather fall faster than, slower than, or the same as the penny?

9. A ball is thrown up in the air with an initial velocity of 1.2 m/sec. If the ball is caught at exactly the same height from which it was thrown, what will its velocity be?

10. Suppose you have two balls of equal mass. Although their masses are equal, they are made of different substances. As a result, the first ball is significantly smaller than the second ball. Which ball has the highest terminal velocity?

PRACTICE PROBLEMS

1. An athlete runs at a constant velocity of 3.0 m/sec. How far will the athlete travel in 25 minutes?

2. A BMW 535I can accelerate from 0.0 to 60.0 mph in 7.9 seconds. What is its acceleration?

3. How far does the BMW travel during the acceleration described in problem #2?

4. A car, traveling at 25 m/sec, stops in 0.103 km. What is the deceleration provided by the brakes?

5. A rocket ship blasts off from rest and accelerates for 2.0×10^3 yards at 9.7×10^4 ft/sec^2. What is the rocket's velocity when it is finished accelerating?

6. In an amusement park, there is a free fall ride that drops riders straight down for 30.0 m. What velocity is the ride traveling at when it reaches the bottom of this straight drop?

7. A physics student drops a rock off of a cliff. If the rock takes 3.5 seconds to reach the bottom of the cliff, how tall is the cliff (in meters)?

8. A parachutist falls from a plane for 5.0 seconds before opening the parachute. How fast is he traveling when he opens his parachute?

9. A child throws a ball up in the air with an initial velocity of 3.9 m/sec. What is the maximum height that the ball reaches?

10. A physics student stands on a ladder 12 feet above the ground and throws a ball up into the air. If it takes the ball 4.2 seconds to hit the ground, what was the ball's initial velocity?

MODULE #3: Two-Dimensional Vectors

Introduction

In the previous two modules, we learned some very valuable ways to analyze motion as well as several equations that allow us to calculate many of the important variables which describe motion. While those concepts and equations are very powerful, they have their limitations. One of the biggest limitations is the fact that they only apply to one-dimensional motion. Even though there are many practical applications for one-dimensional motion, much of the motion that occurs in real life is two-dimensional.

In this module, we will begin to learn how to analyze two-dimensional motion. To make sure you understand what I am talking about, however, let me define two-dimensional motion:

Two-dimensional motion - Motion that occurs in a plane

While two-dimensional motion *is* more difficult to analyze and understand than one-dimensional motion, it's really not that bad. We don't need to learn any new physics, but we do need to learn a new mathematical concept. That's what we'll concentrate on in this module.

Vectors

You should already have learned about vectors as a part of your mathematics education. If you haven't specifically learned about vectors, you should have at least learned about the trigonometric functions (sine, cosine, and tangent) that are used in vector mathematics. While I usually do not like to review math concepts that you should already know, it is valuable to go through a careful review of vectors. Math courses often do not teach this concept in a way that students can apply to physics. So in this module, I will give you a physicist's version of vectors. Even if you think you are fully competent on the subject of vectors, please read through this module carefully and perform all of the problems. If you really know it all, this module should be a breeze!

Now we've already learned a little about vectors. You should remember from Module #1 that vector quantities contain information about direction. In one-dimensional motion, the direction information was contained in the sign of the vector. A positive sign indicated motion in one direction, while a negative sign indicated motion in the opposite direction. In two-dimensional motion, however, direction is much more difficult to keep track of.

Let's suppose you're on a hike. Your map tells you to walk 1.2 miles due east and 4.3 miles due north in order to get to your destination. Once you get there, how would you describe your displacement relative to where you started? Of course, one way would be to simply say that you are 1.2 miles east and 4.3 miles north of your original location. However, there is a better way to describe your position. To illustrate this, let's look at your hike on a Cartesian (kar tee' shun) coordinate plane:

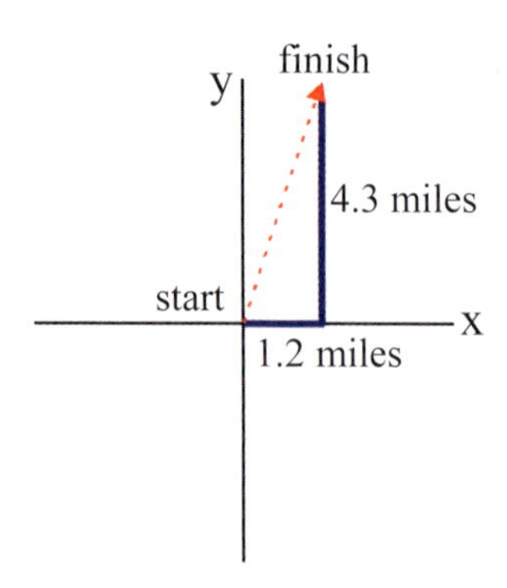

Since we can think of any two-dimensional motion in terms of an x/y graph, we will be using the Cartesian coordinate plane quite a bit. For this particular example, we represent eastern motion as motion along the positive x-axis. Northern motion is along the positive y-axis. If there were southern motion, we would represent it as motion in the negative y direction. Likewise, western motion would be represented as motion in the negative x direction.

Now that we have that down, let's look at the graph. In order to get to your destination, you moved 1.2 miles in the positive x-direction (east) and 4.3 miles in the positive y-direction (north). The position of your destination relative to your starting place is represented by the dashed arrow. The length of that arrow represents the distance between your starting position and your destination. The arrow points out the direction from your starting position to your destination. Notice that this arrow, then, represents both distance and direction. What do you have when you are given both distance and direction? You have displacement. Thus, the arrow represents the displacement vector.

In general, we use arrows to represent two-dimensional vectors. The length of the arrow is called the **magnitude** of the vector, and the arrow points out the direction of the vector. The magnitude of a vector tells us how much. In other words, the magnitude of a displacement vector tells us the distance, while the magnitude of a velocity vector tells us the speed. This is the first important point concerning vectors:

Arrows represent two-dimensional vectors. The length of the arrow is the magnitude of the vector, while the arrow points out the direction.

How would we calculate the length of the vector pictured above? Well, since the x- and y-axes in the Cartesian coordinate system are perpendicular, the two heavy blue lines and the arrow pictured above form a right triangle. Thus, we can use the Pythagorean theorem to calculate the arrow's length:

$$\text{magnitude} = \sqrt{(1.2\text{ miles})^2 + (4.3\text{ miles})^2} = 4.5\text{ miles}$$

Since we know that the magnitude of this vector represents the distance from the starting point to the destination, we can now say that your destination is 4.5 miles away from your starting point.

Okay, so now we know how to tell someone the distance between the starting place and the destination, but how do we communicate direction? We need some way of describing where the arrow points. How do we do that? Here's how:

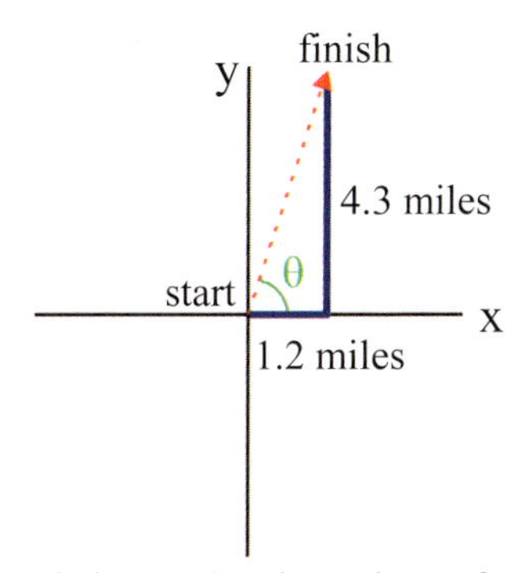

The angle between the positive x-axis and the arrow is unique for each possible direction that the arrow can point. Thus, we can use this angle to describe the direction that the arrow points. This is the second major point concerning two-dimensional vectors:

The angle between the arrow representing a two-dimensional vector and the positive x-axis provides the direction of the vector.

Since the magnitude tells us how much and the angle tells us direction, we now have everything we need to define a two-dimensional vector.

Well, if the angle between the positive x-axis and the arrow represents the direction of the vector, it is important to be able to calculate it. This is where you must know some trigonometry. If you have not had enough trigonometry to know what sine, cosine, and tangent mean, you can go no further in this course, because there is no way I can review all of the information that you need to know. Thus, if you have not already learned the three basic trigonometric functions (sine, cosine, and tangent) and how they are defined on a right triangle, put this course away until you have.

If you have had some trigonometry, you should see how to calculate the value for that angle. Since the x- and y-axes in a Cartesian coordinate system are perpendicular to each other, the triangle formed in the picture above is a right triangle. The trigonometric functions sine, cosine, and tangent are based on a right triangle. In particular, the tangent is defined as follows:

$$\tan(\theta) = \frac{\text{opposite side of the triangle}}{\text{adjacent side of the triangle}} \tag{3.1}$$

Equation (3.1) means that in order to calculate the tangent of the angle, you need to take the length of the leg across from the angle and divide it by the leg that is adjacent to the angle, but not the hypotenuse. For the angle we are interested in, what would that be? Well, the side of the triangle that is across from the angle has a length of 4.3 miles. The arrow is the hypotenuse, and we ignore that when calculating the tangent. Thus, the side that is adjacent to the angle (but is not the hypotenuse) has a length of 1.2 miles. Equation (3.1), then, becomes:

$$\tan(\theta) = \frac{4.3\ \cancel{\text{miles}}}{1.2\ \cancel{\text{miles}}}$$

$$\tan(\theta) = 3.6$$

At this point, we have the tangent of the angle but we do not have the angle itself. From math class, you should have learned that to solve for θ in the above equation, you simply need to take the **inverse tangent** of both sides. When you do this, you get:

$$\theta = \tan^{-1}(3.6) = 74^{\circ}$$

Try this on your calculator to make sure you know how to calculate the inverse tangent. If you do not get 74 degrees, you could have one of two problems. Your calculator may not be in degrees mode. Look at your calculator manual for instructions on changing from degrees to radians to gradients. Make sure it is in degrees mode. If you still do not get the correct answer (74.475889 ignoring significant figures), you probably are not using your calculator properly.

Now we have both the magnitude of the vector (4.5 miles) and its direction (74°). So, to describe your position to someone else, you would say, "I am 4.5 miles away from my starting position, at a heading of 74 degrees." When you say this, a person need not follow your route to get to you. Rather than walking 1.2 miles east and 4.3 miles north (for a total of 5.5 miles), a person could just head out at an angle of 74° and walk 4.5 miles, cutting a full mile off of the total trip.

Adding and Subtracting Two-Dimensional Vectors: The Graphical Approach

So now we know that in two dimensions, vectors are represented by arrows. The magnitude of the vector tells us "how much," while the angle of the vector relative to the positive x-axis tells us the direction. Now that we have this knowledge under our belts, we need to figure out how to add and subtract vectors. After all, in most of our one-dimensional problems, we were adding and subtracting vectors all of the time. We need to do the same thing in two dimensions as well.

It turns out that there are two approaches to adding and subtracting vectors. One is the graphical method, which allows you to visualize the resulting vector. The other is the analytical method, which allows you to get precise values for the magnitude and direction of the result. In physics, we need to be able to do both of these things. We'll start with the easiest of the two: the graphical method.

Let's suppose you are on a hike and you have a set of directions. These directions tell you to travel 1.2 miles at a heading of 60.0 degrees and then turn to a heading of 135 degrees and travel 3.4 miles. Where would your final destination be? Well, notice that you were given two vectors that are two-dimensional. In a Cartesian coordinate plane, your travel would look something like this:

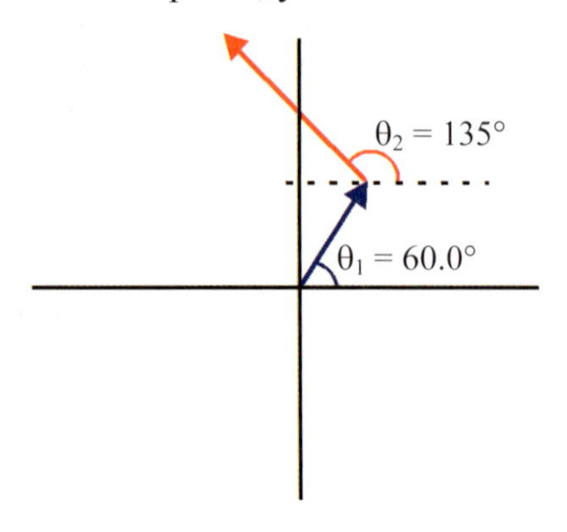

In this picture, the blue arrow represents the first leg of your journey, when you traveled 1.2 miles at a heading of 60 degrees. The red arrow represents the second leg of your journey, when you turned to an angle of 135 degrees and traveled for 3.4 miles. Notice that the direction is *always* defined relative to the positive x-axis. Thus, the 135° angle is defined relative to an imaginary line that runs parallel to the positive x-axis. An angle of 135° relative to that imaginary line is also 135° relative to the positive x-axis. Notice also that since the second leg of your journey was longer than the first leg, the length of the second arrow is larger than the length of the first arrow.

The end of your journey, then, is represented by the tip of the second arrow. After all, the red arrow represents the second leg of your journey, and the tip of the second arrow is the end of that leg. Thus, once you have completed the trip, you are at the tip of the second arrow. What vector describes your final destination? As we already learned, in order to construct a two-dimensional vector, we draw an arrow from the starting point to the finish. Thus, the dashed green arrow below represents your final destination:

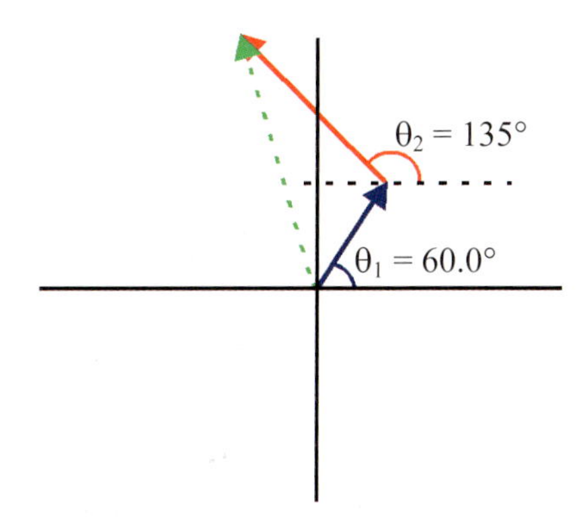

I have just guided you through the process of adding vectors graphically. Notice what I did. I was given two vectors. In order to add them together, I took the first vector and drew it starting at the origin. I then took the beginning of the second vector (which is called the "tail" of the vector) and started it at the end of the first vector (which is called the "head" of the vector). The arrow that then goes from the tail of the first vector to the head of the second vector completes the triangle and represents the sum of the two vectors.

You need to be completely comfortable with adding vectors in this way. Thus, you must first remember the steps we just did:

When adding vectors graphically, take the tail of the second vector and place it at the head of the first vector. The vector that then completes the triangle by starting at the tail of the first and pointing to the head of the second represents the sum of the two vectors.

Now please realize that when you are asked to add vectors, they might not be as conveniently placed as the ones in my example. In fact, in order to add vectors graphically, you must often move the second vector so that its tail is resting on the head of the first vector. That's perfectly fine. You can move a vector anywhere you want, as long as you do not change its magnitude (length) or direction (angle relative to the x-axis). See what I mean by studying the following example problem:

EXAMPLE 3.1

Given vectors A and B as drawn below, draw the vector that represents the sum A + B. Also draw the vector that represents B + A.

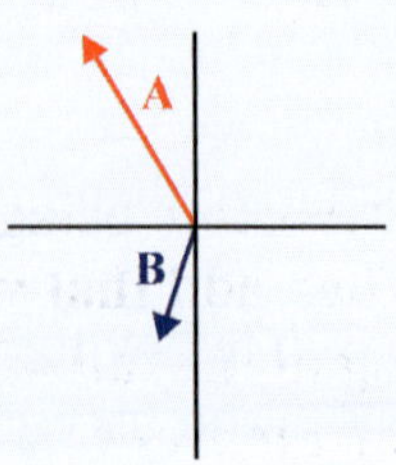

In order to solve this problem, we simply follow the rules established above. Realizing that we can move vector **B** around as long as we do not change its magnitude or direction, I will move it until its tail rests on the head of vector **A**:

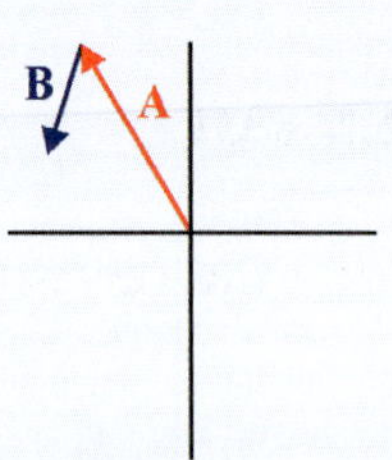

Notice that even though I moved vector **B**, its direction is still the same, as is its length. Now that I have moved it into the proper position, the sum of these two vectors is illustrated by drawing an arrow from the origin to the head of vector **B**:

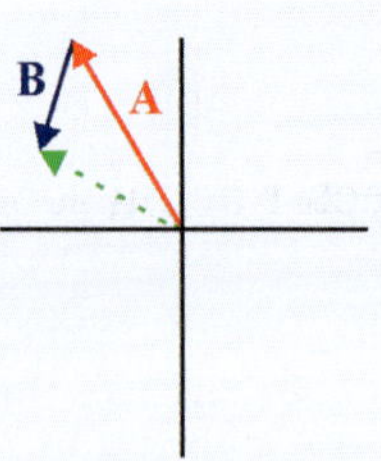

The dashed arrow, then, represents **A** + **B**. To calculate the sum **B** + **A**, we do the same thing, but this time, we move vector **A**, because it is the second vector:

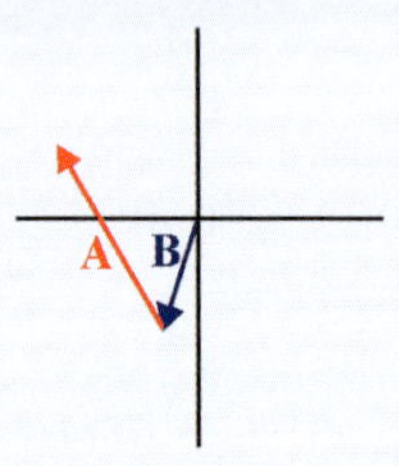

Now, since vector **A** is the second vector this time, the sum is represented by drawing an arrow from the origin to the head of vector **A**:

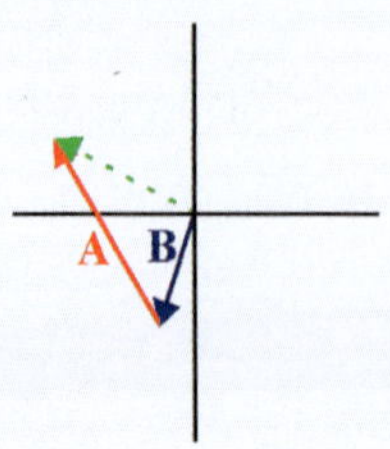

Notice that the dashed vector, which represents **B** + **A**, is the same size and direction as the vector representing **A** + **B**. This is a general rule. Vector addition is commutative, just as scalar addition is.

What about subtracting vectors? Well, it is nearly the same as adding them. When you need to compute **A** - **B**, you do it by taking the vector **A** and adding it to the negative of **B**. How do we determine the negative of a vector? As we learned in one-dimensional motion, the positive or negative sign of a vector denotes direction. Thus, the negative of a given vector is the same size vector, going in precisely the opposite direction.

When subtracting vectors, take the vector being subtracted and make it point in precisely the opposite direction. Then add that vector to the first. The vector that then completes the triangle by starting at the tail of the first and pointing to the head of the second represents the difference between the two vectors.

If that all seems a little confusing, the following example will hopefully clear things up.

EXAMPLE 3.2

Given vectors A and B as drawn below, draw the vector that represents A - B, and draw the vector that represents B - A.

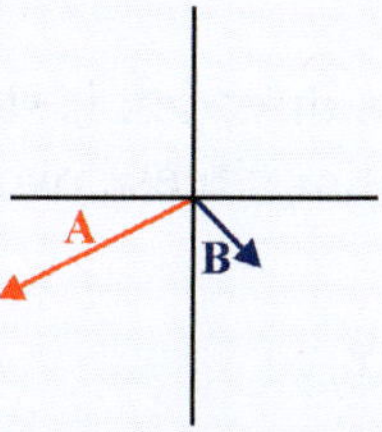

To subtract vectors, we must first take the vector being subtracted and point it in precisely the opposite direction:

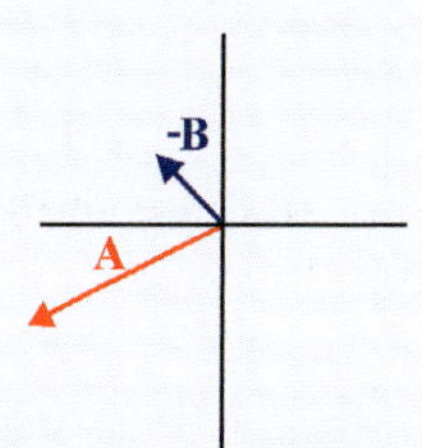

Now that we have **-B**, we simply add it to **A**:

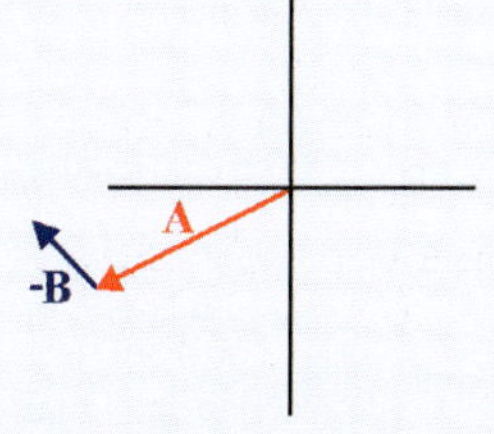

The arrow that is now drawn from the tail of the first vector to the head of the second will represent the difference **A** - **B**:

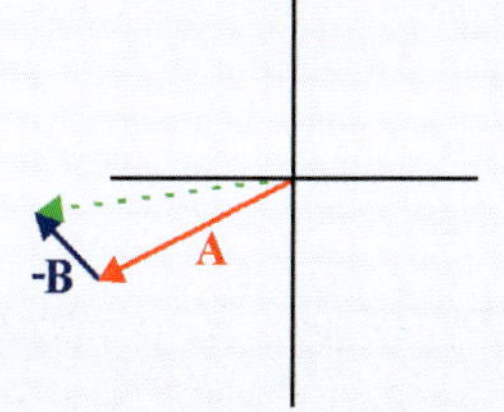

Drawing the vector that represents **B** - **A** is done in precisely the same way, but this time, we take the negative of vector **A**, because that's the vector being subtracted. The end result is as follows:

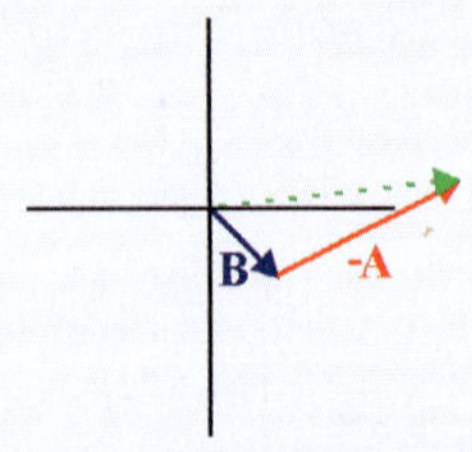

Notice that the dashed arrow in this case (representing **B** - **A**) is different from the previous dashed arrow (representing **A** - **B**). From the two drawings, you should be able to see that these vectors have the same magnitude, but opposite directions. That makes them different vectors, which tells us that even though vector addition is commutative, vector subtraction is not.

Even though these exercises might seem a bit pointless right now, we will see in the next module that the skills I have discussed here are absolutely critical in analyzing two-dimensional situations in physics. Thus, you need to become very good at adding and subtracting vectors in a graphical way. See if you can by solving the "On Your Own" problems that follow.

ON YOUR OWN

Given the vectors below:

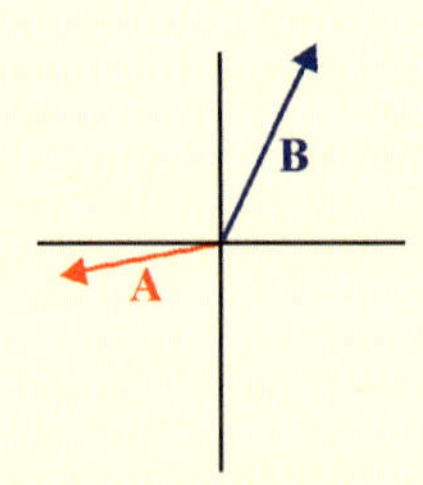

3.1 Draw an arrow that represents the vector **A** + **B**.

3.2 Draw an arrow that represents **B** - **A**.

Vector Components

The graphical method for adding vectors is a useful technique, but it only gives us part of the information that we need to analyze physical situations. After all, the graphical method for adding and subtracting vectors allows us to visualize the vector that results from the addition or subtraction, but it does not allow us to get numbers for the magnitude and direction of the vector. If we need hard and fast numbers to describe the vector in question, we turn to the analytical method.

In order to use the analytical method, however, we need to learn more about two-dimensional vectors. In order to do this, let's go back and look at the very first vector we considered in this module. We said that if you walked 1.2 miles east and 4.3 miles north, your final destination could be represented by the red arrow (vector A) drawn on the next page.

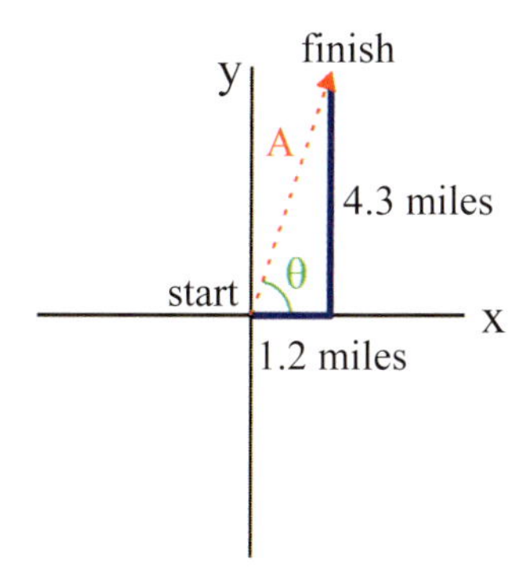

Rather than concentrating on the vector this time, however, let's look at the blue lines in the picture. Remember what these heavy lines mean. They represent how far you walked in each direction to get to the destination. The heavy line that goes along the x-axis is called the **x-component** of vector **A**, because it tells you how far along the x-axis you traveled to get to the destination. We usually label this component "A_x." In the same way, the heavy vertical line is called the **y-component** of vector **A**, because it represents the distance you had to travel in the y-direction. It is usually labeled "A_y."

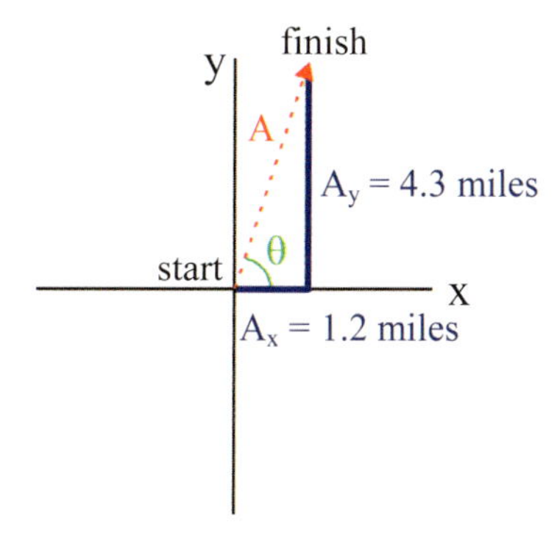

The important thing to realize is that *all two-dimensional vectors* have x and y components, and it turns out that these components are critical in being able to analyze the vector.

All two-dimensional vectors have x- and y-components.

In order to be able to add and subtract vectors using the analytical method, we need to see how these components relate to the overall vector.

Remember what we did with this vector earlier? In order to calculate the magnitude of the vector, we recognized that the triangle was a right triangle and that the vector was the hypotenuse of that triangle. Thus, we could relate the components to the vector's magnitude using the Pythagorean theorem. This is a general rule. If you have the x- and y-components of a vector (we'll call it vector **A**), you can calculate its magnitude with the following equation:

$$\text{magnitude of A} = \sqrt{A_x^{\ 2} + A_y^{\ 2}} \tag{3.2}$$

What about direction? Remember how we did it before? We said that the angle of the vector was given by the inverse tangent of the opposite leg of the triangle divided by the adjacent leg. Well, based on the fact that we must define the angle of a vector relative to the positive x-axis, the y-component of the vector will *always* be the opposite leg of the right triangle, and the x-component will *always* be the adjacent leg. As a result, the following general equation will allow us to calculate the direction of a vector if we have its components:

$$\theta = \tan^{-1}\left(\frac{A_y}{A_x}\right)^{**} \tag{3.3}$$

Notice that I have two asterisks on Equation (3.3). I will explain them in a moment. First, however, I want you to get some hands-on experience using Equations (3.2) and (3.3).

EXPERIMENT 3.1
Vector Components

Note: A sample set of calculations is available in the solutions and tests guide. It is with the solutions to the practice problems.

Supplies:

- Safety goggles
- Modeling clay (Play-Doh® or Silly Putty® will work as well.)
- A pencil (It should be at least 6 inches long and have a sharp point on the end.)
- A wooden ruler (It needs to have a flat end at zero inches.)
- A plain 8.5-in. x 11-in. sheet of paper
- A pen
- A protractor

1. Measure the length of the pencil, from the end of the eraser to the tip of the pencil's point.
2. Place the piece of paper on a flat surface.
3. Make a lump of modeling clay and place it on the paper.
4. Stick the eraser end of the pencil into the lump of clay. The pencil should be at an angle, and the top of the eraser should protrude out the back of the lump, while most of the pencil sticks out the other side.
5. Make a mark to indicate where the end of the eraser touches the paper.
6. Work with the clay so that the pencil remains pointing up at an angle even when you let go of it (See the drawing under step 8) .
7. Place the ruler upright so that its end at zero inches is flat against the paper.
8. Move the ruler so that it is very, very close to the point of the pencil. It should not actually touch the pencil, because that will move the pencil. Just get it as close as possible without actually touching it. In the end, your setup should look something like this:

Illustration by Megan Whitaker

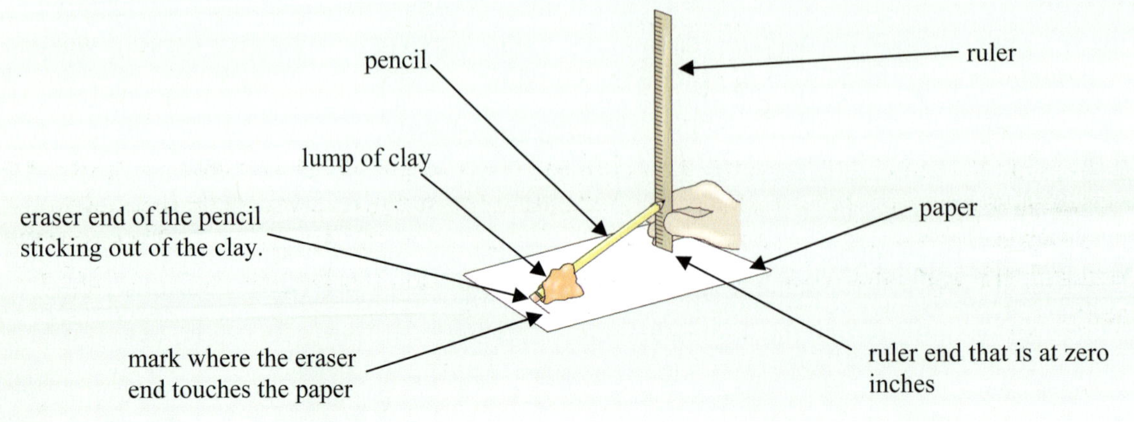

9. Keep the end of the ruler flat against the paper, so that there is a 90° angle between the paper and the ruler.
10. Record the mark on the ruler at which the pencil is pointing. As long as the ruler makes a 90° angle with the paper, this is the height of the angled pencil.
11. Use the pen to mark where the face of the ruler touches the paper.
12. Measure the distance between that mark and the mark that indicates where the eraser end touches the paper. This is the horizontal distance spanned by the pencil.
13. Use the protractor to measure the angle at which the pencil is pointing.
14. What do we have here? Well, the pencil represents a vector. Its length is the magnitude of the vector, and the angle you measured in the previous step is the direction of the vector. In step 10, you measured the vertical distance spanned by the vector. That's its y-component. In step 12, you measured the horizontal distance spanned by the vector. That's its x-component.
15. Use Equation (3.2) and the two components you measured to calculate the magnitude of the vector. Compare the result to the length of the pencil. The numbers should be equal to each other, within about 10%, which is a reasonable error margin for this experiment.
16. Use Equation (3.3) and the two components you measured to calculate the angle at which the vector points. Compare that to the angle you measured with your protractor. Once again, these numbers should be within about 10% of each other.
17. Repeat this experiment with the pencil pointing at a noticeably different angle.
18. Clean up your mess.

Now that you've had some hands-on experience with Equations (3.2) and (3.3), we need to spend a little more time discussing Equation (3.3). Remember, there are asterisks next to Equation (3.3). That's because this equation is a little harder to use than the experiment led you to believe. You see, when you use a calculator to take the inverse tangent of a number, the angle it gives you as the answer is not always defined the same way that we define it. In order to keep all of our vectors consistent, physicists like to always define the angle of a vector starting at the positive x-axis. Your calculator, however, gives you the result of the inverse tangent function relative to the *nearest* x-axis (often called a "reference angle"), which might, indeed, be the negative x-axis. In addition, your calculator can also give you negative angles as a result of the inverse tangent function. Physicists, on the other hand, like to keep their vector angles positive. All of this can lead to some real confusion if we don't spend some time on it right now.

You should remember from algebra that when we look at a Cartesian coordinate plane, we can divide it into four regions:

When taking the inverse tangent of a number, the definition of the angle that your calculator gives you depends on which of these regions the vector is in. If the vector is in region I, the angle that your calculator gives you is defined relative to the positive x-axis, just as physicists define it. Thus, if your vector is in region I, the angle that your calculator gives you for the inverse tangent function will be defined properly.

However, if the vector is in region II of the Cartesian coordinate plane, the angle that your calculator gives you is defined relative to the negative x-axis and is negative. In the Cartesian coordinate system, negative angles mean clockwise rotation while positive angles mean counterclockwise rotation. So, when a vector is in region II, your calculator gives you the number of degrees clockwise from the negative x-axis. Thus, if your vector is in region II, and your calculator gives you a direction of -60°, this is what it means:

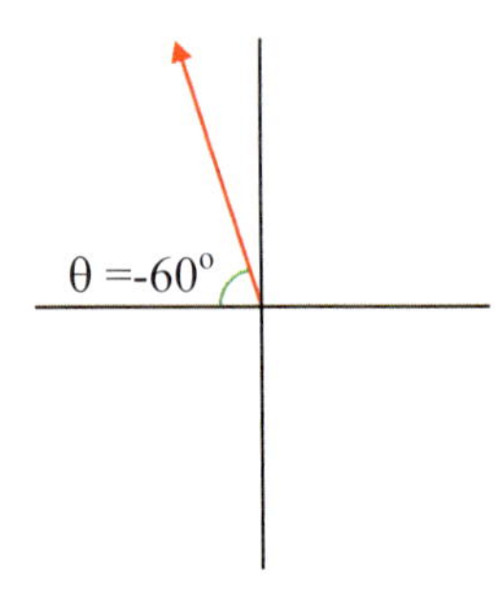

This definition of angle definitely does not fit in with our definition of angle. As a result, we must convert it to our definition. It turns out that in both regions II and III, if you simply add 180 to the angle that your calculator gives you, you will have converted your calculator's answer to one that is consistent with our angle definition. If the vector is in region IV, you must add 360 to the calculator's answer in order to get the properly defined angle.

In summary, Figure 3.1 tells you what you must do in order to change your calculator's answer into a properly defined vector angle, based on the region of the Cartesian coordinate system.

FIGURE 3.1
Converting Reference Angle to Vector Angle*

+ 180°	do nothing
+ 180°	+ 360°

*Please note: The angles in this figure are *exact*. Thus, they have infinite precision.

As the note indicates, these angles are exact. As a result, the 180° is really 180.00000…°. Thus, if your vector is in regions II-IV, you will be adding an infinitely precise number to the angle that you get from your calculator. As a result, the precision of the angle from your calculator will determine the precision of your final answer.

This figure, of course, does you no good if you can't tell what region of the Cartesian coordinate plane your vector is in. Luckily, however, this is not a difficult task. All you have to do is look at the signs on the vector components. If the x-component and y-component are both positive, the vector must be in region I. After all, a positive x-component indicates that you are to the right of the origin, while a positive y-component means you are above the origin. The region that is both to the right of and above the origin is region I. On the other hand, suppose that both components are negative. Since a negative x-component means left of the origin, and a negative y-component means below the origin, you must be in region III, since that is the only region that is to the left of and below the origin. Do you see how to do this? Following the same logic, if the x-component is negative and the y-component is positive, the vector must be in region II. If the x-component is positive and the y-component is negative, however, the vector must be in region IV.

Now if all of this has been a bit confusing, don't worry. Study the next two example problems and solve the "On Your Own" problems that follow. That should clear up any confusion you have regarding vector magnitudes and directions and how they relate to the vector's x- and y-components.

EXAMPLE 3.3

Vector C has the following components: C_x = -1.2 m, C_y = 3.4 m. What are the magnitude and direction of the vector?

The magnitude is easy. We just use Equation (3.2):

$$\text{magnitude} = \sqrt{C_x^{\ 2} + C_y^{\ 2}} = \sqrt{(-1.2\text{ m})^2 + (3.4\text{ m})^2} = 3.6\text{ m}$$

We can get the direction of the vector by using Equation (3.3):

$$\theta = \tan^{-1}\left(\frac{C_y}{C_x}\right) = \tan^{-1}\left(\frac{3.4\ \cancel{\text{m}}}{-1.2\ \cancel{\text{m}}}\right) = \tan^{-1}(-2.833) = -71^\circ$$

In order to make sure you can use your calculator properly, I would suggest going through the calculation above with your own calculator and making sure you get the same answer. If you do not, consult your calculator instructions to make sure you are using the inverse tangent function correctly and that your calculator is in degrees mode.

Of course, this is not necessarily the final answer. Since your calculator does not define angle the same way that we do, we must convert this angle to the properly-defined vector angle. To do this, we must determine the region of the Cartesian coordinate system in which the vector is located. Since the x-component is negative and the y-component is positive, the vector lies to the left of and above the origin. This is region II. In that region, we must add 180° to the calculator's answer to get the properly defined angle:

$$-71^\circ + 180^\circ = 109^\circ$$

We find, then, that the vector has a magnitude of <u>3.6 m and a direction of 109°</u>.

Before we move on to the next problem, I want to discuss significant figures for a moment. When we calculated the magnitude of the vector, we had to square the numbers, add the results, and then take the square root. How do we handle the significant figures here? If we were just taking a square root, we would consider this a multiplication problem and would count significant figures to determine where to round. However, that's not what we are doing here. Since we square, add, and then take the square root, the act of taking the square root negates the act of squaring, at least from the view of significant figures. So, in terms of significant figures, this is an addition problem. Thus, we use decimal place to determine where to round the answer. Since both the x-component and y-component go out to the tenths place, the answer must be reported to the tenths place as well.

How do we deal with significant figures in a trigonometric function like the inverse tangent? From a significant figures point of view, we will treat this like multiplication and division. Please note that determining how errors propagate through trigonometric functions is actually a lot more complex than that, but for the purpose of this text, we will treat significant figures in trigonometric functions as if they were multiplication and division problems. It is a simplification, but it is one that we will live with. Remember, however, that we wait until the end of the equation to actually take care of the significant figures. Thus, when I divided 3.4 m by 1.2 m, I did not worry about significant figures. However, after I got the inverse tangent of that quotient, I looked at the x-component and y-component and noted that they each had two significant figures. That's why I limited my answer to two significant figures. That was not the end of the problem, however. In the final step of the solution, I added -71° to 180°, which you should remember is exact. Since I added two numbers, I had to look at decimal place, and since 71° went out to the ones place and 180 is exact and thus is infinitely precise, the answer had to be reported to the ones place. That's why the final angle was 109°.

I know that this is a lot to deal with right now, but it will get easier. Just remember that calculating the magnitude should be treated as an addition problem from the standpoint of significant figures, while using trigonometric functions should be treated as a multiplication/division problem.

A velocity vector (V) has components V_x = 2.1 m/sec and V_y = -4.2 m/sec. What is the magnitude and direction of the vector?

Now don't get confused because of the units here. I realize that I have discussed vectors totally in terms of displacement, but all that we have discussed is applicable to *any* vector quantity. So don't worry that we're suddenly talking about velocity instead of displacement. Vectors are vectors. So, to get the magnitude of this vector, we use Equation (3.2):

$$\text{magnitude} = \sqrt{V_x^2 + V_y^2} = \sqrt{(2.1\,\frac{\text{m}}{\text{sec}})^2 + (-4.2\,\frac{\text{m}}{\text{sec}})^2} = 4.7\,\frac{\text{m}}{\text{sec}}$$

Note that since both components have their last significant figures in the tenths place, the magnitude must also have its last significant figure in the tenths place.

To find the direction of the vector, we use Equation (3.3) :

$$\theta = \tan^{-1}(\frac{V_y}{V_x}) = \tan^{-1}\left(\frac{-4.2\,\frac{\cancel{\text{m}}}{\cancel{\text{sec}}}}{2.1\,\frac{\cancel{\text{m}}}{\cancel{\text{sec}}}}\right) = \tan^{-1}(-2.00) = -63^{\circ}$$

Since each component has only two significant figures, the angle can have only two significant figures. Also, please note that I did not worry about the significant figures until the end of the equation.

We cannot use the answer we get from this equation until we determine the region of the Cartesian coordinate plane in which the vector is located. This vector has a positive x-component and a negative y-component, which means it is to the right of and below the origin. Thus, the vector is in region IV. This means that we add 360 to the answer that the calculator gives us:

$$-63° + 360° = 297°$$

Note that this is addition and 360° is exact. Since 63° has its last significant figure in the ones place, the final angle must have its last significant figure in the ones place. The vector, therefore, has a magnitude of 4.7 m/sec and a direction of 297°.

ON YOUR OWN

3.3 The velocity vector of a car has an x-component of 23 m/sec and a y-component of 11 m/sec. What are the magnitude and direction of the velocity vector?

3.4 A person walks due west (to the left on a Cartesian coordinate plane) for 1.1 miles. She then walks due south (down on a Cartesian coordinate plane) for 2.0 miles. What is the magnitude and direction of her displacement vector?

Determining a Vector's Components from Its Magnitude and Direction

Believe it or not, there's one more thing we need to learn regarding vectors and their components. We know how to get the magnitude and direction of a vector if we have its components. But how do we go the other way? How do we get a vector's components if we have its magnitude and direction? Well, let's look at our original diagram again:

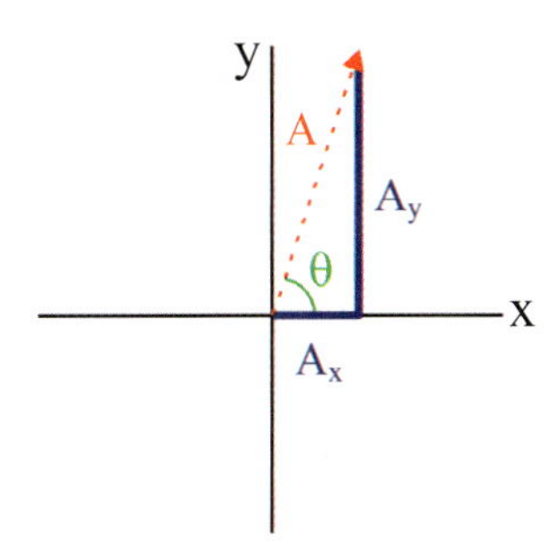

You should remember from trigonometry that the sine and cosine functions are defined as follows:

$$\sin(\theta) = \frac{A_y}{A} \tag{3.4}$$

$$\cos(\theta) = \frac{A_x}{A} \tag{3.5}$$

We can rearrange Equation (3.4) in order to calculate the y-component of the vector:

$$A_y = A \cdot \sin(\theta) \tag{3.6}$$

Similarly, we can rearrange Equation (3.5) to calculate the x-component of the vector:

$$A_x = A \cdot \cos(\theta) \tag{3.7}$$

Based on Equations (3.6) and (3.7), then, to calculate the x-component of a vector, you merely take the magnitude of the vector and multiply by the cosine of the vector's angle. The y-component is calculated in a similar way, except that the sine of the angle is used instead of the cosine. In order for these equations to work, however, the angle must be defined counterclockwise relative to the positive x-axis, as a vector's angle should be defined. Make sure you understand how to use these equations by studying the next example and solving the "On Your Own" problem that follows.

EXAMPLE 3.4

What are the x- and y-components of an acceleration vector that has a magnitude of 3.4 m/sec² and a direction of 230°?

To calculate the x-component of this vector, we simply use Equation (3.7):

$$A_x = A \cdot \cos(\theta)$$

$$A_x = (3.4 \frac{m}{sec^2}) \cdot \cos(230°) = -2.2 \frac{m}{sec^2}$$

The y-component of the vector can be calculated using Equation (3.6):

$$A_y = A \cdot \sin(\theta)$$

$$A_y = (3.4 \frac{m}{sec^2}) \cdot \sin(230°) = -2.6 \frac{m}{sec^2}$$

This tells us that the vector's x-component is -2.2 m/sec², and its y-component is -2.6 m/sec². Note that both components are negative, which means that the vector points below and to the left of the origin. Thus, the vector occupies region III of the Cartesian coordinate plane.

ON YOUR OWN

3.5 An acceleration vector has a magnitude of 55 m/sec² and an angle of 300.0°. What are its x- and y-components?

Adding and Subtracting Two-Dimensional Vectors: The Analytical Approach

Now that we have Equations (3.2), (3.3), (3.6), and (3.7), we can finally learn how to add and subtract vectors using the analytical approach. These equations are important because in order to add and subtract vectors analytically, you must first break them down into their x- and y-components. This requires Equations (3.6) and (3.7). Once you have done that, you can add or subtract the x-components of the vectors just like you add or subtract numbers. The result will be the x-component of the answer. You can do the same to the y-component, getting the y-component of the answer. Once you have both the x- and y-components of the answer, you can then use Equations (3.2) and (3.3) to get the magnitude and direction of the final vector. Study the next two examples to make sure you understand what I'm talking about.

EXAMPLE 3.5

Vector A has a magnitude of 34.0 m/sec and a direction of 45 degrees. Vector B has a magnitude of 20.0 m/sec and a direction of 123 degrees. What is the sum of the vectors?

In this problem, we are asked to add two-dimensional vectors together. We can't simply add the magnitudes of the vectors and then add the directions, because that is just not how two-dimensional vectors work. Since the directions of the two vectors are different, we must add them in a very specific way in order to get the correct answer. Before we do this analytically, however, let's do it graphically. We will simply estimate the direction and relative lengths of the vectors in order to give us some idea of what things look like:

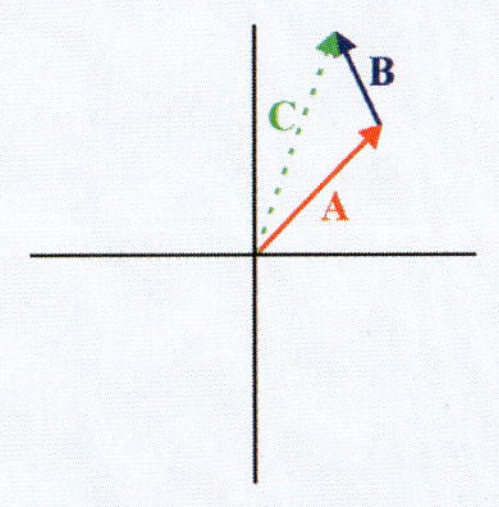

As we learned before, the dashed arrow (vector **C**) gives us the sum. We just did this to get an idea of what the final vector should look like. We didn't really need to do this to get the answer, but it helps us visualize exactly what's going on. Now let's add the vectors analytically. The first step is to break both vectors down into their components using Equations (3.6) and (3.7):

$$A_x = (34.0\,\frac{\text{m}}{\text{sec}}) \cdot \cos(45^\circ) = 24\,\frac{\text{m}}{\text{sec}}$$

$$A_y = (34.0\,\frac{\text{m}}{\text{sec}}) \cdot \sin(45^\circ) = 24\,\frac{\text{m}}{\text{sec}}$$

$$B_x = (20.0\,\frac{\text{m}}{\text{sec}}) \cdot \cos(123^\circ) = -10.9\,\frac{\text{m}}{\text{sec}}$$

$$B_y = (20.0\,\frac{\text{m}}{\text{sec}}) \cdot \sin(123^\circ) = 16.8\,\frac{\text{m}}{\text{sec}}$$

Note that since the angle for vector **A** (45°) has only two significant figures, each component of vector **A** has two significant figures. However, since both the magnitude (20.0 m/sec) and direction (123°) of vector **B** have three significant figures, the components of vector **B** also have three significant figures.

Now that we have the individual components, we can add them like numbers. Let's call the sum vector **C**, as we did in the drawing above. We can get the x-component of vector **C** by just adding the two x-components together:

$$C_x = A_x + B_x = 24\,\frac{\text{m}}{\text{sec}} + -10.9\,\frac{\text{m}}{\text{sec}} = 13\,\frac{\text{m}}{\text{sec}}$$

Please note that we are adding, so we look at decimal place instead of counting significant figures. Since the least precise number (24) has its last significant figure in the ones place, the answer must also have its last significant figure in the ones place.

The y-component of our answer can be calculated by simply adding up the y-components of the two vectors:

$$C_y = A_y + B_y = 24\,\frac{\text{m}}{\text{sec}} + 16.8\,\frac{\text{m}}{\text{sec}} = 41\,\frac{\text{m}}{\text{sec}}$$

Now that we have the components to our answer, we can use Equations (3.2) and (3.3) to give us the magnitude and direction of the answer. To get the magnitude of this vector, we use Equation (3.2):

$$\text{magnitude} = \sqrt{C_x^{\;2} + C_y^{\;2}} = \sqrt{(13\,\frac{\text{m}}{\text{sec}})^2 + (41\,\frac{\text{m}}{\text{sec}})^2} = 43\,\frac{\text{m}}{\text{sec}}$$

As I noted before, when we do a calculation like this, we treat it as an addition problem when it comes to significant figures. Since each of the components have their last significant figure in the ones place, the answer must have its last significant figure in the ones place.

To find the direction of the vector, we use Equation (3.3):

$$\theta = \tan^{-1}\left(\frac{C_y}{C_x}\right) = \tan^{-1}\left(\frac{41\,\frac{\cancel{\text{m}}}{\cancel{\text{sec}}}}{13\,\frac{\cancel{\text{m}}}{\cancel{\text{sec}}}}\right) = 72^{\circ}$$

The two significant figures in 41 and 13 limit the angle to two significant figures. Since the x- and y-components are both positive, the vector is in the first region of the Cartesian coordinate plane. This is consistent with the graphical answer we drew to begin with, and it means that we do not need to do anything to the result of Equation (3.3). Thus, the sum of vectors **A** and **B** has a magnitude of <u>43 m/sec and a direction of 72°</u>.

Following the directions on a map, a Girl Scout walks 3.2 miles at 130°. She then turns and walks 2.3 miles at 40.0 degrees. What is her final position relative to her starting point?

In this two-dimensional problem, we are given two displacement vectors followed by the Girl Scout, and we are asked to come up with her final displacement vector. The final vector must be the sum of the two. In the end, then, we simply have to add these two vectors. Before we do this analytically, let's do it graphically to visualize what's going on here:

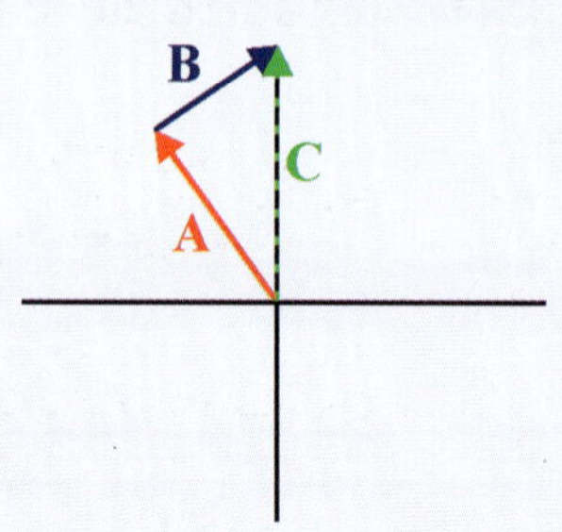

In this diagram, we are calling the first leg of her journey vector **A**, while the second leg is vector **B**. The final displacement vector, then, is vector **C**.

Now that we have some idea of what the final displacement vector looks like, we need to get its magnitude and direction. The first step is to break both vectors down into their components using Equations (3.6) and (3.7):

$$A_x = (3.2 \text{ miles}) \cdot \cos(130^\circ) = -2.1 \text{ miles}$$

$$A_y = (3.2 \text{ miles}) \cdot \sin(130^\circ) = 2.5 \text{ miles}$$

$$B_x = (2.3 \text{ miles}) \cdot \cos(40.0^\circ) = 1.8 \text{ miles}$$

$$B_y = (2.3 \text{ miles}) \cdot \sin(40.0^\circ) = 1.5 \text{ miles}$$

Now that we have the individual components, we can add them like numbers. We can get the x-component of the sum (vector **C**) by just adding the two x-components together:

$$C_x = A_x + B_x = -2.1 \text{ miles} + 1.8 \text{ miles} = -0.3 \text{ miles}$$

Note that since this is addition, the precision of the two x-components limits the precision of the sum. Since both components have their last significant figure in the tenths place, the answer must have its last significant figure in the tenths place.

Similarly, the y-component of our answer can be calculated by simply adding up the y-components of the two vectors:

$$C_y = A_y + B_y = 2.5 \text{ miles} + 1.5 \text{ miles} = 4.0 \text{ miles}$$

Now that we have the components to our answer, we can use Equations (3.2) and (3.3) to give us the magnitude and direction of the answer. To get the magnitude of this vector, we use Equation (3.2):

$$\text{magnitude} = \sqrt{C_x^2 + C_y^2} = \sqrt{(-0.3 \text{ miles})^2 + (4.0 \text{ miles})^2} = 4.0 \text{ miles}$$

Remember, in terms of significant figures, we treat this as an addition problem, so the fact that both components have their last significant figure in the tenths place means that our answer must have its last significant figure in the tenths place as well. To find the direction of the vector, we use Equation (3.3):

$$\theta = \tan^{-1}\left(\frac{C_y}{C_x}\right) = \tan^{-1}\left(\frac{4.0 \text{ \sout{miles}}}{-0.3 \text{ \sout{miles}}}\right) = -90^\circ$$

Note that since -0.3 miles has only one significant figure, the answer can have only one significant figure.

The result of Equation (3.3) is not the proper answer. Since the x-component of vector **C** is negative and the y-component is positive, the vector is in region II of the Cartesian coordinate plane. This means that we need to add 180° to the result of Equation (3.3) in order to get the angle defined properly. Thus, the proper angle is $-90^{\circ} + 180^{\circ} = 90^{\circ}$. So her final displacement is <u>4.0 miles at a direction of 90°</u>. Notice how consistent these values are with the picture we drew at the beginning.

I want to make sure you understand exactly what happened here. In order to add two-dimensional vectors, we broke them down into their components. Think about that for a minute. The x-component represents one of the vector's dimensions while the y-component represents the other dimension. Thus, in order to solve this two-dimensional problem, we broke it down into two one-dimensional problems. Adding the x-components of two vectors is a one-dimensional job, as is adding the y-components of two vectors. This is a fundamental way that we approach two-dimensional problems in physics. When we are faced with such a problem, we typically break it down into two one-dimensional problems. Since one-dimensional problems are relatively easy to solve, this makes the job a whole lot easier. If you can really grasp the idea of breaking a two-dimensional problem down into two one-dimensional problems, the rest of this module and all of the next one will be a piece of cake!

ON YOUR OWN

3.6 Vector **A** has a magnitude of 3.1 m/sec at an angle of 60.0°, and vector **B** has a magnitude of 1.4 m/sec at an angle of 290°. What is the sum of these two vectors?

3.7 A Boy Scout travels 8.2 km with a heading of 120° and then 3.2 km at a heading of 250°. What is his final displacement relative to his starting point?

Applying Vector Addition to Physical Situations

Now that we know the mechanics of how to add and subtract vectors, it is time to see how this technique can be used to analyze physical situations. Let's start with a simple experiment. Suppose you are piloting a plane that must fly from one city to another. Then, the plane must take off again and fly to another city. How would you calculate the overall displacement you experienced as a result of the entire journey? Well, one way you could do it is to add the displacement vector from the first leg of your journey to the displacement vector from the second leg of your journey. If you add the vectors properly, the result will be your overall displacement. Perform the following experiment to see what I mean.

EXPERIMENT 3.2
Vector Addition

Note: A sample set of calculations is available in the solutions and tests guide. It is with the solutions to the practice problems.

Supplies:

- Ruler
- Pencil
- A protractor
- Map given below

1. You will be using the map below in this experiment. It can also be found on the course website. You can photocopy the map from this page, or you can print it from the website.

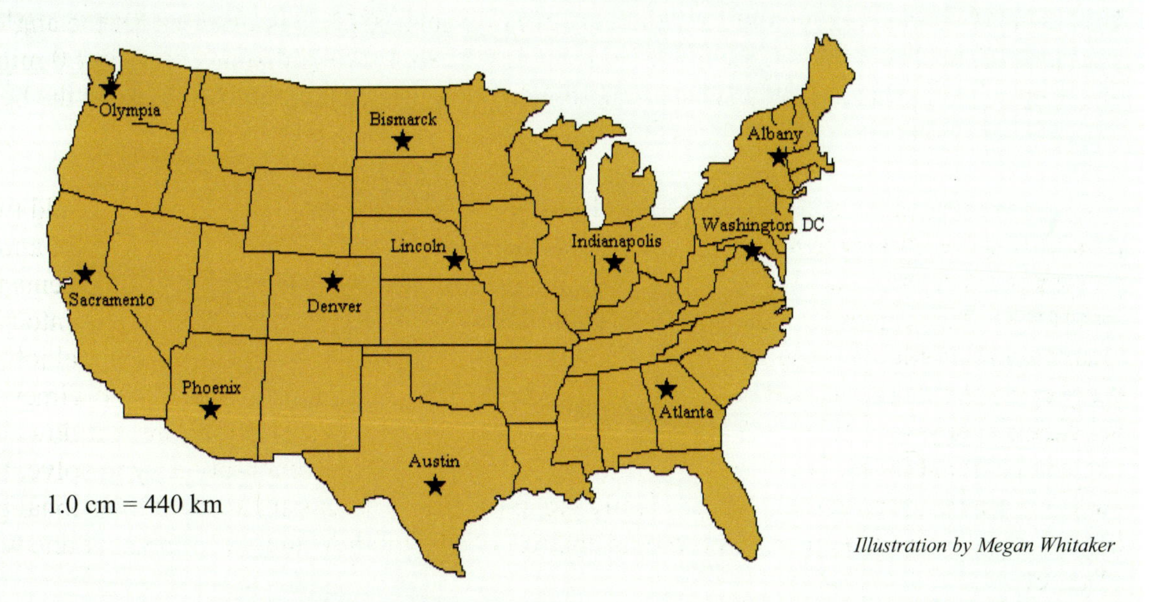

Illustration by Megan Whitaker

2. Note the scale given on the bottom left corner of the map. This will allow you to convert from centimeters on the map into kilometers in real life.
3. Use the ruler to draw an arrow from Phoenix, AZ to Bismarck, ND. This will represent the displacement vector for the first leg of a hypothetical journey.
4. Measure the length of the arrow, and then use the map's scale (mentioned in step 2) to convert from centimeters on the map to kilometers in real life.
5. Use the ruler to draw x- and y-axes at Phoenix. The x-axis should run east/west and the y-axis should run north/south.
6. Use the protractor to measure the angle that the vector makes with the positive x-axis.
7. Use the ruler to draw an arrow from Bismarck, ND to Washington, DC. This will represent the displacement vector for the second leg of the journey.
8. Once again, measure the length of the arrow, and then use the map's scale to convert from centimeters on the map to kilometers in real life.
9. Draw x- and y-axes at Bismarck. The x-axis should run east/west and the y-axis should run north/south.
10. Use the protractor to measure the angle that the vector makes with the positive x-axis. Remember, you need to define the angle counterclockwise from the positive x-axis. As a result, the angle will be greater than 270°.
11. Now that you have the magnitude and direction for each of the two displacement vectors, add the vectors together to get the displacement vector for the entire trip.
12. Now you can check to see how well you added the vectors. Use the ruler to draw an arrow from Phoenix, AZ to Washington, DC.
13. Measure the length of the arrow, and then use the map's scale to convert from centimeters on the map to kilometers in real life.
14. Use the protractor to measure the angle the arrow makes with the positive x-axis at Phoenix.
15. Compare your measurements to the results of the vector addition. Within a reasonable margin of error (5% or so), they should be equal.
16. Repeat the experiment with a two-segment trip of your own choosing.
17. Clean up your mess.

Although displacement calculations such as the ones that you did in the experiment are useful, one of the most important applications of vector addition occurs in the navigation of planes and ships. Suppose, for example, you are piloting an airplane. If you know that your destination is due east of your present location, do you simply point your aircraft due east and fly to your destination? Most likely not. You see, when an airplane is in flight, it is influenced by the wind. To see how airplanes are affected by the wind, examine Figure 3.2.

FIGURE 3.2
The Wind's Effect on an Airplane

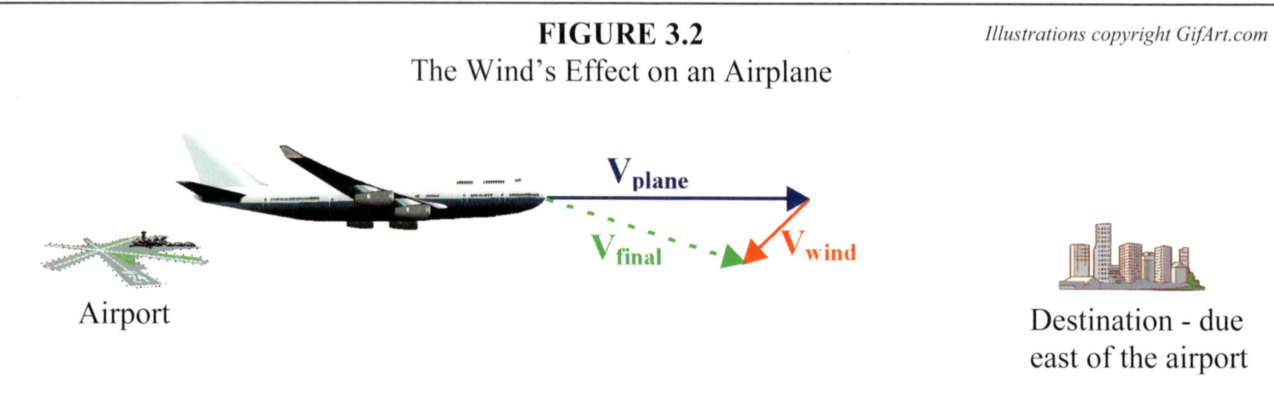

In this drawing, the plane is flying from an airport to a location due east of that airport. While the plane is flying, however, the wind is blowing southwest. The force of the wind pushes on the aircraft, causing the plane to drift in the direction of the wind. This causes the plane to be blown off course. We can see how much off course by graphically adding the vectors as I have done above. The vector $\mathbf{V}_{\mathbf{plane}}$ indicates the velocity of the plane as determined by the direction the pilot points the plane and the speed of the engines. If there were no wind, this would be the plane's final velocity, and it would fly on course to its destination. Because there is wind, however, the wind's velocity ($\mathbf{V}_{\mathbf{wind}}$) adds to the plane's velocity. The resulting vector ($\mathbf{V}_{\mathbf{final}}$) represents the actual velocity with which the plane flies. Notice that with this velocity vector, the plane will end up far south of its destination.

How do planes correct for this? Well, they keep track of wind velocity, and the navigational system calculates what direction and speed the plane must fly so that the sum of the plane's velocity and the wind's velocity will end up being the vector that will keep the plane on course, as shown below:

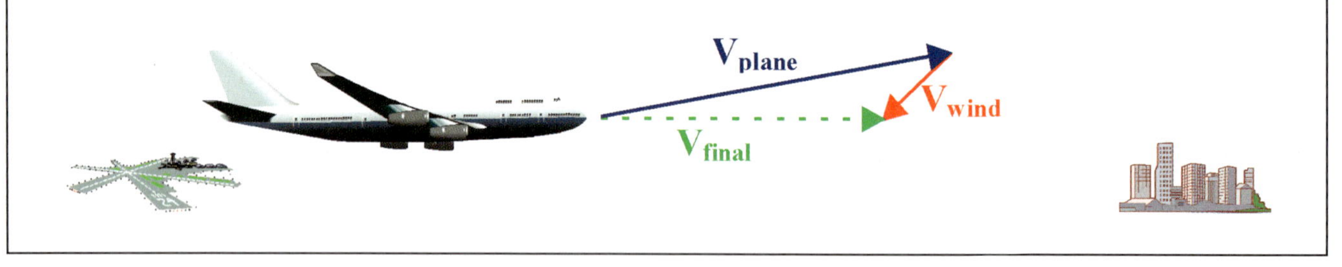

Do you see what happened here? In order to "fight" the wind, the pilot had to point the plane off course. That way, with the wind's velocity added to the plane's velocity, the final velocity vector would be pointing directly towards the destination. It turns out that the same thing happens to boats. When boats float in the water, the velocity of the current adds to the velocity of the boat. In the end, that final velocity is what determines the direction of the boat. See if you can understand how to analyze this kind of problem by studying the following examples.

EXAMPLE 3.6

Illustrations from the MasterClips collection

A pilot steers her plane with a velocity of 125 mph at a heading of 30.0°. If the wind is blowing southwest (225°) at 20.0 mph, what is the plane's final velocity?

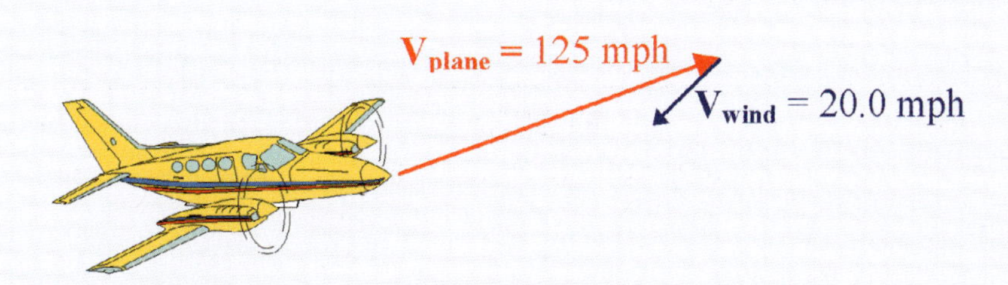

Since the plane is flying, its actual velocity is the vector sum of the plane's velocity and the wind's velocity:

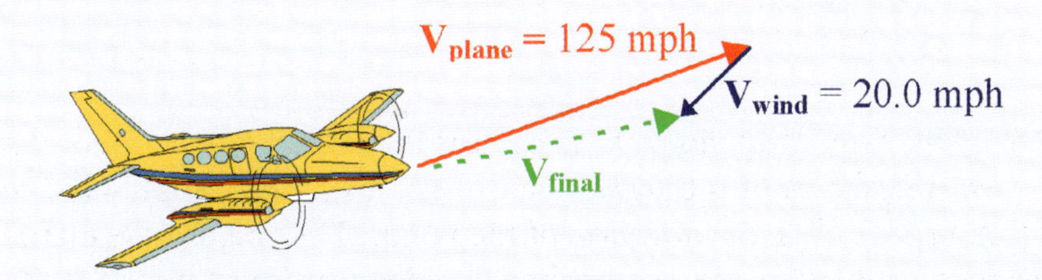

The dashed arrow, then, shows us the actual course of the plane. To determine the exact velocity, however, we must add the vectors analytically. First, we split them up into their components:

$$V_{planex} = (125 \text{ mph}) \cdot \cos(30.0^\circ) = 108 \text{ mph}$$

$$V_{planey} = (125 \text{ mph}) \cdot \sin(30.0^\circ) = 62.5 \text{ mph}$$

$$V_{windx} = (20.0 \text{ mph}) \cdot \cos(225^\circ) = -14.1 \text{ mph}$$

$$V_{windy} = (20.0 \text{ mph}) \cdot \sin(225^\circ) = -14.1 \text{ mph}$$

Now we add the components together to get the components of the final velocity:

$$V_{finalx} = 108 \text{ mph} + -14.1 \text{ mph} = 94 \text{ mph}$$

$$V_{finaly} = 62.5 \text{ mph} + -14.1 \text{ mph} = 48.4 \text{ mph}$$

With these components, we can get the magnitude and direction of the final velocity:

$$\text{magnitude} = \sqrt{V_{finalx}{}^2 + V_{finaly}{}^2} = \sqrt{(94 \text{ mph})^2 + (48.4 \text{ mph})^2} = 106 \text{ mph}$$

To find the direction of the vector, we use Equation (3.3):

$$\theta = \tan^{-1}\left(\frac{V_{finaly}}{V_{finalx}}\right) = \tan^{-1}\left(\frac{48.4\ \cancel{mph}}{94\ \cancel{mph}}\right) = 27^{o}$$

Since both components of the final velocity are positive, the result of Equation (3.3) is the angle of the vector. In the end, then, the final velocity of the plane is 106 mph at an angle of 27°. Notice that these values are different from the heading used by the pilot. That's the effect of the wind.

A man rows a boat across a river. He is trying to reach his campsite directly north of his position. Since he does not know physics, the man points his boat directly north ($\theta = 90.0^{o}$) and rows at 3.0 m/sec. The current of the river is 1.2 m/sec due west ($\theta = 180.0^{o}$). What will the man's actual velocity be?

Illustration by Megan Whitaker

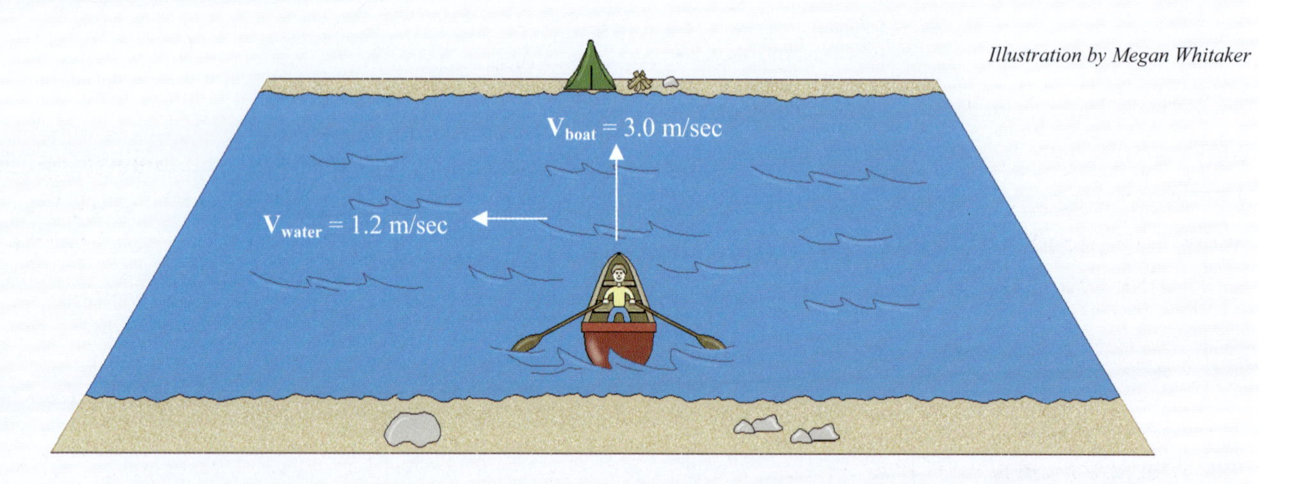

Since the boat will be carried by the river's current, the actual velocity of the boat will be the vector sum of the boat and the current. To do this, we first split them up into their components:

$$V_{boatx} = (3.0\ \frac{m}{sec}) \cdot \cos(90.0^{o}) = 0$$

$$V_{boaty} = (3.0\ \frac{m}{sec}) \cdot \sin(90.0^{o}) = 3.0\ \frac{m}{sec}$$

$$V_{currentx} = (1.2\ \frac{m}{sec}) \cdot \cos(180.0^{o}) = -1.2\ \frac{m}{sec}$$

$$V_{currenty} = (1.2\ \frac{m}{sec}) \cdot \sin(180.0^{o}) = 0$$

Now we add the components together to get the components of the final velocity:

$$V_{finalx} = 0\ \frac{m}{sec} + -1.2\ \frac{m}{sec} = -1.2\ \frac{m}{sec}$$

$$V_{finaly} = 3.0\ \frac{m}{sec} + 0\ \frac{m}{sec} = 3.0\ \frac{m}{sec}$$

With these components, we can get the magnitude and direction of the final velocity:

$$\text{magnitude} = \sqrt{V_{finalx}^2 + V_{finaly}^2} = \sqrt{(-1.2\ \frac{m}{sec})^2 + (3.0\ \frac{m}{sec})^2} = 3.2\ \frac{m}{sec}$$

To find the direction of the vector, we use Equation (3.3):

$$\theta = \tan^{-1}\left(\frac{V_{finaly}}{V_{finalx}}\right) = \tan^{-1}\left(\frac{3.0\ \frac{\cancel{m}}{\cancel{sec}}}{-1.2\ \frac{\cancel{m}}{\cancel{sec}}}\right) = -68^{o}$$

Since the x-component of the vector is negative and the y-component is positive, the vector is in region II of the Cartesian coordinate plane. As a result, we must add 180° to the result of -68°. Thus, the actual velocity of the boat is <u>3.2 m/sec at 112°</u>. This heading, of course, will make the boater end up quite a bit west of the campsite!

ON YOUR OWN

3.8 A boat travels across a wide river. If the boat can travel at a speed of 30.0 mph and its pilot heads in a direction of 120 degrees, while the current's velocity is 5.1 mph at a heading of 315 degrees, what will be the final velocity of the boat?

3.9 A plane heads due east ($\theta = 0^{o}$) at a speed of 300.0 mph. If the wind's velocity is southeast (315°) at 25 mph, what will be the actual heading of the plane?

If you can answer problems like "On Your Own" 3.8 and 3.9, you have all of the skills necessary to continue on to the next module. However, if you are not very sure about problems such as these, try to shore up your skills with the practice problems at the end of the module. If you are still unsure after that, solve the extra practice problems in Appendix B, because you need to be very comfortable with two-dimensional vectors in order to be successful in the next two (or more) modules.

ANSWERS TO THE "ON YOUR OWN" PROBLEMS

3.1 To add these two vectors, we must move the second so that its tail is at the head of the first:

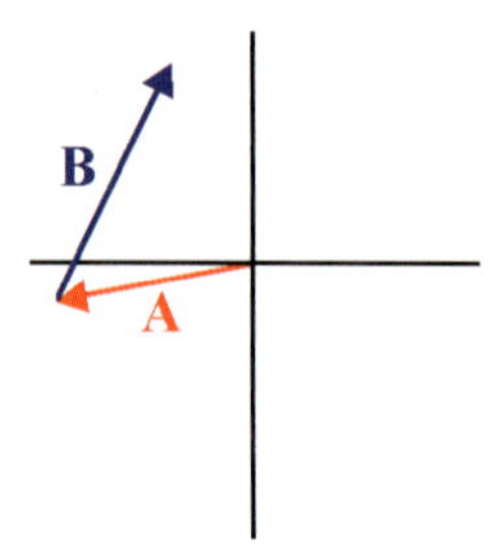

The vector that starts at the tail of the first and points to the head of the second represents the sum:

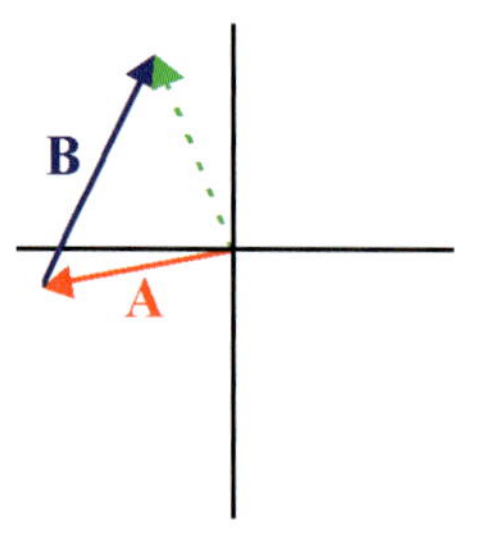

The dashed arrow, then, represents **A** + **B**.

3.2 To subtract **A** from **B**, we must first take the negative of **A**:

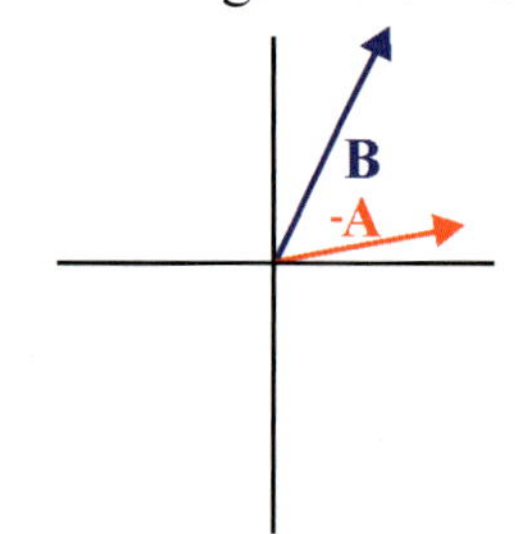

Then we add -**A** to **B**:

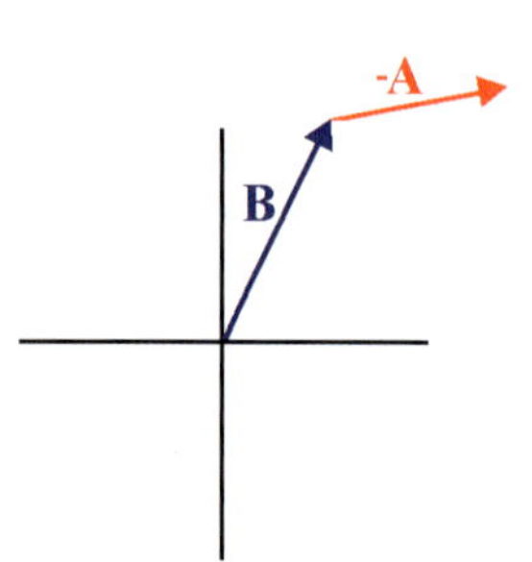

The vector representing the difference is then drawn by starting at the tail of the first vector and pointing to the head of the second:

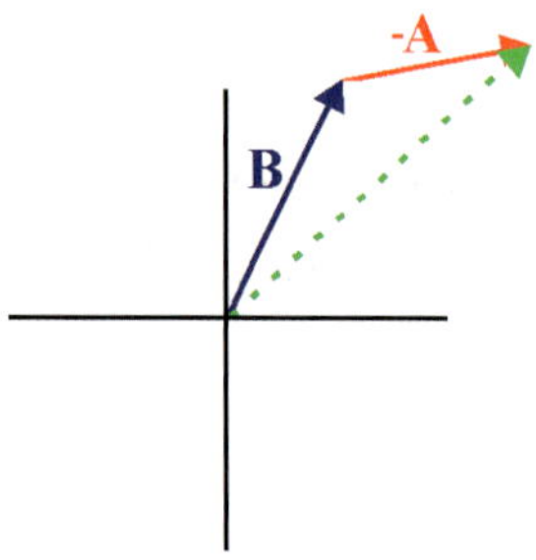

The dashed arrow, then, represents **B** - **A**.

3.3 In this problem, we are given the x- and y-components of a vector and are asked to calculate its magnitude and direction. To get the magnitude, we use Equation (3.2):

$$\text{magnitude} = \sqrt{V_x^{\,2} + V_y^{\,2}} = \sqrt{(23\frac{\text{m}}{\text{sec}})^2 + (11\frac{\text{m}}{\text{sec}})^2} = 25\frac{\text{m}}{\text{sec}}$$

To get the angle, we start with Equation (3.3):

$$\theta = \tan^{-1}(\frac{V_y}{V_x}) = \tan^{-1}(\frac{11\frac{\cancel{\text{m}}}{\cancel{\text{sec}}}}{23\frac{\cancel{\text{m}}}{\cancel{\text{sec}}}}) = 26^{\text{o}}$$

We aren't necessarily finished yet, however. We have to determine the region of the Cartesian coordinate system in which the vector is located. Since both its components are positive, this tells us that the vector is to the right of and above the origin, which means that the vector is in region I. According to our rules, we don't need to do anything to the result of Equation (3.3) when the vector is in region I, so 26° is the proper angle. Thus, the vector has magnitude of 25 m/sec and direction of 26°.

3.4 This problem is just a bit trickier because it does not give the direction of the x- and y-components in terms of plus and minus signs. Instead, it tells us direction explicitly. We need to infer the signs. Since we are told that west is to the left, this tells us that the x-component is negative, because motion to the left is always negative on a Cartesian coordinate plane. Thus, the x-component is -1.1 miles. Similarly, the y-component is -2.0 miles because downward motion is always negative on a Cartesian coordinate plane. Now that we have the components, we can begin to use our equations. To get the magnitude, we use Equation (3.2):

$$\text{magnitude} = \sqrt{A_x^{\,2} + A_y^{\,2}} = \sqrt{(-1.1\text{ miles})^2 + (-2.0\text{ miles})^2} = 2.3\text{ miles}$$

To get the angle, we start with Equation (3.3):

$$\theta = \tan^{-1}(\frac{A_y}{A_x}) = \tan^{-1}(\frac{-2.0\,\cancel{\text{miles}}}{-1.1\,\cancel{\text{miles}}}) = 61^{\text{o}}$$

We aren't necessarily finished yet, however. We have to determine the region of the Cartesian coordinate system in which the vector is located. Since both its components are negative, this tells us that the vector is to the left of and below the origin, which means that the vector is in region III. According to our rules, we add 180° to the result of Equation (3.3) when the vector is in region III, so the angle is 61° + 180° = 241°. Thus, the vector has magnitude of 2.3 miles and direction of 241°.

3.5 This problem is a straightforward example of using Equations (3.6) and (3.7):

$$A_x = (55\ \frac{m}{sec^2}) \cdot \cos(300.0^o) = \underline{28\ \frac{m}{sec^2}}$$

$$A_y = (55\ \frac{m}{sec^2}) \cdot \sin(300.0^o) = \underline{-48\ \frac{m}{sec^2}}$$

Notice that since the x-component is positive and the y-component is negative, the vector must occupy region IV of the Cartesian coordinate plane.

3.6 In this problem, we are asked to add two-dimensional vectors together. Before we do this analytically, let's do it graphically:

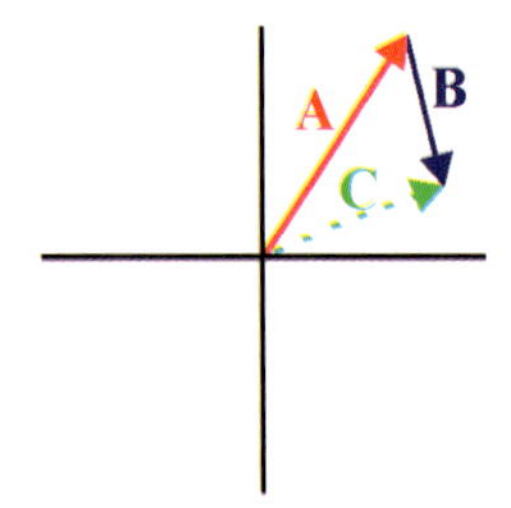

As we learned before, the dashed arrow (vector **C**) gives us the sum.

The first step in adding vectors analytically is to break both vectors down into their components:

$$A_x = (3.1\ \frac{m}{sec}) \cdot \cos(60.0^o) = 1.6\ \frac{m}{sec}$$

$$A_y = (3.1\ \frac{m}{sec}) \cdot \sin(60.0^o) = 2.7\ \frac{m}{sec}$$

$$A_x = (1.4\ \frac{m}{sec}) \cdot \cos(290^o) = 0.48\ \frac{m}{sec}$$

$$A_y = (1.4\ \frac{m}{sec}) \cdot \sin(290^o) = -1.3\ \frac{m}{sec}$$

Now that we have the individual components, we can add them like numbers.

$$C_x = A_x + B_x = 1.6\ \frac{m}{sec} + 0.48\ \frac{m}{sec} = 2.1\ \frac{m}{sec}$$

$$C_y = A_y + B_y = 2.7\ \frac{m}{sec} + -1.3\ \frac{m}{sec} = 1.4\ \frac{m}{sec}$$

Now that we have the components to our answer, we can use Equations (3.2) and (3.3) to give us the magnitude and direction of the answer. To get the magnitude of this vector, we use Equation (3.2):

$$\text{magnitude} = \sqrt{C_x^2 + C_y^2} = \sqrt{(2.1\ \frac{m}{sec})^2 + (1.4\ \frac{m}{sec})^2} = 2.5\ \frac{m}{sec}$$

To find the direction of the vector, we use Equation (3.3):

$$\theta = \tan^{-1}(\frac{C_y}{C_x}) = \tan^{-1}(\frac{1.4\ \frac{\cancel{m}}{\cancel{sec}}}{2.1\ \frac{\cancel{m}}{\cancel{sec}}}) = 34^o$$

Since the x- and y-components are both positive, the vector is in the first region of the Cartesian coordinate plane. This is consistent with the graphical answer we drew to begin with, and it means that we do not need to do anything to the result of Equation (3.3). Thus, the sum of vectors **A** and **B** has a magnitude of <u>2.5 m/sec and a direction of 34°</u>.

3.7 In this two-dimensional problem, we are given two displacement vectors followed by the Boy Scout, and we are asked to come up with his final displacement vector. The final vector must be the sum of the two. In the end, then, we simply have to add these two vectors. Before we do this analytically, let's do it graphically to visualize what's going on here:

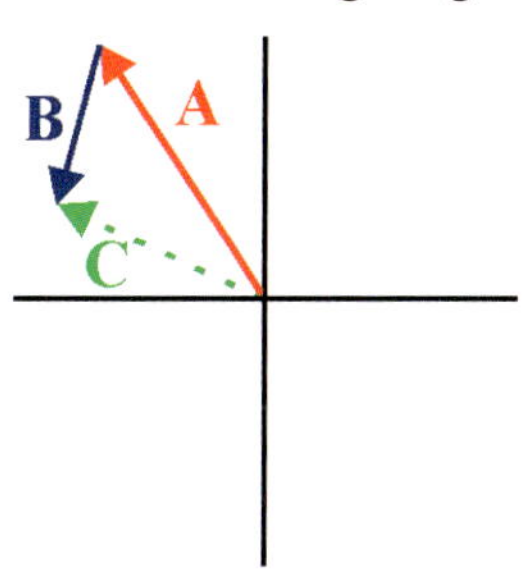

In this diagram, we are calling the first leg of his journey vector **A**, while the second leg is vector **B**. The final displacement vector, then, is vector **C**.

Now that we have some idea of what the final displacement vector looks like, we need to get its magnitude and direction. The first step is to break both vectors down into their components:

$$A_x = (8.2\ km) \cdot \cos(120^o) = -4.1\ km$$

$$A_y = (8.2\ km) \cdot \sin(120^o) = 7.1\ km$$

$$B_x = (3.2\ km) \cdot \cos(250^o) = -1.1\ km$$

$$B_y = (3.2\ km) \cdot \sin(250^o) = -3.0\ km$$

Now that we have the individual components, we can add them like numbers.

$$C_x = A_x + B_x = -4.1 \text{ km} + -1.1 \text{ km} = -5.2 \text{ km}$$

$$C_y = A_y + B_y = 7.1 \text{ km} + -3.0 \text{ km} = 4.1 \text{ km}$$

Now that we have the components to our answer, we can use Equations (3.2) and (3.3) to give us the magnitude and direction of the answer. To get the magnitude of this vector, we use Equation (3.2):

$$\text{magnitude} = \sqrt{C_x^{\,2} + C_y^{\,2}} = \sqrt{(-5.2 \text{ km})^2 + (4.1 \text{ km})^2} = 6.6 \text{ km}$$

To find the direction of the vector, we use Equation (3.3):

$$\theta = \tan^{-1}\left(\frac{C_y}{C_x}\right) = \tan^{-1}\left(\frac{4.1 \cancel{\text{km}}}{-5.2 \cancel{\text{km}}}\right) = -38^{\circ}$$

Since the x-component of vector **C** is negative and the y-component is positive, the vector is in region II of the Cartesian coordinate plane. This means that we need to add 180° to the result of Equation (3.3) in order to get the angle defined properly. Thus, the proper angle is -38° + 180° = 142°. So the Boy Scout's final displacement is <u>6.6 km at a direction of 142°</u>. Notice how consistent these values are with the picture we drew at the beginning.

3.8 Since the boat will be carried by the river's current, the actual velocity of the boat will be the vector sum of the boat and the current. To do this, we first split them up into their components:

$$V_{\text{boatx}} = (30.0 \text{ mph}) \cdot \cos(120^{\circ}) = -15 \text{ mph}$$

$$V_{\text{boaty}} = (30.0 \text{ mph}) \cdot \sin(120^{\circ}) = 26 \text{ mph}$$

$$V_{\text{currentx}} = (5.1 \text{ mph}) \cdot \cos(315^{\circ}) = 3.6 \text{ mph}$$

$$V_{\text{currenty}} = (5.1 \text{ mph}) \cdot \sin(315^{\circ}) = -3.6 \text{ mph}$$

Now we add the components together to get the components of the final velocity:

$$V_{\text{finalx}} = -15 \text{ mph} + 3.6 \text{ mph} = -11 \text{ mph}$$

$$V_{\text{finaly}} = 26 \text{ mph} + -3.6 \text{ mph} = 22 \text{ mph}$$

With these components, we can get the magnitude and direction of the final velocity:

$$\text{magnitude} = \sqrt{V_{\text{finalx}}^{\,2} + V_{\text{finaly}}^{\,2}} = \sqrt{(-11 \text{ mph})^2 + (22 \text{ mph})^2} = 25 \text{ mph}$$

To find the direction of the vector, we use Equation (3.3):

$$\theta = \tan^{-1}\left(\frac{V_{final_y}}{V_{final_x}}\right) = \tan^{-1}\left(\frac{22\ \cancel{mph}}{-11\ \cancel{mph}}\right) = -63^{\circ}$$

Since the x-component is negative and the y-component is positive, the vector is in region II of the Cartesian coordinate plane. We must therefore add 180° to the result of Equation (3.7). The boat's final velocity, then, is 25 mph at 117°.

3.9 The final velocity of the plane will be the vector sum of the plane's velocity and the wind's velocity.

$$V_{planex} = (300.0\text{ mph})\cdot\cos(0^{\circ}) = 300.0\text{ mph}$$

$$V_{planey} = (300.0\text{ mph})\cdot\sin(0^{\circ}) = 0$$

$$V_{windx} = (25\text{ mph})\cdot\cos(315^{\circ}) = 18\text{ mph}$$

$$V_{windy} = (25\text{ mph})\cdot\sin(315^{\circ}) = -18\text{ mph}$$

Now we add the components together to get the components of the final velocity:

$$V_{finalx} = 300.0\text{ mph} + 18\text{ mph} = 318\text{ mph}$$

$$V_{finaly} = 0\text{ mph} + -18\text{ mph} = -18\text{ mph}$$

With these components, we can get the magnitude and direction of the final velocity:

$$\text{Magnitude} = \sqrt{{V_{final_x}}^2 + {V_{final_y}}^2} = \sqrt{(318\text{ mph})^2 + (-18\text{ mph})^2} = 319\text{ mph}$$

To find the direction of the vector, we use Equation (3.3):

$$\theta = \tan^{-1}\left(\frac{V_{final_y}}{V_{final_x}}\right) = \tan^{-1}\left(\frac{-18\ \cancel{mph}}{318\ \cancel{mph}}\right) = -3.2^{\circ}$$

Since the x-component is positive and the y-component is negative, the final vector is in region IV of the Cartesian coordinate plane. Therefore, we must add 360° to the result of Equation (3.7). The final velocity, then, is 319 mph at 356.8°.

REVIEW QUESTIONS

1. We always define the angle of a vector relative to what?

2. For the following observables, tell whether they would be represented by the magnitude or the angle of a vector:

a. Speed b. Heading c. Direction d. Distance

3. Consider the following vector which points straight up the y-axis:

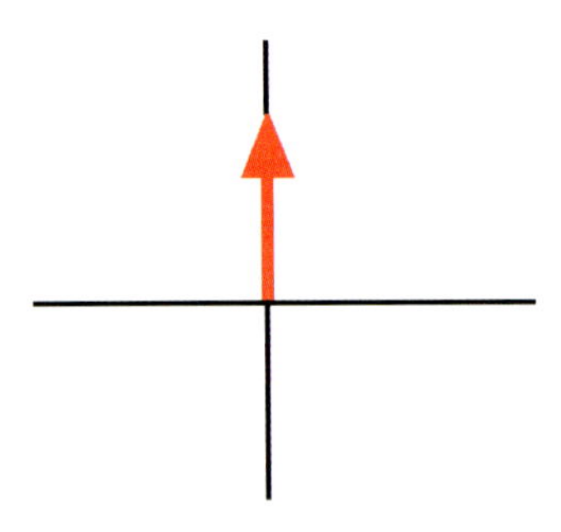

Does this vector have an x-component? Why or why not?

4. If a vector has an x-component that is negative and a y-component that is positive, which region of the Cartesian coordinate plane does it fall in? If we were using Equation (3.3) to calculate the angle of the vector, what would we have to do to ensure that the angle was defined properly?

5. If a vector has an x-component that is negative and a y-component that is negative, which region of the Cartesian coordinate plane does it fall in? If we were using Equation (3.3) to calculate the angle of the vector, what would we have to do to ensure that the angle was defined properly?

6. If a vector has an x-component that is positive and a y-component that is negative, which region of the Cartesian coordinate plane does it fall in? If we were using Equation (3.3) to calculate the angle of the vector, what would we have to do to ensure that the angle was defined properly?

7. Estimate the angle of the following vector:

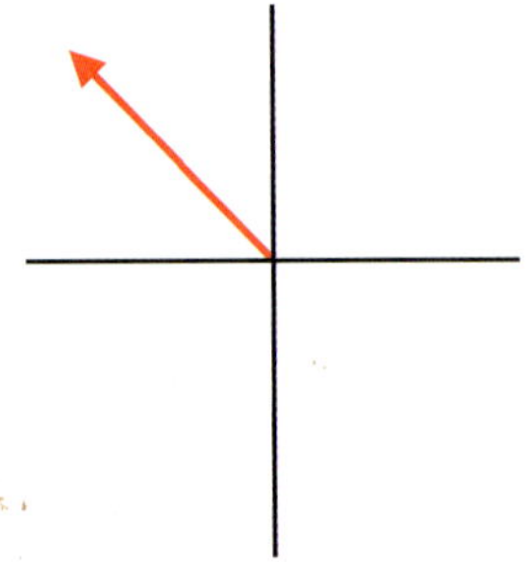

8. Vectors **A** and **B** below represent the accelerations of objects 1 and 2, respectively.

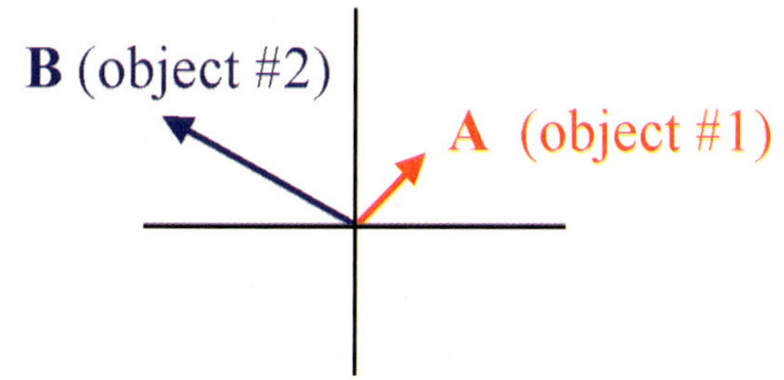

Which object is accelerating more quickly?

9. Vectors **C** and **D** below represent the velocities of objects 1 and 2, respectively, in Problem #8.

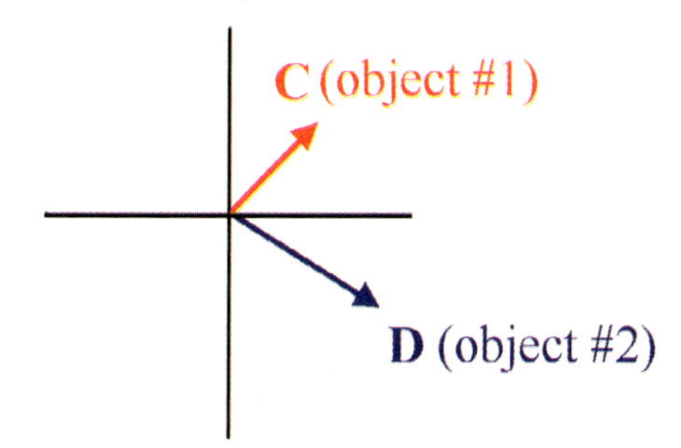

Based on the acceleration vectors in Problem #8 and the velocity vectors in this problem, which object is speeding up?

10. Fill in the blank: When solving two-dimensional problems, physicists usually split them up into ________ _____-_________ problems.

PRACTICE PROBLEMS

1. Graphically add the following vectors:

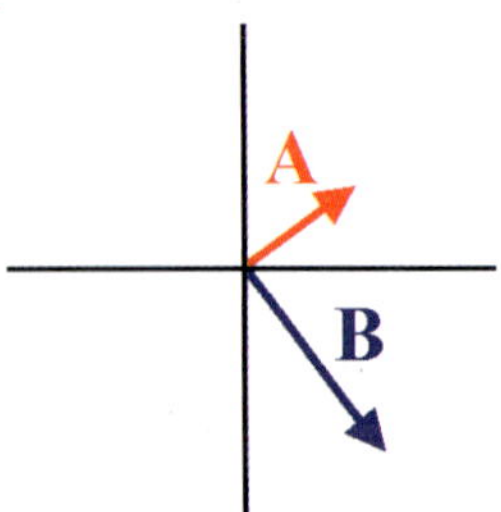

2. Graphically subtract vector **B** from vector **A** in Problem #1.

3. The velocity of a plane is pictured below along with the velocity of the wind through which the plane is flying. Graphically show the final heading of the plane.

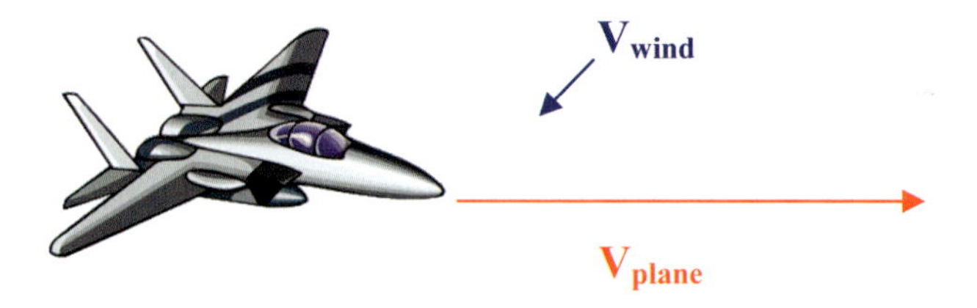

4. Vector **A** has a magnitude of 35.1 m/sec and a direction of 215 degrees. What are its x- and y-components?

5. The velocity vector of a car has an x-component of 17 m/sec and a y-component of -11 m/sec. What are the magnitude and direction of the velocity vector?

6. Vector **A** has a magnitude of 31.1 m and an angle of 160.0°, and vector **B** has a magnitude of 11.4 m and an angle of 290.0°. What is the sum of these two vectors?

7. A Boy Scout travels 1.2 miles with a heading of 150.0° and then 3.2 miles at a heading of 250.0°. What is the Boy Scout's final displacement relative to his starting point?

8. A boat travels across a wide river. If the boat can travel at a speed of 25 mph and its pilot heads in a direction of 30.0°, while the current's velocity is 5.1 mph at a heading of 270.0°, what will be the final velocity of the boat?

9. A plane heads due west (θ = 180.0°) at a speed of 200.0 mph. If the wind's velocity is southeast (315°) at 15 mph, what will be the actual velocity of the plane?

10. A man rows a boat across a river. The man can row at a speed of 3.0 m/sec and heads due east (θ = 0°). If the current's velocity is 1.5 m/sec at a heading of 135°, what will be the final velocity of the boat?

MODULE #4: Motion in Two Dimensions

Introduction

Now that we understand how to use and interpret two-dimensional vectors, it is time to begin analyzing situations in which two-dimensional motion occurs. There are, of course, many more situations in which two-dimensional motion occurs than those in which strictly one-dimensional motion occurs. Remember, however, the "trick" that we learned in the previous module. Whenever we analyze a two-dimensional problem, we should try to split it up into two one-dimensional motion problems. This is a trick that we employ heavily in this module.

Navigation in Two Dimensions

In order to get you started understanding how to see the one-dimensional problem or problems that exist in a two-dimensional situation, you need to study the following example.

EXAMPLE 4.1

A man walks at a constant speed of 1.20 m/sec with a direction of 45°. What will be his displacement in 301 seconds?

When you read this problem, you should picture in your mind a two-dimensional velocity vector:

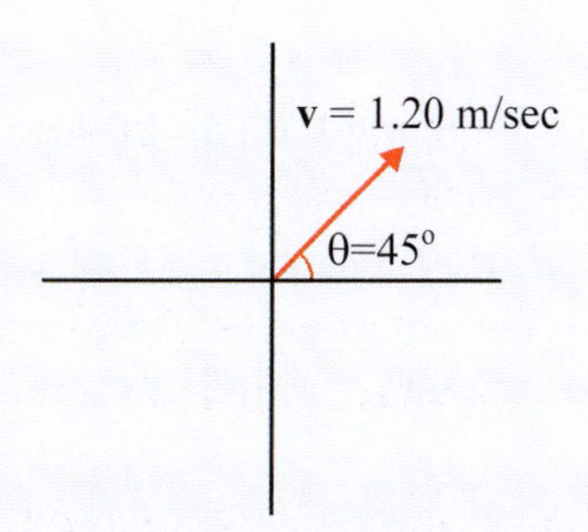

Remember, when we talk about angles in this course, the proper way to define them is counterclockwise from the positive x-axis. Unless I tell you differently, any angle that I give you in a problem will always be defined in that manner.

Now that we have a visual for the velocity vector, we need to calculate what the person's displacement would be after walking for 301 seconds with that velocity. At first, you might think that this is a tough problem because it involves a two-dimensional vector. Remember our "trick," however. In any problem that uses two dimensions, we need to find the one-dimensional problem or problems within it.

Think about how the man in this problem is walking. He is walking *in a straight line*. Sure, that line is pictured by a two-dimensional velocity vector, but nevertheless, he travels straight. He never changes direction. As a result, even though the velocity vector is two-dimensional, this problem is one-dimensional. Thus, we can use our one-dimensional motion equations. In this problem, we are given velocity (1.20 m/sec) and time (301 seconds). Also, since the speed is constant, we know that acceleration is 0. In order to get displacement, then, we just use Equation (2.19):

$$\Delta\mathbf{x} = \mathbf{v}_0 t + \frac{1}{2}\mathbf{a}t^2$$

$$\Delta\mathbf{x} = (1.20\ \frac{\text{m}}{\cancel{\text{sec}}})\cdot(301\ \cancel{\text{sec}}) + \frac{1}{2}\cdot(0)\cdot(301\ \text{sec})^2$$

$$\Delta\mathbf{x} = 361\ \text{m}$$

Now remember, in this problem we were asked to calculate displacement. Thus, we must include a direction in our answer. What direction do we give? Well, the man started walking at an angle of 45° and never deviated from that path. As a result, we know that no matter where he is in his walk, he can always be found at a direction of 45°. Thus, after he has walked 361 meters, he is still at a direction of 45°. The displacement vector, then, is 361 m at 45°.

Now this might seem pretty simple to you. After all, since the person in the example traveled in a straight line, we were able to use our one-dimensional motion equations. In addition, since we've already studied one-dimensional motion, solving the problem was a breeze. Well, that's all true. Things can get a lot more complicated, however. Consider the next example.

EXAMPLE 4.2

An explorer follows directions on a map. She first travels at a constant 0.92 m/sec in a direction of 113° for 2.3 hours. She then turns to a heading of 220° and travels at a constant 0.56 m/sec for 1.2 hours. What is the explorer's final displacement?

The first thing to do here is to picture what's going on:

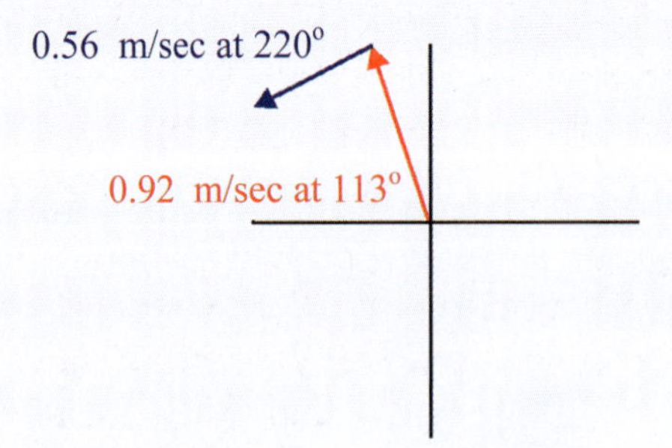

Now that we see what happened in the problem, let's talk about how to solve it. Can we add velocity vectors together? Well, we could, but it wouldn't help much. Remember, the explorer first travels in one direction and *then* travels in another direction. So, unlike the airplane and boat problems we did in the previous module, these velocities do not affect each other. They happened independently, so adding them would tell us nothing.

Instead, we need to look for the one-dimensional problems within this two-dimensional situation. The first leg of the journey is one of the one-dimensional problems. After all, the explorer travels in a straight line during that leg of the journey. She never turns or strays from the straight line. As a result, the first leg of her journey can be solved as a one-dimensional motion problem. We know the velocity (0.92 m/sec), the acceleration (constant velocity means 0 acceleration), and the time (2.3 hours).

The first thing we should notice is that the units we have are not consistent. While the velocity is given in m/sec, the time is given in hours. Thus, we need to convert hours to seconds:

$$\frac{2.3\ \cancel{\text{hr}}}{1} \times \frac{3600\text{ sec}}{1\ \cancel{\text{hr}}} = 8.3 \times 10^3\text{ sec}$$

Now that our units are consistent, we can use Equation (2.19) to figure out the displacement on the first leg of the journey:

$$\Delta\mathbf{x} = \mathbf{v}_0 t + \frac{1}{2}\mathbf{a}t^2$$

$$\Delta\mathbf{x} = (0.92\ \frac{\text{m}}{\cancel{\text{sec}}}) \cdot (8.3 \times 10^3\ \cancel{\text{sec}}) + \frac{1}{2} \cdot (0) \cdot (8.3 \times 10^3\text{ sec})^2$$

$$\Delta\mathbf{x} = 7.6 \times 10^3\text{ m}$$

Is this our answer? Of course not! This was only the first leg of the journey. We did learn something, however. We learned that the first leg of the journey resulted in a displacement of 7.6 x 10^3 m at an angle of 113°. That, then, is the point at which the second leg of the journey starts. If we can figure out the displacement that results from the second leg, we should have enough information to determine the final displacement of the explorer.

Well, we can analyze the second leg of the journey the same way that we analyzed the first leg. We can take the velocity, time, and acceleration given for this leg and use Equation (2.19) to determine the displacement. Just like we did in the first leg, however, we must convert the units of time from hours to seconds. Once we do that, Equation (2.19) looks like this:

$$\Delta\mathbf{x} = \mathbf{v}_0 t + \frac{1}{2}\mathbf{a}t^2$$

$$\Delta\mathbf{x} = (0.56\ \frac{\text{m}}{\cancel{\text{sec}}}) \cdot (4.3 \times 10^3\ \cancel{\text{sec}}) + \frac{1}{2} \cdot (0) \cdot (4.3 \times 10^3\text{ sec})^2$$

$$\Delta\mathbf{x} = 2.4 \times 10^3\text{ m}$$

What do we have now? Well, we know that the first leg of the journey resulted in a displacement of 7.6 x 10^3 m at 113°, while the second leg resulted in a displacement of 2.4 x 10^3 m at 220°. How can we find the total displacement? Just like we did in the previous module, when we have two displacements, we can add them in order to get the final displacement.

To add these two displacements, we must first break the vectors down into their components. To make the notation easier, let's call the displacement vector resulting from the first leg of the journey vector **A**. We will therefore call the second vector **B**, and the final displacement vector will be known as **C**:

$$A_x = (7.6 \times 10^3 \text{ m}) \cdot \cos(113^\circ) = -3.0 \times 10^3 \text{ m}$$

$$A_y = (7.6 \times 10^3 \text{ m}) \cdot \sin(113^\circ) = 7.0 \times 10^3 \text{ m}$$

$$B_x = (2.4 \times 10^3 \text{ m}) \cdot \cos(220^\circ) = -1.8 \times 10^3 \text{ m}$$

$$B_y = (2.4 \times 10^3 \text{ m}) \cdot \sin(220^\circ) = -1.5 \times 10^3 \text{ m}$$

$$C_x = -3.0 \text{ x } 10^3 \text{ m} + -1.8 \text{ x } 10^3 \text{ m} = -4.8 \text{ x } 10^3 \text{ m}$$

$$C_y = 7.0 \text{ x } 10^3 \text{ m} + -1.5 \text{ x } 10^3 \text{ m} = 5.5 \text{ x } 10^3 \text{ m}$$

Now we can determine the magnitude and direction of the final displacement vector:

$$C = \sqrt{C_x^2 + C_y^2} = \sqrt{(-4.8 \times 10^3 \text{ m})^2 + (5.5 \times 10^3 \text{ m})^2} = 7.3 \times 10^3 \text{ m}$$

$$\theta = \tan^{-1}\left(\frac{5.5 \times 10^3 \cancel{\text{m}}}{-4.8 \times 10^3 \cancel{\text{m}}}\right) = -49^\circ$$

Based on our rules regarding Equation (3.3), this means that the angle for the final displacement vector is really 180° + -49° = 131°. In the end, then, the explorer's final displacement vector is <u>7.3 x 10^3 m at an angle of 131°</u>.

Now this last problem was long and complicated, but notice that there are no new skills to learn here. All we had to do was employ the "trick." We realized that in order to get the final displacement vector, we had to break the problem up into two one-dimensional problems. Since the explorer took two different straight routes, we calculated the displacement resulting from each route. Once that was done, it was pretty easy to see that by just adding the displacements, we could get the overall displacement. That's how we need to approach two-dimensional problems. We look for the one-dimensional parts of the problem. Make sure you can master this technique by solving these "On Your Own" problems.

ON YOUR OWN

4.1 An airplane flies with a velocity of 214 mph at a heading of 65.0°, including the effects of the wind. The pilot stays on that course for 2.10 hours and then turns his plane so that it flies with a velocity of 187 mph at a heading of 335°, including the effects of the wind. If the airplane flies on that course for an additional 1.40 hours, what is the plane's final displacement?

4.2 A treasure map says that you can reach the treasure on an island by standing at a certain tree and traveling with a velocity of 1.5 m/sec at an angle of 192° for 45 minutes. It then tells you to turn to a heading of 45° and travel with the same speed for 12 minutes. What is the displacement of the treasure relative to the tree?

Projectile Motion in Two Dimensions

Although the problems that you just solved are long, they should no longer seem complicated. After all, to solve the problems, you just do two one-dimensional motion problems and a little vector addition. Now we will look at a more difficult application of two-dimensional motion: projectile motion.

Suppose you are firing a cannon. You point the barrel at an angle in air and then you fire the cannon. Ignoring air resistance, what does the path of the cannonball look like? Examine Figure 4.1 to find out.

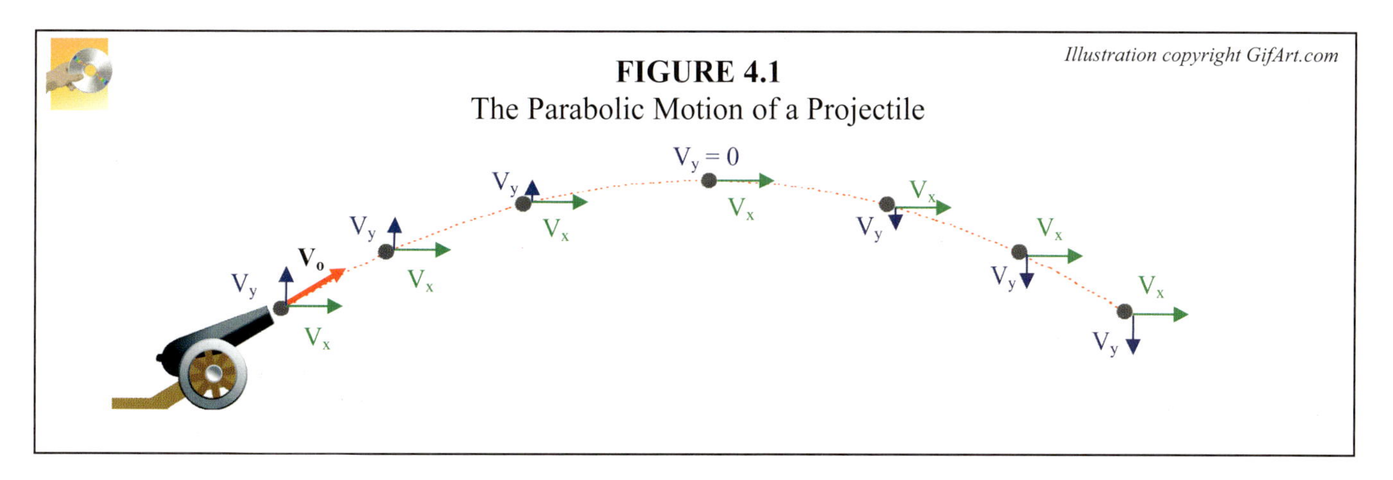

FIGURE 4.1
The Parabolic Motion of a Projectile

When a projectile is fired at an angle relative to the ground, its motion is called **parabolic** (pehr' uh bol' ik), because it follows the curved path of a parabola. At first, this kind of motion might seem pretty complicated. After all, up to this point, we have dealt exclusively with situations where there is motion only in a straight line. Even the two-dimensional problems we did in the previous section were, in effect, two one-dimensional problems. As a result, we have really only dealt with motion in one (or more) straight line(s). The motion of the cannonball in Figure 4.1, however, is curved. There are no straight lines. What do we do?

Well, believe it or not, we still look for the one-dimensional motion within this situation. The one-dimensional motion we look for now, however, resides in the x- and y-components of the vectors. Look at Figure 4.1 for a moment. When the cannonball is first fired from the cannon, it has a velocity vector (**V**) pointed in the direction that the cannon is aimed. This velocity vector, however, can be split into two components: an x-component (V_x) and a y-component (V_y). Rather than trying to describe how the vector **V** changes throughout the course of the cannonball's flight, let's make it easy on ourselves by simply looking at how the x- and y-components change. Since each component is one-dimensional, this simplifies the situation immensely. Even though the situation is simplified, we do not lose any information. After all, when we have the components of a vector, we really have the vector itself. Thus, splitting up this two-dimensional situation into two one-dimensional situations makes everything a lot simpler.

Let's first look at what happens to the x-component of the cannonball's velocity vector. Think about the cannonball once it is in flight. Will there be any change in the x-component of the velocity vector while the cannonball is in flight? No. This answer might surprise you, but if you really concentrate on what's happening to the cannonball, you should be able to understand why. When the cannonball is in the cannon, the air pressure caused by the explosion of the gunpowder pushes on the cannonball. This accelerates the cannonball so that it flies out of the cannon's barrel. After the cannonball leaves the barrel; however, *the cannon can no longer push on the cannonball.* As a result, the cannon stops being important once the ball leaves the barrel.

Well, if the cannon cannot affect the cannonball once it leaves the barrel, does anything else affect it? The obvious answer to that question has to be yes. After all, if there were nothing affecting the cannonball, it would simply travel straight in the direction it was fired. That's not what happens, however. The path of the cannonball curves as the cannonball rises. Then, after a while, the cannonball begins to fall in a similar curved fashion. What makes it do this? *GRAVITY!*

Once the cannonball leaves the barrel, the *only* thing acting on it is gravity. In fact, that's not quite true. Air resistance also affects the ball, but we are neglecting it here. Anyway, gravity only acts in the y-dimension. In other words, gravity can affect the y-component of the cannonball's velocity (because it is pulling the cannonball down), but it cannot affect the x-component of the velocity. As a result, the x-component of the velocity *never changes*. That's what I've tried to illustrate in Figure 4.1. Since nothing is pulling or pushing on the cannonball in the x-dimension, the x-component of the velocity vector stays exactly the same length and points in exactly the same direction throughout the entire flight of the cannonball.

What all this means is that the x-dimension in this situation is very easy to analyze. When the cannonball first comes out of the cannon, it has some initial velocity and is heading at an initial angle. We can use this information to calculate the x-component of the cannonball's velocity vector. Once we have that x-component, we know it never changes. From the time it leaves the cannon to the time it hits the ground, the x-component of its velocity is always the same.

Now the y-component of the cannonball's velocity is not as easy to analyze. Since gravity is constantly pulling the cannonball down, the y-component of the velocity changes. Once again, refer to Figure 4.1 in order to see exactly how it changes. When the cannonball first leaves the cannon's barrel, it has an initial velocity and is headed at the same angle that the cannon is pointed. From that information, we can calculate an initial y-component for the velocity, which is represented by "V_y" in the figure.

Notice what happens to the y-component of the velocity. At first, it is pointed up because the cannon is shooting the ball up in the air. Since gravity is pulling the cannonball down, however, the upwards velocity is decreasing. In the middle of the cannonball's trip, gravity has reduced the y-component of the ball's velocity to zero. All of the time before this happened, the cannonball was going up. Now what happens? Well, gravity is still pulling down on the cannonball, so the y-component of its velocity continues to change. It now starts to point downwards. As a result, the cannonball starts to fall. Since the velocity is now pointed in the same direction as gravity's pull, the y-component of the velocity increases, making the cannonball fall even faster.

See how all of this is illustrated in Figure 4.1. At first, the y-component of the cannonball's velocity is pointed up. However, as the ball continues in its flight, the y-component of the velocity gets smaller. In the precise center of the cannonball's journey, V_y is actually zero. After that, V_y gets more and more negative, causing the cannonball to fall.

The behavior of the x- and y-components of a projectile's velocity is actually the reason that projectiles travel in a parabolic curve. Since the x-component of the velocity never changes, the cannonball travels in the same horizontal direction with the same horizontal speed until it hits the ground. At the same time, however, the projectile's upwards velocity is continually decreasing until it reaches zero. At that point, the projectile begins to fall, gaining downward speed the entire time. If you think about it, the way that these two one-dimensional velocities add together forms a parabolic curve:

Parabolic motion - Motion that occurs when an object moves in two dimensions but has zero acceleration in one of those dimensions and a constant, non-zero acceleration in the other

Before we leave the analysis of the cannonball's motion, let me ask you a couple of questions. What are the values of the x- and y-components of the cannonball's velocity when it reaches its maximum height? Well, the x-component is easy. Since it never changes, the x-component of the cannonball's velocity is the same as the initial x-component. What about the y-component? Remember, we are treating this as two separate one-dimensional problems. What did we say back in Module #2 about the maximum height that a projectile reaches in one dimension? This will always occur when the vertical velocity is zero. That's the point at which the object has stopped rising and is about to start falling. Thus, while the x-component of the cannonball is the same as it always has been, the y-component is zero when the cannonball reaches its maximum height.

The next question is similar. What are the values of the x- and y-components of the cannonball's velocity the instant the cannonball lands? Once again, the x-component will be equal to the initial x-component of the velocity. What about the y-component? Well, in one-dimensional motion (Module #2), we said that a projectile will land with the exact opposite velocity that it started with. Thus, the velocity vector will be the negative of the initial velocity vector. Since the y-component of the cannonball's motion is a one-dimensional problem in and of itself, the answer is the same. The y-component of the cannonball's velocity will be the negative of the initial y-component, assuming that the cannonball lands at the height from which it was launched.

Think about this answer for a minute. The x-component of the cannonball will be the same as when the cannonball was launched. The y-component will have the same magnitude as it had initially, but it will have the opposite direction. What does this tell you about the final *speed* of the cannonball? Well, since the magnitude of each velocity component is the same as the initial velocity components, the magnitude of the overall velocity vector must also be the same as the magnitude of the initial velocity vector. This means, then, that the *speed* of the cannonball is the same when it lands as it was when it was launched. Once again, this assumes that the cannonball lands at the height from which it was launched.

One more question: Where along its flight path does the cannonball reach its maximum height? Well, once again assuming that the cannonball lands at the height from which it was launched, it will reach maximum height at the midpoint of its journey. After all, since the acceleration due to gravity is constant, and since the speed of the cannonball is the same at the beginning and end of its flight, gravity must spend just as much time stopping the cannonball's upward motion as it does accelerating the cannonball back to earth. Thus, the cannonball will reach its maximum height at the midpoint of its flight.

Now if all of this seems a little complicated, don't worry. Parabolic motion is one of the more difficult concepts that we need to cover, so you should be a bit in the dark at this point. To make sure you understand exactly what we just talked about, let me summarize the important points that you need to keep in mind. Please remember that we are neglecting air resistance throughout this discussion.

1. **When a projectile is fired or thrown near the surface of the earth, its path is parabolic.**

2. **The x- and y-components of a projectile's motion can be treated as two separate one-dimensional situations. The x-component of the projectile's velocity will never change once it has been launched. The y-component of the velocity will be affected by gravity, so it will decrease until it reaches zero, at which point it will get more and more negative.**

3. **A projectile's maximum height will be reached when the y-component of its velocity is zero.**

4. **If the projectile lands at the height from which it was launched, it will reach its maximum height at the midpoint of the journey. The final value of V_y will be the negative of the initial value of V_y. This makes its final speed the same as its initial speed.**

If you can remember these facts, then you should have no trouble analyzing the projectile motion situations that I give you in the following example and "On Your Own" problems.

EXAMPLE 4.3

Illustrations copyright GifArt.com

A football player throws a football to his teammate far downfield. Draw the path of the ball as it leaves the quarterback's hands and travels to the receiver's hands. The ball is thrown at an angle of 45.0° relative to the ground and with a speed of 15.0 m/sec. Relative to the height at which it was thrown, what will be the ball's maximum height?

The path of the ball will be parabolic:

Now remember, even though this is a two-dimensional problem, we can break it down into two one-dimensional problems. The question asks what the maximum height of the ball will be. What dimension do we need to consider in order to determine that? We need to look at the y-dimension. To do this, we must first figure out what the y-component of the velocity is:

$$v_y = (15.0\frac{m}{sec}) \cdot \sin(45.0^o) = 10.6\frac{m}{sec}$$

Now we need to see what we have and what we are asked to calculate. We have the initial velocity in the y-dimension, and we also know that at the ball's maximum height, the y-component of its velocity is zero. We also know that gravity affects the ball in the y-dimension, so the acceleration of the ball in the y-dimension is 9.8 m/sec^2. So, we have initial velocity, final velocity, and acceleration. We are asked to calculate displacement. What equation should we use? If you think back to Module #2, you will remember that Equation (2.15) will allow us to solve this problem:

$$\mathbf{v}^2 = \mathbf{v}_o^2 + 2\mathbf{a} \cdot \Delta\mathbf{x}$$

Since we are only concentrating on the y-dimension, we can simply plug the y-dimension numbers into this equation and treat this like a one-dimensional problem. Remember, we must consider direction. If the initial velocity is positive, that means up is positive. Therefore, the acceleration (which is directed down) must be negative.

$$0^2 = (10.6\frac{m}{sec})^2 + 2 \cdot (-9.8\frac{m}{sec^2}) \cdot \Delta\mathbf{x}$$

$$\Delta\mathbf{x} = \frac{(10.6\frac{m}{\cancel{sec}})^2}{2 \cdot (9.8\frac{\cancel{m}}{\cancel{sec^2}})} = 5.7\,m$$

This tells us that the ball's maximum y-displacement (which is, of course, the same thing as its maximum height relative to the height at which it was thrown) is <u>5.7 m</u>.

Do you see what happened here? Even though this is a two-dimensional problem, the last part of the question only asked about one dimension, because it only wanted to know the maximum height. Thus, the two-dimensional problem was really just a one-dimensional problem in disguise!

In the situation depicted above, how long will it take the ball to reach the receiver, who is 23 meters down field?

What dimension are we concerned about in this part of the problem? Well, the receiver is about the same height as the quarterback, since they are standing on a level playing field. Any small differences in the heights of the players will not affect the result of the problem much, so we will simply ignore them. Thus, since both players are on the same level, we can ignore the y-component of the problem. After all, when the ball reaches the receiver's hands, it will have the same height as it did when it left the quarterback's hands. As a result, there is no net change in the y-dimension of the problem. Thus, all we have to look at is the x-dimension.

First, we need to determine the x-component of the ball's velocity:

$$V_x = (15.0\frac{m}{sec}) \cdot \cos(45.0^{\circ}) = 10.6\frac{m}{sec}$$

Now it is time to see what we know and what we are asked to calculate. We are given displacement (23 m), we calculated the initial velocity in this dimension (10.6 m/sec), and we also know that the x-component of the velocity is constant (therefore $\mathbf{a} = 0$). Using this information, we can calculate time with Equation (2.19):

$$\Delta\mathbf{x} = \mathbf{v}_0 t + \frac{1}{2}\mathbf{a}t^2$$

$$23\text{ m} = (10.6\frac{m}{sec}) \cdot t + \frac{1}{2} \cdot 0 \cdot t^2$$

$$t = \frac{23\ \cancel{m}}{10.6\frac{\cancel{m}}{sec}} = 2.2\text{ sec}$$

So we find out from this analysis that the ball will reach the receiver in <u>2.2 seconds</u>. As a sidelight, since we know the total time the ball takes to travel the distance, we also know that the ball must have reached its maximum height in 1.1 seconds, because when a projectile lands at the height from which it was launched, the maximum height is always reached halfway through the journey.

It may seem like I'm harping on this, but bear with me. It is important for you to see what happened in the examples given above. Even though we were clearly dealing with a two-dimensional problem, we broke it down into two one-dimensional problems. Since each question only asked about one dimension, they were easy to answer. They each boiled down to one-dimensional problems. Try to see how we do this by solving the following "On Your Own" problems.

ON YOUR OWN

Illustrations by Megan Whitaker

4.3 A battleship sinks an enemy ship by firing a missile with an initial velocity of 125 m/sec at an angle of 30.0°. How long does it take the missile to reach its maximum height? How long does it take for the missile to reach the enemy ship?

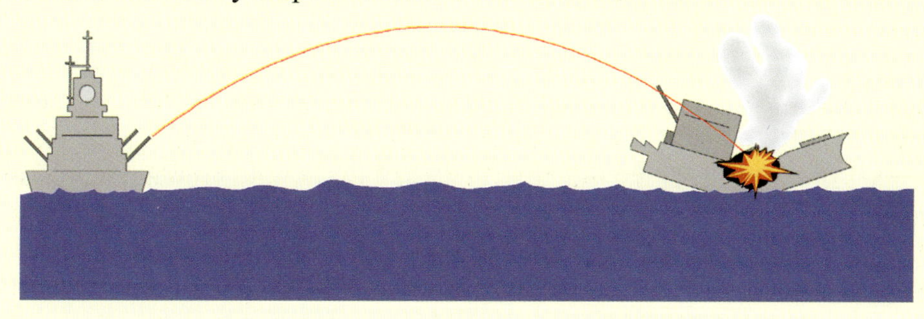

4.4 A basketball player lobs the ball over the heads of her opponents to her teammate down court. The player throws the ball with a velocity of 12.3 m/sec at an angle of 65.0°. If her teammate catches the ball in 2.28 seconds, how far apart are the two players?

The Range Equation

Although the problems that we just did should have been instructive, they really aren't all that useful. It turns out that I gave you more information than you needed in each of the example and "On Your Own" problems presented in the previous section. When we analyze a two-dimensional situation, we can usually learn a piece of information by analyzing one of the dimensions and then use that information to help us analyze something in the other dimension. To see what I'm talking about, let's look at the situation from "On Your Own" Problem 4.3.

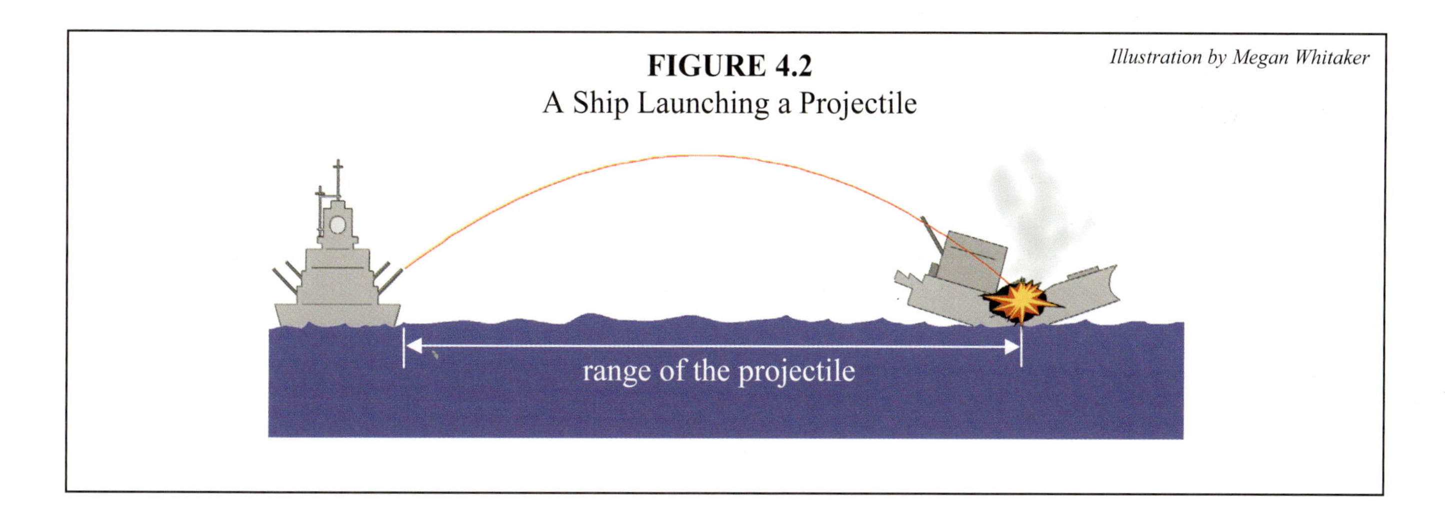

FIGURE 4.2
A Ship Launching a Projectile

Illustration by Megan Whitaker

In this situation, I asked you to calculate the time that it took for the projectile to reach its maximum height. Although that is an interesting piece of information, it's not all that useful to the person firing the projectile. After all, the person firing the projectile needs to know where to aim the cannon so that the projectile will land on the enemy ship. Thus, the really useful piece of information we should seek is the distance that the projectile will travel in the x-dimension. This will tell us whether or not the projectile will hit the enemy ship. When we calculate the distance that the projectile travels in the x-dimension, we usually say that we are calculating the **range** of the projectile.

How can we do this? In "On Your Own" Problem 4.4, we actually did determine the range of a projectile. In that problem, we used the initial velocity and the time to determine the distance that a basketball traveled in the x-dimension. In other words, we determined the range of the basketball. Now, although that was a useful exercise to go through, it wasn't all that applicable to the real world. After all, it is hard for someone firing a projectile (or throwing a basketball) to know how long the projectile will be in the air. The person firing the gun on the ship, for example, only knows the initial velocity of the projectile. Does that mean he or she will never be able to determine where to aim the cannon, since one of the pieces of data required to calculate the range is missing? Of course not!

Think about both of the "On Your Own" problems we have been discussing, and think of them together. In "On Your Own" Problem 4.3, you were able to calculate the time it took for the projectile to reach its maximum height when you were given nothing but the initial velocity. I pointed out in the solution to that problem that if you double your answer, you get the total time that the projectile was in the air. So, given just the initial velocity, you were able to determine the total time it took for the projectile to travel to its destination. What did you do in "On Your Own" Problem 4.4? Given the initial velocity *and the time it took for the ball to reach the destination*, you calculated the ball's range. Now think about it. In "On Your Own" Problem 4.4, I didn't really need to give you the total time that the ball was in the air. You could have determined that yourself, by calculating the time it took for the ball to reach its maximum height. If you doubled that answer, you would have calculated the time that I originally gave to you!

Do you see what I'm talking about here? In order to calculate the range of a projectile, you need to concentrate on the x-dimension. Before you can do that, however, you need the total time it takes for the projectile to reach its destination. You can calculate that by looking at the y-dimension! In other words, you examine the y-dimension first, in order to get a crucial piece of information, and then you analyze the x-dimension. That's how you can determine the range of any projectile when you are given just the initial velocity. I'm going to use this idea to develop a general equation that can be used to calculate the range of any projectile that lands at the same height from which it was fired.

In order to develop a general equation, I am not going to use numbers in this derivation. I am going to assume that someone told me the initial velocity of the projectile, but I am not going to use a number for it. Instead, I will refer to the initial speed as v_o (it is not in bold face type because speed is not a vector) and the angle (remember, speed and angle together make velocity) as θ. Furthermore, I will assume that θ is between 0° and 90°. Using these two symbols for the initial velocity of the projectile, I will now develop a general equation for the range of a projectile.

Since we know the initial velocity, we can determine its y-component:

$$v_{oy} = v_o \cdot \sin(\theta) \qquad (4.1)$$

Now that we know the initial velocity in the y-dimension, we can calculate the time it takes for the projectile to reach its maximum height. After all, we know that the y-component of its velocity will be zero at its maximum height, and we know that the acceleration in the y-dimension is simply due to gravity. In order to keep things completely general, we will refer to the acceleration due to gravity as "g." That way, we can use either 32 ft/sec^2 or 9.8 m/sec^2, depending on the units of the problem. So, we have initial velocity, final velocity, and acceleration. From that information, we need to calculate time. This calls for Equation (2.6):

$$\mathbf{v} = \mathbf{v}_o + \mathbf{a}t \tag{2.6}$$

Now I will substitute in the things I know. Please note that I will list the acceleration as "-g." I will do this because, in the end, I want the *value* for "g" to be positive. However, the acceleration is negative in this situation, so I need to include the negative sign explicitly:

$$0 = v_o \cdot \sin(\theta) + (-g) \cdot t \tag{4.2}$$

Solving for "t" gives us:

$$t = \frac{v_o \cdot \sin(\theta)}{g} \tag{4.3}$$

Since I am assuming that θ is between 0° and 90°, we know that this equation will always be positive, which is good, since time must always be positive.

So now we know how long it takes for the projectile to reach its maximum height. That's not quite what we need to know, however. We really need to know how much total time the projectile is in the air. But remember, when a projectile lands at the same height from which it was fired, it reaches its maximum height exactly halfway along its journey. This means that the total time the projectile is in the air is simply twice the time we just calculated:

$$\text{total time} = 2 \cdot (\text{time to reach maxium height}) \tag{4.4}$$

$$\text{total time} = 2 \cdot \frac{v_o \cdot \sin(\theta)}{g} \tag{4.5}$$

So you see, we didn't need to be told the time that the projectile was in the air. We just had to calculate it from some of the things we knew about the y-dimension of the problem. Now that we have this information, we can move to the x-dimension. We need to know the distance that the projectile travels in the x-dimension. In order to calculate this, we have some data. We just calculated the time it takes for the projectile to travel. We can also calculate the x-component of the initial velocity:

$$v_{ox} = v_o \cdot \cos(\theta) \tag{4.6}$$

Finally, we also know that in the x-dimension, acceleration is zero for any projectile. So, we have time, initial velocity, and acceleration. We need to calculate displacement. Equation (2.19) will do the trick:

$$\Delta\mathbf{x} = \mathbf{v}_o t + \frac{1}{2}\mathbf{a}t^2 \tag{2.19}$$

To use this equation, we need to substitute in the things that we know. The initial velocity (remember, we're in the x-dimension now) is given by Equation (4.6), the acceleration is zero, and the time is given by Equation (4.5). Substituting these things into Equation (2.19) gives us:

$$\Delta \mathbf{x} = [v_o \cdot \cos(\theta)] \cdot [2 \cdot \frac{v_o \cdot \sin(\theta)}{g}] + \frac{1}{2} \cdot (0) \cdot t^2 = \frac{2 \cdot v_o^2 \cdot [\sin(\theta)] \cdot [\cos(\theta)]}{g} \tag{4.7}$$

In the end, then, Equation (4.7) is one way that we can calculate the range of a projectile given just its initial velocity (v_o and θ). I want to do a couple of things to this equation, though, before we're finished. First, rather than calling our answer "$\Delta\mathbf{x}$," I'll call it "range," because that's what we're really solving for here. Second, I want to make the equation a little more compact. In trigonometry, you might have learned the following identity:

$$[\sin(\theta)] \cdot [\cos(\theta)] = \frac{\sin(2\theta)}{2} \tag{4.8}$$

If you didn't learn it, don't worry. It's simply an equation. It says that anywhere we see "$[\sin(\theta)] \cdot [\cos(\theta)]$," we can substitute "$\frac{\sin(2\theta)}{2}$." I will use this substitution in Equation (4.7). These two changes turn our range equation into:

$$\text{range} = \frac{v_o^2 \cdot \sin(2\theta)}{g} \tag{4.9}$$

Thus, if we are given the initial velocity of a projectile, we can determine its range by simply using Equation (4.9). Now remember, I assumed that θ was between 0° and 90° in this derivation, so this equation is valid only for angles within that range.

Now, before I go on to show you an example of how to use this equation, let me get a couple of things straight. First, because I put the negative sign on the acceleration due to gravity in Equation (4.2), you needn't use a negative number for g in this equation. That's already taken into account. Second, *this equation does not apply to all two-dimensional projectiles*. I cannot emphasize this point too strongly. Students see this range equation and think that they can apply it to all situations. However, it doesn't work that way. Since we assumed in the derivation that the projectile is going to land at the height from which it was launched, this equation only works for those kinds of projectiles. If a projectile lands at a height *other than that from which it was launched*, this equation *does not apply!* Please remember these two things:

Gravitational acceleration is always positive in Equation (4.9). Also, Equation (4.9) applies only to projectiles that land at the same height from which they are launched.

I really want to make sure that you understand when you can and cannot use Equation (4.9). Study Figure 4.3. Note that for the situation depicted on the left side of the figure, Equation (4.9) applies. However, for the situation shown on the right side of the figure, you *cannot* use Equation (4.9)!

FIGURE 4.3
Where To Apply Equation (4.9)

Illustration by Megan Whitaker

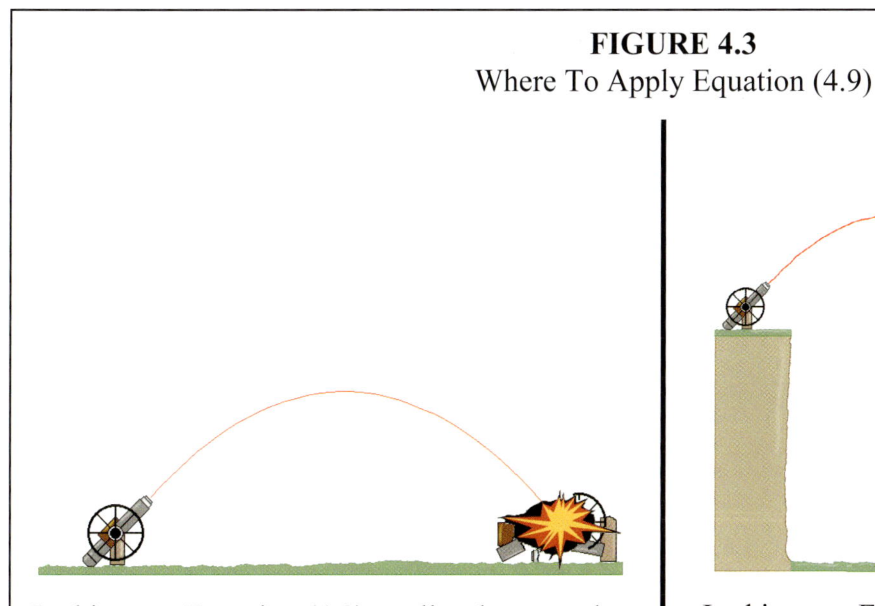

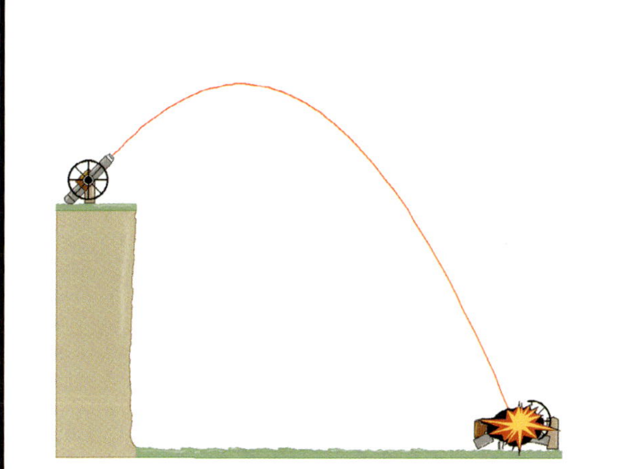

In this case, Equation (4.9) applies, because the projectile lands at approximately the same height as that from which it was launched.

In this case, Equation (4.9) does not apply, because the projectile lands significantly lower than the height from which it was launched.

Now that you see where Equation (4.9) does and does not apply, we can solve some problems using it.

EXAMPLE 4.4

Using a missile launcher, a soldier attempts to destroy an enemy outpost 7,512 meters away from his position. The outpost is at the same elevation as the soldier. If the launcher can fire its missiles with an initial speed of 345 m/sec, at what angle should the soldier aim the launcher?

To solve this problem, we merely need to use Equation (4.9) We are given the range (7,512 m) that the missile must have, and we are also given its initial speed. The only variable left in Equation (4.9) is the angle, so we can solve for it:

$$\text{range} = \frac{v_o^{\,2} \cdot \sin(2\theta)}{g}$$

$$7{,}512 \text{ m} = \frac{(345 \frac{\text{m}}{\text{sec}})^2 \cdot \sin(2\theta)}{9.8 \frac{\text{m}}{\text{sec}^2}}$$

$$\sin(2\theta) = \frac{7{,}512 \cancel{\text{m}} \cdot 9.8 \frac{\cancel{\text{m}}}{\cancel{\text{sec}^2}}}{119025 \frac{\cancel{\text{m}^2}}{\cancel{\text{sec}^2}}}$$

$$\sin(2\theta) = 0.62$$

We've solved for sin (2θ), so we are close to being done. How do we get θ? In order to remove a function from one side of an algebraic equation, we always take the inverse function of both sides. Since the inverse function of sin is $\sin^{-1}$, we can get rid of the sin function by taking the $\sin^{-1}$ of each side. When we do this, we get:

$$2\theta = \sin^{-1}(0.62)$$

$$2\theta = 38^{\circ}$$

$$\theta = 19^{\circ}$$

Thus, the soldier must aim his missile launcher at an angle of <u>19°</u> in order to hit the outpost.

See how well you can use Equation (4.9) by solving the following "On Your Own" problems.

ON YOUR OWN

4.5 A cannon is tilted at an angle of 42° relative to the ground. If the cannon launches its projectile with an initial speed of 150 m/sec, what is the range of the cannonball on level ground?

4.6 A missile is aimed at an angle of 23° relative to the ground. If the missile's target is 5.4×10^4 m away and at the same elevation, at what speed must the missile launch in order to hit it?

4.7 A rifle shoots its bullet with an initial speed of 2.00×10^2 ft/sec. At what angle should a marksman aim his rifle to hit a target that is 5.00×10^2 feet away and at the same height as the rifle?

Now I realize I've harped and harped on the concept that two-dimensional problems can be broken down into two or more one-dimensional problems. In particular, this last section shows how we can treat the y-dimension and x-dimension in a two-dimensional problem as two completely separate situations. Now we'll see how closely you have been listening. I want you to consider a situation and then I will ask you a question. The accuracy of your answer will tell you how closely you have been paying attention to my harping!

Suppose a person holds a pistol in one hand and one of the pistol's bullets in the other. Suppose further that the pistol is pointed level to the ground. In other words, its angle relative to the ground is zero. Let me now ask my question. If the person fires the pistol and drops the bullet at precisely the same instant, which will hit the ground first: the bullet that the person dropped or the bullet that was fired from the gun? Think about it, and then write your answer down on a piece of paper. Once you have done that, perform the following experiment to find out whether or not you were right.

EXPERIMENT 4.1

The Two Dimensions of a Rubber Band's Flight

Supplies:

- Safety goggles
- A rubber band
- A stopwatch
- A person to help you

1. In order to find out whether or not you answered my question correctly, we need to simulate the situation discussed above. Of course, we can't use a pistol, so we will use a rubber band. First, hold the rubber band in front of your face and be ready to drop it. You should grab the rubber band on only one side, so that it hangs down from your fingers.
2. Align the *lowest* portion of the rubber band with your nose. Your helper should hold the stopwatch and be ready to time how long it takes for the rubber band to hit the floor.
3. Have your helper say "go." When he or she says "go," your helper should start the stopwatch and you should immediately let go of the rubber band. Your helper should then stop the watch as soon as *any* part of the rubber band hits the floor.
4. Repeat this four more times, and average the times that your helper measured. When you are finished, you have a relatively accurate measure of how long it takes the rubber band to fall from the height of your nose to the floor.
5. In the second portion of this experiment, you need to practice flipping the rubber band a little. You need to get in a hallway or a large room that is reasonably free of obstruction.
6. Hook one end of the rubber band on the tip of your index finger and pull the other end back with your other hand, so that when your hand releases the rubber band, it will fly in the direction that your index finger points.
7. Practice flipping the rubber band a few times from the height of your nose. It really doesn't matter if the rubber band goes straight. It is more important that it is aimed perfectly horizontally and that it does not hit any objects in its travel. In other words, the initial angle of the rubber band relative to the ground should be zero, and the rubber band's flight should be uninhibited. Also, you needn't launch the rubber band with great speed. If you are in a small room or hallway, pull back gently on the rubber band. Once you think you are pretty good at launching the rubber band horizontally from the height of your nose, proceed to the next step.
8. Aim your rubber band horizontally, and make sure it is as high as the bottom of your nose.
9. Once again have your helper hold the stopwatch and say "go." When that happens, you should release the rubber band and your helper should start the timer. Once again, the helper should stop the timer when he or she sees *any part* of the rubber band hit the ground.
10. Do this a total of five times and average the results. Be sure that the rubber band does not go up at all when it is released. It needs to travel horizontally at first and then down. It should never travel up. If it does, re-do that trial.
11. Compare the time it takes for the launched rubber band to hit the floor to the time it took for the rubber band to simply fall on the floor. Keeping in mind that this experiment is probably accurate to about 10%, how do these two numbers compare?
12. Have your parent / teacher read the experiment answer (found in the *Solutions and Tests Guide* after the answers for the Module #4 test) to determine whether or not your experiment turned out correctly.
13. Put away the stopwatch and the rubber band.

The multimedia CD has a video demonstration that illustrates the point made by the answer your parent / teacher just read to you.

Two-Dimensional Situations in Which You Cannot Use the Range Equation

Now that you are a veritable expert at using the range equation, it is time to analyze situations in which the range equation does not apply. As we learned in the previous section, this would be any situation in which the projectile lands at a height different than that from which it was launched. In these kinds of situations, we cannot use Equation (4.9); however, most of the concepts that we used to derive Equation (4.9) are directly applicable. Examine the following example problem so that you can see the reasoning necessary to analyze such problems.

EXAMPLE 4.5

Cannon illustration copyright GifArt.com
Terrain illustration by Megan Whitaker

A cannon shoots horizontally (θ = 0°) off of a cliff. If the cannon can shoot with an initial speed of 150.0 m/sec and is 100.0 m above the ground, how far will the cannonball be from the cliff when it finally hits the ground?

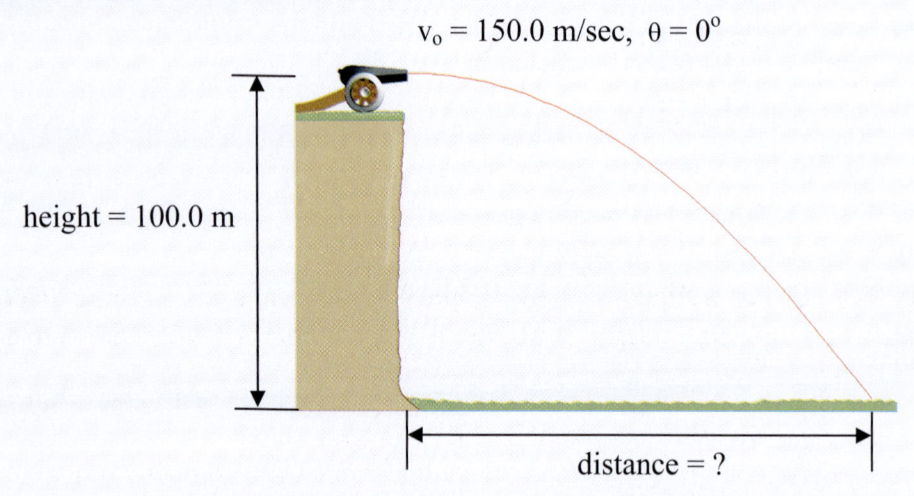

From the picture we have above, it is clear that Equation (4.9) does not apply, because the cannonball lands 100.0 meters below the height from which it was launched. Thus, we must find another way to solve this problem. Let's first determine what we have. Well, we have both the angle and the speed. We therefore can calculate the initial velocity in both the x- and the y-dimensions. We also know the acceleration in each of those dimensions (0 in the x-dimension and -9.8 m/sec^2 in the y-dimension). Finally, we know how far the cannonball will travel in the y-dimension. Using all of this information, we need to figure out how far the cannonball will travel in the x-dimension. How do we approach this?

Well, since we already know the initial velocity and the acceleration in the x-dimension, we could figure out the distance traveled using Equation (2.19) if we knew the time that the cannonball was in the air. Although we are not given the time, we can get that information by looking at the y-dimension. In that dimension, we have displacement, initial velocity, and acceleration. We can use Equation (2.19) and these facts in that dimension to calculate the time. We can then use that time in Equation (2.19) for the x-dimension.

To get things going in the y-dimension, we first have to figure out the y-component of the initial velocity:

$$v_{oy} = (150.0 \frac{m}{sec}) \cdot \sin(0^{\circ}) = 0$$

We also need to determine direction. Let's stick with the convention that a positive sign means upward motion while a negative sign denotes downward motion. With that convention, the displacement is negative (because the cannonball ended up below its starting point), and the acceleration is negative as well. Now we use Equation (2.19):

$$\Delta \mathbf{x} = \mathbf{v}_o t + \frac{1}{2}\mathbf{a}t^2$$

$$-100.0 \text{ m} = (0) \cdot t + \frac{1}{2} \cdot (-9.8 \frac{m}{sec^2}) \cdot t^2$$

$$t^2 = \frac{2 \cdot 100.0 \cancel{m}}{9.8 \frac{\cancel{m}}{sec^2}}$$

$$t = 4.5 \text{ sec}$$

This tells us, then, that the cannonball is in flight for 4.5 seconds. Now that we have this information, we can go back to the x-dimension. We first need to calculate the x-component of the initial velocity:

$$v_{ox} = (150.0 \frac{m}{sec}) \cdot \cos(0^{\circ}) = 150.0 \frac{m}{sec}$$

Now that we've done that, we have all we need to use Equation (2.19) to solve the problem.

$$\Delta \mathbf{x} = \mathbf{v}_o t + \frac{1}{2}\mathbf{a}t^2$$

$$\Delta \mathbf{x} = (150.0 \frac{m}{\cancel{sec}}) \cdot (4.5 \cancel{sec}) + \frac{1}{2} \cdot (0) \cdot (4.5 \text{ sec})^2$$

$$\Delta \mathbf{x} = 680 \text{ m}$$

The cannonball, then, ends up 680 m from the edge of the cliff.

Do you see how I solved this problem? The question asked me to analyze the distance traveled in the x-dimension. I realized, however, that I couldn't immediately solve for that because I did not have enough information to do so. As a result, I went to the other dimension (the y-dimension) to see if I could find the information that I needed there. After calculating the necessary quantity, I then went back to the dimension of interest (the x-dimension) to finish the problem.

In order to give you some hands-on experience separating the x- and y-components of two-dimensional motion, I want you to perform the following experiment. It will allow you to measure the speed of a toy car or the speed at which you can flick a ping-pong ball.

EXPERIMENT 4.2

Measuring the Horizontal Speed of an Object without a Stopwatch

Supplies:

- Safety goggles
- A toy car or a ping-pong ball
- A meterstick
- A person to help you
- A flat table in a room with plenty of space on at least one side of the table

Note: A sample set of calculations is available in the solutions and tests guide. It is with the solutions to the practice problems.

1. Measure the height of the table. Make sure you measure it from the floor to the table top.
2. You are going to roll either the toy car or the ping-pong ball off the table. Your helper needs to find the spot where the toy or ball first hits the floor. Thus, have your helper stand away from the table, in the direction in which you will roll the car or the ball off the table.
3. If you are using a toy car, wind it up and place it on the table, facing the direction in which you want it to roll. If it is not a windup car, just set it on the table and push it as hard as you can. If you are using a ball, place the ball on the table and use your thumb and forefinger to flick the ball as hard as you can so that it will roll off the table.
4. Have your helper spot the point at which the car or ball first hits the floor.
5. Measure the horizontal distance from the edge of the table to the spot where the car hit the floor. The best way to do this is to lay the meterstick on the floor, starting at where the edge of the table would be if it were on the floor. Then, measure the length of a straight line between that point and the point at which the car or ball hit the floor.
6. Believe it or not, you now have all of the information you need to calculate how fast the car or ball was rolling when it hit the edge of the table. Once the car was in the air, the only force acting on it (neglecting air resistance) was gravity. Since gravity pulls objects straight down, it operates only in the y-dimension. What else do we know about the y-dimension? The table top is flat. Thus, when the car or ball rolled off of the table, its initial velocity was completely horizontal. This means that the y-component of its initial velocity was zero. The acceleration in the y-dimension is -9.8 m/sec^2, and the displacement in the y-dimension is the negative (remember, down is negative) of whatever you measured in step 1. Since you have the initial velocity, the acceleration, and the displacement, you can use Equation (2.19) to calculate the time it took for the car or ball to fall.
7. Now let's consider the x-dimension. In the x-dimension, there is nothing pushing or pulling the car once it leaves the table. Thus, in the x-dimension, there is no acceleration. The car is simply traveling along with whatever speed it had when it launched off of the table. The displacement is the value you measured in step 5. Thus, you have the displacement, acceleration, and time (you calculated that in the previous step). You can plug those numbers into Equation (2.19) to calculate the x-component of the velocity as the car or ball left the table. Since *all* of the car's velocity was pointed in the x-dimension when the car hit the end of the table, you have just determined the velocity of the car or ball!
8. Repeat this experiment with another toy or, if you are using the ping-pong ball, switch with your helper and measure the speed at which he or she can flick a ping-pong ball.
9. Clean up your mess.

Just to make sure you really understand how to analyze two-dimensional motion, I want to go through one more example problem for you. It is very similar to the experiment you just did.

EXAMPLE 4.6

Illustration by Megan Whitaker

An archer measures how fast her arrows launch in the following way. She stands 50.0 feet from a tree and aims her bow and arrow horizontally (θ = 0°) at the tree. She then stretches back the bow and releases the arrow so that it embeds into the tree trunk. She then measures how far the arrow dropped during its flight and finds that distance to be 4.12 feet. How fast did the archer launch her arrow?

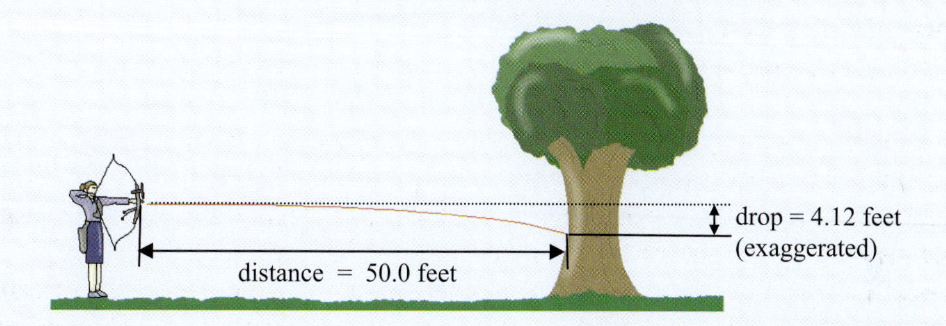

In this problem, of course, we cannot use Equation (4.9), because the arrow lands below the height from which it was launched. We must analyze each dimension, then, and try to figure out what information we can glean from the data given. We are asked to calculate the initial velocity of the arrow. If we look at the initial velocity in each of the two dimensions, we will quickly see which dimension allows us to calculate that quantity:

$$v_{ox} = v_o \cdot \cos(0^o) = v_o$$

$$v_{oy} = v_o \cdot \sin(0^o) = 0$$

Because the arrow is aimed horizontally, the y-component of the velocity is zero. As a result, we cannot calculate the initial velocity using the y-dimension. The x-dimension has a non-zero initial velocity, but we do not have enough information yet. In the x-dimension, we know the displacement (50.0 feet) and acceleration (0). If we could find out the time, we could use Equation (2.19) to determine the initial velocity. Unfortunately, we don't have the time yet. Let's see if we can get that from the y-dimension.

In the y-dimension, we know the initial velocity (0), the acceleration (-32 ft/sec^2), and the displacement (-4.12 ft). That's all we need to calculate the time from Equation (2.19):

$$\Delta \mathbf{x} = \mathbf{v}_o t + \frac{1}{2}\mathbf{a}t^2$$

$$-4.12 \text{ ft} = (0) \cdot t + \frac{1}{2} \cdot (-32 \frac{\text{ft}}{\text{sec}^2}) \cdot t^2$$

$$t^2 = \frac{2 \cdot 4.12 \cancel{\text{ft}}}{32 \frac{\cancel{\text{ft}}}{\text{sec}^2}}$$

$$t = 0.51 \text{ sec}$$

Now we know the time it took the arrow to reach its destination. At this point, we can go back to the x-dimension to finish the problem:

$$\Delta \mathbf{x} = \mathbf{v}_o t + \frac{1}{2}\mathbf{a}t^2$$

$$50.0 \text{ ft} = (v_o) \cdot (0.51 \text{ sec}) + \frac{1}{2} \cdot (0) \cdot (0.51 \text{ sec})^2$$

$$v_o = \frac{50.0 \text{ ft}}{0.51 \text{ sec}}$$

$$v_o = 98 \frac{\text{ft}}{\text{sec}}$$

This means that the archer released her arrow at a speed of 98 ft/sec.

Even though this example was different from Example 4.5, we essentially used the same skills to analyze the situation. Although we had no initial idea of what dimension we were interested in, we quickly found out that the y-dimension could not be used to calculate the initial velocity, because the angle at which the arrow was launched caused no initial velocity in the y-dimension. Once we looked at the x-dimension, however, we realized that we were short some information. We needed the time that the arrow was in flight. So we went to the y-dimension to get that information and then solved the problem in the x-dimension. See if you can perform such problems on your own.

ON YOUR OWN

Illustrations from the MasterClips collection

4.8 A reckless teenager throws a baseball horizontally ($\theta = 0°$) off of the top of his house. The baseball was thrown with a speed of 11.5 m/sec and lands 15.1 m away from the house. Neglecting the height of the boy, how tall is the house?

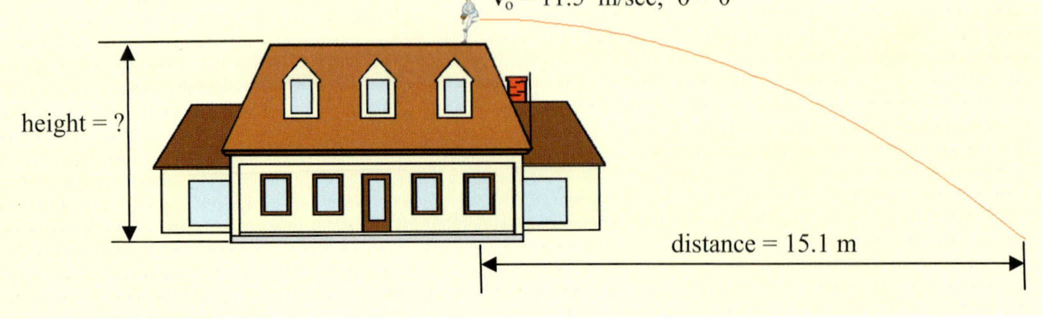

4.9 A bomber is flying 9,112 meters above the ground (relative to its target) and is traveling at a speed of 145 m/sec. When the plane drops its bomb, the bomb will have an initial speed equal to that of the plane and an initial angle of zero degrees relative to the ground. How far away from the target must the plane be when it drops the bomb?

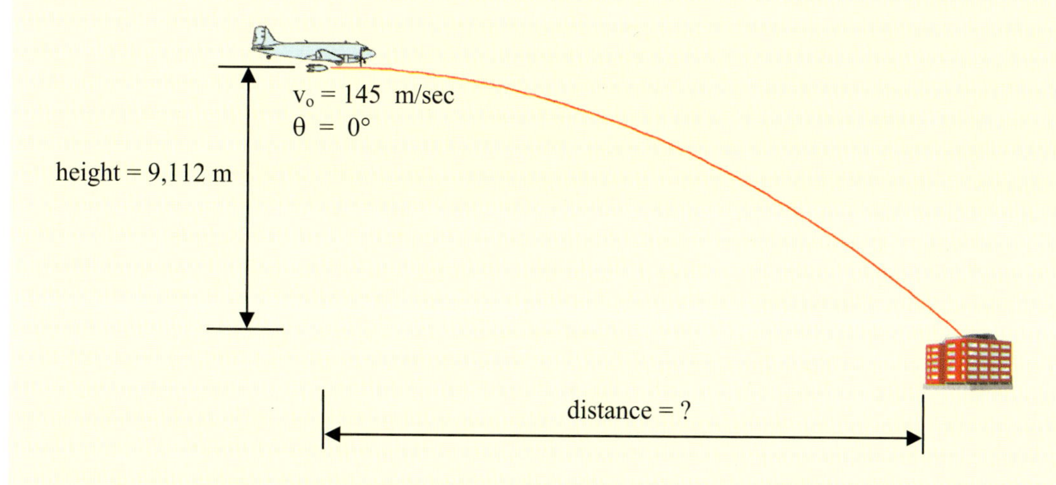

4.10 A marksman aims his gun horizontally (θ = 0°) and fires at a target that stands 81.0 meters away. If the bullet hits the target 0.45 meters below the aim of the gun, what was the initial speed of the bullet ?

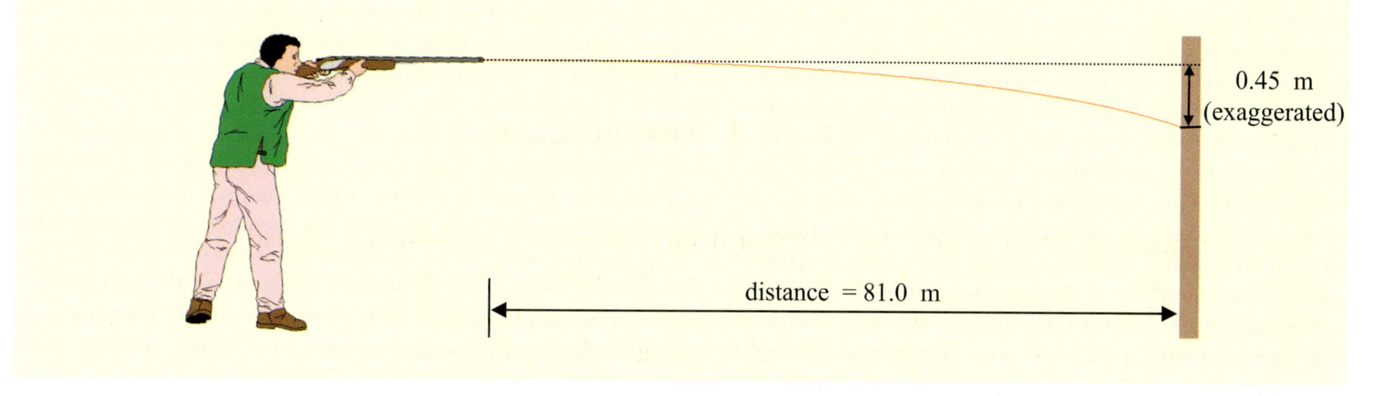

If these problems seemed a bit hard, don't worry about it. This subject is one of the more difficult ones in physics. In fact, aside from some of the problems we will solve in Module #6, these are probably the most difficult problems in the course. Thus, if you're a little unsure of how to solve them and therefore get a lower grade on this module, don't worry. Most of what we do in physics is a bit easier than this. On the other hand, if you were able to solve the last "On Your Own" problems, you probably have a knack for physics!

ANSWERS TO THE "ON YOUR OWN" PROBLEMS

4.1 The first thing to do here is to picture what's going on:

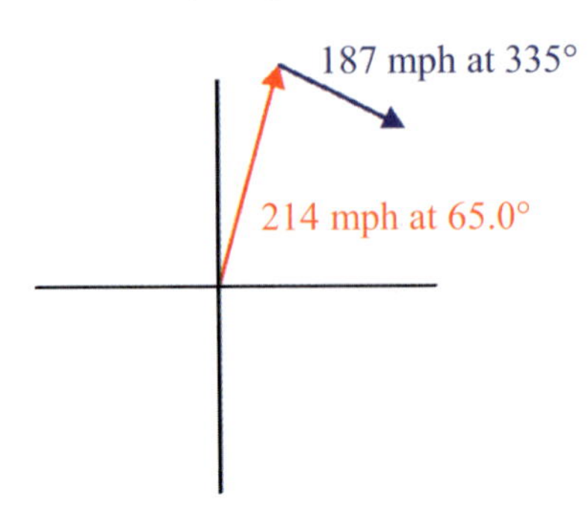

As with all such problems, we need to look for the one-dimensional situations within this two-dimensional problem. Each leg of the journey is an independent one-dimensional problem, because the airplane travels in a straight line in each case. In the first leg of the journey, the velocity is 214 mph, the acceleration is 0 (because the velocity is constant), and the time is 2.10 hours. Equation (2.19), then, looks like this:

$$\Delta\mathbf{x} = \mathbf{v}_o t + \frac{1}{2}\mathbf{a}t^2$$

$$\Delta\mathbf{x} = (214 \frac{\text{miles}}{\cancel{\text{hr}}})\cdot(2.10\,\cancel{\text{hr}}) + \frac{1}{2}\cdot 0\cdot(2.10\text{ hr})^2$$

$$\Delta\mathbf{x} = 449 \text{ miles}$$

Is this our answer? Of course not! This was only the first leg of the journey. We did learn something, however. We learned that the first leg of the journey resulted in a displacement of 449 miles at an angle of 65.0°. That, then, is where the second leg of the journey begins. If we can figure out the displacement that results from this second leg, we should have enough information to determine the final displacement of the airplane.

Well, we can analyze the second leg of the journey the same way that we analyzed the first leg.

$$\Delta\mathbf{x} = \mathbf{v}_o t + \frac{1}{2}\mathbf{a}t^2$$

$$\Delta\mathbf{x} = (187 \frac{\text{miles}}{\cancel{\text{hr}}})\cdot(1.40\,\cancel{\text{hr}}) + \frac{1}{2}\cdot 0\cdot(1.40\text{ hr})^2$$

$$\Delta\mathbf{x} = 262 \text{ miles}$$

What do we have now? Well, we know that the first leg of the journey resulted in a displacement of 449 miles at 65.0°, while the second leg resulted in a displacement of 262 miles at 335°. How can we find the total displacement? We can add them in order to get the final displacement.

To add these two displacements, we must first break the vectors down into their components. To make the notation easier, let's call the displacement vector resulting from the first leg of the journey vector **A**. We will therefore call the second vector **B**, and the final displacement vector will be known as **C**.

$$A_x = (449 \text{ miles}) \cdot \cos(65.0°) = 1.90 \times 10^2 \text{ miles}$$

$$A_y = (449 \text{ miles}) \cdot \sin(65.0°) = 407 \text{ miles}$$

$$B_x = (262 \text{ miles}) \cdot \cos(335°) = 237 \text{ miles}$$

$$B_y = (262 \text{ miles}) \cdot \sin(335°) = -111 \text{ miles}$$

$$C_x = 1.90 \text{ x } 10^2 \text{ miles} + 237 \text{ miles} = 427 \text{ miles}$$

$$C_y = 407 \text{ miles} + -111 \text{ miles} = 296 \text{ miles}$$

Now we can determine the magnitude and direction of the final displacement vector:

$$C = \sqrt{C_x^{\ 2} + C_y^{\ 2}} = \sqrt{(427 \text{ miles})^2 + (296 \text{ miles})^2} = 5.20 \times 10^2 \text{miles}$$

$$\theta = \tan^{-1}\left(\frac{C_y}{C_x}\right) = \tan^{-1}\left(\frac{296 \cancel{\text{ miles}}}{427 \cancel{\text{ miles}}}\right) = 34.7°$$

Based on our rules regarding Equation (3.3), this is the angle for the final displacement vector. In the end, then, the airplane's final displacement vector is <u>5.20 x 10^2 miles at an angle of 34.7°</u>.

4.2 The first thing to do here is to picture what's going on:

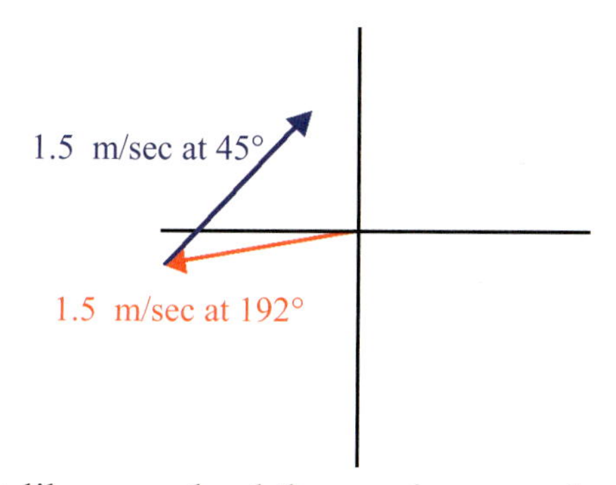

We can solve this problem just like we solved the previous one, but we have to recognize that the units are not consistent. The velocities are given in m/sec while the time is in minutes. Thus, we need to convert minutes to seconds:

$$\frac{45\ \cancel{\text{min}}}{1} \times \frac{60\ \text{sec}}{1\ \cancel{\text{min}}} = 2.7 \times 10^3\ \text{sec}$$

Now that our units are consistent, we can use Equation (2.19) to figure out the displacement on the first leg of the journey:

$$\Delta \mathbf{x} = \mathbf{v}_0 t + \frac{1}{2}\mathbf{a}t^2$$

$$\Delta \mathbf{x} = (1.5\ \frac{\text{m}}{\cancel{\text{sec}}}) \cdot (2.7 \times 10^3\ \cancel{\text{sec}}) + \frac{1}{2} \cdot 0 \cdot (2.7 \times 10^3\ \text{sec})^2$$

$$\Delta \mathbf{x} = 4.1 \times 10^3\ \text{m}$$

We can analyze the second leg of the journey in the same way. Just like we did in the first leg, however, we must convert the units of time from minutes to seconds. Once we do that, Equation (2.19) looks like this:

$$\Delta \mathbf{x} = \mathbf{v}_0 t + \frac{1}{2}\mathbf{a}t^2$$

$$\Delta \mathbf{x} = (1.5\ \frac{\text{m}}{\cancel{\text{sec}}}) \cdot (7.2 \times 10^2\ \cancel{\text{sec}}) + \frac{1}{2} \cdot 0 \cdot (7.2 \times 10^2\ \text{sec})^2$$

$$\Delta \mathbf{x} = 1.1 \times 10^3\ \text{m}$$

Just like we did in the last problem, we can add the two displacements in order to get the final displacement.

To add these two displacements, we must first break the vectors down into their components. To make the notation easier, let's call the displacement vector resulting from the first leg of the journey vector **A**. We will therefore call the second vector **B**, and the final displacement vector will be known as **C**.

$$A_x = (4.1 \times 10^3\ \text{m}) \cdot \cos(192^\circ) = -4.0 \times 10^3\ \text{m}$$

$$A_y = (4.1 \times 10^3\ \text{m}) \cdot \sin(192^\circ) = -8.5 \times 10^2\ \text{m}$$

$$B_x = (1.1 \times 10^3\ \text{m}) \cdot \cos(45^\circ) = 7.8 \times 10^2\ \text{m}$$

$$B_y = (1.1 \times 10^3\ \text{m}) \cdot \sin(45^\circ) = 7.8 \times 10^2\ \text{m}$$

$$C_x = -4.0 \times 10^3 \text{ m} + 7.8 \times 10^2 \text{ m} = -3.2 \times 10^3 \text{ m}$$

$$C_y = -8.5 \times 10^2 \text{ m} + 7.8 \times 10^2 \text{ m} = -70 \text{ m}$$

Look at how the significant figures work out in the y-component of C. Both numbers have their last significant figure in the *tens* place ($0.5 \times 10^2 = 50$, $0.8 \times 10^2 = 80$). Thus, the answer must have its last significant figure in the tens place. Now we can determine the magnitude and direction of the final displacement vector:

$$C = \sqrt{C_x^{\ 2} + C_y^{\ 2}} = \sqrt{(-3.2 \times 10^3 \text{ m})^2 + (-70 \text{ m})^2} = 3.2 \times 10^3 \text{ m}$$

$$\theta = \tan^{-1}\left(\frac{-70 \cancel{\text{m}}}{-3.2 \times 10^3 \cancel{\text{m}}}\right) = 1^\circ$$

Based on our rules regarding Equation (3.3), this means that the angle for the final displacement vector is really $180^\circ + 1^\circ = 181^\circ$. In the end, then, the treasure's displacement vector is <u>3.2×10^3 m at an angle of 181°</u>. Notice once again how the significant figures worked out. The angle from Equation (3.3) was limited to one significant figure because 70 has only one significant figure. However, the 1° was then added to 180°, which is exact. Thus, since 1° has its last significant figure in the ones place, the answer must have its last significant figure in the ones place, which is why the final angle is 181°.

4.3 In solving this problem, we first have to see what dimension we are dealing with. Since the question asks about height, we are at first only interested in the y-dimension. Thus, we should get the y-component of the velocity before we start:

$$v_y = (125 \frac{\text{m}}{\text{sec}}) \cdot \sin(30.0^\circ) = 62.5 \frac{\text{m}}{\text{sec}}$$

Now we have the initial velocity in the dimension of interest. In this dimension, we also know the acceleration (-9.8 m/sec^2), because in the y-dimension, gravity is at work. We know the final velocity as well, because at a projectile's maximum height, the final velocity is zero. With these data, we are asked to calculate time. We therefore must use Equation (2.6):

$$\mathbf{v} = \mathbf{v}_o + \mathbf{a}\mathbf{t}$$

$$0 = 62.5 \frac{\text{m}}{\text{sec}} + (-9.8 \frac{\text{m}}{\text{sec}^2}) \cdot t$$

$$t = \frac{62.5 \frac{\cancel{\text{m}}}{\cancel{\text{sec}}}}{9.8 \frac{\cancel{\text{m}}}{\text{sec}^{\cancel{2}}}} = 6.4 \text{ sec}$$

Notice the signs I used. If we define the upward initial velocity as positive, then the downward acceleration must be negative. Thus, our answer tells us that it takes 6.4 seconds to reach maximum height.

It turns out that this is all the information we need to answer the second part of the problem as well. After all, we know that when the projectile lands at the same height from which it is launched, it reaches maximum height at the halfway point of its journey. Thus, if it takes 6.4 seconds to reach its maximum height, it takes 12.8 seconds to reach the enemy ship.

4.4 This problem is concerned only with the x-dimension. After all, the ball will land at the same height from which it was launched. Thus, we are only interested in the distance that it travels *horizontally*. Since we are only interested in the x-dimension, then, we should figure out what the x-component of the ball's initial velocity is:

$$v_x = (12.3\,\frac{\text{m}}{\text{sec}}) \cdot \cos(65.0^\circ) = 5.20\,\frac{\text{m}}{\text{sec}}$$

So we know the initial velocity in the x-dimension as well as the time that the ball travels. In addition, we know that there is no acceleration in the x-dimension of a projectile's travel. With those three pieces of information, we need to determine displacement. Equation (2.19) seems to be just what we need:

$$\Delta\mathbf{x} = \mathbf{v}_o t + \frac{1}{2}\mathbf{a}t^2$$

$$\Delta\mathbf{x} = (5.20\,\frac{\text{m}}{\cancel{\text{sec}}}) \cdot (2.28\,\cancel{\text{sec}}) + \frac{1}{2} \cdot (0) \cdot (2.28\text{ sec})^2$$

$$\Delta\mathbf{x} = 11.9\text{ m}$$

Thus, the players are 11.9 meters apart from one another.

4.5 This is a simple application of Equation (4.9). We are given the initial speed (150 m/sec) and the angle (42°). From those facts, we are asked to calculate the range of the projectile:

$$\text{range} = \frac{v_o^{\,2} \cdot \sin(2\theta)}{g}$$

$$\text{range} = \frac{(150\frac{\text{m}}{\cancel{\text{sec}}})^2 \cdot \sin(2 \cdot 42^\circ)}{9.8\frac{\cancel{\text{m}}}{\cancel{\text{sec}^2}}} = 2.3 \times 10^3\text{ m}$$

So the cannonball's range is 2.3 x 10^3 m.

4.6 In this use of Equation (4.9), we are given the angle (23°) and the desired range (5.4×10^4 m). We need to determine the necessary initial speed. Thus, we just need to rearrange Equation (4.9) to solve for v_o after we have plugged in the numbers that we know:

$$\text{Range} = \frac{v_o^2 \cdot \sin(2\theta)}{g}$$

$$5.4 \times 10^4 \text{ m} = \frac{v_o^2 \cdot \sin(2 \cdot 23)}{9.8 \frac{\text{m}}{\text{sec}^2}}$$

$$v_o^2 = \frac{5.4 \times 10^4 \text{ m} \cdot 9.8 \frac{\text{m}}{\text{sec}^2}}{0.719}$$

$$v_o = 8.6 \times 10^2 \frac{\text{m}}{\text{sec}}$$

So, in order to hit its target, the missile must have an initial velocity of $\underline{8.6 \times 10^2 \text{ m/sec}}$.

4.7 In this application of Equation (4.9), we are given the initial speed (2.00×10^2 ft/sec) and the desired range (5.00×10^2 feet). With that information, we must calculate the angle at which the rifle should be aimed. To do that, we will plug in our numbers and then rearrange the equation to solve for θ. Before we do this, however, notice that the units are in feet and seconds. This tells us that we need to use 32 ft/sec^2 for the acceleration due to gravity so that the units will be consistent.

$$\text{Range} = \frac{v_o^2 \cdot \sin(2\theta)}{g}$$

$$5.00 \times 10^2 \text{ ft} = \frac{(2.00 \times 10^2 \frac{\text{ft}}{\text{sec}})^2 \cdot \sin(2\theta)}{32 \frac{\text{ft}}{\text{sec}^2}}$$

$$\sin(2\theta) = \frac{5.00 \times 10^2 \cancel{\text{ft}} \cdot 32 \frac{\cancel{\text{ft}}}{\cancel{\text{sec}^2}}}{4.00 \times 10^4 \frac{\cancel{\text{ft}^2}}{\cancel{\text{sec}^2}}}$$

$$\sin 2\theta = 0.40$$

$$2\theta = \sin^{-1}(0.40)$$

$$\theta = 12^{\circ}$$

The marksman, then, must aim his rifle at an angle of 12 degrees.

4.8 In this problem, we want to determine the height of a house. Thus, we are interested in the y-dimension. We can figure out the y-component of the initial velocity, and we also know the acceleration in the y-dimension. Unfortunately, to determine the distance that the ball travels in the y-dimension, we need to figure out the time it takes for the ball to hit the ground. We can do that by looking at the x-dimension.

In that dimension, we can calculate the initial velocity, and we know that the acceleration is zero. We also know that the ball traveled 15.1 m in the x-dimension. With this information, we can use Equation (2.19) to determine the time that the ball was in the air. First, we need to determine the x-component of the initial velocity:

$$v_{ox} = (11.5\ \frac{m}{sec}) \cdot \cos(0^{\circ}) = 11.5\ \frac{m}{sec}$$

Now we can take the data we know for the x-dimension and plug them into Equation (2.19). At that point, the only unknown will be the time, and we can solve for it:

$$\Delta \mathbf{x} = \mathbf{v}_o t + \frac{1}{2}\mathbf{a}t^2$$

$$15.1\ m = (11.5\ \frac{m}{sec}) \cdot t + \frac{1}{2} \cdot (0) \cdot t^2$$

$$t = \frac{15.1\ \cancel{m}}{11.5\ \frac{\cancel{m}}{sec}}$$

$$t = 1.31\ sec$$

Now that we have the time, we can go back to the y-dimension and figure out the height of the building. To do that, we need to determine the initial velocity in the y-dimension:

$$v_{oy} = (11.5\ \frac{m}{sec}) \cdot \sin(0^{\circ}) = 0$$

We can take this initial velocity, the acceleration (-9.8 m/sec^2), and the time (1.31 sec) and plug them into Equation (2.19) to determine the distance that the ball falls in the y-dimension. This is equal to the height of the house:

$$\Delta\mathbf{x} = \mathbf{v}_o t + \frac{1}{2}\mathbf{a}t^2$$

$$\Delta\mathbf{x} = (0)\cdot(1.31\,\text{sec}) + \frac{1}{2}\cdot(-9.8\,\frac{\text{m}}{\cancel{\text{sec}^2}})\cdot(1.31\,\cancel{\text{sec}})^2$$

$$\Delta\mathbf{x} = -8.4\text{ m}$$

Realize that our answer is negative simply because it denotes that the ball traveled *down* for 8.4 m. Since the distance that the ball falls down is the height of the house, then the house is 8.4 m high.

4.9 This is a standard problem that bombers have to do in order to determine when to release their payloads. Because the bomb will have an initial velocity that is equal to that of the plane, the bombers must release their payloads long before they are over their target, or they will miss. In order to solve the problem, we need to determine how far the bomb travels in the x-dimension. Unfortunately, in order to make this calculation, we need to know the time that the bomb is in the air. To find out the time, let's move to the y-dimension.

The initial velocity in the y-dimension is:

$$v_{oy} = (145\,\frac{\text{m}}{\text{sec}})\cdot\sin 0 = 0$$

We also know the acceleration in the y-dimension (-9.8 m/sec^2) as well as the displacement in the y-dimension (-9,112 m). The displacement is negative because the bomb falls down in the y-dimension. We can use this information, along with Equation (2.19), to determine the time that the bomb is in the air:

$$\Delta\mathbf{x} = \mathbf{v}_o t + \frac{1}{2}\mathbf{a}t^2$$

$$-9{,}112\text{ m} = (0)\cdot t + \frac{1}{2}\cdot(-9.8\,\frac{\text{m}}{\text{sec}^2})\cdot t^2$$

$$t^2 = \frac{2\cdot(-9{,}112\,\cancel{\text{m}})}{-9.8\,\frac{\cancel{\text{m}}}{\text{sec}^2}}$$

$$t = 43\text{ sec}$$

Now that we have the time that the bomb takes to fall, we can figure out how far the bomb travels in the x-dimension during that time. To do this, we first have to calculate the x-component of the velocity:

$$v_{ox} = (145\ \frac{m}{sec}) \cdot \cos 0 = 145\ \frac{m}{sec}$$

Now we can use the data from the x-dimension and Equation (2.19) to calculate what we want to know:

$$\Delta\mathbf{x} = \mathbf{v}_o t + \frac{1}{2}\mathbf{a}t^2$$

$$\Delta\mathbf{x} = (145\ \frac{m}{\cancel{sec}}) \cdot (43\ \cancel{sec}) + \frac{1}{2} \cdot (0) \cdot (43\ sec)^2$$

$$\Delta\mathbf{x} = 6.2 \times 10^3\ m$$

Since the bomb will travel that distance in the time that it takes for the bomb to drop, the bomber must release his payload <u>6.2×10^3 m before the target</u>.

4.10 In this problem, of course, we cannot use Equation (4.9), because the bullet lands below the height from which it was launched. We must analyze each dimension, then, and try to figure out what information we can glean from the data given. We are asked to calculate the initial velocity of the bullet. If we look at the initial velocity in each of the two dimensions, we will quickly see which dimension allows us to calculate that quantity:

$$v_{ox} = v_o \cdot \cos(0^o) = v_o$$

$$v_{oy} = v_o \cdot \sin(0^o) = 0$$

So, because the gun is aimed horizontally, the y-component of the velocity is zero. As a result, we can not calculate the initial velocity using the y-dimension. The x-dimension has a non-zero initial velocity, but we do not have enough information yet. In the x-dimension, we know the displacement (81.0 m) and acceleration (0). If we could find out the time, we could use Equation (2.19) to determine the initial velocity. Unfortunately, we don't have the time yet. Let's see if we can get that from the y-dimension.

In the y-dimension, we know the initial velocity (0), the acceleration (-9.8 m/sec^2), and the displacement (-0.45 m). That's all we need to calculate the time from Equation (2.19):

$$\Delta\mathbf{x} = \mathbf{v}_o t + \frac{1}{2}\mathbf{a}t^2$$

$$-0.45\ m = (0) \cdot t + \frac{1}{2} \cdot (-9.8\ \frac{m}{sec^2}) \cdot t^2$$

$$t^2 = \frac{2 \cdot 0.45 \cancel{m}}{9.8 \frac{\cancel{m}}{sec^2}}$$

$$t = 0.30 sec$$

Now we know the time it took the bullet to reach its destination. At this point, we can go back to the x-dimension to finish the problem:

$$\Delta\mathbf{x} = \mathbf{v}_o t + \frac{1}{2}\mathbf{a}t^2$$

$$81.0 \text{ m} = (\mathbf{v}_o) \cdot (0.30 \text{ sec}) + \frac{1}{2} \cdot (0) \cdot (0.30 \text{ sec})^2$$

$$\mathbf{v}_o = \frac{81.0 \text{ m}}{0.30 \text{ sec}}$$

$$\mathbf{v}_o = 2.7 \times 10^2 \frac{\text{m}}{\text{sec}}$$

This means that the bullet has an initial speed of <u>2.7×10^2 m/sec</u>.

REVIEW QUESTIONS

1. Explain how you should look at two-dimensional situations so as to make them as easy as possible to analyze.

2. What do we know about the x- and y-components of a projectile's velocity when the projectile is at its maximum height?

3. A projectile lands at a height equal to the height at which it was launched after a total flight time of 4.0 seconds. How many seconds after launch does it reach its maximum height?

4. A projectile is fired upwards with a speed of 150 m/sec. When it once again reaches the height from which it was launched, what is its speed?

5. What happens to the y-component of a projectile's velocity as the projectile travels through the air?

6. If a projectile is fired in empty space (no planets or stars in the area), will it still follow a parabolic path? Why or why not?

7. In which of the following situations can you use Equation (4.9)?

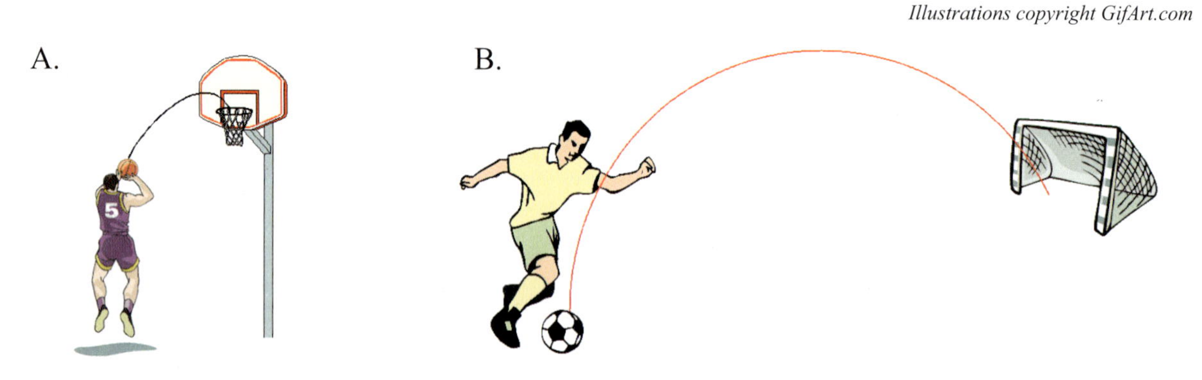

8. A sharpshooter tells a nervous man to hold out a can at arm's length. The sharpshooter says that he will shoot the can out of the man's hand. The sharpshooter realizes, however, that the nervous man will be scared at the sound of the gun and will drop the can as soon as the gun fires. In order to still hit the can, should the sharpshooter aim above, below, or straight at the can? (This is, of course, a hypothetical situation. It would be far too dangerous to actually do this!)

9. In the question above, suppose the man holding the can is not nervous, and the sharpshooter knows that he will hold onto the can even after the gun is fired. Should the sharpshooter aim above, below, or at the can in order to hit it?

10. What have we ignored when solving all of the projectile motion problems that we have done?

PRACTICE PROBLEMS

1. A ship travels 3.20 hours with a velocity of 21.1 mph at $\theta = 123°$, including the effects of the current. The pilot then turns to a heading of 190.0° with a speed of 18.2 mph, including the effects of the current. If the pilot stays on this heading for 1.10 hours, what is the ship's final displacement from its original starting point?

2. A map gives you the following directions:

Travel 45 minutes at a speed of 1.2 m/sec and a heading of 45°. Turn to a heading of 330° and travel at the same speed for 32 minutes.

What will your displacement be if you follow these directions?

3. A ship fires at an on-shore enemy factory, which sits behind a rather large mountain range. The tallest mountain in the range is 550 m high, and it stands roughly halfway between the ship and the factory. In order to hit the factory, the crew determines that the projectiles' initial velocity must be 200.0 m/sec at an angle of 30.0 degrees. That's the only setting that will give the projectiles the range necessary to hit the target. Determine the maximum height of the projectiles to see if there is any chance of hitting the factory.

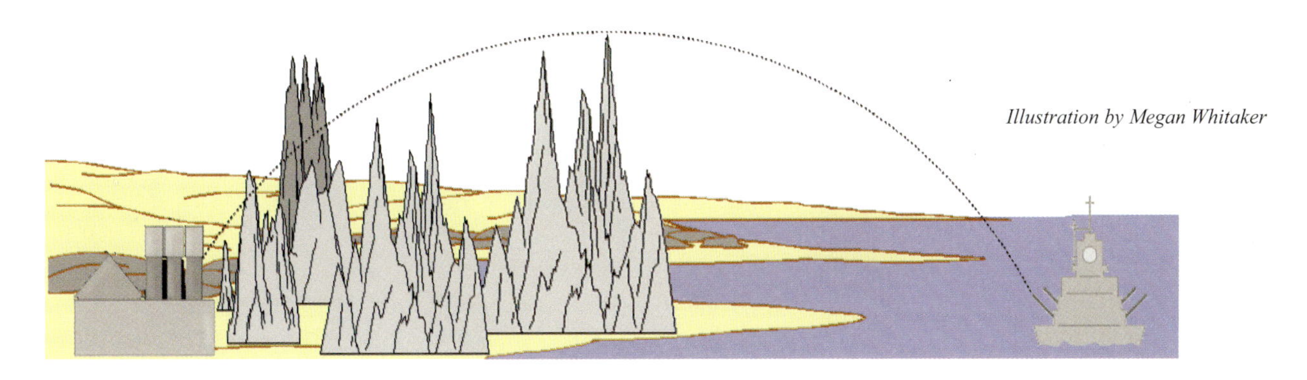

Illustration by Megan Whitaker

4. A soccer player kicks the ball from the ground with an initial velocity of 8.2 ft/sec at an angle of 40.0 degrees. How long will it take for the ball to reach its maximum height? How long will it take the ball to hit the ground again?

5. What is the range of a cannon that fires its cannonballs with an initial velocity of 350.0 ft/sec at an angle of 35 degrees?

** **Practice problems continue on the next page.** **

6. A soccer player wants to kick the ball to another teammate who is 31 meters downfield. To get the maximum range, he kicks it at an angle of 45 degrees. What initial speed must the ball have so that it can reach the intended target?

Illustration from the MasterClips collection

7. A football kicker is kicking off. He wants the ball to have a range of 65 yards (195 feet) so that it will land back on the ground in the end zone. If the kicker can kick the ball with a speed of 81 ft/sec, at what angle must he aim the ball in order to get the desired range?

8. A golfer hits his ball up to a green that is on a hill. If the ball's initial velocity is 75 ft/sec at $\theta = 60.0$ degrees, and it hits the green 94 feet from the golfer, how high is the hill?

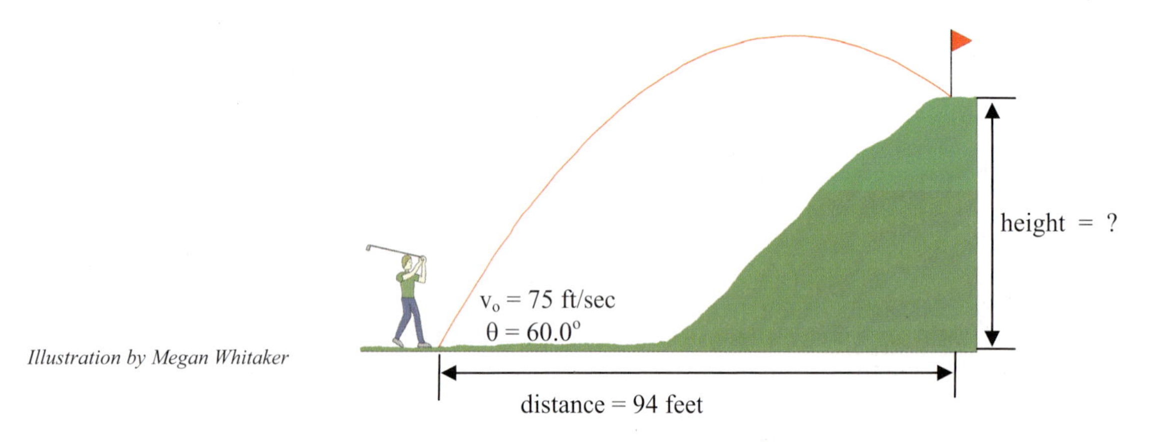

Illustration by Megan Whitaker

9. A pitcher throws a baseball horizontally ($\theta = 0°$) at a wall that is 51 feet away. If the ball falls 2.9 feet while it travels to the wall, at what speed was it originally thrown?

10. A little boy is playing with a toy car and rolls it across a 4.1 feet high table with a speed of 1.2 ft/sec. The toy car rolls off the edge of the table and onto the floor. Assuming that the initial angle of the car's velocity is zero as it rolls off the table, how far away from the table will the car land?

Illustration from the MasterClips collection

MODULE #5: Newton's Laws

Introduction

In the previous four modules, we have discussed units, vectors, and motion. In these discussions, we have concentrated solely on how to describe physical situations. We can predict how long it will take to make a journey, how fast an object will be traveling after a certain amount of time, and where an object will be under certain conditions. Now, of course, all of these things are useful and interesting, but they don't address a fundamental question: **WHY?** You see, the skills we have discussed so far help you understand how and where, but not why. That's what we'll learn in this module, because now we're ready to investigate Newton's Laws. Before we look at the laws themselves, however, we first need to learn a little about the man who wrote them.

Sir Isaac Newton

Sir Isaac Newton is a pivotal figure in the history of science. His portrait and a brief biography are given in Figure 5.1.

Image in the public domain

FIGURE 5.1
Sir Isaac Newton

Sir Isaac Newton was born in England in 1642. From an early age, he was interested in learning about how the world worked, and he devoted his life to performing experiments designed to help him understand creation. He is credited with many, many discoveries. As we will see in this module, he discovered three laws of motion, but that's not all. He developed a theory describing gravity (which you will learn later); he did the famous prism experiment which showed that white light is composed of many colors; and in order to help his scientific investigations, Newton developed a new kind of mathematics that we now call "calculus." Amazingly enough, these three accomplishments were completed in less than 18 months! He also built the first reflecting telescope. Newton was knighted for his accomplishments by Queen Anne in 1705, and he died in 1727.

Clearly, Sir Isaac Newton was a genius. Most physics books discuss this fact, but they do not tell the whole story. In fact, Newton was a devoutly religious man who spent as much time studying the Bible as he did studying science. As *The Columbia History of the World* says, "…at the end of his days he spent more time studying and writing about the prophecies in the Book of Daniel than he did in charting the heavens" (J.A. Garraty and P. Gray, eds., [New York: Harper & Row, 1972], 709) Although not an orthodox Christian (he rejected the Trinity and the divinity of Christ), he held to many standard Christian beliefs. Here is a good quote that sums up Newton's theological views: "There is one God, the Father, ever living, omnipresent, omniscient, almighty, the maker of heaven and earth, and one mediator between God and man, the man Christ Jesus" (Sir Isaac Newton, *Theological*

Manuscripts, Ed. H. McLachlan [Liverpool: Liverpool University Press, 1950], 56). As you can see, he clearly saw God as the almighty Creator, and he saw Christ as the mediator between man and God. At the same time, however, he did not see Christ as divine. Instead, he refers to Christ as a man throughout his theological works. This is part of what made his religious views unorthodox.

Isaac Newton believed that in studying science, he was actually learning about God. In fact, it was his strong belief in God that made him study science. After all, he reasoned, studying science was a way of learning about creation, and learning about creation was a way of learning about God. Of course, Newton also realized that studying creation cannot be the sole means of learning about God. That's why he spent so much time studying the Bible as well. Newton applied his strong mind to interpreting Scripture and wrote many commentaries on passages in the Bible. He was especially drawn to the book of Daniel, as the quote from *The Columbia History of the World* indicates. Clearly, Newton had a great sense of priorities. He recognized the importance of science, but he also realized that learning about God is even more important. Scientists today could learn a real lesson from this brilliant man.

Newton's First Law

Suppose you take a block of wood and slide it across the floor. It eventually comes to a halt, doesn't it? Once it stops, it will stay at rest until you push it again. Observations like this one led an early scientist named Aristotle to conclude that matter "wants" to be at rest, and it will stay at rest until acted on by some outside force. In other words, the "natural" state of matter is for it to be at rest, and the only way to get it to move is to perturb that "natural" state with a force (such as a push). Although Aristotle lived in the fourth century B.C., his reasoning on this and other matters of physics held powerful sway among scientists until the late 1500s.

What happened in the late 1500s? Italian physicist Galileo Galilei began doing experiments that contradicted Aristotle's views. A portrait of Galileo and a brief biography are given below:

Image in the public domain

FIGURE 5.2
Galileo Galilei

Galileo Galilei was born in Italy in 1564. He is famous for challenging many of the scientific beliefs of his time. He did experiments on bodies in motion that clearly contradicted the prevailing views, which were mostly based on the work of Aristotle. Using a telescope that he built based on another person's design, he discovered mountains and craters on the moon and Jupiter's four largest moons. Later, he also discovered sunspots. All of these observations went against the prevailing "earth-centered" view of the solar system, and he became an advocate for the "sun-centered" view of the solar system. This put him at odds with the Roman Catholic church, and the church eventually put him under house arrest for suspicion of heresy. He died in 1642 (the year Newton was born) while still under house arrest. Although at odds with the Roman Catholic church because of his scientific views, he considered himself a devout Roman Catholic.

Galileo showed by experiment that even though the idea that matter "wants" to remain at rest might seem logical, it is wrong. This view fails to account for a "hidden" force, which we call **friction**. Galileo deduced that when two surfaces come into contact with one another (the surface of a wooden block and the surface of a floor, for example), the surfaces grab onto each other, resulting in a force that inhibits motion.

Friction - A force that opposes motion, resulting from the contact of two surfaces

Since friction opposes motion, it will reduce the speed of an object until that object stops. Thus, Galileo concluded that a block of wood sliding across the floor does not stop because it "wants" to stop. It stops because friction slows it down and eventually stops it. If there were no friction between the block and the floor (and if there were no air resistance), the block would slide on forever, never slowing and never stopping.

Galileo's experiments and reasoning led Newton to his first law, which is often called the **Law of Inertia** (in ur' shuh):

Newton's First Law (The Law of Inertia) - An object in motion (or at rest) will tend to stay in motion (or at rest) until it is acted upon by an outside force.

Although this law might sound similar to what Aristotle believed, it is really quite different. Newton's First Law says that matter has no preferred state. It neither "wants" to stay at rest nor to stay in motion. It simply stays in whatever state it finds itself until an outside force causes it to change.

When students read the Law of Inertia, they tend to think "hmm...that makes sense," and then they move on. That's probably your reaction at this point. "Okay," you might think, "that makes sense, but so what?" Well, it turns out that Newton's First Law has many, many physical implications. To see one of these implications, perform the following experiment.

EXPERIMENT 5.1
Inertia

Supplies:

- Someone to help you
- A small beanbag, or any object that does not bounce when dropped
- A sidewalk, driveway, or long, flat yard
- Three rocks

1. Have your helper stand at one end of the driveway, sidewalk, or yard.
2. Walk to the other end of the driveway, sidewalk, or yard with the beanbag in your hand. Hold the beanbag in one hand and keep it as high as your shoulders.
3. Even though it is rather uncomfortable (and even though you'll look at bit silly), run as fast as you can down the driveway, sidewalk, or yard, keeping the beanbag at shoulder height. As soon as you pass your helper, drop the beanbag. Once you have dropped the beanbag, you can slow down and stop.
4. Go back to where the beanbag landed on the ground and mark the place that it landed with one of the rocks.

5. Have your helper stand at the same place and hold the beanbag at the same height that you held it (your shoulder height). Have him or her hold it so that it will be *very* easy for you to knock out of his or her hand.
6. Go back to the other end of the driveway, sidewalk, or yard and run towards your helper again. This time, as you pass him or her, knock the beanbag out of his or her hand. Do not hit it hard. Instead, just barely flick it as you pass. Once again, go back and mark the place where the beanbag landed with a rock.
7. Have your helper stand in the same place and once again hold the beanbag at your shoulder height.
8. Instruct him or her to drop the beanbag as soon as you pass by.
9. Start from the same place you did before and once again start running towards your helper. As you pass by your helper, he or she should drop the beanbag.
10. Mark the spot where the beanbag landed with the last rock.
11. Record the positions of the three rocks.
12. Pick up the rocks and beanbag and return them to where you found them.

What were the relative positions of the rocks? The rock that marked the first time you dropped the beanbag should have been the farthest away from where your helper stood, in the direction that you ran. The reason for this is quite simple. While you were carrying the beanbag, it had the same velocity that you had. When you dropped it, that velocity did not vanish. After all, the Law of Inertia says that a body in motion stays in motion until acted on by an outside force. As a result, when you dropped the beanbag, it continued moving with the same velocity it had while you were holding on to it. At the same time, however, the force of gravity began acting on it. As we learned in the previous module, however, gravity only acts in the vertical direction. The velocity that the beanbag had while you were holding onto it was horizontal; therefore, the force of gravity had no effect on it. As a result, the beanbag continued with both a horizontal and a vertical velocity. Thus, the beanbag moved away from your helper as it fell. This illustrates Newton's First Law.

Depending on how hard you hit the beanbag in the second part of the experiment, the last two rocks you used to mark the landing spot of the beanbag might have been right on top of each other. You see, in the second and third trials of the experiment, the beanbag had no initial velocity, because your helper was holding onto it, and your helper was standing still. As a result, when you knocked it out of his hand (and when he dropped it himself), the beanbag had no initial horizontal velocity. Thus, the beanbag should have dropped straight down. It is possible, however, that when you knocked it out of your helper's hand, you gave it a little bit of horizontal velocity when you hit it. The second rock, therefore, might be a little farther away from your helper than the third rock. Even if it is, however, it should not be nearly as far away from your helper as the first rock was.

So now you should see how inertia works. When you dropped the beanbag, it had your velocity and it continued traveling with your velocity because no outside force altered it. As a result, it traveled away from your helper in the direction that you were running. When you knocked the beanbag out of your helper's hand, and when your helper dropped it, the beanbag had no initial velocity in the direction that you were running, so it dropped straight down. If you remember one of the problems you did in the previous module, you had to calculate how early a bomber had to drop his payload in order to hit his target. The reason the bomber has to drop the bombs early is due to the Law of Inertia.

FIGURE 5.3
Inertia and Military Aircraft

This is a photograph of an F-20 Tigershark dropping Mk82 bombs. While in the plane, the bombs travel along with the plane. Thus, they have the same velocity as the plane. At the instant in which they are dropped from the plane, Newton's First Law says that they will *still* have the same velocity as the plane. If the pilot waited until the airplane was over the target to release the bombs, they would never hit the target. Instead, as they fell, they would still travel in the direction in which the airplane was traveling when they were released. As a result, they would hit the ground far past the target. In order to get the bombs to hit the target, then, the pilot must release the bombs well before the airplane flies over the target. That way, as the bombs fall, they will approach the target. If the pilot drops the bombs at the right time, they will reach the target just as they reach the ground.

So you see that the Law of Inertia explains a lot of situations involving motion. Suppose, for example, you are riding in the front seat of a car. The car is moving along at a good speed when suddenly the driver slams on the brakes. What happens in this situation? Your body lurches forward until the seat belt grabs you and keeps you from going through the windshield. Once again, this happens because of inertia. Since you are riding in the car, your body is traveling with the same velocity as the car. When the brakes are applied, the car stops, but your body continues to move because it still has the original velocity of the car. An outside force (friction between your body and the car seat) does work to slow you down, but it is not strong enough to do the job. Thus, the force applied by the seat belt is the outside force necessary to rid you of the rest of the velocity that the car originally gave you.

Use what you have learned about Newton's First Law to explain the situations described in the following "On Your Own" problems.

ON YOUR OWN

5.1 You have started your Christmas shopping early and are driving on a highway with several packages on the seat next to you. Suddenly, you are forced to accelerate quickly in order to avoid hitting a car that is getting on the highway from an entrance ramp. You notice that, even though the road is level, the packages slide backwards in the seat until they hit the backrest. Why did the packages slide backwards when you accelerated forwards?

5.2 We all know that if you slide a block of wood on ice, it will travel much farther than it does on concrete. Use the Law of Inertia and the concept of friction to explain this fact.

Newton's Second Law

Newton needed more to explain motion than just the Law of Inertia. After all, that very law references the fact that an object's velocity can change if the object is acted on by an outside force. Newton's Second Law explains how this can happen:

Newton's Second Law - When an object is acted on by one or more outside forces, the vector sum of those forces is equal to the mass of the object times the resulting acceleration vector.

In other words, if you push on an object, the amount that it accelerates depends on two things: the mass of the object and the magnitude of the force with which you push. In addition, the direction of the resulting acceleration will be the same as the direction of the force. Newton's Second Law is more often expressed with an equation rather than with words:

$$\Sigma \mathbf{F} = m\mathbf{a} \tag{5.1}$$

If you understand the symbols in Equation (5.1), you can see how the equation is equivalent to the definition given above. To begin with, the "Σ" symbol is the capital Greek letter "sigma." It means "sum of." Thus, "Σ**F**" means "the sum of the forces." The "m" in Equation (5.1) represents the mass of the object in question, and the "**a**" stands for the resulting acceleration. Since both "**F**" and "**a**" are in boldface type, they represent vectors. Equation (5.1), therefore, says "the vector sum of the forces is equal to the mass times the acceleration vector," which is identical to what the above definition says.

Equation (5.1) is probably one of the most fundamental equations in all of physics, so we will spend a *great deal* of time on it. Let's start with the quantities in the equation. You should already be familiar with the concepts of mass and acceleration; however, you probably have little experience with the concept of force. That, then, is where we must start.

What is force? Well, a force is essentially a push or a pull exerted on an object in an effort to change that object's velocity. Equation (5.1) tells us that we can calculate force by multiplying mass and acceleration. Let's think about the units that result when we do this. Remember from my introductory remarks that the standard units we use in science are called **SI** units. The SI unit for mass is kilograms, and the SI unit for acceleration is m/sec^2. When we multiply these two quantities together, we get the unit $\frac{\text{kg}\cdot\text{m}}{\text{sec}^2}$. This rather complicated unit is often referred to as the **Newton**, in honor of Sir Isaac Newton. This is fitting. Since Newton is one of the most important figures in the history of physics, he deserves to have a unit that belongs in one of the most important equations in physics. This unit is something you'll have to remember:

The Newton is the SI unit of force and is defined as a $\frac{\textbf{kg}\cdot\textbf{m}}{\textbf{sec}^2}$

Now, of course, any unit that has a mass unit multiplied by a displacement unit divided by a time unit squared would be considered a force unit. Thus, $\frac{\text{g}\cdot\text{km}}{\text{min}^2}$ would also be a valid force unit; we just don't run into it that often. One force unit that you might run into occasionally is the **dyne**. It is used when you are dealing with smaller objects and forces, and is equivalent to a $\frac{\text{g} \cdot \text{cm}}{\text{sec}^2}$. To give you an idea of how much force is associated with these units, if you were to hold a gallon of water in your hand, it would pull your hand down with a force of 40 Newtons. On the other hand, when a fly lands on your finger, it pushes your finger down with a force of approximately 1,000 dynes. Thus, whereas one Newton is a pretty significant force, one dyne is a very, very small force.

The only thing left to learn about force is why it is a vector quantity. If you think about this, it should make sense. After all, if you push an object, the direction in which it will begin to accelerate depends on the direction in which you push it. If you push left, the object will accelerate to the left. If you push right, the object will accelerate to the right. Thus, if acceleration is a vector quantity, force must be as well. In fact, any acceleration that occurs as a result of a force must be in the same direction as the force.

So now we know about force. It is a vector quantity whose magnitude is usually measured in Newtons. When a force is applied to an object, that object will experience an acceleration in the same direction as the applied force. The magnitude of the acceleration depends on both the magnitude of the force and the mass of the object. Massive objects take a lot of force to achieve even a little acceleration. Objects that have little mass need only a little force to achieve a large acceleration. Since we finally know what force is, we can go back to studying Newton's Second Law.

As with many things in science, at first glance Newton's Second Law seems simple. After all, Equation (5.1) has only three variables. None of them are squared, and the only mathematical function in the equation is multiplication. It doesn't get any easier than this, right? Well, not exactly. Although Newton's Second Law is simple in and of itself, the application of this law can get really complex. While you'll get a taste of the complexity in this module, you'll really see how complicated it can get in the next one!

Right now, we'll start out simple and gradually build the level of complexity. On its surface, Equation (5.1) is, indeed, just an equation that relates three variables. Thus, I can use it to analyze situations such as those presented in the next example and the "On Your Own" problems that follow.

EXAMPLE 5.1

A person pushes a rock with an acceleration of 0.12 m/sec^2 east. If the rock has a mass of 4.2 x 10^4 g, what force is the person exerting to move it?

This is a straightforward application of Equation (5.1). We are given the mass and the acceleration and are asked for the force. Even though the problem does not instruct us to do so, however, I want my final answer to be in Newtons. In order to do this, my mass must be in kg, because a Newton is a $\frac{\text{kg} \cdot \text{m}}{\text{sec}^2}$. Thus, the first thing we have to do is convert the mass to kg:

$$\frac{4.2 \times 10^4 \ \cancel{\text{g}}}{1} \times \frac{1 \text{ kg}}{1{,}000 \ \cancel{\text{g}}} = 42 \text{ kg}$$

Now we just use Equation (5.1):

$$\mathbf{F} = \mathrm{m} \cdot \mathbf{a}$$

$$\mathbf{F} = (42 \text{ kg}) \cdot (0.12 \frac{\text{m}}{\text{sec}^2})$$

$$\mathbf{F} = 5.0 \text{ Newtons}$$

Now remember, the acceleration was directed east. Since I put acceleration into the equation as positive, I implicitly defined east as the positive direction. Since the force ended up positive, that means the force is directed east as well. This makes sense, since the net force and the acceleration always point in the same direction. Thus, the person must exert a force of 5.0 Newtons east in order to accelerate the rock 0.12 m/sec^2 east.

ON YOUR OWN

5.3 A man is pushing a car that has run out of gas. If he pushes with a force of 15 Newtons west, and the car has a mass of 1.23×10^6 grams, what will be the car's acceleration?

5.4 A tennis ball bounces as shown in the diagram on the right.

Illustration copyright GifArt.com

a. How can you tell that a force other than gravity must have acted on the tennis ball?
b. At what point did that force act on the ball?
c. What is the general direction (up, down, right, or left) of that force?
d. What is the general direction (up, down, right, or left) of the acceleration resulting from that force?

Mass and Weight

One important application of Newton's Second Law involves distinguishing mass from weight. In chemistry, you should have learned the difference between mass and weight. Mass is a measure of how much matter is in a object, while weight is a measure of how hard gravity is pulling on an object. One of the principal differences between mass and weight is that while the weight of an object depends on its location (what planet it is on, what altitude it is at), the mass of an object does not depend on its location. For example, if I were to weigh myself on the moon, I would weigh considerably less than I do on earth, because the moon's gravity is significantly weaker than that of the earth. On the other hand, my mass is the same on the moon as it is on the earth.

Another important difference between mass and weight is that mass is a scalar quantity, while weight is a vector quantity. After all, weight is a measure of a *force*: the force with which gravity pulls on an object. Since weight is a measure of force, and since force is a vector quantity, weight is a vector quantity as well. Mass, on the other hand, simply measures the amount of matter in an object. There is no direction attached to the amount of matter in an object, so mass is a scalar quantity.

Given the fact that weight is a measure of the force due to gravity, Equation (5.1) is actually the best means of distinguishing mass from weight. Since force is equal to mass times acceleration, the weight of an object is actually the mass of that object times the acceleration due to gravity. In other words:

$$\mathbf{w} = \mathbf{mg} \tag{5.2}$$

where "m" is the mass of an object, and "**g**" is the acceleration due to gravity. This is the best way to see that weight changes while mass does not. The acceleration due to gravity is different for each planet. As we will see in a later module, it also changes slightly depending on your altitude. Well, if you calculate weight by taking the mass and multiplying by the acceleration due to gravity (which is position-dependent), it follows that weight should be position-dependent as well.

Now think about this for a moment. I just said that weight is a force. We also learned in the previous section that the SI unit for force is the Newton. Thus, the Newton is the SI unit for weight. When you weigh yourself in the metric system, the measurement is done in Newtons. What about the weight unit you are most familiar with, the pound? Well, it's the English unit for force. Remember, you just learned that any displacement unit times any mass unit divided by any time unit squared is a valid unit for force. The standard English unit for mass is the **slug** (no kidding); the standard English unit for displacement is the foot; and the standard unit for time is the second. Thus, the standard English unit for force is the $\frac{\text{slug} \cdot \text{ft}}{\text{sec}^2}$, which is (thankfully) called the **pound**. Thus, when you are using the pound unit, you are talking about a force. When you use the slug unit, you are talking about mass.

This is one of the first places students get confused when dealing with problems that use mass or weight. When physicists refer to an object, they will rarely say things like "an object whose mass is 15 kg." Instead, they will generally say "a 15-kg object." Physicists expect you to realize that since the unit kg was used, the measurement must be the object's mass. In the same way, physicists will not often use the term "weight" when giving you the weight of an object. Rather, they will simply give you the number with a weight unit (pounds or Newtons) and expect you to realize that this means weight. Thus, physicists interchange mass and weight rather freely, expecting you to tell which is which by what unit is used. You must, therefore, become very comfortable recognizing the difference between mass and weight by recognizing the units:

When a measurement is referred to in slugs, grams, or any prefix unit based on grams (mg, kg, etc.), the <u>mass</u> of the object is being reported. When a measurement is given in Newtons, dynes, or pounds, the <u>weight</u> of the object is being reported.

As we work through the different problems in this module, you will see how crucial it is to be able to distinguish between mass and weight using just the units. Let's deal with first things first, however. Study the example problem below in order to get an introduction to the concept of weight.

EXAMPLE 5.2

A man steps on a scale and determines that he weighs 151 pounds, which is equivalent to 672 Newtons. Determine the man's mass in both kg and slugs.

Since we are given the weight and asked to calculate the mass, we will clearly be using Equation (5.2). Since the problem asks us to solve for the mass in both kg and slugs, we will be working in both the English and the metric systems. Let's start with the metric system. The SI unit for weight is the Newton, so we should use 672 Newtons as the weight. To use Equation (5.2), however, we also need to know the acceleration due to gravity. Since we are in the metric system at this time, we will use 9.8 m/sec^2. Now please remember that weight and acceleration are *vectors*. Thus, we need to be worried about *direction*. Weight is the force due to gravity, and since gravity pulls *down*, weight is direction down as well. Thus, by using positive numbers for both the weight and the acceleration due to gravity, we are implicitly defining down as the positive direction.

$$\mathbf{w} = \mathbf{m}\mathbf{g}$$

$$672 \text{ Newtons} = \text{m} \cdot (9.8 \frac{\text{m}}{\text{sec}^2})$$

$$\text{m} = \frac{672 \frac{\text{kg} \cdot \cancel{\text{m}}}{\cancel{\text{sec}^2}}}{9.8 \frac{\cancel{\text{m}}}{\cancel{\text{sec}^2}}} = 69 \text{ kg}$$

Notice that I had to substitute the definition of the Newton when it came time to cancel out the units. You will usually have to do this when solving problems like the ones in this module. In the end, then, the man has a mass of <u>69 kg</u>.

We're only halfway done. We also have to report our answer in slugs. Now, of course, if we knew the conversion factor between kg and slugs, we could simply convert between the two. However, to get you more familiar with Equation (5.2), we will do it the long way. We will solve the equation again, this time using English units. Thus, we will use 151 pounds as the weight, and the acceleration due to gravity will be 32 ft/sec^2, as is appropriate for the English system:

$$\mathbf{w} = \mathbf{m}\mathbf{g}$$

$$151 \text{ lbs} = \text{m} \cdot (32 \frac{\text{ft}}{\text{sec}^2})$$

$$\text{m} = \frac{151 \frac{\text{slug} \cdot \cancel{\text{ft}}}{\cancel{\text{sec}^2}}}{32 \frac{\cancel{\text{ft}}}{\cancel{\text{sec}^2}}} = 4.7 \text{ slugs}$$

This tells us that a mass of 69 kg is the same as a mass of <u>4.7 slugs</u>.

Before we look at another problem, I want to use this example to make you aware of something rather important. It turns out that when we measure the mass of an object we are often, in fact, measuring its *weight* and then doing a calculation like the one that we just did in order to convert that weight into mass. For example, if you took a chemistry class that had a lab component, you probably had to measure the mass of certain reagents. You probably did that by putting the reagents on a scale and reading the mass in grams. It turns out, however, that a scale *does not* measure the mass of an object.

A scale is composed of something (a spring or a flexible needle) that is sensitive to the *force* that it exerts. When an object sits on a scale, gravity tries to pull the object to the center of the earth, but the scale stops that from happening because the spring or needle in the scale pushes back with a force equal to that of gravity. Since gravity pulls the object down and the spring or needle pushes the object up with an equal force, the forces cancel and the object stays still. The scale, however, can read the force exerted by the spring or needle. Thus, since the scale reads *force*, it measures *weight*. How do we read the mass in grams, then? Well, the makers of the scale simply take the weight and divide by the acceleration due to gravity, turning a weight scale into a mass scale.

In order to do that, however, the makers of the scale have to use a set number for the acceleration due to gravity, like 9.8 m/sec^2. Remember, however, that the value for the acceleration due to gravity changes depending on location. On a different planet (Mercury, for example), the acceleration due to gravity is different. Thus, a scale that reads the proper mass while on earth would NOT read the proper mass on Mercury, because in order to convert from what is measured (weight) into mass, the scale must divide by the acceleration due to gravity. If that value is the correct one for the earth, it will not be correct for Mercury.

You don't have to go to Mercury to encounter such a problem, however. We will learn in a later module that the acceleration due to gravity on earth depends slightly on where you are. At the North Pole, for example, the acceleration due to gravity is 9.83 m/sec^2 whereas near the equator, the acceleration due to gravity is 9.78 m/sec^2. Both of these numbers, of course, round to the number that we use (9.8 m/sec^2), but if you are looking for a very precise mass measurement, a scale that reads the proper mass of objects at the North Pole will read slightly too low for the mass of objects when that scale is taken to the equator.

Now that you know a bit about how a scale works, let's go back to seeing how we can apply Equation (5.2).

EXAMPLE 5.3

An object weighs 3.42 x 10^3 pounds on earth. How many pounds will it weigh on the moon? (The acceleration due to gravity on the moon is 5.3 ft/sec^2.)

In order to solve this problem, we have to realize that even though the weight of an object changes when it travels from the earth to the moon, its mass does not. Thus, if we convert from weight to mass, we will have something that does not change between the earth and the moon. We can use Equation (5.2) to do this:

$$\mathbf{w = mg}$$

$$3.42 \times 10^3 \text{ lbs} = \text{m} \cdot (32 \frac{\text{ft}}{\text{sec}^2})$$

$$\text{m} = \frac{3.42 \times 10^3 \ \frac{\text{slug} \cdot \cancel{\text{ft}}}{\cancel{\text{sec}^2}}}{32 \ \frac{\cancel{\text{ft}}}{\cancel{\text{sec}^2}}} = 1.1 \times 10^2 \text{ slugs}$$

So on both the moon *and* the earth, the object has a mass of 110 slugs. The problem does not ask for mass, however. It asks for the weight of the object on the moon. Since we have mass and the acceleration due to gravity on the moon, we can use Equation (5.2) again to determine the weight:

$$\mathbf{w = mg}$$

$$\mathbf{w} = (1.1 \times 10^2 \text{ slugs}) \cdot (5.3 \frac{\text{ft}}{\text{sec}^2}) = 5.8 \times 10^2 \ \frac{\text{slug} \cdot \text{ft}}{\text{sec}^2} = 5.8 \times 10^2 \text{ lbs}$$

Thus, an object that weighs 3.42×10^3 pounds on earth weighs <u>5.8×10^2 pounds</u> on the moon.

ON YOUR OWN

5.5 What is the weight (on earth) of a 13.5-gram object? Give your answer in Newtons.

5.6 A Martian rock weighs 15 Newtons on Mars. How much will it weigh on Earth? (The acceleration due to gravity on Mars is 3.7 m/sec^2.)

The Normal Force

At this time, I need to elaborate on a point that I passed over rather quickly a moment ago. When discussing how we measure mass and weight, I mentioned that in order to stop an object from being pulled to the center of the earth, a scale will push up against an object with the same force that gravity is using to pull down on the object. This cancels out the gravitational force and keeps the object from falling to the center of the earth. This is, in fact, what the scale actually measures. It cannot measure the force with which gravity is pulling an object down. However, it can measure the force with which it must push back to counteract that gravitational force. As long as the scale and object are at rest (or moving with a constant velocity), the two forces will be equal, however, so the magnitude of the force with which the scale pushes back will be equal to the magnitude of the gravitational force acting on the object. Examine Figure 5.4 so that you have a better understanding of this.

FIGURE 5.4

Think about a man standing on a floor. Why doesn't the man just float up off of the floor? Well, any elementary student can tell you it's because of gravity. Gravity pulls on the man, giving him weight. This pull from gravity keeps the man's feet firmly on the ground. I could represent the force due to gravity with an arrow pointing down, because the force of gravity is pulling him down. I will say that motion downwards is negative motion; thus, this is a negative force:

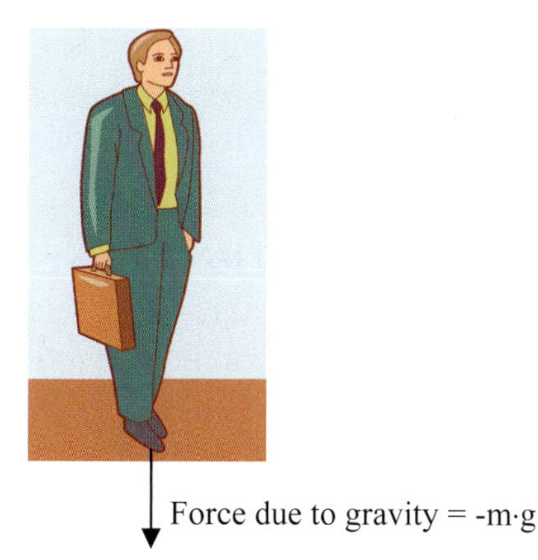

Now wait a minute. Newton's Second Law says that when a force is applied to an object, the object accelerates. Does this mean that the man will accelerate and crash through the floor? Of course not. The reason that the man does not start accelerating through the floor is that the floor pushes back on the man. It pushes with exactly the same force that gravity is using, but it pushes in the opposite direction; thus, it is a positive force:

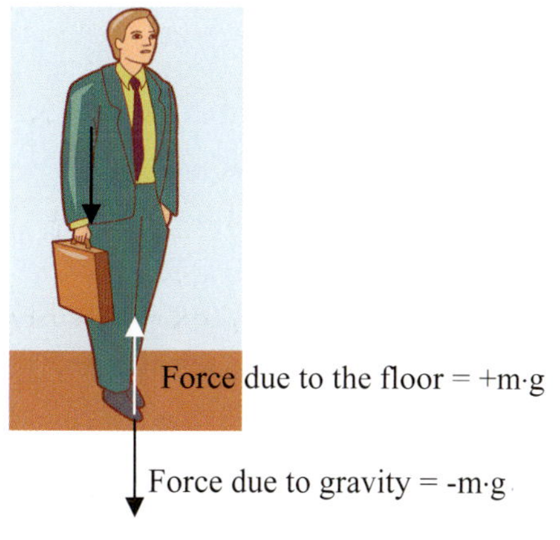

Newton's Second Law says that the acceleration of an object depends on the *sum of the forces applied to the object*. In other words, in order to determine the acceleration, we need to add these forces. Both forces are pointed in the y-dimension, so this is really a one-dimensional situation. As a result, we can simply add the two forces:

$$\Sigma\mathbf{F} = m \cdot g + - m \cdot g = 0$$

Thus, the man does not accelerate because the force with which the floor pushes up on him cancels out the force with which gravity is pulling down on him. Since the sum of the forces is zero, the man's acceleration is zero as well.

So we see that the reason we stand on a surface and do not go accelerating through the surface to the center of the earth is due to the fact that any surface upon which we are standing fights gravity by applying an equal but opposite force on us. This cancels out the effect of gravity and allows us to stand (or sit or kneel or whatever) on the surface without continuing to fall towards the center of the earth. The force that the surface applies will always be perpendicular to the surface at the point where contact occurs. Thus, we call this the **normal force**, because in geometry, the word "normal" means "perpendicular":

Normal force - A force that results from the contact of two bodies and is perpendicular to the surface of contact

The normal force figures prominently in the upcoming discussion on friction; thus, we need to make sure we understand how to use it.

EXAMPLE 5.4

A 75-kg man stands on a level floor. What is the normal force that the floor exerts on the man?

According to what we just learned, the floor should push up on the man with the same force that gravity is using to pull him down. We learned before that the force due to gravity is the same as a person's weight. Since we were given the man's mass (we know it's mass because its units are kg), we can use Equation (5.2) to determine the weight:

$$\mathbf{w} = m \cdot \mathbf{g}$$

$$\mathbf{w} = (75 \text{ kg}) \cdot (9.8 \frac{\text{m}}{\text{sec}^2}) = 740 \frac{\text{kg} \cdot \text{m}}{\text{sec}^2} = 740 \text{ Newtons}$$

Now remember, gravity accelerates an object down. Since I used a positive number for **g**, that means I have defined downward motion as positive. The normal force exerted by the floor will be equal to and opposite of the force exerted on it by gravity, so the normal force is -740 Newtons. Since the downward direction has been defined as positive, the upward direction must be negative, so the normal force is 740 Newtons straight up. This means that gravity pulls on the man with a force of 740 Newtons and the floor pushes back with a force of -740 Newtons. The sum of the forces on the man is zero; thus, he doesn't move up or down.

Now, suppose the wood that makes up the floor in the example above was rotten and too weak to exert a normal force with a magnitude of 740 Newtons. At that point, there would be a greater force pulling the man down than pushing him up. This would cause him to accelerate downwards. He would crash through the floor and continue to fall until he encountered a surface that would be able to push with a normal force of 740 Newtons straight up. Only then would he stop falling.

ON YOUR OWN

5.7 A 3.8-slug woman stands on a step. What is the normal force exerted by the step?

Friction

Now that you are familiar with Newton's Second Law, the difference between mass and weight, and the concept of a normal force, it is time to talk about friction. I've mentioned friction before, but always in a vague way. It is now time to really study the concept of friction so that you have a firm grasp on the subject. First, we need to know why friction exists. Friction is a phenomenon brought about by the fact that matter is made up of atoms or molecules. Examine Figure 5.5 to see how this fact brings about friction:

Illustration by Megan Whitaker

FIGURE 5.5

Why Friction Exists

Suppose a box is on the floor. From experience, we know that if someone were to give the box a quick shove, it might slide across the floor a little, but it would eventually come to a halt. Newton's First Law says that once the box is put in motion by the shove, it should stay in motion until acted on by an outside force. Since the box stops sliding, we know that there must be an outside force. I have already explained that the outside force which stops the box is friction. The question is, why does friction exist? Well, let's picture the box sitting on the floor:

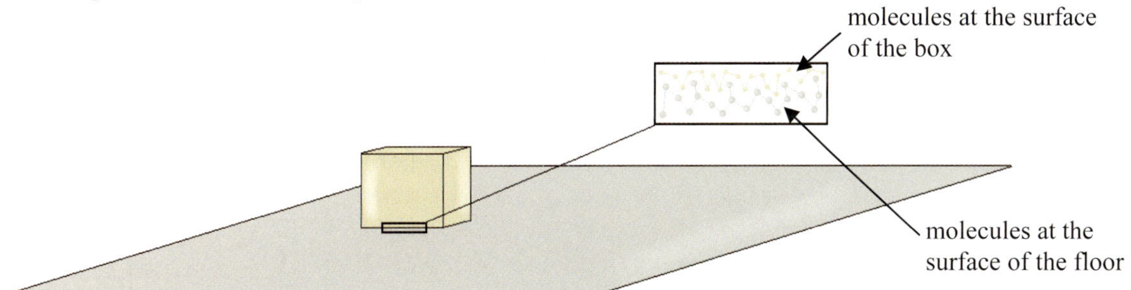

From our point of view, the surface of the box and the surface of the floor are both pretty smooth. However, suppose we could magnify a portion of the box and the floor where the two surfaces contact one another. If it were possible to magnify that portion so much that we could see things on the atomic scale, we would see something like what is drawn in the inset picture above. You see, on the atomic scale, there are no really smooth surfaces. All surfaces have grooves and bumps in them. These grooves and bumps determine how close the molecules (or atoms) of one surface can get to the molecules (or atoms) of the other surface. This is important, as these molecules (or atoms) are attracted to one another, and the closer they can get to one another, the more they can attract each other. The more they attract each other, the more they will be able to resist motion.

With this picture in your head, you should be able to understand a great deal about friction. Think about the box in the picture above. While it is stationary, the molecules that make up the box have nestled as close as possible to the molecules that make up the floor. If I come along and try to push the box, the attraction of these molecules will fight the motion of the box. If I push the box with a tiny force, the attraction between the molecules will be able to fight the force that I apply, and the box will not move. As I apply more force, the attraction between the molecules will fight even harder, trying to keep the molecules close to one another. If I continue to increase the force that I apply to the box, I will eventually push with a force greater than the molecular attraction, and the box will begin to move.

This is, of course, why friction opposes motion. The molecules of each surface are attracted to one another. Thus, they would like to stay together. If the box moves, then the molecules of the box move away from the molecules to which they were near. The mutual attraction of the molecules fights that motion.

Think a little harder about Figure 5.5. Suppose the box were sitting still on a floor. In order to get it moving at all, you would have to apply enough force to overcome the force with which the molecules of the box and floor attract one another. Since the box has been sitting there, the bumps and grooves have "settled" a little, nestling the molecules close together. This makes the force with which the molecules attract one another large. Once the box gets going, however, the bumps and grooves really don't have a chance to settle into each other, because they move past each other before they have a chance to line up well. As a result, the molecules do not interact as strongly, and they can't attract each other with as strong a force as they did when the box was stationary.

What does this all mean? It means that the frictional force between two surfaces is different when the surfaces are at rest relative to one another compared to when they are in motion relative to one another. Thus, we typically split friction into two distinct types:

Kinetic (kuh net' ik) friction - Friction that opposes motion once the motion has already started

Static friction - Friction that opposes the initiation of motion

Now that you know some basics about friction, I want you to perform an experiment that will illustrate some of what you have just learned as well as provide a little more detail about the nature of the frictional force.

EXPERIMENT 5.2
The Frictional Force

Supplies:

- Safety goggles
- A plastic tub (like the kind that holds margarine) with lid
- A board that is at least a meter long and wider than the plastic tub listed above
- A rubber band (Thin rubber bands work better than thick ones.)
- Scissors
- Aluminum foil
- Liquid soap (Dish detergent or body wash will work as well.)
- A washcloth
- A ruler
- Sand or kitty litter

1. Lay the board down on a flat surface.
2. Cut the rubber band so that it can be stretched out as one straight piece of rubber.
3. Fill the plastic tub about one-third full of sand or kitty litter.
4. Lay one end of the rubber band over the lip of the plastic tub so that at least one-half of an inch of the rubber band extends into the tub.

5. Put the lid on the plastic tub, right over the rubber band. Even though the rubber band is between the lip of the tub and the lid, you should still be able to get the lid closed. This will anchor the end of the rubber band to the tub.
6. Put the tub on the board and pull the rubber band so that it is taut but not stretched. In other words, just straighten out the rubber band so that it doesn't bow in the middle, but do not stretch the rubber band.
7. Lay the ruler on the board so that the 1-cm mark is lined up with the fingers you are using to pull on the rubber band. The numbers on the ruler's scale should increase in the direction that you are pulling on the rubber band.
8. **Very slowly**, begin pulling on the rubber band. As you pull, it will stretch. You can measure the length that it stretches by looking down at the fingers you are using to pull on the rubber band. You will see the ruler right underneath, allowing you to measure the stretch of the rubber band.
9. Continue to pull slowly until the tub begins to move. Record the distance that the rubber band had stretched the instant before the tub began to move.
10. Reposition the tub so that it is in approximately the same place as it was before, make the rubber band taut again, and repeat steps 7-9. Do this four more times, so that you have a total of five measurements for the stretch of the rubber band the instant prior to the tub moving. Average your results.
11. Open the tub and add more sand or kitty litter, so that it is approximately two-thirds full.
12. Repeat steps 4-10, so that you have five measurements of the distance that the rubber band stretched in order to get the tub moving when it is two-thirds full of sand or kitty litter. Average your results.
13. Open the tub and add more sand or kitty litter, so that it is full.
14. Repeat steps 4-10, so that you have five measurements of the distance that the rubber band stretched in order to get the tub moving when it is full of sand or kitty litter. Average your results.
15. Remove the ruler from the board and put the tub back to its original position.
16. Stretch the rubber band again, but as the tub begins to move, try to continue pulling on the rubber band so that the tub moves across the board at a reasonably constant speed. Note the difference in how much the rubber band is stretched while the tub is moving as compared to how much it was stretched in order to get the tub moving. This is hard, and all I want you to do is make a general assessment of the situation, so don't worry if you have trouble with it. Just try to make a general statement that compares the amount the rubber band is stretched while the tub is moving with a constant speed compared to the amount the rubber band was stretched in order to get it moving.
17. Squirt some liquid soap on the board, and use the washcloth to spread the soap around. It need not cover the entire board. It simply needs to cover the part of the board where you place the tub at the beginning of each trial and extend a few centimeters down the board so that when the tub begins to slide, it will still slide on the soap.
18. Repeat steps 6-10 so that you have five measurements of how far the rubber band stretched in order to get the tub to move on the soapy part of the board. Average your results.
19. Cover the board with aluminum foil. Wrap it around the ends of the board so that it will remain stationary, and stretch it as much as possible to reduce the wrinkles in the foil. The foil need only cover the same portion of the board that the soap covers.
20. Repeat steps 6-10 so that you have five measurements of how far the rubber band stretched in order to get the tub to move on the aluminum foil. Average your results.
21. Clean up your mess.

There are several things that we can learn from this experiment. First, however, you need to understand that the amount the rubber band stretched is actually proportional to the force with which the rubber band pulled on the tub. You will learn more about this in a later module, when you study Hooke's Law. For now, just realize that the farther the rubber band stretched, the more force it was exerting on the tub.

Okay, then, let's think about what happened in the very first part of the experiment. As you stretched the rubber band a little, it began pulling on the tub, but the tub did not move. Obviously, this is because of friction. Since the tub did not move, the friction you were working with at that point was static friction. Think about what this means from the perspective of Newton's Second Law. Since the tub did not move, its acceleration was zero. That means the sum of the forces acting on the tub was zero, right? Well, the only two forces acting on the tub were the pull of the rubber band and static friction. This means that the static frictional force was equal in magnitude to the pull of the rubber band, but it was opposite in direction.

Now, when you stretched the rubber band a little more, the tub still did not move. That means the sum of the forces was still zero, right? However, since you stretched the rubber band more, the force of the rubber band's pull increased, right? What does that tell you about the force that static friction exerted? It increased as well. This tells you something very important about the nature of the static frictional force: it can vary, depending on the force that you apply to the object you are trying to move. Suppose, for the sake of argument, that when you stretched the rubber band a little, it exerted a force of 1 Newton. Since the tub did not move, the static frictional force must have exerted a force of -1 Newton. However, suppose that when you stretched the rubber band a bit more, the force doubled to 2 Newtons. Since the tub still did not move, the static frictional force must have increased to -2 Newtons.

What happened as you continued to stretch the rubber band? Eventually, the tub moved. What does that tell us about the nature of the static frictional force? It tells us that there is a maximum value for the static frictional force. The static frictional force can increase to fight the force trying to cause motion, but at some point, it reaches a maximum value. After that, it cannot increase, and any increase in the force trying to cause motion will result in the object starting to move. This is an important conclusion:

The static frictional force can increase to counteract an increasing force which attempts to cause motion. However, for a given object and surface upon which it sits, there is a maximum value that the static frictional force can reach. If a force greater than the maximum static frictional force is applied to an object, the object will move.

That's the first conclusion from our experiment.

I actually want to skip the results of the next part of the experiment (the part where you increased the amount of sand or kitty litter in the tub) for now. I will come back to it in the next section of this module. For right now, I want to move on to the part where you compared how far the rubber band stretched in order to get the tub moving to the amount that it stretched when you kept the tub moving at a constant speed. You should have noticed that the rubber band stretched less when you were keeping the tub moving at a constant speed as compared to the stretch required to get the tub moving initially.

Why should you have gotten that result? Well, think about the situation in terms of Newton's Second Law again. When you moved the tub across the board at a constant speed, the acceleration was still zero. Remember, acceleration is the *change* in velocity. If the speed and direction were both constant, the velocity was constant and thus the acceleration was *zero*. Based on Newton's Second Law, then, the sum of the forces was *still zero*, because even though the tub was moving, the acceleration was zero. Thus, the force with which the rubber band was pulling was still being counteracted by friction. However, since the tub was moving, the friction that was in effect was *kinetic friction*, not static friction.

Think about what this means. The amount that the rubber band stretched in order to get the tub moving is a measure of the maximum *static* frictional force between the tub and the board. However, the stretch of the rubber band once the tub was moving at a constant speed is a measure of the *kinetic* friction that the tub experienced. Thus, since the rubber band stretched less while the tub was moving at a constant speed as compared to the amount that it stretched to get the tub moving, the kinetic frictional force was *less* than the maximum static frictional force. This is an important point to remember:

The kinetic frictional force between a given object and the surface upon which it moves is less than the maximum static frictional force between the object and the surface.

That's the second conclusion from our experiment.

In the next part of the experiment, you added soap to the surface of the board and used the stretch of the rubber band to measure the maximum static frictional force between the tub and the soapy board. You might have been surprised to find that the rubber band had to stretch *farther* to move the tub on the soapy board, indicating that the maximum static frictional force between the tub and the soapy board is greater than the maximum static frictional force between the tub and the dry board. Generally, we think of soap as being "slippery," so you probably thought that a soapy board would be more slippery (and thus have less friction) than a dry board. The experiment should have indicated otherwise.

Remember, the force of friction depends on the attraction between the molecules of the two surfaces as well as how close those molecules can get to one another. The molecules in the soap actually resulted in *more* attraction between the surfaces, so the result was that the maximum static frictional force increased. Now please understand that this result is strongly dependent on the materials used. Had you been using a metal block on a metal surface, the soap would have *decreased* the frictional force. So soap *can* reduce friction between some surfaces. However, it can increase friction between other surfaces.

Your results with the aluminum foil probably indicated that the maximum static frictional force between the aluminum foil and the tub was less than the maximum static frictional force between the tub and the dry board. These results, then, lead to another conclusion from our experiment:

The frictional force between two surfaces is strongly dependent on the molecules which make up those surfaces.

An Equation for the Frictional Force

Now let's return to the second part of Experiment 5.2. In that portion of the experiment, you added more sand or kitty litter to the tub and measured how that affected the stretch of the rubber band required to get the tub moving. Your results probably did not surprise you: the more sand or kitty litter, the greater the stretch required to initiate motion.

We can understand this rather easily. By increasing the amount of sand or kitty litter in the tub, you were increasing the total mass of the object. This, of course, means that the weight of the object increased as well. Now remember, when an object rests on a surface, the surface exerts a normal force on the object. In the case of a level surface, that normal force is equal in magnitude to the weight and opposite in direction. Thus, by increasing the mass of the object, you also increased the normal force exerted by the board on the tub.

Based on this knowledge, what do your results tell you? They tell you that the frictional force between an object and a surface depends on the normal force. The greater the normal force, the greater the frictional force. This should make sense based on what we know about friction. After all, the greater the normal force, the more the surface pushes against the object. This will result in the molecules of the surface and the molecules of the object getting closer, which will enhance the attraction between the molecules of the surface and those of the object. That will increase the friction.

We now know two things that affect the strength of the frictional force between an object and a surface: the nature of the object and the surface as well as the normal force that the surface exerts on the object. We can take these two factors and put them into an equation as follows:

$$f = \mu F_n \qquad (5.3)$$

In this equation, "f" stands for the magnitude of the frictional force while "F_n" represents the normal force. The symbol "μ" (the lower-case Greek letter "mu") is called the **coefficient of friction**. This coefficient takes into account the nature of the object and the surface. If a surface holds on to an object well, the coefficient of friction is large. If the surface does not hold on to an object well, the coefficient of friction is small. So you see that Equation (5.3) takes into account both the nature of the object and the surface as well as the strength of the normal force.

Notice in Equation (5.3) that there are no vectors. It turns out that even though the frictional force and the normal force are both vectors, their directional information does not apply in Equation (5.3). We will therefore only use this equation to calculate the *magnitude* of the frictional force. We will use our own reasoning to determine the *direction* of the frictional force. This is rather easy to do, because we know that friction always opposes motion. Therefore, the frictional force will always point opposite the motion or attempted motion.

Now before we actually use Equation (5.3), there is one more thing we need to tie into it. We learned earlier that there are two kinds of friction: static and kinetic. How do we take account of this fact in terms of Equation (5.3)? We do this by having two separate coefficients of friction for a given surface: a **coefficient of static friction (μ_s)** and a **coefficient of kinetic friction (μ_k)**. If an object is stationary on a surface, we will use the surface's coefficient of static friction to calculate the magnitude of the frictional force. Please note that as we define things here in this course, using the coefficient of static friction in Equation (5.3) will calculate the *maximum* static frictional force. If the object is already moving, we will use its coefficient of kinetic friction to calculate the kinetic frictional force.

There is no way for you to know the values of the coefficients of friction for a given surface. That's something which must be determined by experiment (an experiment you will do in the next module). Thus, I must either give you the coefficients of friction that you need or provide you a way to calculate them. There is one thing that you *do* know about the coefficients of friction, however. You know that the coefficient of static friction is larger than the coefficient of kinetic friction. This should be pretty obvious. After all, we learned previously that the maximum static frictional force is larger than the kinetic frictional force. Thus, the coefficient of static friction (which allows us to calculate the maximum static frictional force) must be larger than the coefficient of kinetic friction (which allows us to calculate the kinetic frictional force).

For a given object and surface, the coefficient of static friction (μ_s) is greater than the coefficient of kinetic friction (μ_k).

So how do we use all of this new information to help us analyze physical situations? Study the following examples to see how we blend this new information with Newton's Second Law to analyze realistic situations.

EXAMPLE 5.5

Illustrations copyright GifArt.com

A 300.0-kg log rests on the ground (μ_s = 0.74, μ_k = 0.45). What is the frictional force that a pulling truck would have to overcome in order to get the log to move?

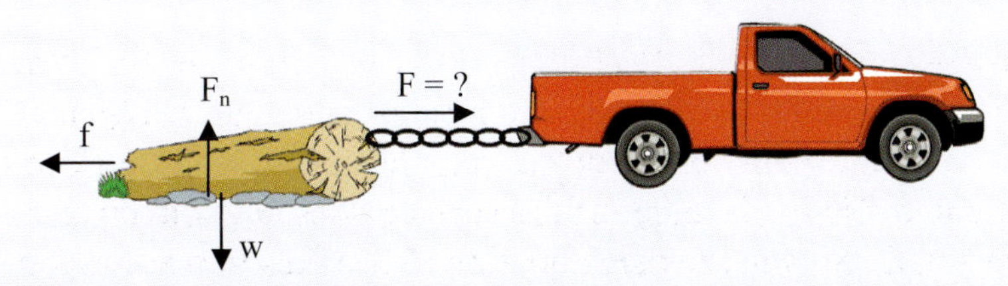

In order to solve any problem that deals with Newton's Second Law, we must analyze all of the forces involved in the problem. The first force that comes to mind is the force that we must solve for. If the truck is pulling, it is obviously exerting a force. I have used the arrow labeled "F=?" to represent that force. There are many other forces at work here, however, and we cannot solve the problem until we recognize all of them. The next most obvious force is gravity. I have represented it with the arrow labeled "w" (for weight). As with anything sitting on a surface, the surface is also exerting a normal force on the log. This force is labeled "F_n." Finally, the thing that makes the log hard to move is the frictional force (labeled "f") that opposes motion. These are all of the forces acting on the log.

The problem asks us to determine the frictional force that the truck must overcome. Therefore, we must calculate f. That's not so bad, because we have a formula [Equation (5.3)] for it. There's only one problem. Equation (5.3) needs the normal force, and we don't have that yet. Our first job, then, is to calculate the normal force. Well, since the normal force will simply counteract the force of gravity (which is the weight), we can just calculate the weight of the log and say that the magnitude of the normal force is the same.

$$\mathbf{w} = \mathbf{mg}$$

$$\mathbf{w} = (300.0 \text{ kg}) \cdot (9.8 \frac{\text{m}}{\text{sec}^2})$$

$$\mathbf{w} = 2.9 \times 10^3 \frac{\text{kg} \cdot \text{m}}{\text{sec}^2} = 2.9 \times 10^3 \text{ Newtons}$$

Now that we have the weight, we know that the normal force has the same magnitude. We can therefore use this value in Equation (5.3). Before we use the equation, however, we also need to determine which coefficient of friction to use. Since the log is stationary and the truck is trying to get it to start moving, we have to use the coefficient of static friction:

$$f = \mu F_n$$

$$f = (0.74) \cdot (2.9 \times 10^3 \text{ Newtons})$$

$$f = 2.1 \times 10^3 \text{ Newtons}$$

This tells us that the truck must overcome a frictional force of 2.1x 10^3 Newtons in order to get the log moving.

A man is pushing a 150.0-kg desk across a floor. The frictional coefficients between the desk and the floor are $\mu_s = 0.45$ and $\mu_k = 0.35$. If the desk is already moving, how much force must the man exert to keep the desk accelerating at 0.30 m/sec^2?

In this problem, we are given the mass of the desk and the desired acceleration and asked to calculate how much force must be applied. This looks a lot like another problem we did in the first section that discussed Newton's Second Law. The difference is that now we must take friction into account. Well, the way to do this is to look at the situation and see what forces are acting on the desk. Newton's Second Law says that the *sum of the forces* will equal the mass times the acceleration.

What forces are acting on the desk? Well, the man is pushing, so he is exerting a force. I used the arrow labeled "F = ?" to represent that force. Another force acting on the desk is gravity, which pulls the desk straight down. I used the arrow labeled "w" (for weight) to indicate the force due to gravity. Also, the surface fights gravity by pushing back with a normal force, represented by the label "F_n." Finally, friction is at work as well. Since it always opposes motion, it acts opposite to the direction that the man is pushing. This is represented by the arrow labeled "f."

Well, now we have all of the forces acting in the situation. What in the world do we do with them all? The first thing to remember is that we can completely separate the vertical from the horizontal direction. In the horizontal direction, we know that the forces at work are the force that the man is pushing with (F) and the frictional force (f). Newton's Second Law says that if we add them, we can set them equal to the mass times the acceleration. Let's remember, however, that the forces are pointed in opposite directions; thus, one of them will be negative. I will call motion to the right positive, indicating that the frictional force is negative. Newton's Second Law, then, looks like this:

$$F - f = ma$$

$$F - f = (150.0 \text{ kg}) \cdot (0.30 \frac{\text{m}}{\text{sec}^2})$$

$$F - f = 45 \text{ Newtons}$$

Notice that there are no bold-faced letters here. That's because we are dealing only with the magnitudes of the force and acceleration vectors. Because of this, I need to put the directions in with signs as a part of the equation. That's why I subtracted the friction force from the force of the man. This explicitly tells us that the two forces are opposite each other. We will be using this kind of reasoning throughout the next two modules, so you need to get used to it.

The equation we ended up with would be a pretty easy equation to solve if we knew what the frictional force is. Well, we can find that out. Remember, the frictional force can be calculated using the coefficient of friction (which was given) and the normal force. Thus, we need to calculate the normal force and then use it to calculate the frictional force. Calculating the normal force is easy. We just calculate the weight. The magnitude of the normal force is identical:

$$\mathbf{w} = m\mathbf{g}$$

$$\mathbf{w} = (150.0 \text{ kg}) \cdot (9.8 \frac{\text{m}}{\text{sec}^2}) = 1.5 \times 10^3 \text{ Newtons}$$

This means that the magnitude of the normal force is also 1.5×10^3 Newtons. We can now calculate the frictional force. Which coefficient do we use, however? Since the desk is already moving, we use the coefficient of kinetic friction:

$$f = \mu F_n$$

$$f = (0.35) \cdot (1.5 \times 10^3 \text{ Newtons}) = 530 \text{ Newtons}$$

Now we can go back to our original equation and solve for F:

$$F - f = 45 \text{ Newtons}$$

$$F - 530 \text{ Newtons} = 45 \text{ Newtons}$$

$$F = 580 \text{ Newtons}$$

In the end, then, in order to overcome friction and still give the desk an acceleration of 0.30 m/sec^2, the man has to push the desk with a force of 580 Newtons. Notice how the significant figures worked out in this case. To solve for F, I had to add 530 to 45. The rule of addition and subtraction says to report your answer to the same decimal place as the least precise number in the problem. Since 530 has its last significant figure in the tens place, I had to report my answer to the tens place.

A man is pushing a 2.2-kg box across a floor at a constant velocity of 2.1 m/sec. If the man exerts a force of 15 Newtons to keep the box going, what is the coefficient of kinetic friction between the box and the floor?

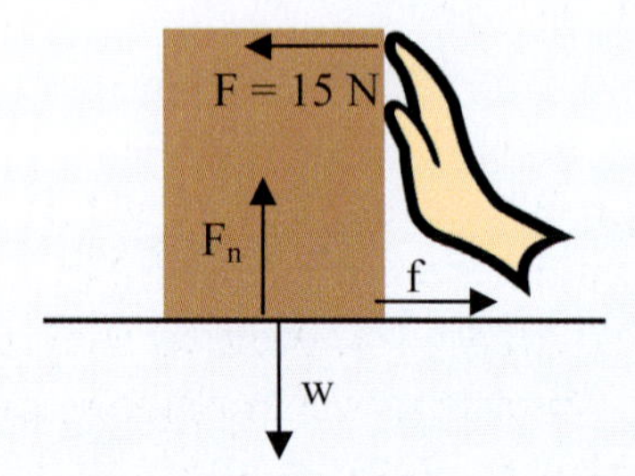

Once again, before we analyze any situation like this, we must keep track of all forces that act on the box. The man is exerting a force (F = 15 N); gravity is exerting a force (w); the floor exerts a normal force (F_n); and friction exerts a force that opposes the motion (f). According to Newton's Second Law, the sum of the forces exerted on the box equals the box's mass times its acceleration. What is the box's acceleration? Well, it is moving at a constant velocity. What does that tell us about acceleration? It is *zero*. Thus, in this case, the sum of the forces acting on the box must equal zero.

In order to solve Newton's Second Law in the horizontal dimension, we will need to know something about friction. This means we need to know the normal force. Thus, we might as well get that out of the way. To get the normal force, we simply have to calculate the box's weight:

$$\mathbf{w} = \text{m}\mathbf{g}$$

$$w = (2.2 \text{ kg}) \cdot (9.8 \frac{\text{m}}{\text{sec}^2}) = 22 \text{ Newtons}$$

Since this is the weight, we know that the normal force also has a magnitude of 22 Newtons. Now we can look in the horizontal dimension. In this dimension, the sum of the forces must be zero because the box has no acceleration. Just as we did in the previous problem, however, we will need to take care of the directions explicitly in the problem. Thus, since the frictional force points to the right, we will call it positive. This makes the force from the hand negative:

$$f - F = ma$$

$$f - 15 \text{ Newtons} = 0$$

$$f = 15 \text{ Newtons}$$

So we know that the frictional force is 15 Newtons. To determine μ_k, then, we simply need to use Equation (5.3):

$$f = \mu_k F_n$$

$$15 \text{ Newtons} = \mu_k \cdot (22 \text{ Newtons})$$

$$\mu_k = 0.68$$

The coefficient of kinetic friction, then, is 0.68.

Now these final two example problems might have seemed really complicated, but they aren't if you sit and think about them. All you have to do is find all of the forces. Gravity will almost always be one of the forces. The normal force will also be there as long as the object is in contact with a surface (usually the ground or floor). There is usually someone or something applying a force. Finally, you may or may not need to take friction into account. How will you know whether or not to take it into account? Unless I use the term "frictionless" or unless I specifically tell you to ignore friction, you must take it into account.

Once you have all of the forces, you usually use the forces in the vertical dimension to calculate the normal force so that you can calculate friction. You then sum up the forces in the horizontal dimension and set that sum equal to the mass times the acceleration. The only real trick here is to make sure you put the signs associated with the forces right into the equation, like I did in the last two examples. Try to cement this skill into your mind with the following "On Your Own" problems.

ON YOUR OWN

5.8 A man is pushing on a 12.0-slug rock. What is the maximum frictional force that will keep the rock from moving? ($\mu_s = 0.65$, $\mu_k = 0.33$)

5.9 A student starts to push a 22-kg pile of books across a desk. If the pile of books is at rest, what frictional force must be overcome to get the books moving? ($\mu_s = 0.45$, $\mu_k = 0.23$)

5.10 A man pushes a 6,675-Newton rock with a force of 95 Newtons. If the rock is already moving, what will its acceleration be? ($\mu_s = 0.15$, $\mu_k = 0.011$)

5.11 A child pushes a 2.3-kg toy along the floor at a constant velocity. What force is the child exerting? ($\mu_s = 0.25$, $\mu_k = 0.13$)

Illustrations copyright GifArt.com

Newton's Third Law

You never thought we'd get to the last law, did you? Well, you'll be very happy to know that Newton's Third Law does not require nearly as much discussion! Stated simply, Newton's Third Law is easy to remember:

Newton's Third Law - For every action, there is an equal and opposite reaction.

This law is most easily understood by illustration. Suppose you are pushing against a wall:

Illustration copyright GifArt.com

When you push the wall, it actually pushes back at you! In fact, it pushes back with equal strength. This is the equal and opposite reaction that Newton's Third Law talks about.

This may seem pretty simple to you, but there are some pretty profound consequences to Newton's Third Law. Suppose, for example, you drop a ball. What happens? Well, it falls to the earth. According to Newton's Third Law, every action must have an equal and opposite reaction. If the action is that the ball falls to the earth, what is the equal and opposite reaction? Believe it or not, the *earth climbs up towards the ball!* That's right! The ball does not just fall; the earth actually moves up to reach the ball. Now if this sounds a little bizarre, remember one thing. The equal and opposite part of Newton's Third Law applies to the *forces* involved. Thus, the earth exerts a force (gravity) that accelerates the ball to it. The ball exerts an equal and opposite force causing the earth to rise and meet the ball. Think about it, though. The forces are equal, but the masses are not. The earth is millions of times more massive than the ball, which means that the acceleration of the ball towards the earth is millions of times faster than the acceleration of the earth to the ball. Thus, the ball will move a lot, and the earth will only move a little. Nevertheless, it is important to understand that the earth does move a little in this situation.

It is very important to realize that the equal and opposite forces talked about in Newton's Third Law do not act on the *same* object. If that were the case, then there would never be any motion. After all, if equal and opposite forces act on the same object, they cancel each other out and the resulting force would be zero. Instead, the equal and opposite forces discussed in Newton's Third Law affect *different* objects. In our first example, you exerted a force on the wall. The equal and opposite force was exerted not on the wall, but on you. Thus, the two forces acted on two completely different objects. In the second example, the earth exerted a force on the ball, causing it to accelerate. The ball then caused an equal and opposite force on the earth, causing it to accelerate. Once again, the two forces worked on two *separate* objects. One force worked on the earth, the other on the ball. Both were able to accelerate, since the forces did not cancel, because they acted on different objects.

One more example should help. Suppose you jump onto a trampoline. When you hit the trampoline, you exert a force on the trampoline. What does Newton's Third Law say will happen? The trampoline will exert an equal but opposite force on you. How can you tell that this is happening? Well, the trampoline's surface bends. This tells you that you are exerting a force on the trampoline. How do you know that the trampoline is exerting a force back? You start to slow down, stop, and then accelerate in a completely different direction. You could have never done that if a force had not acted on you. That force was the equal and opposite force required in Newton's Third Law. Thus, you exerted a force on the trampoline (causing the surface to bow), and the trampoline exerted a force right back on you (causing you to accelerate in a different direction). See if you can use Newton's Third Law to answer the following "On Your Own" problem:

ON YOUR OWN

5.12 A tennis ball hits a tennis racquet. What two forces (besides gravity) exist in this situation? What are the visible ramifications of these two forces?

As you review this module, pay close attention to all of our discussions of Newton's Second Law. The next module is devoted exclusively to more applications of this very important law, so you need to have a firm grasp of everything related to it in this module!

ANSWERS TO THE "ON YOUR OWN" PROBLEMS

5.1 While the car was traveling at a relatively constant velocity, the packages were simply moving along with the car at the same velocity. However, as the car began to accelerate quickly, the packages were still moving with the old velocity. In order for them to speed up, the Law of Inertia says that an outside force had to act on them. This outside force was the friction between the car seat and the packages. Unfortunately, the frictional force was not strong enough to change the velocity of the packages as quickly as the car's velocity was changing. As a result, the packages moved with a smaller velocity than the car. Thus, the car started moving ahead of the packages, making the packages look like they were moving backwards. Once the packages hit the back of the seat, a new force (the push from the seat) was added to the frictional force. This provided enough force to accelerate the packages as fast as the car, so the packages stopped sliding. In principle, if the packages had been *very* heavy and the seat back had been weak, even the force of the seat back pushing on the packages would not have been enough to accelerate them. If that had happened, the seat back would have been broken down by the packages, and they would have continued to slide backwards. That's what inertia is all about.

5.2 When you slide a block of wood across any surface, the Law of Inertia says that it will continue to slide at the velocity you gave it until acted on by an outside force. The outside force that will slow the block down and eventually stop it is friction. Since friction opposes motion, the frictional force will work to remove all velocity from the wood. Well, the less friction that there is, the longer it will take for this to happen. When you slide a block of wood on ice, the slippery nature of the ice reduces the friction. Thus, when a block of wood slides on ice, it experiences less friction than when it slides on a sidewalk. With less friction on the ice, the block will slide a lot farther, because it takes longer for friction to remove the velocity with which the block of wood is traveling.

5.3 In this problem, we are given force and mass and asked to calculate acceleration. Thus, this is a simple application of Equation (5.1). The problem, however, is that our units are not consistent. A Newton is a $\frac{\text{kg} \cdot \text{m}}{\text{sec}^2}$, but the mass is in g. Thus, we need to convert from g to kg:

$$\frac{1.23 \times 10^6 \,\cancel{\text{g}}}{1} \times \frac{1 \text{ kg}}{1{,}000 \,\cancel{\text{g}}} = 1{,}230 \text{ kg}$$

Now we can use Equation (5.1):

$$\mathbf{F} = m\mathbf{a}$$

$$15 \text{ Newtons} = (1{,}230 \text{ kg}) \cdot (\mathbf{a})$$

$$\mathbf{a} = \frac{15 \frac{\cancel{\text{kg}} \cdot \text{m}}{\text{sec}^2}}{1{,}230 \,\cancel{\text{kg}}} = 0.012 \frac{\text{m}}{\text{sec}^2}$$

Notice that in order to properly cancel the units, I had to use the definition of the Newton in the last line of the equation. This is one reason you need to know what a Newton is. In the end, then, a 15-

Newton force (pretty considerable) does not accelerate the car very quickly, only 0.012 m/sec^2. Since the acceleration has the same sign as the force, you know that they are pointed in the same direction.

5.4 a. The tennis ball changed direction, which means its velocity changed. Thus, it experienced acceleration, which means a force must have acted on it.

b. The force acted on the ball when its velocity changed, which is when it hit the ground.

c. The velocity of the ball was heading down and to the right. After the bounce, it was headed up and to the right. Thus, the force must have been directed up.

d. Force and acceleration point in the same direction, so the acceleration is pointed up.

5.5 Since the units used are grams, we know that we have been given the mass of an object. We must use Equation (5.2) to change that into Newtons. Before we can do that, however, because a Newton is defined as a $\frac{\text{kg}\cdot\text{m}}{\text{sec}^2}$, we must have our mass in kg:

$$\frac{13.5\ \cancel{\text{g}}}{1} \times \frac{1\ \text{kg}}{1{,}000\ \cancel{\text{g}}} = 0.0135\ \text{kg}$$

Now we can use Equation (5.2):

$$\mathbf{w} = \text{m}\mathbf{g}$$

$$\mathbf{w} = (0.0135\ \text{kg})\cdot(9.8\frac{\text{m}}{\text{sec}^2}) = 0.13\frac{\text{kg}\cdot\text{m}}{\text{sec}^2} = \underline{0.13\ \text{Newtons}}$$

5.6 When an object moves from Mars to the earth, its weight changes but its mass does not. Thus, we must figure out mass so we know something that is consistent between the two locations. We can use Equation (5.2) to do this, but let's make sure we know what acceleration due to gravity we need. Since we know the weight on Mars, we must use the Martian acceleration due to gravity (3.7 m/sec^2):

$$\mathbf{w} = \text{m}\mathbf{g}$$

$$15\text{Newtons} = \text{m}\cdot(3.7\frac{\text{m}}{\text{sec}^2})$$

$$\text{m} = \frac{15\frac{\text{kg}\cdot\cancel{\text{m}}}{\cancel{\text{sec}^2}}}{3.7\frac{\cancel{\text{m}}}{\cancel{\text{sec}^2}}} = 4.1\ \text{kg}$$

So we now know that the rock has a mass of 4.1 kg, which remains the same no matter where the rock is. To calculate the weight on earth, then, we take this unchanging mass and multiply by the acceleration due to gravity, in accordance with Equation (5.2):

$$\mathbf{w} = \mathrm{mg}$$

$$\mathbf{w} = (4.1\text{ kg}) \cdot (9.8\frac{\text{m}}{\text{sec}^2}) = 4.0 \times 10^1 \frac{\text{kg} \cdot \text{m}}{\text{sec}^2} = 4.0 \times 10^1 \text{ Newtons}$$

A rock that weighs 15 Newtons on Mars, then, weighs $\underline{4.0 \times 10^1 \text{ Newtons}}$ on earth. The answer must be in scientific notation, since it must have two significant figures.

5.7 We are given the woman's mass (we know it is mass because the unit is slug), so we can use Equation (5.2) to get weight. Since the weight is the same as the force due to gravity, we can use it to determine the normal force:

$$\mathbf{w} = \mathrm{mg}$$

$$\mathbf{w} = (3.8\text{ slugs}) \cdot (32\,\frac{\text{ft}}{\text{sec}^2}) = 120\,\frac{\text{slug} \cdot \text{ft}}{\text{sec}^2} = 120 \text{ pounds}$$

This tells us that gravity exerts a 120-pound force down on the woman, causing the step to exert a normal force of <u>120 pounds up</u>.

5.8 To determine the frictional force, we need to use Equation (5.3). This equation, however, uses the normal force. Thus, we first have to calculate the normal force. Since the normal force counteracts weight, we need to just calculate the weight and realize that the normal force has the same magnitude:

$$\mathbf{w} = \mathrm{mg}$$

$$\mathbf{w} = (12.0\text{ slugs}) \cdot (32\frac{\text{ft}}{\text{sec}^2}) = 380\,\frac{\text{slug} \cdot \text{ft}}{\text{sec}^2} = 380 \text{ pounds}$$

Now that we have the weight, we know that the normal force has the same magnitude. We can therefore use this value and the coefficient of static friction (which resists the initiation of motion) in Equation (5.3):

$$f = \mu F_n$$

$$f = (0.65) \cdot (380\text{ pounds}) = \underline{250 \text{ pounds}}$$

5.9 What forces are at play here? Well, there is gravity, pulling the books down, and the normal force, pushing up. The student is applying a force to get the books to move, and friction is opposing the motion. Thus, the situation looks like this:

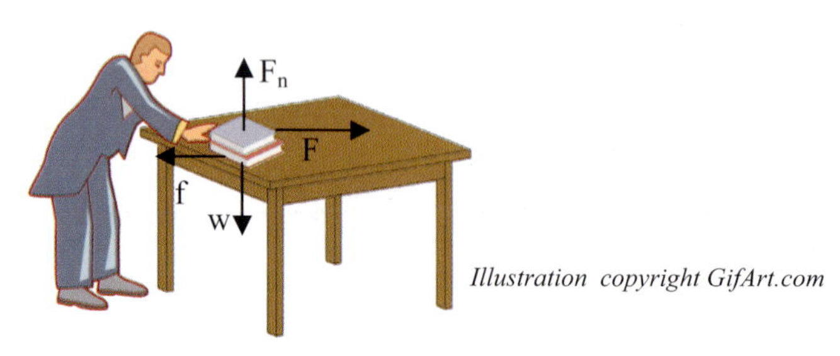

Illustration copyright GifArt.com

To figure out the frictional force, we must first determine the normal force. Luckily, that's easy:

$$\mathbf{w} = m\mathbf{g}$$

$$\mathbf{w} = (22\text{ kg})\cdot(9.8\frac{\text{m}}{\text{sec}^2}) = 220\frac{\text{kg}\cdot\text{m}}{\text{sec}^2} = 220\text{ Newtons}$$

Now that we have the weight, we know that the normal force has the same magnitude. We can therefore use this value and the coefficient of static friction (because the books are not moving) in Equation (5.3):

$$f = \mu F_n$$

$$f = (0.45)\cdot(220\text{ Newtons}) = 99\text{ Newtons}$$

The frictional force that must be overcome, then, is 99 Newtons. Now remember, static frictional force *increases* with the increasing force used against it. Thus, this is the maximum static frictional force. If the student pushes with 50 Newtons, the static frictional force will push back with 50 Newtons. However, the maximum force with which static friction can push is 99 Newtons. That, then, is the force which must be overcome.

5.10 The first thing we need to do is look at all of the forces in the problem:

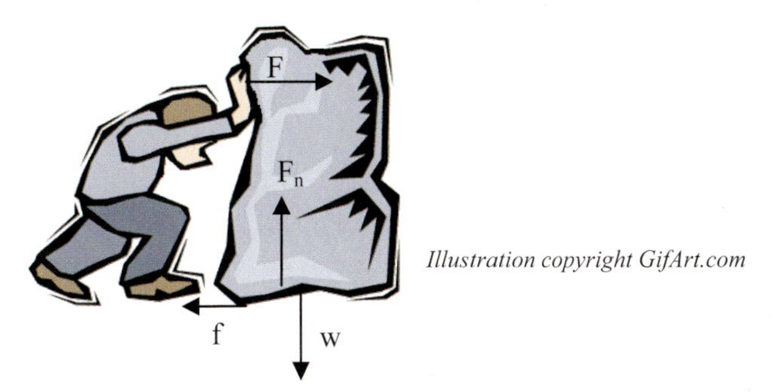

Illustration copyright GifArt.com

The man pushes with a force (F); gravity pulls down with a force (w); the road pushes against gravity with the normal force (F_n); and friction opposes the motion (f). To determine the acceleration (which will be in the horizontal dimension), we must look in the horizontal dimension. However, we will need to know something about friction, and for that we need the normal force, so let's start in the vertical dimension.

In the vertical dimension, we can determine the normal force's magnitude by simply calculating the weight. We needn't calculate it in this case, however, because it is given. We know that the number given is weight because it has the unit of Newtons. Thus, the weight is 6,675 Newtons, indicating that the normal force has a magnitude of 6,675 Newtons.

Now we can look in the horizontal dimension. The applied force (F) and friction (f) are working on the rock. Since the applied force is pointed to the right, I will call it positive. This makes friction negative. I must explicitly put these signs into my equation, since the variables in the equation represent only magnitudes:

$$F - f = ma$$

To use this equation, though, I need to determine a few things. First, I need to determine the force due to friction. I can do that:

$$f = \mu F_n$$

$$f = (0.011) \cdot (6{,}675 \text{ Newtons}) = 73 \text{ Newtons}$$

Notice that I used the coefficient of kinetic friction, since the rock is already moving. I still need to calculate one more thing: I need to determine the rock's mass. Since the weight was given earlier, I need to use Equation (5.2) to determine the mass. You must be very careful to look at the unit and determine what you have and what you need. Since the unit is Newtons, we have the weight. However, we need the mass as well, so we must convert Newtons to kg:

$$\mathbf{w} = m\mathbf{g}$$

$$6{,}675 \text{ Newtons} = m \cdot (9.8 \frac{\text{m}}{\text{sec}^2})$$

$$m = \frac{6{,}675 \frac{\text{kg} \cdot \cancel{\text{m}}}{\cancel{\text{sec}^2}}}{9.8 \frac{\cancel{\text{m}}}{\cancel{\text{sec}^2}}} = 680 \text{ kg}$$

Now we can go back to our original equation and solve for a:

$$F - f = ma$$

$$95 \text{ Newtons} - 73 \text{ Newtons} = (680 \text{ kg}) \cdot a$$

$$a = \frac{22 \frac{\cancel{\text{kg}} \cdot \text{m}}{\text{sec}^2}}{6.8 \times 10^2 \cancel{\text{kg}}} = 0.032 \frac{\text{m}}{\text{sec}^2}$$

In then end, then, the rock accelerates at a mere <u>0.032 m/sec²</u>.

5.11 Once again, before we analyze any situation like this, we must keep track of all forces. The child is exerting a force (F); gravity is exerting a force (w); the floor exerts a normal force (F_n); and friction exerts a force that opposes the motion (f).

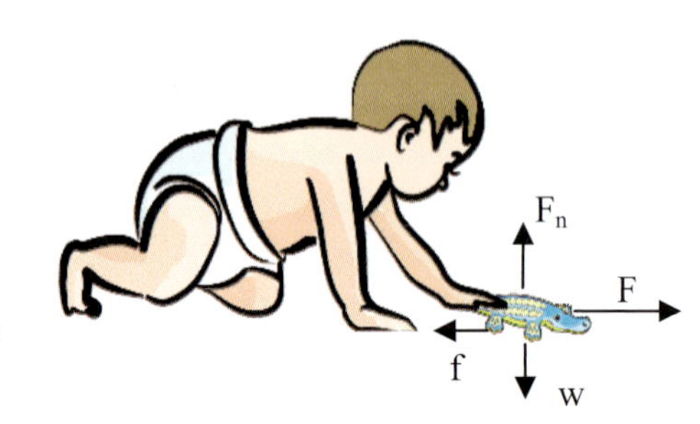

Illustration copyright GifArt.com

According to Newton's Second Law, the sum of the forces exerted on the toy equals the toy's mass times its acceleration. What is the toy's acceleration? Well, it is moving at a constant velocity. What does that tell us about acceleration? It is *zero*. Thus, in this case, the sum of the forces acting on the toy must equal zero.

In order to solve Newton's Second Law in the horizontal dimension, we will need to know something about friction. This means we need to know the normal force. Thus, we might as well get that out of the way. To get the normal force, we simply have to calculate the toy's weight:

$$\mathbf{w} = m\mathbf{g}$$

$$\mathbf{w} = (2.3\,\text{kg})\cdot(9.8\frac{\text{m}}{\text{sec}^2}) = 23\,\frac{\text{kg}\cdot\text{m}}{\text{sec}^2} = 23\text{ Newtons}$$

Since this is the weight, we know that the normal force also has a magnitude of 23 Newtons. Now we need to calculate the frictional force using Equation (5.3). In this equation, we will use the coefficient of kinetic friction, since the toy is already moving:

$$f = \mu_k F_n$$

$$f = (0.13)\cdot(23\text{ Newtons}) = 3.0\text{ Newtons}$$

Now we can look in the horizontal dimension. In this dimension, the sum of the forces must be zero because the toy has no acceleration. Just as we did in the last problem, we will need to take care of the directions explicitly in the problem. Thus, since the frictional force points to the left, we will call it negative. This makes the force from the child positive:

$$F - f = ma$$

$$F - 3.0\text{ Newtons} = 0$$

$$F = 3.0\text{ Newtons}$$

The child is therefore pushing with a force of 3.0 Newtons.

5.12 The racquet exerts a force on the ball, and the ball exerts an equal and opposite force on the racquet. You can tell the racquet exerts a force on the ball because the ball accelerates in an opposite direction. This can only happen if a force is acting on it. You can also tell that the ball exerts a force on the racquet because the racquet strings bow back in reaction to it.

REVIEW QUESTIONS

1. State Newton's three laws in your own words.

2. Why will friction always be present when two surfaces touch each other?

3. A man is riding a horse at a quick gallop. Suddenly, the horse stops dead in its tracks. When this happens, the rider flies out of the saddle and lands on the ground. Does the rider fly forward (over the horse's head) or backwards (over the horse's rump)? Use Newton's First Law to explain your answer.

4. A student applies the same force to two objects. If the first is 1,000 times more massive than the second, which accelerates more quickly? How much more quickly?

5. Which of the following are legitimate force units?

$$\text{pound, slug,} \frac{\text{kg} \cdot \text{m}}{\text{sec}}, \frac{\text{slug} \cdot \text{mile}}{\text{hr}^2}$$

6. A physics student measures an object and writes down a value of 13.1 Newtons. What was the student measuring: the object's mass or its weight?

7. A physicist uses the same scale to measure the mass of an object at two different locations: Death Valley and the top of Mount Everest. Is the measurement accurate in both locations? Why or why not?

8. What is a normal force? Why is it important in physics?

9. A physics student is measuring friction coefficients. For one surface, he measures the coefficients but neglects to write down which is the static coefficient and which is the kinetic one. If the numbers are 0.23 and 0.54, which is which?

10. When a tire rolls on a road, it pushes against the road. According to Newton's Third Law, what must happen in response to this push?

PRACTICE PROBLEMS

1. A father is teaching his young daughter to ice skate by pulling her across the ice with a short rope. If the daughter (whose mass is 25 kg) starts to accelerate at 0.35 m/sec^2, with what force is the father pulling her? (Assume that the ice is frictionless.)

2. A 501-pound rock has fallen onto a frozen pond that you want to use as an ice-skating rink. If you push the rock with a force of 16 pounds, how fast will the rock accelerate? (You can assume that there is no friction between the rock and the ice.)

3. In Problem #2, I asked you to assume that there is no friction between the rock and the ice. You do, however, have to assume that there is friction between your feet (or whatever is on them) and the ice. Why?

4. A rock has a mass of 523 grams. What is its weight in Newtons?

5. If a box of cargo bound for space exploration weighs 1,234 pounds on earth, what will its weight be on the moon? (The acceleration due to gravity on the moon is 5.3 ft/sec^2.)

6. If an astronaut weighs 296 Newtons on Mercury (where the acceleration due to gravity is 3.95 m/sec^2), how much does she weigh on earth?

7. A 745-kg box is sliding across a floor (μ_s = 0.45, μ_k=0.32). What is the frictional force between the box and the floor?

8. While driving down a country road, a man has to stop because a 314-pound rock has fallen off of a hill and has landed in the middle of the road (μ_s = 0.45, μ_k=0.32). What is the magnitude of the frictional force the man must overcome to get the rock to move?

Illustration copyright GifArt.com

9. A construction worker is sliding a 74-kg box of bricks across a wooden board (μ_s = 0.26, μ_k=0.12). If the man pushes with a force of 92 Newtons, how fast will the box accelerate? Make the calculation for when the box is already moving.

Illustration copyright GifArt.com

10. A young boy found a large box that he can use to make a fort. He put some other "treasures" he found into the box, so that its weight is 11 pounds. He tied it to the back of his bike and is pulling it down the road ($\mu_s = 0.45$, $\mu_k = 0.32$). With what force must he pull the box in order to keep it moving at a constant velocity?

Illustration from Arts and Letters Express

Cartoon by Speartoons

MODULE #6: Applications of Newton's Second Law

Introduction

In the previous module, we looked at all three of Newton's Laws of Motion. Admittedly, we concentrated mostly on Newton's Second Law, but that's because it's so important. Well, guess what? We're going to spend *almost all of this module* on Newton's Second Law. Before we do this, however, I want to make a strong point. Even though I have said several times that Newton's Second Law is very important, I don't want to belittle the other two laws that we have learned. The reason I consider Newton's Second Law so important is because it is so *useful*. Without the other two laws, we would not understand motion, so they are also important. Because Newton's Second Law is applicable to many problems, however, it gets the most attention.

You might be wondering what in the world is left to discuss about Newton's Second Law after all we studied in the previous module. Well, one thing we have not done is analyze situations in which there are many forces at play. Also, we have not discussed how to apply this law to situations in which the forces at play are not aligned in the horizontal or vertical directions. Finally, we also need to learn a little bit about how to apply this important law to extended objects which are in rotational motion. That's what lies ahead of us in this module.

Translational Equilibrium

Newton's Second Law states that the sum of the forces on an object is equal to the mass of that object times its acceleration. What happens, then, when the sum of the forces on an object is zero? Newton's Second Law says that the object's acceleration must be zero as well. For example, if you push on an object while someone else pushes with the same force in the opposite direction, the sum of the forces on the object will be zero, because the force of your push will be canceled out by the force of the other person's push in the opposite direction. When something like this happens, the object will not accelerate in any direction. When an object is in such a state, we say that it is in **translational** (trans lay' shun uhl) **equilibrium** (ee kwuh lib' ree uhm).

<u>Translational equilibrium</u> - An object is said to be in translational equilibrium when the sum of the forces acting on it is equal to zero.

Let's think for a moment about what can happen when an object is in translational equilibrium. In this situation, the object has no acceleration. Does this mean that the object cannot move? Well, not exactly. An object certainly *will* have an acceleration of zero if it is at rest, but that's not the only situation in which an object has zero acceleration. If an object moves at a constant velocity, it also has zero acceleration. Thus, when an object is in translational equilibrium, it can either be at rest or moving with a constant velocity. As a result, we usually distinguish between these two types of equilibrium:

<u>Static equilibrium</u> - When an object is at rest, it is said to be in static equilibrium.

<u>Dynamic equilibrium</u> - When an object moves with a constant velocity, it is said to be in dynamic equilibrium.

In this module, we will concentrate on static equilibrium.

Let's look at a simple example of static equilibrium. Suppose you hang a lamp from the ceiling with a piece of rope:

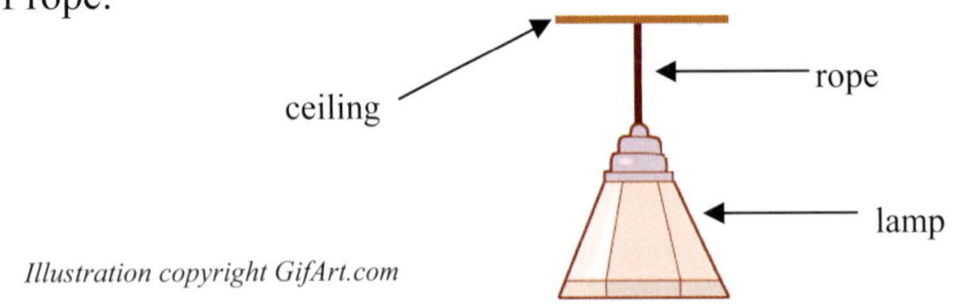

Unless the rope suddenly breaks, the lamp is clearly in static equilibrium. Let's look at the forces to see what keeps it in static equilibrium. One force that is certainly at work is gravity, which pulls the object down. If we call the mass of this object "m," the force of gravity is equal to m·g:

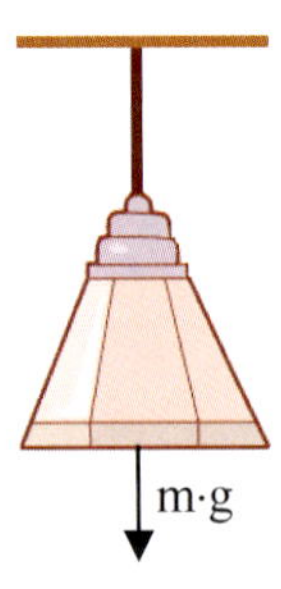

Now there must be another force at work here. After all, if gravity were the only force at play, the object would accelerate downwards. Where is this second force, then? It must work *on the object*, because the only way that the force of gravity can be canceled is by another force that acts *on the object* in the opposite direction.

Although exactly what this force is may be a bit of a mystery to you, I'm sure you know where it is coming from. The rope is holding the object. Thus, the rope is fighting gravity. How does it do this? It uses **tension** (ten' shun).

Tension - The force from a tight string, rope, or chain. This force is directed away from the object to which the string, rope, or chain is anchored.

So, the weight of the object stretches the rope to which it is attached. The rope responds by pulling in the opposite direction. Tension (which we will label as "T") is the force behind that pull:

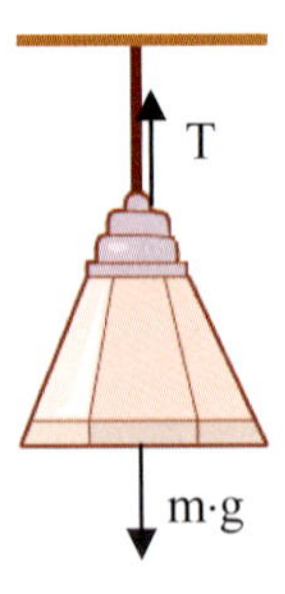

Now, since the object is at rest, we know that the sum of the forces must equal zero. The forces are aligned in the vertical dimension, so we can add them directly, as long as we take into account their directions. The tension is directed upward, so we will say it is the positive force. This makes the gravitational force (weight) negative. As we did in the previous module, we will put the directions into Equation (5.1) explicitly:

$$T - m \cdot g = 0$$

Presumably, we know the mass of this object. Also, we know g; it's just the acceleration due to gravity. Thus, we can solve for the only thing that we might not know, the tension in the rope:

$$T = m \cdot g$$

So we see that the tension in the rope (in this case) is equal to the weight of the object. It is directed opposite to the force of gravity, so it cancels the gravitational force.

Now you might think that this was a lot of work to determine a rather trivial point. After all, the force supplied by the rope *has* to be equal to the weight of the object, or the object would start accelerating. Why bother to go through all of this rigmarole then? Well, it turns out that this can get *really* complicated in two-dimensional situations. Take a look at the example below to see what I mean.

EXAMPLE 6.1

Illustrations from Arts and Letters Express

A 50.0-Newton painting is hung from a ceiling with two strings, according to the diagram below. What is the tension in each string?

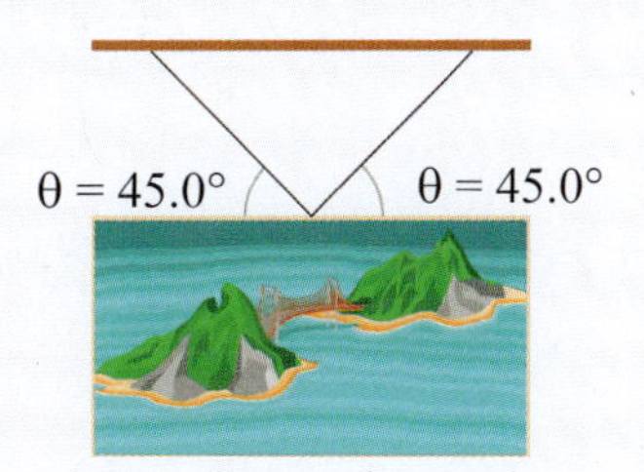

As we learned in the previous module, the way to solve problems that involve multiple forces is to make sure we know where those forces are and how they are directed. First, there is gravity pulling the picture down. The strings fight gravity by pulling up on the picture. The tension in each string pulls up, and it is obviously directed along the strings. There is no friction in this situation, because the surface of the painting is not in contact with any other surfaces, as far as we can tell.

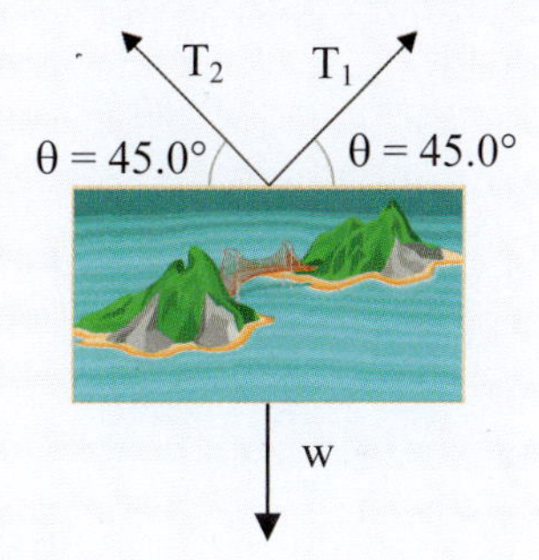

Notice that I called the tension from one string "T_1" while the other string's tension is "T_2." This is because we have to treat each string separately. We don't know yet whether or not the strings pull with equal force, so for right now, we have to assume that they do not.

So what are we faced with here? We know the weight of the picture, because it was given as 50.0 Newtons. We know that this is weight and not mass because the unit "Newton" can only mean weight. Thus, we know one of the three forces, and we have to find the other two. How can we do that? Well, the picture is in static equilibrium, so the sum of the forces on it must equal zero. Since the forces are not aligned in one dimension, however, we cannot simply add them together. We have to use our knowledge of (oh no!) two-dimensional vectors.

When we add two-dimensional vectors together, we split them up into their x- and y-components and add those components separately. Let's start with the tension in the first string. The diagram tells us that the string is directed 45° from the picture, so we can get the x- and y-components using Equations (3.6) and (3.7):

$$T_{1x} = T_1 \cdot \cos(45.0^\circ) = 0.707 \cdot T_1$$

$$T_{1y} = T_1 \cdot \sin(45.0^\circ) = 0.707 \cdot T_1$$

When we analyze the force of tension from the second string, however, we need to be a little careful. Do we use 45° for the vector's angle? No we don't. Remember, in order to keep the signs correct, we *always* define vector angle counterclockwise from the *positive x-axis*. Well, because of where the angle is drawn, the angle given plus the angle from the positive x-axis must add up to 180°. If the angle given is 45.0°, the angle from the positive x-axis is 135.0°. Now we can split the vector up into its components:

$$T_{2x} = T_2 \cdot \cos(135.0^\circ) = -0.7071 \cdot T_2$$

$$T_{2y} = T_2 \cdot \sin(135.0^\circ) = 0.7071 \cdot T_2$$

Finally, we can split the weight vector into its x- and y-components. In order to do this, however, we need its angle. Since it is pointing down, it is aligned along the negative y axis, which is 270.0° away from the positive x-axis. Now realize that we will not call weight negative here. This is because when you split two-dimensional vectors into components, the definition of angle takes care of the negative and positive signs. Thus, all of our forces will be positive until they are multiplied by the appropriate trigonometric functions. At *that* point, direction is automatically taken into account by the trigonometric function:

$$w_x = (50.0 \text{ Newtons}) \cdot \cos(270.0^\circ) = 0$$

$$w_y = (50.0 \text{ Newtons}) \cdot \sin(270.0^\circ) = -50.0 \text{ Newtons}$$

See what I mean? We end up with a negative weight because the sin of 270.0° is negative. Always remember, the only time you put in the negatives and positives on your own is when you are not using the trigonometric functions to split a vector into its components. Under those circumstances, you must think for yourself about the direction of the vector. When you use the trigonometric functions, however, they do the thinking for you, as long as you have defined the angle properly.

Now that we have all vectors split up into their components, we can add those components together and set them equal to zero. This will allow us to solve the problem. Let's start with the x-components. According to the definition of static equilibrium, the sum of the forces in the x-direction must equal zero:

$$\Sigma F_x = 0$$

$$0.707 \cdot T_1 - 0.7071 \cdot T_2 + 0 = 0$$

$$T_2 = (1.00) \cdot T_1$$

Okay, this doesn't give us our answer, but it helps out. This tells us that the tensions in the string are the same. Now let's look at the y-components:

$$\Sigma F_y = 0$$

$$0.707 \cdot T_1 + 0.7071 \cdot T_2 - 50.0\,\text{Newtons} = 0$$

Now you might think that we can't solve this equation because it has two unknowns in it. However, the first equation we got (from the x-components) tells us that $T_2 = (1.00) \cdot T_1$. I can use that fact to substitute "$(1.00) \cdot T_1$" in for "T_2" in the equation above:

$$0.707 \cdot T_1 + 0.7071 \cdot [(1.00) \cdot T_1] - 50.0\,\text{Newtons} = 0$$

$$1.414 \cdot T_1 = 50.0\text{ Newtons}$$

$$T_1 = 35.4\text{ Newtons}$$

Since we have already learned that $T_2 = (1.00) \cdot T_1$, this tells us that the tension in each string is 35.4 Newtons.

Let's think about this result for a moment. If only one string were holding up the picture, what would its tension be? We learned in the first part of this module that it would be the same as the weight of the object it is holding up. Thus, if only one string were holding the picture up, the tension in it would be 50.0 Newtons. Since there are two strings, the tension in each is less than that. Why isn't the tension in each string half of what it would be with one string? Well, it would be, if both strings were pulling straight up. Since they are pulling at an angle, however, they also pull in the horizontal direction a little. One string pulls one way in the horizontal, while the other pulls the other way. Since the strings must fight both gravity *and each other in the horizontal dimension*, they must have more tension than they would if they were both pulling straight up.

This seems like a long, drawn-out problem, but that's mostly because there are two-dimensional vectors involved and, as a result, we have to split them up into their components, resulting in two separate problems: one for the x-dimension and one for the y-dimension. If you look at the thinking involved, however, these problems aren't all that hard. First, you identify all of the forces in the problem, just as we did in the problems from the previous chapter. After that, you split the two-dimensional force vectors into their components and add the components in each dimension separately. This will result in two equations. You can use those two equations to solve the problem. Let's see how these problems can get a little harder.

EXAMPLE 6.2

Illustrations from Arts and Letters Express

A 5.00-kg flower pot is suspended from two strings as shown below. What is the tension in each string?

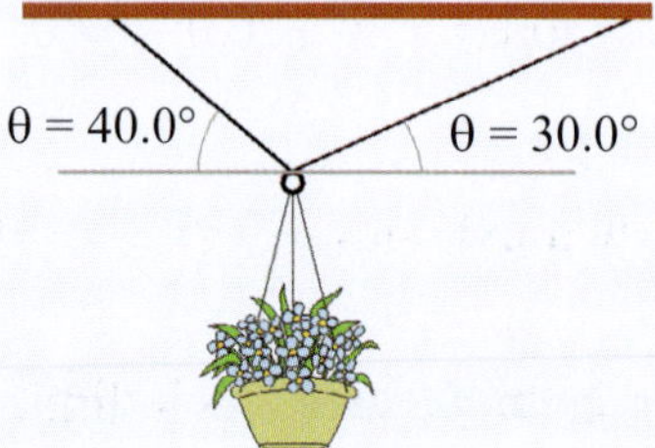

The first thing that we must do is figure out all of the forces that are acting on the object of interest: the flower pot. Gravity is pulling it down, of course, and the tension in each of the strings is fighting gravity's pull. These forces, however, are not pointed in the same direction. Gravity is pulling straight down while the tension in each string must be directed along that string. In the end, then, the forces look like this:

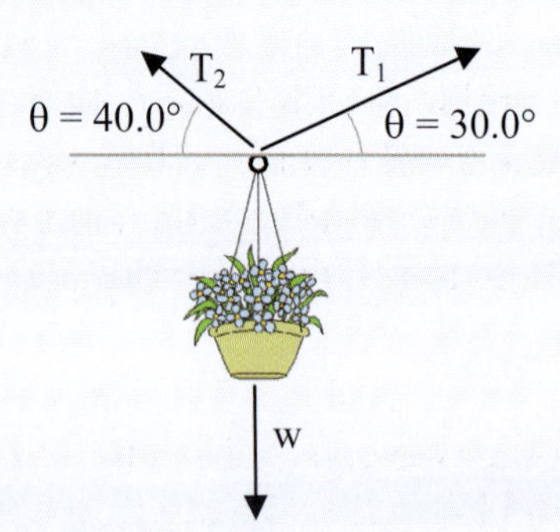

Our final goal is to add these forces together and set them equal to zero, because the flowerpot is in static equilibrium. Since the forces are not all directed in the same dimension, we must first break them down into their components before we can add them together. When breaking vectors down into their components, however, we must be very careful about how we define the angle. Angle must always be defined counterclockwise from the positive x-axis. This is, indeed, the way the angle is defined for T_1, so finding its components is an easy task.

$$T_{1x} = T_1 \cdot \cos(30.0^\circ) = 0.866 \cdot T_1$$

$$T_{1y} = T_1 \cdot \sin(30.0^\circ) = 0.500 \cdot T_1$$

To find the components of T_2, we must redefine the angle. The angle currently is 40.0° from the negative x-axis. In order to define it from the positive x-axis, the angle becomes 140.0°. That's the value we will use when we break T_2 into its components:

$$T_{2x} = T_2 \cdot \cos(140.0^\circ) = -0.7660 \cdot T_2$$

$$T_{2y} = T_2 \cdot \sin(140.0^\circ) = 0.6428 \cdot T_2$$

Finally, the weight vector must be split into its components. What is the weight equal to? According to Equation (5.2), the weight is mass times the acceleration due to gravity. The mass is given (5.00 kg) and, since it is the SI unit for mass, we will use the SI unit for the acceleration due to gravity, 9.8 m/sec^2. This makes the weight 49 Newtons. Also, since the vector points straight down, its angle is 270.0°. Remember, we don't worry about the sign of this vector. The trigonometric functions will take care of that.

$$w_x = (49 \text{ Newtons}) \cdot \cos(270.0^\circ) = 0$$

$$w_y = (49 \text{ Newtons}) \cdot \sin(270.0^\circ) = -49 \text{ Newtons}$$

Now we are finally ready to add the forces together in each dimension and set them equal to zero. Let's start in the x-dimension:

$$0.866 \cdot T_1 + -0.7660 \cdot T_2 = 0$$

This equation isn't all that useful, however. We have two unknowns. Well, let's see if the y-dimension gives us any ideas:

$$0.500 \cdot T_1 + 0.6428 \cdot T_2 - 49 \text{ Newtons} = 0$$

Hmm. It seems that we have the same problem with this equation. There are two unknowns. How do we solve the problem? We use a mathematical technique that you should have learned in algebra. When you have two separate equations, you can solve them simultaneously for two unknowns. If this example doesn't jog your memory on how to use this technique, go back to your algebra book and review it.

The way you simultaneously solve two equations is to take the first equation and solve for one variable in terms of the other. For example, I will take my first equation and solve for T_2:

$$0.866 \cdot T_1 - 0.7660 \cdot T_2 = 0$$

$$0.866 \cdot T_1 = 0.7660 \cdot T_2$$

$$T_2 = \frac{0.866}{0.7660} \cdot T_1 = 1.13 \cdot T_1$$

This doesn't tell us exactly what T_2 is, but it does allow us to do something. Since $1.13 \cdot T_1$ is the same as T_2, we can substitute it in for T_2 in our second equation:

$$0.500 \cdot T_1 + 0.6428 \cdot T_2 - 49 \text{ Newtons} = 0$$

$$0.500 \cdot T_1 + 0.6428 \cdot (1.13 \cdot T_1) - 49 \text{ Newtons} = 0$$

Now we've reduced this to an equation with only one variable. We can therefore solve for T_1:

$$0.500 \cdot T_1 + 0.6428 \cdot (1.13 \cdot T_1) - 49 \text{ Newtons} = 0$$

$$1.226 \cdot T_1 = 49 \text{ Newtons}$$

$$T_1 = 4.0 \times 10^1 \text{ Newtons}$$

This tells us that the tension in the first string is 4.0×10^1 Newtons. Note that we have to use scientific notation with this answer, since we need two significant figures. Had we reported the answer as 40 Newtons, there would only be one significant figure.

What about the second string? Well, we have an equation above that has T_2 in terms of T_1. Now that we have T_1, we can use that equation to determine T_2:

$$T_2 = 1.13 \cdot T_1 = 1.13 \cdot (4.0 \times 10^1 \text{ Newtons}) = 45 \text{ Newtons}$$

Thus, the tension in the second string is 45 Newtons. Notice once again that the strings have more total tension than the weight of the flowerpot. That's because not only are the strings pulling against gravity (in the y-dimension), but they are also pulling against each other (in the x-dimension).

Do you see why this kind of problem is more difficult that those we covered in the previous module? It's not because the *physics* is harder. We're still just splitting vectors up into their components, adding those components together, and setting them equal to zero. The difficulty comes in the math. We had to simultaneously solve two equations for two unknowns. Unfortunately, you will have to get used to this. Many problems in physics involve that technique for getting to the answer. Thus, if you are a little shaky at simultaneously solving algebraic equations, you had better do some mathematics review before you proceed any further!

ON YOUR OWN

6.1 Two farmers are trying to pull a mule into a barn. The mule does not want to go, so she is pulling back against the farmers with a force of 2.00×10^3 Newtons, including the effects of friction. If the farmers are pulling as illustrated below and the mule does not move, what is the tension in their ropes?

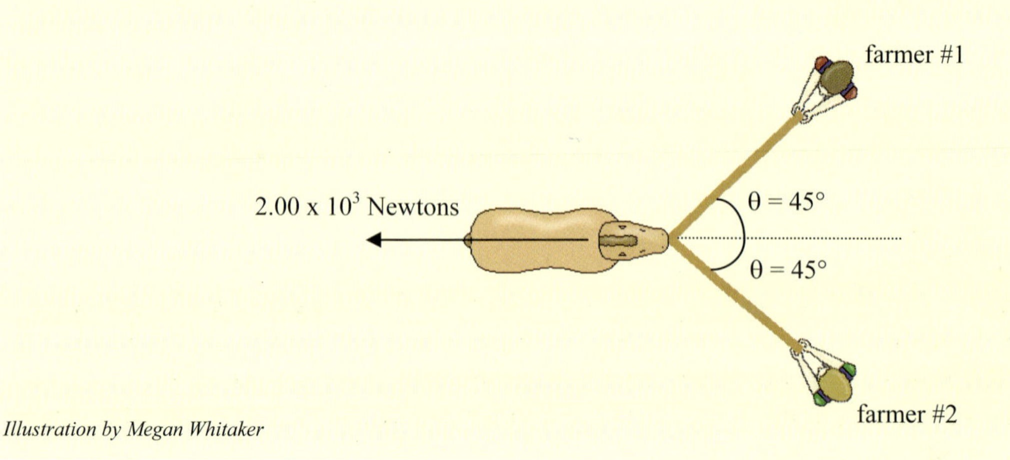

Illustration by Megan Whitaker

6.2 A 112.1-kg sign is hung from two strings as shown below. What is the tension in each string?

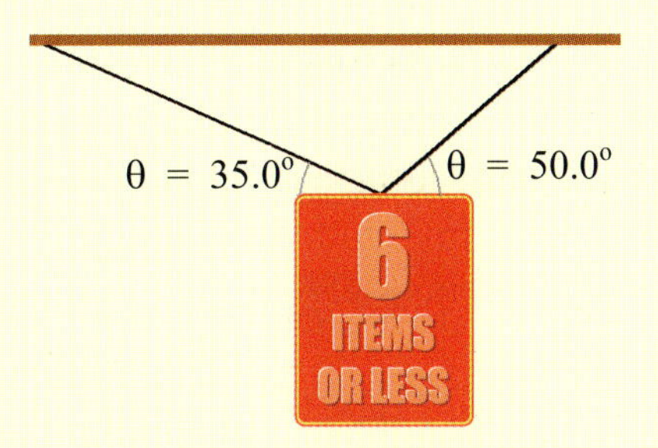

Illustration copyright GifArt.com

6.3 A fisherman is displaying his prize catch by hanging it as illustrated below. The horizontal string in the hanging apparatus is weak and will only be able to stand a tension of 13.0 pounds before it breaks. What is the maximum weight for a fish to hang on this system?

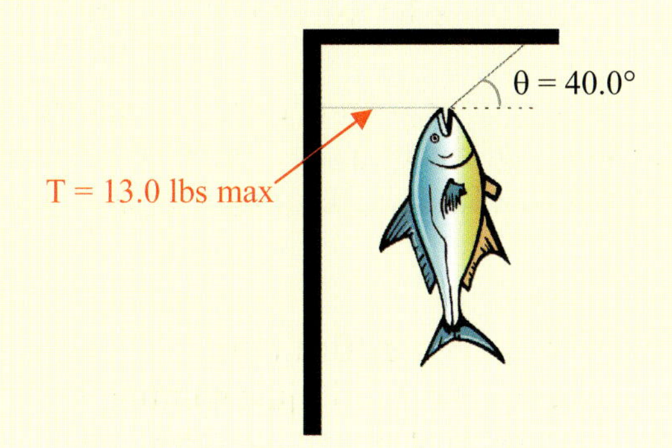

Illustration copyright GifArt.com

Translational Equilibrium and Measuring Weight

In the previous module, I told you how a scale measures your weight by measuring the normal force that the scale exerts to counteract the gravitational pull of the earth. In general, this works well. However, there are situations in which a scale will not give you an accurate reading for your weight. Perform the following experiment to see what I mean.

EXPERIMENT 6.1
Measuring Acceleration in an Elevator

Note: A sample set of calculations is available in the solutions and tests guide. It is with the solutions to the practice problems.

Supplies:

- A building with an elevator (In other words, it's time for a field trip!)
- A bathroom scale (A digital scale may not work. It needs to be a scale that responds quickly.)

1. Get into the elevator and put the scale on the floor of the elevator.
2. While the elevator is still, read your weight from the scale.
3. Push the button for the topmost floor, and watch the scale from the time the elevator starts moving until the time that it stops. Record the maximum and minimum weights that the scale reads. As the elevator stops, the scale reading might vary wildly because the elevator shakes. If so, ignore it.
4. Once the elevator has reached the top floor, push the button for the ground floor and take the same readings again as the elevator travels down.
5. Don't forget to say hello to anyone who is traveling with you on the elevator!

What did you see in the experiment? Well, you should have seen that when the elevator began to travel upwards, the reading from the scale *increased* to something heavier than your actual weight. Most likely, the scale returned to reading your actual weight while it was in transit to the top floor, but it began reading a *lighter* weight when the elevator slowed down to stop at the top floor. Once the elevator had stopped, however, the scale read your actual weight again. On the way down, you should have seen the opposite effect: the scale initially read lighter than your actual weight; then it read your actual weight; then it read more than your actual weight; and then, once the elevator had stopped, it read your actual weight again.

How can we understand these results? Remember, the scale reads the normal force with which it pushes. Study the figure below to see how that can change:

Illustration by Megan Whitaker

FIGURE 6.1
Measuring Weight on an Elevator

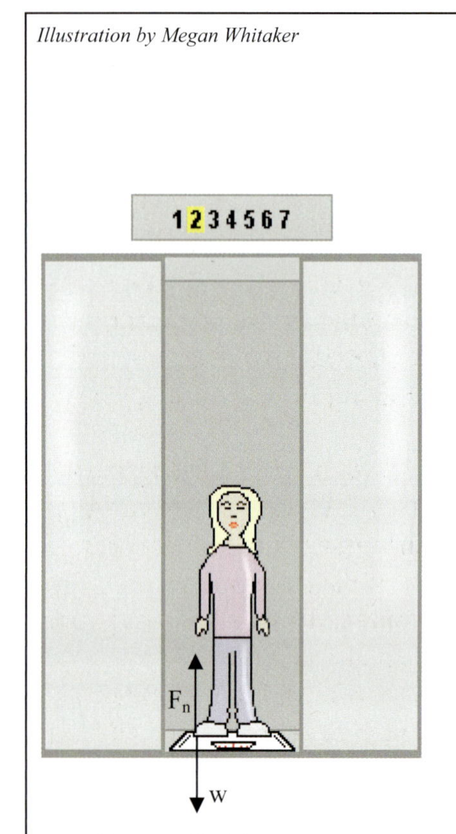

Consider the forces acting on the girl in the elevator. Gravity pulls her down (w) while the scale pushes her up with a normal force (F_n). The sum of those two forces equals the mass of the girl times her acceleration:

$$F_n - w = m\mathbf{a} \qquad (6.1)$$

Now remember, the scale reads F_n. If we solve this equation for F_n, we get:

$$F_n = m\mathbf{a} + w \qquad (6.2)$$

When the elevator is still (or moving with a constant velocity), the girl's acceleration is zero. As a result, $F_n = w$, and the scale reads the girl's weight. However, once the elevator begins to accelerate, the girl accelerates with it. Thus, the normal force (what the scale reads) is the *sum* of the girl's weight plus the girl's mass times her acceleration. Notice that since I defined weight as negative, that defines the downward direction as negative. Thus, if the elevator accelerates downwards, "**a**" must be negative.

So you see that because the scale actually measures the normal force it exerts rather than the weight of the object upon it, a scale must be in translational equilibrium in order for the scale to read the weight accurately.

Let's apply this knowledge to the experiment. When the elevator in the experiment was stopped, its acceleration and your acceleration were zero. As a result, the normal force that the scale exerted on you was equal to the gravitational force being exerted on you, and the scale measured your weight. However, once the elevator began accelerating upward, the reading on the scale increased. Why? Look at Equation (6.2). Since the elevator was accelerating upward, the "**a**" in Equation (6.2) was positive. Thus, the normal force exerted by the scale was equal to your weight *plus* your mass times the elevator's acceleration. Thus, the scale read a value greater than your weight.

As the elevator traveled upwards, it probably reached a point at which it was traveling at a constant velocity. At that point, your acceleration was equal to zero (you were in dynamic translational equilibrium), so the scale went back to measuring your true weight. However, as the elevator neared the top floor, it had to slow down. At that point, the elevator and you were accelerating, but your acceleration was negative (opposed to the upward motion of the elevator). Thus, the "**a**" in Equation (6.2) was negative, so the scale read a weight *lower* than your actual weight. Once the elevator did finally stop, your acceleration was once again zero, so the scale measured your weight.

Now consider what happened in the next part of the experiment. When the elevator began to travel down, it accelerated in that direction. Thus, the "**a**" in Equation (6.2) was negative, and the scale read something lower than your weight. Once again, the elevator probably started traveling at a constant velocity at some point, so your acceleration reduced to zero, and the scale measured your actual weight. However, as the elevator neared the bottom floor, it had to start accelerating upwards to slow the descent of the elevator. As a result, "**a**" in Equation (6.2) was positive, and the scale read something larger than your actual weight.

As I warned you in the experiment, at the very end of each journey, the reading might have varied wildly up and down for a moment. Why was that? Well, as an elevator gets ready to stop, it makes sure that its floor is level with the building's floor. Sometimes, it stops short of that level or passes it, so it has to make quick adjustments to get to the proper level. This results in a quick succession of up and down accelerations, which causes the scale to vary wildly until the elevator reaches the proper level. I told you to ignore these readings because the acceleration changes so quickly in those situations that it is too hard to get any kind of accurate reading from the scale.

Now that you know why the scale's reading changed during the course of your elevator trips, use Equation (6.2) to calculate the maximum accelerations experienced by the elevator. You can use the maximum weight in the first part of the experiment to calculate the maximum upward acceleration as the elevator rose. You can use the minimum weight from the first part of the experiment to calculate the maximum downward acceleration as the elevator stopped at the top floor. Similarly, use the minimum weight in the second part of the experiment to calculate the maximum downward acceleration the elevator had as it moved to the bottom floor, and use the maximum weight in the second part of the experiment to calculate the maximum upward acceleration of the elevator as it stopped at the ground floor.

Now please realize as you make these calculations that you need to know both your weight (w) and your mass (m). The reading of the scale when the elevator was stationary is your weight. To get your mass, you will have to take that reading and divide by the acceleration due to gravity. Since the scale probably read your weight in pounds, use 32 ft/sec^2 for the acceleration due to gravity. That will make your mass come out in slugs. As is the case with all experiments that require calculations, sample calculations for this experiment can be found in the Solutions and Tests guide, after the practice problem solutions for this module.

In the end, then, a scale will measure the weight of an object only when it is in translational equilibrium. This is just another consequence of Newton's Second Law. Of course, this consequence applies to situations other than a scale. Try the following "On Your Own" problem to make sure you really understand this kind of situation.

Illustration by Megan Whitaker

ON YOUR OWN

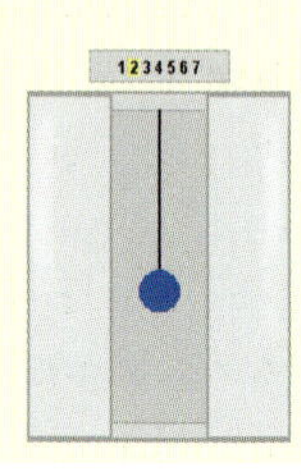

6.4 A spherical mass (50.0 kg) hangs from a string in an elevator as illustrated in the drawing to the right. If the tension in the string is 460 Newtons, what is the acceleration (magnitude *and* direction) of the elevator? From this information, is it possible to tell whether the elevator is moving up or down? Why or why not?

Rotational Motion and Torque

In the course so far, we have concentrated on only one type of motion: motion in a straight line. We have either dealt with situations that involve motion along a line, motion that can be split up into a few straight-line segments, or motion that can be "easily" analyzed by breaking it down into its horizontal and vertical components. In all of those situations, we reduced the problem down to straight-line motion. Now it's time to move beyond that.

When an object moves from one point to another (like we have studied up to this point), physicists say that it is in **translational motion**. On the other hand, suppose an object moves in a circular or elliptical pattern so that it keeps passing the same point over and over again. In that case, we say that the object is in **rotational motion**. The distinction between these two types of motion is important.

Translational motion - Motion from one point to another which does not involve repeatedly passing the same point in space

Rotational motion - Motion around a central axis such that an object could repeatedly pass the same point in space relative to that axis

To understand our definition of rotational motion, you need to understand what an **axis of rotation** is. Suppose you have a toy airplane tied to a string. If you twirl that airplane around your head, it is clearly in rotational motion. The plane passes the same point over and over again as it moves in a circle around your head. When studying this motion, we would say that the point at which your hand holds the string is the axis of rotation, because it is at the center of the rotational motion. If an object is moving in a circle, we say that the center of the circle is the axis of rotation. In other words, whatever point in space is at the center of the rotational motion is what we would call the axis of rotation.

You need to understand the difference between rotational and translational motion, so I want to spend a little bit of time on it. Suppose you walk from your house to a friend's house. What kind of motion are you exhibiting? Translational motion, of course. What about a baton being twirled by a drum majorette? The tips of the baton keep passing the same point in space over and over again. The place where she holds the baton is the axis of rotation, so the baton is clearly in rotational motion. It is important to realize, however, that the two types of motion are not mutually exclusive.

Think about the tires on a car that is traveling down the street. The tires themselves do not repeatedly pass the same point in space. Instead, they travel from one place to another. Thus, they are in translational motion. At the same time, however, the tires are spinning around their centers. The axis of rotation is the center of the tire, and the surface of the tire keeps passing the same point (relative to that axis) over and over again. Thus, the tires are also in rotational motion. This is a classic case of an object having both rotational and translational motion.

Why is the distinction between translational and rotational motion important? Well, the means by which translational acceleration occurs is different than the means by which rotational acceleration occurs. What causes translational acceleration? You should know this by now! A force will cause translational acceleration. If you want to increase the velocity of an object, for example, you give it a push or a pull. The force of the push or the pull causes acceleration, which alters the object's velocity. Rotational acceleration isn't that simple, however. Perform the following experiment to understand why.

Illustrations by Megan Whitaker

EXPERIMENT 6.2

What Causes Rotational Acceleration?

Supplies:

- Safety goggles
- A large nut and bolt. The nut should be at least 1/4 inch in diameter and the bolt should fit the nut.
- A board or something that has a hole big enough for the bolt to fit through
- Either one wrench that has a long handle or three wrenches that are of different lengths but nevertheless all fit the nut
- Another wrench or a pair of pliers that can hold onto the head of the bolt

1. Push the bolt through the hole in the board and then screw the nut onto the bolt with your finger. As you are screwing the nut on, think about the motion. The nut is clearly exhibiting rotational motion. What is the axis of rotation? Well, the nut is spinning around the center of the bolt. Thus, that's the axis of rotation.
2. Tighten the nut (not too tight) with your fingers; do not use the wrenches yet.
3. Now, let's suppose you want to unscrew the nut. In order to do this, you will have to induce the nut into rotational motion. Try to do this by pushing on the center of the bolt as if you are trying to push the bolt back out of the hole:

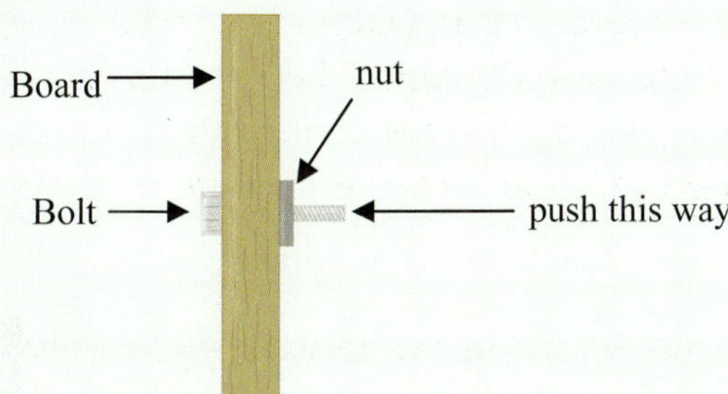

   Does anything happen? Of course not! The nut holds the bolt in place and will not let it move in response to the force that you are applying.
4. Try placing your thumb on top of the nut, directly above the bolt. Now push straight down on the nut as illustrated below:

Push straight down.

5. Does this cause the nut to unscrew? No, it does not. You don't want the nut to move down; you want it to move in a circle.
6. Unscrew the nut the way that you know it should be unscrewed. Notice what you had to do. You had to apply a force on the edge of the nut, perpendicular to the bolt, as illustrated below:

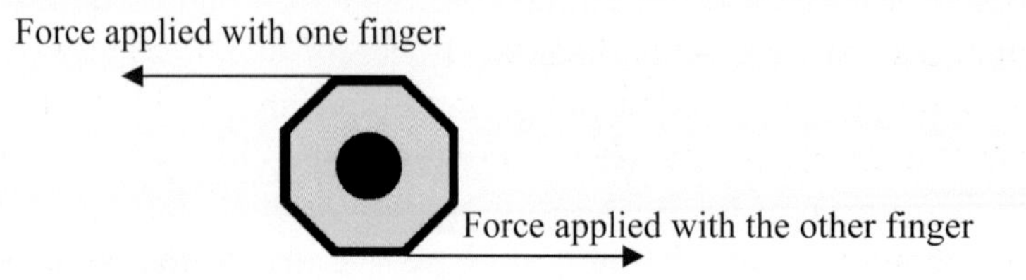

That's what it takes to make the nut unscrew.

7. Screw the nut back onto the bolt until it is once again finger tight.
8. Use your pliers to hold onto the head of the bolt. Next, take your largest wrench, fit it on the nut, and grasp the wrench at the edge of the handle. Now, tighten the nut as hard as you can. Be sure to make it as tight as possible, always grasping the wrench at the end of the handle.
9. Once the nut is tight, take the wrench off of the nut and fit your smallest wrench onto the nut. If you are just using one long wrench, stop grasping the wrench at the end of the handle, and grasp it as close to the nut as possible. Now try to unscrew the nut. You probably can't.
10. Try using the wrench with the medium-sized handle. If you are using one long wrench, grasp it about halfway down the handle. Try unscrewing the nut again. If you really tightened it as hard as you could at first, you *still* won't be able to unscrew the nut.
11. Use your largest wrench again and grasp it at the end of the handle to unscrew the nut. You should be able to unscrew it this time.
12. Clean up your mess.

What did we learn from this experiment? Well, the purpose of the experiment was to take a nut that was not rotating and make it start rotating around the center of the bolt. We learned that in order to do this, we couldn't apply the force just anywhere. When you pushed on the center of the bolt (the axis of rotation), the nut would not unscrew. Compare this to translational motion. When you are trying to push a box, you can push on it anywhere, and the box will accelerate in the direction of the applied force. Not so in rotational motion. If you apply a force at the axis of rotation, you will not cause rotational acceleration. In the case of the nut, there was no rotational motion, and there continued to be no rotational motion after the force was applied. Thus, there was no change in the rotational motion (no rotational acceleration).

We also learned that the direction of the applied force makes a big difference. When you pushed straight down on the nut, it did not unscrew. On the other hand, when you applied a force at the edge of the nut and perpendicular to the bolt, the nut unscrewed. Once again, compare this to translational motion. If you are trying to move a box, the direction of the force does not affect whether or not the box accelerates, as long as the box is not obstructed in any way. The direction of the force does affect the *direction* of the acceleration, but no matter which way you push on the box, it will accelerate.

To cause rotational acceleration, however, a force must be directed perpendicular to an imaginary line from the axis of rotation to the point of the applied force. This imaginary line is called the **lever arm**.

Lever arm - The length of an imaginary line drawn from the axis of rotation to the point at which the force is being applied

Finally, we learned that rotational acceleration is easier to induce the farther the applied force is away from the axis of rotation. After all, the only way you could unscrew the nut was to use your largest wrench and grasp it at the end of the handle. When you used smaller wrenches (or grasped the wrench closer to the nut), you could not get the nut to move. That's because the applied force is more effective at inducing rotational acceleration the farther it is from the axis of rotation.

How do we sum up all of these facts? We do it with the concept of **torque**.

Torque - The tendency of a force to cause rotational acceleration. The magnitude of the torque is equal to the length of the lever arm times the component of the force that is applied perpendicular to it.

The definition of torque can be restated mathematically as follows:

$$\tau = F_{\perp} \cdot r \tag{6.3}$$

In this equation, the "τ" is the lowercase Greek letter "tau." It is used in physics to represent torque. The symbol "$F_{\perp}$" represents the component of the applied force that is perpendicular to the lever arm, and "r" stands for the length of the lever arm. If you're not quite sure what all of this means, look at the diagram shown below.

Illustration by Megan Whitaker

FIGURE 6.2
The Definition of Torque

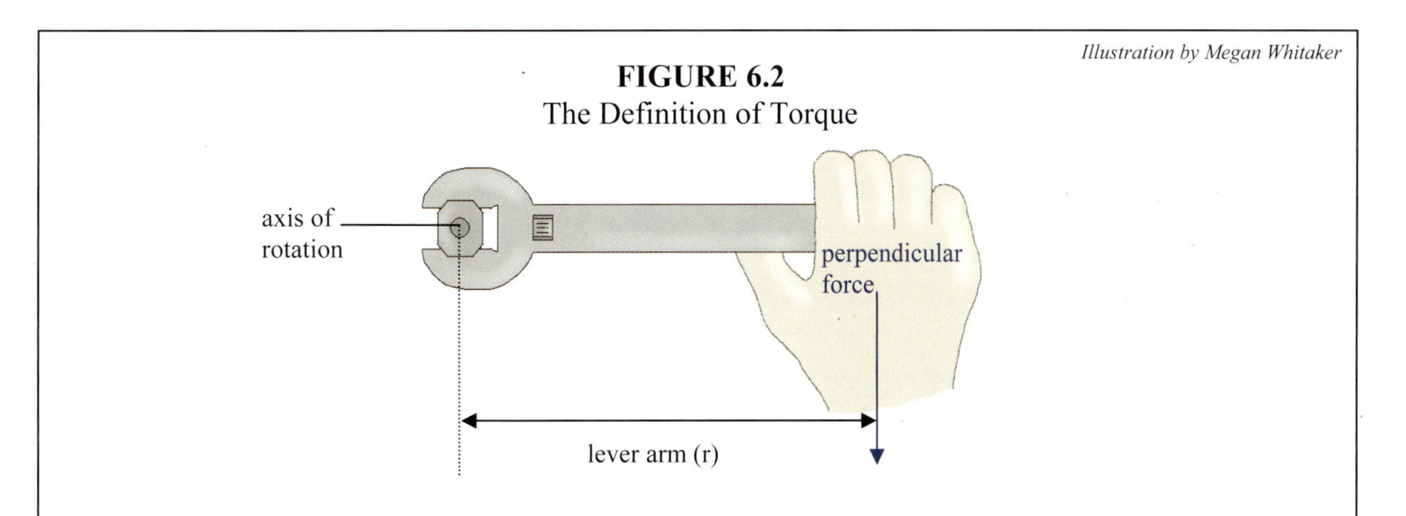

When determining the torque applied in a situation, you need to look at several things. First, you need to find the axis of rotation. Once you've found that, you need to determine how far away from that axis the force is being applied. That distance is the lever arm, r. Also, you have to determine whether or not the force being applied is perpendicular to the lever arm. If it is not, then you must break it down into its components and find out what part of the force is directed perpendicular to the lever arm. Only that component of the force counts towards the torque. Once you have determined these things, you take the perpendicular component of the force and multiply it by the length of the lever arm. The resulting number is the magnitude of the torque.

Well, now that we know a little bit about torque, what is it all about? Actually, torque is the rotational equivalent of force. When we are trying to cause *translational* acceleration, we use a force. The more force that's applied, the more translational acceleration will occur. When trying to analyze *rotational* acceleration, however, we do not look at force. Instead, we examine the torque. The more torque that is applied to an object, the more rotational acceleration will occur. This is an important concept:

The magnitude of the force applied to an object determines the amount of translational acceleration that will occur. The amount of torque applied to an object determines the amount of rotational acceleration that will occur.

What unit will the magnitude of the torque have? In order to get torque, you take the force applied (Newtons) and multiply by the distance of the lever arm (meters). In the end, then, the SI unit for torque is Newton·m. See how we calculate torque by examining the following example problem.

EXAMPLE 6.3

A plumber is trying to unscrew one pipe from another with a 6.00-inch (15.2 cm) wrench. He pulls straight down on the wrench with all of his might (a force of 653 Newtons), but he cannot unscrew the pipe. He pulls a 12.0-inch (30.5 cm) wrench out and tries again, using the same force. This time the pipe unscrews quite readily. Examine the two torques applied in order to see why. In each case, assume that the plumber grasps the wrench at the very end and that the applied force is perpendicular to the lever arm. Also, the length of a wrench is given from the axis of rotation of the object in its grasp, so the length of each wrench is the length of the lever arm.

When trying to get something to unscrew, you are clearly trying to start rotational motion. Thus, you need rotational acceleration, and you therefore must examine the torque applied. The plumber applied a force straight down, which we will assume was perpendicular to the wrench. Also, we are told that the length of each wrench is, in fact, the length of the lever arm. Thus, we simply take the magnitude of the applied force times the length of the wrench in order to get torque. The units should be Newtons and meters, however, so we need to convert from cm to m. I will assume that you know how to do that by now. Using Equation (6.3), then, the torque applied in the first case is:

$$\tau = F_{\perp} \cdot r = (653 \text{ Newtons}) \cdot (0.152 \text{ m}) = 99.3 \text{ Newton} \cdot \text{m}$$

By itself, this number doesn't mean a whole lot to us. However, if we calculate the torque applied in the second case, we will be able to compare one torque to another, and that will make a little more sense. In the second case, the torque applied is:

$$\tau = F_{\perp} \cdot r = (653 \text{ Newtons}) \cdot (0.305 \text{ m}) = 199 \text{ Newton} \cdot \text{m}$$

Now we can see why the pipe unscrewed in the second case. Notice that although the force remained the same, the torque applied to the pipe was twice as much when the plumber used the longer wrench. With twice as much torque, you have a better chance of overcoming friction and starting the rotational motion. Thus, the pipe moved readily with the longer wrench because the torque exerted was much greater.

If you look in a plumber's tool box, you will see some really long wrenches. This is because pipes are notoriously hard to unscrew, and therefore the lever arm of pipe wrenches must be long in order to provide a torque sufficient to unscrew them. Make sure you understand the concept of torque by solving the following "On Your Own" problem.

ON YOUR OWN

6.5 A father and daughter are trying to change a tire on the car. The father attempts to unscrew the lug nuts with a small (12.3 cm) wrench and is capable of applying a 612-Newton perpendicular force. The daughter, on the other hand, can only deliver a 214-Newton perpendicular force. If she chooses a longer wrench (41.1 cm), is she or her father more likely to get the nuts unscrewed?

Now before we move on and really apply the concept of torque, I want to make you aware of one thing. As I have said before, the only force that can be counted as a part of torque is that component of the force which is perpendicular to the lever arm. Thus, in the picture below:

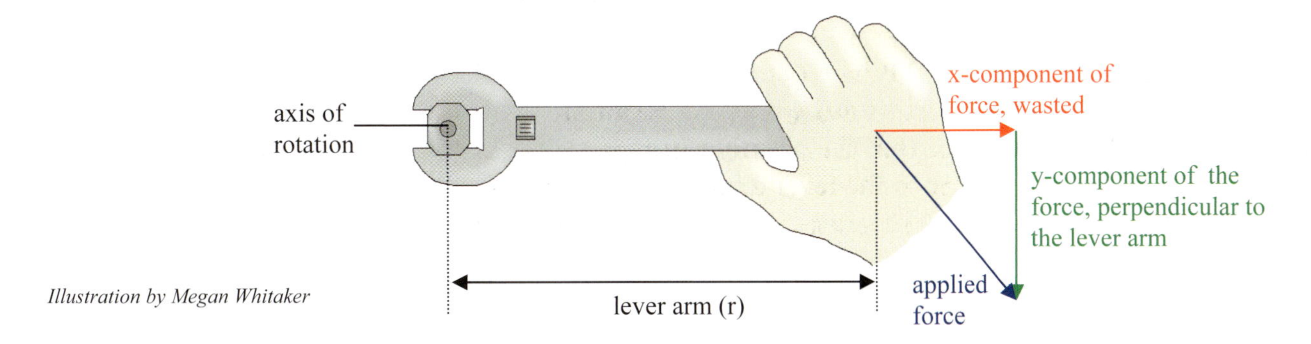

The person pulling on the wrench is wasting some of the applied force. Since only the component of the force that is perpendicular to the lever arm can contribute to the torque, only the y-component of this force can be counted as part of the torque. The x-component of the force is wasted.

Now although it is important for you to see this effect, do not worry that you will ever be required to calculate the component of the force that is perpendicular to the lever arm when you are doing torque problems in this course. For all problems that involve torque in this course, we will always assume that the applied force is perpendicular to the lever arm. Since that is the most efficient way to generate a torque anyway, it makes the most sense to work with forces that are always applied perpendicular to the lever arm. Even though you don't have to worry about this concept while *solving problems*, don't forget it. I guarantee you a conceptual question on the test which involves noticing that only the component of a force perpendicular to the lever arm counts towards the applied torque!

Rotational Equilibrium

In the first part of this module, we studied translational equilibrium, which governs any situation in which the sum of the forces on an object equals zero. We will now apply the same principles to the study of rotational motion. Suppose you had a rod lying on the ground:

If the rod is not moving, we say that it is in static equilibrium. This is because there are two forces working on it, each canceling out the other. Gravity pulls the rod down, while the ground applies an equal force (the normal force) up. These two forces cancel out, so the sum of the forces is zero. Thus, the rod is in *translational* equilibrium. Suppose, however, you place a fulcrum underneath the rod, but not at its center:

Will the rod stay balanced like that? Of course not. It will tilt and fall over until its edge hits the ground:

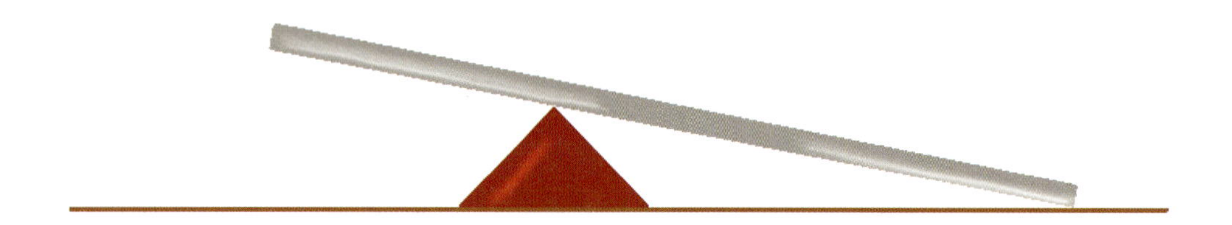

Why can't the rod stay balanced? To answer that question, think about what kind of motion is involved when the rod tilts and falls. Is it translational motion? No. The rod is still in translational equilibrium when the fulcrum is placed underneath it. This is because the same two forces are still at play. Gravity still pulls down on the rod and the fulcrum still exerts an equal normal force that opposes gravity. The rod, when it tilts and falls, is not experiencing translational motion. It is experiencing *rotational* motion. The fulcrum is the rod's axis of rotation, and the rod rotates around that point. This rotation is, of course, cut short when the rod hits the ground. Nevertheless, while the rod is tilting and falling, it is experiencing *rotational* motion.

The rod cannot stay balanced because it is not in **rotational equilibrium**. Why isn't it in rotational equilibrium? Well, the sum of the *forces* acting on the rod does, indeed, equal zero, but the sum of the *torques* acting on the rod is not zero. When the fulcrum is placed under the rod, the rod has the opportunity to rotate around the axis of rotation formed by the fulcrum. If the fulcrum is not placed at the center of the rod, gravity has a longer lever arm on one side of the rod than it does on the other. As a result, the side of the rod farthest from the fulcrum has more torque acting on it than the other side of the rod. Thus, the rod tilts (starts rotational motion) towards the side that has more torque.

In addition to translational equilibrium, then, we have to learn about rotational equilibrium:

Rotational equilibrium - The state in which the sum of the torques acting on an object is zero

All illustrations on this page are by Megan Whitaker.

In other words, if it is possible for an object to experience rotational acceleration, it will, unless the sum of the torques acting on it turns out to be zero. I hope you see the correlation here. To achieve *translational equilibrium*, the sum of the *forces* acting on an object must be zero. To achieve *rotational equilibrium*, however, the sum of the *torques* acting on an object must be zero. Study the next example to see how we apply the concept of torque to rotational equilibrium.

EXAMPLE 6.4

Illustration by Megan Whitaker

Two children are playing on a seesaw. They get a little tired rocking back and forth , so they decide to balance the seesaw. One child weighs 201 Newtons and sits 1.30 m from the center of the seesaw. The other finds that he must be 1.00 m from the center of the seesaw to achieve balance. What does he weigh?

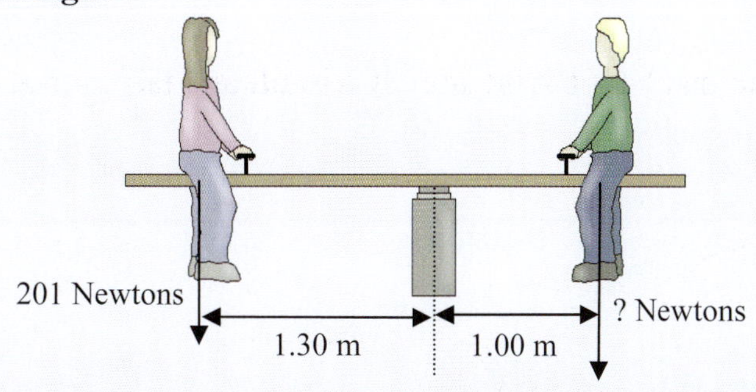

To achieve rotational equilibrium, the sum of the torques on an object must be zero. The seesaw is the object in question, because that's what the kids are trying to balance. In order to determine torque, we must first determine the axis of rotation. Where is that? Well, the seesaw wants to rotate around the fulcrum, so that's the axis of rotation. To calculate torque, we must take the force applied perpendicular to the lever arm and multiply it by the length of the lever arm. Well, what forces come into play? That's where this gets a little tricky.

First, we will always ignore the force of gravity acting on the seesaw alone. We do this because compared to the torques exerted by the children, the torque exerted by the seesaw using its own weight is very small. If you take advanced physics, you will concentrate much more on the effects of the weight of the seesaw, but for right now we will ignore it. Thus, two forces we need to consider are the weights of the two children. The weight of one child is given, and the weight of the other is what we need to calculate. We are given the distances from the axis of rotation to the point at which the weight of each child is pushing down on the seesaw, so to calculate torque, we simply multiply the weight by the distance.

There is one very important thing we have to consider, though. We must determine the *direction* of each torque. Torque is a vector, just like force. Equation (6.3) does not use boldface type to indicate this, because it is used to get the *magnitude* of the torque. Just like we do with force when we do not have trigonometric functions, we must determine the direction of the torque ourselves. We do this with positive and negative signs, using the following rule: torques that cause clockwise motion are negative while torques that cause counterclockwise motion are positive.

Since we have both the weight and the lever arm for the force exerted on the seesaw by the girl, let's start with her. If the boy were not on the seesaw, which way would it rotate with just the girl sitting on it? Without the boy on the other side, the seesaw would tilt towards the side that has the girl on it. That kind of motion is counterclockwise, because the rotation of the seesaw as it falls goes opposite the motion of a clock. Thus, we would say that the torque exerted by the girl is positive:

$$\tau = (201 \text{ Newtons})\cdot(1.30 \text{ m}) = 261 \text{ Newton}\cdot\text{m}$$

Now what about the boy? Without the girl on the seesaw, the boy's torque would cause it to tilt the other way, making clockwise motion. Thus, the boy's torque is negative. We don't know what the value for the torque is, but we could come up with an equation for it:

$$\tau = -(\text{w})\cdot(1.00 \text{ m})$$

where "w" represents the boy's weight. The condition for balancing is that the seesaw must be in rotational equilibrium. For that to be the case, the sum of the torques must be zero. Thus, we can add these two torques and set them equal to zero:

$$261 \text{ Newton}\cdot\text{m} - (\text{w})\cdot(1.00 \text{ m}) = 0$$

$$\text{w} = \frac{261 \text{ Newton}\cdot\cancel{\text{m}}}{1.00\ \cancel{\text{m}}} = 261 \text{ Newtons}$$

Thus, the weight of the boy is 261 Newtons.

This probably seemed like a complicated problem, but let's go back and dissect it. First, we ignored the weight of the seesaw. This makes the problem easier because there is one less force to consider. In these rotational equilibrium problems, we will *always* ignore the weight of the thing that is rotating. Second, we had to determine the axis of rotation so that we could determine the lever arm lengths for the remaining forces. That wasn't so bad. Next, we had to calculate the torques acting on the seesaw, making sure to assign the positive or negative signs based on this rule:

Torques that cause clockwise motion are considered negative torques, while torques that cause counterclockwise motion are considered positive torques.

Once we assigned the positive and negative signs, then we added the torques and set them equal to zero. This allowed us to solve for the unknown in the problem. Now try this yourself.

Illustration by Megan Whitaker

ON YOUR OWN

6.6 Two children play on a seesaw. One weighs 221 Newtons and sits 1.12 meters away from the seesaw's fulcrum. If the other weighs 171 Newtons, how far from the fulcrum must she sit in order to balance the seesaw?

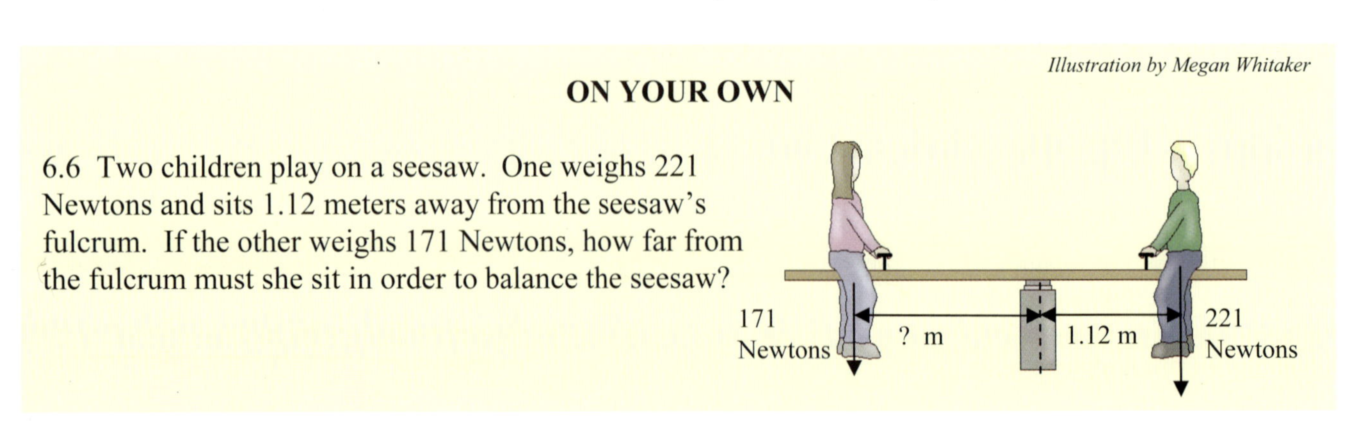

6.7 In a physics experiment, a bar is balanced on a fulcrum. A 3.0-kg mass is placed on the bar, 45 cm away from the fulcrum. If another mass is placed on the other side of the fulcrum, 75 cm away, what is its mass?

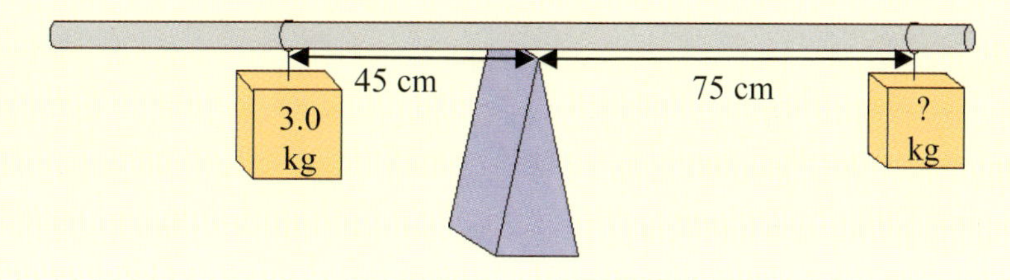

Objects on an Inclined Surface

When dealing with Newton's Second Law, we often run into situations in which objects are resting on a surface, but that surface is not horizontal. Instead, it is inclined. It turns out that to really analyze situations such as these, there is a trick you can employ in order to make things much easier. You see, in the problems that we did above, we always separated the vectors into their horizontal and vertical components. Well, when things are hanging straight down or resting on a horizontal surface, this makes sense. However, what if an object is on an inclined surface? Consider, for example, a skier going down a hill:

Illustration copyright GifArt.com

Does it still make sense to work in the horizontal and vertical dimensions? No, it doesn't.

As we learned back in Module #3, we can split vectors up into *any* two components, as long as those components are perpendicular to each other. Well, when dealing with objects on an inclined surface, the best dimensions to consider are the ones in which something actually happens. For example, the skier in the illustration above does not move horizontally. Instead, he moves parallel to the inclined surface. That dimension is physically interesting, so it makes sense to look at the dimension that runs parallel to the surface of the incline. Also, if we are interested in friction, we have to determine the normal force. By definition, the normal force is perpendicular to the surface that the skier is on. Thus, the dimension perpendicular to the surface is also physically interesting.

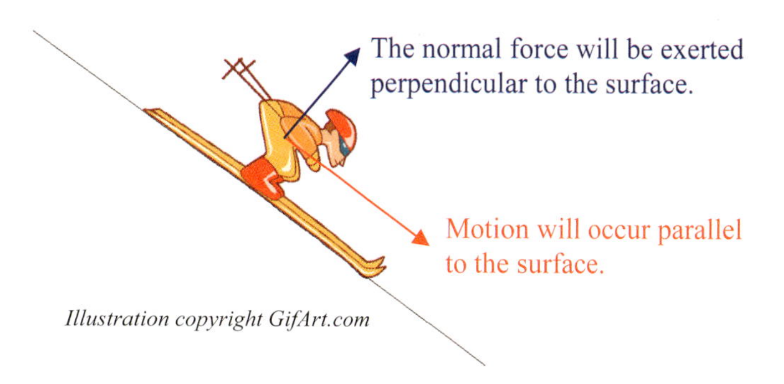

Illustration copyright GifArt.com

Notice that these two dimensions are perpendicular to each other. Thus, we can split up any vectors into components that exist in these two dimensions. As a result, the components that we split the vectors into will actually be physically useful.

Well, now that we know *what* to do, *how* do we do it? How do we split vectors into components that are parallel and perpendicular to an inclined surface? Actually, the geometry required to do that is rather complicated. The result of that complicated geometry is rather simple, however. Rather than doing the complicated geometry, then, I will simply give you the rather easy result. Suppose I wanted to calculate the components of the gravitational force that run parallel and perpendicular to the inclined surface. Figure 6.3 shows you how to do that.

FIGURE 6.3

Determining the Parallel and Perpendicular Components of Gravity on an Inclined Surface

If the incline of a surface is determined by the angle θ, then the components of an object's weight that run parallel and perpendicular to the surface are:

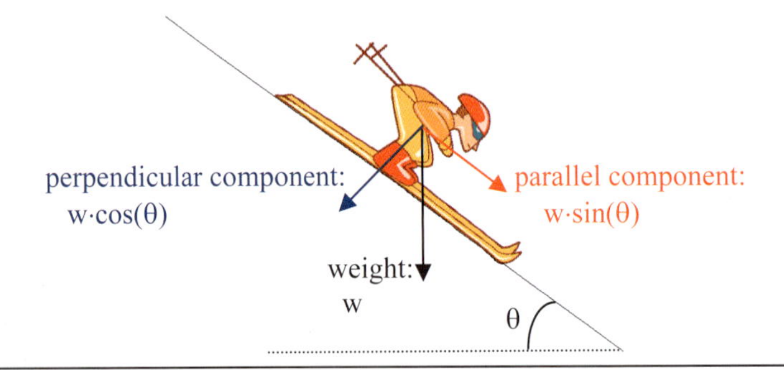

In then end, then, when studying an object that rests on an inclined surface, it is best to split the force of gravity up into components that run parallel to and perpendicular to the surface, because those are the dimensions in which all of the action takes place. To do this, use the following rule:

The component of an object's gravitational force that runs parallel to an inclined surface is equal to the weight of the object times the sine of the incline angle. The gravitational component perpendicular to the surface is equal to the weight of the object times the cosine of the angle.

So how in the world does all of this apply to real situations? Perform the following experiment to see.

EXPERIMENT 6.3

Measuring a Coefficient of Static Friction

Note: A sample set of calculations is available in the solutions and tests guide. It is with the solutions to the practice problems.

Supplies:

- Safety goggles
- A board that is at least half a meter long
- A meter stick
- A block of wood or metal that can fit on the surface of the board. If you don't have a block of metal or wood, try a small cardboard box.

1. Lay the board flat on the floor and measure its length.
2. Put the block on top of the board, somewhere near the middle.
3. Slowly raise one end of the board while keeping the other end fixed. Watch the block carefully. The very instant that you see it start to fall, stop raising the end of the board, but do not lower it at all. Hold it right where it was when the block started to move.
4. Measure the height of the end of the board from the floor.
5. Do this five times, each time recording the height that you raised the board to in order to get the block to move.
6. Average your results.
7. Clean up your mess.

Believe it or not, you just measured the coefficient of static friction that exists between the board and the block. Why? Well, first let's look at what you did. You raised one end of the board so that it formed an incline:

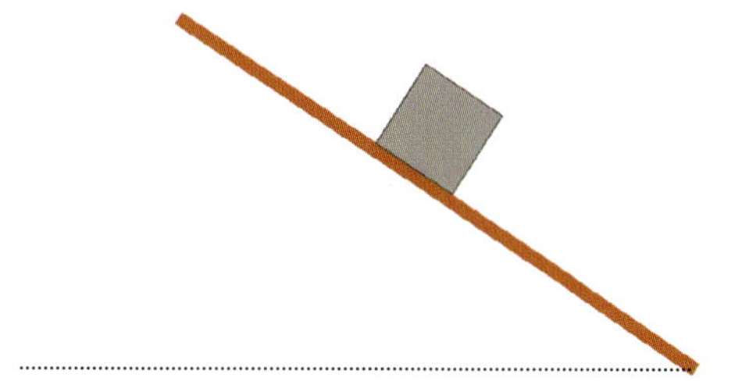

When the board is tilted like that, the force of gravity on the block can be split into two components, one perpendicular to the incline, and one parallel to the incline.

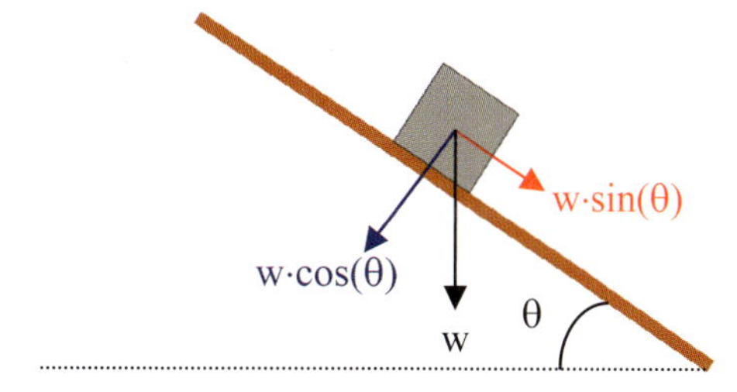

Other than the weight of the block, the only forces acting on the block are friction (which is trying to oppose the motion of the block) and the normal force (which is exerted perpendicular to the surface of the board). This means that the force diagram looks like this:

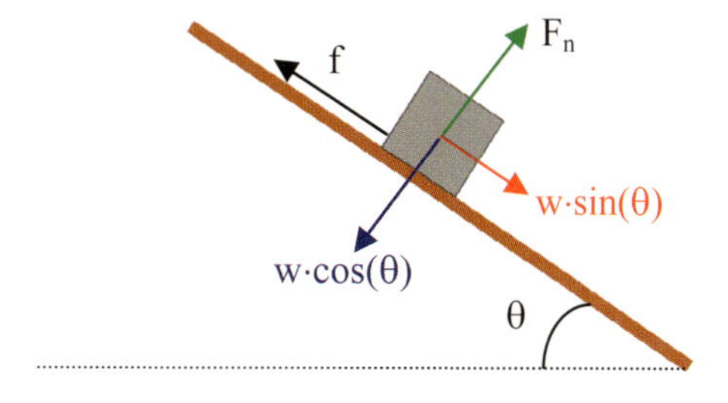

Now, in order to analyze the motion of the block, we are going to have to get a handle on the frictional force. To do this, we need to calculate the normal force and then use Equation (5.3) to determine the friction. What is the normal force? Well, to find that out, we examine the dimension that is perpendicular to the surface. In that dimension, the normal force pushes upwards (so we will call it positive), and the perpendicular component of the gravitational force is pushing downwards (so we will call it negative). There is no motion in that dimension, so those two forces must add up to zero:

$$F_n + -w \cdot \cos(\theta) = 0$$

$$F_n = w \cdot \cos(\theta)$$

This, then, allows us to determine friction. In this experiment, we are examining when the block just starts moving, so the friction we see the block overcoming is static friction. Thus, the frictional force is:

$$f = \mu_s \cdot F_n$$

$$f = \mu_s \cdot w \cdot \cos(\theta)$$

Now that we have an equation for the frictional force, we can go back to the dimension parallel to the incline and sum up the forces. We will say that the parallel component of the block's weight pushes in the positive direction, which means that the frictional force is negative. Now, while the box stayed still, the sum of the forces had to be zero. Thus, at the point that you stopped raising the board, the sum of the forces had just grown to a hair above zero. To a good approximation, however, we can say that if you stopped the board soon enough, you were just at the point where the maximum static frictional force was at play, and the forces just barely added to zero. Thus, at the point where you stopped the board:

$$w \cdot \sin(\theta) - f = 0$$

$$w \cdot \sin(\theta) - \mu_s \cdot w \cdot \cos(\theta) = 0$$

Let's take this equation and solve for μ_s:

$$\mu_s = \frac{\cancel{w} \cdot \sin(\theta)}{\cancel{w} \cdot \cos(\theta)} = \tan(\theta)$$

In case you are wondering, $\sin(\theta)$ divided by $\cos(\theta)$ equals $\tan(\theta)$. In the end, then, this really complicated discussion leads to a very simple fact. At the point where you stopped the board, you can calculate the coefficient of static friction by simply taking the tangent of the incline angle.

You might be wondering if you measured the incline angle in this experiment. Yes, you did. You measured the length of the board, which is the hypotenuse of the incline. You also measured the height of the side of the incline opposite the angle. Thus, if you take the height that you measured and divide by the length of the board, you have the sine of the incline angle. Mathematically:

$$\theta = \sin^{-1}\left(\frac{\text{height}}{\text{length}}\right)$$

So use this equation and your data to calculate the angle of the incline. Once you have done that, you can take the tangent of that angle. This is the coefficient of static friction. It should be less than one.

So what did you learn from all of this? Well, one thing you probably learned is that problems involving an inclined surface are really tough! That's certainly true. I hope, however, that you learned a little something else. When objects are on an incline, using the parallel and perpendicular components of the gravitational force is the best way to solve the problem. When we do that, we end up with two dimensions that have all of the interesting action in them. Study the following example and solve the "On Your Own" problems that come after in order to better understand motion on an inclined surface.

EXAMPLE 6.5

A 35-Newton block slides down a board that is inclined at 30.0°. If the coefficient of kinetic friction between the block and the board is 0.12, what is the block's acceleration?

To solve any Newton's Second Law problem, we must first determine what forces are at play. First, the force of gravity acting on the block has a parallel and perpendicular component. In addition, friction opposes the motion, and there is a normal force exerted by the board. The force diagram, then, looks like this:

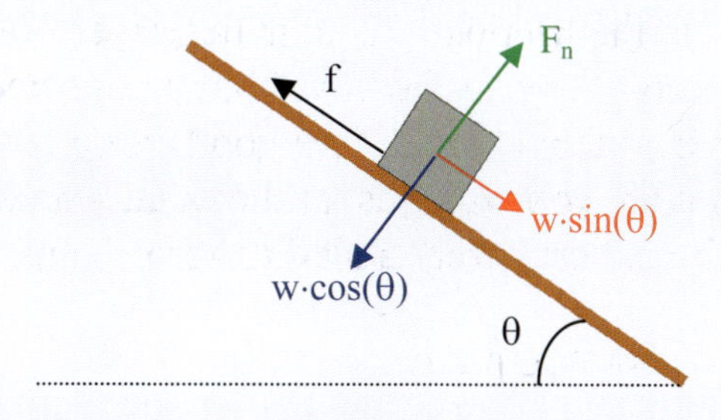

To analyze the motion, we will have to determine friction, which requires knowing the normal force, so we might as well start there.

In the perpendicular dimension, the normal force is positive and the perpendicular component of the gravity is negative. Now don't get confused here. I told you earlier that you do not use positives and negatives when you have the trigonometric functions. That's only true when you are working with the normal horizontal and vertical dimensions. When we work with the dimensions that are perpendicular and parallel to the incline, we need to use positive and negative signs. Thus, in the perpendicular dimension, the forces add up to zero as follows:

$$F_n + -w \cdot \cos(\theta) = 0$$

$$F_n = w \cdot \cos(\theta)$$

$$F_n = (35 \text{ Newtons}) \cdot \cos(30.0°) = 3.0 \times 10^1 \text{ Newtons}$$

Notice that since we needed two significant figures in our answer, we had to use scientific notation in order to make the zero significant. Now we can use this and Equation (5.3) to calculate the frictional force:

$$f = \mu_k \cdot F_n$$

$$f = (0.12)\cdot(3.0 \times 10^1 \text{ Newtons}) = 3.6 \text{ Newtons}$$

Now that we have the frictional force, we can concentrate on the parallel dimension. In that dimension, the block is accelerating, and we are asked to determine the magnitude of that acceleration. Since the block is accelerating in the dimension parallel to the incline, the sum of the forces parallel to the incline is not zero. Instead, the sum of the forces equals mass times acceleration. We were not given the mass of the block; we were given its weight. Thus, we need to convert to mass. I assume you can do that. The answer is 3.6 kg.

Let's keep the same direction definitions we used in the analysis of the experiment. We will say that motion down the board is positive, so the component of the gravitational force parallel to the board is positive while friction is negative:

$$w \cdot \sin(\theta) - f = m\mathbf{a}$$

$$(35 \text{ Newtons}) \cdot \sin(30.0^\circ) - 3.6 \text{ Newtons} = (3.6 \text{ kg}) \cdot \mathbf{a}$$

$$\mathbf{a} = \frac{14 \text{ Newtons}}{3.6 \text{ kg}} = \frac{14 \frac{\cancel{\text{kg}} \cdot \text{m}}{\text{sec}^2}}{3.6 \cancel{\text{kg}}} = 3.9 \frac{\text{m}}{\text{sec}^2}$$

So the block falls <u>down the incline (remember, we defined down the incline as positive) at an acceleration of 3.9 m/sec^2</u>.

ON YOUR OWN

6.8 A 45-kg block slides down an incline that is tilted at an angle of 45 degrees. If the coefficient of kinetic friction between the incline and the block is 0.30, what is the acceleration of the block?

6.9 A 10.0-kg block slides down an incline at a constant velocity. If the incline is tilted at an angle of 22 degrees, what is the coefficient of kinetic friction between the block and the incline?

Applying Newton's Second Law to More Than One Object at a Time

So far, when we have applied Newton's Second Law to a situation, we have ended up applying it to only one object. For example, when we had the problem with two farmers pulling on a mule, there were, in principle, three objects in the problem: two farmers and one mule. In the end, however, we were only interested in the motion of the mule, so we applied Newton's Second Law only once, to the mule. In this last section of the module, we will learn how to apply Newton's Second Law to *multiple* objects in a single problem. The best way to see how this is done is to study an example.

EXAMPLE 6.6

Two toy cars (3.5 kg and 4.5 kg) are tied together with a piece of string. If a child pulls the lead car with a 3.2 Newton force, what acceleration will the two cars experience? What will be the tension in the string? You may ignore friction in this problem.

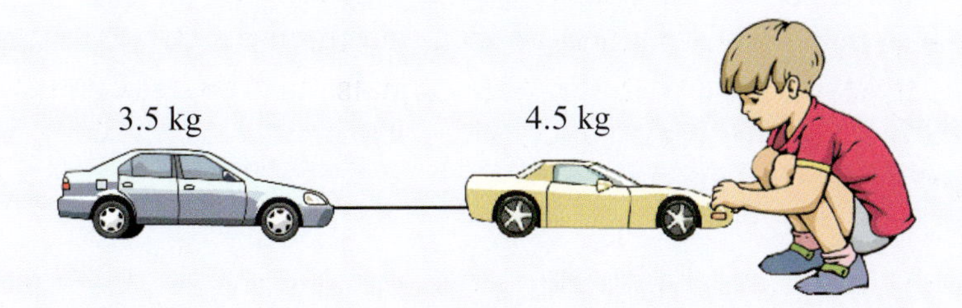

As with all Newton's Second Law problems, we must first analyze all forces acting on the objects of interest. Since, in the end, we are only interested in calculating the acceleration of the two cars, those are the objects of interest. The lead car has the force of the child pulling on it. It also has gravity pulling it down and the normal force pushing it back up. There is one more force, however. The string that attaches the lead car to the rear car actually pulls back on the lead car. Thus, the tension in the string also acts on the lead car. The rear car has gravity and the normal force working on it, but it also has the tension in the string pulling on it. Thus, the forces look something like this:

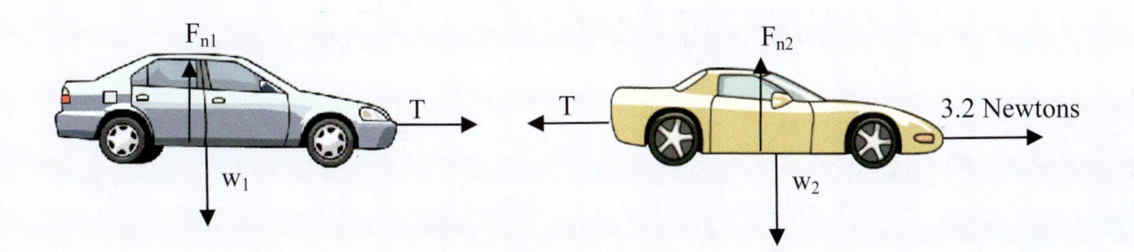

This may look like a *really* complicated force diagram, but it's not bad if we look at each car individually. Since the rear car has fewer forces acting on it, let's start with the rear car:

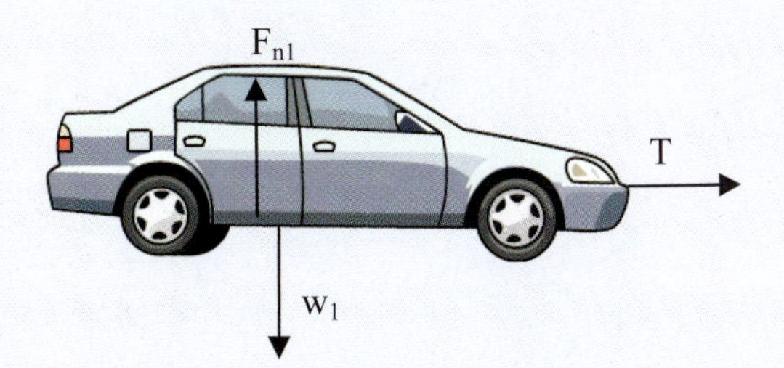

According to Newton's Law, we can sum up all of the forces in a dimension and set them equal to the mass of the object times the object's acceleration. Well, the only motion that occurs is in the horizontal dimension, so let's start there. In the horizontal dimension, the only force present is the tension in the string. Since the tension in the string is pulling to the right, it is a positive force. Thus, Newton's Second Law for the horizontal dimension would say:

$$\Sigma \mathbf{F} = \mathrm{ma}$$

$$T = (3.5\ \text{kg})\cdot\mathbf{a}$$

If we knew the magnitude of the tension in the string, we would be able to solve for acceleration and finish the problem. Unfortunately, we don't know how much tension is in the string. In fact, that's one of the things we need to solve for. So we need to move on.

What do we look at next? Do we go to the vertical dimension, or do we move on to the next car? Well, in these kinds of problems, what do we usually learn from the vertical dimension? We usually learn the normal force so that we can calculate friction. In this problem, however, we are told to ignore friction. Thus, rather than looking at the vertical dimension, let's move to the other car.

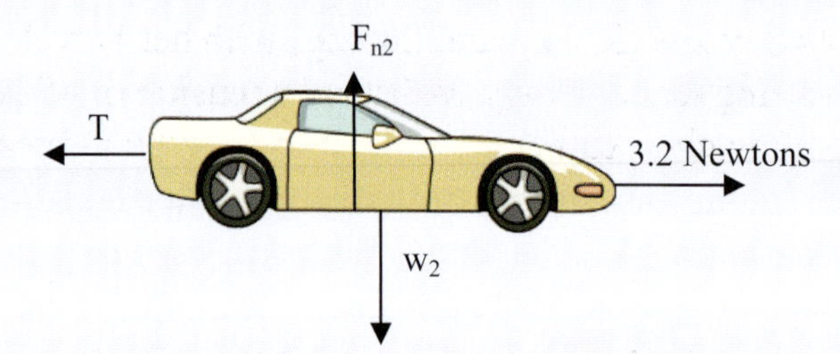

We can apply Newton's Second Law to the horizontal dimension in this case as well. Since the 3.2 Newton force pulls to the right, it is positive. The tension, however, pulls to the left and is therefore negative:

$$\Sigma\mathbf{F} = m\mathbf{a}$$

$$3.2\ \text{Newtons} + -T = (4.5\ \text{kg})\cdot\mathbf{a}$$

We seem to be stuck here as well. We once again have one equation and two unknowns. Wait a minute, though. The other car's equation ended up telling us that T = (3.5 kg)·**a**. If we use that expression for T in this equation, we are left with only one unknown:

$$3.2\ \text{Newtons} + -(3.5)\cdot\mathbf{a} = (4.5\ \text{kg})\cdot\mathbf{a}$$

Now we can use this equation to solve for **a**:

$$3.2\ \text{Newtons} - (3.5\,\text{kg})\cdot\mathbf{a} = (4.5\ \text{kg})\cdot\mathbf{a}$$

$$\mathbf{a} = \frac{3.2\ \text{Newtons}}{8.0\ \text{kg}} = \frac{3.2\ \frac{\cancel{\text{kg}}\cdot\text{m}}{\text{sec}^2}}{8.0\ \cancel{\text{kg}}} = 0.40\ \frac{\text{m}}{\text{sec}^2}$$

So we see that the cars will accelerate to the right (remember, since the acceleration is positive it means to the right) at <u>0.40 m/sec^2</u>. We're not done yet, however. We also have to calculate the tension in the string. How do we do that? Well, the first car's equation related the tension in the string and the acceleration. Let's use that relationship:

$$T = (3.5\ \text{kg})\cdot\mathbf{a} = (3.5\ \text{kg})\cdot(0.40\ \text{m/sec}^2) = 1.4\ \text{Newtons}$$

So we learn not only the acceleration (0.40 m/sec^2), but also the <u>tension in the string, 1.4 Newtons</u>.

This was a difficult problem, so it is worthwhile to go back and review exactly how we solved it. First, we determined all of the forces acting on the objects of interest in the problem. Then, in order to make things easier, we split up the objects and considered each one individually. Each object provided us with an equation. When we solved the equations simultaneously, we ended up getting the answers that we needed. That's how we apply Newton's Second Law to multiple objects. We consider each object separately. Each object gives us one equation. If we take all of the equations and solve them simultaneously, we can solve the problem. Try this on your own.

Illustration from Arts and Letters Express

ON YOUR OWN

6.10 A girl is dragging two boxes (322 kg and 245 kg) with her bicycle. The two boxes are attached to each other with a thin string that can only withstand a tension of 45 Newtons before it breaks. What is the maximum force that the child can exert with her bike without breaking the string? Ignore friction in this problem.

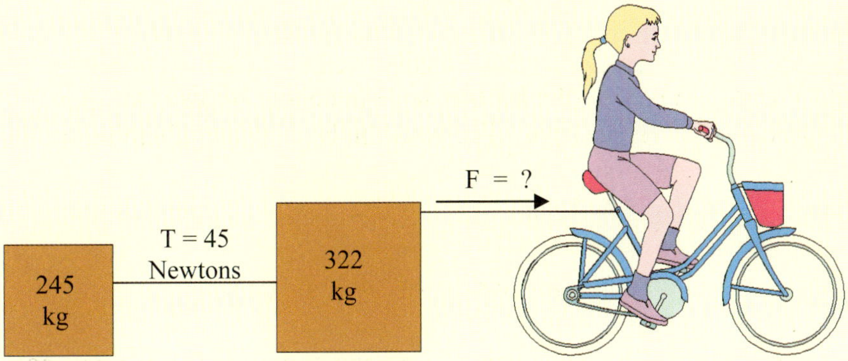

245 kg

T = 45 Newtons

322 kg

F = ?

If this module seemed really rough, don't sweat it too much. Many students find this material to be the hardest in the course. So, even if your grade on this module is a little low, never fear. Other modules are easier, and they will bring your grade up. If, on the other hand, you can really understand the material in this module, you probably have a knack for physics!

ANSWERS TO THE "ON YOUR OWN" PROBLEMS

6.1 First, we need to see all of the relevant forces in the problem. We are told that the mule is not moving, so that means it is in static equilibrium. We therefore must look at all of the forces acting on the mule. The mule pulls horizontally ($\theta = 180.0^{\circ}$) against the desired motion. Friction does as well. The problem states that friction and the mule combine for a total of 2.00×10^3 Newtons, so we will consider them one force. Also, each of the farmers is pulling on the mule. Since the farmers are not directly touching the mule, however, it does not feel *their* force. Instead, it feels the tension from the ropes. In the end, then, the force diagram looks like this:

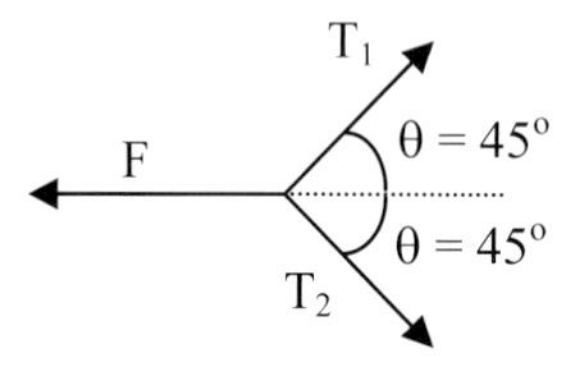

The first thing we are faced with doing is splitting all of the vectors up into their components. This should be second nature to you by now, so I will just list the equations after I remind you that the second rope's angle must be defined from the positive x-axis. Thus, the angle is 315°. Also, we don't need to put in any signs for these forces; the trigonometric functions will take care of that. The components of the vectors, then, come out to be:

$$T_{1x} = T_1 \cdot \cos(45^{\circ}) = 0.71 \cdot T_1$$

$$T_{1y} = T_1 \cdot \sin(45^{\circ}) = 0.71 \cdot T_1$$

$$T_{2x} = T_2 \cdot \cos(315^{\circ}) = 0.707 \cdot T_2$$

$$T_{2y} = T_2 \cdot \sin(315^{\circ}) = -0.707 \cdot T_2$$

$$F_x = (2.00 \times 10^3 \text{ Newtons}) \cdot \cos(180.0^{\circ}) = -2.00 \times 10^3 \text{ Newtons}$$

$$F_y = (2.00 \times 10^3 \text{ Newtons}) \cdot \sin(180.0^{\circ}) = 0$$

We add the forces in the x-dimension and set them equal to zero:

$$0.71 \cdot T_1 + 0.707 \cdot T_2 - 2.00 \times 10^3 \text{ Newtons} = 0$$

This equation doesn't tell us all that much, so let's see if the y-dimension can help us out any:

$$0.71 \cdot T_1 + -0.707 \cdot T_2 + 0 = 0$$

$$T_1 = (1.0) \cdot T_2$$

Based on this equation, we can substitute "(1.0)·T_1" in for "T_2" in our other equation:

$$0.71 \cdot T_1 + 0.707 \cdot T_1 - 2.00 \times 10^3 \text{ Newtons} = 0$$

$$1.42 \cdot T_1 = 2.00 \times 10^3 \text{ Newtons}$$

$$T_1 = 1.41 \times 10^3 \text{ Newtons}$$

Since $T_1 = (1.0) \cdot T_2$, we know that both ropes have a tension of 1.41×10^3 Newtons.

6.2 The first thing that we need to do is diagram the forces. It should be clear that the force diagram looks like this:

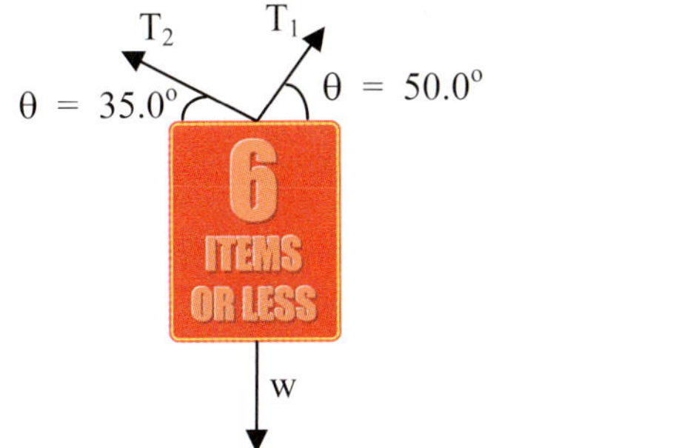

Illustration copyright GifArt.com

The problem gives us the mass (112.1 kg), which you should be able to easily convert to weight (1.1×10^3 Newtons). Now we just need to split all of the forces into their components. Once again, realize that T_2's angle must change to conform to the standard vector angle:

$$T_{1x} = T_1 \cdot \cos(50.0^\circ) = 0.643 \cdot T_1$$

$$T_{1y} = T_1 \cdot \sin(50.0^\circ) = 0.766 \cdot T_1$$

$$T_{2x} = T_2 \cdot \cos(145.0^\circ) = -0.8192 \cdot T_2$$

$$T_{2y} = T_2 \cdot \sin(145.0^\circ) = 0.5736 \cdot T_2$$

$$w_x = (1.1 \times 10^3 \text{ Newtons}) \cdot \cos(270.0^\circ) = 0$$

$$w_y = (1.1 \times 10^3 \text{ Newtons}) \cdot \sin(270.0^\circ) = -1.1 \times 10^3 \text{ Newtons}$$

Now we sum up the forces and set them equal to zero in each dimension. First, the x-dimension:

$$0.643 \cdot T_1 + -0.8192 \cdot T_2 = 0$$

Then the y-dimension:

$$0.766 \cdot T_1 + 0.5736 \cdot T_2 + -1.1 \times 10^3 \text{ Newtons} = 0$$

So we have a two-equation, two-unknown math problem here. I will use the first equation to solve for T_2 in terms of T_1:

$$0.643 \cdot T_1 + -0.8192 \cdot T_2 = 0$$

$$T_2 = 0.785 \cdot T_1$$

Now that I have an alternate expression for T_2, I will plug it into the second equation:

$$0.766 \cdot T_1 + 0.5736 \cdot T_2 + -1.1 \text{ x } 10^3 \text{ Newtons} = 0$$

$$0.766 \cdot T_1 + 0.5736 \cdot (0.785 \cdot T_1) + -1.1 \text{ x } 10^3 \text{ Newtons} = 0$$

$$1.216 \cdot T_1 = 1.1 \text{ x } 10^3 \text{ Newtons}$$

$$T_1 = 9.0 \text{ x } 10^2 \text{ Newtons}$$

This tells us that the tension in the first rope is $9.0 \text{ x } 10^2$ Newtons. Now that we have T_1, we can use our alternate expression for T_2 to calculate its value:

$$T_2 = 0.785 \cdot (9.0 \text{ x } 10^2 \text{ Newtons})$$

$$T_2 = 7.1 \text{ x } 10^2 \text{ Newtons}$$

Thus, the tension in the second rope is $7.1 \text{ x } 10^2$ Newtons.

6.3 The fish is in static equilibrium, and will remain so as long as the tension in the horizontal string does not exceed 13.0 pounds. We will calculate the maximum weight of the fish by setting that string's tension to 13.0 pounds and seeing what weight fish can remain in static equilibrium under that condition. If a fish is heavier than that, then the tension in the string will obviously be greater and the string will break. Thus, we must sum up all of the forces acting on the fish and set them equal to zero. The force diagram looks like this:

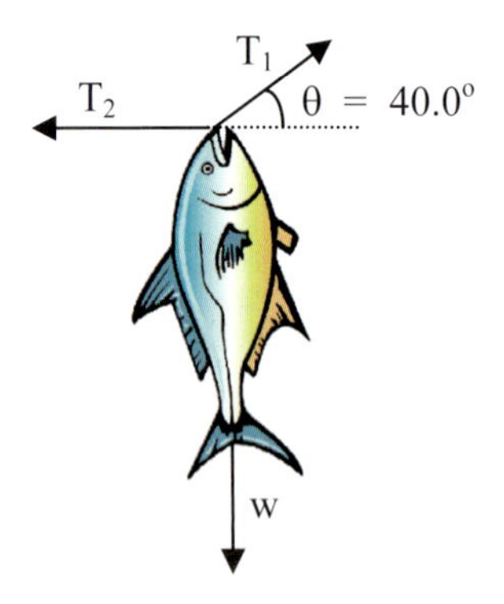

Illustration copyright GifArt.com

Now we need to get all of the forces broken down into their components:

$$T_{1x} = T_1 \cdot \cos(40.0^\circ) = 0.766 \cdot T_1$$

$$T_{1y} = T_1 \cdot \sin(40.0^\circ) = 0.643 \cdot T_1$$

$$T_{2x} = (13.0 \text{ lbs}) \cdot \cos(180.0^\circ) = -13.0 \text{ pounds}$$

$$T_{2y} = (13.0 \text{ lbs}) \cdot \sin(180.0^\circ) = 0$$

$$w_x = w \cdot \cos(270.0^\circ) = 0$$

$$w_y = w \cdot \sin(270.0^\circ) = -w$$

Summing up the forces in the x-dimension:

$$0.766 \cdot T_1 - 13.0 \text{ lbs} + 0 = 0$$

$$T_1 = 17.0 \text{ lbs}$$

So the tension in the other string will be 17.0 pounds if the tension in the horizontal string is 13.0 pounds. This isn't the answer, but it will help us immensely when we sum up the forces in the y-dimension:

$$0.643 \cdot T_1 + 0 + -w = 0$$

$$0.643 \cdot (17.0 \text{ lbs}) = w$$

$$w = 10.9 \text{ lbs}$$

As long as the fish is under 10.9 pounds, the horizontal string will not break.

6.4 There are only two forces acting on the mass: gravity pulls it down and the tension in the string pulls it up. Therefore, the sum of those two forces must equal the mass times the acceleration:

$$T - w = m\mathbf{a}$$

We know the tension and the mass; and since we know the mass, we can calculate the weight. That means **a** is the only variable:

$$460 \text{ Newtons} - (50.0 \text{ kg}) \cdot (9.8 \frac{\text{m}}{\text{sec}^2}) = (50.0 \text{ kg}) \cdot \mathbf{a}$$

$$\mathbf{a} = \frac{460 \text{ Newtons} - 490 \text{ Newtons}}{50.0 \text{ kg}} = \frac{-30 \frac{\cancel{\text{kg}} \cdot \text{m}}{\text{sec}^2}}{50.0 \cancel{\text{kg}}} = -0.6 \frac{\text{m}}{\text{sec}^2}$$

Since we defined down as negative, the elevator is accelerating downward at 0.6 m/sec^2. We cannot determine which way the elevator is moving, however. It might be moving upward. Since acceleration is negative, that means it would be slowing down. However, it might also be moving downward. If so, it is speeding up.

6.5 When you are unscrewing a nut, you are trying to produce rotational acceleration. Thus, the person applying the most torque is the one that will, most likely, unscrew the nut. The father can apply the most force, but we need to use Equation (6.3) to determine the torque that he can exert. To get the units right, however, we need to convert cm to meters. I will assume that you know how to do that by now:

$$\tau = F_{\perp} \cdot r$$

$$\tau = (612 \text{ Newtons}) \cdot (0.123 \text{ m}) = 75.3 \text{ Newton} \cdot \text{m}$$

So the father can apply 75.3 Newton·m of torque with his wrench. Now let's see how much torque the daughter can apply:

$$\tau = F_{\perp} \cdot r$$

$$\tau = (214 \text{ Newtons}) \cdot (0.411 \text{ m}) = 88.0 \text{ Newton} \cdot \text{m}$$

This tells us that despite the fact that the father is stronger, the daughter can apply more torque because the wrench that she is using has a longer lever arm. Thus, the daughter is more likely to get the nut unscrewed. Of course, the most effective scenario would be for the father (who can apply the most force) to use the daughter's wrench (which has the longest lever arm). That would be the best way to ensure that the nuts get unscrewed.

6.6 The first thing we need to do is determine the axis of rotation. That's easy. It's the fulcrum. Now that we have that out of the way, we can concentrate on the torques involved. As always, we will ignore the force of gravity due to the weight of the seesaw. That makes the problem easier. Now there are only two torques, one from each child. The boy weighs 221 Newtons and exerts that force 1.12 meters away from the fulcrum. If the girl were not present, the boy's torque would cause the seesaw to rotate clockwise. Thus, his torque is negative. We can use Equation (6.3) to determine the magnitude:

$$\tau = -(221 \text{ Newtons}) \cdot (1.12 \text{ m}) = -248 \text{ Newton} \cdot \text{m}$$

Now we can determine the girl's torque. Since her torque by itself would cause counterclockwise motion, it is positive. Since we do not know the length of the lever arm, we will call it "x."

$$\tau = (171 \text{ Newtons}) \cdot (x)$$

To achieve balance, these torques must sum up to zero:

$$-248 \text{ Newton} \cdot \text{m} + (171 \text{ Newtons}) \cdot x = 0$$

$$x = \frac{248 \cancel{\text{Newton}} \cdot \text{meters}}{171 \cancel{\text{Newtons}}} = 1.45 \text{ meters}$$

The girl, then, must sit 1.45 meters away from the fulcrum.

6.7 The problem gives us the mass of the objects. In order to calculate torque, however, we need force. Thus, we need to convert 3.0 kg into Newtons. I assume you know how to do this by now (the answer is 29 Newtons). Also, the distances are given in cm. They really need to be converted to meters for the calculation of torque. I will also assume that you know how to do that.

Now we can set up the problem. We ignore the weight of the bar, so the only torques come from the masses. The 3.0 kg mass would cause counterclockwise motion if the other mass were not present, so its torque is positive:

$$\tau = (29 \text{ Newtons})\cdot(0.45 \text{ m}) = 13 \text{ Newton}\cdot\text{m}$$

The other mass exerts a negative torque, and we will call its weight "w":

$$\tau = -(\text{w})\cdot(0.75 \text{ m})$$

To balance the bar, the torques must sum up to zero:

$$13 \text{ Newton}\cdot\text{m} - (\text{w})\cdot(0.75 \text{ m}) = 0$$

$$\text{w} = \frac{13 \text{ Newton}\cdot\cancel{\text{m}}}{0.75\ \cancel{\text{m}}} = 17 \text{ Newtons}$$

Is that our answer? Not quite. You see, the problem asks for the mass, but we have found the weight. Thus, we need to divide by the acceleration due to gravity to find the mass of the object, which is 1.7 kg.

6.8 In this problem, there are four forces at play. The gravitational force has a parallel and a perpendicular component, the incline exerts a normal force, and friction opposes the motion. The force diagram looks like this:

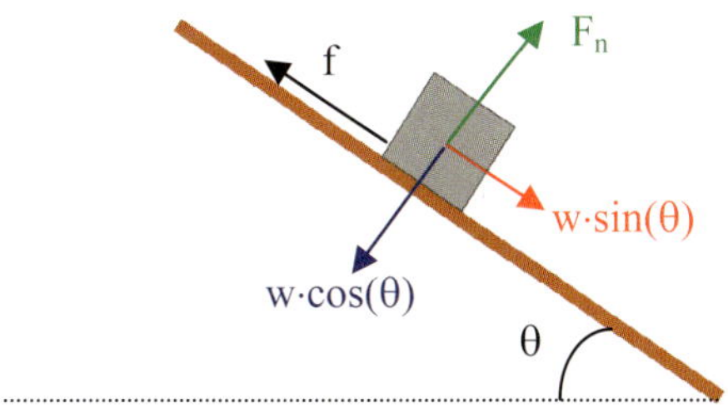

We will need to determine friction, so let's get that out of the way. In the perpendicular dimension, we will call the normal force positive and the perpendicular component to gravity negative. Since there is no movement in this dimension, these forces add up to zero. To get the gravitational force, we will need the weight. The problem gives only mass, so we must calculate weight, which is 4.4×10^2 Newtons:

$$F_n + -\text{w}\cdot\cos(\theta) = 0$$

$$F_n = (4.4 \times 10^2 \text{ Newtons})\cdot\cos(45°) = 3.1 \times 10^2 \text{ Newtons}$$

This allows us to calculate the frictional force:

$$f = \mu_k\cdot F_n$$

$$f = (0.30) \cdot (3.1 \times 10^2 \text{ Newtons}) = 93 \text{ Newtons}$$

Now that we have the frictional force, we can turn our attention to the parallel dimension. In that dimension, we will call the frictional force negative and the parallel component of the gravitational force positive:

$$w \cdot \sin(\theta) - f = m\mathbf{a}$$

$$(4.4 \times 10^2 \text{ Newtons}) \cdot \sin(45^\circ) - 93 \text{ Newtons} = (45 \text{ kg}) \cdot \mathbf{a}$$

$$\mathbf{a} = \frac{2.2 \times 10^2 \frac{\cancel{\text{kg}} \cdot \text{m}}{\text{sec}^2}}{45 \cancel{\text{kg}}} = 4.9 \frac{\text{m}}{\text{sec}^2}$$

The block therefore falls down the incline (down the incline was defined to be positive) at an acceleration of 4.9 m/sec^2.

6.9 This is the same kind of problem as the previous one, so the diagram is the same:

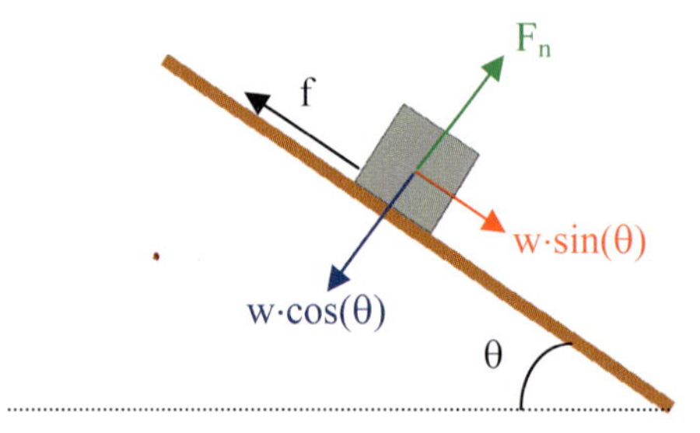

This time, however, we need to determine the frictional coefficient. Since the bulk of the information about friction comes from the perpendicular dimension, let's start with the parallel dimension and see what we can figure out. The block slides at a constant velocity. This means that the acceleration is zero. Thus, we are in dynamic equilibrium and the forces must add up to zero. Keeping the same directional definitions and realizing that we must convert mass to weight, we get:

$$w \cdot \sin(\theta) - f = 0$$

$$f = (98 \text{ Newtons}) \cdot \sin(22^\circ) = 37 \text{ Newtons}$$

Well, now we know the frictional force. If we can calculate the normal force, getting μ will be a snap. To determine the normal force, we look at the perpendicular dimension:

$$F_n + -w \cdot \cos(\theta) = 0$$

$$F_n = (98 \text{ Newtons}) \cdot \cos(22^\circ) = 91 \text{ Newtons}$$

Now we can calculate the coefficient of friction:

$$f = \mu_k \cdot F_n$$

$$37 \text{ Newtons} = \mu_k \cdot (91 \text{ Newtons})$$

$$\mu_k = \frac{37 \text{ ~~Newtons~~}}{91 \text{ ~~Newtons~~}} = 0.41$$

This gives us a coefficient of friction that is 0.41. However, there is a shortcut. If you realized that this problem was the kinetic analog of Experiment 6.3, you could have just taken the tangent of the angle and gotten the coefficient as well. Rounding errors cause the answer I give to be one one-hundredth off of that answer, but they are really the same.

6.10 In this problem, we need to see how much force the girl pulls with when the tension in the string is 45 Newtons. We have two separate boxes, so we need to consider them separately. The box in the rear has the following force diagram:

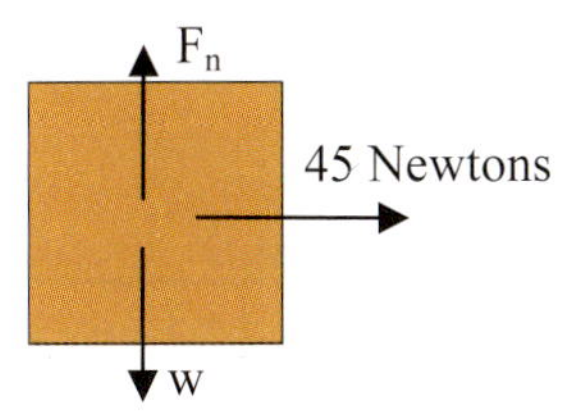

The sum of the forces in the horizontal dimension equals the mass times the acceleration. Thus,

$$\Sigma \mathbf{F} = m \cdot \mathbf{a}$$

$$45 \text{ Newtons} = (245 \text{ kg}) \cdot \mathbf{a}$$

$$\mathbf{a} = 0.18 \text{ m/sec}^2$$

Great, we've determined the acceleration, but that's not what we're looking for. Oh well, let's look at the other box:

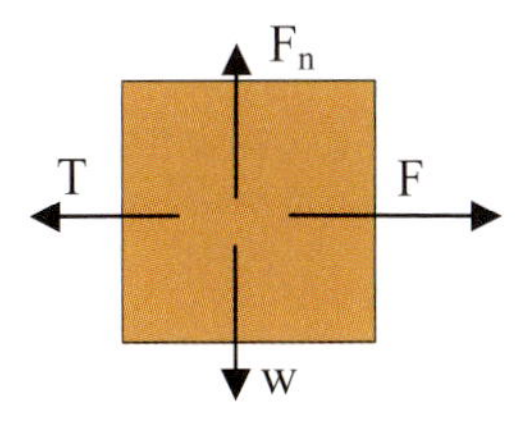

In the horizontal dimension, the forces must sum up to the mass times the acceleration. We know the acceleration, however, so we can just plug it in!

$$F + -45 \text{ Newtons} = (322 \text{ kg}) \cdot (0.18 \frac{\text{m}}{\text{sec}^2})$$

$$F = 103 \text{ Newtons}$$

The girl can pull with a force of 103 Newtons without the string breaking.

REVIEW QUESTIONS

1. What is required for an object to be in translational equilibrium?

2. Distinguish between dynamic equilibrium and static equilibrium.

3. What is tension in physics? With what unit is it measured?

4. Which of the following situations will result in the most tension in the two strings?

5. What is the difference between torque and force?

6. What is required for an object to be in rotational equilibrium?

7. Which of the following situations results in more torque being applied? Assume the forces are equal.

Illustrations by Megan Whitaker

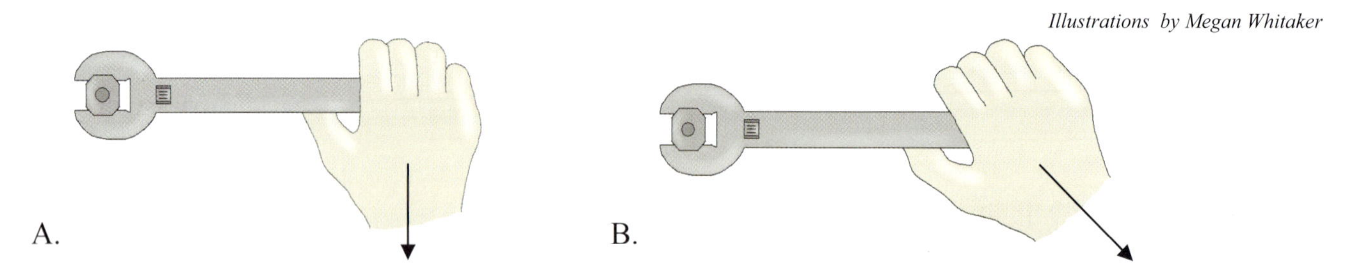

8. A mechanic tries with all of his might to unscrew a nut. No matter how hard he tries, however, he just can't do it. What would you suggest the mechanic do in order to get the nut unscrewed?

9. Before cars came out with power steering, a big car required a much bigger steering wheel than a small car. Why?

10. When analyzing motion on an incline, why do we split the forces up into components that are parallel and perpendicular to the incline?

PRACTICE PROBLEMS

1. A 675-Newton acrobat walks on a tightrope. The woman's weight makes the tightrope sag, so that it looks like this:

Illustration from the MasterClips collection

What is the tension of the rope on each side of the acrobat?

2. A 15-kg sign hangs from the ceiling by two strings as pictured below:

Illustration copyright GifArt.com

What is the tension in each string?

3. A 75-kg man weighs himself with a scale on an elevator.

a. What weight does the scale read when the elevator is stopped?
b. What weight does the scale read when the elevator is rising at a constant speed?
c. What weight does the scale read when the elevator is traveling down and speeding up with an acceleration of 0.95 m/sec^2?
d. What weight does the scale read when the elevator is traveling up and slowing down with an acceleration of 0.95 m/sec^2?

4. If a mechanic is exerting 97 Newton·m of torque on a nut and is applying a 415-Newton force perpendicular to the wrench, how far away from the center of the nut is he grasping the wrench?

5. Two children balance on a seesaw. If the boy weighs 352 Newtons and sits 1.12 meters from the fulcrum, how much does the girl weigh if she must sit 1.34 meters from the fulcrum to achieve balance?

6. The captain of a ship holds the steering wheel and exerts 211 Newtons of force with each hand as shown below. How much total torque is he exerting on the ship's steering wheel?

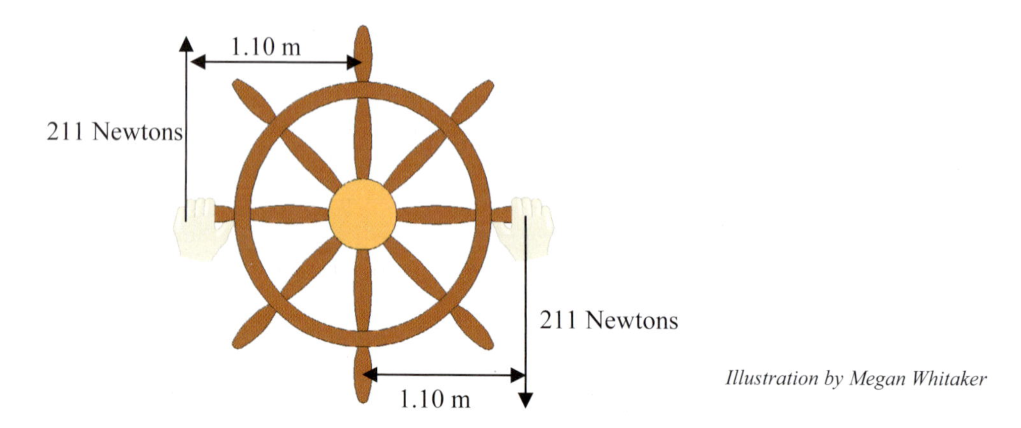

Illustration by Megan Whitaker

7. A physics experiment is done in which three masses are used to balance a bar on a fulcrum. Two masses (21 kg and 12 kg) are placed on one side of the fulcrum, 12 cm and 21 cm away, respectively. Where must the third mass (18 kg) be placed in order to make the bar balance?

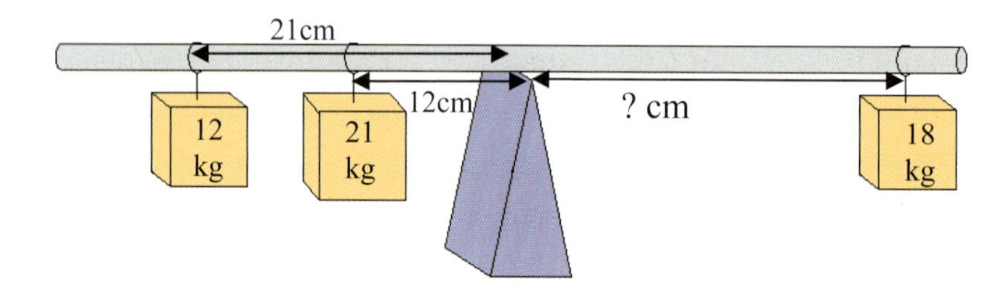

8. A 30.1-kg block of wood slides down an incline angled at 23 degrees. If friction is ignored, what is the block's acceleration?

9. Consider the block sliding down the plane as discussed in Problem #8. If you take friction into account ($\mu_k = 0.35$), what is the block's acceleration?

10. Two toy boats (0.45 kg and 0.32 kg) are tied together with a string. A child sits in the bathtub and pulls the lead boat with a force of 3.1 Newtons. Ignoring friction, what are the acceleration of this two-boat system and the tension in the string that connects the boats?

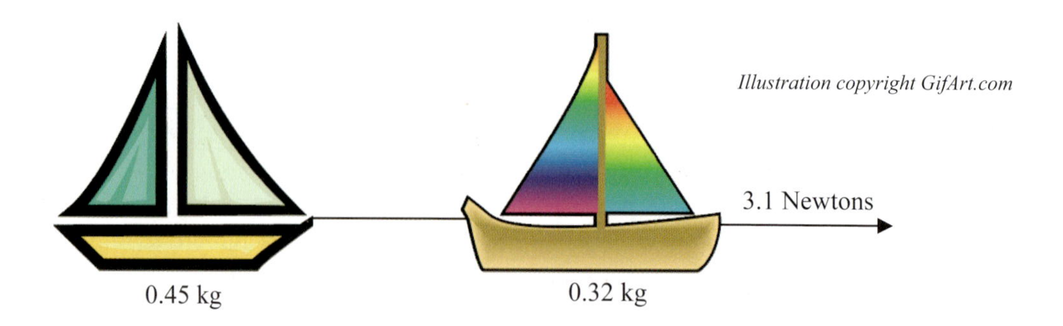

Illustration copyright GifArt.com

MODULE #7: Uniform Circular Motion and Gravity

Introduction

Before we go any further, I want you to breathe a sigh of relief. Why? Well, these next two modules contain *absolutely no trigonometry!!!!* Yes, it turns out that we're done with the really intense trigonometry for a while. Most students love to hear this, so I thought I would bring it to your attention right away. Now that I've taken care of that, let's move on to learning more physics.

In the previous module, we learned a bit about rotational motion. However, we concentrated on extended objects rotating around a fixed point. When we analyzed the balance of a seesaw, for example, the extended object was the seesaw, and it rotated around the fixed point of the fulcrum. When we analyzed using a wrench to unscrew a nut, the extended object was the wrench, and it was rotating around the center of the bolt.

In this module, I want to concentrate on another kind of rotational motion: uniform circular motion. Think about an object traveling in a circle. That object is exhibiting rotational motion as defined in the previous module, because it passes the same point over and over again as it goes around and around in a circle. It turns out that not only is this interesting motion, but it also leads directly to a study of our solar system. Why? Well, let's get through the basics first. Then you will see.

Uniform Circular Motion

I want to start our discussion by considering a particle that is moving in a circle at constant speed:

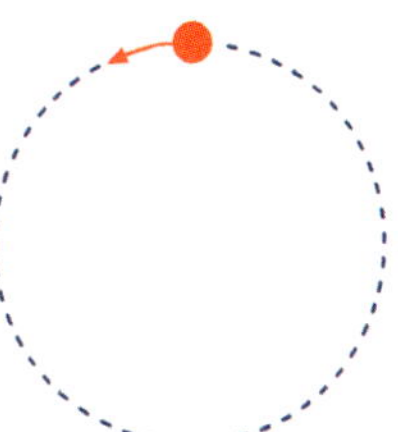

This type of motion is called is called **uniform circular motion**. Because the object is moving at a constant speed, we know that the sum of the torques acting on it is zero. After all, if a net torque were acting on the object, it would be experiencing rotational acceleration. That would mean that the speed of its rotational motion would be changing. Since the speed of the rotational motion is not changing, there must be no net torque acting on the object.

Now let me ask you a question. Think very carefully about it. If the object is moving around the circle at a constant speed, does that mean its acceleration is zero? You might think that the answer to this question is rather obvious, but it is not. The answer is, in fact, "no." All objects that exhibit uniform circular motion do, indeed, have a non-zero acceleration. How is this possible? If the object neither slows down or speeds up, how can it have acceleration?

Think about it. Acceleration is the change in an object's *velocity*. What is velocity? It is a vector quantity whose magnitude is the speed of the object and whose direction is the same as the direction of travel. Even though the speed (the magnitude of the velocity vector) does not change for an object in uniform circular motion, the direction does. In order to move in the circle, the object must continually turn. This means that the object is continually changing direction. Thus, the object's *velocity* is continually changing, because the direction of the object is changing. In order for the velocity of an object to change, the object must have acceleration. Thus, an object that exhibits uniform circular motion has a non-zero acceleration!

The first thing that we must do, then, is figure out how the velocity and acceleration of the object interact in order to keep it moving in a circle at a constant speed. In order to understand this, take a look at the following figure.

FIGURE 7.1
Velocity and Acceleration Vectors in Uniform Circular Motion

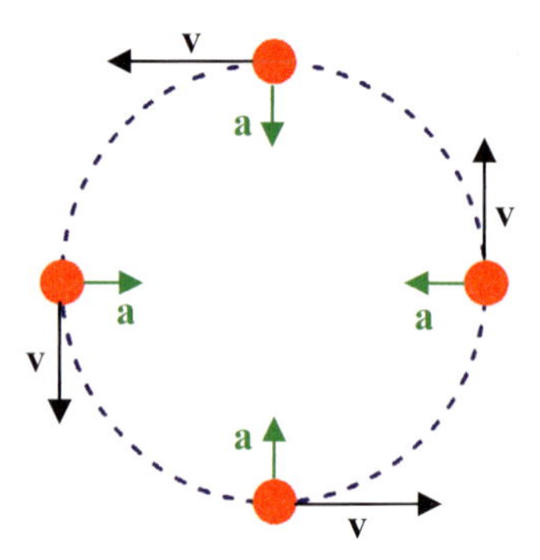

When an object is in uniform circular motion, the velocity vector is always tangent to the circle in which the object is traveling. The acceleration, however, is always directed into the circle, along the radius.

Newton's Second Law tells us that wherever there is acceleration, there must be a force causing the acceleration. Where is that force? Well, Newton's Second Law also tells us that the force vector is always pointed in the same direction as the acceleration. Thus, the force vector is also pointed towards the center of the circle, as shown in Figure 7.2.

FIGURE 7.2
Velocity and Force Vectors in Uniform Circular Motion

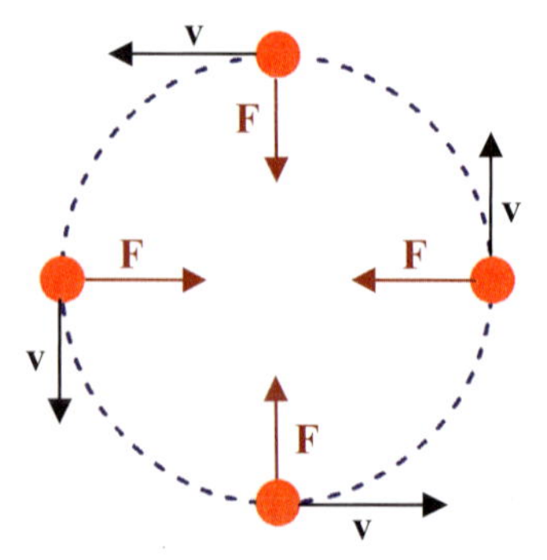

In uniform circular motion, the force vector is pointed in the same direction as the acceleration vector, towards the axis of rotation along the radius of the circle.

Before we go on to discuss the force and acceleration vectors in detail, let me point out something. We know that there is no net *torque* acting on an object in uniform circular motion. However, as Figure 7.2 illustrates, there is a net *force* acting on the object. This is an example of a situation in which a force acts on an object but the force causes no torque. You should see why this is the case. After all, the force vector is parallel to the lever arm (the imaginary line from the axis of rotation to the force). Thus, there is no component of the force which is perpendicular to the lever arm. As a result, this force exerts no torque. Therefore, even though the sum of the forces on the object is not zero (and the object is therefore not in translational equilibrium), the sum of the torques is zero (and the object is therefore in rotational equilibrium).

Centripetal Force and Centripetal Acceleration

Since we know that there is a force and therefore an acceleration associated with uniform circular motion, we had better learn a little more about them. First of all, physicists have a special name for the force and acceleration associated with circular motion: **centripetal** (sen trip' uh tul) **force** and **centripetal acceleration**.

Centripetal force - The force necessary to make an object move in a circle. It is directed towards the center of the circle.

Centripetal acceleration - The acceleration caused by centripetal force

Notice that the definition of centripetal force mentions nothing about *uniform* circular motion. It simply says that centripetal force is necessary for circular motion of any kind. Don't let this fact slip by you. When we are speaking about uniform circular motion, centripetal force is the *only* force acting on the object in question. However, centripetal force is necessary for all circular motion. Thus, if an object is moving in a circle but not at a constant speed, there is still a centripetal force acting on the object. In that case, however, the centripetal force will not be the only force acting on the object. There must also be some other force present which is changing the speed of the object. This is an important point:

Centripetal force is necessary for *any* kind of circular motion.

Now that we know its name, I want you to perform the following experiment so that you understand some of the properties of the centripetal force.

EXPERIMENT 7.1
Centripetal Force

Note: A sample set of calculations is available in the solutions and tests guide. It is with the solutions to the practice problems.

Supplies:
- Safety goggles
- A pen that can be disassembled
- String
- A ruler
- Four washers
- One washer that is a bit smaller than the four washers used above
- A marker
- A stopwatch

1. Disassemble the pen. You want to use the tube that makes up the body of the pen for this experiment.
2. Cut a length of string approximately one meter long.
3. Feed the string through the tube that made up the body of the pen.
4. Tie two of the larger washers to one end of the string. If the tube from the pen tapers, make sure these washers are tied to the end of the string that comes out of the tapered side of the tube
5. Tie the smaller washer on the other side of the string. In the end, your setup should look like this:

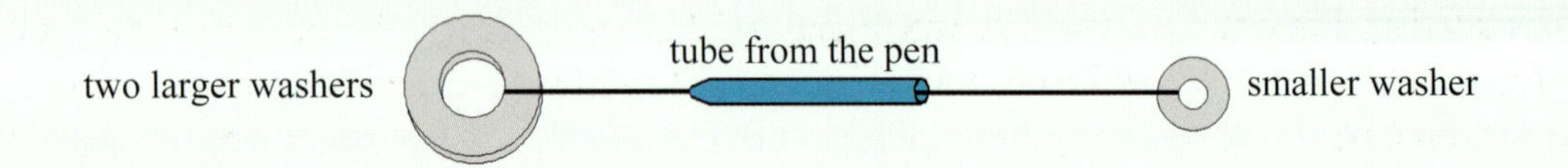

6. Use the marker to make a mark on the string 48 cm from the center of the smaller washer.
7. Grasp the tube of the pen and start twirling the smaller washer around in a circle. Keep the circle above your head so that you don't hit yourself with the washer. Try to get the washer to twirl with a constant speed, and do not touch the string in any way. The only thing you should be touching is the tube from the pen. Get used to maintaining this motion, because you are going to have to do some pretty difficult things with this setup.
8. Once you are used to twirling the small washer at constant speed, adjust the string so that the mark you made is just visible over the top of the tube. You can do this by using your free hand to raise or lower the two larger washers until you see the mark. Then, you can slowly let go of the two washers. Vary the speed with which you twirl until you can keep the small washer twirling at a constant speed while keeping the mark on the string just above the top of the tube. By keeping the mark right above the tube, you are ensuring that the small washer is orbiting the tube with a radius of 48 cm. Remember, once you get the motion started, you should be able to keep the washer twirling at a constant rate and keep the mark just above the tube *without touching the string at all.* Get used to doing this for a while.
9. Okay, now here comes the hard part. Hold the stopwatch in your other hand and start the washer twirling again, making sure that the mark on the string is just above the tube and making sure that you are not touching the string.
10. When you are ready, use the stopwatch to measure the time it takes for the washer to twirl around the tube 20 times. Do this five times and average the result.
11. Since the washer was traveling in a circle, you can calculate the distance it traveled in 20 revolutions. Take the radius of the motion (0.48 m) and multiply by 2π. That gives you the distance around the circle. Then, multiply by 20. That gives you the distance the washer traveled in 20 revolutions. Then, divide that number by the average of the times that you measured. Now you have the average speed of the small washer in m/sec.
12. Take another large washer and tie it to the end of the string that already has two washers on it. That way, there are now three large washers on the end of the string.
13. Repeat steps 9-11.
14. Take the last large washer and tie it to the end of the string that already has three washers on it. That way, there are now four large washers on the end of the string.
15. Repeat steps 9-11.
16. Take all but one of the large washers off the string, so that the string has only one large washer tied to it.
17. Repeat steps 9-11.

18. Lay the setup down and make three more marks on the string. Make one mark 36 cm from the center of the small washer; make another that is 24 cm from the center of the small washer; and then make one that is 12 cm from the center of the small washer.
19. Tie another large washer on the end of the string so that there are now two large washers there.
20. Repeat steps 9-11, this time keeping the mark that is 36 cm away from the center of the washer at the top of the tube. You will need to use a radius of 0.36 m in your calculation of the speed.
21. Repeat steps 9-11, this time keeping the mark that is 24 cm away from the center of the washer at the top of the tube. You will need to use a radius of 0.24 m in your calculation of the speed.
22. Repeat steps 9-11, this time keeping the mark that is 12 cm away from the center of the washer at the top of the tube. You will need to use a radius of 0.12 m in your calculation of the speed.
23. Clean up your mess.

Are you dizzy from all of that twirling? I hope not, because now we need to see what this experiment meant. In the first part of the experiment (steps 9-17), you measured the speed of the small washer twirling in a circle of fixed radius while varying the number of larger washers pulling down on the string. What were you varying by changing the number of large washers? You were varying the *centripetal force*. Think about it. The tension on the string pulled on the small washer, and the direction of that pull was towards the tube, which was at the center of the washer's motion. Thus, the tension in the string was exerting a *centripetal force*. The more washers you added, the larger the tension in the string, so the larger the centripetal force.

Graph the data you collected in the first part of the experiment. Put the number of large washers on the x-axis and the speed of the smaller washer on the y-axis. You should see that the smallest speed was achieved when one washer was used, and as the number of washers was increased, the speed increased as well. What does this tell you? It tells you that when the radius of the motion is kept constant, the speed of the object increases with increasing centripetal force.

Graph the data you collected in the second part of the experiment. This time, put the radius of motion on the x-axis and the speed on the y-axis. Remember that you can use the first speed you measured in the first part of the experiment on the graph as well. After all, the first speed you measured was for a radius of motion of 48 cm and for two large washers on the string. That's the same number of washers used in the second part of the experiment, so that piece of data is consistent with the second part of the experiment. What do you see from the graph? In this case, you should see that as the radius of the motion increased, the speed of the washer increased as well. Now remember, in this part of the experiment, you kept the centripetal force constant, because you had two large washers on the string for each trial. Thus, these data tell us that when we keep the centripetal force constant and increase the radius, the speed of the object must increase.

Those are two very important properties of centripetal force. How can we sum them up? Well, there is a formula that relates the speed of an object in circular motion and the centripetal force needed to keep that object in circular motion:

$$F_c = \frac{mv^2}{r} \tag{7.1}$$

In this equation, "F_c" represents the centripetal force, "m" stands for the mass of the object that is moving, "v" represents the speed of the object (remember, when "v" is not bold we are talking about the magnitude of the velocity vector), and "r" is the radius of the circle in which the object is traveling.

Notice first how well your experimental results agree with the formula. If you hold "r" constant and increase "F_c," what must happen to the speed? In order to keep the two sides equal, "v" must increase, as you saw in the experiment. Also, what happens when you hold "F_c" constant and increase "r?" In order for the two sides of the equation to equal one another, "v" must increase as well. Once again, this is consistent with your experimental results.

Notice also how the units work out in this equation. When mass is multiplied by v^2, the units become $\frac{\text{kg}\cdot\text{m}^2}{\text{sec}^2}$. When that unit is divided by the radius (in meters), the unit becomes $\frac{\text{kg}\cdot\text{m}}{\text{sec}^2}$, which is the same as a Newton. That's the unit we expect for force.

As Newton's Second Law tells us, now that we have the force, calculating the acceleration is trivial. Force is equal to mass times acceleration, so acceleration is the same as force divided by mass. Thus, if I take Equation (7.1) and divide by the mass of the object, I get the acceleration:

$$a_c = \frac{v^2}{r} \tag{7.2}$$

Notice that in both Equation (7.1) and Equation (7.2), I have no boldface type. This is because although both force and acceleration are vector quantities, these equations only allow us to calculate their magnitudes. We have already learned, however, that their directions will always be the same. Centripetal force and acceleration vectors always point to the center of the circle. Thus, we can use these two equations to calculate the magnitude of the centripetal force or the centripetal acceleration vectors, but we will always need to think about the direction on our own. Luckily, however, this is easy, because these vectors will always point towards the center of the circle. Let's look at an example problem to see how we can use these equations to determine centripetal acceleration and force vectors.

EXAMPLE 7.1

A 151-g toy truck travels on a circular track whose radius is 1.1 meters. If the truck travels at a constant speed of 2.1 meters per second, what are the magnitudes of the centripetal force and acceleration? Draw the centripetal force vector.

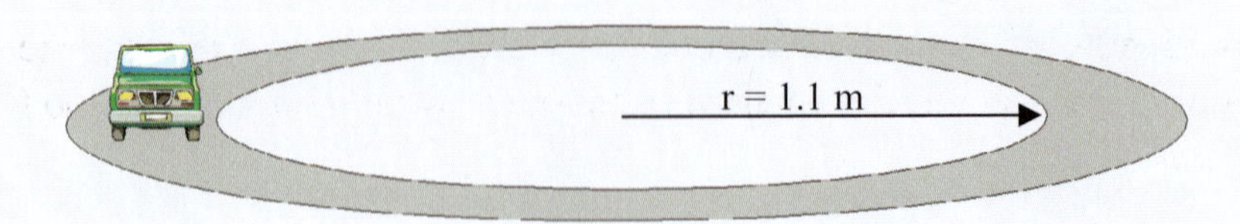

To calculate the magnitudes of the centripetal force and acceleration, we simply use Equations (7.1) and (7.2). When we do that, however, we must convert the mass of the truck to kg, because that's the SI unit for mass:

$$F_c = \frac{mv^2}{r}$$

$$F_c = \frac{(0.151\text{ kg})\cdot(2.1\ \frac{\text{m}}{\text{sec}})^2}{1.1\text{ m}} = \frac{(0.151\text{ kg})\cdot(4.41\ \frac{\text{m}^{\not{2}}}{\text{sec}^2})}{1.1\ \not{\text{m}}} = 0.61\ \frac{\text{kg}\cdot\text{m}}{\text{sec}^2} = \underline{0.61\text{ Newtons}}$$

$$a_c = \frac{v^2}{r}$$

$$a_c = \frac{(2.1\ \frac{\text{m}}{\text{sec}})^2}{1.1\text{ m}} = \frac{(4.41\ \frac{\text{m}^{\not{2}}}{\text{sec}^2})}{1.1\ \not{\text{m}}} = \underline{4.0\ \frac{\text{m}}{\text{sec}^2}}$$

These are the magnitudes of the acceleration and force vectors. Notice that I waited until the end of each equation to determine the number of significant figures to use. Since 2.1 has only two significant figures, our answer is limited to two significant figures. However, I did not worry about that until the end of each equation.

To draw in the force vector, we simply need to remember that centripetal force is always directed to the center of the circle:

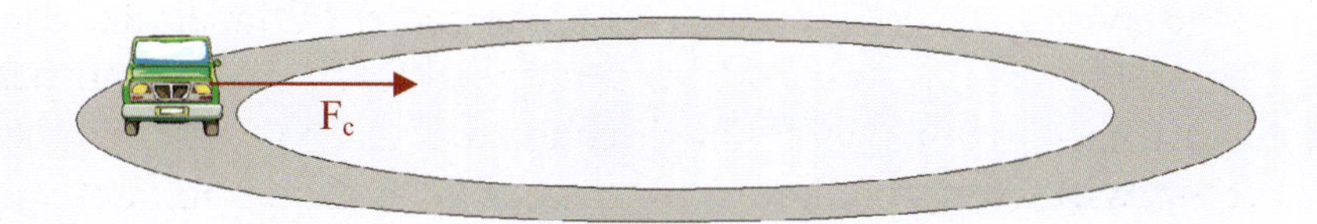

Now see if you can solve the following "On Your Own" problems.

ON YOUR OWN

7.1 A 1.234-kg object travels in a circle at a constant speed of 12.1 m/sec. If the centripetal force applied to the object is 4.67 Newtons, what is the radius of the circle in which the object is traveling?

7.2 Consider an object moving in a circle at a constant speed. If the centripetal force on the object were to suddenly double but the speed remained unchanged, what would happen to the radius of the circle in which it is traveling?

The Source of Centripetal Force

By now you should know that subjects in physics are not as simple as they might first appear. Although determining centripetal force and acceleration vectors is a rather easy task, situations that involve uniform circular motion can become pretty complex. Before we tackle a couple of problems involving these more complex situations, let's talk a little bit about what *causes* centripetal force and acceleration. For example, suppose a boy is twirling a toy airplane on a string at a constant rate:

Illustration copyright GifArt.com

We know that there must be a centripetal force acting on the plane, or it would not be moving in a circle. The question is, however, what *causes* the centripetal force? Well, you might think that the boy does. Ultimately, that is true, because the plane would never travel in a circle without the boy there. However, the boy is not touching the plane. Therefore, he cannot exert a force directly on the plane. What is exerting the force directly on the plane? The string is. The tension in the string exerts a force on the plane that is pointed directly into the center of the circle. That's the *source* of the centripetal force:

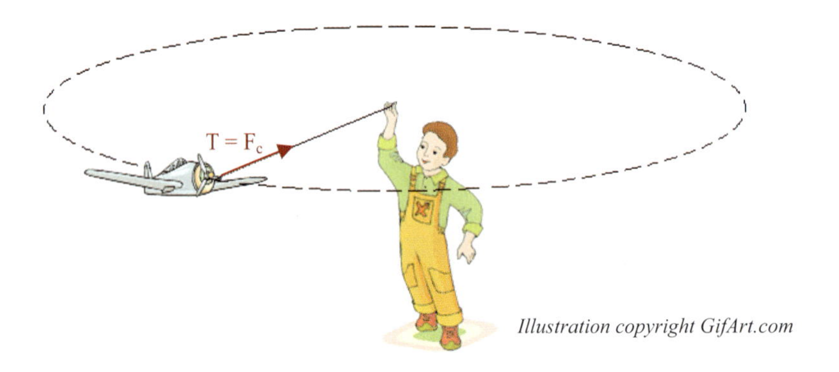

Illustration copyright GifArt.com

This is the same as what happened in the experiment. The tension in the string provided the centripetal force which caused the smaller washer to travel in a circle.

Let's consider another example. Suppose you are riding your bike in a circle:

Illustration from Arts and Letters Express

Once again, we know that there must be a centripetal force, but what is supplying that force? Well, ultimately, you are because the bike would not travel in a circle without your control; however, you cannot be the direct supplier of the centripetal force. After all, you are on the bike. You can turn the wheels or tilt the bike one way or another. Neither of these actions, however, will result in a force that is directed precisely towards the center of the circle. It turns out that friction is supplying the centripetal force in this case:

Illustration from Arts and Letters Express

Yes, friction acts on the tires. When combined with you tilting on the bike and steering the handlebars, the force friction exerts is directed precisely towards the center of the circle, making it the centripetal force.

Do you see what I am driving at here? In order to have circular motion, there must be centripetal force. It is important to understand what supplies that force, because often this information drives to the heart of the situation. For example, consider the situation of the boy twirling the airplane in a circle. Suppose the centripetal force needed to keep the plane moving in the circle was 10 Newtons. Suppose, however, that the string could only withstand a tension of 5 Newtons. What would happen? Well, since the centripetal force needed is supplied by the tension in the string, the string would end up breaking because it could withstand only half of the tension needed. Thus, the string would break and the plane would go flying off. Consider the case of you riding your bike. What if the area you were riding on was very slippery and there wasn't much friction? You would not be able to keep your bike moving in a circle because friction would not be able to provide the needed centripetal force.

So you see that determining *what supplies* the centripetal force in a situation is almost as important as being able to determine the magnitude and direction of the centripetal force and acceleration vectors. Study the following examples and then solve the "On Your Own" problems afterwards to make sure you understand this concept.

EXAMPLE 7.2

Illustrations copyright GifArt.com

One track and field event is called the hammer toss. In this event, the competitors twirl a mass on a rope and then release the rope, causing the mass to fly off in the direction that it was released. Suppose a competitor is twirling a 12-kg mass on the end of a 0.65 meter rope. If the tension on the rope is 2.1 x 10^3 Newtons, how fast is the mass moving?

As we discussed earlier, the tension in the rope is what supplies the centripetal force. Thus, since the tension is 2.1 x 10^3 Newtons, we know that the centripetal force is 2.1x10^3 Newtons. Now that we've made that connection, the rest of the problem is rather easy. After all, we know the centripetal force, the radius, and the mass. We need to find the speed. Equation (7.1) relates all of these quantities, so we just need to set up the equation and solve it:

$$F_c = \frac{mv^2}{r}$$

$$2.1 \times 10^3 \text{ Newtons} = \frac{(12 \text{ kg}) \cdot v^2}{0.65 \text{ m}}$$

$$v^2 = \frac{(2.1 \times 10^3 \frac{\cancel{\text{kg}} \cdot \text{m}}{\text{sec}^2}) \cdot (0.65 \text{ m})}{12 \cancel{\text{kg}}}$$

$$v = 11 \frac{\text{m}}{\text{sec}}$$

The mass is twirling around at a speed of 11 m/sec.

A car is traveling down a road which curves. The curve's radius of curvature is 15.1 meters and the coefficient of friction between the car's tires and the road is 0.34. With what maximum speed can the car travel around the curve?

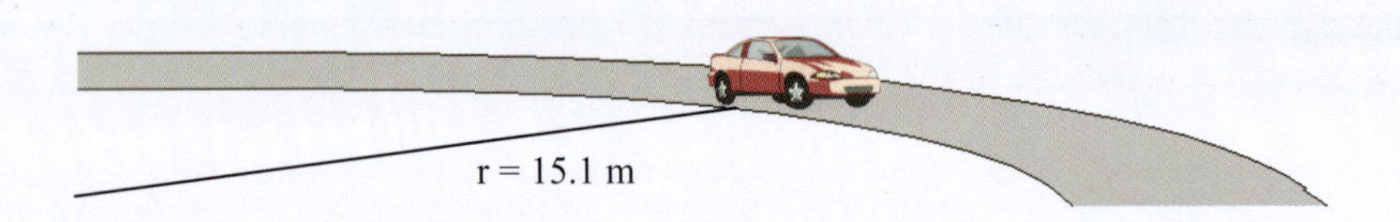

In order to solve this problem, we need to realize a few things. First, when we say "radius of curvature," we mean that the curve is an arc of a circle. If the circle were complete, it would have a radius equal to the radius of curvature. Thus, while the car is traveling on the curve, it is just as if the car were traveling a circle of that radius. Also, since the car is traveling in a circle, there must be a centripetal force. Just like the bicycle case we discussed above, friction is supplying that force.

Let's think about this for a moment. Suppose we knew the frictional force between the tires and the road. What would that tell us? Well, it would tell us how much force friction *could* supply as centripetal force. If the car needed less centripetal force than friction can supply, everything would be fine. The car would stay on the curve. Suppose, however, the car needed more centripetal force than friction could supply. What would happen then? Well, at that point, there would not be enough centripetal force, and the car would slide off the curve. So, if we could calculate the frictional force, we could say that it is the maximum centripetal force available. We could then use Equation (7.1) to solve for the maximum speed.

The problem, of course, is that we cannot calculate the frictional force. After all, to calculate the frictional force, we need the coefficient of friction and the normal force. Well, when an object is flat on the ground, the normal force is equal to the weight of the object. Since we are not given the mass of the car, we cannot calculate the weight. Even though we can't get a *number* for the frictional force, let's go ahead and get an expression for it, by using "m" in place of the mass of the car:

$$f = \mu F_n$$

$$f = (0.34) \cdot (m) \cdot (9.8 \frac{\text{meters}}{\text{sec}^2})$$

$$f = 3.3 \frac{\text{meters}}{\text{sec}^2} \cdot (m)$$

Now, even though we don't have a number for the frictional force, let's go ahead and put the expression we have for it into Equation (7.1). Remember our strategy. The expression we have for the frictional force represents the maximum possible centripetal force available. If we put that into Equation (7.1), we can use it to calculate the maximum possible speed:

$$F_c = \frac{mv^2}{r}$$

$$3.3 \frac{\text{meters}}{\text{sec}^2} \cdot (m) = \frac{m \cdot v^2}{15.1 \text{ meters}}$$

Do you see what happens now? Since mass appears on both sides of the equation, it cancels out. Thus, it doesn't matter that we don't know the mass of the car, because it cancels out of the final equation. Now we can simply solve for v:

$$3.3 \frac{\text{meters}}{\text{sec}^2} \cdot (\cancel{m}) = \frac{\cancel{m} \cdot v^2}{15.1 \text{ meters}}$$

$$v^2 = (3.3 \frac{\text{meters}}{\text{sec}^2}) \cdot (15.1 \text{ meters})$$

$$v = 7.1 \frac{\text{meters}}{\text{sec}}$$

So we see that as long as the car is traveling with a speed of 7.1 m/sec or less, it will stay on the curve.

Before we leave this example problem, we need to discuss a couple of things. First, let's review how we solved the problem. Even though we didn't have enough information to get a number for the frictional force, we said that if we could calculate friction, that would represent the maximum centripetal force available. Thus, we calculated an expression for frictional force and went ahead and used it. In then end, the unknown part of that expression dropped out, and we were able to solve for the speed of the car. This is sometimes what we must do to solve a problem. If we know that a particular quantity is necessary to solve a problem, we should not fret if we can't calculate it completely. We should calculate it as much as we can and hope that the unknowns left will cancel out as the problem unfolds.

Second, look at the *meaning* of this problem. We can learn two things from it. One thing we can learn is that the reason cars slide out of turns is because there is not enough friction on the road. If the speed of a car is high, then a large amount of frictional force is necessary to keep it moving along a curve. If friction cannot supply the necessary centripetal force, the car cannot stay on the curve and it slides away. This is why cars must take curves slowly when the road is slick. A slow speed reduces the centripetal force needed, and even a slick road will have a little friction, keeping the car on the curve.

The other thing we see from this problem is that the mass of the car does not affect whether or not it can stay on the curve. Even though a more massive car will have a greater frictional force, that doesn't make it any easier for the car to stay on the road. This is because a larger mass requires a greater centripetal force to keep it moving in a circle. Thus, the gain in friction is offset by the fact that a greater centripetal force is necessary. This is an important point:

On a flat curve, the mass of the car does not affect the speed at which it can stay on the curve.

This is a good thing, because different cars have different masses, and it would be difficult to build safe road curves for all of those different car masses. Since the mass is irrelevant, though, road engineers needn't worry about that.

ON YOUR OWN

7.3 A child twirls a 314-g toy airplane around his head. If the 67-cm string attached to the plane can withstand a tension of 18.1 Newtons, what is the maximum speed with which the toy plane can travel?

7.4 A car travels around a curve in the road. If the radius of curvature is 12.1 meters and the car travels at a speed of 21 m/sec, what is the minimum coefficient of friction necessary to keep the car on the curve?

A Fictional Force

Before we move on, I want to take a minute to make sure you do not get the concept of centripetal force confused with the concept of "centrifugal (sen trif' you gul) force." It is important to distinguish between the two, because while centripetal force is a valid concept in physics, *centrifugal force is not*! Although you might have heard about centrifugal force (indeed, it is even in the dictionary), *it is not a real force*.

The concept of a centrifugal force is a result of poor physical analysis of certain situations. For example, think about traveling down the road in a car. Suppose there are some books on the car's dashboard. If the car turns a corner at a high rate of speed, what will happen to the books? They will begin to move across the dashboard, won't they? The books were stationary on the dashboard; then they suddenly started moving when the car made a turn. Thus, a person might conclude that the books must have experienced some sort of force in order for them to start moving like that. The person might call this a "centrifugal force."

It turns out that such an analysis is wrong. There is a force at play in this situation, but it is not a "centrifugal" force. The force is simply friction, and the motion of the books is a result of the fact that the frictional force is not very strong. Remember Newton's First Law. The books are traveling along with the car. Thus, they have the same velocity as the car. When the car makes a quick turn, the books will continue to travel with their *previous* velocity until acted on by a force. The friction between the books and the dashboard is the force that acts. It pushes the books, changing their velocity so that the books continue to travel with the car.

What happens, however, when the friction between the dashboard and the books is not strong enough to change the velocity of the books to match the new velocity of the car? The books will begin sliding along the dashboard, in the general direction of the car's previous velocity. That's why the books move. They are not acted on by "centrifugal" force. They are simply obeying Newton's First Law, and friction is not powerful enough to give them the acceleration they need to travel with the car.

It is easy to get centrifugal force and centripetal force confused, since they both are associated with circular motion. For example, chemists and biologists often use a machine called a centrifuge (sen' trih fuj), which is pictured below.

Photo copyright Getty Images

FIGURE 7.3
A Centrifuge

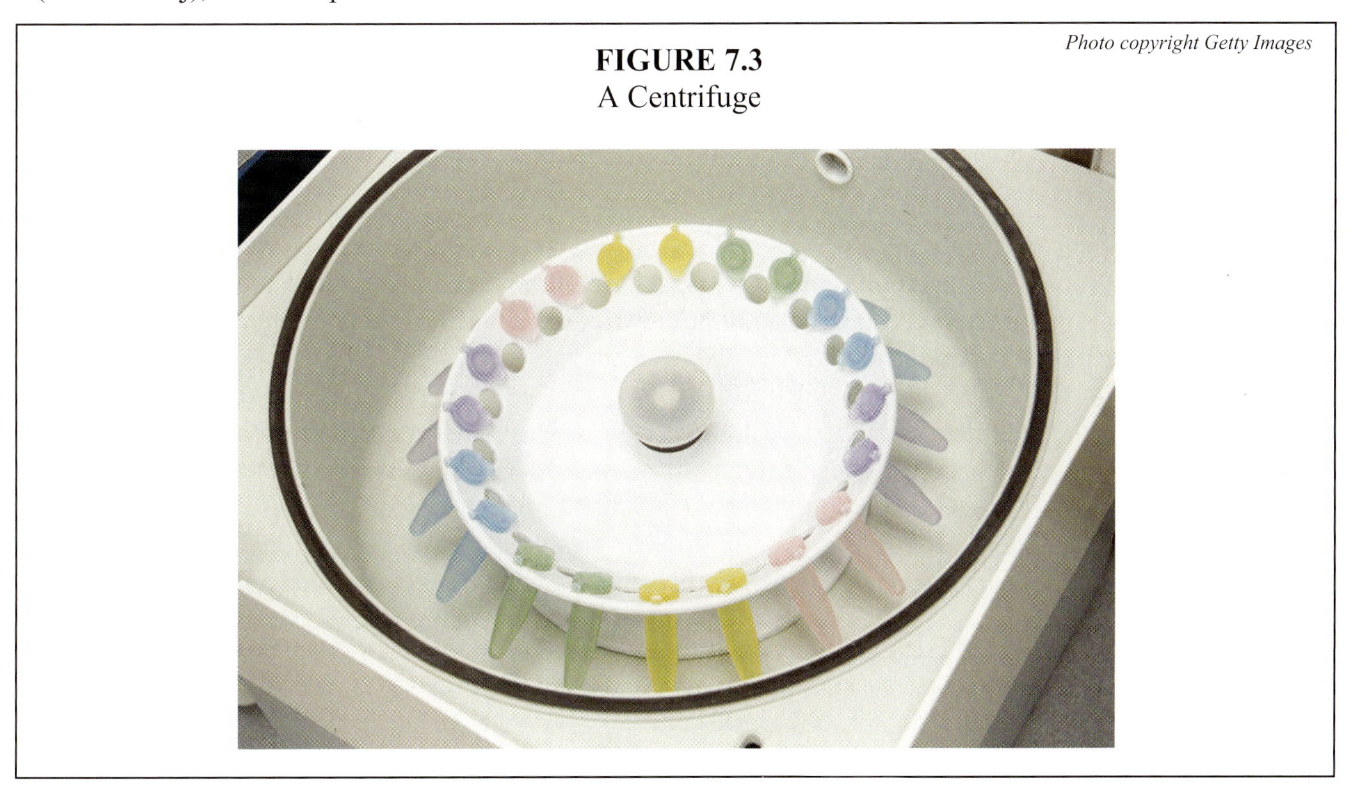

The plastic vials you see in the centrifuge are often filled with a liquid that has a solid suspended in it. When the centrifuge is turned on, the vials begin spinning around in a circle at high speed. When the spinning stops, the solid is separated from the liquid. Not only can a centrifuge separate solids from liquids, but it can also separate liquids of different densities. The cellular components of blood, for example, can be separated from blood plasma using a centrifuge.

Once again, however, a centrifuge does not separate solids from liquids and cellular components from blood plasma because of "centrifugal force." It does so simply because the components in the vials must obey Newton's First Law. As the vials spin in a circle, their contents initially continue to travel straight, along the tangent of the circle. This continues until a force is exerted to change their velocity. The only way this can happen is for the vials to exert a force on their contents, accelerating them so that they continue to travel in a circle. The more dense substances require more force (**F** = m**a**) to achieve the same acceleration, so they typically continue to move straight until they hit the edge of the vial. At that point, the vial can exert enough force to accelerate the substance so that it continues to follow the circular motion. Thus, even a centrifuge does not use "centrifugal force," as there is really no such thing!

Gravity

Now that we've had some practice identifying the supplier of centripetal force in circular motion, it is time to look at an important supplier of centripetal force: gravity. We all know something about gravity; it makes things fall and keeps us from flying off the earth. But what is it really? Well, we don't know exactly. We know that when an object has mass, it attracts other masses to itself. That's what we call gravity:

Gravity - The attractive force that exists between all objects which have mass

Even though we know that all objects which have mass are attracted to all other objects which have mass by the force known as gravity, we don't really know *why*.

There are currently two theories as to why massive objects attract one another. One theory states that all objects exchange tiny particles called "gravitons." All masses like to exchange these particles, and they try to get as close as possible to each other in order to facilitate such exchange. We see this as an attractive force and call it "gravity." The nice aspect of this theory is that physicists think that they know what causes positive and negative charges to be attracted to one another as well as north and south poles of magnets to be attracted to one another. It is through the exchange of tiny packets of light called **photons**. Thus, since we are fairly sure that the exchange of photons is what causes the force between charges and magnets, it would make sense that gravity would be caused by the exchange of particles as well. Although this sounds like a nice theory, we cannot seem to detect gravitons. Many proponents of this theory point out that gravitons should be very difficult to detect, so they are not dissuaded against their theory simply because we have not yet detected any.

The other theory was proposed by Albert Einstein and is called the General Theory of Relativity. This theory states that mass actually deforms space and time in its vicinity. This deformity tends to make a "hole" in space and time, and any masses near the hole will fall down the hole, towards the mass that caused the deformation. In this view, then, gravity is not really a force. It is a result of how mass affects space and time. It turns out that there is a lot of evidence to support this view, and most physicists consider Einstein's General Theory of Relativity to be the best explanation of the "force" that we call gravity.

Now if you didn't quite understand the last two paragraphs, don't worry. I only shared these two theories with you to show that although we have learned a lot about gravity, we still have a lot to learn. Thus, there is still a lot of exciting research going on in physics which tries to uncover its secrets!

Let's stop talking about the things that we don't know and start learning the things that we do know about gravity. First of all, Sir Isaac Newton developed an equation that he called the **Law of Universal Gravitation:**

$$F_g = \frac{Gm_1m_2}{r^2} \qquad (7.3)$$

What does this law mean? First, let's examine the term "universal." According to Newton, every object in the universe attracts every other object in the universe according to Equation (7.3). Thus, this law applies to *every object in the universe*.

What do the terms in this equation mean? The first term, "F_g," represents the force due to gravity. The "G" on the opposite side of the equation is called the **universal gravitational constant**. This is another fundamental constant whose value never changes. No matter where you are in the universe, this constant has a value of $6.67 \times 10^{-11} \frac{\text{Newton} \cdot \text{m}^2}{\text{kg}^2}$. The symbols "$m_1$" and "$m_2$" in Equation (7.3) represent the masses of the objects that are attracting one another. Finally, "r" stands for the distance between the centers of the two objects.

So, the first thing we see about the force due to gravity is that it gets stronger the closer two objects are to each other. If two masses are far from each other, "r" is large and thus their gravitational attraction is relatively weak. As they get closer to one another, "r" decreases, increasing the gravitational attraction between them. Also, the masses of both objects are important in determining the gravitational attraction that they have towards one another. Both masses must be taken into account before we can determine their mutual gravitational attraction. Study the following example in order to see how we apply Equation (7.3).

EXAMPLE 7.3

The centers of two masses (11 kg and 15 kg) are 1.2 meters apart from one another. What is the gravitational force between them? (G = $6.67 \times 10^{-11} \frac{\text{Newton} \cdot \text{m}^2}{\text{kg}^2}$)

This is a straightforward application of Equation (7.3). We know the universal gravitational constant, the mass of each object, and the distance between them. We can just plug these values into Equation (7.3) and come up with a value for the gravitational force:

$$F_g = \frac{Gm_1m_2}{r^2}$$

$$F_g = \frac{(6.67 \times 10^{-11}\ \frac{\text{Newton} \cdot \cancel{\text{m}^2}}{\cancel{\text{kg}^2}}) \cdot (11\ \cancel{\text{kg}}) \cdot (15\ \cancel{\text{kg}})}{(1.2\ \cancel{\text{m}})^2} = 7.6 \times 10^{-9}\ \text{Newtons}$$

Now you see why the universal gravitational constant has such strange units. With those units, the gravitational force calculated using Equation (7.3) ends up with the unit Newtons, as it should. So the mutual gravitational attraction between these two objects is only $\underline{7.6 \times 10^{-9}\ \text{Newtons}}$ when they are 1.2 meters apart.

What does this tell us about gravity? Well, it's a relatively weak force. The attractive force that we just calculated between two relatively heavy objects not too far apart from each other is really quite small. Thus, as forces go, gravity is pretty weak. Of course, the more massive things are, the more powerful gravity becomes. Overall, however, it takes a *lot* of mass to make a large gravitational force.

Now remember what Newton's Law of Universal Gravitation says. It says that every particle in the universe attracts every other particle in the universe with a gravitational force. Thus, if I put two objects on a table (a salt shaker and a pepper shaker, for example), they attract one another with a mutual gravitational force. When we see salt and pepper shakers next to each other on a table, then, why don't we see them inching towards one another? After all, if there is an attractive force between them, they should move towards each other, right?

Well, that's true only if the force acting to pull them together is greater than the static frictional force trying to keep them from moving. As we can see from the number we calculated in Example 7.3, gravity is a rather weak force. For most objects in our day-to-day life, the gravitational force that exists between them is far weaker than the static frictional force holding the objects stationary. As a result, even though everything on your dinner table is attracted to everything else on your dinner table, the gravitational attractions are so small that friction wins out, and the objects don't move towards one another.

All of this reasoning suddenly changes if one (or both) of the objects in question is very massive. Consider, for example, the case when one of the objects in question is the earth. The earth has a mass of 5.98×10^{24} kg. That's a pretty substantial mass! Because of this large mass, the earth exerts a pretty hefty gravitational force on virtually any object in its vicinity, as the following example demonstrates.

EXAMPLE 7.4

What is the gravitational attraction between a 1.0-kg object 1.2 meters above the earth's surface and the earth itself? (G = $6.67 \times 10^{-11}\ \frac{\text{Newton} \cdot \text{m}^2}{\text{kg}^2}$, mass of earth = 5.98×10^{24} kg, radius of earth = 6.38×10^6 m)

This is another straightforward application of Equation (7.3). We know the universal gravitational constant and the mass of each object. The only problem is what to use for "r." Remember, in Equation (7.3), "r" stands for the distance between the centers of the two objects. Well, if you examine the sketch below, you will see what to use for "r."

Illustration copyright GifArt.com

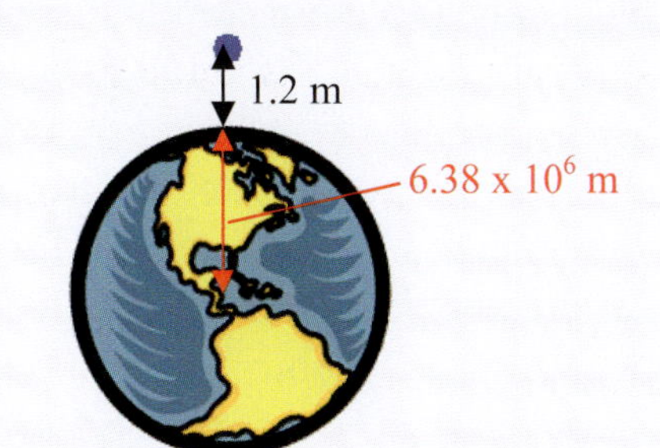

NOTE: This figure is not drawn to scale. The height of the object is overemphasized to make the illustration clear.

The distance between the centers of the two objects, then, is 6.38 x 10^6 m + 1.2 m. Now what does that equal? Remember, when adding numbers, we must report our answer to the same decimal place as the least precise number in the problem. Well, 6.38 x 10^6 m has its last significant figure in the ten thousands place (0.01 x 10^6 = 10,000), while 1.2 has its last significant figure in the tenths place. Thus, we must report our answer to the ten thousands place. This means that 1.2 m is simply not large enough to make a significant difference when added to 6.38 x 10^6. In the end, then, r = 6.38 x 10^6 m.

Now that we have the value for "r," we plug our numbers into Equation (7.3) and come up with a value for the gravitational force:

$$F_g = \frac{Gm_1m_2}{r^2}$$

$$F_g = \frac{(6.67 \times 10^{-11}\ \frac{\text{Newton} \cdot \cancel{m^2}}{\cancel{kg^2}}) \cdot (1.0\ \cancel{kg}) \cdot (5.98 \times 10^{24}\ \cancel{kg})}{(6.38 \times 10^6\ \cancel{m})^2} = 9.8 \text{ Newtons}$$

So the mutual gravitational attraction between these two objects is 9.8 Newtons.

Now the real reason I went through that example was not to show you how much stronger the gravitational force is when a massive object like earth is involved. I wanted you to see how to determine "r" when one of the objects in the situation is rather large. Since "r" is defined as the distance between the centers of the objects in question, you must make sure that the distance you use is defined that way. Since the distance in the problem was given from the earth's surface, the radius of the earth had to be added in order to properly define "r."

You are now ready to learn that I lied to you way back in Module #2. Are you surprised? Sometimes I have to lie a little in order not to confuse you. I promise you, however, that if I lie to you, I will either tell you that I am lying or clear the lie up in a later module. In this case, I am doing the latter. I hope you still remember the acceleration due to gravity. What is it? It's 9.8 m/sec^2, right? Well, not exactly. That's where the lie comes in. You see, the value we have learned for the acceleration due to gravity is just an approximation.

Think about it. The acceleration due to gravity exists because of the gravitational force that the earth exerts on all objects in its vicinity. Well, according to Equation (7.3), the gravitational force between the earth and any other object depends on the distance between them. If the object is far from the earth, the gravitational force is weaker than if the object is close to earth. Why, then, did we say that *regardless* of where the object was in relationship to the earth's surface, the acceleration due to gravity was always 9.8 m/sec^2?

Think about the previous example. When the object was 1.2 meters above the earth's surface, we had to calculate "r" by adding the height (1.2 meters) to the radius of the earth (6.38×10^6 meters). As I discussed in the example, 1.2 meters is so small compared to 6.38×10^6 meters, that, in the end, it does not affect the answer. Let's suppose the object was 1,000 meters (a pretty hefty distance) above the earth's surface. What would "r" be then? It would be 6.38×10^6 m + 1,000 m. To the proper significant figures, this is still 6.38×10^6 meters. Thus, whenever the object is reasonably close to the earth's surface, the radius of the earth is large enough to overwhelm the effect of the object's position relative to the surface of the earth. Thus, for all practical purposes, the distance between the center of the earth and any object anywhere near the earth's surface is 6.38×10^6 meters. What happens when I use Equation (7.3) to calculate the gravitational attraction between any object whose mass is "m" and the earth, using this value for "r?"

$$F_g = \frac{Gm_1m_2}{r^2}$$

$$F_g = \frac{(6.67 \times 10^{-11}\ \frac{\text{Newton} \cdot \cancel{\text{meter}^2}}{\cancel{\text{kg}^2}}) \cdot (\text{m}\ \cancel{\text{kg}}) \cdot (5.98 \times 10^{24}\ \cancel{\text{kg}})}{(6.38 \times 10^6\ \cancel{\text{meters}})^2} = \text{m} \cdot (9.8)\ \text{Newtons} = \text{mg Newtons}$$

According to Newton's Second Law, this force is equal to the mass of the object times the object's acceleration. In other words, 9.8 m/sec^2 is the acceleration due to gravity, as long as the distance between the object and the surface of the earth is small compared to the radius of the earth.

The point of this whole discussion is that the acceleration due to gravity that we learned way back in Module #2 is an approximation. It is a good approximation, but it is still an approximation. To be perfectly accurate, we would have to say that the acceleration due to gravity changes depending on where you are in relationship to the earth's surface. However, since most heights that we deal with are small compared to the radius of the earth, we can ignore this effect and say that, for all practical purposes, the acceleration due to gravity is reasonably constant anywhere near the surface of the earth.

ON YOUR OWN

7.5 What is the gravitational force between the moon and the earth? (mass of earth = 5.98×10^{24} kg, mass of moon = 7.36×10^{22} kg, radius of earth = 6.38×10^6 m, radius of moon = 1.74×10^6 m, distance between surface of earth and surface of moon = 3.76×10^8 m, G = $6.67 \times 10^{-11}\ \frac{\text{Newton} \cdot \text{m}^2}{\text{kg}^2}$).

Circular Motion Terminology

Before we go any further in our discussion of gravity, we need to get some terminology out of the way. It turns out that when studying circular motion, physicists have developed some special terms that condense the conversation. To make it easy on ourselves in a little while, we need to learn these terms now.

When an object travels in a circle, it continually passes by the same point over and over again. After all, that's the nature of circular motion. Well, suppose you were sitting at a particular point on a circle and an object passed you by. After a while, as you stayed at that point, you would see the object pass by you again, and then again, and then again. If the object travels at a constant speed, the time you need to wait before you see the object again will always be the same. In other words, if the object passes by you and then passes by you again two minutes later, you will see that object pass by you every two minutes as long as the object is in uniform circular motion. This time interval is called the **period** of the object's motion:

Period (T) - The time it takes for an object in uniform circular motion to travel through one full circle

Thus, if you see the object pass by you every two minutes, you can infer that the object has a period of two minutes.

Let's look at things from another perspective now. Let's assume that you are watching something that moves very quickly around a circle. It moves so quickly, in fact, that it's hard to time how long it takes for the object to move around the circle, because it is much less than one second. Is there something else we can measure that tells us something about the circular motion of the object? Well, suppose rather than trying to measure the time it takes to move around the circle, you instead counted the number of times each second that it passed you by. If you did that, you would be measuring its **frequency**.

Frequency (f) - The number of times per second an object in uniform circular motion travels around the circle

The unit for frequency is $\frac{1}{\text{sec}}$ which is often abbreviated "Hz," for "Hertz." The name for this unit was chosen to honor Heinrich Rudolf Hertz, a German physicist who greatly increased our understanding of electricity and magnetism. So, if you are watching an object in uniform circular motion, and that object passes you by five times every second, you could say that its frequency is 5 $\frac{1}{\text{sec}}$, or 5 Hz.

Now if you think about it, there is a relationship between frequency and period. If the period is large, it takes the object a long time to travel around the circle. This means that the frequency will be small. On the other hand, if the period is small, the object travels around the circle very quickly. This means that you will probably see it pass you by several times a second, making the frequency large. Thus, frequency and period are inversely related to each other. When frequency is large, period is small, and vice versa. Mathematically, we could say:

$$f = \frac{1}{T} \tag{7.4}$$

See how we use these definitions and this equation by studying the following example.

EXAMPLE 7.5

A hyperactive dog runs in a circle. The circle has a circumference of 2.2 meters, and the dog runs at a constant speed of 1.6 m/sec. What are the period and frequency of the dog's motion?

In order to make a full circle, the dog must travel 2.2 meters. If it is running with a speed of 1.6 meters per second, we can easily calculate how long it takes to make one full circle:

$$\text{distance} = \text{rate} \cdot \text{time}$$

$$2.2\ \text{m} = 1.6\frac{\text{m}}{\text{sec}} \cdot \text{t}$$

$$\text{t} = \frac{2.2\ \cancel{\text{m}}}{1.6\ \frac{\cancel{\text{m}}}{\text{sec}}} = 1.4\ \text{sec}$$

Since the definition of period is the time it takes to make one full circle, we can say that the period is 1.4 seconds. Now that we have period, we can determine the frequency using Equation (7.4):

$$f = \frac{1}{T}$$

$$f = \frac{1}{1.4\ \text{sec}} = 0.71\ \frac{1}{\text{sec}}$$

Thus, the dog's frequency is 0.71 Hz.

ON YOUR OWN

7.6 A child's train moves on a circular track. If the train has a constant speed of 1.8 m/sec and the radius of the circular track is 0.81 m, what is the train's period and frequency?

Gravity and the Motion of Planets

Now that we've covered the basics on gravity and we have circular motion terminology down pat, it is finally time to focus on the planets and their motion. Let's start with a figure.

Planet illustrations courtesy of NASA

FIGURE 7.4
A Schematic Representation of Our Solar System

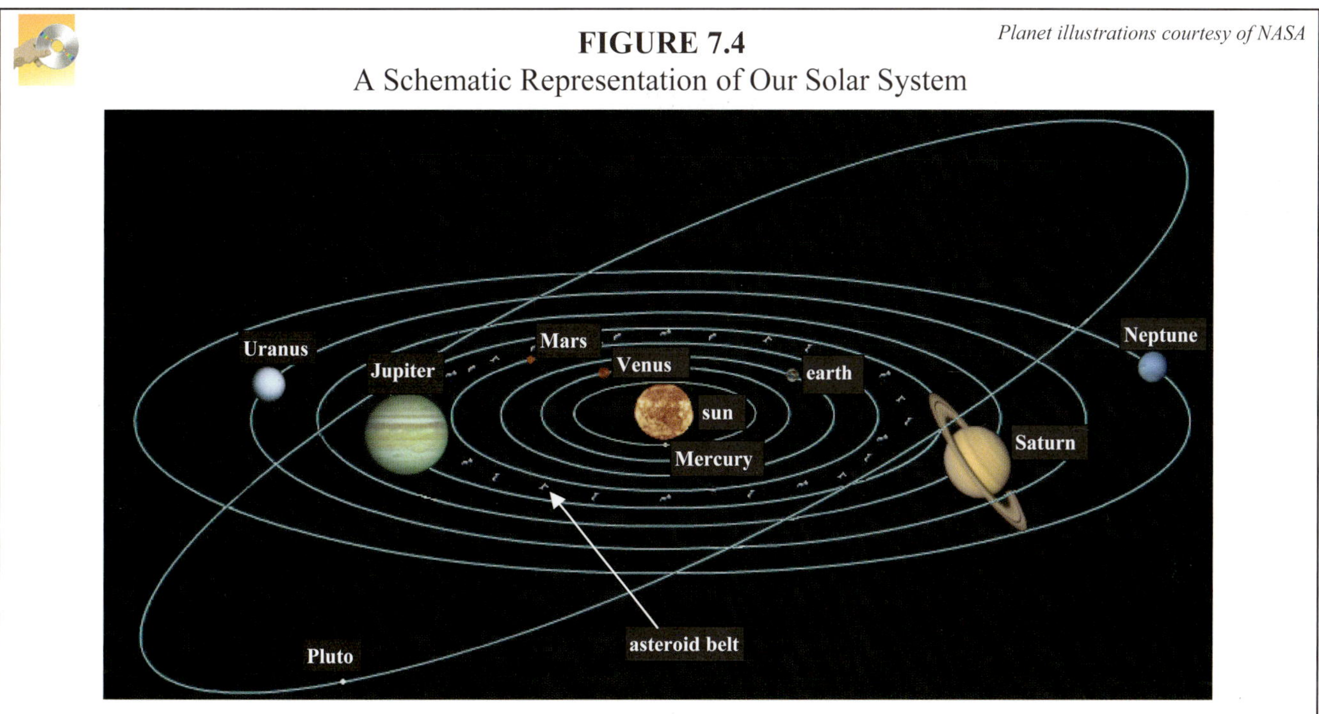

The sun is at the center of the solar system. Orbiting around the sun in nearly circular orbits are, in order, Mercury, Venus, earth, Mars, Jupiter, Saturn, Uranus, Neptune, and Pluto. Between Mars and Jupiter there is a belt of asteroids that orbit the sun. This belt provides a convenient distinction between the "inner planets" (Mercury, Venus, earth, and Mars) and the "outer planets" (Jupiter, Saturn, Uranus, Neptune, and Pluto). Please note that the planets are drawn roughly to scale, but the sun and the orbits are not. The sun is the largest object in the solar system, and drawing the orbits to scale would reduce the clarity of the illustration. Also, on August 24, 2006, the International Astronomical Union officially decided Pluto is no longer a planet. Check the course website I discussed in the "Student Notes" section at the beginning of the book to learn more about this change.

I'm sure that you've seen this before, but you've probably never realized why the planets orbit the sun in the way that they do. Of course, it looks nice, and it is also probably the most orderly arrangement for the planets, but that's not an explanation as to *why*. With the physics that you've just learned, you should now see why. After all, the sun exerts a gravitational force on each of the planets. This gravitational force pulls directly on each planet, towards the center of the sun. What do we call a force pulling on an object directly towards the center of a circle? We call it a centripetal force.

You see, the sun's gravitational attraction supplies a centripetal force to each planet, causing it to move in a circle. That's why the planets orbit the way that they do. Now it turns out that these orbits are not perfect circles. They are actually ellipses. However, these ellipses are nearly circular, so in this course, we will treat them as circles. If you take advanced physics later on, you will learn more about the elliptical nature of planetary orbits.

Now that we know *why* the planets move in nearly circular orbits, what can we learn about them? Well, we can learn a great deal, actually. Consider this next example:

EXAMPLE 7.6

How fast does Mars travel in its orbit around the sun? (mass of sun = 2.0×10^{30} kg, radius of Mars's orbit = 2.3×10^{11} m, G = 6.67×10^{-11} $\frac{\text{Newton} \cdot \text{m}^2}{\text{kg}^2}$).

To solve any problem like this, we first have to realize that the reason Mars orbits the sun in the first place is because the gravitational attraction between the sun and Mars provides the centripetal acceleration necessary to keep Mars in its nearly circular path. Thus, we could say that the force due to gravity (F_g) is equal to the centripetal force (F_c):

$$F_g = F_c$$

We needn't stop there, however. We have equations that allow us to develop expressions for each of those forces. Equation (7.1) relates centripetal force to radius and speed. Since I want to learn Mars' speed, this will be a good equation to substitute into the expression above. Also, Equation (7.3) relates the gravitational force to the mass of the planets and the radius of Mars's orbit. This will also be a good equation to use. So, if we use Equations (7.1) and (7.3) to substitute into the equation above, we get the following:

$$\frac{G \cdot m_{Mars} \cdot m_{sun}}{r^2} = \frac{m_{Mars} \cdot v^2}{r}$$

Notice what we did in our substitution. When we used Equation (7.3) to substitute for F_g, we explicitly wrote "m_{sun}" for the mass of the sun and "m_{Mars}" for the mass of Mars. We did this because we know that those are the two masses in this situation. Now look at the other side of the equation. When we used Equation (7.1) to substitute in for F_c we also had a mass term in the equation. Why did we label that mass term as "m_{Mars}?" Well, the "m" in Equation (7.1) refers to the mass of the thing that is moving in a circle. Which is moving in the circle, Mars or the sun? Mars is, of course. Thus, we use the mass of Mars in Equation (7.1).

Do you see what happens when we use this notation? The term "m_{Mars}" appears in the numerator of each fraction on both sides of the equation. Thus, it cancels. Also, there is an "r" term that appears in the denominator of each fraction on both sides of the equation. This allows us to cancel an "r" from both sides. In the end, then, our expression reduces to:

$$\frac{G \cdot \cancel{m_{Mars}} \cdot m_{sun}}{r^{\cancel{2}}} = \frac{\cancel{m_{Mars}} \cdot v^2}{\cancel{r}}$$

$$\frac{G \cdot m_{sun}}{r} = v^2$$

Look at what we have now. We have an equation for the speed of Mars, which is exactly what we wanted to solve for. All we have to do now is plug the given information into this equation and we will get our answer:

$$v^2 = \frac{(6.67 \times 10^{-11}\ \frac{\text{Newton} \cdot \text{m}^2}{\text{kg}^2}) \cdot (2.0 \times 10^{30}\ \text{kg})}{2.3 \times 10^{11}\ \text{m}} = \frac{(6.67 \times 10^{-11}\ \frac{\frac{\cancel{\text{kg}} \cdot \text{m}}{\text{sec}^2} \cdot \text{m}^2}{\cancel{\text{kg}^2}}) \cdot (2.0 \times 10^{30}\ \cancel{\text{kg}})}{2.3 \times 10^{11}\ \cancel{\text{m}}}$$

$$v = 2.4 \times 10^4\ \frac{\text{m}}{\text{sec}}$$

So, Mars travels a whopping $\underline{2.4 \text{ x } 10^4 \text{ m/sec } (5.4 \text{ x } 10^4 \text{ mph})}$ in its orbit!

Given the speed calculated in the previous problem, what is Mars's period of rotation around the sun?

Remember what period means. It means the time it takes to travel one full circle. Well, we know the speed; and, with the radius of the orbit given above, we can calculate the distance around the orbit. That will tell us how long it takes to travel one circle. First, let's calculate the distance around the orbit:

$$\text{circumference} = 2\pi \cdot (2.3 \times 10^{11}\ \text{m}) = 1.4 \times 10^{12}\ \text{m}$$

This is the distance around the orbit. Now that we have distance and speed, we can calculate time:

$$\text{distance} = \text{rate} \cdot \text{time}$$

$$\text{time} = \frac{\text{distance}}{\text{rate}} = \frac{1.4 \times 10^{12}\ \cancel{\text{m}}}{2.4 \times 10^4\ \frac{\cancel{\text{m}}}{\text{sec}}} = 5.8 \times 10^7\ \text{sec} = 1.8\ \text{years}$$

So the period of Mars's orbit is $\underline{5.8 \text{ x } 10^7 \text{ sec or } 1.8 \text{ years}}$.

See how much we can learn once we know about gravity and centripetal force? Actually, we can learn even more. Notice that so far in this module, I have given you the mass of the sun, the mass of earth, and the mass of the moon. How do we know these numbers? No one has gone out and put these planets on a scale! So how do we know their mass? Study the next example to find out.

EXAMPLE 7.7

We all know that the orbital period of the earth is one year ($3.2 \text{ x } 10^7$ seconds). Using this fact and the fact that the radius of earth's orbit around the sun is $1.51 \text{ x } 10^{11}$ meters, calculate the mass of the sun. ($G = 6.67 \text{ x } 10^{-11}\ \frac{\text{Newton} \cdot \text{m}^2}{\text{kg}^2}$).

Now, of course, we already have the answer to this problem. It was given in the last example. The purpose of this example is to show you how such data is determined. Where do we begin? The same place we always do in circular motion problems. We know that there is a centripetal force. In this case, the centripetal force is supplied by gravity, thus:

$$F_g = F_c$$

Equations (7.1) and (7.3) give us expressions for these forces, so we will substitute them in. In this case, the earth is the object that is in circular motion, so the mass term in Equation (7.1) is the mass of the earth. Thus, our expression looks like this:

$$\frac{G \cdot \cancel{m}_{earth} \cdot m_{sun}}{r^{\cancel{2}}} = \frac{\cancel{m}_{earth} \cdot v^2}{\cancel{r}}$$

$$\frac{G \cdot m_{sun}}{r} = v^2$$

In this problem, we are trying to find the mass of the sun. We know the values for "G" and "r," but to solve this equation for mass, we need to find "v." How do we do that? Well, we know the radius of earth's orbit, so with that information, we can calculate the distance the earth travels in one orbit. We also know the time is takes to make that orbit. If we divide distance by time, what do we have? We have "v."

$$\text{circumference} = 2\pi \cdot (1.51 \times 10^{11}\ \text{m}) = 9.49 \times 10^{11}\ \text{m}$$

Well, there's the distance. What time do I divide by? I'll be using "v" in an equation that uses Newtons as a unit. The unit Newton has seconds in it. Thus, to make the units consistent, I will have to use the time in seconds:

$$v = \frac{9.49 \times 10^{11}\ \text{m}}{3.2 \times 10^{7}\ \text{sec}} = 3.0 \times 10^4\ \frac{\text{m}}{\text{sec}}$$

So, along the way we see that the earth travels a blazing 3.0 x 10^4 m/sec (6.7 x 10^4 mph) in its orbit. Now that we know "v," we can plug it into our equation and solve for "m_{sun}":

$$\frac{G \cdot m_{sun}}{r} = v^2$$

$$\frac{(6.67 \times 10^{-11}\ \frac{\text{Newton} \cdot \text{meter}^2}{\text{kg}^2}) \cdot m_{sun}}{1.51 \times 10^{11}\ \text{meters}} = (3.0 \times 10^4\ \frac{\text{meters}}{\text{sec}})^2$$

$$m_{sun} = \frac{(9.0 \times 10^8\ \frac{\cancel{\text{meters}^2}}{\cancel{\text{sec}^2}}) \cdot 1.51 \times 10^{11}\ \cancel{\text{meters}}}{(6.67 \times 10^{-11}\ \frac{\frac{\text{kg} \cdot \cancel{\text{meter}}}{\cancel{\text{sec}^2}} \cdot \cancel{\text{meter}^2}}{\text{kg}^{\cancel{2}}})} = \underline{2.0 \times 10^{30}\ \text{kg}}$$

So, we measure the mass of the heavenly bodies by doing gravitational calculations that utilize our gravity and centripetal force equations. A pretty impressive feat, isn't it?

Since this example dealt with the orbit of the earth around the sun, I thought I would point out something very interesting about the radius of earth's orbit. The sun is earth's main energy source. Without that energy, life could not exist. Did you know, however, that with too much of the sun's energy, life would also cease to exist? You see, one of the many things that the energy from the sun does is heat up the earth. If the earth were just a few percent farther from the sun, it would not be warm enough to support life. On the other hand, if the earth were just a few percent closer to the sun, it would be too hot to support life. Isn't it great that the earth happens to be in *exactly* the right orbit to make life possible? This is just one more piece of evidence that science gives us to show that the earth was designed by a very intelligent designer!

Now the neat thing about the concepts and equations we have discussed so far is that they are applicable to more things than just planets and their orbits. These kinds of calculations can be done in any situation that uses gravity as a centripetal force. So, not only can you calculate information about the planets, the moon, and the sun, but you can also learn about any satellite that orbits a planet. For example, these same equations and concepts are used by television, armed forces, and communication engineers to analyze the motion of the satellites that they place in orbit around the earth.

Speaking of satellites, how do we put them in orbit? Suppose I am a cellular phone company and I want to place a satellite in orbit around the earth in order to facilitate my communications business. How do my engineers do it? Well, they first have to deploy the satellite into space. Usually, they simply attach it to a rocket. The rocket lifts the satellite straight into space. Eventually, the rocket is jettisoned, and then a thruster gives a velocity to the satellite that is tangential to the desired orbit. Since gravity pulls the satellite towards the center of the earth, and since the velocity imparted to the satellite by the thrusters is perpendicular to this force, circular motion is the result. In order to determine the altitude at which the satellite should be placed and what velocity it must be given, the engineers do calculations much like the ones we have done here.

These kinds of calculations are also used to measure the masses of many objects outside of our solar system. For example, most astronomers are convinced that a very massive, dense object sits at the center of our galaxy, which is called the "Milky Way." This object's mass has been calculated based on the speed at which stars near the center of the galaxy orbit around the center. Based on this mass and the size (which has been estimated by other means), astrophysicists say that this object is so dense that even light cannot escape its strong gravitational pull. Such an object is called a **black hole**, because it cannot be seen by telescopes. After all, telescopes collect the light that comes from planets and stars. Since light cannot escape a black hole, there is no light for a telescope to collect, and thus it cannot be seen. Black holes were predicted to exist based on equations from Einstein's **General Theory of Relativity**, and observations like these indicate that such objects do, indeed, exist. If you take advanced physics, you will learn more about black holes.

Just to give you an idea of what I am talking about, Figure 7.5 shows you a picture of the Milky Way galaxy. This picture was taken by the Goddard Space Center's Cosmic Background Explorer (COBE) in its orbit around the earth. Since the COBE is actually in the Milky Way galaxy, the best it can do is take an "edge-on" picture of the galaxy, which is what is shown in the figure. Also, it is important to realize that this picture was produced using infrared light, which we cannot detect with our eyes. Thus, the light you see in the picture is not actually the light that COBE detected. It is a computer-generated visible representation of infrared light.

Image courtesy of NASA

FIGURE 7.5
The Milky Way Galaxy

This is an edge-on shot of the Milky Way galaxy in infrared light. Note the bright bulge at the center of the galaxy. The light that you see in the bulge comes from stars that orbit the galaxy's center. Calculations indicate that these stars are orbiting a black hole.

Well, you've now scratched the surface of the fascinating world of gravity, the planets, satellites, and the galaxy. Of course, there is a lot more to learn about all of this, but we must move on to other areas of physics. If you find some of these topics interesting, I would encourage you to investigate the subject more in detail. Check out a book on astronomy from the library. You'll be fascinated by what you read! For right now, however, solve the "On Your Own" problems below so that you will master what we have covered here.

ON YOUR OWN

$$(G = 6.67 \times 10^{-11} \frac{\text{Newton} \cdot \text{m}^2}{\text{kg}^2})$$

7.7 Using the facts that the mass of the sun is 2.0×10^{30} kg and the radius of Pluto's orbit is 5.9×10^{12} m, calculate Pluto's orbital period.

7.8 The orbital period of the moon is 27.3 days (2.36×10^{6} sec). If its orbital radius is 3.8×10^{8} m, what is the mass of the earth?

7.9 A satellite engineer wishes to place a satellite in orbit around the earth (mass = 5.98×10^{24} kg). If the engineer wishes her satellite to have a circular orbit with a radius of 2.1×10^{7} m, what speed must the satellite have in orbit?

7.10 A satellite orbits the earth (mass = 5.98×10^{24} kg) with a velocity of 6.6×10^{3} m/sec. What is its altitude? (radius of the earth = 6.38×10^{6} m)

ANSWERS TO THE "ON YOUR OWN" PROBLEMS

7.1 We can use Equation (7.1) to relate centripetal force to the radius of the circular path that the object travels:

$$F_c = \frac{mv^2}{r}$$

$$4.67 \text{ Newtons} = \frac{(1.234 \text{ kg}) \cdot (12.1 \frac{\text{m}}{\text{sec}})^2}{r}$$

Now we just have to rearrange the equation to solve for r:

$$r = \frac{1.234 \text{ kg} \cdot (12.1 \frac{\text{m}}{\text{sec}})^2}{4.67 \text{ Newtons}} = \frac{1.234 \cancel{\text{kg}} \cdot (146.41 \frac{\text{m}^{\cancel{2}}}{\cancel{\text{sec}^2}})}{4.67 \frac{\cancel{\text{kg}} \cdot \cancel{\text{m}}}{\cancel{\text{sec}^2}}} = 38.7 \text{ m}$$

The object travels in a circle that has a radius of 38.7 m.

7.2 The centripetal force, radius, and velocity are all related by Equation (7.1). According to that equation, the centripetal force is inversely proportional to the radius of the circle. Thus, if centripetal force doubles and the speed does not change, the radius is cut in half.

7.3 Since the tension in the string is supplying the centripetal force, the circular motion cannot require a centripetal force of more than 18.1 Newtons. The faster the airplane twirls, the more centripetal force it will require to continue moving in a circle. Thus, if we put 18.1 Newtons into Equation (7.1), the speed we calculate with the equation will be the maximum possible speed.

Before we do that, however, we need to look at the units. The radius is given in cm and the mass is given in grams. The tension, however, is given in Newtons, the SI unit for force. Thus, I will need to convert the mass and radius to SI units as well. I assume that you know how to do that.

$$F_c = \frac{mv^2}{r}$$

$$18.1 \text{ Newtons} = \frac{(0.314 \text{ kg}) \cdot v^2}{0.67 \text{ m}}$$

$$v^2 = \frac{(18.1 \frac{\cancel{\text{kg}} \cdot \text{m}}{\text{sec}^2}) \cdot (0.67 \text{ m})}{0.314 \cancel{\text{kg}}}$$

$$v = 6.2 \frac{\text{m}}{\text{sec}}$$

7.4 As we did in the example problem, we must first develop an expression for the friction that exists between the tires and the road. The force of friction depends on the normal force (which for flat surfaces is just the weight of the car) times the coefficient of friction. We know neither the mass nor the coefficient, so our expression is not very enlightening:

$$f = \mu \cdot F_n$$

$$f = (\mu)\cdot(m)\cdot(9.8\frac{\text{meters}}{\text{sec}^2}) = (9.8\frac{\text{meters}}{\text{sec}^2})\cdot(\mu)\cdot(m)$$

Now, even though we don't have a number for the frictional force, let's go ahead and put the expression we have for it into Equation (7.1):

$$F_c = \frac{m \cdot v^2}{r}$$

$$9.8\frac{\text{meters}}{\text{sec}^2}\cdot(\mu)\cdot(m) = \frac{m\cdot(21\frac{\text{meters}}{\text{sec}})^2}{12.1\text{ meters}}$$

Since mass appears on both sides of the equation, it cancels out. Thus, it doesn't matter that we don't know the mass of the car, because it cancels out of the final equation. Now we can simply solve for μ:

$$9.8\frac{\text{meters}}{\text{sec}^2}\cdot(\mu)\cdot(\cancel{m}) = \frac{\cancel{m}\cdot(21\frac{\text{meters}}{\text{sec}})^2}{12.1\text{ meters}}$$

$$\mu = \frac{441\frac{\cancel{\text{meters}^2}}{\cancel{\text{sec}^2}}}{(12.1\ \cancel{\text{meters}})\cdot 9.8\frac{\cancel{\text{meters}}}{\cancel{\text{sec}^2}}} = \underline{3.7}$$

Since coefficients of friction are usually less than one, we find that this car had better slow down, because at this speed, friction will not be able to supply the centripetal force necessary to keep the car on the curve.

7.5 This is a simple application of Equation (7.3), provided we get the value for "r" correct. To make sure we know what we are doing, look at the drawing on the right. In order to get the distance between the *centers* of these two objects, we must add the radii to the distance between the surfaces. Thus:

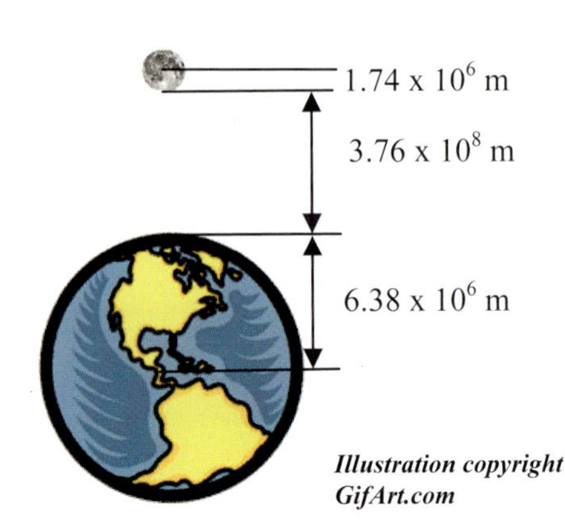

Illustration copyright GifArt.com

$$r = 6.38 \times 10^6\text{ m} + 3.76 \times 10^8\text{ m} + 1.74 \times 10^6\text{ m} = 3.84 \times 10^8\text{ m}$$

Now that we have "r," we can use Equation (7.3):

$$F_g = \frac{Gm_1m_2}{r^2}$$

$$F_g = \frac{(6.67 \times 10^{-11} \frac{\text{Newton} \cdot \cancel{m^2}}{\cancel{kg^2}}) \cdot (7.36 \times 10^{22} \cancel{kg}) \cdot (5.98 \times 10^{24} \cancel{kg})}{(3.84 \times 10^8 \cancel{m})^2} = \underline{1.99 \times 10^{20} \text{ Newtons}}$$

7.6 To determine the period, we must determine how long it takes for the train to travel one full circle. Well, we know its speed, so we only need to know the distance that the train travels when it makes a circle. We know the radius of the circle, and you should remember from geometry that the circumference of a circle is 2π times its radius. Thus:

$$\text{distance} = 2\pi \cdot (0.81 \text{ m}) = 5.1 \text{ m}$$

Now that we know the distance that the train travels to complete one full circle, calculating the time it takes for that to happen is trivial:

$$\text{distance} = \text{rate} \cdot \text{time}$$

$$5.1 \text{ m} = 1.8 \frac{\text{m}}{\text{sec}} \cdot \text{t}$$

$$\text{t} = \frac{5.1 \cancel{m}}{1.8 \frac{\cancel{m}}{\text{sec}}} = 2.8 \text{sec}$$

So the period is <u>2.8 sec</u>. To get frequency, we use Equation (7.4):

$$f = \frac{1}{T}$$

$$f = \frac{1}{2.8 \text{ sec}} = 0.36 \frac{1}{\text{sec}}$$

The frequency, then, is <u>0.36 Hz</u>.

7.7 As we do in all problems such as this, we must set the gravitational force equal to the centripetal force. Since Pluto is traveling in a what is almost a circle, the mass term in Equation (7.1) refers to the mass of Pluto.

$$F_g = F_c$$

$$\frac{G \cdot m_{sun} \cdot \cancel{m_{Pluto}}}{r^{\cancel{2}}} = \frac{\cancel{m_{Pluto}} \cdot v^2}{\cancel{r}}$$

$$\frac{(6.67 \times 10^{-11} \frac{\text{Newton} \cdot \text{m}^{\not 2}}{\text{kg}^{\not 2}}) \cdot (2.0 \times 10^{30} \not{\text{kg}})}{5.9 \times 10^{12} \not{\text{m}}} = \text{v}^2$$

$$\text{v} = 4.8 \times 10^3 \ \frac{\text{m}}{\text{sec}}$$

This, of course, is not our answer. To calculate period, we need to determine how far Pluto travels in one orbit. Given the orbital radius, this is no problem:

$$\text{circumference} = 2\pi \cdot (5.9 \times 10^{12} \ \text{m}) = 3.7 \times 10^{13} \ \text{m}$$

That's the distance traveled in one orbit. Now we can calculate the period:

$$\text{distance} = \text{rate} \cdot \text{time}$$

$$\text{time} = \frac{3.7 \times 10^{13} \not{\text{m}}}{4.8 \times 10^3 \ \frac{\not{\text{m}}}{\text{sec}}} = 7.7 \times 10^9 \ \text{sec}$$

This tells us that it takes Pluto 7.7 x 10^9 sec (2.4 x 10^2 years) to make one orbit around the sun.

7.8 We know that the orbital period of the moon is 2.36 x 10^6 seconds. Since we know that our equations do not use period, we might as well convert this to speed right away:

$$\text{circumference} = 2\pi \cdot \text{r} = 2 \cdot \pi \cdot (3.8 \times 10^8 \ \text{m}) = 2.4 \times 10^9 \ \text{m}$$

$$\text{speed} = \frac{2.4 \times 10^9 \ \text{m}}{2.36 \times 10^6 \ \text{sec}} = 1.0 \times 10^3 \ \frac{\text{m}}{\text{sec}}$$

Now that we have speed, we can set the gravitational force equal to the centripetal force and solve for the mass of the earth:

$$\text{F}_\text{g} = \text{F}_\text{c}$$

$$\frac{\text{G} \cdot \text{m}_\text{earth} \cdot \not{\text{m}}_\text{moon}}{\text{r}^2} = \frac{\not{\text{m}}_\text{moon} \cdot \text{v}^2}{\text{r}}$$

$$\frac{(6.67 \times 10^{-11} \frac{\text{Newton} \cdot \text{m}^2}{\text{kg}^2}) \cdot (\text{m}_\text{earth})}{3.8 \times 10^8 \ \text{meters}} = (1.0 \times 10^3 \ \frac{\text{m}}{\text{sec}})^2$$

$$m_{earth} = 5.7 \times 10^{24}\ \text{kg}$$

This answer is slightly different from the number that we have been using for the mass of the earth, due to rounding errors and the fact that the moon's orbit is not a perfect circle. Nevertheless, according to this calculation, the mass of the earth is $\underline{5.7 \times 10^{24}\ \text{kg}}$.

7.9 In order to get the satellite into the proper orbit, the thrusters must give it the exact velocity necessary to match the desired radius. To get the velocity, we simply set the gravitational force equal to the centripetal force, as we always have:

$$F_g = F_c$$

$$\frac{G \cdot m_{earth} \cdot \cancel{m_{satellite}}}{r^{\cancel{2}}} = \frac{\cancel{m_{satellite}} \cdot v^2}{\cancel{r}}$$

$$\frac{(6.67 \times 10^{-11}\ \frac{\text{Newton} \cdot \text{m}^2}{\text{kg}^2}) \cdot (5.98 \times 10^{24}\ \text{kg})}{2.1 \times 10^7\ \text{m}} = v^2$$

$$v = 4.4 \times 10^3\ \frac{\text{m}}{\text{sec}}$$

The thrusters must give the satellite a velocity of $\underline{4.4 \times 10^3\ \text{m/sec}}$ in order to keep it in the desired orbit.

7.10 The first step to solving this problem is the same as we have done in the three previous problems:

$$F_g = F_c$$

$$\frac{G \cdot m_{earth} \cdot \cancel{m_{satellite}}}{r^2} = \frac{\cancel{m_{satellite}} \cdot v^2}{\cancel{r}}$$

$$\frac{(6.67 \times 10^{-11}\ \frac{\text{Newton} \cdot \text{m}^2}{\text{kg}^2}) \cdot (5.98 \times 10^{24}\ \text{kg})}{r} = (6.6 \times 10^3\ \frac{\text{m}}{\text{sec}})^2$$

$$r = 9.2 \times 10^6\ \text{m}$$

Now this is not the answer. The question asked for the altitude of the satellite. What does that mean? It means the distance above the earth's surface. The radius that we just solved for is the distance from the earth's center. To get the altitude, we must subtract the radius of the earth from the orbital radius:

$$\text{altitude} = 9.2 \times 10^6\ \text{m} - 6.38 \times 10^6\ \text{m} = \underline{2.8 \times 10^6\ \text{m}}$$

REVIEW QUESTIONS

1. An object moves at constant speed. Is the acceleration necessarily zero? Why or why not?

2. An object moves from point A to point B at constant speed along the path drawn below. Draw its velocity and acceleration vectors. Don't worry about magnitude; I am interested only in direction.

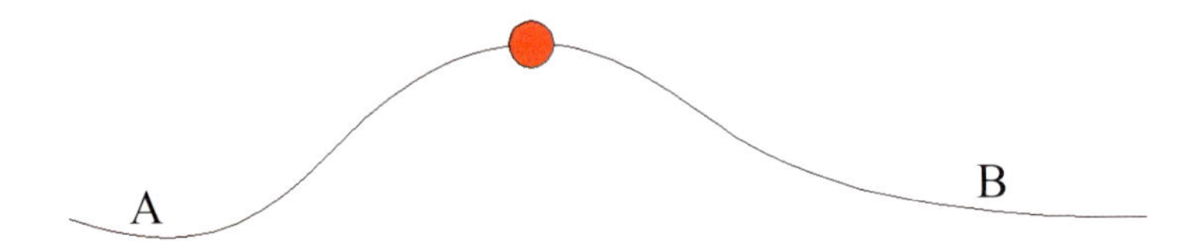

3. Given the force and velocity vectors drawn in the two situations below, which will result in uniform circular motion?

4. Consider an object moving uniformly in a circle. If the speed of the object suddenly doubles, but the radius of the motion does not change, what new centripetal acceleration is needed to keep the object moving in a circle?

5. A student places two objects on a table only a few centimeters apart. The student says that since the objects do not move towards one another, there must be no gravitational force between the two objects. Is the student correct? Why or why not?

6. Newton's Third Law states that for every action there is an equal and opposite reaction. Suppose two massive objects sit close to one another. The first object exerts a gravitational force on the second object. What is the equal and opposite force that exists in reaction to this gravitational force?

7. You have already been told that the acceleration due to gravity is not really constant near the surface of the earth. It varies a bit depending on where you are. Compare the acceleration due to gravity at sea level to that at the tip of Pike's Peak. Where do you expect the acceleration due to gravity to be greater?

8. Two objects are in close proximity to one another. If the distance between them is suddenly doubled, what happens to the strength of their mutual gravitational attraction?

9. An annoying fly travels around your head in a circle. If the fly suddenly increases its speed but stays in the same circle, does the fly's orbital period increase or decrease? What about its orbital frequency?

10. In the spin cycle of an automatic washing machine, the clothes are spun around inside the machine at a high rate of speed. This removes a large amount of the water that has soaked into the clothes. Given the fact that the walls of the washing machine have small holes in them, explain the physics behind why this works.

PRACTICE PROBLEMS

$$(G = 6.67 \times 10^{-11} \frac{\text{Newton} \cdot \text{m}^2}{\text{kg}^2})$$

1. An airplane pilot nose-dives and then pulls his 455-kg plane upwards, forming a semicircle in the air. If the radius of this semicircle is 112.2 m and the airplane's speed is a constant 73.3 m/sec, what is the centripetal force on the plane?

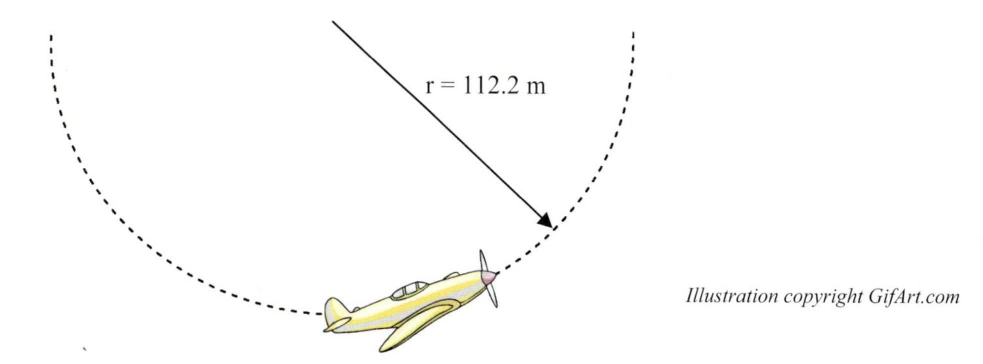

Illustration copyright GifArt.com

2. A popular amusement park ride is made up of a hollow cylinder that spins. People stand against the outside wall of the cylinder and, while it is spinning at its fastest speed, the floor falls away. The people do not fall through the hole, however, because they are "stuck" to the wall of the cylinder. This is because the cylinder's walls push against the people, exerting a centripetal force on them. The friction that is then created by the wall pressing against the people holds them up, keeping them from falling, despite the fact that there is no floor for them to stand on. If such a ride has a radius of 5.1 meters and spins with a speed of 15.1 m/sec, how much centripetal acceleration do the riders experience?

Illustration by Megan Whitaker

3. A child attaches a 34.5-cm string to a 15.0-g ball and begins twirling it around his head. He twirls it faster and faster. Eventually, when the ball reaches a speed of 12.1 m/sec, the string breaks. What is the maximum tension that the string can withstand?

4. A car is traveling down a slippery road ($\mu = 0.22$). What is the maximum speed at which the car can take a flat curve whose radius of curvature is 13.4 m?

5. A child watches his toy train moving around a circular track. He finds out that the train makes 10.0 completes trips around the track every 1.0 minute. What is the train's frequency? What is its period?

6. Two objects (m_1 = 15.0 kg, m_2 = 25.0 kg) are 45 cm apart from one another. What is the gravitational attraction between them?

7. How long does it take Saturn to make one orbit around the sun? (mass of the sun = 2.0×10^{30} kg, orbital radius of Saturn = 1.4×10^{12} m).

8. The orbital period of Io, one of Jupiter's moons, is 1.77 days (1.53×10^5 sec). If Io orbits Jupiter at a radius of 4.22×10^8 meters, what is the mass of Jupiter?

9. A satellite orbits the earth (mass = 5.98×10^{24} kg) with a speed of 1,123 m/sec. What is its orbital radius?

10. A satellite orbits the earth at an altitude of 2,314 km. What is the satellite's orbital period? (radius of earth = 6.38×10^6 m, mass of earth = 5.98×10^{24} kg).

Cartoon by Speartoons

MODULE #8: Work and Energy

Introduction

In previous modules of this course, we spent a lot of time discussing motion. We used our one-dimensional motion equations to describe and understand both one- and two-dimensional motion. We then used Newton's laws to better understand *why* motion occurred according to our equations. Finally, we even studied circular motion and how it applies to gravity.

It is now time to discuss something that affects all of these subjects: energy. Without energy, none of the things that we have discussed so far would ever occur. As a result, the fundamental concepts involved in understanding energy are some of the most important concepts that you will ever learn in physics. Indeed, the fundamental ideas that scientists have developed about energy have probably done more to help our understanding of creation than any other set of ideas. Obviously, then, it is important for any physics student to explore these concepts extensively.

The Definitions of Work and Energy

If you have already taken a good chemistry course, some of the things discussed in this section will be review. Nevertheless, there is probably a lot here that you haven't seen before, so pay close attention. First, if we are going to talk about energy, we had better define it:

<u>Energy</u> - The ability to do work

This might seem like a silly definition. After all, what is work? Well, physicists define work as follows:

<u>Work</u> - The product of the displacement of an object and the component of the applied force that is parallel to the displacement

If you think about it, this definition of work is much more restrictive than is our everyday usage of the word. For example, suppose you were trying to move a large rock in your yard. You push and push on the rock with all of your might, but the rock does not move. In the end, you give up, totally exhausted. Have you done any work on the rock? You might think that you have, but by the definition given above, you really haven't! You see, in order for work to occur, there must be displacement. In addition, the work that is actually done is calculated by determining the product of the displacement and the component of the force that is parallel to the displacement. In pushing on the rock, you were, indeed, applying a force, but the rock did not move. Thus, there was no displacement. As a result, you actually did *no work* on the rock!

Now remember, energy is the ability to do work. Thus, energy is the ability to apply a force which results in a displacement that has a component parallel to the force. So, if you push a rock with a force, and if the rock moves so that at least a component of its displacement is parallel to the force, energy has been expended. You need to remember, however, that energy is the *ability* to do these things. In order for energy to exist in a situation, the application of the force and the displacement do not have to actually occur. The *ability* for the application of the force and the resulting displacement is what matters. This is actually a very important point, and we will come back to it in a moment.

The Mathematical Definition of Work

Although I have already given you the definition of work, I want to show you how we define it mathematically. After all, physics does us little good if we cannot take the concepts that we learn and apply them to specific situations. That's what mathematics allows us to do.

$$W = F_{\parallel} \cdot \Delta x \qquad (8.1)$$

In this equation, "W" represents work , "Δx" stands for the magnitude of the displacement vector, and "$F_{\parallel}$" is used to indicate the magnitude of the component of the force vector that is parallel to the displacement vector. Please note that this equation assumes the force applied is constant throughout the displacement and that the displacement occurs in a straight line. If the force is not constant, or if the motion is not in a straight line, work can still be calculated, but you cannot use this equation.

To understand why Equation (8.1) uses only that portion of the applied force that is parallel to the motion, consider Figure 8.1:

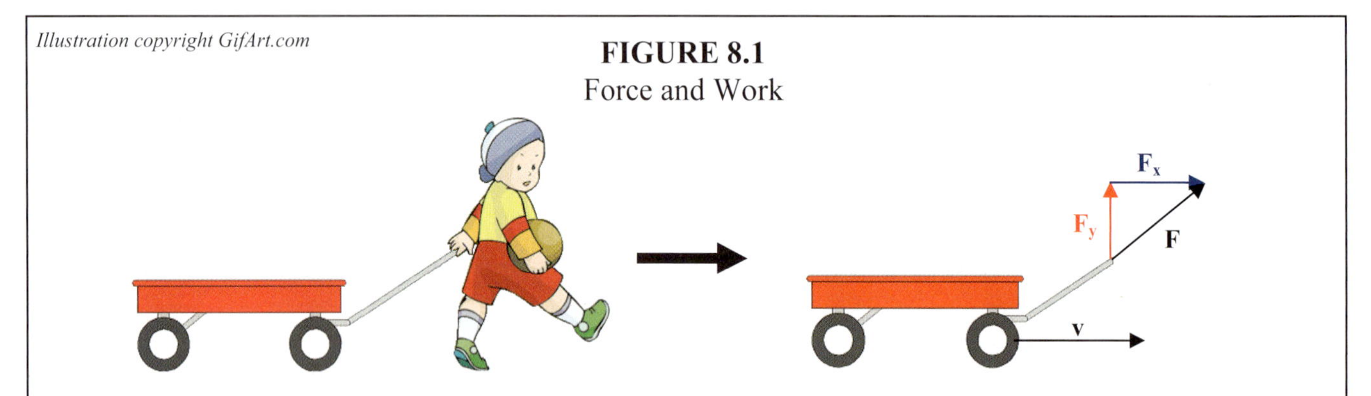

FIGURE 8.1
Force and Work

In the left side of this figure, a boy is pulling a wagon. If we remove the picture of the boy (right side of the figure) and just look at the vectors involved, we see several things. First, the wagon travels in a straight, horizontal line, as indicated by the horizontal velocity vector (**v**). The force that the boy is applying (**F**) goes in the same direction as the wagon's handle. As a result, his pulling force is not parallel with the motion. Since the force is a two-dimensional vector, we can split it into vertical (F_y) and horizontal (F_x) components. The horizontal component is parallel to the motion and, as you can see, is the only portion of the force that contributes to the motion. In contrast, there is no motion in the perpendicular direction. In other words, the wagon is not moving upwards. In the end, the perpendicular portion of the force fights gravity. Since it is not strong enough to overcome gravity, the wagon does not move up. Remember, when a force is applied but there is no displacement, there is no work. Thus, the perpendicular component of the boy's pulling force is wasted. It causes no displacement, and therefore it accomplishes no work. That's why only the portion of the force parallel to the displacement is considered when calculating the work done by that force.

So, when you are faced with a situation in which you must calculate the work done by a force, you can use Equation (8.1), as long as you use only the component of force which is parallel to the motion. How would you do that? Well, you can take any force vector and use trigonometry to split it into its components. Then, you can find the component that is parallel to the motion and plug it into

Equation (8.1). Even though such an exercise should be old hat to you, I won't require you to do it in this module. After all, in Module #7, I promised that you would see no trigonometry in this module. As a result, in all situations that I present in this module, the force will be parallel to the motion. Nevertheless, I do expect you to remember that only the parallel component of the force counts towards the work, and I expect you to understand why!

Now that I've gone through the concepts behind Equation (8.1), it's time to get into the heart of the mathematics. First, let's talk about the units involved. The SI unit for force is the Newton, and the SI unit for distance is the meter. Thus, when I calculate work, I take Newtons and multiply by meters. This gives me a $\text{Newton} \cdot \text{m}$. In honor of James Prescott Joule (jew' uhl), a brilliant British physicist (and a devout Christian) who did pioneering research in the study of energy, we call this unit the "Joule" (abbreviated as "J"). A portrait and brief biography of James Prescott Joule are given below.

Image from the AIP Emilio Segrè Visual Archives, Physics Today Collection

FIGURE 8.2
James Prescott Joule

James Prescott Joule was born in Lancashire, England in 1818. He was one of the greatest physicists of his day and is well known for his studies of both electricity and energy. He developed a law (now known as "Joule's Law") that describes the heat produced in an electrical circuit. He also studied the behavior of gases, which led him to conclude that the temperature of a gas falls when it expands without doing work. This is the principle by which refrigerators and air conditioners work. Finally, he showed how mechanical energy can be converted into heat and developed a mathematical relationship between the two. Joule was a devout Christian who saw science as a means of understanding God. He said, "After the knowledge of, and obedience to, the will of God, the next aim must be to know something of His attributes of wisdom, power and goodness as evidenced by His handiwork...It is evident that an acquaintance with natural laws means no less than an acquaintance with the mind of God therein expressed."(J.G. Crowther, *British Scientists of the Nineteenth Century* [London, K. Paul, Trench, Trubner & Co., Ltd., 1935], 138]

Since the Joule is a metric unit, we can add prefixes to it. Thus, kiloJoules, centiJoules, and megaJoules are all legitimate energy units as well. Also, you must remember that the Joule is simply the SI unit for work. In fact, any force unit (pound, Newton, dyne, etc.) multiplied by any distance unit (inches, cm, feet, etc.) is a valid energy unit. The standard energy unit in the English measurement system is $\text{ft} \cdot \text{lbs}$. Now that you see the units associated with work, it's time to look at a simple example and an "On Your Own" problem.

EXAMPLE 8.1

A father is teaching his daughter to ride a bike. He pushes her horizontally with a force of 15.1 Newtons. If the bike travels in the same direction for 0.213 km, how much work has he done?

The father is pushing his daughter in the same direction that she is moving, thus, as I promised, the force is parallel to the motion. So, we can plug it directly into Equation (8.1). Before we do that, however, we need to convert from km to meters so that our answer will be in Joules. After we do that, the equation looks like this:

$$W = F_{\parallel} \cdot \Delta x$$

$$W = (15.1 \text{ Newtons}) \cdot (213 \text{ m}) = 3.22 \times 10^3 \text{ Newton} \cdot \text{m} = 3.22 \times 10^3 \text{ J}$$

The father does 3.22×10^3 Joules of work while pushing his daughter. At this point, that number doesn't mean much to you. However, as we move along in this module, you'll come to appreciate just how much work that really is. There are a few more concepts we need to get through first, however.

ON YOUR OWN

8.1 A gardener is expanding her garden. In order to do this, she must push a heavy rock out of the way. She pushes the rock straight in the direction that it travels, moving it 451 cm. If she does a total of 34.5 Joules of work in the process, how much force did she apply?

Kinetic and Potential Energy

Now that we have learned an equation for work, it is time to learn some equations for energy. Before we do that, however, we need to learn a couple more definitions. Remember that I said energy exists in a situation simply if there is an *ability* to do the work. The work need not be done in order for the energy to exist. Even though this is the case, there is clearly a big difference between actually performing work and simply having the ability to do so. After all, when you have a job, your employer hires you because you have the *ability* to do the work, but you won't get paid until you actually *do* the work!

In physics, we also make a distinction between the ability to do work and actually doing the work. If energy exists in a situation but is not used to perform work, we call it **potential energy**. If the work is actually being performed, then motion occurs (remember: work must include motion). When a system is in motion, we call the energy associated with that motion **kinetic energy**. Another way to think about this distinction is that potential energy means energy that is stored up, while kinetic energy is energy in motion. That's the way we will define things:

Potential energy - Energy that is stored, ready to do work

Kinetic energy - Energy in motion

The distinction and interactions between kinetic and potential energy make up the bulk of the material in this module, so it is important that you grasp these concepts.

Suppose a rock is on the edge of a cliff. The rock is just sitting there, doing nothing. Is there energy in this situation? You might be tempted to say "no," because the rock is not doing anything. Nevertheless, you would be wrong. After all, what would happen if the rock were given a tiny push off the cliff? If that happened, the force of gravity would start pulling the rock, causing it to fall. In this situation, a force (gravity) would be applied in the direction of the displacement (the rock falls down and the force of gravity is pointed down). That's work! Thus, even though the rock is just sitting there, since it has the potential to fall, we say there is potential energy in this situation. When the rock actually falls, however, it starts moving. When that happens, we would say there is kinetic energy in the situation, because the rock is in motion. Do you see the distinction?

Since potential energy and kinetic energy are different, it should not be surprising to you that their equations are different as well:

$$PE = mgh \tag{8.2}$$

$$KE = \frac{1}{2}mv^2 \tag{8.3}$$

In these equations, "PE" means potential energy, whereas "KE" means kinetic energy; "h" is height, "g" represents the acceleration due to gravity, and "v" is speed. In both equations, "m" stands for the mass of the object in question. Let's consider each of these equations individually.

In Equation (8.2), we use the height of the object. What does that mean? Suppose you are done studying (yeah!!!) and you set your book down on the desk, leave your chair so that it is right under the book, and walk away. If you are asked to determine the height of the book, you most likely will measure the height from the floor to the book. You might then plug this height into Equation (8.2) to determine the potential energy in this situation. Nevertheless, if the book falls, it will probably hit the chair, stopping there. Thus, the ability of gravity to do work is really limited by the chair, because the book cannot fall any farther than the chair. For the purposes of determining the potential energy, then, you might just want to measure the height from the chair to the book. If you do that, however, the potential energy you calculate will be different. Which one is right?

It turns out that they both are, because potential energy is a *relative* quantity. In other words, potential energy must be calculated in relation to something else. Thus, there is a certain amount of potential energy if we consider the possibility of the book falling onto the chair, and there is a larger amount of potential energy if we consider the possibility of the book falling on the floor. When you calculate potential energy, then, you must determine what you are measuring it relative to. If you are ever unsure, you can always calculate the potential energy relative to the lowest spot to which an object can fall. You'll see how this works in the next example.

Before we get to the example, however, let's take a look at the units. In Equation (8.2), you multiply mass (usually measured in kg) by the acceleration due to gravity (usually in m/sec^2) and by height (usually in meters). The resulting unit is $\frac{kg \cdot m^2}{sec^2}$. Notice, however, that since a Newton is a

$\frac{\text{kg} \cdot \text{m}}{\text{sec}^2}$, this reduces to a Newton · m. Does this look familiar? It should, because it's a Joule. Now if you think about it, this should make perfect sense. After all, since energy is the ability to do work, the units for work and energy should be the same.

In fact, when you work on an object, you are actually adding energy to or taking energy away from the situation, depending on what you are doing. If, for example, you shove a toy car that is at rest, it rolls. That's because your shove performs work on the car. That work gives kinetic energy to the car, allowing the car to move. If you perform 5 Joules of work on the car, you will add 5 Joules of kinetic energy to it. In the same way, if you catch a baseball, it stops in your hand. This is because when the ball hits your hand, you begin to apply a force that opposes the motion. Your hand moves back a little in the process of catching the ball, so that force is being applied through a displacement. In other words, you are doing work. That work removes energy from the ball. If the kinetic energy of the baseball is 200 Joules, you must do 200 Joules of work to get the ball to stop. Thus, work and energy have a one-to-one correspondence. If you do 1 Joule of work, energy either increases or decreases by 1 Joule. If you don't quite understand this concept, don't worry; we'll come back to it soon. For right now, see if you have the concept of potential energy down by studying this example and solving the "On Your Own" problem that follows.

EXAMPLE 8.2

A ball (m = 34 g) rests on a staircase. If it is on the third step from the floor and each step is 45 cm high, calculate the ball's potential energy relative to both the second step and the floor.

This is a rather easy problem, as long as we get the units right. In order to come up with Joules, we need the mass in kg (0.034 kg) and the height of each stair in meters (0.45 m). Now that we have that out of the way, we can use Equation (8.2). Since the ball is on the third step, the second step is right below it. Each step is 0.45 m above the previous one, so the ball rests 0.45 m above the second step. Thus, Equation (8.2) becomes:

$$PE = mgh$$

$$PE = (0.034\ \text{kg}) \cdot (9.8\ \frac{\text{m}}{\text{sec}^2}) \cdot (0.45\ \text{m}) = 0.15\ \frac{\text{kg} \cdot \text{m}^2}{\text{sec}^2} = 0.15\ \text{J}$$

Relative to the second step, then, the ball's potential energy is <u>0.15 J</u>. What about relative to the floor? Well, since the ball is on the third step and each step is 0.45 m high, the ball is 1.35 m above the floor. Equation (8.2), then, becomes:

$$PE = mgh$$

$$PE = (0.034\ \text{kg}) \cdot (9.8\ \frac{\text{m}}{\text{sec}^2}) \cdot (1.35\ \text{m}) = 0.45\ \frac{\text{kg} \cdot \text{m}^2}{\text{sec}^2} = 0.45\ \text{J}$$

Thus, relative to the floor, the ball has a potential energy of <u>0.45 J</u>.

ON YOUR OWN

8.2 A roller coaster car (m = 365 kg) is at the top of the highest hill on the track. If the potential energy is 7.72×10^5 J relative to the ground and 6.67×10^4 J relative to the next hill the car encounters, what are the heights of the two hills ?

Now that you have a working knowledge of potential energy, let's move our attention to kinetic energy. As I said before, kinetic energy is energy in motion. Mathematically, we can calculate that energy with Equation (8.3). Once again, before we look at an example using this equation, let's examine the units. The SI unit for mass is kg, while the SI unit for speed is m/sec. If we multiply mass by the speed squared, we get $\frac{\text{kg} \cdot \text{m}^2}{\text{sec}^2}$. As we have learned already, this is the same as a Joule. Thus, as you might expect, the SI unit for kinetic energy is also the Joule.

EXAMPLE 8.3

A 6,675-Newton car speeds down the highway at 37 m/sec. What is its kinetic energy?

This is a straightforward application of Equation (8.3), as long as we watch our units. First, Equation (8.3) uses mass. Were we given mass? No, we were given weight, because the unit is Newtons. We will have to therefore use Equation (5.2) to calculate the mass:

$$w = mg$$

$$6{,}675 \text{ Newtons} = m \cdot (9.8 \frac{\text{m}}{\text{sec}^2})$$

$$m = \frac{6{,}675 \frac{\text{kg} \cdot \cancel{\text{m}}}{\cancel{\text{sec}^2}}}{9.8 \frac{\cancel{\text{m}}}{\cancel{\text{sec}^2}}} = 6.8 \times 10^2 \text{ kg}$$

Now that we have the mass, we can use Equation (8.3):

$$KE = \frac{1}{2}mv^2 = \frac{1}{2} \cdot (6.8 \times 10^2 \text{ kg}) \cdot (37 \frac{\text{m}}{\text{sec}})^2 = 4.7 \times 10^5 \frac{\text{kg} \cdot \text{m}^2}{\text{sec}^2} = 4.7 \times 10^5 \text{ J}$$

The car, therefore, is moving with $\underline{4.7 \times 10^5 \text{ J}}$ of kinetic energy. Once again, this number probably doesn't mean a whole lot to you, but after a while, you'll gain some perspective.

ON YOUR OWN

8.3 The engines of a rocket ship (m = 1.34×10^4 kg) give the ship a kinetic energy of 4.5×10^{10} J. How fast is the rocket ship traveling?

The First Law of Thermodynamics

The previous section gave you an introduction to the separate concepts of kinetic energy and potential energy. Although these concepts are useful by themselves, they become very, very powerful when you realize that they are connected through the First Law of Thermodynamics (ther' moh dye nam' iks). If you took a good chemistry course, you should have already been introduced to this law. Nevertheless, it is such an important law that I will review it now. The First Law itself is rather straightforward:

The First Law of Thermodynamics- Energy cannot be created or destroyed. It can only change form.

This important law tells us that there is a certain amount of energy in the universe. There is no way to add to this amount of energy or subtract from it. From the time of creation, the total amount of energy in the universe is fixed.

While it is impossible to add to or take away from the total amount of energy in the universe, that energy is free to change form. What do I mean when I say that energy can "change form?" Well, suppose you boil a pot of water on a stove. There is energy in the stove element (whether it is a gas flame or a hot electrical coil). That energy is in the form of heat. The heat warms up the water. Once the water gets hot enough, it boils. As you should have learned in chemistry, when a liquid boils, its molecules become gaseous, increasing their motion. Thus, when you boil a pot of water, energy in the form of heat (from the stove) changes into the energy contained in the motion of the gas molecules. In other words, energy changes form, from heat into motion. There are many such ways in which energy can change form.

For our purposes in this section of the module, the most important way that energy can change form is when it changes from potential energy into kinetic energy. For example, consider a rock at the edge of a cliff. I have already mentioned that even though the rock is just sitting there, energy exists in the form of potential energy. If I knew the mass of the rock and the height of the cliff, I could even use Equation (8.2) to calculate the potential energy in the situation. Now, suppose you give the rock a little shove. What will happen? The rock will begin falling, of course. At that point, there is kinetic energy, because the rock is moving. Where does that kinetic energy come from? According to the First Law of Thermodynamics, it couldn't have been created. So where does it come from?

Well, if you think about it, when the rock begins to fall, its height decreases. When its height decreases, what happens to its potential energy? The potential energy decreases. While this happens, however, the rock accelerates, so the rock's speed increases. What does that do to the kinetic energy? According to Equation (8.3), the kinetic energy increases when speed increases. So, as the rock falls, its potential energy decreases while its kinetic energy increases. This tells us that energy is changing form from potential energy to kinetic energy. As the rock's height decreases, potential energy decreases, which causes a corresponding increase in the kinetic energy. That increase in kinetic energy increases the speed of the rock.

In order to follow the First Law of Thermodynamics, there must be a one-to-one relationship between the potential energy decrease and the kinetic energy increase. In other words, as the rock falls, each Joule of potential energy lost must immediately be converted into a Joule of kinetic energy. That way, no energy will be created or destroyed. Suppose the height of the rock indicates that there are 100 Joules of potential energy in the situation. As the rock falls, potential energy will decrease and kinetic energy will increase, but the total amount of energy cannot change (according to the First Law of Thermodynamics). Thus, when it has fallen halfway down the cliff, its height will be half as much, so the potential energy will be half as much as well. Thus, when it is halfway down, there will be 50 Joules of potential energy. Since any potential energy lost must have been converted to kinetic energy, there will also be 50 Joules of kinetic energy.

Do you see what's going on here? There is a certain amount of energy in the situation. That energy can change from potential to kinetic, but the *total* amount of energy must remain the same. Mathematically, we could say the following:

$$\text{TE} = \text{PE} + \text{KE} \tag{8.4}$$

where "TE" stands for the total energy of the rock, "PE" represents the rock's potential energy at any given moment, and "KE" is the kinetic energy of the rock at that same moment. This equation is incredibly useful (as we will see in the examples that follow), and is often called the "conservation of energy equation" because it shows that as energy transforms from potential to kinetic (or vice versa), the total amount of energy cannot change.

Before I show you how Equation (8.4) can be used to analyze situations, I need to make a quick point about what "total energy" means in Equation (8.4). Please realize that you can analyze many different types of energy associated with a rock. The molecules that make up the rock are in random motion, for example, even when the rock itself is still. That is a form of kinetic energy. The molecules in the rock are held together with chemical bonds, which are a form of potential energy. Neither of these forms of energy are a part of Equation (8.4). As a result, the "total energy" described in Equation (8.4) is not the sum total of *all* energy in the rock. Instead, the "total energy" expressed in Equation (8.4) is the energy related to the linear motion of the rock itself. This is often called **mechanical energy**, and we will discuss this a bit more in an upcoming section. Thus, you could say that Equation (8.4) allows you to calculate the total *mechanical* energy of the rock. Now that you understand the terms in Equation (8.4), it is time to use it in analyzing physical situations.

EXAMPLE 8.4

A rock is at rest on top of a 121-m high cliff. If the rock falls off the edge of the cliff, how fast will it be traveling the instant before it hits the ground? Ignore air resistance.

In order to solve this problem, let's think about the energy involved. When the rock is at rest on top of the cliff, how much kinetic energy does it have? Well, if it's just sitting there, its speed is zero. According to Equation (8.3), this means that its kinetic energy is zero as well. The potential energy, however, is not zero, because the rock is on a cliff. Thus, the rock has the potential to fall and allow gravity to work on it. We can use Equation (8.2) to calculate its potential energy:

$$\text{PE} = \text{mgh}$$

$$\text{PE} = \text{m} \cdot (9.8\frac{\text{m}}{\text{sec}^2}) \cdot (121 \text{ meters})$$

We don't have the mass of the rock, so we cannot get a number for the potential energy. Is that a problem? Not really. Even though we do not have a *number* for the potential energy, we have an *expression* for it, and as we have seen in previous modules, sometimes the unknowns in an expression drop out when we use them. Thus, for right now, let's be satisfied with what we have.

Now, according to Equation (8.4), we can calculate the total amount of energy available to the rock by adding its potential and kinetic energy together. Since the rock is not moving, the kinetic energy is zero:

$$\text{TE} = \text{PE} + \text{KE}$$

$$\text{TE} = \text{m} \cdot (9.8 \frac{\text{meters}}{\text{sec}^2}) \cdot (121 \text{ meters}) + 0$$

$$\text{TE} = \text{m} \cdot (9.8 \frac{\text{meters}}{\text{sec}^2}) \cdot (121 \text{ meters})$$

This tells us that no matter where the rock is, the total energy will be given by the expression above. Thus, at the instant before it hits the ground, it will still have a total energy given by the expression above. Let's think about that for a moment. At the instant before the rock hits the ground, its height relative to the ground is zero. Thus, the potential energy will be zero. As a result, the expression for total energy at that point will be:

$$\text{TE} = \text{PE} + \text{KE} = 0 + \text{KE}$$

$$\text{TE} = \text{KE}$$

Now remember, we have an expression for the total energy. We determined that previously, and we know that it cannot change. We also have Equation (8.3), which relates the kinetic energy to the mass and speed. Let's use those two expressions in the equation we just determined:

$$\text{TE} = \text{KE}$$

$$\text{m} \cdot (9.8 \frac{\text{meters}}{\text{sec}^2}) \cdot (121 \text{ meters}) = \frac{1}{2} \text{mv}^2$$

Look what happens. Mass appears on both sides of the equation, so it cancels. Thus, we did not need to know the mass, because it eventually drops out of the equations! Now we can solve for speed:

$$\cancel{\text{m}} \cdot (9.8 \frac{\text{meters}}{\text{sec}^2}) \cdot (121 \text{ meters}) = \frac{1}{2} \cancel{\text{m}}\text{v}^2$$

$$\text{v} = \sqrt{2 \cdot (9.8 \frac{\text{meters}}{\text{sec}^2}) \cdot (121 \text{ meters})} = 49 \frac{\text{m}}{\text{sec}}$$

The instant before the rock hits the ground, then, it will be traveling with a speed of 49 m/sec.

Do you see what these equations and concepts allow us to do? Using relatively simple energy arguments, we can calculate the speed of an object as it falls. That's a pretty powerful tool. "But wait a minute," you might be thinking, "we already learned to solve problems like this, in Module #2. What's so powerful about this method, then?" Well, remember the major limitations of the concepts we learned in Modules #1-#4. In order to solve a problem, it had to be composed (or broken up into sections that were composed) solely of linear motion. In addition, the acceleration had to be constant in order for most of our equations to apply. The concepts that we have just learned apply to *all* situations, regardless of whether or not the motion is linear and regardless of whether or not the acceleration is constant. Consider the next example.

Illustration by Megan Whitaker

EXAMPLE 8.5

A roller coaster car is pulled up a hill that is 113 meters high. It stops at the top of the hill for a second and then begins to roll down the hill. It accelerates to the bottom of the hill (which is still 22 meters above the ground), and then it begins to climb another hill which is 67 meters high (see figure below). Ignoring friction, what is the car's speed when it reaches the bottom of the first hill and when it reaches the top of the next hill?

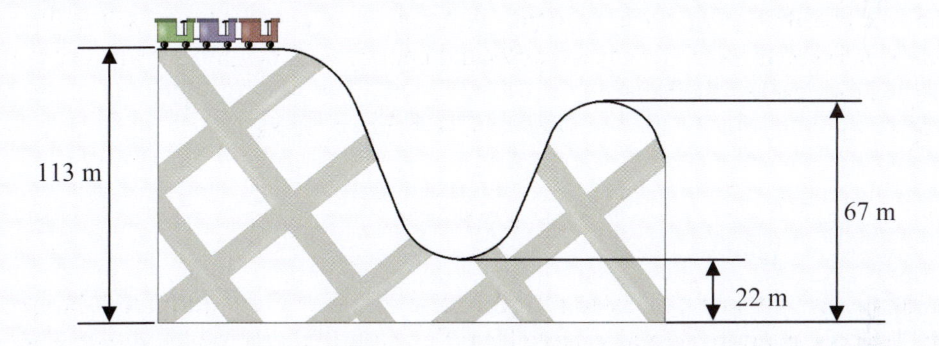

The only way to solve a problem like this is to think in terms of the energy involved. We don't know how far the car travels and, since the angle of incline is not constant, we cannot calculate the car's acceleration. Nevertheless, using our energy concepts and equations, we can solve this problem. First, we have to figure out the total energy. The easiest way to do this is to find a place where either the potential energy or kinetic energy is zero. That way, we only have to use one equation to calculate the total energy. In this case, we know that the car rests on top of the first hill for a moment without moving. At that point, the kinetic energy is zero, so all of the energy must be potential:

$$\text{PE} = \text{mgh}$$

$$\text{PE} = \text{m}\cdot(9.8\frac{\text{meters}}{\text{sec}^2})\cdot(113\text{ meters})$$

Now, once again, we do not have a number for potential energy, but we do have an expression. That should be good enough. Since there is no kinetic energy at this point, the total energy is just:

$$\text{TE} = \text{PE} + \text{KE} = \text{m}\cdot(9.8\frac{\text{meters}}{\text{sec}^2})\cdot(113\text{ meters}) + 0$$

$$TE = m \cdot (9.8\frac{\text{meters}}{\text{sec}^2}) \cdot (113 \text{ meters})$$

Thus, throughout the car's entire trip, the total energy must equal the expression given above. So, even when the car is at the bottom of the hill, the total energy is equal to the expression given above. However, at that point, the car is still above the ground (at a height of 22 meters), and thus there is still potential energy. At the same time, however, it is moving, which means that there is also kinetic energy. The sum of these two energies must equal the expression given above.

Since we know the height, we can calculate an expression for the potential energy at the bottom of the hill:

$$PE = mgh$$

$$PE = m \cdot (9.8\frac{\text{meters}}{\text{sec}^2}) \cdot (22 \text{ meters})$$

We also know an equation for kinetic energy:

$$KE = \frac{1}{2}mv^2$$

Now remember, the total energy at any time must equal the kinetic energy plus the potential energy. Well, we have an equation for the total energy at any time (the equation at the top of the page), and we now have equations for the potential energy and kinetic energy at the bottom of the first hill. Thus:

$$TE = PE + KE$$

$$m \cdot (9.8\frac{\text{meters}}{\text{sec}^2}) \cdot (113 \text{ meters}) = m \cdot (9.8\frac{\text{meters}}{\text{sec}^2}) \cdot (22 \text{ meters}) + \frac{1}{2}mv^2$$

Now look what happens. The mass appears in every term of the equation, so it cancels out. Once again, then, we did not need numbers for the total energy, potential energy, and kinetic energy. When we put our expressions for TE, PE, and KE into Equation (8.4), all unknowns except speed (which we need to calculate) dropped out!

$$\cancel{m} \cdot (9.8\frac{\text{meters}}{\text{sec}^2}) \cdot (113 \text{ meters}) = \cancel{m} \cdot (9.8\frac{\text{meters}}{\text{sec}^2}) \cdot (22 \text{ meters}) + \frac{1}{2}\cancel{m}v^2$$

$$\frac{1}{2}v^2 = (9.8\frac{\text{meters}}{\text{sec}^2}) \cdot (113 \text{ meters}) - (9.8\frac{\text{meters}}{\text{sec}^2}) \cdot (22 \text{ meters}) = 890\ \frac{\text{meters}^2}{\text{sec}^2}$$

$$v = \sqrt{2 \cdot 890\ \frac{\text{meters}^2}{\text{sec}^2}} = 42\ \frac{\text{m}}{\text{sec}}$$

At the bottom of the first hill, then, the car is moving at 42 m/sec.

We can calculate the speed of the car when it reaches the top of the next hill in the same way. After all, when it reaches the top of the next hill, it still must have a total energy given by the expression we calculated previously:

$$TE = m \cdot (9.8\frac{\text{meters}}{\text{sec}^2}) \cdot (113 \text{ meters})$$

Remember, the total energy cannot change, so this expression must be true everywhere, including at the top of thc next hill. Since we know that TE = PE + KE, all we need to do is get expressions for PE and KE at the top of the next hill, and we will be all set. We can use the height of the hill to get an expression for the potential energy:

$$PE = mgh$$

$$PE = m \cdot (9.8\frac{\text{meters}}{\text{sec}^2}) \cdot (67 \text{ meters})$$

We also have an equation for kinetic energy :

$$KE = \frac{1}{2}mv^2$$

Now we just set our expression for TE equal to the sum of the expressions we have for PE and KE:

$$TE = PE + KE$$

$$m \cdot (9.8\frac{\text{meters}}{\text{sec}^2}) \cdot (113 \text{ meters}) = m \cdot (9.8\frac{\text{meters}}{\text{sec}^2}) \cdot (67 \text{ meters}) + \frac{1}{2}mv^2$$

Once again, the unknown mass drops out. That way, we can simply solve for v:

$$\cancel{m} \cdot (9.8\frac{\text{meters}}{\text{sec}^2}) \cdot (113 \text{ meters}) = \cancel{m} \cdot (9.8\frac{\text{meters}}{\text{sec}^2}) \cdot (67 \text{ meters}) + \frac{1}{2}\cancel{m}v^2$$

$$\frac{1}{2}v^2 = (9.8\frac{\text{meters}}{\text{sec}^2}) \cdot (113 \text{ meters}) - (9.8\frac{\text{meters}}{\text{sec}^2}) \cdot (67 \text{ meters}) = 450\ \frac{\text{meters}^2}{\text{sec}^2}$$

$$v = \sqrt{2 \cdot 450\ \frac{\text{meters}^2}{\text{sec}^2}} = 3.0 \times 10^1\ \frac{\text{m}}{\text{sec}}$$

At the top of the next hill, then, the car is moving with a speed of 3.0×10^1 m/sec. Note that we must use scientific notation to express this answer, as the rules for significant figures tell us that the answer must have two significant figures. The only way to express an answer of 30 with two significant figures is to use scientific notation.

Now this example problem may have seemed really long, and the math was probably a little difficult. However, think about what happened here. Since we know that the total energy in the system cannot change, once we develop an expression for the total energy, we know that it stays the same no matter what goes on in the problem. Thus, our first job was to determine an expression for the total energy of the system. We chose to do that when the roller coaster was not moving. That way, the kinetic energy was zero. Thus, the total energy was the same as the potential energy. Even though we could not get a *number* for the total energy, we could get a mathematical expression for it. At that point, we realized that we could use that expression for the total energy *at any other point in the problem.*

Thus, we simply developed expressions for the potential and kinetic energies at each point in the problem. Since total energy always equals potential energy plus kinetic energy, we simply set our expression for the total energy equal to our expression for the kinetic energy plus our expression for the potential energy. At that point, the annoying unknown mass dropped out, and the only unknown left was the speed, which was what we were asked to calculate. See if you can do this kind of reasoning yourself by solving the following "On Your Own" problems.

ON YOUR OWN

8.4 A biker pedals hard to ride his bike to the top of a 21-meter hill. He rests momentarily at the top, exhausted. He then decides to let his bike coast down the hill, so that he does not have to pedal. He is having so much fun coasting that he decides not to pedal even when his bike begins climbing the next hill. If that hill is 12 meters high, what will the biker's speed be at the top of that hill? Ignore friction.

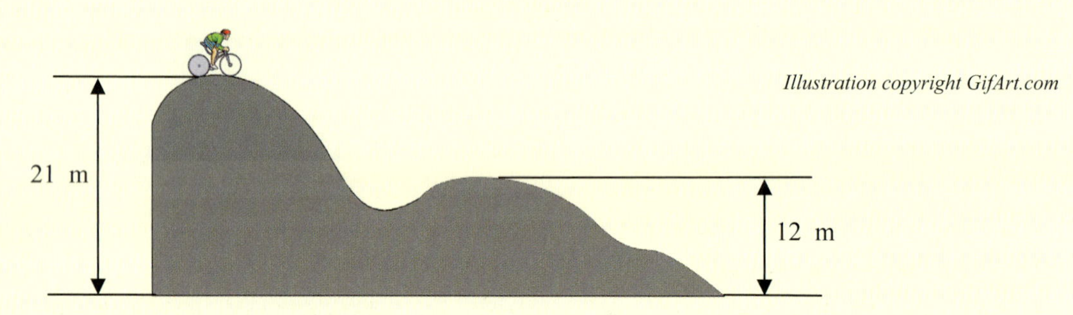

Illustration copyright GifArt.com

8.5 A teenager shoots a target with his BB gun. The target is mounted on a post so that it is 25 meters higher than the tip of the gun. If the gun shoots the BB with an initial speed of 99 m/sec, at what speed does the BB hit the target? Ignore air resistance.

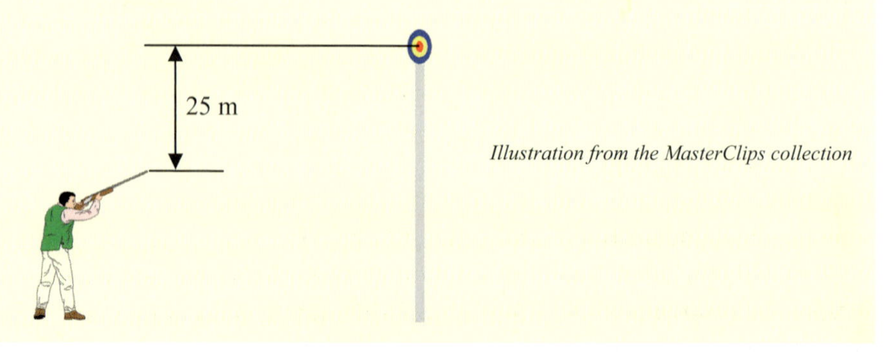

Illustration from the MasterClips collection

Before we move on to another section, I want you to be aware of one important consequence of Equation (8.4). The best way to learn this consequence is through experiment.

EXPERIMENT 8.1
Energy in a Pendulum

Supplies:

- Safety goggles
- A piece of string at least 25 cm long
- A ruler
- A desk or table
- Some tape
- Some books
- A mass (a ball or a nut, for example) to hang on the string. It should be heavy enough to keep the string taut when it hangs at the end.

1. Attach the mass to the string. You can either tie it to the string or attach it with tape; it really doesn't matter.
2. Lay the ruler on the edge of a desk or table so that the ruler sticks out over the edge.
3. Place the books on top of the ruler, at the end that is on the desk. This will allow the ruler to stay on the desk.
4. Attach the other end of the string to the ruler on the side that is sticking out over the edge of the desk, so that the mass hangs down on the string. You have just constructed a pendulum (pen' juh lum). It should look something like the figure below:

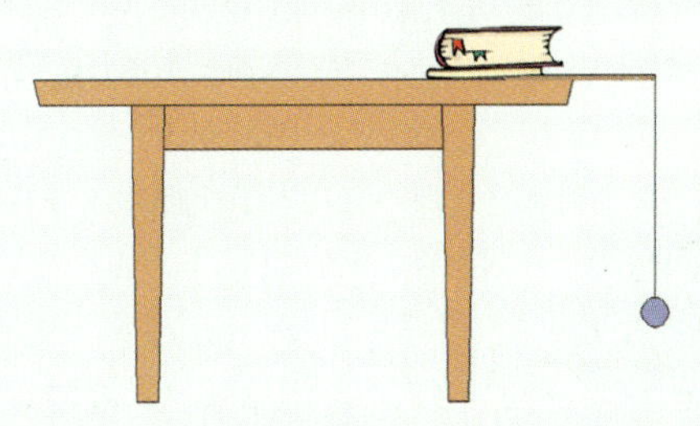

5. Once you have your pendulum set up, kneel down and grasp the mass in your hand. Pull it up so that it touches your nose (see diagram below) and release the mass. Do not give it any kind of shove. Just release it and keep perfectly still. Do not allow your head to move. Watch the mass as it swings back and forth a few times. Record what happens.

6. Clean up your mess.

What happened in the experiment? If you didn't give the mass any kind of shove when you released it, the mass should have swung away from your face and, eventually, it should have swung back. As long as you kept your head still, the mass should have come very close to (and perhaps even slightly touched) your nose. Each time the mass swung back to your face, it should not have swung quite as high as the time before. Thus, the mass stayed farther away from your nose with each swing.

Why did this happen? Well, let's think about the energy involved for a moment. When you lifted the mass and held it at your nose, you added potential energy to the system, because the mass was higher than its equilibrium position. As you were grasping the mass, it was not moving, so there was no kinetic energy. Thus, the total energy was the same as the potential energy. When you released the mass, gravity started pulling it down. This reduced its height, reducing the potential energy. At the same time, it began accelerating, increasing the kinetic energy. When the mass reached the point at which it was hanging straight down, there was a lot of kinetic energy and no potential energy relative to its equilibrium position. Once it passed that point, however, it began to climb again. This increased the potential energy and decreased the kinetic energy. The mass, therefore, started to slow down. Eventually, it climbed high enough to convert all of the kinetic energy back into potential energy. At that point, the mass stopped.

Once the mass stopped, however, gravity began to pull it down again, and the whole process started over. This made the mass swing back to your face. Why did it stop when it reached your nose, though? Why didn't it continue and whack your nose really hard? Think about it. When you initially raised the mass, you raised it to your nose's height. That gave potential energy to the system. Since the mass was not moving at the time, it had no kinetic energy, so the mass's total energy was the same as its potential energy. As the mass began to swing back and forth, it started changing potential energy into kinetic energy and vice versa. When the mass was falling, it converted potential energy into kinetic energy. When the mass was climbing, it changed kinetic energy into potential energy.

So, when the mass began climbing towards your face again, it started converting kinetic energy into potential energy. This couldn't go on indefinitely, however. Once *all* of the energy was converted to potential energy, the mass would stop. Where did that happen? Right where you released it to begin with. The height at which you released the mass determined the potential energy. Since it had no kinetic energy at that point, the height also determined the total energy. As a result, the mass could never climb higher than that height, because there was no more energy to convert to potential energy. Thus, no matter how fast the mass seemed to be approaching your face, you didn't need to flinch or move, because the First Law of Thermodynamics, coupled with Equation (8.4), tells us that there is no way the mass could ever really hit you. It simply would never have enough energy to climb higher than the point at which you released it.

Now, of course, if you supplied it with a little extra energy by giving it a shove when you released it, then it would be able to come back and crack you in the nose. That's because shoving it would give it kinetic energy *in addition* to the potential energy. Thus, as it began climbing, it would have extra energy to convert to potential energy, allowing it to climb higher and smack into your nose.

There is at least one question that should come to your mind. As the pendulum continued to swing back and forth, it never climbed as high on the second swing as it did on the first swing. Each time the mass started climbing towards your face, it started falling again at a lower height. Why did that happen? As you might expect, it's because of friction. There is friction between the molecules in the air and the mass on the pendulum. The friction takes energy away from the mass; thus, each time the mass begins to climb, it has less energy to convert to potential energy, so it cannot climb as high as it did before.

Wait a minute, though? Doesn't the First Law of Thermodynamics tell us that you cannot destroy energy? If friction takes energy away from the motion of the mass, where does it go? That's what you'll learn in the next section.

Friction, Work, and Energy

So far, we've talked a lot about converting potential and kinetic energy back and forth, but we haven't talked a lot about other forms of energy and how systems can convert between them. There are many forms that energy can take. Some are listed below:

Mechanical energy - Energy associated with the movement (or potential movement) of objects

Chemical energy - Energy associated with the chemical bonds of a molecule

Electrical energy - Energy associated with the motion (or potential motion) of charged particles

Heat - Energy that is transferred from one object to another as a result of a difference in temperature

With the exception of heat, all of these forms of energy can be either kinetic or potential. For example, a rock at the edge of a cliff is an example of potential mechanical energy. If that energy is converted to kinetic form, the rock will move. Thus, the potential energy is associated with moving the rock downwards. Since it is associated with potential movement, we say it is potential mechanical energy. When the rock actually does fall, the kinetic energy it has is kinetic mechanical energy. In the same way, when a molecule exists, there is energy stored up in its chemical bonds. Thus, we say it has potential chemical energy. When that molecule participates in a chemical reaction, the energy is released. At that point, we say it is kinetic chemical energy. Finally, a battery has electrical energy stored in it, so we say it has potential electrical energy. When you hook it up to your radio and turn the radio on, that energy is released in the form of electrons moving down wires. That's kinetic electrical energy.

Heat is the only one of the above forms that must always be kinetic. You should see that the definition of heat requires this. After all, if energy is transferred from one object to another, *something* must be moving. Therefore, heat must be kinetic energy. If you had a good chemistry course, you should have learned about heat.

Now that you have seen a few forms of energy, let's talk a little about how energy is converted from one form to another. Suppose you push a toy car across a floor. When you give the toy car a shove, you've given it some kinetic energy. How did you do that? You did it by a series of energy conversions. First, you ate some food. The potential chemical energy stored in the molecules of food was sent to the cells in your body. When you decided to give the car a shove, your nervous system cells converted some of that chemical potential energy into kinetic electrical energy so that your nerves could tell your arm muscles to contract. Once your arm muscles received the electrical signal from your nerves, they converted some chemical potential energy into kinetic mechanical energy, moving your arms. This allowed you to work on the car. When you work on an object, its total energy changes. By working on the car, you gave it extra energy. That increased the car's kinetic mechanical energy, and thus the car started to move.

Do you see what happened here? At no point in the process was energy created. It was transformed in several different ways, however. It started out as chemical energy, was transformed into electrical energy, and was then transformed into mechanical energy. You might wonder where the potential chemical energy in the food came from. Well, plants get most of their energy from the sun.

Animals get their energy from the plants (and other animals) they eat. Where does the energy from the sun come from? It comes from nuclear reactions that occur in its core. Where did the energy in the nuclei taking part in the reactions come from? These kinds of questions can go on forever.

All of modern science clearly tells us that energy cannot be created by any *natural* process. That's the First Law of Thermodynamics. This law is probably the most trusted law in all of science. Thus, scientists are as sure about this as they can be about anything. Nevertheless, we know that there is plenty of energy in the universe. If energy cannot be created by any natural process, where did all of the energy in the universe come from? It seems rather clear that if *natural* processes cannot create energy, this energy must have been created by a *supernatural* process. This is just one more piece of evidence that science gives us for the existence of a supernatural creator.

Now that we've discussed how energy can be transformed, I hope that you can begin to see where energy goes in situations such as Experiment 8.1. I have already said that the reason the mass's maximum height kept decreasing as the pendulum kept swinging back and forth is that friction reduced the mass's total energy. The question still remains, however: where does that energy go? Well, friction slows an object down because the molecules in the object rub up against the molecules of the surface or, in the case of the pendulum, the air. When you rub one thing against another thing, what happens? They both heat up. So friction removes mechanical kinetic energy from a moving object, transforming it into heat.

We can do more than just determine that friction converts mechanical kinetic energy into heat. We can actually determine how much mechanical kinetic energy can be transformed. After all, we have already learned that in order to affect the energy of an object, you simply have to work on it. If you apply a force in the direction of motion, you are working *with* the motion of the object. This increases the mechanical energy. Thus, when you shove a toy car, the work that you do in the shove is added to the kinetic energy of the car. If you do 10 Joules of work, the kinetic energy will raise by 10 Joules. In the same way, if you apply a force opposed to the direction of motion, you are working *against* the motion of the object. This removes kinetic energy, usually transforming it into heat. This is what friction does. It works against the motion of the object. Every Joule of work that friction does removes a Joule of kinetic energy.

In other words, suppose a pendulum starts out with a total energy of 50 Joules. If, in the course of a few swings, air resistance does 10 Joules of work, the total energy of the pendulum reduces to 40 Joules. The 10 Joules of energy lost by the object is transformed into heat, slightly increasing the temperature of the pendulum and the surrounding air. See if you understand this reasoning by following the next example.

EXAMPLE 8.6

A 230-gram toy car is pushed so that it rolls across a level floor. If it starts with a kinetic energy of 0.46 Joules, how fast will it be going when it gets to the other side of the room, after friction has done 0.30 Joules of work?

The car starts out with 0.46 Joules of kinetic energy. We could use Equation (8.3) to calculate what speed the car starts with, but that won't really help us. After all, we need to know the speed of the car *after* friction has worked on it. Well, any work that friction does reduces the total energy. Thus, we can calculate the total energy after friction has worked:

$$TE_{after} = TE_{before} - W_{friction}$$

What's the total energy of the car before friction worked? Well, the floor is level, so we can say that potential energy is zero everywhere. Thus, the total energy at the beginning is the same as the kinetic energy at the beginning:

$$TE_{after} = 0.46 \text{ J} - 0.30 \text{ J} = 0.16 \text{ J}$$

Since the potential energy is zero everywhere, the total energy is equal to the kinetic energy. This tells us that kinetic energy afterwards is also 0.16 J. We can now use Equation (8.3) to calculate the speed of the car:

$$KE = \frac{1}{2}mv^2$$

$$0.16 \text{ J} = \frac{1}{2} \cdot (0.230 \text{ kg}) \cdot v^2$$

$$v = \sqrt{\frac{2 \cdot 0.16 \frac{\cancel{kg} \cdot m^2}{sec^2}}{0.230 \cancel{kg}}} = 1.2 \frac{m}{sec}$$

Notice that to show you how the units work out, I changed the Joule unit into $\frac{kg \cdot m^2}{sec^2}$, which is the same as a Newton·meter, which is the same as a Joule. After friction did its work, then, the car had a speed of <u>1.2 m/sec</u>.

Now that you've seen this reasoning in an example problem, I want you to get some hands-on experience with it.

EXPERIMENT 8.2

Estimating the Work Done by Friction

Note: A sample set of calculations is available in the solutions and tests guide. It is with the solutions to the practice problems.

Supplies:

- Safety goggles
- The inner cardboard tube from a roll of paper towels
- Scissors
- A golf ball or marble
- Some tape
- A ruler
- Some books
- Someone to help you
- A desk, table, or counter that you can tape things to. Make sure the finish won't come off with tape, or that your parents don't care if it does. There should be plenty of space on at least one side.

1. Use the scissors to cut the cardboard tube in half lengthwise. That way, each half of the tube will be as long as the original tube.
2. Take one of the halves of the tube and tape one end securely to the table, approximately 10 cm from the edge of the table. Choose an edge that has plenty of space beyond it, as you will be rolling the ball off of the edge, much like you did in Experiment 4.2.
3. Stack the books underneath the end of the tube that is not taped down. This will turn the tube into an incline. The top of the inclined tube should be somewhere between 12 and 18 cm above the surface of the table.
4. Try to make the point at which the tube is taped to the table as smooth as possible. You want your ball to roll down this incline, onto the table, and over the edge of the table. The smoother you make the transition from the tube to the table, the better your experiment will work. Your setup should look something like this:

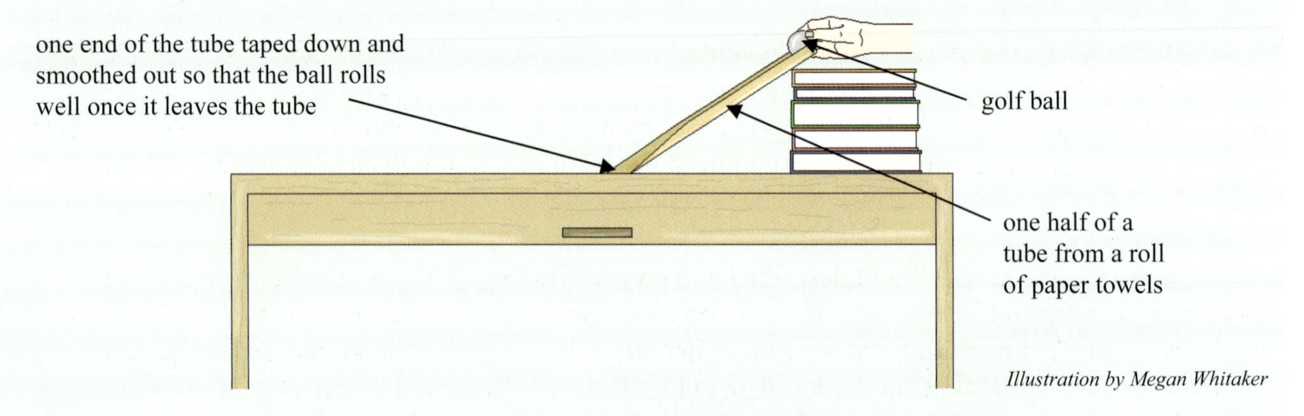

Illustration by Megan Whitaker

5. Measure the height of the top of the incline relative to the surface of the table.
6. Hold the ball at the top of the incline. Have your helper stand so that he can easily see where the ball hits the floor when it rolls off the table.
7. Release the ball without giving it any kind of shove. Have your helper mark where the ball hits the floor.
8. Measure the distance from the edge of the table to the point at which the ball hit the floor.
9. Repeat steps 6-8 four more times, so that you have five total measurements of the distance between the edge of the table and the spot where the ball hit the floor.
10. Average your five results.
11. Measure the height of the tabletop relative to the floor.
12. Clean up your mess.

Now let's see what we can learn from the experiment. When you held the ball at the top of the incline formed by the tube, there was potential energy in the system. Since you held the ball still, there was no kinetic energy, so the total energy of the system was just the potential energy. When you let go of the ball, potential energy began being converted into kinetic energy as the ball rolled down the incline. When the ball reached the end of the incline, relative to the table, the potential energy was zero. Thus, at that point, the total energy of the system was equal to the kinetic energy of the ball. Well, since the total energy of the system cannot change, the potential energy of the ball at the top of the incline will equal the kinetic energy of the ball at the bottom of the incline.

Now that should sound familiar. We used that reasoning in Example 8.4. Here's where I have to throw you a curve ball. It turns out that the experiment is more complicated than Example 8.4, because the ball is *rolling*. Since the ball is rolling, it is in *rotational motion* as well as translational motion. As it rolled down the incline, *some* of the potential energy was converted to kinetic energy of rotation and the rest was converted to kinetic energy of translation. Using the equations that we have developed so far, we cannot account for how much energy went into translational motion and how much went into rotational motion. However, I can give you an equation that will. *For the specific situation of a sphere rolling without slipping*, the kinetic energy is given by:

$$KE = \frac{7}{10}mv^2$$

Don't worry about how I got that equation, and don't ever use it again in this course. You are using it specifically for this experiment and nothing else. If you take advanced physics, you will learn how I got this equation. For right now, you will just use it for this experiment.

If you take the equation for the potential energy when the ball was at the top of the incline and set it equal to the equation for kinetic energy that I just gave you, the mass of the ball will cancel. Since you measured the height of the incline in step 5 (be sure to convert to meters), the only unknown left will be the speed of the ball at the bottom of the incline. Thus, you can solve for it. Call that speed "v_1."

Now in the analysis I just described, we ignored friction. Thus, it was not as realistic an analysis as we would like to have. It would be nice if we could determine how much work friction did on the ball. Well, it turns out that we can. After all, v_1 is the speed that the ball should have at the bottom of the incline if friction did no work at all. However, we can also calculate (to a reasonable approximation) the speed that the ball *actually* had when it reached the bottom of the incline.

How do we do that? Well, remember Experiment 4.2. In that experiment, we measured the speed of a toy by measuring how far it traveled when it rolled off of a table using our one-dimensional motion equations. First, we need to determine how much time it took for the ball to reach the floor once it rolled off the table. That's easy. You measured the height of the table in step 11. Use that height and the fact that the acceleration due to gravity is 9.8 m/sec^2 in Equation (2.19) to determine how long it took the ball to fall in the vertical dimension. Remember, since the ball rolled horizontally off of the table, its initial vertical velocity was zero.

Now that you have the time from the vertical dimension, divide that into the horizontal distance that the ball traveled before hitting the floor (the average you computed in step 10). That gives you the horizontal speed of the ball when it left the table. Since it had no vertical component to its velocity at that point, then the horizontal speed of the ball is the total speed of the ball. Label that speed "v_2."

Now compare v_1 and v_2. You should see that v_2 (the measured speed of the ball when it left the table) is smaller than v_1 (the calculated speed of the ball). Why? As the ball rolled, friction slowed it down. Thus, the measured speed of the ball (v_2) turned out to be less than the calculated speed of the ball (v_1).

We can get a really good idea about how much work friction did in this situation by comparing the squares of the two speeds. After all, the kinetic energy depends on the mass and the square of the speed. Square each speed, then divide v_2^2 by v_1^2. What does that represent? Think about it. Suppose I calculated the kinetic energy based on v_2 and then divided it by the kinetic energy based on v_1. Since the mass is the same in each case, it would cancel. Thus, the ratio of v_2^2 to v_1^2 is the same as the ratio of the *kinetic energies* corresponding to v_2 and v_1. Since v_1 was calculated assuming that there was no friction, and since v_2 is a more realistic value for the speed which includes the effects of friction, the ratio of v_2^2 to v_1^2 tells you the fraction of energy that remained after friction did its work. When I did this experiment, for example, the ratio of v_2^2 to v_1^2 was 0.79. This means that the work due to friction removed 21% of the energy from the system, so that only 79% of the total energy remained.

Now don't get lost in all of the math here. As is the case with all experiments that require calculations, there are sample calculations for this experiment after the solutions to the practice problems for this module in the *Solutions and Tests Guide*. If you are having trouble with the calculations, look there for help. What I want you to do now is concentrate on the *reasoning* we used. We calculated the speed of the ball assuming no friction was present. That led to v_1. We then used our one-dimensional motion equations to actually measure the speed of the ball. That led to v_2. We noted that v_2 was smaller than v_1 because of friction. Then, we calculated the ratio of v_2^2 to v_1^2, which is the same as the ratio of the measured kinetic energy to the kinetic energy that was calculated ignoring friction. The ratio, then, was a good estimate of the fraction of energy that remained after friction worked on the ball. Thus, it can be used to approximate the work done by friction.

ON YOUR OWN

8.6 A child shoves his block (m = 0.125 kg) down a sidewalk. If the block starts sliding at 2.5 m/sec and eventually stops, how much work did friction do the entire time that the block was sliding? Assume that the sidewalk is level.

That problem wasn't very tough, was it? It can, of course, get a lot harder. For example, Equation (8.1) allows us to determine the work done by a force over a given displacement. Thus, if we know the frictional coefficient of a surface and the normal force exerted by the surface on an object, we do not need to be told the work friction does; we can calculate it. That adds another level of difficulty to these kinds of problems. Before we look at a couple of problems like this, however, let's be sure that we can actually apply Equation (8.1) to friction.

EXAMPLE 8.7

A driver slams on the brakes. His car (m = 1,234 kg) skids down the road for 61 m. If the coefficient of kinetic friction of the road is 0.45, how much work does friction do during the course of the car's skid? (Assume that the road is level.)

Equation (8.1) allows us to calculate the work done by a force, provided that we know both the force and the displacement and that they are parallel to one another. The displacement is given and, since we have the mass of the car and the coefficient of kinetic friction between the road and the car, we can use Equation (5.3) to calculate the force due to friction:

$$f = \mu F_n$$

Since the car is level on the road, the normal force is simply the weight $(m \cdot g)$ of the car. Thus, the frictional force is:

$$f = \mu_k F_n$$

$$f = (0.45) \cdot (1{,}234 \text{ kg}) \cdot (9.8 \frac{\text{m}}{\text{sec}^2}) = 5.4 \times 10^3 \text{ Newtons}$$

Now that we know the frictional force, we can simply plug it into Equation (8.1) and calculate the work done:

$$W = F_{\parallel} \cdot \Delta x$$

$$W = (5.4 \times 10^3 \text{ Newtons}) \cdot (61 \text{ m}) = 3.3 \times 10^5 \text{ J}$$

So friction does a whopping $\underline{3.3 \times 10^5 \text{ J}}$ of work on the car during the skid.

Make sure you can solve this kind of problem on your own.

ON YOUR OWN

8.7 Friction does 4.5 J of work on a 430-g box that slides along a level sidewalk. If the work is done while the box slides 2.5 meters, what is the coefficient of kinetic friction between the box and the sidewalk?

We can now try to put these concepts together into a more difficult problem. Study the example and solve the "On Your Own" problems that follow to make sure you understand how this is done.

EXAMPLE 8.8

A student shoves a 2.15-kg book and watches it slide to the other end of the table. If the book starts out with a speed of 0.95 m/sec, how far will it travel before it stops? Assume that the table is level. (μ_k = 0.34)

You should see that in giving us the mass and initial speed of the book, the problem is, in fact, giving us the book's initial kinetic energy. Also, since the table is level, we can assume that the book's potential energy is zero as long as it stays on the table. Thus, all of the book's energy is kinetic. We can calculate exactly how much energy that is:

$$KE = \frac{1}{2}mv^2 = \frac{1}{2} \cdot (2.15 \text{ kg}) \cdot (0.95 \frac{\text{m}}{\text{sec}})^2 = 0.97 \text{ J}$$

When it stops, the book will no longer have any kinetic energy. Since the book's potential energy is also zero, this means that the book has zero total energy. Where did that energy go? Friction converted it to heat. This tells us that friction must have done 0.97 J worth of work. Since we know the mass of the book (which gives us the normal force) and the coefficient of kinetic friction, we can calculate how much force friction used to perform this work:

$$f = \mu_k F_n$$

$$f = (0.34) \cdot (2.15 \text{ kg}) \cdot (9.8 \frac{\text{m}}{\text{sec}^2}) = 7.2 \text{ Newtons}$$

With this force, we can use Equation (8.1) to calculate the distance it would take to do 0.97 J of work. That corresponds to the distance that the book travels before stopping:

$$W = F_{\parallel} \cdot \Delta x$$

$$0.97 \text{ J} = (7.2 \text{ Newtons}) \cdot \Delta x$$

$$\Delta x = \frac{0.97 \cancel{\text{Newton}} \cdot \text{m}}{7.2 \cancel{\text{Newtons}}} = 0.13 \text{ m}$$

The book, therefore, travels only 0.13 m (13 cm) before stopping.

ON YOUR OWN

8.8 A girl kicks a 155-g rock down a sidewalk.. If the rock starts out with a speed of 3.2 m/sec, how far will it slide down the sidewalk before it stops? Assume that the sidewalk is level. ($\mu_k = 0.44$)

8.9 A 2.3-kg wooden block slides across a carpentry shop's floor. If the floor is level and the block stops after traveling 3.5 m, what was its initial speed? ($\mu_k = 0.52$)

Energy and Power

Although we now know a great deal about energy, there is one concept we have ignored up until this point: **power**. The concept of power tells us how much energy is used during a certain time interval. For example, suppose your parents told you to go out and mow the lawn. If you approached the project in a leisurely way and took several hours to finish the job, would you be very tired? Probably not. After all, you could push the mower slowly, take rests in between, and even sit down to have a cool drink. In the end, you would have wasted a lot of time, but you wouldn't be incredibly worn out. Suppose, on the other hand, you got it done in only 30 minutes. You would be a lot more tired at that point, right? After all, in order to get the job done in such a short time, you wouldn't have any time to rest. You would also have to push the mower a lot faster, which can get tiring.

Now let me ask you a question. In which case would you have done more work? The answer is "neither." The amount of work necessary to get the job done is independent of how much time it took. You need to push the lawn mower a certain distance in order to get the job done. Since you push the lawn mower at a relatively constant rate, there is little acceleration involved during the vast majority of the job. Thus, the force you apply in pushing the lawn mower is essentially the force necessary to overcome friction. This force will be pretty much the same regardless of whether you push the mower quickly or slowly. Thus, in both cases, you apply essentially the same force over the majority of the distance. The distance is the same in both cases. Thus, since the force is essentially the same and the distance is exactly the same, the work (which is force times distance) is essentially the same in both cases.

Why, then, are you so much more tired when you do the job in 30 minutes? Well, you expend the same amount of energy, but in a much smaller time interval. The energy you expend is used up quickly, and that makes you tired. This is the concept of power. Power measures the amount of energy expended per second. When you use a lot of power, you tire quickly.

Power - The amount of energy expended per second

We could restate this definition mathematically as follows:

$$P = \frac{\Delta W}{\Delta t} \tag{8.5}$$

where "P" represents power, "ΔW" is the work done within a certain time interval, and "Δt" is the time interval. Look at that equation for a moment. What does it look like? If we replaced "W" with distance, we would have the equation for speed, right? What is speed? Speed is the *time rate of change* in the distance. Thus, power is the *time rate of change* of the work done.

What units are associated with power? Well, according to Equation (8.5), to get power we take work done in a certain time interval (measured in Joules) and divide by time (measured in seconds), which gives us $\frac{\text{J}}{\text{sec}}$ as a power unit. This unit is named the "Watt," in honor of James Watt, a Scottish engineer whose work on steam engines in the late 1700s made steam power practical. Watt was not the inventor of the steam engine, but he made improvements to the design which helped make steam power practical. Realize, of course, that the Watt is the SI unit for power, but any energy unit divided by any time unit is a legitimate power unit.

The Watt should be familiar to you. When you buy light bulbs, what determines their brightness? The number of Watts at which the light bulb is rated. This number simply tells you how many Joules of energy that the light bulb burns every second that it is on. So, a 100-Watt light bulb burns 100 Joules of energy every second, while a 40-Watt light bulb burns 40 Joules of energy every second.

As far as we are concerned, what's the difference between a 40-Watt light bulb and a 100-Watt light bulb? The brightness. A 40-Watt light bulb seems quite dim in comparison to a 100-Watt light bulb. This is another excellent example of what power really means. After all, when you screw a light bulb into a socket, you are screwing it into an energy source. Whether you screw the 40-Watt light bulb or the 100-Watt light bulb into the socket, each bulb has access to exactly the same amount of energy. Since the 100-Watt light bulb burns that energy much faster, however, it is a lot brighter. That's the difference between power and energy. Both bulbs have access to exactly the same amount of energy. Since one bulb burns that energy faster, however, it is brighter. Now please realize that there are many factors that contribute to the brightness of a light bulb. Thus, two light bulbs with equal Watt ratings might be of differing brightness because of other factors. However, if all other factors are the same, the higher the Watt rating on a light bulb, the brighter it will be.

How much power exists in a Watt? Well, let's put it in terms with which we are familiar. An average teenage male burns 3,000 food Calories each day. As you should have learned in chemistry, a food Calorie is, in fact 1,000 chemistry calories, and a chemistry calorie is a unit of energy that equals 4.18 Joules. Thus, in units with which we are familiar, the average teenage male burns 1.25×10^7 Joules in a day, which is equal to 8.64×10^4 seconds. Thus, the average power used by the average teenage male is:

$$P = \frac{\Delta W}{\Delta t} = \frac{1.25 \times 10^7 \text{ J}}{8.64 \times 10^4 \text{ sec}} = 145 \text{ Watts}$$

This tells us that 1.5 100-Watt light bulbs consume just a little more power than the average teenage male! This should give you an appreciation of how much energy you consume when you turn on a light bulb. For every second that a 100-Watt light bulb is on, it is burning 69% of the energy that the average teenage male burns!

Now that you have an appreciation for how much power exists in a Watt, see if you can apply this concept to an example and an "On Your Own" problem.

EXAMPLE 8.9

A woman pushes a rock 4.3 meters. If she exerts 54 Newtons of force over 2.10 minutes to accomplish this task, what would be her average power output?

According to Equation (8.5), power is the work done during a certain time interval divided by time. We can calculate the work from the force and distance:

$$W = F_{\parallel} \cdot \Delta x = (54 \text{ Newtons}) \cdot (4.3 \text{ m}) = 2.3 \times 10^2 \text{ Newton} \cdot \text{m} = 2.3 \times 10^2 \text{ J}$$

This work was done over 2.10 minutes. After converting to seconds, we can simply plug the numbers into Equation (8.5):

$$P = \frac{\Delta W}{\Delta t} = \frac{2.3 \times 10^2 \text{ J}}{126 \text{ sec}} = 1.8 \text{ Watts}$$

Compared to the power of a light bulb, 1.8 Watts is nothing!

ON YOUR OWN

8.10 The engine in a machine is rated at 125 Watts. If this machine uses a force of 55 Newtons to move an object 351 meters, how long will it take the machine to complete the task?

ANSWERS TO THE "ON YOUR OWN" PROBLEMS

8.1 In this problem, we need to plug the distance (after we convert it to meters) and the work into Equation (8.1) and solve for the force. Since the force was applied in the same direction as the motion, the force that we solve for will be the total force:

$$W = F_{||} \cdot \Delta x$$

$$34.5 \text{ J} = (F) \cdot (4.51 \text{ m})$$

$$F = \frac{34.5 \text{ Newton} \cdot \cancel{\text{m}}}{4.51 \cancel{\text{m}}} = 7.65 \text{ Newtons}$$

Notice that in order to get the units to work out, I renamed the Joule to the $\text{Newton} \cdot \text{m}$.

8.2 For this problem, we are given the potential energy and the mass, and we are asked to calculate the height. This is a straightforward application of Equation (8.2). The potential energy relative to the ground will give us the height of the first hill:

$$PE = mgh$$

$$7.72 \times 10^5 \text{ J} = (365 \text{ kg}) \cdot (9.8 \frac{\text{m}}{\text{sec}^2}) \cdot h$$

$$h = \frac{7.72 \times 10^5 \frac{\cancel{\text{kg}} \cdot \text{m}^{\cancel{2}}}{\cancel{\text{sec}^2}}}{(365 \cancel{\text{kg}}) \cdot (9.8 \frac{\cancel{\text{m}}}{\cancel{\text{sec}^2}})} = 2.2 \times 10^2 \text{ m}$$

Notice that I changed Joules into $\frac{\text{kg} \cdot \text{m}^2}{\text{sec}^2}$ to make sure that the units worked out. The first hill, then, is <u>2.2×10^2 m</u> high. To get the height of the second hill, however, we have to think. The potential energy relative to the ground told us how high it was above the ground. That is the same as the height of the hill. The potential energy relative to the second hill, however, merely tells us how high *above* the second hill the car is. It does not tell us the height of the second hill. Thus, when we use Equation (8.2), we are not calculating the height of the second hill. Instead, we are calculating how high above the second hill the car sits. Doing that gives us the following answer:

$$PE = mgh$$

$$6.67 \times 10^4 \text{ J} = (365 \text{ kg}) \cdot (9.8 \frac{\text{m}}{\text{sec}^2}) \cdot h$$

$$h = \frac{6.67 \times 10^4 \frac{\cancel{kg} \cdot m^{\cancel{2}}}{\cancel{sec^2}}}{(365 \cancel{kg}) \cdot (9.8 \frac{\cancel{m}}{\cancel{sec^2}})} = 19 \text{ m}$$

Now what does this answer mean? Well, we already know that the first hill is 2.2 x 10^2 m above the ground. The calculation that we just did tells us that when the car on the first hill, it is only 19 m above the next hill. Thus, the height of the second hill must be 2.2 x 10^2 m - 19 m, or 2.0 x 10^2 m.

8.3 Since we know the rocket's kinetic energy and mass, solving for the velocity is a straightforward application of Equation (8.3):

$$KE = \frac{1}{2}mv^2$$

$$4.5 \times 10^{10} \text{ J} = \frac{1}{2} \cdot (1.34 \times 10^4 \text{ kg}) \cdot v^2$$

$$v = \sqrt{\frac{2 \cdot (4.5 \times 10^{10} \text{ J})}{1.34 \times 10^4 \text{ kg}}} = \sqrt{\frac{2 \cdot (4.5 \times 10^{10} \frac{\cancel{kg} \cdot m^2}{sec^2})}{1.34 \times 10^4 \cancel{kg}}} = 2.6 \times 10^3 \frac{m}{sec}$$

The rocket, then, travels at 2.6 x 10^3 m/sec.

8.4 This problem is much like the roller coaster problem in the example. When the biker rests at the top of the first hill, he has no kinetic energy, because he is not moving. Thus, all of his energy is potential. Since we know the height of the hill, we can develop an expression for that potential energy:

$$PE = mgh$$

$$PE = m \cdot (9.8 \frac{m}{sec^2}) \cdot (21 \text{ meters})$$

Now remember, this is also the *total* energy that the biker has, because he has no kinetic energy at the top of the hill. Since the biker never pedals, the total energy of the bike does not change. As a result, when the biker hits the top of the next hill, the total energy is still given by:

$$TE = m \cdot (9.8 \frac{m}{sec^2}) \cdot (21 \text{ meters})$$

At the top of the next hill, the biker is still moving. At this point, then, he also has kinetic energy. All we know is that the sum of the potential energy and kinetic energy must be equal to the expression given above:

$$TE = PE + KE = mgh + \frac{1}{2}mv^2$$

Think about what we have. We have an expression for TE. We also have the height of the hill that the biker is currently on, and we know "g." Let's stick those things into this equation:

$$TE = mgh + \frac{1}{2}mv^2$$

$$m \cdot (9.8\frac{m}{sec^2}) \cdot (21 \text{ meters}) = m \cdot (9.8\frac{\text{meters}}{sec^2}) \cdot (12 \text{ meters}) + \frac{1}{2}mv^2$$

Notice what happens now. The mass drops out. That leaves "v" as the only unknown, so we can solve for it.

$$\cancel{m} \cdot (9.8\frac{\text{meters}}{sec^2}) \cdot (21 \text{ meters}) = \cancel{m} \cdot (9.8\frac{\text{meters}}{sec^2}) \cdot (12 \text{ meters}) + \frac{1}{2}\cancel{m}v^2$$

$$\frac{1}{2}v^2 = (9.8\frac{\text{meters}}{sec^2}) \cdot (21 \text{ meters}) - (9.8\frac{\text{meters}}{sec^2}) \cdot (12 \text{ meters}) = 90\frac{\text{meters}^2}{sec^2}$$

$$v = \sqrt{2 \cdot 90\frac{\text{meters}^2}{sec^2}} = 10\frac{m}{sec}$$

At the top of the second hill, the biker is traveling at 10 m/sec. The significant figures are worth discussing here. Typically, we do not round for significant figures until the end of an equation, but we are mixing mathematical procedures here. First we have to multiply, and then we have to subtract. Thus, we need to change our significant figures rules in the middle of the equation. As a result, we need to deal with significant figures as we go. When we multiply 9.8 and 21, we get 210 (to two significant figures). When we multiply 9.8 times 12, we get 120 (to two significant figures). When we subtract 120 from 210, we are restricted by the rule of addition and subtraction to the tens place. As a result, we have 90. Then, when we multiply by two and take the square root, we are left with only one significant figure, since 90 has only one. That's why the answer is 10 m/sec and not 13 m/sec.

8.5 This problem is a little different from the example, but it uses exactly the same concepts. First of all, we are interested in the BB. It starts from the tip of the gun and travels through the air to the target, 25 m above the gun's tip. Now, although we don't know how high the gun's tip is from the ground, we don't need that information to use potential energy in this problem. As I said earlier, potential energy is *relative*. Just as in Example 8.2 where we define the potential energy of a ball on a staircase relative to both the floor and the second step, we will define potential energy in this problem in a relative manner. Since the lowest height that the BB ever reaches is the tip of the gun, we could say that the BB's height (and therefore the potential energy) is zero at that point. Thus, when the BB is at the tip of the gun, all of the energy is kinetic, for which we can develop an expression:

$$KE = \frac{1}{2}mv^2 = \frac{1}{2} \cdot (m) \cdot (99\frac{\text{meters}}{sec})^2$$

Since we have defined the BB's initial height as zero, it has no potential energy. The expression above, then, is also the expression for the BB's total energy:

$$TE = \frac{1}{2} \cdot (m) \cdot (99 \frac{\text{meters}}{\text{sec}})^2$$

Now remember, total energy is equal to kinetic energy plus potential energy. Thus, at any point in the BB's trip:

$$TE = PE + KE$$

$$\frac{1}{2} \cdot (m) \cdot (99 \frac{\text{meters}}{\text{sec}})^2 = mgh + \frac{1}{2} mv^2$$

When the BB is at the target, its height relative to the tip of the gun (the point at which we said potential energy was zero) is 25 m. Thus, it is 25 m above zero potential energy, so h = 25 m. We also know "g," so we can put values for "h" and "g" in the equation above:

$$TE = PE + KE$$

$$\frac{1}{2} \cdot (m) \cdot (99 \frac{\text{meters}}{\text{sec}})^2 = m \cdot (9.8 \frac{\text{meters}}{\text{sec}^2}) \cdot (25 \text{ meters}) + \frac{1}{2} mv^2$$

Now the mass drops out, leaving the speed as the only variable:

$$\frac{1}{2} \cdot (\not{m}) \cdot (99 \frac{\text{meters}}{\text{sec}})^2 = \not{m} \cdot (9.8 \frac{\text{meters}}{\text{sec}^2}) \cdot (25 \text{ meters}) + \frac{1}{2} \not{m} v^2$$

$$\frac{1}{2} v^2 = \frac{1}{2} \cdot (99 \frac{\text{meters}}{\text{sec}})^2 - (9.8 \frac{\text{meters}}{\text{sec}^2}) \cdot (25 \text{ meters}) = 4700 \frac{\text{meters}^2}{\text{sec}^2}$$

$$v = \sqrt{2 \cdot 4700 \frac{\text{meters}^2}{\text{sec}^2}} = 97 \frac{\text{m}}{\text{sec}}$$

In order to reach the target, then, the bullet lost some speed because it had to convert kinetic energy to potential energy. It hits the target at 97 m/sec, rather than the 99 m/sec at which it was shot.

8.6 Since we can assume that the sidewalk is level, we can ignore potential energy, saying it is equal to zero during the entire time. If that's the case, then all energy is kinetic energy. If the block starts sliding at 2.5 m/sec, we can calculate how much total energy it has at the beginning by just using Equation (8.3):

$$KE = \frac{1}{2} mv^2$$

$$KE = \frac{1}{2} \cdot (0.125 \text{ kg}) \cdot (2.5 \frac{\text{m}}{\text{sec}})^2 = 0.39 \text{ J}$$

Now, when the block stops sliding, how much kinetic energy does it have? When an object has no speed, it has no kinetic energy. Since potential energy is always zero in this problem, we can say that the block has lost all 0.39 J of its energy because of friction. Thus, if friction took all of its energy away, friction must have done 0.39 J of work.

8.7 In this problem, we have the box's mass, which will give us the normal force. Based on Equation (5.3), if we know the normal force and the frictional force, we can calculate the coefficient of friction. Thus, we need to determine the frictional force. Well, we know how much work the frictional force did (4.5 J), and we know the displacement over which the work was done (2.5 m). Thus, we can use Equation (8.1) to determine the frictional force:

$$W = F_{\parallel} \cdot \Delta x$$

$$4.5 \text{ J} = (F) \cdot (2.5 \text{ m})$$

$$F = \frac{4.5 \text{ Newton} \cdot \cancel{\text{m}}}{2.5 \cancel{\text{m}}} = 1.8 \text{ Newtons}$$

This, then, is the frictional force. Since the sidewalk is level, the normal force is just the weight of the box $(m \cdot g)$. We must, of course, make sure that the mass is in kg. Plugging these things into Equation (5.3):

$$f = \mu F_n$$

$$1.8 \text{ Newtons} = \mu \cdot (0.430 \text{ kg}) \cdot (9.8 \frac{\text{m}}{\text{sec}^2})$$

$$\mu = \frac{1.8 \frac{\cancel{\text{kg}} \cdot \cancel{\text{m}}}{\cancel{\text{sec}^2}}}{(0.430 \cancel{\text{kg}}) \cdot (9.8 \frac{\cancel{\text{m}}}{\cancel{\text{sec}^2}})} = 0.43$$

The coefficient of kinetic friction, then, is 0.43.

8.8 Given the mass and the initial speed, we have the rock's initial kinetic energy. Since the sidewalk is level, we can say that this is also its total energy:

$$KE = \frac{1}{2} \cdot m \cdot v^2$$

$$KE = \frac{1}{2} \cdot (0.155\,kg) \cdot (3.2\,\frac{m}{sec})^2 = 0.79\,J$$

In order for the rock to stop, friction must convert all of this energy into heat. To do that, it must perform 0.79 J of work. Since we have the coefficient of friction and the rock's mass, we can determine the force with which friction works:

$$f = \mu F_n$$

$$f = (0.44) \cdot (0.155\text{ kg}) \cdot (9.8\,\frac{m}{sec^2}) = 0.67\text{ Newtons}$$

Now we can use Equation (8.1) to determine over what distance friction must work in order to convert 0.79 J of kinetic energy into heat:

$$W = F_{||} \cdot \Delta x$$

$$0.79\text{ J} = (0.67\text{ Newtons}) \cdot \Delta x$$

$$\Delta x = \frac{0.79\,\cancel{\text{Newton}} \cdot m}{0.67\,\cancel{\text{Newtons}}} = 1.2\text{ m}$$

Thus, the rock travels 1.2 m before coming to a halt.

8.9 Regardless of the block's initial speed, in order for it to stop, friction must remove all of its kinetic energy. Since the floor is level, this also corresponds to its total energy. Thus, if we can determine how much work friction did, we can say that this work must equal the block's initial kinetic energy. Well, we can start by calculating the frictional force:

$$f = \mu F_n$$

$$f = (0.52) \cdot (2.3\text{ kg}) \cdot (9.8\,\frac{m}{sec^2}) = 12\text{ Newtons}$$

Now that we know the frictional force, we can calculate how much work friction did:

$$W = F_{||} \cdot \Delta x$$

$$W = (12\text{ Newtons}) \cdot (3.5\text{ m}) = 42\text{ J}$$

If friction did 42 J of work, this tells us that the block initially had 42 J of kinetic energy. Now we can calculate the block's initial speed:

$$KE = \frac{1}{2}mv^2$$

$$42 \text{ J} = \frac{1}{2} \cdot (2.3 \text{ kg}) \cdot v^2$$

$$v = \sqrt{\frac{2 \cdot 42 \text{ J}}{2.3 \text{ kg}}} = 6.0 \frac{\text{m}}{\text{sec}}$$

The block, therefore, had an initial speed of 6.0 m/sec.

8.10 Power is work divided by time. The force and distance given will tell us how much work needs to be done. Given the power, then, we can calculate the time it takes to get that work done. Let's start by calculating the work done:

$$W = F_{\parallel} \cdot \Delta x$$

$$W = (55 \text{ Newtons}) \cdot (351 \text{ m}) = 1.9 \times 10^4 \text{ J}$$

Now that we know how much work is to be done, we can use Equation (8.5) to determine how much time it takes to do that much work:

$$P = \frac{\Delta W}{\Delta t}$$

$$125 \text{ Watts} = \frac{1.9 \times 10^4 \text{ J}}{\Delta t}$$

$$\Delta t = \frac{1.9 \times 10^4 \cancel{\text{J}}}{125 \frac{\cancel{\text{J}}}{\text{sec}}} = 1.5 \times 10^2 \text{ sec}$$

The machine, then, will take 1.5×10^2 sec to get the job done.

REVIEW QUESTIONS

1. A secret agent is locked in a room. He pushes and pushes against the door but cannot open it. Finally, he falls to the floor exhausted. Has he done any work on the door? Why or why not?

2. Two men attempt to move a desk through the same displacement using the same magnitude force (F). The first ties a rope to the desk and pulls on it; the second pushes. Which man will be able to do the most work? Refer to the diagram below:

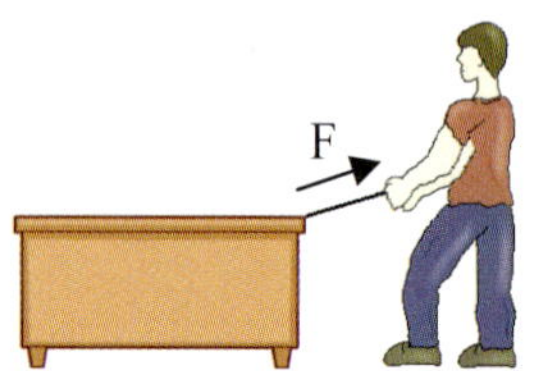

Illustrations of the desk and the man on the right copyright GifArt.com

Illustration of the man on the left by Megan Whitaker

3. Which of the following are legitimate units for energy?

$$\frac{J}{s},\ \text{inch}\cdot\text{lbs},\ \frac{\text{kg}\cdot\text{m}}{\text{sec}^2},\ \text{Newton}\cdot\text{m}^2,\ \text{Newton}\cdot\text{km},$$

4. In each of the cases below, indicate whether the energy is kinetic or potential:

   a. The energy in a tank full of gasoline
   b. The energy in a car traveling down a road
   c. The energy in electricity moving through a wire
   d. The energy in a ripe apple hanging in a tree

5. A rock is at rest on the top of a cliff. Suddenly, a small breeze pushes it off the cliff. Where is the potential energy the greatest? Where is the kinetic energy the greatest? At what point in its fall is the kinetic energy equal to the potential energy?

6. When you lift a heavy box and place it on top of a table, you perform work on the box. After you put the box on the table, it remains stationary. We have already learned that when you work on an object, you either add to or take away from its energy. In this case, did you add to or take away from the energy of the box? Was it potential or kinetic energy?

7. Two boxes slide down a level sidewalk. The first is heavier than the second; other than that, they are identical. If they start with equal speeds, which will slide the farthest?

8. What is the difference between energy and power?

9. Two light bulbs are in two different lamps. Although they are both the same type of light bulb, the first is significantly dimmer than the second. Which light bulb consumes more power?

10. Which of the following are legitimate units for power?

$$\text{kiloWatts},\ \frac{\text{N}}{\text{sec}},\ \text{J}\cdot\text{sec},\ \frac{\text{kiloJoules}}{\text{year}}$$

PRACTICE PROBLEMS

1. A construction worker drags a box across a floor. If the frictional force between the floor and the box is 12.2 Newtons, how much work does friction do as the box moves 11.5 meters?

2. A 567-kg car is traveling down a hilly road. If the car is at the top of a 15.1-meter hill and is moving with a speed of 19.1 m/sec, what is the potential energy relative to the bottom of the hill? What is the kinetic energy? What is the total energy?

3. A roller coaster car comes to a halt at the top of a 75-m hill. If the car starts to roll down the hill, what will its speed be when it reaches the top of the next hill, which is 40.0 meters high?

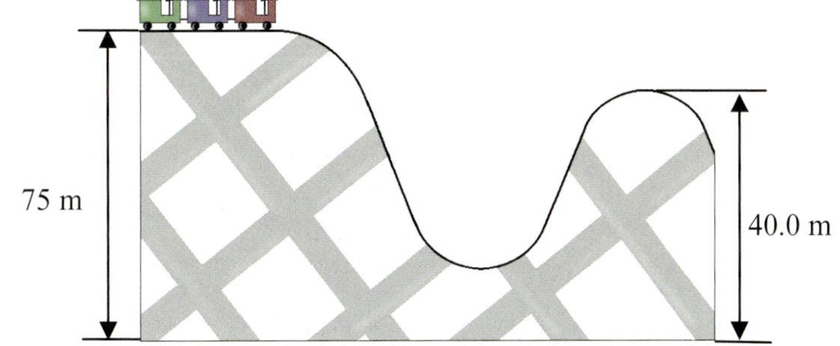

Illustration by Megan Whitaker

4. A car travels down a hilly road. At the top of a 11.5 meter hill, it is traveling with a speed of 21.4 m/sec, and the driver decides to let the car coast down the hill and up the next one. If the next hill is 5.1 meters high, how fast will the car be traveling at the top of that hill?

Illustration copyright GifArt.com

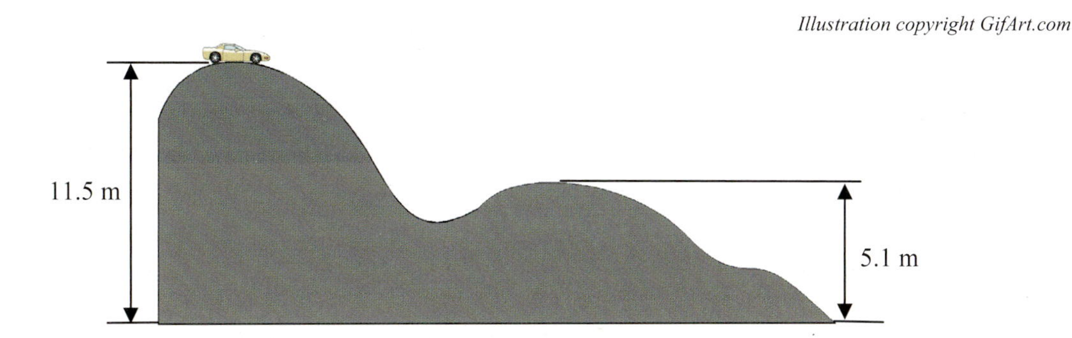

5. The mass on a pendulum is pulled up so that it is 45 cm above the point at which it hangs straight down (its equilibrium position). If it is released from this position, how fast will it be traveling when it reaches its equilibrium position?

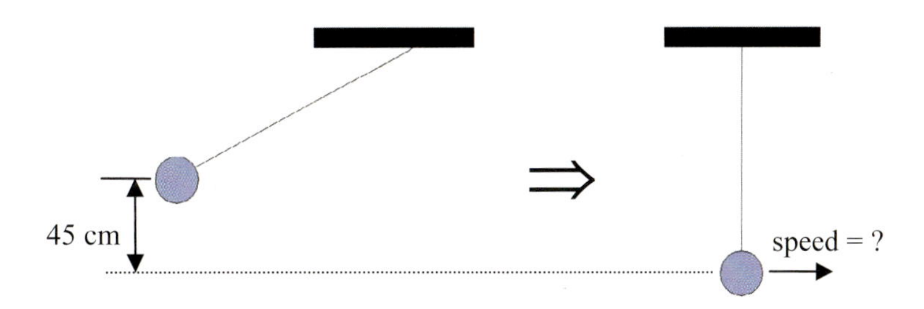

6. A 124-gram block is shoved across the floor with an initial speed of 3.1 m/sec. When the block finally comes to a halt, how much work has friction performed?

7. A 1.1-kg package is at the top of a 1.2-meter high slope. When the package slides to the bottom of the slope, its speed is 2.1 m/sec. How much work did friction do?

8. A student gives a 534-gram book a hard shove across the table. If the coefficient of kinetic friction between the book and the table is 0.45 and the book's initial speed is 4.1 m/sec, how far will the book slide before coming to a halt?

9. A janitor pushes a box 11.1 meters across the floor in 2.40 minutes. If the janitor exerts 67 Newtons of force to do the job, how much power has he exerted?

10. Suppose you were able to convert the energy used by light bulbs into kinetic mechanical energy in order to move objects. If a 101-Watt light bulb shone for 1.0 hours, how much force could be generated over a distance of 25 meters ?

MODULE #9: Momentum

Introduction

To start this module, let me ask you a question. If you were going to be in an automobile accident, which would you rather collide with: a compact car or an 18-wheel semitrailer? Would you prefer that it be traveling slowly or quickly? The answers are rather obvious, aren't they? Clearly, a compact car will not cause nearly as much damage in an accident as will an 18-wheel semitrailer. We would also rather get into an accident involving low speeds than high speeds. Why? Intuitively we just know that the lighter an object is and the slower it is traveling, the less damage it causes in a collision. This intuitive knowledge is based on a fundamental concept in physics: momentum. We will study that concept extensively in this module.

Definition of Momentum

When an object is in motion, it has momentum. Momentum is a vector quantity, and its definition is best expressed mathematically:

$$\mathbf{p} = \mathrm{m}\mathbf{v} \tag{9.1}$$

In this definition, "**p**" represents the momentum vector, "m" stands for the mass of the object, and "**v**" is the object's velocity vector. The first thing that we ought to notice about this equation is that the momentum is calculated by taking a scalar (non-vector) quantity and multiplying it by the velocity vector. When a scalar multiplies a vector, no direction change occurs. As a result, this definition tells us that the momentum vector is always pointed in the same direction as the velocity vector.

Now, what about the units? In order to get momentum, we take mass (the SI unit is kg) and multiply by velocity (the SI unit is $\frac{\text{m}}{\text{sec}}$). The resulting SI unit is $\frac{\text{kg}\cdot\text{m}}{\text{sec}}$. Unlike many of the other complicated units in physics, this does not have an abbreviation or another name. It is just a $\frac{\text{kg}\cdot\text{m}}{\text{sec}}$. I do, of course, have a proposition to solve this problem. In the tradition of taking complicated units and renaming them to honor famous physicists, I suggest that we call this unit the "Wile." I'm not sure that my proposition will ever be taken seriously, but you can't blame a guy for trying, can you?

Remember, since momentum is a vector quantity, just providing momentum as a number attached to a unit isn't enough. You must also give a direction. Based on our previous discussion, that direction will always be the same as the direction of the velocity vector. Since we haven't dealt with vectors in a while, let me remind you of what that means. Sometimes, directions like west, north, northwest, etc. are given. If this is the case, we use those directions with our answer. If the motion we are examining is one-dimensional motion, however, a sign often tells us the direction. The standard convention is that horizontal motion to the right and vertical motion up are considered positive, while horizontal motion to the left and vertical motion down are considered negative. If the motion is two-dimensional, we usually denote direction with an angle.

Let's make sure that you understand what Equation (9.1) means and how to use it by studying a quick example and then solving an "On Your Own" problem.

EXAMPLE 9.1

A 345-kg car travels with a velocity of 19.1 m/sec southwest. What is its momentum?

This is a simple application of Equation (9.1). We are given both the mass and the velocity (all in SI units), and we just need to plug them into the equation:

$$\mathbf{p} = \text{mv}$$

$$\mathbf{p} = (345 \text{ kg}) \cdot (19.1 \frac{\text{m}}{\text{sec}}) = 6.59 \times 10^3 \frac{\text{kg} \cdot \text{m}}{\text{sec}}$$

Since momentum always points in the same direction as the velocity, the car's momentum is 6.59×10^3 Wiles, uh, I mean $\frac{\text{kg} \cdot \text{m}}{\text{sec}}$ southwest. This number probably means nothing to you right now, but don't worry; you'll get a feel for momentum as this module continues.

ON YOUR OWN

9.1 A 35-Newton rock rolls down a hill. At the bottom of the hill, it is rolling horizontally with a velocity of -13.2 m/sec. What is its momentum at that point?

So, we know the units for momentum and the fact that momentum is a vector quantity. We can also calculate momentum given mass and velocity. That's all well and good, but what *is* momentum? That's a tough question to answer. Newton referred to momentum as "quantity of motion." In other words, the larger the momentum, the more total "motion" exists. I tend to view momentum as a highly useful mathematical concept that allows us to better analyze physical situations. As a way of illustrating what I mean, the next section shows how the mathematical concept of momentum allows us to analyze situations in which the velocity of an object changes.

Impulse

Suppose you threw a wad of paper at a window. Would the window break? Probably not. What if you threw a baseball at the window with the same velocity as the paper wad? The window would most likely break, provided that the velocity was reasonably large. What's the difference between the paper wad and the baseball? In this case, the major difference is the mass of the two objects. Since the mass of the baseball is larger than that of the paper wad, its momentum (mass times velocity) is greater. This momentum is what determines whether or not the window will break. How does momentum determine that? Well, let's go back to Newton's Second Law in order to find out.

When an object strikes the glass in a window, the window impedes the progress of the object. This is true whether the object is a paper wad, a falling leaf, a molecule of oxygen in the air, or a baseball. In order for the window to stop an object, it must change that object's velocity. This means that there must be acceleration involved. Newton's Second Law says that wherever there is an acceleration, there must be a force causing that acceleration:

$$\mathbf{F} = m\mathbf{a} \tag{5.1}$$

This tells us that the window exerts a force on each object, attempting to change that object's velocity.

What does Newton's Third Law tell us? It says that for every applied force, there is an equal and opposite force. In this case, the object striking the glass applies an equal but opposite force on the window. That force is also given by Equation (5.1). Now remember, acceleration is defined as the change in velocity over a particular time interval:

$$\mathbf{a} = \frac{\Delta\mathbf{v}}{\Delta t} \tag{1.3}$$

Plugging this definition of acceleration into Equation (5.1):

$$\mathbf{F} = m\frac{\Delta\mathbf{v}}{\Delta t} \tag{9.2}$$

The change in velocity is simply the difference between the final and initial velocities:

$$\mathbf{F} = m\frac{(\mathbf{v}_f - \mathbf{v}_o)}{\Delta t} \tag{9.3}$$

Distributing the mass gives us:

$$\mathbf{F} = \frac{(m\mathbf{v}_f - m\mathbf{v}_o)}{\Delta t} \tag{9.4}$$

Look at this equation for a moment. We have mass times the final velocity minus mass times the initial velocity. Equation (9.1) tells us that mass times velocity is momentum. Thus, Equation (9.4) becomes:

$$\mathbf{F} = \frac{(\mathbf{p}_f - \mathbf{p}_o)}{\Delta t} \tag{9.5}$$

Final momentum minus initial momentum, however, is simply the change in momentum:

$$\mathbf{F} = \frac{\Delta\mathbf{p}}{\Delta t} \tag{9.6}$$

This important equation is an expression of Newton's Second Law. It tells us that in order for an object's momentum to change, there must be a force applied. The strength of that force and the time interval over which the force is applied will determine how much the object's momentum changes. It also tells us that the direction of the applied force must be the same as that of the change in momentum. As an historical note, this is actually the way that Newton expressed his second law. The use of Equation (5.1) to express Newton's law came later.

Equation (9.6) tells us why a paper wad thrown at a window will not break the glass, but a baseball thrown with the same velocity will. Since the baseball has a larger mass, its momentum is larger. As a result, when the window impedes the progress of the baseball, it must apply a significantly larger force than it did when it impeded the progress of the paper wad. Newton's Third Law tells us that both the baseball and the paper wad will apply equal and opposite forces on the glass. However, since the glass must apply more force to the baseball than to the paper wad, the baseball (according to Newton's Third Law) applies more force on the glass. This larger force is enough to break the glass.

Since the change in an object's momentum is a rather fundamental quantity to examine when studying motion, physicists often rearrange Equation (9.6) to solve for it:

$$\Delta \mathbf{p} = \mathbf{F}\Delta t \qquad (9.7)$$

The quantity $\mathbf{F}\Delta t$ is often called "**impulse**." Thus, physicists often say that change in momentum which an object experiences is equal to the impulse imparted to the object. It should be obvious that the units for impulse are the same as the units for momentum, since Equation (9.7) says that the change in momentum is equal to impulse.

The concept of impulse is most useful when you deal with forces that are applied over a short time interval. These forces are often called "**impulsive forces**" to emphasize the fact that impulse plays a critical role. For example, consider the case of a baseball player hitting a baseball with a bat. The bat hits the ball and stays in contact with the ball for a short time interval. During that time, the bat exerts a force on the ball. The product of that force and the time interval over which the ball and the bat are in contact is the impulse imparted to the ball by the bat. This impulse will stop the ball from traveling towards the batter and will force it to travel in the other direction. In addition, the batter tries to increase the new velocity of the ball as much as possible, because (as we know from our study of projectile motion) the faster it travels, the farther it goes. These direction and velocity changes create a change in momentum.

Based on this information, then, what two things can be done to increase the velocity (and therefore the range) of the ball as it leaves the bat? Obviously, the larger the force that the bat applies to the ball, the faster (and therefore farther) the ball will travel. Thus, the baseball player needs to hit the ball as hard as he can. The other thing that the player can do, however, is to keep the ball and bat in contact as long as possible. The longer they are in contact, the more impulse is imparted, because Δt in Equation (9.7) is larger. This results in a larger change in momentum.

How does a baseball player keep the ball and bat in contact for a longer time interval? The player "follows through" on the swing. Rather than just stopping the swing when the player feels the bat hit the ball, he continues on with the swing. Thus, as the ball changes direction and begins traveling away from the player, the bat will keep up with the ball for a while, increasing the time that the bat and ball are in contact. Follow-through is an important part of a baseball player's swing and is often the skill that separates good players from bad ones. Of course, the concept of follow-through applies to any sport in which you are trying to change the momentum of an object to achieve a large velocity. Thus, tennis players, golf players, and hockey players must all have good follow-through in their swings as well.

Perform the following experiment to get a better feel for the concept of impulse.

EXPERIMENT 9.1
Egg Drop

Note: This experiment can be modified to make a fun and interesting contest. Read the note on p. 309 to see what I mean.

Supplies:
- Safety goggles
- Two helpers (wearing old clothes or large aprons)
- Old clothes or an apron for yourself
- Two eggs
- A reasonably large board or some other hard, flat surface such as a big cookie sheet
- A fitted bed sheet

1. Go outside or somewhere that a mess will be easy to clean up, because you will end up breaking one of the eggs in this experiment.
2. Have your helpers stand in front of you, holding the board waist high, parallel to the ground.
3. Hold one of the eggs in your hand, and stretch out your arm so that your hand is above the board.
4. Throw the egg straight up in the air, so that when it falls, it hits the board. What happens? The egg breaks, splattering both you and your helpers!
5. Have your helpers put the board away and hold the bed sheet in exactly the same position as they held the board. They should grasp the bed sheet firmly and hold it so that the surface is flat and parallel to the ground. They should not go out of their way to stretch the bed sheet tightly. They should simply make sure that its surface is flat.
6. Take the other egg and do exactly the same thing: throw it up in the air so that it lands on the bed sheet. What happens this time? The egg did not break, did it?
7. Repeat step 6, being careful to watch what happens as the egg hits the sheet.
8. Record all of your observations.
9. Clean up your mess.

Why didn't the egg break when it landed on the bed sheet? You might be tempted to say that the bed sheet is softer. Well, that's true, but that's not the real reason. When an egg lands on a soft surface, it is less likely to break, but you still haven't determined *why* this is the case. Remember your observations in step 7. As you carefully watched the egg land on the sheet, you should have seen the sheet flex downward, moving with the egg. *This* is why the egg did not break.

Think about the concept of impulse. When the egg hit either the board or the sheet, it had a certain amount of momentum, because it had both mass and velocity. Both the board and the sheet needed to stop the egg. In order to do that, the egg's velocity had to be changed to zero, which means its momentum had to change to zero. In order to change the egg's momentum, both the board and the sheet applied an impulse. Since the eggs had roughly the same momentum in each case, the board and sheet applied roughly the same impulse. Since the board couldn't flex and move with the egg like the sheet could, the board applied this impulse in a very short time interval. The sheet, on the other hand, had more time to stop the egg, because it flexed, "stretching out" the time it took to stop the egg.

Since the board delivered the impulse over a short time interval, the force it applied was large. After all, since impulse is defined as $\mathbf{F}\cdot\Delta t$, if Δt is small, $\mathbf{F}$ must be large. In the case of the bed sheet, however, the time interval (Δt) was larger, so in order to deliver the same impulse, the force ($\mathbf{F}$) was much smaller. Since the bed sheet exerted a smaller force while stopping the egg, it did not break the egg. This is, in fact, why soft objects are used to cushion falls. If you fall off a building onto a cement sidewalk, the cement sidewalk must stop you in a very short time. Thus, it must apply a lot of force.

This force will be sufficient to break your bones and maybe even kill you. If, on the other hand, you land on a pile of hay, the hay will flex and bend, increasing the time interval over which it must stop you. This makes the force that the hay must exert rather small, most likely small enough so that you suffer no real damage.

The concept of impulse is an important one in physics, so you might want to spend some extra time on it. At the end of this module (p. 309), there is a note on how to make a contest out of the concept of impulse. You might want to do a contest like this the next time you get together with some fellow physics students. For right now, however, I want to make sure you understand impulse by applying the concept in order to solve problems.

EXAMPLE 9.2

In order to get the necessary range on her drives, a golfer must hit the ball (m = 0.0459 kg) off the tee so that its initial velocity is 30.0 m/sec at an angle of 45 degrees. If the ball and the club are in contact for 0.120 seconds, with what force must she hit the ball?

When the ball sits on the tee, its velocity is zero. Thus, according to Equation (9.1), its momentum is zero as well. When the ball leaves the club, it is supposed to have a velocity of 30.0 m/sec directed at an angle of 45 degrees. This means its momentum will be:

$$\mathbf{p} = m\mathbf{v}$$

$$\mathbf{p} = (0.0459 \text{ kg}) \cdot (30.0 \frac{\text{m}}{\text{sec}}) = 1.38 \frac{\text{kg} \cdot \text{m}}{\text{sec}}$$

The direction of momentum and velocity are the same, meaning that the momentum is also directed at an angle of 45 degrees. Thus, a change in momentum has occurred. To calculate the change, we take the final momentum minus the initial momentum. This is easy to do, because the initial momentum was zero. The change in momentum, then, is simply equal to the final momentum. Now that we have both the change in momentum and the time interval over which this change occurs, we can use Equation (9.6) to determine the force. Since we know that the force must have the same direction as the change in momentum, we will just ignore direction until the end, and then attach it to the force when we are done.

$$\mathbf{F} = \frac{\Delta \mathbf{p}}{\Delta t}$$

$$\mathbf{F} = \frac{1.38 \frac{\text{kg} \cdot \text{m}}{\text{sec}}}{0.120 \text{ sec}} = 11.5 \frac{\text{kg} \cdot \text{m}}{\text{sec}^2} = 11.5 \text{ Newtons}$$

The force that needs to be applied is <u>11.5 Newtons at an angle of 45 degrees</u>.

Before we move on to the next example problem, I want to point out something. We used our concepts of momentum and impulse to figure out the force exerted by the golf club on the ball. Now think about what we learned in the previous module. The ball had no kinetic energy while it sat on the tee, but it had kinetic energy when it left the tee. Where did the energy come from? The golf club worked on it. Given the mass and final velocity (which we know), we could calculate the amount of energy that the ball was given, which would be equal to the work done by the club. Well, since we could calculate the work done by the club, and since we just figured out the force exerted by the club, we could use Equation (9.1) to determine the distance over which the club did the work. Since the work must be done when the ball and club are in contact, this would tell us how far the ball and club traveled together during the swing. Finally, since we have that distance and the time interval over which it occurred, we could calculate the average speed of the club during that time, which would tell us how fast the golfer swung the club. Are you starting to see how powerful the physics you have learned can be?

A baseball player catches a 145.0-gram ball that is traveling with a speed of 22.1 m/sec. If the player can exert a force of 55 Newtons, how long will it take to stop the ball?

There are two things you should notice very quickly in this problem. First, the mass is not given in SI units. We will need to convert it to kg before we use it. Secondly, notice that you are not given *velocity*; you are only given the speed. That's okay, though. We know that the ball travels in a straight line. Thus, we can simply denote direction by sign. I will say that motion in the original direction of the ball's travel is positive motion. Thus, the velocity of the ball is positive 22.1 m/sec.

With those two issues out of the way, we can now solve the problem. First, in order to stop the ball, the player must change the ball's velocity from 22.1 m/sec to zero. This means that a change in momentum will occur. The initial momentum is:

$$\mathbf{p} = m\mathbf{v}$$

$$\mathbf{p} = (0.1450 \text{ kg}) \cdot (22.1 \frac{\text{m}}{\text{sec}}) = 3.20 \frac{\text{kg} \cdot \text{m}}{\text{sec}}$$

When the ball stops, its momentum will equal zero. The change in momentum, then, will be:

$$\Delta\mathbf{p} = \mathbf{p}_{\text{final}} - \mathbf{p}_{\text{initial}} = 0 \frac{\text{kg} \cdot \text{m}}{\text{sec}} - 3.20 \frac{\text{kg} \cdot \text{m}}{\text{sec}} = -3.20 \frac{\text{kg} \cdot \text{m}}{\text{sec}}$$

What does that negative sign mean? Remember, signs in physics generally denote direction. The negative sign here means that the change in momentum is in the negative direction. Since the initial velocity and momentum of the ball were denoted as positive, this simply means that the change in momentum is in the opposite direction as compared to the initial momentum. This should make sense. If the ball stops, the change in momentum must oppose the initial momentum.

Now that we have the change in momentum, we can use Equation (9.6) to determine the time interval over which the momentum changed. However, we do have to think about something. The problem says that the player can exert a force of 55 Newtons. However, in what direction will that force be pointed? It will be pointed *opposite* the initial velocity of the ball. In other words, *the force will be -55 Newtons*. After all, the initial motion of the ball is positive, and the player wants to stop the ball. As a result, he will have to apply a force in the opposite direction, which has been defined as negative. Thus, Equation (9.6) becomes:

$$\mathbf{F} = \frac{\Delta p}{\Delta t}$$

$$-55\,\text{Newtons} = \frac{-3.20\,\frac{\text{kg} \cdot \text{m}}{\text{sec}}}{\Delta t}$$

$$\Delta t = \frac{-3.20\,\frac{\text{kg} \cdot \text{m}}{\text{sec}}}{-55\,\frac{\text{kg} \cdot \text{m}}{\text{sec}^2}} = 0.058\ \text{sec}$$

It takes 0.058 seconds for the baseball player to stop the ball.

ON YOUR OWN

9.2 A baseball (m = 145 grams) thrown by a pitching machine is clocked by radar with a speed of 45 m/sec. If the roller wheels of the pitching machine exert 43 Newtons of force on the baseball as it is launched, how long did the baseball stay in contact with them?

9.3 A tennis ball (m = 56.7 grams) travels with a horizontal velocity of 23.1 m/sec. A tennis player takes her racquet and hits the ball, returning it with a horizontal velocity of -29.7 m/sec. If the ball and racquet were in contact for 0.21 seconds, how much force did the tennis player exert?

The Conservation of Momentum

We are not yet done exploiting the concept of momentum. It turns out that when we state Newton's Second Law in terms of momentum (as we did in Equation 9.6), we can learn another useful fact. Consider a system in which there are no external forces at play. For example, suppose you were watching two objects moving freely (without friction, without being pushed, etc.) on a collision course with each other. If there are really no forces acting on the objects, then the force in Equation (9.6) would be zero. This would make Equation (9.6) look like:

$$0 = \frac{\Delta \mathbf{p}}{\Delta t} \tag{9.8}$$

If we now multiply both sides of the equation by Δt, we get:

$$\Delta \mathbf{p} = 0 \tag{9.9}$$

Do you see the significance of Equation (9.9)? If $\Delta\mathbf{p} = 0$, we know that *total momentum of the system cannot change*. Even though the two objects collide with each other, their total momentum must stay the same. This means that, like energy, momentum must be conserved in a system which has no forces working on it. In fact, we can make an even stronger statement than that. In order for Equation (9.9) to be true, we only need the force term of Equation (9.6) to equal zero. This can happen even when there are forces at work, as long as those forces cancel each other out. The fact that $\Delta\mathbf{p} = 0$ under such conditions is often called the **Law of Momentum Conservation**:

Law of Momentum Conservation - When the sum of the forces working on a system is zero, the total momentum in the system cannot change.

Even though I just said that the Law of Momentum Conservation is like the First Law of Thermodynamics, there is one very big difference. The First Law of Thermodynamics has no exceptions. Energy is always conserved, no matter what. This law contains a large caveat: in order for momentum to be conserved, the sum of the forces acting on a system must be zero. If friction, for example, is working on an object, then that object's momentum will not be conserved. If you are pushing on an object, that object's momentum will not be conserved. In order for the momentum to be conserved, the sum of the forces working on the system of interest must equal zero.

Now, if you think about it, there are, in fact, very few systems in which this is true. After all, every real situation has friction. If you are standing on a floor, there is friction between your feet and the floor. If an object falls through the air, there is air resistance, another form of friction. Even if the object is in outer space, there are still *some* molecules to bounce off of it, and that causes friction. Thus, in the end, the systems to which the Law of Momentum Conservation truly applies are really quite few and far between.

Why do we bother to study it, then? Well, even though friction applies to all systems, sometimes the effect of friction is so small that it can be ignored. For example, suppose you were standing on a smooth sheet of ice instead of a floor. The friction in that situation is very low and can probably be ignored. Also, for most dense objects moving at reasonable speeds, air resistance is pretty small and can probably be ignored. So, even though the Law of Momentum Conservation very rarely applies to real situations, sometimes the outside forces acting on a system can be so small that they can be ignored, allowing us to *approximate* the fact that the Law of Momentum Conservation applies.

In order to truly study a physical situation, you will need to get a feel for whether or not you can use the Law of Momentum Conservation. In some problems, I will simply tell you whether or not it applies. In other problems, I may require you to figure it out for yourself. In order to get some practice in the latter situation, study the examples and solve the "On Your Own" problems that follow.

EXAMPLE 9.3

A tennis player hits a tennis ball with her racquet. Does the Law of Momentum Conservation apply to the ball and racquet? Why or why not?

In order to determine whether or not the Law of Momentum Conservation applies to a physical situation, the best thing to do is try to imagine what happens. In this case, that's easy. The ball will hit the racquet, and the racquet will cause the ball to start traveling in the other direction. Now, before the ball hits the racquet, what is the momentum of each object under consideration? Well, the problem asks us about the ball and the racquet. Since the woman is swinging the racquet, it has momentum in the direction of the swing. The ball, however, has momentum in the opposite direction, because it is approaching the racquet.

After the ball hits the racquet, what happens to the momentum of each object? Well, in hitting the ball, the racquet may have slowed down a bit, but it will generally continue to move in the direction of the swing. Thus, its momentum may have changed a little, but not much. The ball, on the other hand, has a huge change in momentum. It was traveling *towards* the racquet. After being hit, however, it travels *away* from the racquet. Thus, the momentum has changed direction entirely. Since the racquet's momentum changed little (if at all), and the ball's momentum changed substantially, the total momentum of the system changed. The Law of Momentum Conservation, therefore, does not apply. Why doesn't it apply? Well, there is an outside force. The hand and arm of the player holding the racquet apply a force on the racquet; thus, the sum of the forces acting on the system is not zero.

A couple is performing an ice skating routine. The man stands still, and the woman leaps into his arms. Does the Law of Momentum Conservation apply to the ice skaters? Why or why not?

Unlike the previous example, it might be hard for you to predict what will happen in this case. If you cannot predict what will happen, you need to examine the forces acting on the system. The problem asks about the skaters, so we must examine all forces acting on the skaters. Now remember, we need to worry only about the forces acting *on* the system. Any forces acting *within* the system are irrelevant. This means that although the skaters themselves are exerting all sorts of forces (the woman must exert force to jump and the man must exert force to catch her), they are irrelevant to the question. We need only concern ourselves with the forces acting on the skaters. What forces act on the two skaters? Is anyone else pushing or pulling on them? No. Gravity does pull them down, but the normal force exerted by the ice cancels gravity, making the sum of those two forces zero. If the sum is zero, then momentum can still be conserved. There is only one other force working on them: friction. However, since they are on ice, the friction is rather small. Thus, the Law of Momentum Conservation does apply because the only outside force not canceled out is friction and, in this case, it is small enough to be ignored.

Now that we've answered the question, can you figure out what will happen? Since momentum is conserved, the total momentum of the skaters must be the same before and after the man catches the woman. Well, before the woman is caught, she is traveling towards him. Thus, she has momentum towards the man. The man has no momentum, because he is standing still. Thus, the total momentum of the skaters is the same as that of the woman. When the man catches the woman, then, the momentum must be the same. How will this happen? Both the man and the woman will start moving in the same direction that the woman jumped. This ensures that the skaters still have momentum and that the direction of the momentum is the same as before the woman was caught. When the man catches the woman, then, he and the woman will being moving together, in the same direction as the woman's original motion.

Not only the direction of the momentum vector must be conserved, however. The magnitude must also be conserved. Well, before the catch was made, the total momentum was just the woman's mass times her velocity. Now that the woman and the man are traveling together, the total mass has increased. To keep the magnitude of the momentum the same, then, the total velocity must decrease. The conclusion, then, is that when the man catches the woman, they will both start traveling in the same direction that the woman was traveling, but at a lower velocity than that of the woman by herself. If this discussion is a little above your head right now, don't worry. We'll do a concrete example in a bit, and that will help clear things up.

ON YOUR OWN

9.4 An astronaut is floating in space, near his ship. He has a ball in his hands and is essentially motionless. He suddenly throws the ball. Does the Law of Momentum Conservation apply to the astronaut and the ball? Why or why not?

9.5 A toy car is rolling along a floor. It suddenly crashes into the wall, coming to a complete halt. Does the Law of Momentum Conservation apply to the car and the wall? Why or why not?

Before we leave this section, I want you to get some "hands-on" experience with momentum conservation.

EXPERIMENT 9.2
Momentum and Energy Conservation

Supplies:

- Safety Goggles
- Four ping-pong balls
- Some thread
- Some tape
- A ruler or meterstick
- Several books

1. Make two piles of books. The piles should each be 15-20 cm tall and of equal height.
2. Pull the book piles about 26 cm apart so that the ruler or meterstick can span the gap in between them, much like a bridge.
3. Cut the thread so you have four pieces that are roughly as long as the book piles are high.
4. Use the tape to attach the end of one of the pieces of thread to one of the ping-pong balls.
5. Use the tape to attach the other end of that piece of thread to the ruler, so that the ping-pong ball hangs down from the ruler. The ping-pong ball should be at least 4 cm above the desk or floor upon which the books sit.
6. Repeat steps 4-5 so that another ping-pong ball hangs from the ruler. It needs to hang at exactly the same height as the first ping-pong ball, and it should be just barely touching the first ping-pong ball. In the end, your setup should look something like this:

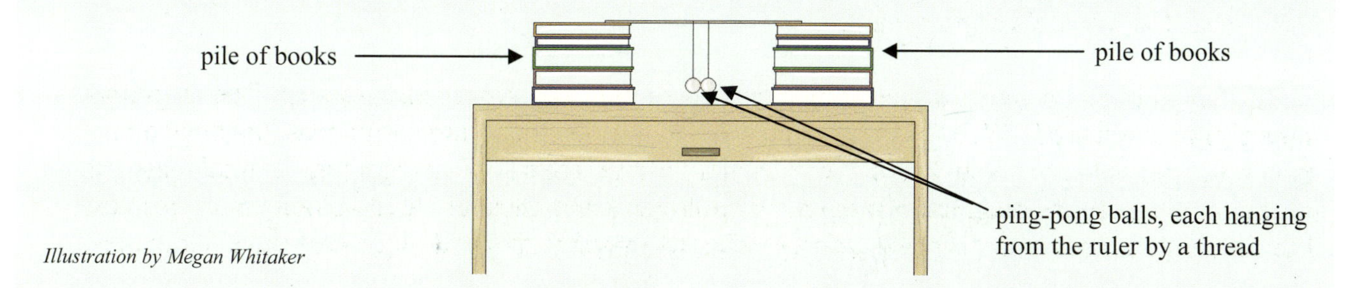

Illustration by Megan Whitaker

7. Lift one ping-pong ball directly away from the other, in the direction of the nearest pile of books, so that the angle between the ruler and the thread is about 45 degrees. Make sure that the thread holding it to the ruler stays taut.
8. Release the ping-pong ball so that it will fall, colliding head on with the other ping-pong ball. Note what happens.
9. Make sure that both balls are still and repeat steps 7 and 8, this time using the other ping-pong ball.
10. Repeat steps 4-5 again, so that there is now a third ping-pong ball hanging from the ruler. Remember, this ping-pong ball must hang at the same height as the others, and it must be barely touching the ball to which it is closest.
11. Lift one of the ping-pong balls on the end directly away from the other two balls, in the direction of the nearest pile of books. Make sure that the thread holding it to the ruler stays taut.
12. Release the ping-pong ball so that it will fall, colliding head-on with the first ping-pong ball in the line of two ping-pong balls that are hanging still. Note what happens.
13. Make sure all of the ping-pong balls are still again.
14. Lift two of the ping-pong balls at once. Make sure they continue to touch one another and are in line with each other, and make sure both of their threads stay taut.
15. Release both of the ping-pong balls together, so that they collide with the one remaining ping-pong ball that is still hanging straight down. Note what happens.
16. Repeat steps 4-5 again, so that there is now a fourth ping-pong ball hanging from the ruler. Remember, this ping-pong ball must hang at the same height as the others, and it must be barely touching the ball to which it is closest.
17. Repeat steps 11-15, each time noting what happens.
18. You can continue to play with this system and explore many different combinations of collisions.
19. Clean up your mess.

What happened in the experiment? Well, in the first part of the experiment when you had only two balls hanging from the ruler, you should have seen that when the moving ping-pong ball collided with the stationary one, the moving ball stopped, and the stationary one suddenly started moving upwards. It then reached a maximum height and began falling, eventually colliding with the first ping-pong ball. It then stopped, and the ping-pong ball you originally lifted began to move upwards again. This kind of motion probably died away quickly and probably got more erratic as time went on.

How can we understand this behavior? Well, the instant before the collision, the moving ping-pong ball had a certain amount of momentum. If we neglect air resistance, the sum of the forces acting on the ping-pong balls is zero, so the conservation of momentum applies. After the collision, then, the momentum still had to be the same. Since the ping-pong balls were of roughly equal mass, this happened by the first ball stopping and the second ball starting to move with the same velocity as the first. Thus, the momentum before the collision was m**v**, where "m" is the mass of the first ping-pong ball and "**v**" is its velocity. After the collision, the momentum had to be the same, so the second ping-pong ball had to move with the same velocity as the first.

As time went on, air resistance as well as some friction in the collision reduced the kinetic energy, so the motion eventually died away. However, for the first couple of collisions, you should have seen the moving ball come to rest and the stationary ball begin moving. The motion of the balls may have become a bit erratic as time went on, because the balls probably developed a bit of rotational motion since they did not collide perfectly head-on. This perturbed the motion of the balls, causing them to sway in directions other than the original direction of motion.

What happened next? In the second part of the experiment, you had three balls hanging from the ruler. When you lifted one ball and then let it go so that it would collide with the other two balls, you should have seen the ball come to a halt, the middle ball stay stationary, and the ball on the other end start to move up. Once again, this is the result of momentum conservation. The ball that was originally moving had momentum of m**v**. Since the momentum after the collision had to still be m**v**, only one of the two stationary balls began to move. That way, the mass was the same. Thus, the velocity would also have to be the same as well.

"Now wait a minute," you might be saying, "why did only one ball move?" After all, momentum would also be conserved if *both* stationary balls began to move with half of the original velocity. After all, the momentum before the collision was m**v**. If both stationary balls moved, the mass would be twice as much (2m), but if each ball had half the velocity (**v**/2), the final momentum would still be (2m)·(**v**/2), which is m**v**. Thus, that would conserve momentum as well. Why didn't that happen?

Well, if both stationary balls moved with a velocity of **v**/2, momentum would be conserved, *but energy would not*. Remember, the First Law of Thermodynamics says that energy cannot be created or destroyed. However, if both stationary balls moved with a velocity of **v**/2, kinetic energy would have to be destroyed. Think about it. Before the collision, the kinetic energy of the system was:

$$KE = \frac{1}{2}mv^2$$

If both stationary balls moved after the collision with a velocity of **v**/2, momentum would be conserved, but the total kinetic energy (the sum of each ball's kinetic energy) would be:

$$KE = \frac{1}{2}m(\frac{v}{2})^2 + \frac{1}{2}m(\frac{v}{2})^2 = m(\frac{v}{2})^2 = \frac{1}{4}mv^2$$

Note then that the kinetic energy would have to be reduced by a factor of two! Energy cannot be destroyed, so when the first ball collided with the second ball, there was no way for both momentum and energy to be conserved unless only one of the stationary balls began to move.

What happened when you lifted two balls and let them collide with the remaining stationary ball? In that case, one of the moving balls became stationary after the collision, but the second moving ball continued to move, along with the other ball that was originally stationary. To conserve both energy and momentum, two balls had to be traveling with the speed of the original two balls. Since there was only one stationary ball before the collision, one of the balls that was originally moving had to continue to move.

When you finally had four balls in the system, you should have noticed similar behavior. The number of balls moving after the collision was always the same as the number of balls moving before the collision. Thus, if you picked one ball up and let it collide with the remaining three, only one ball (the ball on the other end) would start moving after the collision. If you picked up two balls and let them collide with the other two stationary balls, those two previously-stationary balls would begin to move, while the originally-moving balls would remain still. The motion resulting from the collisions, then, is simply the result of the fact that both momentum and energy had to be conserved.

You may have recognized this setup. You can often find a system like this sitting on an office desk. It is what executives call a "desk toy." Usually, the balls are metal, and they generally each hang by two strings rather than one. The two strings constrain the motion of the balls better than one string, making sure they fall so that the collisions between the balls is head-on. Catalogs, websites, and stores that cater to executives often sell these desk toys. Now you know the physics behind how they work.

The Mathematics of Momentum Conservation

Now that you have a bit more understanding of when the Law of Momentum Conservation can be applied and how it works, it's time to see how you can use mathematics to analyze situations in which the Law of Momentum Conservation applies. The best way for you to see how to do this is by example.

EXAMPLE 9.4

A 4.5×10^5-kg cannon is mounted on wheels that are free to roll. The cannon fires a cannonball ($m = 1.2 \times 10^4$ grams) with a horizontal velocity of 145 m/sec. Assuming that friction is small enough to be ignored, what happens to the cannon after the shot is fired?

In this problem, our system is the cannon and the ball. Since we are told that friction is small enough to be ignored and that the wheels are free to roll, we know that there are essentially no forces other than gravity working on our system. Gravity, however, is canceled by the normal force, so the Law of Momentum Conservation does apply. In order to use this law, we simply have to determine the total momentum before the shot was fired and make sure that the total momentum after the shot was fired is the same.

The total momentum before the shot was fired is easy. Before it is fired, the cannonball is motionless inside the cannon. Its velocity is zero and, therefore, so is its momentum. In the same way, the cannon sits still, so its velocity and momentum are both zero. Thus, the total momentum of the system before the shot was fired is zero. In order for the Law of Momentum Conservation to hold, then, the total momentum after the shot was fired must also be zero.

You might think that it's impossible for the total momentum after the shot was fired to be zero. After all, as soon as the cannonball is fired, it has velocity and therefore momentum. How, then, can the total momentum of the system be zero? Well, remember that momentum is a vector. Thus, the total momentum can still be zero as long as the cannon has a momentum which cancels out the momentum of the cannonball. Let's see how that happens.

Since we know that the total momentum of both the cannonball and the cannon must be the same before and after the shot was fired, we can write down this equation:

$$(m_{cannon}\mathbf{v}_{cannon} + m_{cannonball}\mathbf{v}_{cannonball})_{before} = (m_{cannon}\mathbf{v}_{cannon} + m_{cannonball}\mathbf{v}_{cannonball})_{after}$$

This is simply another way of stating Equation (9.9). As I said before, the momenta (plural of momentum) of both the cannon and the cannonball before the shot was fired were zero. In addition, we know the masses of each object (the mass of the cannonball must be converted to kg) and the velocity of the cannonball after the shot was fired. Plugging them into our equation gives us:

$$0 + 0 = (4.5 \times 10^5 \text{ kg}) \cdot \mathbf{v}_{\text{cannon}} + (12 \text{ kg}) \cdot (145 \frac{\text{m}}{\text{sec}})$$

In this equation, we have only one variable: the velocity of the cannon. We can therefore solve for it:

$$0 = (4.5 \times 10^5 \text{ kg}) \cdot \mathbf{v}_{\text{cannon}} + (1.7 \times 10^3 \frac{\text{kg} \cdot \text{m}}{\text{sec}})$$

$$\mathbf{v}_{\text{cannon}} = \frac{-1.7 \times 10^3 \frac{\cancel{\text{kg}} \cdot \text{m}}{\text{sec}}}{4.5 \times 10^5 \cancel{\text{kg}}} = -0.0038 \frac{\text{m}}{\text{sec}}$$

What does this tell us? In response to shooting the cannonball, the cannon starts to move. Since the velocity is negative (and we know that signs in physics tell us direction), we know that it travels in the opposite direction compared to the cannonball (whose velocity was given as positive). Thus, we can say that <u>the cannon travels with a velocity of 0.0038 m/sec opposite the cannonball</u>.

Do you see how we applied the Law of Momentum Conservation to this problem? We determined what the system was and added the momentum of each member of the system together for the situation described at the beginning of the problem. We called this the total momentum "before." We then did the same thing to the situation described at the end of the problem, and we called that the total momentum "after." If the Law of Conservation of Momentum applies, we know that these two sums must be the same, so we set them equal to each other. Once we plugged in everything that we knew, there was only one unknown, and we were able to solve for it. You'll get a chance to see a couple more examples in a minute, but I want to point something out first.

It turns out that the velocity we determined in this problem is such an important facet of physics that there is a special name for it. We call it **recoil velocity**.

<u>Recoil velocity</u> - The velocity that an object develops in response to launching another object, which is a result of the Law of Momentum Conservation

In other words, whenever a gun fires a bullet, in order to conserve momentum, the gun must recoil with a momentum opposite that of the bullet. Now, of course, unless the gun is allowed to move freely (as our cannon was), it won't move very far. Instead, the person holding the gun (or the braces on a mounted gun) will exert a force to stop the gun. This force stops the recoil. However, until that force is applied, the gun will recoil backwards. If you've ever shot a gun, this recoil produced the "kick" that you felt. If a person shooting a gun is not prepared for the gun's recoil, it can be a big shock!

Guns are not the only things that have recoil velocities, however. Think of the astronaut situation described in "On Your Own" Problem 9.4. In that problem, the astronaut was motionless.

When he "launched" the ball, however, he developed a recoil velocity in the opposite direction, in order to conserve momentum. Thus, any object that launches another object will develop a recoil velocity, so long as there is nothing to hold it in place.

A very practical use of the Law of Momentum Conservation is in the launching of rockets and missiles. Study Figure 9.1 to see how momentum conservation is responsible for the launching of a rocket.

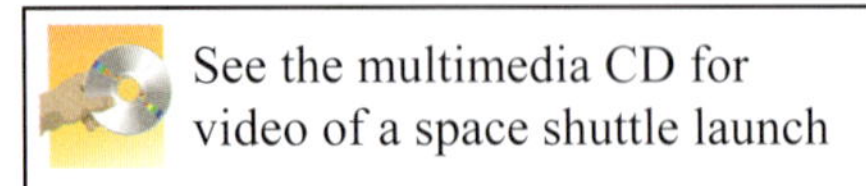

See the multimedia CD for video of a space shuttle launch

Photo courtesy of NASA

FIGURE 9.1
Launching a Rocket

When a rocket is launched, the products of the fuel's combustion speed out the bottom of the rocket. Since the momentum before the combustion began was zero, the momentum afterwards must be zero as well. As a result, the rocket recoils in the other direction so that its momentum cancels the momentum of the combustion products. Even though the rocket is significantly more massive than the molecules that make up the combustion products, there are billions and billions of molecules, each with a large velocity. As a result, the total momentum of all these rocket fuel combustion products is substantial. Thus, the recoil velocity of the rocket is substantial as well.

So, unlike what many people think, rocket engines do not "push" against the ground in order to launch a rocket. The launching of a rocket is a simple result of momentum conservation. There are, of course, many other applications of momentum conservation. Study the following example problems to see more:

EXAMPLE 9.5

Illustrations by Megan Whitaker

In an automobile accident, two cars slam into each other head on. The first (m = 675 kg) is moving with a velocity of 19.3 m/sec while the other (m = 895 kg) has a velocity of -22.1 m/sec. The force of the collision is so great that the two cars stick together, moving as one mass. Assuming that friction between the road and the cars is small enough to be ignored, what is the velocity of the two cars locked together?

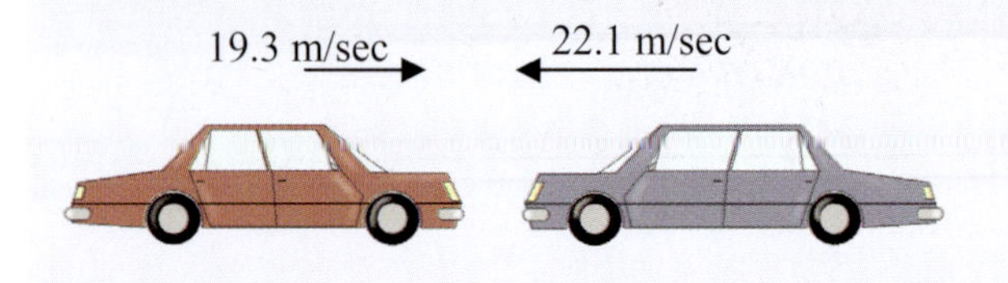

In this collision, we have two independently moving cars at first. When the cars collide, however, they stick together and cannot separate. This means that they now can be treated as one single object. Thus, we start out with two objects, one of mass 675 kg and the other of mass 895 kg, but we end up with only one object of mass 1.570 x 10^3 kg (895 kg + 675 kg). Since we are told to ignore the friction between the road and the cars, and since the force due to gravity working on the cars is canceled by the normal force of the road pushing against the cars, we can say that momentum is conserved. Thus, we can say that the total momentum before the collision must be the same as the total momentum after the collision:

$$(m_{car1}\mathbf{v}_{car1} + m_{car2}\mathbf{v}_{car2})_{before} = (m_{car1} + m_{car2})\mathbf{v}_{two-car\ system}$$

We know everything in this equation except the velocity of the two-car system, so we can solve for it:

$$(675\text{ kg})\cdot(19.3\frac{\text{m}}{\text{sec}}) + (895\text{ kg})\cdot(-22.1\frac{\text{m}}{\text{sec}}) = (675\text{ kg} + 895\text{ kg})\cdot v_{two-carsystem}$$

$$-6.8\times10^3\ \frac{\text{kg}\cdot\text{m}}{\text{sec}} = (1.570\times10^3\text{ kg})\cdot v_{two-carsystem}$$

$$v_{two-carsystem} = \frac{-6.8\times10^3\ \frac{\cancel{\text{kg}}\cdot\text{m}}{\text{sec}}}{1.570\times10^3\ \cancel{\text{kg}}} = -4.3\ \frac{\text{m}}{\text{sec}}$$

Since the velocity is negative, that means that the two-car pileup travels with a velocity of 4.3 m/sec in the direction that the 895-kg car was traveling.

An empty truck (m = 1.2 x 10^3 kg) slowly (v = 2.2 m/sec) moves under a grain hopper. The hopper dumps grain into the truck until it is full. If the hopper puts 7.4 x 10^2 kg of grain into the truck and the driver allows the truck to coast while it is under the hopper, what will the truck's velocity be when it is full? Assume that friction can be neglected.

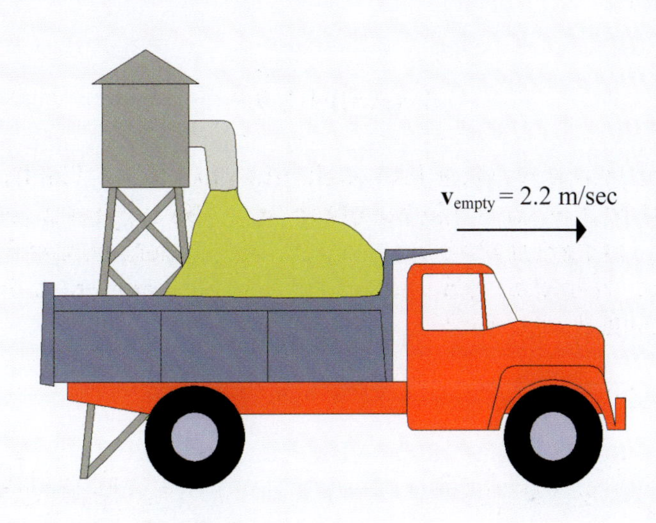

In this problem, the object under consideration (the truck) changes its mass. Since mass is a part of momentum, this affects the truck's momentum. We were told that friction can be ignored, and we know that gravity is countered by the normal force. We also know that the driver is letting the truck coast, so the brakes and the engine are not exerting forces. Thus, we know that momentum is conserved. As a result, we can develop the following equation:

$$m_{empty}\mathbf{v}_{empty} = m_{full}\mathbf{v}_{full}$$

We are told the mass and velocity of the truck when it is empty. In addition, we are also told what mass of grain makes the truck full. Thus, we can say that this mass of grain plus the mass of the empty truck is the mass of the truck when it is full. The only thing that we don't know is the velocity of the truck when it is full, so we can solve for it.

$$(1.2 \times 10^3 \text{ kg}) \cdot (2.2 \frac{\text{m}}{\text{sec}}) = (1.2 \times 10^3 \text{ kg} + 7.4 \times 10^2 \text{ kg}) \cdot \mathbf{v}_{full}$$

$$\mathbf{v}_{full} = \frac{2.6 \times 10^3 \frac{\cancel{\text{kg}} \cdot \text{m}}{\text{sec}}}{1.9 \times 10^3 \cancel{\text{kg}}} = 1.4 \frac{\text{m}}{\text{sec}}$$

The truck, because of the added mass, slows to a velocity of <u>1.4 m/sec</u> in order to conserve momentum. Of course, if the driver used the accelerator or the brakes, this calculation would cease to be correct, because an outside force would make the Law of Momentum Conservation not applicable to this problem.

I hope that you are beginning to see how to determine when and how to use the Law of Momentum Conservation. Try a few problems on your own to see if you get the hang of it.

ON YOUR OWN

9.6 What is the recoil velocity of a 1.21-kg rifle that shoots 25-gram bullets with a velocity of 195 m/sec?

9.7 A 75-kg astronaut is making repairs on his ship in space and accidentally loses his grip. He floats away from his ship and, to his horror, finds that the tie line which supposedly attached him to his ship has come undone. He floats motionless in space, several meters away from his ship. Being a good physicist, he realizes that if he throws the wrench (m = 1.2 kg) he is holding away from the ship, the Law of Momentum Conservation says that he should begin moving towards his ship. If he can throw the wrench with a velocity of 10.1 m/sec away from his ship, at what velocity will he travel back to his ship?

9.8 Two figure skaters are performing a routine. The man (m = 75.0 kg) travels towards the woman at 1.2 m/sec, and the woman (m = 60.0 kg) jumps toward him with a velocity of -3.1 m/sec. After the man catches the woman, at what velocity will the two skaters be traveling?

Angular Momentum

In Module #6, we found that slightly different concepts apply in circular motion as compared to linear motion. For example, in linear motion, force is used to accelerate an object. In order to cause acceleration in circular motion, however, you need to use torque instead of force. Well, it's the same story when we examine momentum in circular motion. When an object moves on a circular path, it is more informative to study its **angular momentum** than to study its linear momentum.

Now before we go into a detailed discussion of angular momentum, you might be curious why it is called "angular momentum" instead of "circular momentum" or something like that. In order to see why this is the case, take a look at Figure 9.2.

FIGURE 9.2
An Object Moving in a Circle

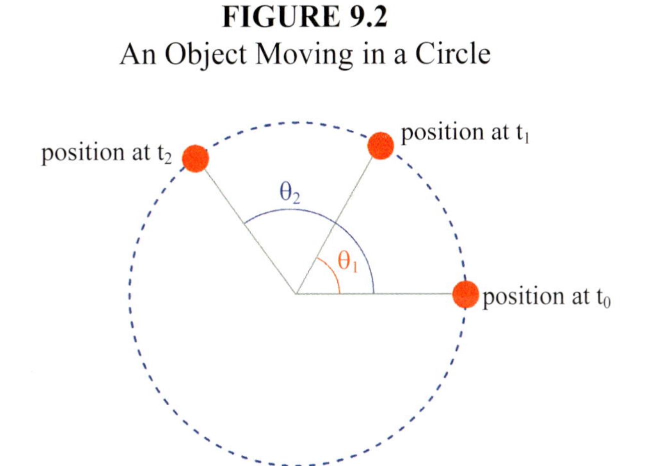

When an object is in rotational motion, the most convenient way to keep track of its position is by the angle swept out between the radius of the circle that points to its initial location and the radius that points to its current location. In the picture above, the object starts at a certain position when the motion begins. This is shown by the position at t_0. After a certain time, it reaches another position on the circle (the position at t_1). If you draw a line from the center of the circle to the initial position and another line from the center of the circle to the position at t_1, you define an angle (θ_1). At a later time, the object reaches a new position (position at t_2). Now the line drawn from the center of the circle to the initial position and the line drawn from the center of the circle to the position at t_2 define another angle (θ_2). This angle is unique for all points on the circle, so as long as the object continues to move around the circle, it is a convenient measure of the object's position.

As you see from the figure, then, the angle continues to change as the object moves around the circle. As a result, we can define something called the **angular velocity**:

Angular velocity - The rate at which the position angle of an object changes in rotational motion

In mathematical terms, we could define the angular velocity this way:

$$\omega = \frac{\Delta\theta}{\Delta t} \tag{9.10}$$

Does this look familiar to you? It should. It looks much like Equation (1.1), the definition of linear velocity. When studying rotational motion, we often look at the object's angular velocity instead of its linear velocity.

Now we really won't be using Equation (9.10) at all, I just wanted to show it to you in order to show you why the term "angular" appears in "angular momentum." Since we can describe rotational motion in terms of angular velocity, we choose to call the momentum associated with such motion "angular momentum."

You must remember that even though angular velocity is a unique physical quantity dealing with rotational motion, it can be *related* to the physical quantities for linear motion. Just as torque (the rotational version of force) can be related to force, angular momentum can be related to linear momentum. As we learned in Module #6, torque is equal to the force times the distance between the axis of rotation and the point at which the force is applied. In the same way, the angular momentum of an object traveling in a circle is equal to the linear momentum ($m \cdot v$) times the radius of the circle:

$$L = mvr \tag{9.11}$$

In this equation, "L" stands for the angular momentum, "m" is the object's mass, "v" represents the object's speed, and "r" is the radius of the circle.

Now the first thing that you might notice about this equation is that it has no vector quantities. Angular momentum is, in fact, a vector quantity, but the vector nature of angular momentum is beyond the scope of this course. If you take advanced physics, you will learn how to determine the direction of the angular momentum vector. For this module, we will only consider the magnitude of the angular momentum vector. This makes things a little easier. You also should note that the units for angular momentum are different from the units for linear momentum. Linear momentum has the units $\frac{\text{kg} \cdot \text{m}}{\text{sec}}$ (alternatively known as the "Wile"). When you take this momentum and multiply by the radius of a circle (as Equation 9.11 says you must in order to get angular momentum), you get units of $\frac{\text{kg} \cdot \text{m}^2}{\text{sec}}$. Those are the units for angular momentum.

Before we go on and study angular momentum in a bit more detail, I need to warn you about one thing. Angular momentum, angular velocity, and torque are all rather difficult subjects. In this course, we barely scratch the surface. As a result, some of the information I will present to you on these subjects is a little simplistic. This is one of those cases. What you will see here is a very limited view of angular momentum. It is, nevertheless, instructive.

Now that we have that out of the way, we can work a little with this concept. One thing that linear and angular momentum have in common is that they both are conserved, under the right circumstances. We have already learned that if the sum of the forces on a system is zero, the linear momentum in that system is conserved. There is a similar law for angular momentum:

<u>Law of Angular Momentum Conservation</u> - If the sum of the torques on a system is equal to zero, the angular momentum never changes.

This law is almost identical to the Law of Momentum Conservation, but the term "force" is replaced by "torque." This should make sense, because torque replaces force when dealing with rotational motion.

What can we learn from this law? Consider, for example, the fact that cats always seem to land on their feet. How do they do that? If a cat is falling and is upside down, the cat must turn over so that its feet are pointing down. The problem is, how does the cat do so when there is nothing to grab on to? The cat does so by moving its tail in a circle. You see, before the cat starts to move its tail, it is experiencing no rotational motion. As a result, its angular momentum is zero. When the cat begins to move its tail in a circle, however, the tail suddenly has angular momentum. In order for the angular momentum to be conserved, something else must rotate in the opposite direction. This will provide an angular momentum that will cancel that of the tail. What starts to rotate? The cat's body, of course. When the cat's tail rotates in one direction, the body rotates in the other direction. This is how a cat controls its body so that it will always land on its feet. In principle, any animal with a tail should be able to do this, but cats seem to have a really strong instinct for it.

Another example of the conservation of angular momentum can be found in figure skating. If you've ever watched figure skaters, you've probably seen them spin in place. If you watch closely, you will see that when they stretch their arms out, the speed at which they spin decreases. When they bring their arms in, the speed at which they spin increases. Once again, this is due to the conservation of angular momentum. When a skater spins, she has angular momentum. The angular momentum of her hand, for example, could be calculated by taking the mass of her hand times the speed of her hand times the radius, which would be the distance from the center of the skater's body to her hand.

Now, if the skater moves that hand closer to her body, what happens? The radius used to calculate the angular momentum decreases. Because angular momentum must be conserved, something else must change in order to keep the angular momentum the same. Well, the only other quantities in Equation (9.11) are the mass and speed of the hand. Since the hand's mass cannot change, its speed must. If the radius decreases, the speed must increase in order to keep angular momentum the same. Thus, the skater spins faster when she pulls her hands in towards her body.

In the same way, if the skater moves her hand farther away from her body, the radius of the hand's motion increases. In order to keep the angular momentum the same, then, the speed of the hand must decrease. Thus, the skater spins more slowly as she extends her hands away from her body.

In some cases, we can, in fact, use Equation (9.11) to apply mathematics to the analysis of a situation involving rotational motion. Study the example and then solve the "On Your Own" problems that follow to see exactly how this is done.

EXAMPLE 9.6

A child's toy is made by tying a ball to a string and passing the string through a hollow tube. The tube is then twirled to get the ball moving in a circle. Once the ball is moving in a circle, the tube is held still and the string is pulled down through the tube, so that the radius of the circle decreases (see figure below).

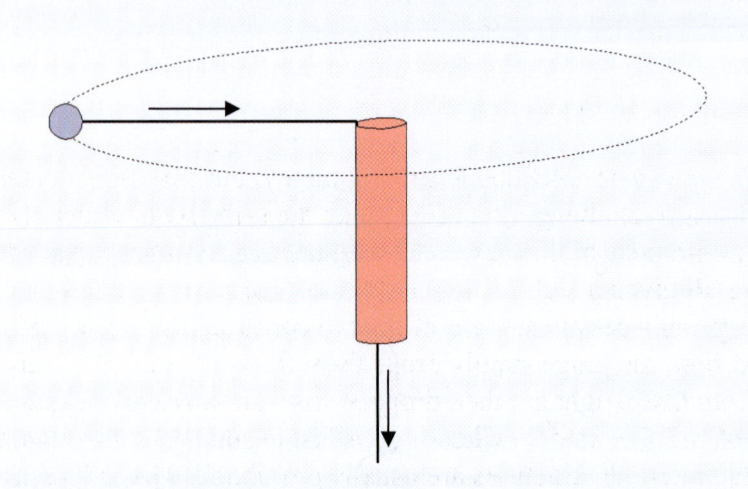

Suppose the initial radius of motion is 34 cm, and the initial speed of the ball is 3.2 m/sec. If the string is pulled so that the radius of motion decreases to 12 cm, what will be the new speed of the ball?

Once the ball starts spinning, it has angular momentum. When the string is pulled, the radius of the motion changes, but the angular momentum cannot change, due to the Law of Angular Momentum Conservation. As a result, the angular momentum before the radius changes must be the same as the angular momentum after it changes:

$$(mvr)_{before} = (mvr)_{after}$$

Since the mass of the ball does not change, it is the same on both sides of the equation, so it cancels out. We know everything else except the speed afterwards, so we can solve for it:

$$\cancel{m} \cdot (3.2\ \frac{\text{meters}}{\text{sec}}) \cdot (0.34\ \text{meters}) = \cancel{m}v_{after} \cdot (0.12\ \text{meters})$$

$$v_{after} = \frac{(3.2\ \frac{\text{meters}}{\text{sec}}) \cdot (0.34\ \cancel{\text{meters}})}{0.12\ \cancel{\text{meters}}} = 9.1\ \frac{\text{meters}}{\text{sec}}$$

The ball speeds up to <u>9.1 meters/sec</u>.

ON YOUR OWN

9.9 In a cruel laboratory experiment, a mouse is placed near the center of a plate that is spinning freely. In an effort to get off of the plate, the mouse begins to crawl towards the plate's edge. What will happen to the speed at which the plate is spinning as the mouse crawls in that direction?

9.10 A toy similar to the one in the example is used to get a ball twirling at a speed of 7.5 m/sec and radius of 22 cm. If the string is adjusted such that the ball travels at 3.2 m/sec, what will be its radius of motion?

**NOTE: Experiment 9.1 can be modified to make a messy, but enjoyable contest, called an "egg drop" contest. The idea is to challenge students to take the same set of materials and make a case that holds a single egg. The purpose of this case is to keep the egg from breaking when it is dropped. Typically, students are all given a certain number of sheets of paper, a certain amount of tape, and a few tissues. They are given a certain amount of time to construct their case, and then a judge begins dropping the eggs. At first, the judge simply drops them from waist height. Any eggs that survive after being dropped from that height are then dropped from an increased height. This keeps going until all but one egg has broken. The student whose egg is last to survive is the winner. In order to keep the focus on impulse, parachute-like structures are usually prohibited. Given that prohibition, the student who comes up with the structure that maximizes the time interval over which the egg comes to a halt will be the winner. I have seen students using just 6 sheets of paper, two tissues, and 25 cm of tape come up with cases that protect an egg from a drop of 41 feet onto cement!

Cartoon by Speartoons

ANSWERS TO THE "ON YOUR OWN" PROBLEMS

9.1 This problem is a simple application of Equation (9.1), as long as we get the units right. We are given the *weight* of the rock (the unit is Newton), not the mass. Thus, we must first convert from weight to mass:

$$\mathbf{w} = m\mathbf{g}$$

$$35 \text{ Newtons} = m \cdot (9.8 \frac{\text{meters}}{\text{sec}^2})$$

$$m = \frac{35 \frac{\text{kg} \cdot \cancel{\text{meters}}}{\cancel{\text{sec}^2}}}{9.8 \frac{\cancel{\text{meters}}}{\cancel{\text{sec}^2}}} = 3.6 \text{ kg}$$

Now that we have the mass, we can use Equation (9.1):

$$\mathbf{p} = m\mathbf{v}$$

$$\mathbf{p} = (3.6 \text{ kg}) \cdot (-13.2 \frac{\text{m}}{\text{sec}}) = -48 \frac{\text{kg} \cdot \text{m}}{\text{sec}}$$

The rock has a momentum of <u>-48 Wiles…I mean… $\frac{\text{kg} \cdot \text{m}}{\text{sec}}$</u>.

9.2 In a pitching machine, the roller wheels exert a force on the ball. At first, the speed (and therefore the momentum) of the baseball is 0. As the baseball rolls through the roller wheels, a force is exerted on it, accelerating the ball. In this case, once the baseball was released from the wheels, the momentum was:

$$\mathbf{p} = m\mathbf{v}$$

$$p = (0.145 \text{ kg}) \cdot (45 \frac{\text{m}}{\text{sec}}) = 6.5 \frac{\text{kg} \cdot \text{m}}{\text{sec}}$$

Since only the speed (not the velocity) was given, I am defining motion in the direction of the throw to be positive. That allows me to turn the speed into the velocity. Notice that the mass had to be converted from grams to kg in order for us to calculate the momentum.

Now we know the change in momentum. If the baseball initially had a momentum of 0 and then was pitched with a momentum of 6.5 $\frac{\text{kg} \cdot \text{m}}{\text{sec}}$, the change in momentum (final minus initial) is 6.5 $\frac{\text{kg} \cdot \text{m}}{\text{sec}}$. Now we know the change in momentum as well as the force applied, so we can calculate the time interval:

$$\mathbf{F} = \frac{\Delta\mathbf{p}}{\Delta t}$$

$$43\,\text{Newtons} = \frac{6.5\,\frac{\text{kg}\cdot\text{m}}{\text{sec}}}{\Delta t}$$

$$\Delta t = \frac{6.5\,\frac{\cancel{\text{kg}\cdot\text{m}}}{\cancel{\text{sec}}}}{43\,\frac{\cancel{\text{kg}\cdot\text{m}}}{\text{sec}^{\cancel{2}}}} = 0.15\text{ sec}$$

The ball is in contact with the roller wheels for <u>0.15 seconds</u>.

9.3 In this problem, the tennis ball has non-zero momentum both before and after the action has occurred. To calculate the change in momentum, we must calculate each momentum individually and then take their difference. Notice that direction is given (the velocities are horizontal, and one is negative while the other is positive), so we will have to keep in mind that we are dealing with direction throughout this problem. Also notice that once again we need to convert grams to kg in order to calculate the momenta:

$$\mathbf{p} = m\mathbf{v}$$

$$\mathbf{p}_{\text{initial}} = (0.0567\text{ kg})\cdot(23.1\frac{\text{m}}{\text{sec}}) = 1.31\,\frac{\text{kg}\cdot\text{m}}{\text{sec}}$$

$$\mathbf{p} = m\mathbf{v}$$

$$\mathbf{p}_{\text{final}} = (0.0567\text{ kg})\cdot(-29.7\,\frac{\text{m}}{\text{sec}}) = -1.68\,\frac{\text{kg}\cdot\text{m}}{\text{sec}}$$

The change in momentum can now be calculated:

$$\Delta\mathbf{p} = \mathbf{p}_{\text{final}} - \mathbf{p}_{\text{initial}} = -1.68\,\frac{\text{kg}\cdot\text{m}}{\text{sec}} - 1.31\,\frac{\text{kg}\cdot\text{m}}{\text{sec}} = -2.99\,\frac{\text{kg}\cdot\text{m}}{\text{sec}}$$

The negative sign denotes direction, and we had better keep it with the number so that we know we are keeping track of direction. Now that we have the change in momentum, we can combine that with the time interval (0.21 seconds) and Equation (9.6) to give us the force used:

$$\mathbf{F} = \frac{\Delta \mathbf{p}}{\Delta t}$$

$$\mathbf{F} = \frac{-2.99\,\frac{\text{kg}\cdot\text{m}}{\text{sec}}}{0.21\,\text{sec}} = -14\,\frac{\text{kg}\cdot\text{m}}{\text{sec}^2} = -14 \text{ Newtons}$$

The negative sign in our answer tells us that the force must be in the same direction as the new velocity of the ball, because the new velocity of the ball is negative as well. This should make sense. The only way that the ball would change direction is if the force were applied in the new direction. The tennis player, then, exerted a force of -14 Newtons.

9.4 The astronaut and the ball are defined in the problem as the system of interest. Since it may not be clear to you what will happen, we must consider the forces acting on the astronaut and ball. Even though the astronaut exerts a force on the ball to throw it, that is irrelevant. Remember, any forces exerted *by* a member of the system are irrelevant. No gravity is at play if the astronaut is floating. Even though there are a *few* molecules in space to rub against the astronaut and the ball, that kind of friction is negligible. Thus, the Law of Momentum Conservation does apply because the only force acting on the system is friction, and that force is negligible.

9.5 In this situation, it is easy to determine what happens. The problem says that the car and the wall make up the system of interest. The car initially has momentum towards the wall. The wall stands still, so its momentum is zero. Before the collision, then, the total momentum in the system is that of the car. Afterwards, the wall is still stationary, but the car is as well. Therefore, both momenta are zero. If this is the case, then, momentum was definitely not conserved. Thus, the Law of Momentum Conservation does not apply. Why not? At first you might be tempted to say that the wall exerts a force on the car. While that is true, it is irrelevant. After all, the wall is a part of the system. We ignore all forces exerted by a member of the system. Where is the force, then? Well, the wall must be held up by something. It is attached to the floor and ceiling of the house. These exert a force on the wall, keeping it in place. *That's* the outside force which prevents momentum conservation.

9.6 In order to determine the recoil velocity, we must calculate the total momentum of the system both before and after the rifle was fired. Since the Law of Momentum Conservation says that they must equal each other, we can build the following equation:

$$(m_{\text{rifle}}\mathbf{v}_{\text{rifle}} + m_{\text{bullet}}\mathbf{v}_{\text{bullet}})_{\text{before}} = (m_{\text{rifle}}\mathbf{v}_{\text{rifle}} + m_{\text{bullet}}\mathbf{v}_{\text{bullet}})_{\text{after}}$$

Since both the rifle and the bullet are stationary, both momenta are equal to zero before the shot. We have all of the rest of the information in the equation except for the velocity of the rifle after the shot was fired. We can therefore solve for it.

$$0 + 0 = (1.21\ \text{kg}) \cdot \mathbf{v}_{\text{rifle}} + (0.025\ \text{kg}) \cdot (195\ \frac{\text{m}}{\text{sec}})$$

$$0 = (1.21\ \text{kg}) \cdot \mathbf{v}_{\text{rifle}} + 4.9\ \frac{\text{kg} \cdot \text{m}}{\text{sec}}$$

$$\mathbf{v}_{\text{rifle}} = \frac{-4.9\ \frac{\cancel{\text{kg}} \cdot \text{m}}{\text{sec}}}{1.21\ \cancel{\text{kg}}} = -4.0\ \frac{\text{m}}{\text{sec}}$$

Notice two things. First, we had to change the units on the bullet's mass from grams to kg. Also, the negative sign simply indicates that the rifle travels in the opposite direction as compared to the bullet. Thus, its recoil velocity is <u>4.0 m/sec away from the bullet</u>.

9.7 Since there are no external forces acting on our system (the astronaut and the wrench), we can simply calculate the momenta before and after the throw and set them equal to each other.

$$(m_{\text{wrench}}\mathbf{v}_{\text{wrench}} + m_{\text{astronaut}}\mathbf{v}_{\text{astronaut}})_{\text{before}} = (m_{\text{wrench}}\mathbf{v}_{\text{wrench}} + m_{\text{astronaut}}\mathbf{v}_{\text{astronaut}})_{\text{after}}$$

The astronaut and wrench float motionless, so their initial velocities (and therefore their initial momenta) are zero.

$$0 + 0 = (75\ \text{kg}) \cdot \mathbf{v}_{\text{astronaut}} + (1.2\ \text{kg}) \cdot (10.1\ \frac{\text{m}}{\text{sec}})$$

$$0 = (75\ \text{kg}) \cdot \mathbf{v}_{\text{astronaut}} + 12\ \frac{\text{kg} \cdot \text{m}}{\text{sec}}$$

$$\mathbf{v}_{\text{astronaut}} = \frac{-12\ \frac{\cancel{\text{kg}} \cdot \text{m}}{\text{sec}}}{75\ \cancel{\text{kg}}} = -0.16\ \frac{\text{m}}{\text{sec}}$$

The negative sign simply means that the astronaut travels in the opposite direction as the wrench, which is towards his ship. So, even though he won't break any speed records, he will travel <u>towards his ship at 0.16 m/sec</u>.

9.8 In this problem, we have two independently moving skaters at first. When the man catches the woman, however, they will stay together. This means that they now can be treated as one single object. Since we can ignore the friction on ice, and since the force due to gravity working on the skaters is canceled by the normal force of the ice pushing against them, we can say that momentum is conserved. Thus, we can say that the total momentum before the catch must be the same as the total momentum afterwards:

$$(m_{\text{man}}\mathbf{v}_{\text{man}} + m_{\text{woman}}\mathbf{v}_{\text{woman}})_{\text{before}} = (m_{\text{man}} + m_{\text{woman}})\mathbf{v}_{\text{both}}$$

We know everything in this equation except the velocity of the skaters when they travel together, so we can solve for it:

$$(75.0\text{ kg})\cdot(1.2\ \frac{\text{m}}{\text{sec}}) + (60.0\text{ kg})\cdot(-3.1\ \frac{\text{m}}{\text{sec}}) = (75.0\text{ kg} + 60.0\text{ kg})\cdot \mathbf{v}_{\text{both}}$$

$$-1.0\times10^{2}\ \frac{\text{kg}\cdot\text{m}}{\text{sec}} = (135.0\text{ kg})\cdot \mathbf{v}_{\text{both}}$$

$$\mathbf{v}_{\text{both}} = \frac{-1.0\times10^{2}\ \frac{\cancel{\text{kg}}\cdot\text{m}}{\text{sec}}}{135.0\ \cancel{\text{kg}}} = -0.74\ \frac{\text{m}}{\text{sec}}$$

Since the velocity is negative, that means that the two skaters travel with a velocity of 0.74 m/sec in the direction that the woman was traveling. The significant figures are worth examining here. When you evaluate the left side of the equation, you are mixing multiplication with subtraction. Thus, you need to deal with significant figures in the multiplication so that you can switch rules for the subtraction. The multiplication results are limited to two significant figures, so you have 9.0×10^1 - 190. This works out to -100. However, since 9.0×10^1 has its last significant figure in the ones place ($0.1 \times 10^1 = 1$) and 190 has its last significant figure in the tens place, the answer must have its last significant figure in the tens place. That's why the left side of the equation becomes -1.0×10^2.

9.9 When the mouse starts to move towards the edge of the plate, the radius of its circular motion increases. In order to conserve angular momentum, the speed of the rotational motion must therefore decrease. Since the plate is causing the mouse's rotational motion, the plate will spin more slowly the closer the mouse gets to the edge of the plate.

9.10 Once the ball starts spinning, it has angular momentum. When the string is adjusted, the radius of the motion changes, but the angular momentum cannot change, due to the Law of Angular Momentum Conservation. As a result, the angular momentum before the radius changes must be the same as the angular momentum after it changes:

$$(\text{mvr})_{\text{before}} = (\text{mvr})_{\text{after}}$$

Since the mass of the ball does not change, it is the same on both sides of the equation, so it cancels out. We know everything except the radius afterwards, so we can solve for it:

$$\cancel{\text{m}}\cdot(7.5\ \frac{\text{meters}}{\text{sec}})\cdot(0.22\text{ meters}) = \cancel{\text{m}}\cdot(3.2\ \frac{\text{m}}{\text{sec}})\cdot \text{r}_{\text{after}}$$

$$\text{v}_{\text{after}} = \frac{(7.5\ \frac{\cancel{\text{meters}}}{\cancel{\text{sec}}})\cdot(0.22\text{ meters})}{3.2\ \frac{\cancel{\text{meters}}}{\cancel{\text{sec}}}} = 0.52\text{ meters}$$

The ball's new radius of motion is 0.52 meters, or 52 cm.

REVIEW QUESTIONS

1. An object's velocity vector points 45 degrees northeast. What is the direction of its momentum vector?

2. Which of the following are legitimate units for momentum?

$$\frac{\text{kg} \cdot \text{m}}{\text{sec}^2}, \frac{\text{g} \cdot \text{km}}{\text{min}}, \frac{\text{ft} \cdot \text{slug}}{\text{hr}}, \frac{\text{g} \cdot \text{m}}{\text{sec}^2}$$

3. Under what conditions can a compact car have the same momentum as a larger, heavier car?

4. A golf ball and a lump of soft clay are both thrown at a wall. They each have the same mass and are thrown at the same velocity, and they each stop dead when they hit the wall. On which object does the wall exert the most impulse?

5. In the situation from question #4, on which object does the wall exert the most force?

6. A cat and mouse are both running with equal kinetic energy. Do they necessarily have the same momentum? Why or why not?

7. Under what conditions is momentum conserved?

8. A marksman fires two guns. They each have approximately the same mass and fire their bullets at roughly the same initial velocities. The first fires heavy bullets, while the second fires lighter ones. Which gun has the greater recoil velocity?

9. Explain in your own words how a rocket is launched.

10. Under what conditions is angular momentum conserved?

PRACTICE PROBLEMS

1. A 45-Newton child runs down a road. If the child's momentum is $-9.6 \frac{\text{kg} \cdot \text{m}}{\text{sec}}$, what is the child's velocity?

2. What is the momentum of a 25.1-g rifle bullet traveling with a velocity of 351 m/sec at 45 degrees northeast?

3. A 46-gram golf ball is hit from a tee so that it has a speed of 55 m/sec. If the ball and club were in contact for 0.090 seconds, what force did the club exert on the ball?

4. A soft lump of clay (m = 456 grams) is thrown at a tree with a speed of 5.9 m/sec. How long does it take to stop the clay if the tree exerts a force of 4.0 Newtons?

5. A 975-kg gun on a ship fires a 85-kg shell with a horizontal velocity of 345 m/sec. What is the recoil velocity of the gun?

6. A 78-kg ice skater at rest throws a 6.0-kg bowling ball with a horizontal velocity of -3.0 m/sec. What is the resulting velocity of the ice skater?

7. A 0.20-kg model railroad car moving with a horizontal velocity of 0.24 m/sec is struck from behind by a 0.42-kg car traveling at 0.45 m/sec. If the two cars stick together when they collide, what will be their final velocity?

8. A 895-kg freight car coasts at 3.1 m/sec. A crane drops a 365-kg junk automobile into the freight car. What is the freight car's new speed?

9. An object has an angular momentum of $4.5 \frac{\text{kg} \cdot \text{m}^2}{\text{sec}}$. If the object's mass is 0.25 kg and it travels with a speed of 3.5 m/sec, what is its radius of motion?

10. A rock is twirled at the end of a string with a speed of 5.6 m/sec and an orbital radius of 34 cm. If the string is suddenly let out so that the rock travels with a speed of 3.2 m/sec, what is its new orbital radius?

MODULE #10: Periodic Motion

Introduction

We've spent a lot of time in this course learning about different types of motion. One type of motion that we've touched on but not studied in detail is **periodic motion**.

Periodic motion - Motion that repeats itself regularly

Why do I say that we've already touched on periodic motion? Well, we've studied circular motion. When an object moves in a circle, it passes the same points over and over again. In uniform circular motion (which we studied in great detail), the object passes those points with a constant frequency. Although we really didn't classify it as such, uniform circular motion is an example of periodic motion.

Uniform circular motion is not the only type of periodic motion, however. Off the top of your head, you can probably think about a lot of examples of periodic motion. Consider the pendulum that swings back and forth on a grandfather clock. Its motion repeats over and over again. A heart expands and contracts at regular intervals in order to pump blood. When a guitar string is plucked, it vibrates back and forth at regular intervals. All of these are examples of periodic motion. As with many things in physics, however, to really understand the phenomenon of periodic motion, we must learn something else first.

Hooke's Law

Robert Hooke, an English scientist who lived in the late seventeenth century, is well known for many accomplishments. In 1665, he wrote a book called *Micrographia* which contained some of the first drawings of the tiny things seen under a microscope. In fact, Hooke coined the term "cell." In this module, however, we won't concentrate on his impressive accomplishments in the field of biology. Instead, we need to look at a law that he developed when he studied the behavior of springs. To understand this law, perform the following experiment.

EXPERIMENT 10.1
Hooke's Law

Supplies:
- Safety goggles
- 2 paper clips
- A Ziploc® plastic bag
- A few heavy books
- Some sand, kitty litter, or fine gravel
- A spring with loops on each end (These are available at any hardware store. You need one that is 3 to 5 cm long. You should be able to stretch it to about 10 to 15 cm with your hands. There is a test at the beginning of the experiment to let you know whether or not your spring is acceptable.)
- A mass scale (You can get these at any supermarket, as dieters use them to weigh their food. It should probably have a range of 0 - 1,000 grams, but a smaller range will work.)
- Two rulers (One of these needn't be a ruler. It just needs to be a long, flat object from which you can hang the spring.)

1. Take one of the rulers (or the long, flat object) and lay it on the table so that it extends off the table's edge.
2. Pile the books on the table to hold it there.
3. Take one paper clip and pull it apart so that it is straight. Fasten it to one of the loops on the end of the spring, and wrap its other end around the ruler so that the spring hangs there securely.
4. Take the other paper clip and make an "S" out of it.
5. Use one hook on the "S" to hang the paper clip off the bottom of the spring.
6. Use the other hook on the "S" to punch through both sides of the Ziploc® bag, right under the zipper. This allows the bag to hang off the end of the spring as pictured below:

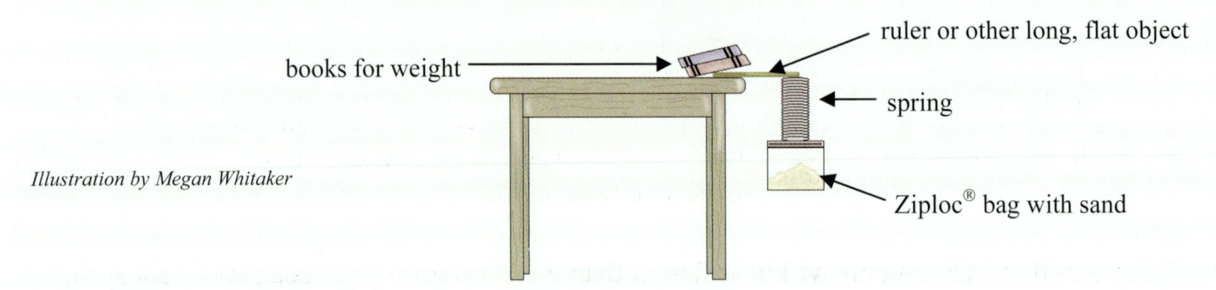

Illustration by Megan Whitaker

7. Now that you have the experiment set up, it is time to test your spring to make sure it is acceptable. Use your other ruler to measure the length of the spring with just the empty bag hanging on it.
8. Fill the bag with 650 grams of sand. If your mass scale does not have a large enough range to read 650 grams, fill the bag in increments until it has a total of 650 grams of sand in it. For example, if your scale reads from 0 to 500 grams, use it to measure 500 grams of sand and add that to the bag. Then, add 150 more grams of sand to the bag.
9. Hang the bag on the spring, and measure the length of the spring again. In order for this experiment to work, the length you measured with the bag full of sand hanging on the spring must be double (or more) the length you measured when the empty bag hung on the spring. If it is significantly less than double, then you need a looser spring.
10. Take the bag off the spring. If the spring goes back to its original length, then it is an acceptable spring. If its does not got back to its original length, you need a tighter spring.
11. Take 100 grams of sand out of the bag so that it has 550 grams of sand in it. The best way to do this is to pour the sand out of the bag and onto the scale. When the scale reads 100 grams, you know that you've gotten the job done.
12. Once you've reduced the mass of sand in the bag to 550 grams, hang the bag on the spring and measure its length again.
13. Repeat steps 11-12 four more times so that you have a total of 6 measurements of the length of the spring with 650, 550, 450, 350, 250, and 150 grams of sand hanging from it.
14. Clean up your mess.

In order to interpret the results of your experiment, you will need to do some math and then make a graph. First, take each of the masses in grams and convert them to kg by dividing by 1,000. Next, multiply each mass in kg by 9.8 m/sec^2. This turns the mass (kg) into weight (Newtons). To stay in SI units for the entire analysis, you also have to convert the length of the spring from cm to m by dividing by 100.

Now I want you to make a graph of your data with the weight on the y-axis and the length of the spring on the x-axis. It should look something like this:

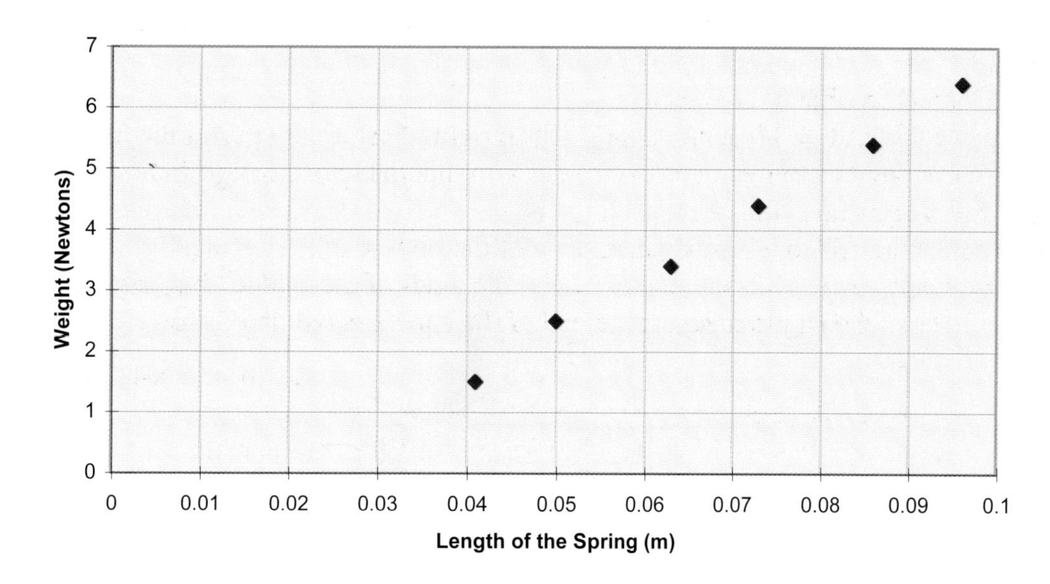

Of course, your length measurements will be different than mine, because you used a different spring. Your weights, however, should be identical, because I told you what masses (and therefore what weights) to use.

Notice that the data tend to form a straight line. True, the points do not line up perfectly, but that's due to experimental error. It seems clear, however, that the basic trend in the data is linear. Since this is the case, you should draw the line that best fits your data. How do you do that? Well, take a ruler and place it on the graph. Now, play with the alignment of the ruler until the edge has as many data points above it as below it, then draw a line on that edge. The line needn't really touch *any* of the points, as long as it has as many data points above it as below it. Your line should look something like this:

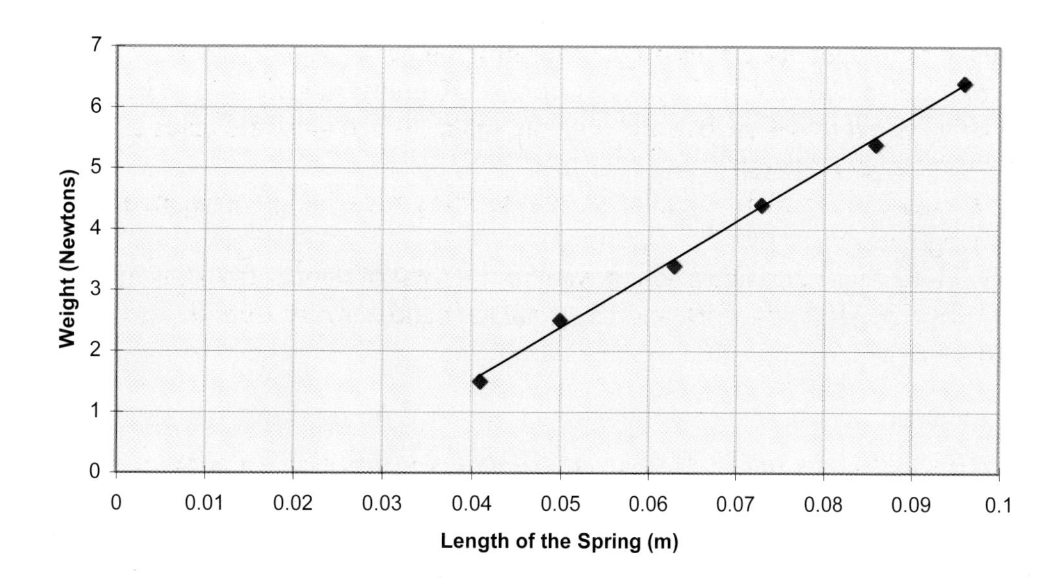

This is something we do in physics quite often. If the data points in a graph seem to follow a line, we draw the line that best fits the data, assuming that if there were no experimental error, all of the data points would fall on this line. This line, then, describes the relationship between the weight of sand we hung on the spring and the length to which the spring stretched.

There's one more thing I want you to do with this graph. I want you to calculate the slope of the line. You should remember that the slope of the line is defined as the rise (Δy) divided by the run (Δx). In order to calculate the slope, you need to choose any two points *that are on the line.* These points needn't be data points, because you are calculating the slope of the *line*, not the *data.* From the graph, read the x- and y-values of each point. Then take the difference between the y-values and divide it by the difference between the x-values. The result, which will have units of Newtons/meter, will be the slope of the line.

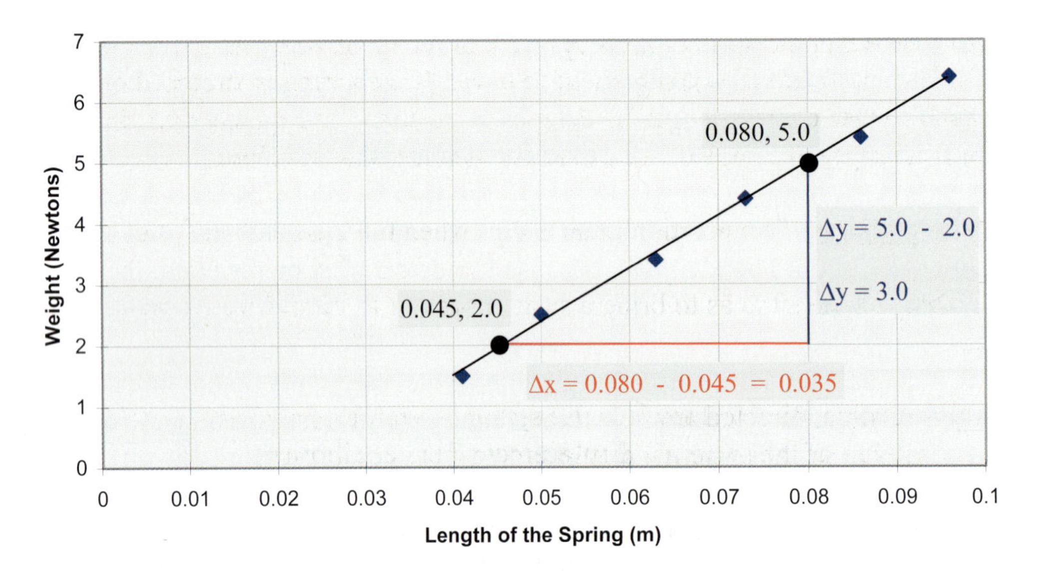

$$\text{slope} = \frac{\text{rise}}{\text{run}} = \frac{3.0\text{ Newtons}}{0.035\text{ m}} = 86\,\frac{\text{Newtons}}{\text{m}}$$

Your slope, of course, will probably be different, since your data are different than mine. This slope turns out to be a very important quantity in physics, as you will soon see.

Experiments such as the one that you just performed led Robert Hooke to determine this rather simple equation for the relationship between the force applied by a spring (**F**) and the displacement through which it stretches (Δ**x**):

$$\mathbf{F} = -k\Delta\mathbf{x} \tag{10.1}$$

While this equation is often referred to as "Hooke's Law," it is, in fact, only one small facet of Hooke's Law. The entire law, however, is really beyond the scope of this course. As a result, we will concentrate only on this small part of Robert Hooke's studies.

Now, what does this equation tell us? Well, think back to the experiment you did. When you hung a bag of sand on the end of the spring, gravity pulled down on the bag. In order to keep the bag hanging in the air, the spring had to pull up on the bag with an equal force. As you added weight to the bag, the spring began to stretch. In addition, the force with which the spring pulled up on the bag increased as well. Your graph shows that the force increased linearly with the displacement over which the spring stretched. That's what Equation (10.1) says. It says that the force and the distance stretched are linearly proportional to each other. The "k" in Equation (10.1) is the proportionality constant. We'll talk more about that in a moment.

Why is the negative sign in the equation? When the spring stretched in the experiment, the bag moved downwards. The spring, however, was pulling up on the bag. Thus, the negative sign tells us that the force which the spring applies will always be in the opposite direction of the spring's displacement. In other words, if I pull down on a spring, it will pull back in the upwards direction. On the other hand, if I compress a spring by pushing up on it, the spring will push downwards, trying to fight back.

Why is the force that the spring exerts opposite of the displacement? In the case of a spring, there is an "optimal" length that the spring wants to have. If the spring is stretched or compressed, it tries to restore itself to that optimal length by applying a force in the opposite direction. We have a name for the optimal length of a spring; we call it the **equilibrium position**:

Equilibrium position - The position of an object when there are no net forces acting on it

When a force is always applied so as to bring a system back to its equilibrium position, we call it a **restoring force**:

Restoring force - A force, directed towards the system's equilibrium position, which is applied as a result of the system's displacement from equilibrium

In other words, when there are no forces pulling or pushing a spring, it stays in its equilibrium position. Any displacement from that position (either by stretching or compressing) will result in a restoring force that opposes the displacement, pointing back towards the equilibrium position. The strength of the restoring force is directly proportional to the displacement of the spring. That's what Equation (10.1) tells us.

Now, if you're a little confused at this point, don't worry. A concrete example should clear everything up. Before we do that, however, we need to talk about the **spring constant**, which is given by "k" in Equation (10.1). First (as usual), we need to worry about its units. The SI unit for force is the Newton, and the SI unit for displacement is the meter. In Equation (10.1), after we multiply "k" by "$\Delta\mathbf{x}$," we are supposed to get an answer whose units are consistent with force. In other words, when the spring constant is multiplied by meters, the resulting unit must be Newtons. Thus, the units for "k" must be $\frac{\text{Newtons}}{\text{meter}}$. That way, when "k" multiplies "$\Delta\mathbf{x}$," the meters cancel, leaving Newtons. As always, however, these are not the only possible units for the spring constant. Any force unit divided by any displacement unit would give us a valid spring constant unit.

What does the spring constant tell us? Well, think about Equation (10.1). If "k" is small, the equation will give a small force. This tells us that when the spring constant is small, it takes only a small force to stretch or compress the spring. On the other hand, if "k" is large, a large force will result from Equation (10.1). This tells us that when the spring constant is large, a large force is necessary to stretch the spring. Thus, the value for the spring constant tells us about the strength of the spring. When the spring constant is small, the spring is easy to stretch or compress. When the spring constant is large, the spring is very strong, making it hard to stretch or compress.

Now look back at your experiment. At the end, I asked you to calculate the slope of the line that best fits your data. What does that slope represent? Well, you graphed the force versus the distance that the spring stretched. If you recall from algebra, the equation of a line is:

$$y = mx + b \tag{10.2}$$

where "m" is the slope of the line and "b" is the y-intercept. Now, suppose the y-intercept is zero. Equation (10.2) would reduce to:

$$y = mx \tag{10.3}$$

Does this look familiar? Suppose you substituted "**F**" for y and "Δx" for "x." This equation would look a lot like Equation (10.1), with "k" in the place of "m." In other words, Equation (10.1) is actually the equation of a line, with the spring constant equal to the slope of the line. This tells us that the slope you calculated in the experiment is the spring constant of the spring that you used!

In order to make this all a little more understandable, study the following example and answer the "On Your Own" problems that come afterwards.

EXAMPLE 10.1

Illustration by Megan Whitaker

A student hangs a spring on the ceiling with no weight on it. She then makes a mark on the wall to indicate the position of the spring's end. After that, she hangs a 451-kg object on the spring. She makes another mark to indicate the new position of the spring's end. If this mark is 2.51 cm lower than the first mark, what is the spring constant?

This problem describes the following situation:

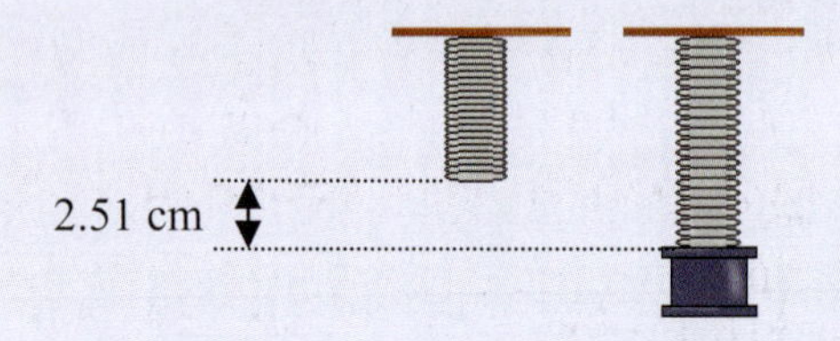

Before the object was hung on the spring, the spring was at its equilibrium position. When the 451-kg object was hung there, it stretched an additional 2.51 cm in length. Thus, the weight of the object caused the spring to experience a displacement of 2.51 cm. This means that $\Delta\mathbf{x}$ in Equation (10.1) is -2.51 cm. I say that $\Delta\mathbf{x}$ is negative because the spring stretched downwards, which we usually define as the negative direction. Also, in order to stop the object from falling, the spring must exert a force equal to but opposite the weight of the spring. Thus, when the system stops moving, we know that the force of gravity due to the object is being canceled by an equal and opposite force, which is **F** in Equation (10.1). We can calculate the weight of the object, which is the force due to gravity acting on the object:

$$\mathbf{w} = m\mathbf{g}$$

$$\mathbf{w} = (451 \text{ kg})\cdot(-9.8 \frac{\text{m}}{\text{sec}^2}) = -4.4 \times 10^3 \text{ Newtons}$$

The force is negative because we have defined the downward direction as negative. Thus, **g** is negative, making **w** negative as well. However, this is *not* **F** in Equation (10.1). The variable **F** refers to the *restoring force* exerted *by the spring*. Well, as I said before, this will be equal to but opposite of the force exerted by gravity; therefore, $\mathbf{F} = 4.4 \times 10^3$ Newtons. Make sure that you understand this reasoning. The **F** in Equation (10.1) is the force exerted by the spring. Since the weight of the object pulls *down* on the spring, the spring reacts by exerting an equal but opposite force (**F**). That makes **F** a positive force.

Now that we have both **F** and **Δx**, we can use Equation (10.1) to solve for k. Keep in mind, however, that in order to get k in SI units, we need to convert cm to m!

$$\mathbf{F} = -\text{k}\mathbf{\Delta x}$$

$$4.4 \times 10^3 \text{ Newtons} = -\text{k}\cdot(-0.0251\text{m})$$

$$\text{k} = \frac{4.4 \times 10^3 \text{ Newtons}}{0.0251\text{m}} = 1.8 \times 10^5 \frac{\text{Newtons}}{\text{m}}$$

This tells us that the spring constant is <u>1.8 x 10^5 Newtons/meter</u>. Now this number probably doesn't mean much to you, but compare it to the slope that you calculated in your experiment. I have already told you that the slope you measured is the spring constant of the spring that you used in the experiment. How does it compare to this spring constant? Your spring's constant should be *significantly* smaller than the one determined here. What does this tell us? Because this spring constant is significantly bigger than the one for your spring, we can say that the spring in this example is *significantly stronger* than the one that you used in the experiment.

What we see from this example, then, is that a simple experiment, along with Equation (10.1), can tell us a lot about the strength of a spring. Solve the following "On Your Own" problems to make sure you understand how to use Equation (10.1).

ON YOUR OWN

10.1 When a student hangs a 25.0-g object on a spring, it stretches 3.45 cm down from its equilibrium position. What is the spring constant of the spring? Is it weaker or stronger than the spring in Example 10.1?

10.2 A physicist hangs a 412-g object on a spring whose spring constant is 11.91 Newtons/meter. How far does the spring stretch?

Uniform Circular Motion: An Example of Periodic Motion

As I mentioned before, uniform circular motion is an example of periodic motion. After all, if an object is moving in a circle at a constant rate, it will pass by the same points again and again at regular intervals. By examining this type of motion a little more closely, we can learn a technique that will help us analyze other forms of periodic motion.

Consider an object moving in a circle.

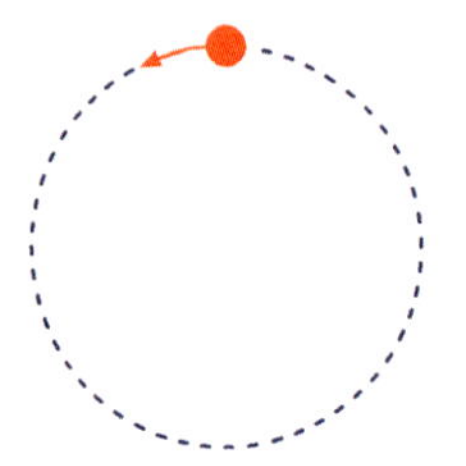

In Module #7, we learned that if the object has a constant speed, the time it takes to make one revolution around the circle is constant. We called that time the *period* of the object's motion (measured in seconds), and we abbreviated it with a "T." Now think for a moment. How far does the object travel over the course of one period? Well, it makes one trip around the circle. If the radius of the circle is "r," the distance around the circle is $2\pi r$. Thus, in one period, the object moves $2\pi r$. So, we know the time (T) for the object to make one revolution, and the distance ($2\pi r$) the object travels in that time. If we divide the two, we get the speed of the object:

$$v = \frac{2\pi r}{T} \tag{10.4}$$

Now remember, even though this speed is constant (because we are talking about uniform circular motion), the object does experience acceleration. This acceleration is called the centripetal acceleration, and is given by Equation (7.2):

$$a_c = \frac{v^2}{r} \tag{7.2}$$

If we use the expression for "v" given by Equation (10.4), we can substitute it into Equation (7.2) to come up with:

$$a_c = \frac{\left(\frac{2\pi r}{T}\right)^2}{r} \tag{10.5}$$

$$a_c = \frac{4\pi^2 r}{T^2} \tag{10.6}$$

Equation (10.6), then, gives us a relationship between the centripetal acceleration that an object in uniform circular motion experiences and the period of its motion. What good is that? Not much, really. In order to determine an object's period, you must know its velocity, and Equation (7.2) gives us an easy way to determine centripetal acceleration if we have velocity. Why did I bother to go through this, then, if the result isn't all that useful? Well, *by itself* the equation isn't all that useful, but wait until you see what we do with it in a little while!

The Mass / Spring System

In this section, we are going to study a system that exhibits a special kind of periodic motion. In order to really understand this system, perform the following experiment.

EXPERIMENT 10.2
The Characteristics of a Mass / Spring System

Supplies:
- Safety goggles
- The spring you used in Experiment 10.1
- 2 paper clips
- A Ziploc® plastic bag
- Some sand, kitty litter, or fine gravel
- A mass scale
- A few heavy books
- Two rulers (One of these needn't be a ruler. It just needs to be a long, flat object from which you can hang the spring.)
- A stopwatch

1. Make sure you take good notes during this experiment (as you should for all experiments), because we will use the results of this experiment later on.
2. Set up this experiment just as you did Experiment 10.1. Put some sand in the bag and measure the mass. Continue to add or remove sand until the measured mass is within 25 grams of 200 g. Write down the mass, because you will use it later on in the module.
3. Align your ruler with the bag so that the top of the bag lines up with the 50.0 cm mark.
4. Pull straight down on the bag so that it is 5.0 cm below the 50.0 cm mark, and then release the bag.
5. Observe the motion of the system. The bag bounces up and down at regular intervals. Although the bouncing motion starts to decrease after a while, the time it takes for the bag to move up and back down really stays the same until the bouncing motion essentially dies out. Notice also that the bag never travels more than 5.0 cm below or 5.0 cm above the 50.0 cm mark. In other words, the distance that you pulled it from its equilibrium position determines how far it will bounce up and down. If you pulled the bag straight down and released it without moving the bag, it should only move up and down. If the bag starts swinging back and forth, try this again until you get used to pulling it straight down and releasing it without moving it.
6. Now that you've seen the basics of this motion, you need to study it in a little more detail. In order to do this, you will measure the period of the object's motion. Steady the bag so that it is still again, and line up the bag so that its top is even with the 50.0 cm mark on the ruler.
7. Pull the bag straight down for 5.0 cm, and release the bag.
8. Start your watch when you release the bag and then begin counting the bounces that the bag makes. When the bag travels all of the way up and then back down to the point where you released it, count this as one bounce. Continue to count until the bag has bounced up and down a total of 10 times. At that point, stop the watch.

9. The time on your watch represents how long it takes for the bag to bounce 10 times. If you divide that number by 10, it tells you how long it takes for the spring to make one bounce up and down. That's the period of the bag's motion. Write down the period. Why did I have you measure the time it took to make 10 bounces? If we wanted to measure the bag's period, why didn't we just measure the time it took for one bounce? Well, in measuring only one bounce, there is a lot of room for error. You might not start or stop the watch at precisely the right instants. If, however, you measure the time it takes for 10 bounces, the importance of such errors is reduced. Thus, the period you measured this way is more accurate than the period you would measure by just timing one bounce. If you timed 100 bounces, your measurement would be even more accurate, but that would get a little dull.
10. Repeat steps 6-9, this time pulling the bag down 10.0 cm before releasing it. Measure the period of the bag's motion under these conditions.
11. Repeat steps 6-9 one more time, measuring the period of the bag's motion when you pull it down 15.0 cm before releasing it.
12. Clean up your mess.

What did we learn from this experiment? Well, first we learned that the motion exhibited by a mass on a spring is periodic. We also learned something else. Compare the three periods that you measured in the experiment. What do you see? Within experimental error (usually about 10%), these numbers should be the same. Does that surprise you? No matter how far you pulled the bag down before releasing it, the period stayed the same. Why?

Think about what you observed. The farther you pulled the bag down before releasing it, the farther it traveled up and back down. Even though it traveled farther, it took the same amount of time for the bag to bounce up and down because the farther the bag was pulled down before it was released, the *faster* it traveled once it was released. The fact that it had to travel farther to bounce up and down is canceled out by the fact that it travels that distance faster. As a result, the period of the bag's motion is independent of how far you pulled it down before you released it.

Before we go on, we need to get some terminology out of the way. In the experiment, you noticed that the bag never traveled any farther from its equilibrium position than the original distance from which you released it. When you released it 10.0 cm below its equilibrium position, it never bounced more than 10.0 cm below or above that equilibrium position. This maximum distance that the bag moves from its equilibrium position is called the **amplitude** (am' pluh tood) of the bag's motion:

Amplitude - The maximum distance away from equilibrium that an object in periodic motion travels

In other words, when you released the bag 10.0 cm below its equilibrium position, the amplitude of its motion was 10.0 cm.

The motion that you observed in the experiment is actually a special kind of periodic motion. When the period of an object's motion is independent of its amplitude (as Experiment 10.2 showed was the case for a mass / spring system), we say that it is experiencing **simple harmonic** (har mahn' ik) **motion**.

Simple harmonic motion - Periodic motion whose period is independent of its amplitude

In a moment, you will see a mathematical derivation that shows *why* the period of motion is independent of amplitude in simple harmonic motion. Before you see that, however, I want you to think a little more deeply about the motion of the mass / spring system that you studied in Experiment 10.2.

During the experiment, did you notice that the speed of the bag continually changed throughout its motion? When you held the bag, its speed was, of course, zero. However, when you let it go, it started moving slowly, but it quickly sped up. However, after the bag passed through its equilibrium position, it actually began to slow down. It slowed down until its speed reached zero. Then, the bag began moving in the other direction, so that it could return to the position from which you released it. At that point, the whole process started over again.

How do we understand this complex motion? Well, when you pulled the bag away from its equilibrium position, the spring began to exert a restoring force in the direction opposite that in which you were pulling. Since you pulled the bag down, the restoring force pulled the bag up. That's why when you released the bag, it started accelerating upwards. As Hooke's Law tells us, however, the strength of the restoring force depends on the displacement from equilibrium. Thus, as the bag began to move up, it got closer to its equilibrium position, so the restoring force decreased. This, of course, caused the acceleration to decrease. Now, even though the *acceleration* of the bag decreased, the acceleration was still in the direction of the velocity, so the velocity (and therefore the speed) kept increasing. So the bag's speed increased as it traveled up, but the acceleration decreased.

When the bag reached its equilibrium position, however, things began to change. At the equilibrium position, the restoring force was equal to zero. This means that the acceleration was also zero. Even though the acceleration was zero, the bag's velocity was not. The bag had been accelerating, so it had a large velocity. As soon as the bag passed the equilibrium position, however, things changed. At that point, its displacement was *above* the equilibrium position. According to Hooke's Law, then, the spring began pushing *down* on the bag, which means that the bag began accelerating downwards. Well, its velocity was still directed upwards at that point, so, as a result, the bag's acceleration was opposite its velocity. Thus, the bag began to slow down. The bag continued to slow down until it reached the same displacement *above* the equilibrium position as the displacement you gave it when you pulled it *below* the equilibrium position. At that point, the bag's acceleration had reduced its velocity to zero. The spring, however, was still pushing down on the bag, so the bag began accelerating downwards.

Do you see what's happening here? Because the restoring force is always pointed in the opposite direction from the displacement from equilibrium, the bag speeds up until it reaches the equilibrium position. This means that the mass's maximum speed will always occur at the equilibrium position. After it passes that point, however, it begins to slow down until its speed is zero. When its speed is zero, it has traveled a full amplitude and is therefore as far away from the equilibrium position as it will ever get. This means that at the amplitudes, the restoring force (and therefore the acceleration) is at its maximum. After that, the bag turns around and begins the journey in the opposite direction. This complex motion is due to the nature of the restoring force that exists in a mass / spring system. It is important that you understand the major aspects of this motion, so I will reiterate them.

- **In a mass / spring system, the maximum acceleration occurs at each amplitude of the motion.**

- **In a mass / spring system, the maximum speed occurs at the equilibrium position.**

- **In a mass / spring system, the acceleration is zero at the equilibrium position.**

- **In a mass / spring system, the speed is zero at each amplitude of the motion.**

Make sure you remember these characteristics and understand why they exist.

The Mathematics of the Mass / Spring System

Now that we've discussed the concepts behind the simple harmonic motion exhibited by the mass / spring system, it's time to add the math. To do this, however, I need to ask you to use your imagination. Previously, we developed Equation (10.6), which allows us to calculate the acceleration of an object moving uniformly in a circle. It turns out that this equation applies to the mass / spring system as well, but I want to try to show you why. Suppose you had an object moving in a circle. Suppose further that you put a movie screen behind this object and shone a flashlight on it, so that the object's shadow hit the movie screen. What would you see? Look at Figure 10.1.

FIGURE 10.1

The Motion of A Mass / Spring System is a One-Dimensional Projection of Circular Motion

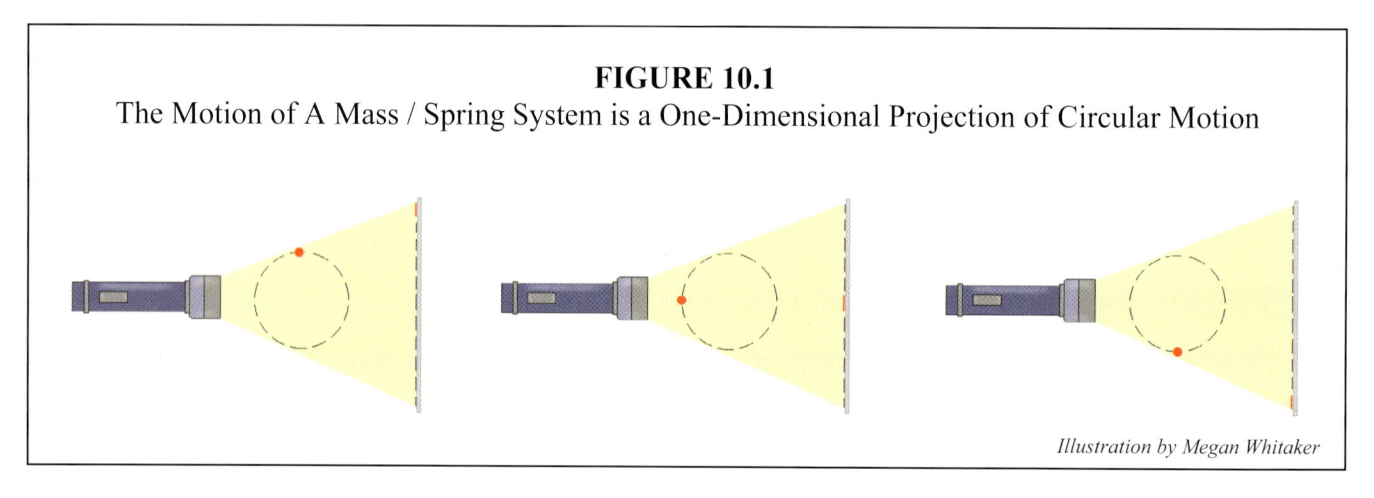

Illustration by Megan Whitaker

An object that moves in a circle is moving in two dimensions. However, if you cause it to cast the correct kind of shadow, that two-dimensional motion can be projected into one dimension. If, for example, you shine a flashlight on the object so that the light strikes the circle directly on the edge, not either one of the sides, the shadow cast by the circle will look like a straight line. The object moving in that circle, then, will appear to move up and down on a straight line, rather than around and around in a circle. The mass / spring system moves just like the shadow in the figure does. While the object moves halfway around the circle, the object's shadow starts at the top of the line and moves down to the bottom. As the object moves around the other half of the circle, its shadow moves from the bottom of the line back to the top. Thus, as the object is moving in a circle, its shadow is moving up and down. Furthermore, in the time that it takes the object to make one full revolution around the circle, its shadow has moved down and then back up. Thus, the *period* of the shadow's motion is the same as the *period* of the object's motion. In the end, then, we can say that a mass / spring system is actually just the one-dimensional equivalent of uniform circular motion.

Although this might be hard to visualize (even with the aid of a drawing), think about it this way. The very act of casting a shadow results in losing one dimension. If, for example, a tree casts a shadow, the shadow is flat. It has width and height but no depth. Thus, the shadow is a two-dimensional representation of the three-dimensional tree. In this case, we are creating the shadow of a two-dimensional situation. Since the process of casting a shadow destroys a dimension, we are turning a two-dimensional situation into a one-dimensional situation. This makes the circle cast a shadow that is a straight line.

If the figure doesn't make sense to you, don't worry about it. Just take my word that the motion of a mass / spring system is the one-dimensional equivalent of uniform circular motion. As a result, Equation (10.6), which we derived a while ago, applies to the mass / spring system, as long as we make an adjustment to one of the variables. Let's look at that equation again:

$$a_c = \frac{4\pi^2 r}{T^2} \qquad (10.6)$$

Because the mass / spring system's motion is not two-dimensional, it cannot have a radius. What would the radius's one-dimensional equivalent be? Well, when the circle casts its shadow, it forms a straight line. Starting from the center of the line, the distance up to the edge is the same as the length of the circle's radius. The length of the line from the center to the lower edge is also the same as the radius of the circle. What do we call the displacement from the center of the mass / spring's motion (the equilibrium position) to the top of its motion? We call it the amplitude. Thus, the "r" in Equation (10.6) can be replaced with an "A" for "amplitude." In addition, since the object is not moving in a circle, its acceleration is not centripetal anymore. As a result, I will drop the "c" in "a_c."

$$a = \frac{4\pi^2 A}{T^2} \qquad (10.7)$$

Equation (10.7), then, gives us a relationship between the acceleration of a mass / spring system, the amplitude of its motion, and the period. What good does that do us? Well, Hooke's Law can help us get an alternative expression for the mass's acceleration. Remember, Hooke's Law was given by Equation (10.1). This equation, however, included vectors. Thus, it took care of direction as well. In this derivation, I'm not worried about direction, so I will get rid of the vectors. This also allows me to get rid of the negative sign, since it denotes direction. Thus, the scalar equivalent of Equation (10.1) is:

$$F = k\Delta x \qquad (10.8)$$

Remember, however, that Newton's Second Law tells us that force is equal to mass times acceleration. Thus, we can substitute "ma" for "F" in Equation (10.8). Also, let's suppose I want to analyze the object when it has moved all the way out to the amplitude of its motion. In that case, $\Delta x = A$. I will make that substitution as well:

$$ma = kA \qquad (10.9)$$

Now we can take this equation and solve for acceleration:

$$a = \frac{kA}{m} \tag{10.10}$$

Let's take this expression for acceleration and plug it back into Equation (10.7):

$$\frac{kA}{m} = \frac{4\pi^2 A}{T^2} \tag{10.10a}$$

Finally, we can rearrange this equation to solve for the period (T). Notice that the amplitude drops out because it exists in the numerator on each side of the equation. After rearranging, we get the following equation:

$$T = 2\pi\sqrt{\frac{m}{k}} \tag{10.11}$$

Think about what this equation tells us. It tells that the period of motion in a mass / spring system depends *only* on two things: the strength of the spring (given by k) and the mass of the object (represented by m). This is why, in Experiment 10.2, the period of the mass's motion was independent of the amplitude. Since amplitude dropped out in our derivation, it does not affect the period in any way.

Since I've drawn a box around this equation, you can bet that it's pretty important. As is always the case, I don't require you to really understand all of the steps that I took to get to the equation. Instead, I want you to realize what the equation means and (as I'll demonstrate in a minute) how to use it. I do need to point out a couple of important things about the derivation, however. Firstly, we were able to perform this derivation because we recognized that the up-and-down motion of the mass / spring system was the one-dimensional equivalent of uniform circular motion. That allowed us to get started. Secondly, we used Hooke's Law in the derivation. It turns out that this was a really critical step. The reason that the amplitude canceled was that it also appeared in Hooke's Law. If the amplitude had not been in the law, or if it had been squared or cubed, it would not have canceled. This is actually a general result. If a system is subjected to a restoring force that follows Hooke's Law (is linearly proportional to the displacement from equilibrium), then it will always exhibit simple harmonic motion. This is, in fact, the only way that a system can exhibit simple harmonic motion.

Thus, even though I don't expect you to understand the derivation of Equation (10.11), I do expect you to remember these two critical things:

The motion of a mass / spring system is the one-dimensional equivalent of uniform circular motion.

and

When a system is subject to a restoring force that is linearly proportional to the displacement from equilibrium, the system exhibits simple harmonic motion.

We will be coming back to the second principle, so don't forget it.

Now let's go on to using the equation. First, think about the two experiments that you did in this module. In Experiment 10.1, you measured the spring constant of the spring you were using. Then, in Experiment 10.2, you measured the period of a mass / spring system that utilized that spring. You also measured the mass in Experiment 10.2. Thus, you measured m, k, and T for your mass / spring system. Let's see how closely your results follow Equation (10.11).

MORE ANALYSIS OF EXPERIMENT 10.2
"Testing" Equation (10.11)

Note: A sample set of calculations is available in the solutions and tests guide. It is with the solutions to the practice problems.

Supplies:
- Laboratory notebook

1. Look at your laboratory notebook entry for Experiment 10.1 and find the spring constant that you measured. That's the value for k. Make sure it is in Newtons/m.
2. Look at your laboratory notebook entry for Experiment 10.2 and find the mass of the bag of sand that you used. That's the value for m.
3. Look at your laboratory notebook entry for Experiment 10.2 and find the three different periods you measured for the mass / spring system. Average them. That average is the value for T.
4. Divide m by k and take the square root.
5. Multiply the result by 2π. You now have the result of the right-hand side of Equation (10.11) for your mass / spring system .
6. Compare this result to your average value for T. Within experimental error (probably about 10%), they should be equal. That's what Equation (10.11) says.

Now, of course, if your value for T does not equal what you calculated for the right side of Equation (10.11), it does not mean that the equation is wrong. Most likely, experimental error is to blame, as Equation (10.11) has been confirmed by many careful experiments. Let's continue learning how to use Equation (10.11) by doing some problems.

EXAMPLE 10.2

A 345-g mass is connected to a spring (k = 35 Newtons/meter). What is the period of the spring's motion?

This is a straightforward application of Equation (10.11). We are given m and k, and we are asked to calculate T. Just remember, however, that the mass must be in kg. So I need to convert g to kg before I use the equation:

$$T = 2\pi\sqrt{\frac{0.345\text{ kg}}{35\frac{\text{Newtons}}{\text{m}}}} = 2\pi\sqrt{\frac{0.345\,\cancel{\text{kg}}}{35\frac{\frac{\cancel{\text{kg}}\cdot\cancel{\text{m}}}{\text{sec}^2}}{\cancel{\text{m}}}}} = 0.62\text{ sec}$$

The mass / spring system, then, has a period of <u>0.62 sec</u>. Notice how the units worked out so that the period ends up in seconds.

Pretty simple, huh? Well, it can get a little more complicated than this, but not by much. There is one thing I can do to make these problems a little more complicated. I can give you the spring constant in a more difficult manner. Rather than just giving you the value for "k," I can give you enough information to calculate it. See if you understand what I mean by studying the following example.

EXAMPLE 10.3

A student hangs a spring from the ceiling. She then hangs a 34.5-kg mass on it, and the spring stretches 59.1 cm. After the spring has stretched to its new equilibrium position, the student pulls the mass down an additional 35.0 cm and releases it. What will be the period of the resulting motion?

Do you see what I have done here? I gave you the mass outright, but there is nothing about the spring constant in this problem. There is, however, enough information to calculate it. After all, I tell you that when the mass is hung on the spring, the spring stretches 59.1 cm. Given the mass and the distance that the spring stretched, we can calculate the spring constant using Hooke's Law, just like we did in Example 10.1. First, we realize that in order to use Hooke's Law, we must determine the weight that is exerting a force on the spring.

$$\mathbf{w} = \mathrm{m}\mathbf{g}$$

$$\mathrm{w} = (34.5\ \mathrm{kg})\cdot(-9.8\ \frac{\mathrm{m}}{\mathrm{sec}^2}) = -\,3.4\times10^2\ \mathrm{Newtons}$$

This is not **F** in Equation (10.1), however. The variable **F** refers to the *restoring force* exerted *by the spring*. Well, this will be equal to but opposite of the force exerted by gravity, therefore, **F** $= 3.4 \times 10^2$ Newtons. Now that we have both **F** and Δ**x**, we can use Equation (10.1) to solve for k. Keep in mind that since the student pulls the mass down, Δ**x** is negative. Also, remember that we must convert cm into m:

$$\mathbf{F} = -\,\mathrm{k}\Delta\mathbf{x}$$

$$3.4\times10^2\ \mathrm{Newtons} = -\,\mathrm{k}\cdot(-\,0.591\mathrm{m})$$

$$\mathrm{k} = \frac{3.4\times10^2\ \mathrm{Newtons}}{0.591\ \mathrm{m}} = 5.8\times10^2\ \frac{\mathrm{Newtons}}{\mathrm{m}}$$

Now that we have the spring constant, we can determine the period:

$$\mathrm{T} = 2\pi\sqrt{\frac{34.5\ \mathrm{kg}}{5.8\times10^2\ \frac{\mathrm{Newtons}}{\mathrm{m}}}} = 2\pi\sqrt{\frac{34.5\ \cancel{\mathrm{kg}}}{5.8\times10^2\ \frac{\frac{\cancel{\mathrm{kg}}\cdot\cancel{\mathrm{m}}}{\mathrm{sec}^2}}{\cancel{\mathrm{m}}}}} = 1.5\ \mathrm{sec}$$

This tells us that the mass bounces up and down on the spring with a period of <u>1.5 sec</u>.

See if you really understand how to use Equation (10.11) by solving the following "On Your Own" problems.

ON YOUR OWN

10.3 An 11.2-kg mass is attached to a spring. When displaced from equilibrium, it bounces up and down with a period of 2.3 seconds. What is the spring constant ?

10.4 When it supports a 351-g mass, a certain spring stretches 12.1 cm. If the mass is pulled down 5.1 cm from this new equilibrium position and then released, what will its period be?

10.5 What would be the period of the mass in problem (10.4) if it were pulled 10.2 cm away from its new equilibrium position instead of just 5.1 cm?

Potential Energy in a Mass / Spring System

Before we leave the mass / spring system, there is one more thing you should know. When a spring is stretched or compressed, it exerts a restoring force, trying to get back to its equilibrium position. Since this force is capable of doing work (for example, moving the mass back and forth), the system has potential energy. The farther the spring is stretched or compressed from equilibrium, the more force it can exert, hence, the more potential energy the system must have. If you think back to the motion of the mass / spring system that you observed in Experiment 10.2, this should make sense.

When you pulled down on the bag in the experiment, you stretched the spring. This gave the system some potential energy. That energy was ready to do work, as soon as you released the bag. Once you did that, the bag started to move. That's because the potential energy in the spring began to be converted to kinetic energy, making the bag move. As more and more potential energy got converted to kinetic energy, the bag moved faster and faster. This continued until the bag reached its equilibrium position. At that point, it was completely out of potential energy. All of the potential energy that you gave the system had been converted to kinetic energy.

Even though, at this point, the spring didn't want to move any more, the bag had other ideas. It was moving rapidly because of all the kinetic energy. As a result, it continued to move, compressing the spring. In order to compress the spring, however, kinetic energy had to be converted into potential energy. As this happened, the bag slowed down, because the system was losing kinetic energy. This continued to happen until the system converted all of its kinetic energy to potential energy and stopped. Of course, with all of the potential energy now in the system, the bag started moving in the opposite direction, converting the potential energy back into the kinetic energy. This process can be represented by the graph in Figure 10.2.

FIGURE 10.2
Kinetic and Potential Energy in a Mass / Spring System

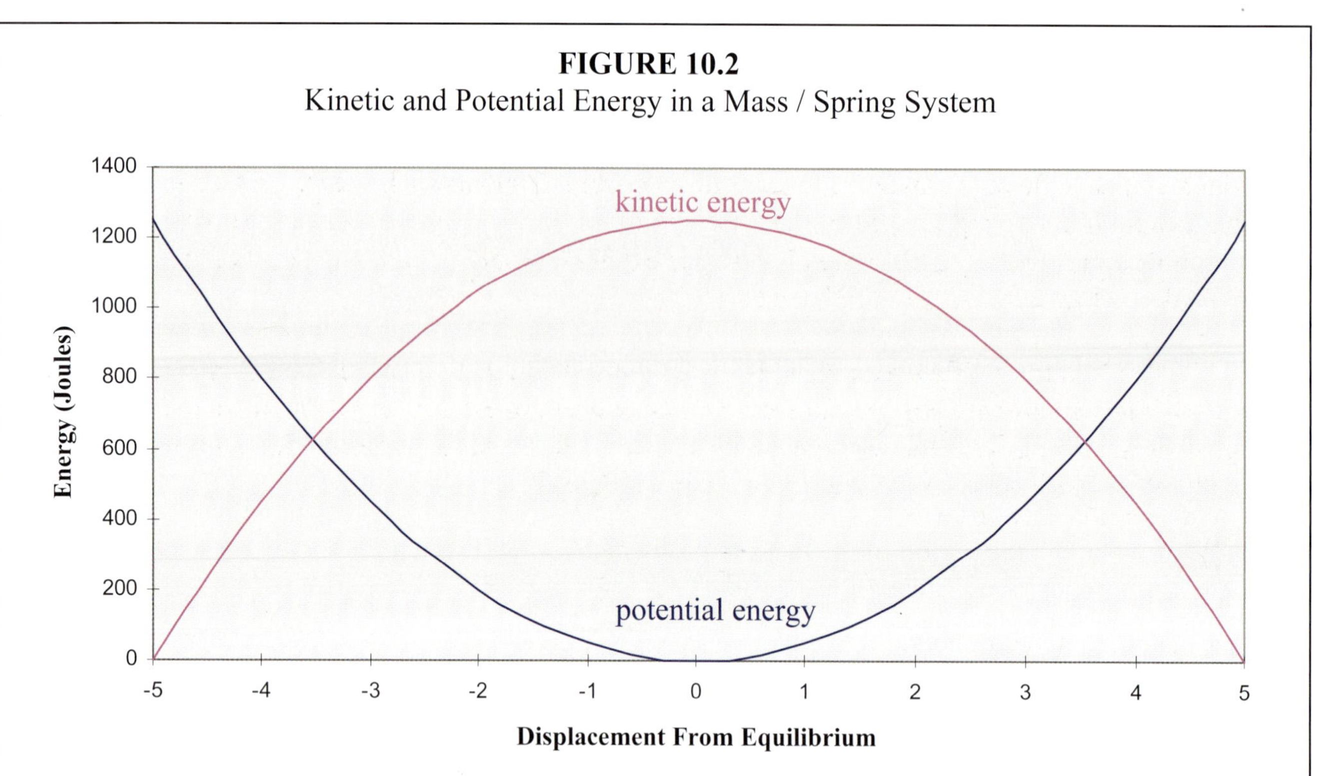

This graph shows the relationship between kinetic and potential energy in a mass / spring system. When the mass is pulled out to the amplitude (either edge of the graph), it is stationary. Thus, the kinetic energy is zero and the potential energy is at its maximum. On the other hand, when the mass is at the equilibrium point (the middle of the graph), all of the potential energy has been converted to kinetic energy. Thus, at this point, the potential energy is zero and the kinetic energy is at its maximum. Also, the First Law of Thermodynamics tells us that the sum of the potential and kinetic energies throughout the course of its journey must always be the same, as long as no work is being done on the system. At any point on the graph, then, the sum of the potential energy and kinetic energy remains the same.

This graph, then, shows us a couple of things about the motion of a mass / spring system that we need to remember:

At the amplitude of its motion, a mass / spring system has no kinetic energy (and therefore zero speed) and its maximum potential energy.

and

At the equilibrium point of its motion, a mass / spring system has no potential energy and its maximum kinetic energy (and therefore its maximum speed).

These facts will help us in the next step of our analysis.

Now you might wonder how I got numbers for the energy on the y-axis of the graph. Well, it turns out that there is a reasonably simple formula that relates a spring's potential energy and its displacement from equilibrium. Although the derivation is a bit too long for this course, the equation itself is very straightforward:

$$PE = \frac{1}{2}k(\Delta x)^2 \qquad (10.12)$$

Although you have no idea where this equation came from, it should make sense. It tells us that the potential energy in a spring depends on the strength of the spring (given by k) and the displacement from equilibrium (given by Δx). When I solve the example problems, you will see that the units work out to Joules, which is the right unit for energy.

Before I go through the examples, I need to remind you that whenever we talk about potential energy, we must always keep kinetic energy in mind. After all, the sum of potential energy and kinetic energy does not change, as long as there is no work being done on the system. Thus, we need to remember the equation for kinetic energy given in Module #8:

$$KE = \frac{1}{2}mv^2 \qquad (8.3)$$

We'll be using this in a moment.

There's one last thing that I need to tell you. Since the sum of the kinetic energy and potential energy does not change, we actually know what that sum must equal. After all, when the mass is displaced all the way out to the amplitude, its speed is zero. This means that there is no kinetic energy in the system. As a result, all of its energy is potential. Well, Equation (10.12) gives us a way to calculate that potential energy. When the mass is at the amplitude of its motion, $\Delta \mathbf{x} = A$. Since there is only potential energy at that point, if we plug in A for x in Equation (10.12), we will end up having the total energy of the spring at any time. Thus, we can say:

$$KE_{\text{mass/spring}} + PE_{\text{mass/spring}} = \frac{1}{2}kA^2 \qquad (10.13)$$

Let's see how to use Equations (10.12, 8.3, and 10.13) in analyzing the motion of a mass / spring system.

EXAMPLE 10.4

In stretching an 11.2-kg mass on a spring, a physicist does 35 Joules of work. What is the spring constant, if this work stretches the spring 14.1 cm? If the spring were let go, what is the maximum kinetic energy it could have? What speed would the mass have at that point?

I ask a lot of questions, don't I? Even though there are a lot of questions, the answers are not too hard to find. First, when a person does work on a system, he transfers energy to that system. Since the physicist stretched the spring, he is taking the work and converting it to potential energy. As a result, the system gets 35 Joules of potential energy. We can now use Equation (10.12) to calculate the spring constant, given the potential energy and the displacement. Remember, however, that I need to convert the distance to meters first.

$$PE = \frac{1}{2}k(\Delta x)^2$$

$$35\text{ Joules} = \frac{1}{2}\text{k}\cdot(0.141\text{ m})^2$$

$$\text{k} = \frac{2\cdot(35\text{ Newton}\cdot\cancel{\text{m}})}{0.0199\text{ m}^{\cancel{2}}} = 3.5\times10^3\ \frac{\text{Newtons}}{\text{m}}$$

Notice that to make sure the units worked out, I renamed "Joules" to "Newton·m," since the two units are equivalent.

Now that we have the spring constant, we can move on. When will the kinetic energy reach a maximum? It will be when the potential energy is zero. Since the spring is stretched 14.1 cm, that will be its amplitude. So, if we used Equation (10.13) and set PE to zero, we will be calculating the maximum KE:

$$\text{KE}_{\text{mass/spring}} + \text{PE}_{\text{mass/spring}} = \frac{1}{2}\text{kA}^2$$

$$\text{KE}_{\text{max}} + 0 = \frac{1}{2}\cdot(3.5\times10^3\ \frac{\text{Newton}}{\text{m}})\cdot(0.141\text{ m})^2$$

$$\text{KE}_{\text{max}} = 35\text{ Newton}\cdot\text{m} = 35\text{ Joules}$$

Finally, we can use Equation (8.3) to convert this kinetic energy into speed to answer the last question:

$$\text{KE} = \frac{1}{2}\text{mv}^2$$

$$35\text{ Joules} = \frac{1}{2}\cdot(11.2\text{ kg})\cdot\text{v}^2$$

$$\text{v} = \sqrt{\frac{2\cdot35\frac{\cancel{\text{kg}}\cdot\text{m}^2}{\text{sec}^2}}{11.2\ \cancel{\text{kg}}}} = 2.5\ \frac{\text{m}}{\text{sec}}$$

Notice that in order to get the units to work out, I needed to rename "Joules" to "$\frac{\text{kg}\cdot\text{m}^2}{\text{sec}^2}$." These calculations tell us, then, that <u>the spring constant is 3.5 x 10^3 Newtons/m; the maximum kinetic energy is 35 J; and the speed at that point is 2.5 m/sec</u>.

A 23-g mass is placed on a spring (k = 46 Newtons/m). If the spring is stretched 12.3 cm and released, what will the mass's speed be when it is 8.0 cm from its equilibrium position?

Seems like an awfully tough problem, doesn't it? Well, it would be, if we hadn't already discussed the conservation of energy problems like we did in Module #8. In order to find the mass's speed, all we need to do is find its kinetic energy. How do we do that? Well, Equation (10.13) gives us a way to relate KE to PE, k, and A. We know A, because we are told how far the spring is stretched before the mass is released. We are given k, so that's taken care of. If we could only determine PE, we would be all set. It turns out that we can. After all, we are told how far away the mass is from its equilibrium position, and that's Δx in Equation (10.12). Using this and the spring constant, we can calculate the potential energy at that point. Remember, since we need to know the PE when the mass is 8.0 cm from its equilibrium position, that's the value we use for Δx, once we have converted to meters.

$$PE = \frac{1}{2}k(\Delta x)^2$$

$$PE = \frac{1}{2} \cdot (46\ \frac{\text{Newtons}}{\text{m}}) \cdot (0.080\ \text{m})^2$$

$$PE = 0.15\ \text{Newton} \cdot \text{m} = 0.15\ \text{Joules}$$

This, then, is the potential energy of the mass for the point at which we want to know the speed. Now we can use Equation (10.13). In this equation, we need to use the amplitude, which is the maximum distance that the mass is from equilibrium. That's 12.3 cm, which we must convert to meters.

$$KE_{\text{mass/spring}} + PE_{\text{mass/spring}} = \frac{1}{2}kA^2$$

$$KE_{\text{mass/spring}} + 0.15\ \text{Joules} = \frac{1}{2} \cdot (46\ \frac{\text{Newton}}{\text{m}}) \cdot (0.123\,\text{m})^2$$

$$KE_{\text{mass/spring}} = 0.20\ \text{Joules}$$

Now that we have the kinetic energy, we can finally get the speed.

$$KE = \frac{1}{2}mv^2$$

$$0.20\ \text{Joules} = \frac{1}{2} \cdot (0.023\ \text{kg}) \cdot v^2$$

$$v = \sqrt{\frac{2 \cdot 0.20 \frac{\cancel{\text{kg}} \cdot \text{m}^2}{\text{sec}^2}}{0.023\ \cancel{\text{kg}}}} = 4.2\ \frac{\text{m}}{\text{sec}}$$

Eight cm from its equilibrium position, the mass will be traveling with a speed of 4.2 m/sec.

These two examples were long problems, but they weren't all that hard. You should be very familiar with the potential energy / kinetic energy concepts from your work in Module #8. The only real difference here is that we have a new equation for potential energy. Cement these skills into your head with the following "On Your Own" problems.

ON YOUR OWN

10.6 How much work is required to stretch a strong spring (k = 3412 Newtons/m) 34.5 cm?

10.7 A 61.2-kg mass is placed on a spring and set into simple harmonic motion by pulling it 23.1 cm from its equilibrium position. If the maximum speed of the mass is 3.4 m/sec, what is the spring constant?

10.8 A 54-g mass is placed on a spring (k = 15 Newtons/m). If it is displaced 5.0 cm from its equilibrium position, how fast will it be traveling when it is only 2.5 cm away from equilibrium?

The Simple Pendulum

Another example of simple harmonic motion is given by the simple pendulum. In Experiment 8.1, you constructed a pendulum, but you did not look at it from a perspective of simple harmonic motion. We will do that now. When a mass is hung by a string or rope and then displaced from equilibrium, it will swing from side to side in a regular manner. This is the motion of a simple pendulum. What causes this motion? Examine Figure 10.3.

FIGURE 10.3
The Simple Pendulum

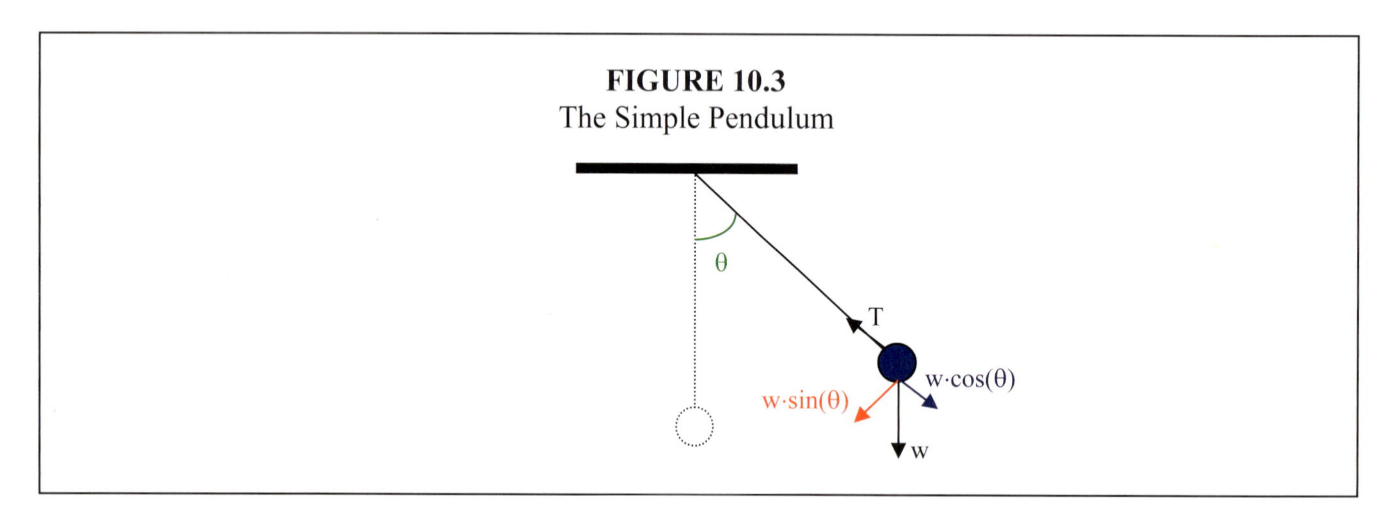

In a simple pendulum, the system would rest in such a way as to have the mass hanging straight down from the string, as illustrated by the dotted drawing in the figure. When hanging straight down, the tension in the string is equal to and opposite of the force due to gravity. This makes the net force zero, which is the definition of equilibrium, so we say that this is the equilibrium position of the pendulum.

When the mass is lifted away from that equilibrium position, the tension in the string can only oppose a component of the gravitational force. The gravitational force (w) points straight down. However, like any vector, this force vector can be split into perpendicular components. If we split the weight vector into a component that is parallel to the string, this component is equal to $w \cdot \cos(\theta)$, where θ is defined as illustrated in the figure. This component of the weight is canceled by the tension in the string (T), which pulls in the opposite direction.

Look at the other component of the weight vector. It is perpendicular to the string and has a magnitude of $w \cdot \sin(\theta)$. This component has nothing to oppose it, so the mass will accelerate back towards the equilibrium position. When the mass reaches the equilibrium position, the net force is once again zero. However, the mass already has significant velocity, so it begins to move away from equilibrium in the other direction. At that point, gravity starts trying to pull it back. Eventually, the mass slows to a speed of zero, and gravity causes it to move the other way. Thus, the mass swings back and forth in a regular pattern.

What you need to notice in the drawing is that the $w \cdot \sin(\theta)$ component of the gravitational force is always pointed towards the equilibrium position. What do we call a force that always tries to push the system back into equilibrium? We call it a restoring force. Thus, the $w \cdot \sin(\theta)$ component of gravity is a restoring force for the simple pendulum.

We have learned that when the restoring force in a system is linearly proportional to the displacement away from equilibrium, the system exhibits simple harmonic motion. It turns out that the restoring force in a simple pendulum is like that. How? Well, you need to remember a couple of things in order to see why.

First, remember that there are two units with which I can measure an angle. The typical unit is degrees, but often in mathematics we use the unit of **radians** to measure an angle. You may or may not have learned in mathematics that when an angle is relatively small and is measured in radians, the sine of the angle is essentially equal to the angle itself. In other words:

$$\sin\theta \approx \theta \qquad \text{when } \theta \text{ is small} \tag{10.14}$$

Even though this is just an approximation (that's what the "$\approx$" means), it is really quite good as long as the angle is less than $\pi/10$ (or about 20 degrees). Thus, as long as the angle in Figure 10.3 is less than 20 degrees, the restoring force in a simple pendulum is essentially:

$$F_r = w \cdot \theta \tag{10.15}$$

where "F_r" stands for the restoring force.

In geometry, you should have learned that the arc length of a circular arc can be calculated using

$$s = r\theta \tag{10.16}$$

where "s" is the length of the arc swept out by the angle θ, and "r" is the radius of the circle. Well, in the case of a simple pendulum, "s" would represent the displacement away from equilibrium (which we call "Δx"), and "r" would simply be the length of the string (which I will call "ℓ"). Thus, for a simple pendulum, this same equation turns into:

$$\Delta x = \ell\theta \tag{10.17}$$

Solving for θ gives us:

$$\theta = \frac{\Delta x}{\ell} \tag{10.18}$$

Finally, if we put this expression for θ back into Equation (10.15):

$$F_r = \frac{w}{\ell} \cdot \Delta x \tag{10.19}$$

What does this equation tell us? Well, "w" and "ℓ" are always constant for a given pendulum, so the only variable on the right-hand side of the equation is "Δx." Thus, the restoring force in a simple pendulum is linearly proportional to the displacement from equilibrium.

Now if you didn't really follow all of that, don't worry. You need only remember two things from the derivation. First, you need to remember that the restoring force in a simple pendulum is linearly proportional to the displacement from equilibrium. Thus, a simple pendulum exhibits simple harmonic motion. Secondly, you need to remember that this result is, in fact, only an approximation. In Equation (10.14), a trigonometric approximation was used. Since this approximation is only good for small angles, the conclusion is only good for small angles. Thus, a simple pendulum can only exhibit simple harmonic motion when the angle of displacement from equilibrium is small (less than 20 degrees). These are important facts.

When the angle of displacement for a simple pendulum is small (less than about 20°), the pendulum exhibits simple harmonic motion.

So what? Well, we know that in simple harmonic motion, the period is independent of the displacement from equilibrium. In a pendulum, therefore, as long as the angle of displacement is under 20 degrees, it doesn't matter what its value is. Regardless of the value from 0 to 20 degrees, the period of a pendulum is always the same. This has a tremendously useful application.

Have you ever seen an old clock? Under the face of the clock, there is a long rod with a mass on the end. The mass travels back and forth, making the "tick-tock" of the clock. You see, since the period of a pendulum's motion is constant regardless of its angle of displacement (as long as that angle is small), it can be used to keep time. In many old clocks, each time the pendulum swung back or forth, the clock advanced one second. The "tick" (when the pendulum swung one way), then, caused the clock to advance a second, and the "tock" (when the pendulum swung the other way) advanced it another second. Since the period was independent of the displacement from equilibrium, it didn't matter how far the pendulum was pulled from equilibrium to get it started. It would keep the same time.

Now although I could derive the equation which allows us to calculate the period of a pendulum, I figure you are rather tired of derivations at this point. I will therefore just give you the final result:

$$T = 2\pi\sqrt{\frac{\ell}{g}} \tag{10.20}$$

In this equation, "T" stands for period, "ℓ" represents the length of the string on the pendulum, and "g" is the acceleration due to gravity. Notice that displacement does not appear in the equation, as we expect for simple harmonic motion.

Looking at Equation (10.20), the period of a pendulum really only depends on two things: what planet you happen to be on (g is different for different planets) and the length of the string which makes up the pendulum. So, if you want to make a pendulum that has a certain period, you really just have to change its length. The longer the pendulum, the longer its period. To see how easy this equation is to use, study the example that follows.

EXAMPLE 10.5

What is the length of a pendulum on a standard grandfather clock?

In a standard grandfather clock, each swing must take a full second. Since the period is based on how long it takes for the pendulum to swing both back and forth, the period of a grandfather clock's pendulum is, in fact, 2.00 seconds. Thus, this question simply asks us to determine the length of a pendulum whose period is 2.00 seconds. Since I assume that this pendulum will be used on planet earth, we can say that $g = 9.8 \text{ m/sec}^2$.

$$T = 2\pi\sqrt{\frac{\ell}{g}}$$

$$2.00 \text{ sec} = 2\pi\sqrt{\frac{\ell}{9.8\frac{\text{m}}{\text{sec}^2}}}$$

$$4.00 \text{ sec}^2 = 4\pi^2\left(\frac{\ell}{9.8\frac{\text{m}}{\text{sec}^2}}\right)$$

$$\ell = \frac{(4.00\ \cancel{\text{sec}^2})\cdot\left(9.8\frac{\text{m}}{\cancel{\text{sec}^2}}\right)}{4\pi^2} = 0.99 \text{ m}$$

Notice how the units worked out to give us a length unit. The pendulum on a standard grandfather clock, therefore, must be 0.99 m in length. Now it is possible for a clock to have a pendulum with a different length. For example, suppose a clock counted a second as the complete back and forth motion of the pendulum. For that clock, the period of the pendulum would need to be 1.00 second, and you can use Equation (10.20) to see that such a clock would have a pendulum that is 0.25 m long.

Since we are on the subject of clocks, I thought you might want to know how we measure the passage of time these days. After all, we do not use grandfather clocks and other pendulum-based clocks to keep time anymore. Even though the period of a pendulum is a good means of keeping time, there are obvious drawbacks. Air resistance and friction reduces the energy of the pendulum with each swing. Thus, the pendulum must be given extra energy from time to time. This is usually accomplished by "winding up" the clock. When you wind up a clock, a mass rises, storing potential energy. As time goes on, the mass slowly falls, moving gears that add energy to the pendulum, restoring what air resistance and friction removed. Although this certainly helps fix the problems of friction and air resistance, it is annoying to keep winding up a clock, and there are other problems associated with pendulum-based clocks. Thus, as technology improved, so did the keeping of time.

By the 1970s, the marvelous quartz watch was introduced. It originally cost about $500 in the United States, but today, quartz watches are incredibly cheap. A quartz watch works by applying a voltage to a sample of quartz, usually shaped like a bar or a tuning fork. The quartz actually oscillates (ah' suh lates) in response to the voltage, and the period of that oscillation is quite regular, usually 32,768 oscillations per second. The clock counts those oscillations, and once the quartz has oscillated 32,768 times, it advances the time by one second. Most wristwatches, wall clocks, alarm clocks, and desk clocks that we use today are quartz clocks.

Even though quartz clocks are marvels of technology, they are not perfect at keeping time. The actual frequency at which the quartz oscillates is affected a bit by the surrounding temperature and other factors, so the frequency is not always 32,768 Hz. As a result, quartz clocks tend to "lose" or "gain" a few seconds each month. In order to improve our ability to mark the passage of time, then, technology has continued to seek new ways to improve clocks.

Today's most accurate clocks are called "atomic clocks." These clocks (which are *not* radioactive) make use of the fact that electrons in atoms have characteristic frequencies at which they absorb energy. For example, most atomic clocks use cesium atoms, which can absorb energy at a frequency of 9,192,631,770 Hz. This characteristic frequency is not nearly as sensitive to temperature and other effects, so it stays quite constant. Every 9,192,631,770 oscillations of the energy source used to excite the cesium atoms advances the clock by one second, making a very accurate timekeeping device. Currently, the most accurate atomic clock, called "NIST-F1," is so accurate that it is estimated that it would lose only one second every 20 million years, if we were around that long to measure it! The official time in the United States is kept by an atomic clock at the National Institute of Standards and Technology (NIST) in Boulder, Colorado.

Although the ways in which we keep time are incredibly interesting, I am not going to test you on this information. I told it to you because I find it interesting. For the purpose of this course, you need only understand the simple pendulum and how to use Equation (10.20). Make sure you can do that by solving the following "On Your Own" problems.

ON YOUR OWN

10.9 What is the period of a pendulum whose length is 2.5 ft?

10.10 If a pendulum has a period of 0.45 seconds, what is its length?

ANSWERS TO THE "ON YOUR OWN" PROBLEMS

10.1 Although this problem is worded a bit differently, you should see that it is really the same as the example. We are given the mass of the object, from which we can calculate the weight of the object, which is the force due to gravity. Note that we must first convert g to kg in order to get our answer in Newtons:

$$\mathbf{w} = m\mathbf{g}$$

$$\mathbf{w} = (0.0250\ \text{kg})\cdot(-9.8\ \frac{\text{m}}{\text{sec}^2}) = -\ 0.25\ \text{Newtons}$$

Thus, gravity pulls the spring with a force of -0.25 Newtons. In order to keep the object from falling, the spring exerts a restoring force equal to but opposite of the force due to gravity. Thus, the spring exerts a force of 0.25 Newtons. This is **F** in Equation (10.1).

When the object hangs on the spring, it stretches downwards by 3.45 cm. Thus, the displacement ($\Delta\mathbf{x}$) is -0.0345 m, after converting to SI units. Using the force calculated above and this displacement, we can use Equation (10.1) to determine the spring constant:

$$\mathbf{F} = -\ k\Delta\mathbf{x}$$

$$0.25\ \text{Newtons} = -\ k\cdot(-0.0345\ \text{m})$$

$$k = \frac{0.25\ \text{Newtons}}{0.0345\ \text{m}} = 7.2\ \frac{\text{Newtons}}{\text{m}}$$

The spring constant, therefore, is 7.2 Newtons/meter. Since this is significantly smaller than the spring constant calculated in the example, this is a weaker spring.

10.2 This is, of course, another problem using Equation (10.1), but this time, we are solving for $\Delta\mathbf{x}$, the displacement of the spring. We are given the object's mass, so we can calculate its weight, once we convert g to kg:

$$\mathbf{w} = m\mathbf{g}$$

$$\mathbf{w} = (0.412\ \text{kg})\cdot(-9.8\ \frac{\text{m}}{\text{sec}^2}) = -\ 4.0\ \text{Newtons}$$

The force exerted by gravity is negative (because gravity pulls down), but the force exerted by the spring (**F**) is opposite that, so **F** = 4.0 Newtons. We are given k, so the only thing we have to do is solve for $\Delta\mathbf{x}$:

$$\mathbf{F} = k\Delta\mathbf{x}$$

$$4.0\ \text{Newtons} = -\ (11.91\ \frac{\text{Newtons}}{\text{m}})\cdot(\Delta\mathbf{x})$$

$$\Delta \mathbf{x} = -\frac{4.0 \text{ ~~Newtons~~}}{11.91 \frac{\text{~~Newtons~~}}{\text{m}}} = -0.34 \text{ m}$$

The negative sign simply means that the spring stretched downwards, which makes sense since gravity pulls downwards. The spring, therefore, stretched 0.34 m downwards.

10.3 In this problem, we are given the mass and the period, and we are asked to calculate the spring constant. This just requires a little rearrangement of Equation (10.11):

$$T = 2\pi\sqrt{\frac{m}{k}}$$

$$2.3 \text{ sec} = 2\pi\sqrt{\frac{11.2 \text{ kg}}{k}}$$

$$(2.3 \text{ sec})^2 = \left(2\pi\sqrt{\frac{11.2 \text{ kg}}{k}}\right)^2$$

$$(2.3 \text{ sec})^2 = 4\pi^2\left(\frac{11.2 \text{ kg}}{k}\right)$$

$$k = 4\pi^2\left(\frac{11.2 \text{ kg}}{(2.3 \text{ sec})^2}\right) = 84 \frac{\text{kg}}{\text{sec}^2}$$

Since a kg/sec^2 is the same as a Newton/m, the spring constant is 84 Newtons/m.

10.4 In this problem, we are not given the spring constant. There is, however, enough information to calculate it. After all, I tell you that when the mass is supported by the spring, the spring stretches 12.1 cm. Given the mass and the distance that the spring stretched, we can calculate the spring constant using Hooke's Law. Notice that we must convert to meters and Newtons, though, to allow the units to work out:

$$\mathbf{w} = \text{mg}$$

$$\mathbf{w} = (0.351 \text{ kg}) \cdot (-9.8 \frac{\text{m}}{\text{sec}^2}) = -3.4 \text{ Newtons}$$

The force due to gravity acting on the object, then, is -3.4 Newtons. This is not **F** in Equation (10.1), however. The variable **F** refers to the *restoring force* exerted *by the spring*. Well, this will be equal to but opposite of the force exerted by gravity, therefore, **F** = 3.4 Newtons. Now that we have both **F** and $\Delta\mathbf{x}$, we can use Equation (10.1) to solve for k. Keep in mind that since the mass is pulled down, $\Delta\mathbf{x}$ is negative. Also, we must convert from cm to m so that the units work out.

$$\mathbf{F} = -\,k\Delta\mathbf{x}$$

$$3.4 \text{ Newtons} = -\,k\cdot(-0.121\,\text{m})$$

$$k = \frac{3.4 \text{ Newtons}}{0.121 \text{ m}} = 28\,\frac{\text{Newtons}}{\text{m}}$$

Now that we have the spring constant, we can determine the period:

$$T = 2\pi\sqrt{\frac{0.351 \text{ kg}}{28\,\frac{\text{Newtons}}{\text{m}}}} = 0.70 \text{ sec}$$

This tells us that the mass bounces up and down on the spring with a period of 0.70 sec. What about the 5.1 cm? Why didn't I use it? Well, the amplitude of the mass's motion does not appear in our equations. In other words, it was extraneous information designed to confuse you.

10.5 Since the amplitude of the mass's motion does not affect its period, the mass's period will be exactly the same, 0.70 sec.

10.6 When you work on an object, you are changing its energy. In stretching a spring, you are giving it potential energy. Thus, the amount of work done stretching the spring is equal to the potential energy given to the spring. In the problem, then, the work required to stretch the spring will be equal to the spring's potential energy. Given k and Δx, calculating that is a simple matter:

$$PE = \frac{1}{2}k(\Delta x)^2$$

$$PE = \frac{1}{2}\cdot(3412\,\frac{\text{Newtons}}{\text{m}})\cdot(0.345 \text{ m})^2$$

$$PE = 203 \text{ Newton}\cdot\text{m} = 203 \text{ Joules}$$

Therefore, it takes 203 Joules of work to stretch the spring 34.5 cm.

10.7 In giving us the maximum speed of the mass, this problem is actually telling us the maximum kinetic energy:

$$KE = \frac{1}{2}mv^2$$

$$KE_{max} = \frac{1}{2} \cdot (61.2 \text{ kg}) \cdot (3.4 \frac{m}{sec})^2 = 350 \text{ Joules}$$

That's the maximum kinetic energy. When kinetic energy is at its maximum, however, potential energy is zero. Thus, if we use Equation (10.13) and set PE = 0, we can use the value we just calculated for KE and relate that to k and A:

$$KE_{spring} + PE_{spring} = \frac{1}{2} \cdot k \cdot A^2$$

$$350 \text{ Joules} + 0 = \frac{1}{2} \cdot k \cdot (0.231 \text{ m})^2$$

$$k = 1.3 \times 10^4 \frac{\text{Newtons}}{m}$$

The spring constant is <u>1.3×10^4 Newtons/m</u>.

10.8 In order to find the mass's speed, all we need to do is find its kinetic energy. Equation (10.13) gives us a way to relate KE to PE, k, and A. We know A, because we are told how far the spring is stretched before the mass is released. We are given k, so that's taken care of. If we could only determine PE, we would be all set. Well, we can. After all, we are told how far away the mass is from its equilibrium position, and that's Δx in Equation (10.12). Using this and the spring constant, we can calculate PE:

$$PE = \frac{1}{2}k(\Delta x)^2$$

$$PE = \frac{1}{2} \cdot (15 \frac{\text{Newtons}}{m}) \cdot (0.025 \text{ m})^2$$

$$PE = 0.0047 \text{ Netwon} \cdot m = 0.0047 \text{ Joules}$$

Now we can use Equation (10.13):

$$KE_{spring} + PE_{spring} = \frac{1}{2}kA^2$$

$$KE_{spring} + 0.0047 \text{ Joules} = \frac{1}{2} \cdot (15 \frac{\text{Newton}}{m}) \cdot (0.050 \text{ m})^2$$

$$KE_{spring} = 0.014 \text{ Joules}$$

Now that we have the kinetic energy, we can finally get the speed:

$$KE = \frac{1}{2}mv^2$$

$$0.014 \text{ Joules} = \frac{1}{2} \cdot (0.054 \text{ kg}) \cdot v^2$$

$$v = \sqrt{\frac{2 \cdot 0.014 \frac{\cancel{kg} \cdot m^2}{sec^2}}{0.054 \cancel{kg}}} = 0.72 \frac{m}{sec}$$

At a distance of 2.5 cm from its equilibrium position, then, the mass will be traveling with a speed of 0.72 m/sec.

10.9 This is a simple application of Equation (10.20). Notice, however, that the length of the pendulum is given in feet. In order to get units straight, we either need to convert feet into meters, or we could simply use the English value for g. Either way, we ensure that the units are consistent. I will choose to do the latter:

$$T = 2\pi\sqrt{\frac{\ell}{g}}$$

$$T = 2\pi\sqrt{\frac{2.5 \cancel{ft}}{32 \frac{\cancel{ft}}{sec^2}}} = 1.8 \text{ sec}$$

This pendulum has a period of 1.8 sec.

10.10 In this problem, we are given the period, and we know the acceleration due to gravity. Figuring out the length is a simple matter:

$$T = 2\pi\sqrt{\frac{\ell}{g}}$$

$$0.45 \text{ sec} = 2\pi\sqrt{\frac{\ell}{9.8 \frac{m}{sec^2}}}$$

$$(0.45\ \text{sec})^2 = 4\pi^2 \cdot \left(\frac{\ell}{9.8\ \frac{\text{m}}{\text{sec}^2}} \right)$$

$$\ell = \frac{(0.45\ \cancel{\text{sec}})^{\cancel{2}} \cdot \left(9.8\ \frac{\text{m}}{\cancel{\text{sec}^2}} \right)}{4\pi^2} = 0.050\ \text{m}$$

This pendulum, then, has a length of 0.050 m or 5.0 cm.

Cartoon by Speartoons

REVIEW QUESTIONS

1. A mass / spring system is displaced from equilibrium by 5.0 cm and its period of motion is determined to be 3.0 seconds. What will the period be if it is displaced from equilibrium by 15.0 cm?

2. What is the definition of a restoring force?

3. A mass / spring system is displaced from equilibrium by 34.1 cm. What is the amplitude of its motion?

4. What condition must be met in order for a system to exhibit simple harmonic motion?

5. In a mass / spring system, where does the system reach its maximum speed?

6. In a mass / spring system, where is the mass subjected to the greatest force?

7. In a mass / spring system, where does the system reach its maximum potential energy?

8. A heavy box is dropped on a spring, and the maximum compression of the spring is measured before the spring pushes the box back up. If the box is dropped from a greater height, what will happen to the maximum compression of the spring?

9. Which of the following situations will result in simple harmonic motion after the mass is released?

10. If a pendulum's period is measured on earth and on the moon, where will its period be the longest?

PRACTICE PROBLEMS

1. A hanging scale uses the fact that the length a spring stretches is directly proportional to the weight of the object suspended from it. If a 10.0-Newton bag of vegetables causes the spring to stretch 5.6 cm, what is the spring constant on the scale?

2. A dart gun consists of a horizontal spring with k = 52 Newtons/m that is compressed 4.3 cm when the gun is cocked. What is the initial acceleration of a 75-g dart when the gun is fired?

3. A 50.0-g puppet bounces on the end of a loose (k = 1.5 Newtons/m) spring. What is the period of its motion?

4. A 0.400-kg mass is attached to a spring and set into simple harmonic motion. If its period is 1.7 seconds, what is the spring constant?

5. A mass on a spring (k = 30.0 Newtons/m) vibrates back and forth with a period of 1.1 seconds. What is the object's mass?

6. How much work does it take to compress a spring (k = 78.1 Newtons/m) 51.1 cm?

7. What is the maximum speed of a 34-g object bouncing on a spring (k = 11.1 Newtons/m) with an amplitude of 3.5 cm?

8. A 1.1-kg object is hung from a spring (k = 23.4 Newtons/m). If it is displaced 14.2 cm from equilibrium and then released, what will its speed be when it is only 5.0 cm from equilibrium?

9. What is the period of a 3.98-meter pendulum?

10. If a pendulum has a period of 3.1 seconds, how long is it?

MODULE #11: Waves

Introduction

In the previous ten modules, we have studied systems and situations that were relatively easy to visualize. After all, we have all observed objects in linear motion, objects in circular motion, and objects in periodic motion. In this module, however, we move away from systems that are easy to visualize. That's because in this module we will concentrate on sound and light. Although we have plenty of everyday experience with both of these things, we really have no visual images in our minds to represent them. Indeed, how does one see sound? How does one draw a picture of light? You see, we don't have visual images in our mind for these things.

Because of this fact, some of the concepts that we discuss here might seem a little vague. If so, don't worry about it. There are aspects of these concepts that are confusing even to physicists. Nevertheless, there are many things about sound and light that you should be able to grasp, and I will try to concentrate on them.

Waves

In order to understand light and sound, you first need to know a little bit about waves. If you had a good chemistry course, part of this should be review. Nevertheless, please read it all carefully to make sure that you really understand it. Now we're all familiar with waves on one level or another. However, to make sure that we're all starting from the same place, let's define a wave:

Wave - A disturbance that propagates in a medium

This definition has two very important implications that need to be discussed.

First, a wave is a disturbance in a medium. This means, of course, that a wave must have a medium in which to travel. Ocean waves, for example, travel in water. Thus, the medium is water, but what is the wave itself? The wave is a *disturbance* in the water. Consider, for example, a still pond. The surface of the water is so still that the whole pond looks like it is glass. Are there waves in the pond? No, of course not. Now, let's imagine that you throw a rock into the pond. What happens? The nice, still surface of the pond is suddenly disturbed by a wave that moves away from the rock. Okay, so now there is a wave in the pond. What is that wave? It is a *disturbance* in the water. That's what I mean when I say that a wave is a disturbance in a medium. In this case, water is the medium, and the wave is a disturbance in water.

Second, a disturbance in a medium is not necessarily a wave. To be a wave, that disturbance must propagate (move) in the medium. The imaginary wave that we just created by throwing an imaginary rock into an imaginary pond moves away from the point at which the rock struck the water. Thus, the disturbance in the water moves (propagates) away from the rock. The water wave, then, is a disturbance that propagates in the medium of water, away from the rock.

Please note that throughout this discussion, I have used the term "wave" to represent all of the ripples that have come from the rock. This is a point at which the common usage of the term "wave" differs from the physical term. Generally, if we see a series of ripples coming from a rock dropped in water, we would call the ripples "waves." However, to a physicist, all of those ripples represent one wave, because they are all the result of a single disturbance. Since the *disturbance* is the wave, one disturbance means one wave. As a result, all of the ripples that come from the rock are the crests and troughs of a single wave.

Okay, now that we know what a wave is, let's examine some details. Consider the imaginary wave we just made in the imaginary pond. From a sideways view, such a wave might look something like this:

FIGURE 11.1
A Transverse Wave

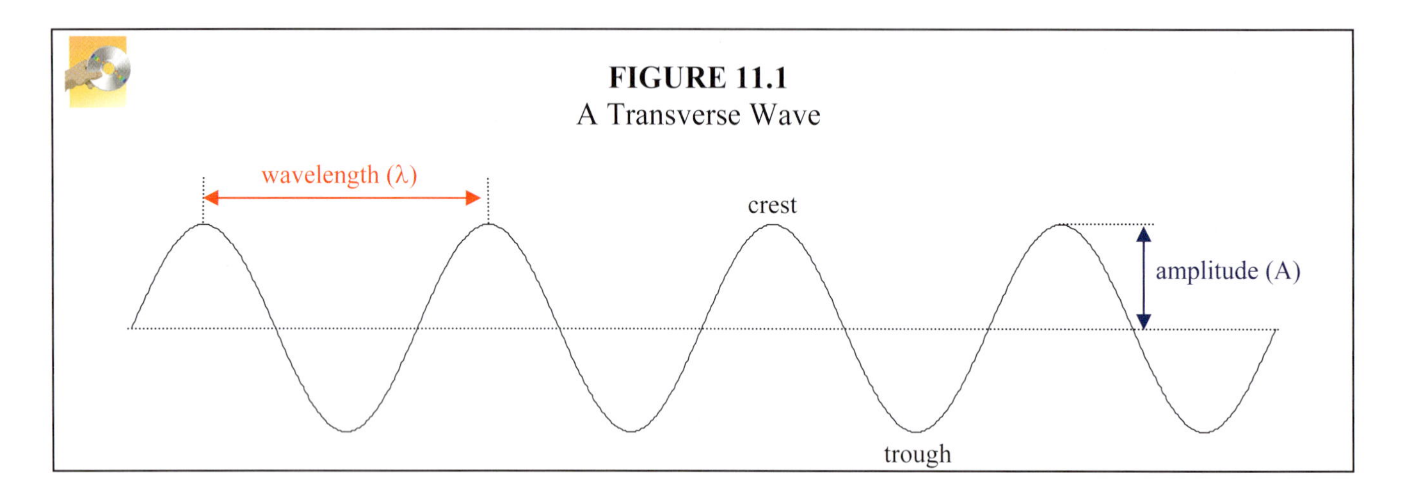

In a wave, the distance between the crests (or the distance between the troughs) is called the **wavelength (λ)** of the wave, while the height of the wave is the called the **amplitude (A)**. Another important characteristic of a wave not shown in Figure 11.1 is **frequency (f)**. The frequency of a wave tells you how many waves hit a certain point every second. Frequency and wavelength are related to one another through the speed of the wave:

$$f = \frac{v}{\lambda} \tag{11.1}$$

In this equation, "f" represents the frequency of the wave, "λ" (the Greek letter "lambda") is the wavelength, and "v" stands for the wave's speed (no boldface means it's not a vector).

First of all, we should examine the units of this equation. Speed is measured in m/sec, while wavelength, since it is a distance, is measured in meters. If I divide m/sec by m, what do I get? I get 1/sec. That's the unit for frequency we learned in Module #7. Remember that this unit is usually called "Hertz" and is abbreviated "Hz." Secondly, you should notice that frequency and wavelength are inversely proportional. In other words, when the wavelength is large, the frequency is small. Alternatively, when the wavelength is small, the frequency is large.

In a moment, we'll use Equation (11.1) to analyze some waves, but first, we need to make a distinction between two different types of waves. The wave pictured in Figure 11.1 is called a **transverse** (trans vurs') **wave**.

<u>Transverse wave</u> - A wave whose propagation is perpendicular to its oscillation

That's a mouthful, isn't it? Actually, this definition is rather simple once you get past the twenty-dollar words. The propagation of a wave is the direction of travel. For example, the waves in an ocean travel towards the shore. That's the direction of the wave's propagation. The oscillation of a wave refers to how its medium moves. Waves in an ocean cause the water to heave in an up-and-down direction. So, an ocean wave propagates horizontally (towards the shore), but it causes the ocean to oscillate vertically (up and down). In this case, the propagation is perpendicular to the oscillation. That's the kind of wave pictured in Figure 11.1.

There is another type of wave, however. Some of the waves that we see in creation are **longitudinal** (lahn' juh tood' nul) **waves**.

Longitudinal wave - A wave whose propagation is parallel to its oscillation

The best way to picture a longitudinal wave is to get a Slinky® and stretch it out on the floor. Hold one end of the Slinky® still, and then start moving the other end back and forth. What will it look like? It will look something like Figure 11.2.

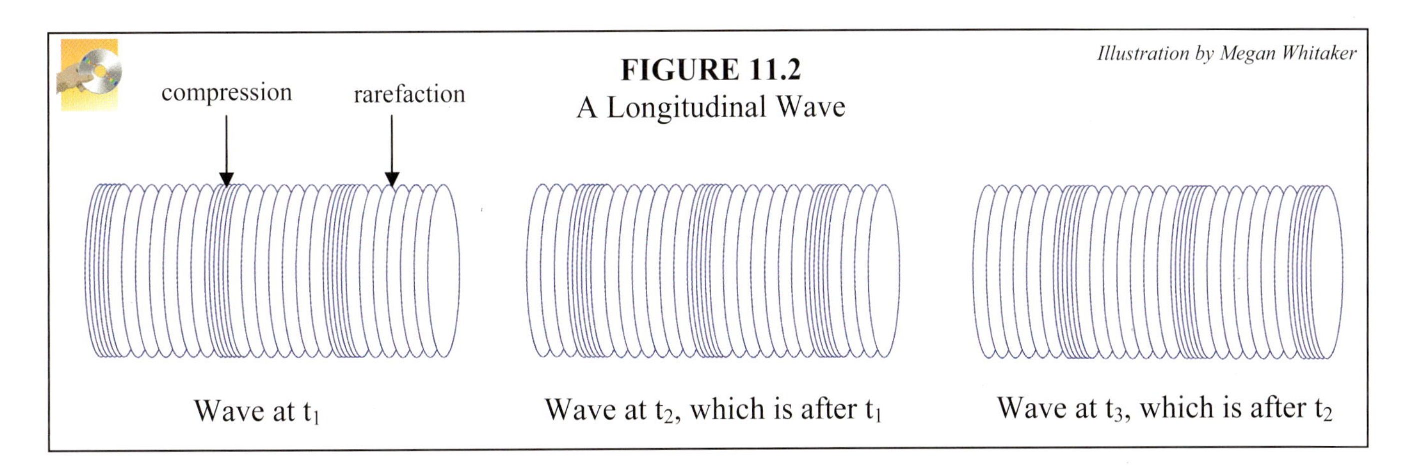

FIGURE 11.2
A Longitudinal Wave

In this figure, you see the Slinky® at three different times. Notice that parts of it are bunched up. Those regions are called **compressions**, and they are the crests of the wave. The parts of the Slinky® that are spread out are called **rarefactions**, and they are the troughs of the wave. The distance between the crests (or the troughs) is the wavelength of the wave.

Now notice what happens to the crests of the wave as time progresses. At t_2, the crests have all moved to the right a bit. At t_3, the crests have moved even farther to the right. Thus, this wave propagates horizontally to the right. The speed at which the crests move to the right would be the speed of the wave. Note that in this case, the medium (the Slinky®) oscillates back and forth in the horizontal dimension. Thus, in this wave, both the propagation and the oscillation are horizontal. Thus, this is a longitudinal wave, because the propagation and oscillation are parallel.

It turns out that while light is considered a transverse wave, sound is a longitudinal wave. Although these two wave types are fundamentally different, Equation (11.1) applies equally to both. Thus, you can use it to calculate the frequency of either kind of wave, given the speed and the wavelength. We will study light in considerable depth in a moment. First, however, let's take a look at sound.

The Physical Nature of Sound

As I mentioned above, sound is a longitudinal wave. The medium through which this wave travels is air. Figure 11.3 illustrates this.

Illustration from the MasterClips collection

FIGURE 11.3
Sound Waves

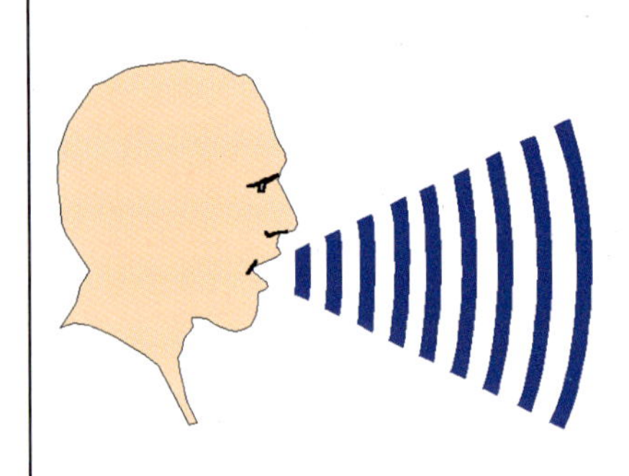

In order for something to make a sound, air must be pushed outwards from the source. When air is pushed, it clumps together while it is being pushed, forming a compression. When the pushing ceases, a rarefaction forms. If the air is pushed outwards like this at regular intervals, the result is a series of compressions (crests) followed by rarefactions (troughs), moving away from the source. This forms a longitudinal wave.

How does a wave like this make sound? To understand that, you first have to know some of the basic anatomy of the ear.

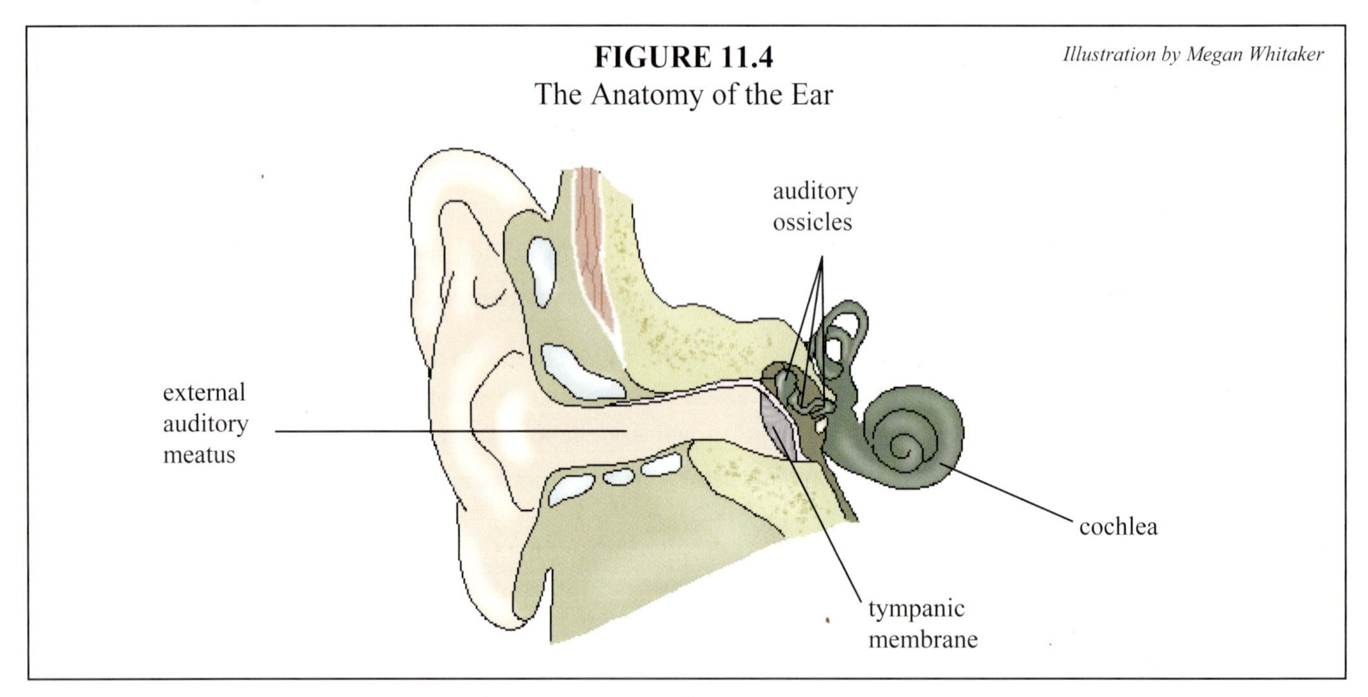

FIGURE 11.4
The Anatomy of the Ear

Illustration by Megan Whitaker

Now don't worry. I am not going to test you on the anatomy of the ear. I just wanted to show you a drawing and point out the major components related to our sense of hearing.

Sound waves travel into our ear through the external auditory meatus (mee aye' tus). The sound waves then hit the tympanic (tim pan' ik) membrane, which is also known as the ear drum. When the compressions of the wave hit the ear drum, it is pushed in. When the rarefactions follow, the ear drum is pulled back out. Thus, the ear drum converts the vibrations in the air to vibrations of the membrane. The vibration of the ear drum is then picked up by the auditory ossicles, which are the three smallest bones in the human body. Those bones vibrate with the ear drum, passing the vibration to the entrance of the cochlea (kok' lee uh). The cochlea is a fluid-filled organ, and the fluid begins to vibrate in response to the vibration of the ossicles. The vibration of the fluid is picked up by tiny hair cells located on the organ of hearing, which is called the spiral organ. Those hair cells generate nerve impulses based on the vibrations that they experience, and those impulses are sent to your brain, which interprets them as sound.

The bottom line, then, is that the compressions and rarefactions in a sound wave cause the ear drum to vibrate. Those vibrations are passed into the organ of hearing, where they are converted to nerve impulses that are interpreted by the brain. This interpretation involves many factors, including the **pitch** of the sound and the **volume**. The frequency of the sound wave is the primary factor in determining the pitch. The higher the frequency, the higher the pitch. Since wavelength and frequency are inversely proportional, we could also say that the smaller the wavelength, the higher the pitch.

What determines the volume of the sounds you hear? That's primarily determined by the amplitude of the wave. The larger the amplitude, the farther in and out the ear drum oscillates. This increases the amplitude of the fluid vibrations in the cochlea, and that amplitude is interpreted by your brain as volume. That's why really loud noises can break a person's ear drum. If the waves have a really large amplitude, the ear drum tries to move really far in and out. The ear drum, however, is a membrane of tissue that is tightly stretched. If the tissue tries to move really far in either direction, it can actually tear apart. That's what we call breaking an ear drum.

If you're having a hard time picturing sound waves (or longitudinal waves in general) don't worry. Just remember these three things:

1. **Sound waves are longitudinal.**
2. **The frequency (or wavelength) is the primary factor that determines the pitch.**
3. **The amplitude determines the volume.**

Now I want you to do an experiment that will give you some experience with the second and third facts listed above.

EXPERIMENT 11.1

Frequency and Volume of Sound Waves

Supplies:

- Safety goggles
- 16 inches of copper pipe, ½-inch diameter preferred (This is available at any hardware store. A good hardware store will even cut it for you. You need two 6-inch pieces and one 4-inch piece.)
- Freezer
- Hacksaw or pipe cutter (You won't need these if the store cuts the pipe for you.)
- File (Depending on the way the pipes are cut, you may not need this.)
- Hot tap water
- Large bowl
- Warm gloves

1. If the store did not cut the pipe for you, cut it into two 6-inch pieces and one 4-inch piece.
2. If the ends of the pipe pieces are rough, file them down to make them smooth.
3. Hold one of the 6-inch pieces of pipe in your hand. Use a finger from the other hand to plug the bottom of the pipe.
4. Purse your lips and blow into the pipe. You will have to play with how you blow into the pipe for a while, but eventually, by blowing into the pipe, you should be able to produce a nice, clear tone. It is very important to completely plug the bottom of the pipe with your finger so that no air can escape. That will result in the best tone for the experiment.
5. Once you have figured out how to blow into the pipe to make a nice clear tone, vary the force with which you blow, and note any differences in both the volume of the tone and its pitch.
6. Repeat steps 3-5 with the other 6-inch piece of pipe. It should sound the same as the first.
7. Repeat steps 3-5 with the 4-inch pipe. Note any difference between the sound you make with the 4-inch pipe and the sound you made with the 6-inch pipe. Feel free to go back to the 6-inch pipe so that you can remember what it sounded like.
8. Take one of the 6-inch pipes and put it in the freezer for 15 minutes.
9. Run the tap water until it is as hot as it will get.
10. Once the tap water is really hot, fill the bowl with the hot tap water.
11. Place the other 6-inch piece of pipe into the hot tap water.
12. After the pipe in the freezer has been in there for 15 minutes, put on your gloves and take it out of the freezer.
13. Quickly repeat steps 3-4 with the pipe from the freezer and note the pitch.
14. Quickly remove your gloves and pull the other pipe out of the hot tap water.
15. Once again, quickly repeat steps 3-4 with the pipe from the hot tap water. Note the difference in pitch between the two pipes
16. If you did not hear any difference in pitch between the two pipes, try again, and this time, make sure that you change pipes quickly. Please note that the gloves are just to protect your hand from the cold pipe. If the gloves are slowing you down, see if you can stand to hold the pipe from the freezer with your bare hands.
17. Clean up your mess.

What did you learn in the experiment? Well, first you should have noted that the pitch of the sound coming from the pipe was not affected by the force with which you blew. That's because pitch is primarily determined by the frequency of the wave, and the frequency of the wave depends on the wavelength and the speed of the wave. The length of the pipe fixed the wavelength produced in the air inside the pipe, and that (along with the speed of sound) fixed the frequency. Thus, no matter how hard you blew, the pitch was the same. However, the harder you blew, the louder the sound became. That's because the harder you blew, the stronger the compressions of the air in the wave, which means the larger the amplitude of the sound wave. Thus, amplitude determines the volume of the sound produced.

When you compared the pitch from the two 6-inch pipes, you should have found that the pitches were the same. Once again, that's because the length of the pipes fixed the wavelength of the wave produced in the pipe, which (along with the speed of sound) fixed the frequency. However, when you compared the pitch of the sound produced in the 6-inch pipe to the pitch of the sound produced in the 4-inch pipe, you should have noticed that the longer pipe produced the lower pitch.

That's because the longer pipe produced a wave with a longer wavelength. A longer wavelength means a lower frequency.

So far, then, the experiment has demonstrated two of the three facts that you need to know about sound waves: the pitch of the sound you hear is determined primarily by the frequency of the sound wave, and the volume is determined by the amplitude of the wave.

What about the last part of the experiment? What did it demonstrate? Remember, frequency and wavelength are related by Equation (11.1):

$$f = \frac{v}{\lambda}$$

In the discussion so far, I have assumed that the speed of sound was the same in all of the pipes. Thus, equal wavelengths produced equal frequencies. However, in the last part of the experiment, something changed.

What did you hear in the last part of the experiment? You should have heard that the pitch of the sound produced by the pipe that was soaked in hot water (let's call it the "hot pipe") was *higher* than the pitch of the sound produced by the pipe that had been in the freezer (let's call it the "cold pipe"). Now remember, the two pipes were of equal length, so the wavelength of the sound wave produced was the same in each pipe. If the pitch (frequency) of the sound wave produced in the hot pipe was higher, what does that tell us about the speed of the sound wave in that pipe? It tells us that the speed was higher. After all, the pipes produced sound waves of equal wavelength. According to Equation (11.1), the only way that their frequencies could be different is if the speed of the waves was different. Since the frequency of the sound wave in the hot pipe was higher than the frequency of the sound in the cold pipe, we can deduce that the speed of sound was higher in the hot pipe. Thus, the higher the temperature, the faster the speed of sound.

This fact should make sense to you if you have already had chemistry. In chemistry, you should have been taught that the faster the molecules and atoms in a substance travel, the warmer the substance is. Since a sound wave is made by moving clumps of air around, the warmer the air, the faster these clumps should move. There is a rather simple equation that relates the temperature of the air and the speed of sound waves traveling through it:

$$v = (331.5 + 0.606 \cdot T)\ \frac{m}{sec} \qquad (11.2)$$

In this equation, "v" is the speed of sound and "T" is the temperature of the air in °C.

Now that we have a wealth of information related to the physical nature of sound, we can do a little bit of analysis of sound waves. Study the following example and solve the "On Your Own" problems to get an idea of how to use the facts that you have learned.

EXAMPLE 11.1

When someone sings a "standard A," the frequency of the sound wave produced is 440.0 Hz. If the temperature of the air is 25 °C, what is the wavelength of the sound wave?

This is a straightforward application of Equations (11.1) and (11.2). To determine the wavelength, we must use Equation (11.1). Before we can do that, however, we need to determine the speed of the waves. For that, we use Equation (11.2):

$$v = (331.5 + 0.606 \cdot T)\,\frac{m}{sec}$$

$$v = [331.5 + 0.606 \cdot (25)]\,\frac{m}{sec} = 347\,\frac{m}{sec}$$

Notice that I did not put the "°C" unit into the equation. That's because this equation is designed to be used without units. As long as you keep temperature in degrees Celsius, the resulting number will always be in "m/sec." Thus, in this particular equation, do not put the units in. Just remember that temperature must be in degrees Celsius. Also, note the significant figures. Since we are mixing addition with multiplication, we first must do the multiplication and round to two significant figures. This gives us 15. Then, when we add 15 to 331.5, we are limited to reporting our answer to the ones place, since 15 has its last significant figure in the ones place. Now that we have speed, we can use Equation (11.1):

$$f = \frac{v}{\lambda}$$

$$440.0\,\frac{1}{sec} = \frac{347\,\frac{m}{sec}}{\lambda}$$

$$\lambda = \frac{347\,\frac{m}{\cancel{sec}}}{440.0\,\frac{1}{\cancel{sec}}} = 0.789\text{ m}$$

Notice that in order to get the units to work out, I renamed "Hz" to "1/sec." What does this wavelength tell us? Well, remember that sound waves are formed by compressions and rarefactions of air. This wavelength tells us that if we were actually able to see the compressions, they would be <u>0.789 m</u> apart.

ON YOUR OWN

11.1 What is the frequency of a sound wave whose wavelength is 1.0 m? Assume that the air has a temperature of 15.0 °C. Is the pitch of this sound higher or lower than a "standard A" (440.0 Hz)?

11.2 If a sound wave of frequency 501 Hz has a wavelength of 0.697 m, what is the temperature of the air through which the wave is traveling?

The Doppler Effect

The fact that the pitch of the sounds you hear is related to the frequency of the sound waves leads to an interesting phenomenon known as the **Doppler effect**. Named after Austrian physicist Christian Doppler, it is best explained with a figure. Consider a car that is traveling down the street. The driver is your friend, and when he sees you standing on the sidewalk, he honks his horn in one steady blast. What do you hear?

FIGURE 11.5
The Doppler Effect

Illustration from the MasterClips collection

The horn from a car produces a sound wave with a constant frequency. When the car moves, however, the horn emits the wave as the car travels. Thus, after it has emitted one crest, it moves forward and emits the next. This causes the crests to be bunched up in front of the car and stretched out behind the car.

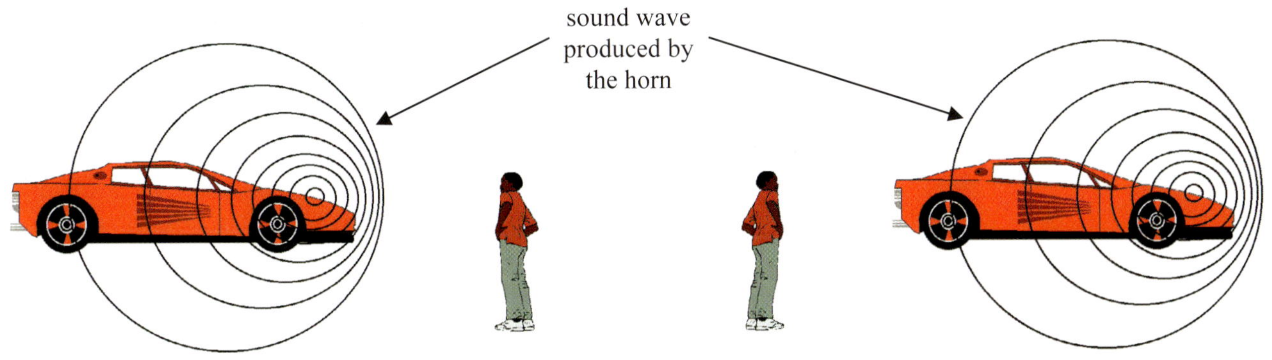

Car Moving Towards You

Since the sound wave crests get bunched up in front of the car, the wavelength seems shorter, so the frequency is higher. If the car is heading towards you, then, you will hear a pitch that is higher than the true pitch of the horn.

Car Moving Away From You

When the car passes you, the sound wave crests that reach your ears are the ones traveling behind the car. These crests are stretched out, which gives them a longer wavelength and thus a lower frequency. This results in a pitch that is lower than that of the true pitch.

The Doppler effect, then, is a result of the wave nature of sound. When a sound-emitting object is moving towards you, the crests and troughs of the sound wave get pushed closer to one another. This decreases the wavelength and therefore increases the frequency of the sound that your ears hear. This results in a pitch that is higher than the pitch you would hear if the object were stationary. Once the object begins moving away from you, the crests and troughs get pulled farther away from one another. This increases the wavelength and therefore decreases the frequency of the sound that hits your ears. As a result, the pitch is lower than what it would be if the object were stationary.

Of course, you can produce the same effect by moving towards or away from a fixed object which is emitting a sound wave. If the object is stationary, its sound wave crests do not get "bunched up" or "stretched out" like those of a moving object. However, if you move towards the object, you will start encountering its crests at a high frequency, because the relative velocity between you and the crests has increased. Thus, moving towards an object generates the same Doppler effect that you experience when a moving object moves towards you. In the same way, if you run away from a stationary object that is emitting a sound, the sound wave crests will hit you with a *lower* frequency, because the relative velocity between you and the crests has decreased. Thus, moving away from a fixed object results in a similar Doppler effect as when a moving object is traveling away from you. Get some experience with the Doppler effect by performing the following experiment.

EXPERIMENT 11.2
The Doppler Effect

Supplies:

- A car with a horn and a parent to drive the car
- The street you live on or a country road
- A bicycle

1. If you have irritable neighbors, you might want to perform this experiment on an isolated road somewhere away from town. If your neighbors are patient or not home, you can do this on the street where you live.
2. Have your parent drive the car to one end of the street.
3. Stand on the sidewalk or in your yard near the street in front of your house. **Be careful!!!! Do not get too close to the street!**
4. Have your parent drive down the street towards your house with a speed of at least 20 miles per hour, preferably faster. A few seconds before the car reaches your house, have your parent blow the horn in a single blast until he has completely passed the house. Note the pitch of the horn before the car passes you and after the car passes you.
5. Have your parent repeat the procedure, this time from the other direction.
6. Have your parent park the car on the street.
7. Tell your parent that you are going to ride towards the car on your bicycle. Tell him that when you get near the car, he should blow the horn in a long, steady blast until you have passed the car and are moving away from it.
8. Get on the bicycle. If the street is ***completely free of other cars***, ride it on the street. Otherwise, ride it on the sidewalk.
9. Ride the bicycle as fast as safety will allow. Note the pitch of the horn as you ride towards and away from the car.
10. Apologize to anyone you have annoyed, and thank your parent.

What did you hear in your experiment? Well, in the first part of the experiment, the pitch of the horn should have sounded higher as the car traveled towards you and lower as it traveled away from you. That's the Doppler effect. The crests and troughs of the horn's sound wave were bunched up as the car approached you, increasing the frequency. They were spread out as the car sped away from you, decreasing the frequency.

The results of the second part of the experiment were probably not nearly as conclusive. If you do not have a good ear for pitch, you might not have noticed much of a difference between the pitch of the horn as you traveled towards the car and away from the car. Why? Didn't the Doppler effect work? Of course it did. However, the speed at which you could pedal the bicycle was pretty small. Thus, the change in frequency produced by the Doppler effect was also small. As a result, if you do not have a good ear for pitch, you might not have noticed any effect. If you do have a good ear for pitch, you should have noticed that the pitch of the horn sounded higher as you traveled towards the car (because you encountered the crests and troughs of the wave more frequently) and lower as you traveled away from the car (because you encountered the crests and troughs of the wave less frequently).

The Doppler effect is reasonably easy to calculate. It depends only on the true frequency of the sound wave (in other words, the pitch you would hear if neither you nor the source of the sound were moving), the speed of the observer, and the speed of the sound's source:

$$f_{observed} = \left(\frac{v_{sound} \pm v_{observer}}{v_{sound} \pm v_{source}}\right) \cdot f_{true} \qquad (11.3)$$

In this equation, "$f_{observed}$" is the frequency that the observer actually hears, while "f_{true}" is the frequency that the observer would hear if the source of the sound and the observer were stationary. There are three speeds in the equation: "v_{sound}" is the speed of sound as given by Equation (11.2), "$v_{observer}$" is the speed of the observer, and "v_{source}" is the speed of the source that is emitting the sound.

Notice first what happens to this equation if the speed of both the observer and the source are zero. When that is the case, the speed of sound cancels, and $f_{observed}$ equals f_{true}. This should make sense. If the source and observer are not moving, there is no Doppler effect, and the observer will hear the actual frequency at which the sound was emitted.

Now let's look at the "±" signs in the equation. What do they mean? They mean that sometimes you add, and sometimes you subtract. Of course, in order to use this equation, you will have to know when to add and when to subtract. How do you know? All you have to do is reason it out. For example, suppose the source that is emitting the sound is moving. That means you will either need to add it to or subtract it from the speed of sound in the denominator of the equation. If it is moving *away from* the observer, you must use a *plus* in the denominator of the equation. After all, when the source moves away from the observer, the frequency observed will be *lower* than the true frequency. The way you will get an observed frequency lower than the true frequency is if the number in the denominator is increased. Thus, you use the plus sign so that the denominator gets bigger. In the same way, suppose the source is moving *towards* the observer. If that's the case, you expect a *larger* frequency than the true frequency. Thus, you need the fraction to become greater in value. Since you can only deal with the denominator of the equation, the only way you can do that is to *decrease* the value of the denominator. That will increase the value of the fraction. To decrease the denominator of the fraction, you will need to *subtract* the speed of the source. Thus, adding or subtracting depends on the motion of the source. If the source moves *towards* the observer, you *subtract* its speed so that the observed frequency is larger than the true frequency. If it moves away from the observer, you add its speed, so that the observed frequency is lower than the true frequency.

You must use similar reasoning if the observer is moving. If the observer moves *towards* the source, then you expect a *higher* frequency. Since $v_{observer}$ appears in the numerator of the fraction, you will have to *add* it to the speed of sound. That way, the fraction gets bigger, which causes the observed frequency to be higher than the true frequency. If the observer is moving *away from* the source, you expect a *lower* frequency. Thus, you will need to *subtract* the speed of the observer from the speed of sound. This will make the fraction smaller, which will make the observed frequency lower than the true frequency.

Make sure you understand this by studying the example problems below and solving the "On Your Own" problems that follow.

EXAMPLE 11.2

A horn emits a sound with a frequency of 355 Hz when the car is at rest. If a car is traveling (v = 21.0 m/sec) towards a stationary person and the driver beeps the horn, what frequency will the person hear? Assume that the speed of sound in this case is 345 m/sec.

This is an application of Equation (11.3). The person is not moving, so $v_{observer}$ is zero. Since the car is traveling towards the observer, he or she will hear a frequency that is *higher* than the horn's true frequency. Thus, we need the fraction in Equation (11.3) to get larger. The speed of the source appears in the denominator of the fraction. To make a fraction larger by changing the denominator, the denominator must *decrease*. Thus, we must subtract the speed of the car from the speed of sound:

$$f_{observed} = \left(\frac{v_{sound} \pm v_{observer}}{v_{sound} - v_{source}}\right) \cdot f_{true} = \left(\frac{345\frac{m}{sec} \pm 0}{345\frac{m}{sec} - 21.0\frac{m}{sec}}\right) \cdot 355\text{Hz} = \left(\frac{345\frac{\cancel{m}}{\cancel{sec}}}{324\frac{\cancel{m}}{\cancel{sec}}}\right) \cdot 355\text{ Hz} = 378\text{ Hz}$$

If the car had been stationary, then, the person would have heard the horn's frequency as 355 Hz. However, since the car was traveling towards the person, she heard its frequency as 378 Hz.

A town has a stationary emergency siren that warns the people in the town of danger. It emits a steady tone at 401 Hz. If a man in a car is traveling away from the siren at 34.0 m/sec when the siren goes off, what frequency does he hear? The speed of sound in this situation is 345 m/sec.

Once again, this is an application of Equation (11.3). Since the siren is stationary, v_{source} is zero. However, the person in the car is moving. Since he is moving away from the siren, we expect the observed frequency to be lower. In Equation (11.3), the speed of the observer is in the numerator of the fraction. Since we expect a lower frequency, we need to reduce the value of the fraction in the equation. The only way we can do that is to lower the value of the numerator, which means we need to subtract $v_{observer}$:

$$f_{observed} = \left(\frac{v_{sound} - v_{observer}}{v_{sound} \pm v_{source}}\right) \cdot f_{true} = \left(\frac{345\frac{m}{sec} - 34.0\frac{m}{sec}}{345\frac{m}{sec} \pm 0}\right) \cdot 401\text{ Hz} = \left(\frac{311\frac{\cancel{m}}{\cancel{sec}}}{345\frac{\cancel{m}}{\cancel{sec}}}\right) \cdot 401\text{Hz} = 361\text{ Hz}$$

ON YOUR OWN

11.3 A man has just said goodbye to his friend, who has boarded a train. He stands and watches the train as it travels away from the station at 35.0 m/sec. If the train blows its horn (frequency = 411 Hz when the train is stationary), what is the frequency that the man will hear? (T = 25.0 °C)

11.4 A woman is driving towards a factory to pick up her husband after work. As she travels towards the factory, the factory starts to blow the whistle indicating the end of the work day. If the whistle's frequency is 494 Hz and the woman hears it as 511 Hz, how fast is the woman driving? (T = 25.0 °C)

Sound Waves in Substances Other Than Air

We are used to thinking about sound as it travels through air, but it is important to realize that sound waves can travel through any medium that can be compressed. After all, a sound wave is simply a series of compressions and rarefactions that form a longitudinal wave. If a substance can be compressed, then, a longitudinal wave can form in it, and therefore sound can travel through it.

If you have ever had to ask people in the next room to be quiet because you are trying to study, for example, you know that sound waves can travel through walls. In addition, people who make their living spying on other people use the fact that sound can travel through glass to help them listen to conversations that they otherwise would not be able to hear. If a spy is trying to find out what is being said in a room to which he does not have access, he can shine a laser beam at one of the room's windows. A portion of the laser beam will reflect off of the window back to a sensor, which will detect fine movements in the beam. As sound waves from the conversation inside the room travel through the glass, the glass will vibrate back and forth as a result of the compressions and rarefactions in the wave. This will cause the beam to "wiggle," and the sensor can analyze that wiggle and actually reconstruct the sound waves that caused it. This allows the spy to actually hear what is being said in the room!

Table 11.1 lists the speed of sound in some substances with which you are familiar.

TABLE 11.1
The Speed of Sound in Certain Substances

Substance	Speed of Sound	Substance	Speed of Sound
Air (20 °C)	343 m/sec	Lead	1,322 m/sec
Methyl Alcohol (25 °C)	1,143 m/sec	Aluminum	5,100 m/sec
Freshwater (25 °C)	1,493 m/sec	Iron	5,130 m/sec
Wood (oak) along the fiber	3,850 m/sec	Copper	3,560 m/sec

Note that sound travels more quickly in the liquids listed than in air, and it travels most quickly in the solids listed. You might be tempted to say that the data indicate that the speed of sound in a substance increases with increasing density. Surprisingly, however, that is not correct. It turns out that the speed of sound actually decreases with increasing density of a substance!

Why, then, does sound travel faster in the solids listed than in the liquids listed? Because the speed of sound in a substance also depends on how easily the substance is compressed. The easier it is to compress the substance, the *slower* sound travels in it. Thus, since air is easy to compress compared to most liquids, sound travels more slowly in air than in most liquids. In the same way, since most liquids are more easily compressed than most solids, sound travels more slowly in most liquids than in most solids. It turns out that the variation in compressibility of different substances is much greater than the variation in their densities, so in general, when you look at the speed of sound in a substance, you generally see the effect of the compressibility of the substance more than the effect of the density. As a result, sound generally travels the slowest in gases, more quickly in liquids, and even more quickly in solids.

Sound Waves Beyond the Ear's Ability to Hear

Before we move on to the topic of light waves, I want to make it clear that there is a difference between the waves that produce sound when they hit an ear and the sound itself. Remember, what we "hear" as sound is actually an electrical response translated by your brain. We do not hear the waves themselves. Your ear drum and the various mechanisms to which it is attached transmits the frequency and amplitude of the wave to your brain, and your brain interprets them as sound.

The human ear, though elegantly designed, cannot detect all frequencies of sound waves. In general, human ears are sensitive to waves whose frequency is between 20 Hz and 20,000 Hz. Waves with these frequencies are called **sonic** (sahn' ik) waves. Waves with frequencies higher than 20,000 Hz are called **ultrasonic** (uhl' truh sahn' ik) waves, and waves with frequencies below 20 Hz are called **infrasonic** (in' fruh sahn' ik) waves. The only difference between these types of waves is their frequencies. Nevertheless, only the sonic waves produce what we hear as sound.

Now, even though we cannot hear ultrasonic waves, they are incredibly useful. Because of their high frequencies, ultrasonic waves have small wavelengths. Their small wavelengths make it very easy to control and manipulate them. One application that uses these easy-to-control waves is the ultrasonic ruler. This device is a small box that emits ultrasonic waves. The circuitry in the device measures the temperature of the air and determines the speed of the waves. When a wave hits a wall or other obstruction, part of the wave is transmitted through the obstruction, and part of the wave is reflected back to the device. The part that is reflected back is detected by the ultrasonic ruler, and the time it took to travel to the obstruction and back is measured by the circuitry. The ultrasonic ruler then uses the speed of the sound wave and the time it measured to calculate the distance from the device to the obstruction. Thus, a person simply holds the device and points it to a wall, and the device determines the distance to that wall.

A more popular application of ultrasonic waves is their use as a medical imaging tool. When a longitudinal wave hits an obstruction, part of the wave is reflected, and part is transmitted. The reflected and transmitted waves are the same frequency, but they each have lower amplitudes than the original wave. If an ultrasonic wave is directed at a human body, a portion of the wave gets transmitted into the body. As the wave travels through the body, it will continue to travel until it hits another obstruction. At that point, a portion of the wave will be reflected, and a portion will be transmitted. If wave sensors are tuned to detect the portion of the wave that was transmitted through the body but reflected back by the first obstruction encountered in the body, the detectors can use the same principles that the ultrasonic ruler uses to determine the distance to the obstruction within the body.

If several such waves are directed across a large area in the human body, this procedure can determine the general shape of the obstruction within the body. The most popular application of this is used for pregnant mothers. Using this technique, the ultrasonic imager can produce the general shape of a fetus in the mother's body. An example of such an image is shown in Figure 11.6.

Image Courtesy of Strong Memorial Hospital

FIGURE 11.6
An Ultrasonic Image of a Human Fetus

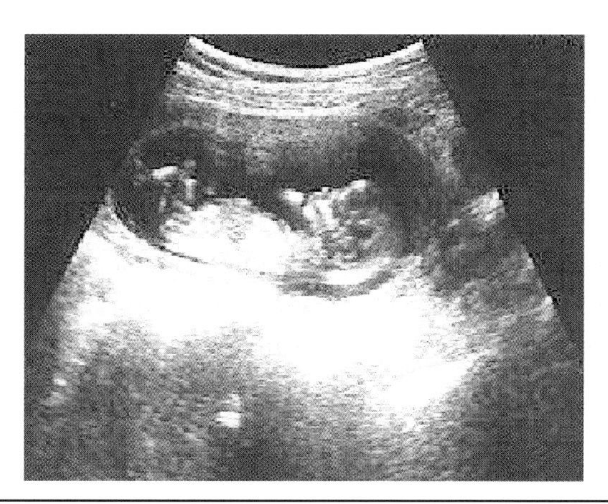

The Speed of Light

Before I tell you about the physical nature of light, I want to spend some time on the speed at which light travels. This is a big difference between sound and light. Even though sound travels really fast, light travels significantly faster. The speed of light in air (to two significant figures) is 3.0×10^8 m/sec and is independent of the temperature of the air. It turns out that this is a rather important number, so you will have to memorize it:

The speed of light in air (to two significant figures) is 3.0×10^8 m/sec.

This is a huge number, so let's put it in a little perspective. If you were able to drive your car around the circumference of the earth, you might be able to average 60 miles per hour. If you kept up that average 24 hours a day, it would take you about 17 days to go around the world. Sound waves, however, can travel the same distance in just over a day, assuming the air stays near a temperature of 25 $^{\circ}$C. Compare this to the 0.13 seconds it would take light to travel that far!

The speed of light never changes as long as the medium in which it is traveling doesn't change. If light waves travel through a medium other than air, however, their speed is different. In general, the more dense the medium, the slower the light waves move. This is another important fact.

The speed of light waves changes depending on the medium through which they travel. Generally speaking, as the medium gets more dense, light waves slow down.

Actually, there are several factors that influence the speed at which light travels through a medium. Thus, the relationship given above is not true in all cases. However, it is a good "rule of thumb." In the next section of this module, you will learn more about the nature of light itself. For right now, however, I want to show you one very important thing that we can learn from the difference between the speed of light as it travels through air and the speed of sound as it travels through air.

EXAMPLE 11.3

A physicist is watching a thunderstorm. She sees a flash of lightning and then hears a thunderclap 1.00 second later. If the air temperature is a cool 15 °C, how far away from the physicist was the lightning formed?

To solve this problem, you first have to realize that thunder is the sound that accompanies a lightning flash. Lightning occurs when friction in a cloud creates an electrical imbalance in some region of the atmosphere. When the electrical imbalance gets large enough, a spark is generated, causing electricity to travel through the region, wiping away the imbalance. The flash of lightning is the light from the spark, and thunder is the sound of the spark. Thus, if you were unfortunate enough to be right at the point where the spark formed, you would see the flash and hear the thunder at precisely the same instant. If you are far away from the spark, however, the light will reach your eyes before the sound reaches your ears, because the light travels so much faster than sound. The difference between these two speeds can allow us to calculate how far away the lightning was formed.

First of all, we will assume that it takes essentially no time for the light to reach your eyes. After all, you just learned that light can travel all the way around the world in about 0.13 seconds. Thus, the time it would take light to travel a few miles (which is, at most, the distance we are interested in) is so small that it is irrelevant. Therefore, we can assume that the lightning was formed at the instant in which the physicist sees it. The time delay that the physicist observes, then, is simply the time it took for the sound to travel from the point at which the lightning was created to the physicist. Since sound waves travel at a constant speed we can simply say:

$$\text{distance} = \text{speed} \times \text{time}$$

To use this equation, though, we need to know the speed of sound. This can be determined by Equation (11.2):

$$v = (331.5 + 0.606 \cdot T)\,\frac{\text{m}}{\text{sec}} = [331.5 + 0.606 \cdot (15)]\,\frac{\text{m}}{\text{sec}} = 340.6\,\frac{\text{m}}{\text{sec}}$$

Now we can calculate the distance that the sound traveled:

$$\text{distance} = \text{speed} \times \text{time}$$

$$\text{distance} = (340.6\,\frac{\text{m}}{\cancel{\text{sec}}}) \times (1.00\,\cancel{\text{sec}}) = 341\text{ m}$$

The lightning, therefore, was formed 341 m away from the physicist.

By measuring the time between when we see a lightning flash and when we hear the thunder, then, we can determine how far away the lightning was formed. Because the speed of sound depends only slightly on temperature, and because 341 meters is about 1/5 of a mile, when you are watching a thunderstorm, you can use as a general rule of thumb that for every second of delay between a lightning flash and the thunder, the lightning was formed about 1/5 of a mile away. Of course, you *cannot* use this rule of thumb when solving problems in this course! When you are working on the "On Your Own" problems, practice problems, and tests, I want you to use the more precise method outlined above.

ON YOUR OWN

11.5 If you hear thunder 2.5 seconds after you see a lightning flash, how far away was the lightning formed? (Assume a temperature of 20.0 °C.)

11.6 Suppose you are watching thc start of a race. The official raises his starter's pistol into the air and fires. You see the cloud of smoke rising from the pistol, indicating that it has fired, but you do not hear the sound of the pistol right away. If you are 402 meters from the starter and the temperature is 25 °C, how long will it take you to hear the sound of the pistol firing?

Light as a Wave

Now it's time to discuss the physical nature of light. It turns out that light is a bit more difficult to understand than sound, but it is important for you to get a good idea of what light really is. First, let's start with the fact that there is solid evidence that light is, indeed, a wave. One of the most important experiments to demonstrate this was originally performed by Thomas Young in the early 1800s. In this experiment, a beam of light hit a screen with a tiny slit in it. This produced a thin beam of light. That beam of light then hit another screen with two tiny slits in it. These two slits were very close to one another. The light that passed through these slits then hit a screen with no slits in it. Young looked at this screen to see what pattern of light formed there.

Now, before I tell you what Young saw in his experiment, I want you to try to guess what he saw. Remember, a thin beam of light hit a screen with two small slits in it. Once the light passed through the slits, it hit a screen with no slits in it, and Young looked at the screen. What did he see?

You might be tempted to say that Young saw two bright spots on the screen: one bright spot resulted from the light passing through one of the slits, and the other bright spot resulted from light passing through the other slit. Although that's probably the result you would expect, it is not the result that Young saw!

He did, indeed, see bright spots on the screen, *but there were a lot more than just two*. Instead, there was a whole series of bright spots separated by dark spots. Despite the fact that there were only two slits through which light could pass before it hit the last screen, Young saw several bright and dark spots on the screen. How can we understand such a surprising result? The only way is to assume that light is a wave.

If we assume that light is a wave, we can picture what happened in Young's experiment. Please study Figure 11.7.

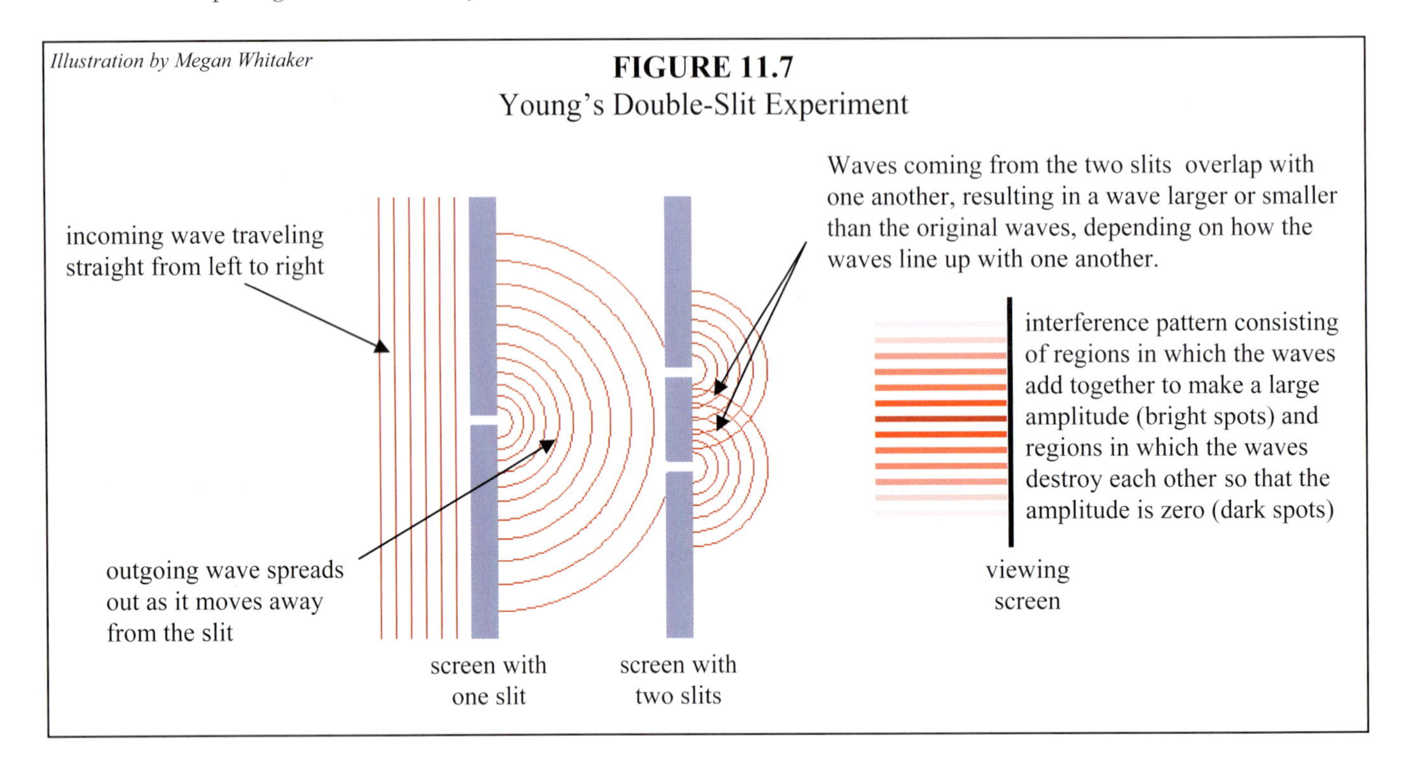

Prior to hitting the screen with the single slit, the light wave was traveling straight from the left side of the figure to the right side of the figure. Once the wave hit the slit, however, it "bent" around the slit and started radiating outwards from the slits. This phenomenon is called **diffraction** (dif rak' shun), and all waves behave this way when they travel through small openings or encounter obstacles.

<u>Diffraction</u> - The spreading of waves around an obstacle

In the experiment, diffraction makes the slit act like an individual light source whose light radiates outward from the slit.

The wave that radiated outward from the slit then hit the screen that contained two slits. Once again, the wave diffracted, bending around each slit. This makes each of the two slits behave like an independent light source. Unlike most individual light sources, however, the light from the two slits is **coherent**, which means that the spatial relationship between the crests and troughs of one wave compared to those of the other wave does not change.

Now think about this for a moment. Two waves are spreading out from two different points in space. Once they start to overlap, what will happen? Well, if two crests happen to overlap, they will add together so that there is a crest that is twice as large as the crest from either of the individual waves. If two troughs overlap, there will be a trough that is twice as deep as the trough from either of the individual waves. What happens if a crest from one wave overlaps with a trough from the other? They will cancel each other out, so that there is no wave at all.

This phenomenon is called **wave interference**, and it can occur whenever two waves overlap with one another. Consider, for example, two speedboats speeding across a lake parallel to one another. Each boat creates a wave that spreads out sideways from the boat. As the waves travel away from their respective boats, they eventually overlap. The crests of each wave can add together to make really large crests; the troughs can add together to make really deep troughs; and the waves can overlap so that their troughs and crests cancel one another out, resulting in no wave at all. As a point of terminology, when the waves overlap so that their crests and troughs add together to make larger crests and troughs, it is called **constructive interference**, because the waves add together to construct a bigger wave. When the waves overlap so that their crests and troughs cancel one another out, it is called **destructive interference**, because the waves destroy one another.

The multimedia CD has an animation that illustrates constructive and destructive interference.

Now you are ready to understand the result of Young's experiment. As the two light waves spread out from the two slits in the experiment, they began to overlap with one another. When the crests (or troughs) of one wave lined up with the crests (or troughs) of the other wave, constructive interference occurred, and a wave of large amplitude was produced. This made a bright spot of light on the viewing screen (illustrated by the red bars in the figure). When the crest of one wave lined up with the trough of another, destructive interference occurred, and the result was no wave. Since there was no light wave, the result was a dark spot (the white spaces in between the red bars in the figure). The fact that light exhibits constructive and destructive interference, then, provides strong evidence that light is a wave.

Now of course, if light is a wave, there is a question that must be answered: *What does light oscillate?* After all, a wave must oscillate a medium. Ocean waves oscillate water. Sound waves oscillate air (or whatever substance through which sound travels). So what does light oscillate? It does not oscillate air, because it travels through the vacuum of space. Sound cannot travel through space because there is nothing for it to oscillate. Light, however, can, as evidenced by the fact that we see light from stars and light reflected off planets. The question remains: *what* does light oscillate?

That question plagued physicists through the turn of the century. For a long time, physicists thought that there was a substance called the **ether** which permeated space, the earth, and anywhere else light traveled. Light travels through space, so the ether must be in space. Light travels through air, so the ether must be in air as well.

In 1887, however, two physicists, A. A. Michelson and E. W. Morley, devised an experiment to measure some physical characteristics of the ether; to their surprise, the experiment indicated that there probably is no such thing as the ether. Several physicists tried to explain their way around that experiment, but the explanations seemed rather desperate. In the end, the Michelson-Morley experiment threw physicists right back to where they were before. They could not answer a fundamental question: "What does light oscillate?"

In 1873, the great physicist James Clerk Maxwell had already derived a series of equations which today we call **Maxwell's Equations**. Although the implications of his equations were not understood until later, they were recognized as important because they described in mathematical detail the relationship between electric and magnetic phenomena. The equations indicate that electric and magnetic phenomena are related and that the relationship involves a wave. The equations also predict the speed of that wave, and the speed predicted is equal to the *measured* speed of light. Now please understand how important this is. Maxwell's equations, which unify electric and magnetic forces into a single **electromagnetic force**, specify that this force is mediated by a wave whose speed is a direct result of the mathematics. That speed is equal to the measured speed of light (3.0×10^8 m/sec), which is typically abbreviated as "c."

What does this tell us? It took physicists a while to realize the implication, but Maxwell's equations tell us that light is an ***electromagnetic wave***. What does that mean? It means that light waves are oscillating electric and magnetic fields. You really don't know what electric and magnetic fields are right now, but you will learn about them in upcoming modules. For right now, you can think about electric and magnetic fields as fields that have the ability to apply electric and magnetic forces. Light waves travel by oscillating these fields, with the electric field being perpendicular to the magnetic field, as shown in Figure 11.8.

FIGURE 11.8
The Generally Accepted View of a Light Wave

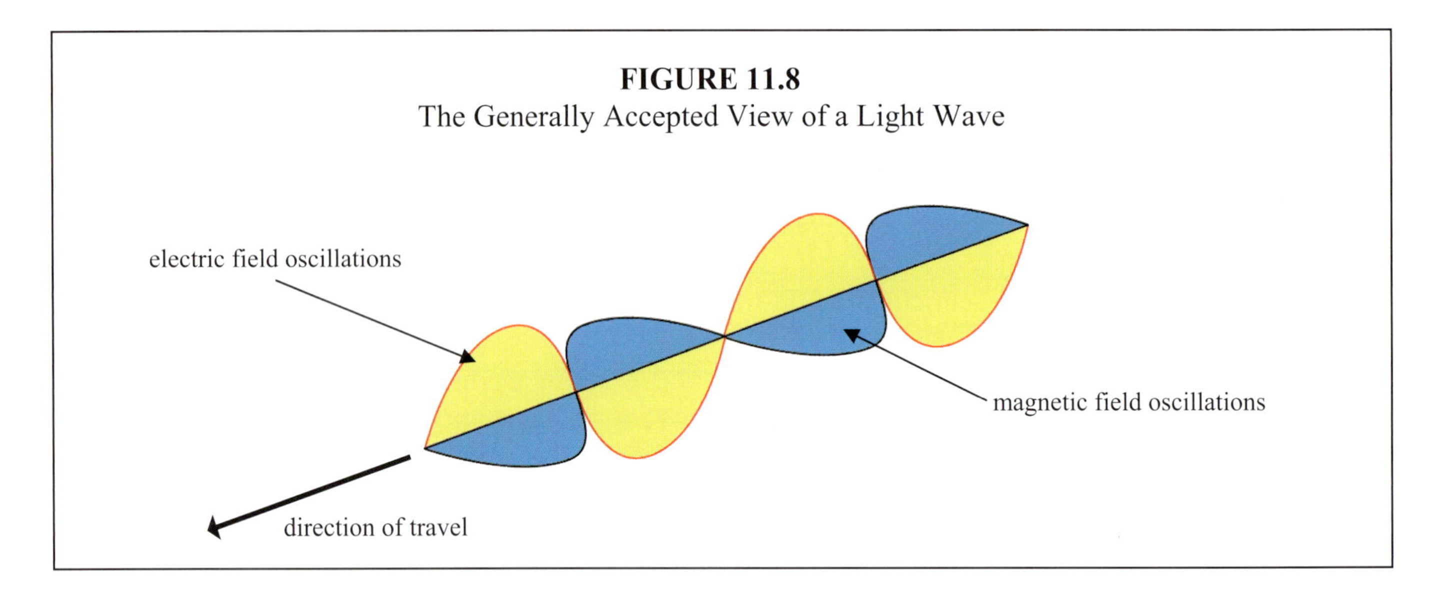

Of course, like any other wave, an electromagnetic wave can have a wide range of wavelengths. As a result, there are many wavelengths in the **electromagnetic spectrum**. Thus, the broader term for light is simply "electromagnetic wave," and the light that we can see (visible light) is only a tiny range of the wavelengths (400-700 nm) in the electromagnetic spectrum, as illustrated in Figure 11.9.

Illustrations from the MasterClips collection

FIGURE 11.9
The Electromagnetic Spectrum

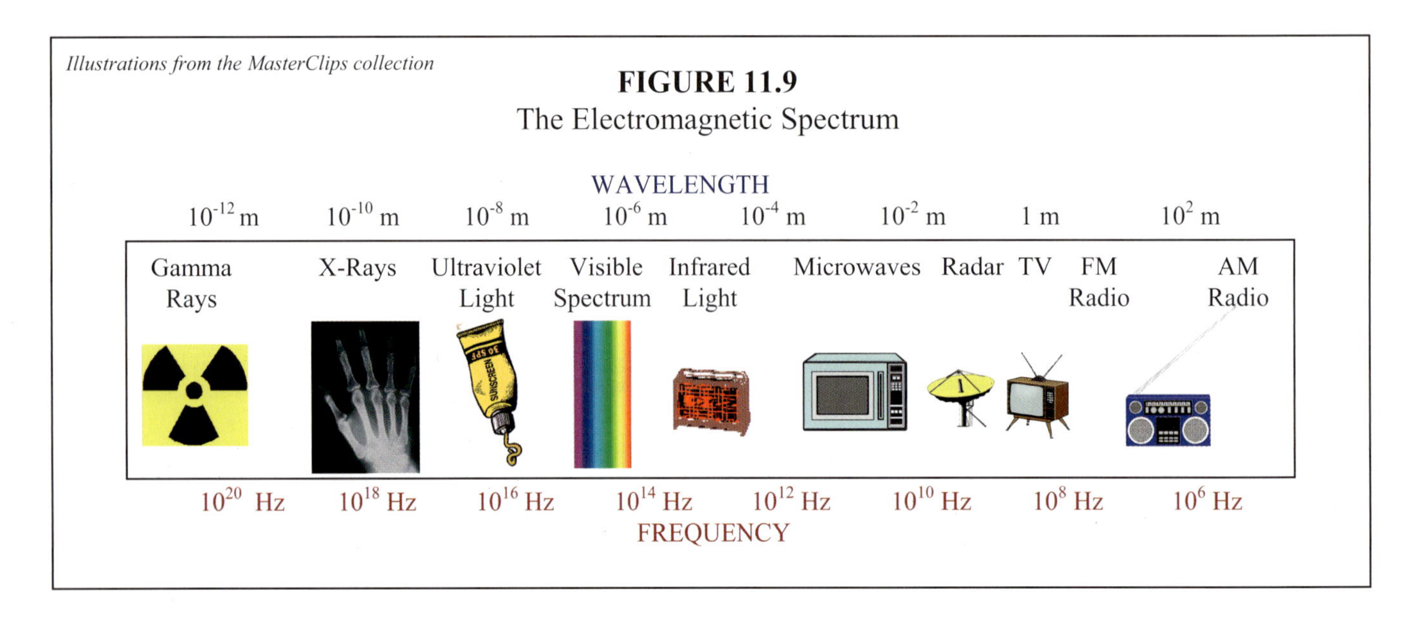

In our day-to-day experience, of course, the tiny portion of the electromagnetic spectrum that we notice is the visible part of the spectrum, so we should discuss it for a moment.

One nice way to look at the visible part of the electromagnetic spectrum is to examine a rainbow. In physical science, you should have learned that tiny droplets of water suspended in the air after a rainfall can separate the light coming from the sun according to its different wavelengths. You will learn more about this in the next module. When that happens, you see the individual colors that make up the white light.

FIGURE 11.10
The Visible Spectrum

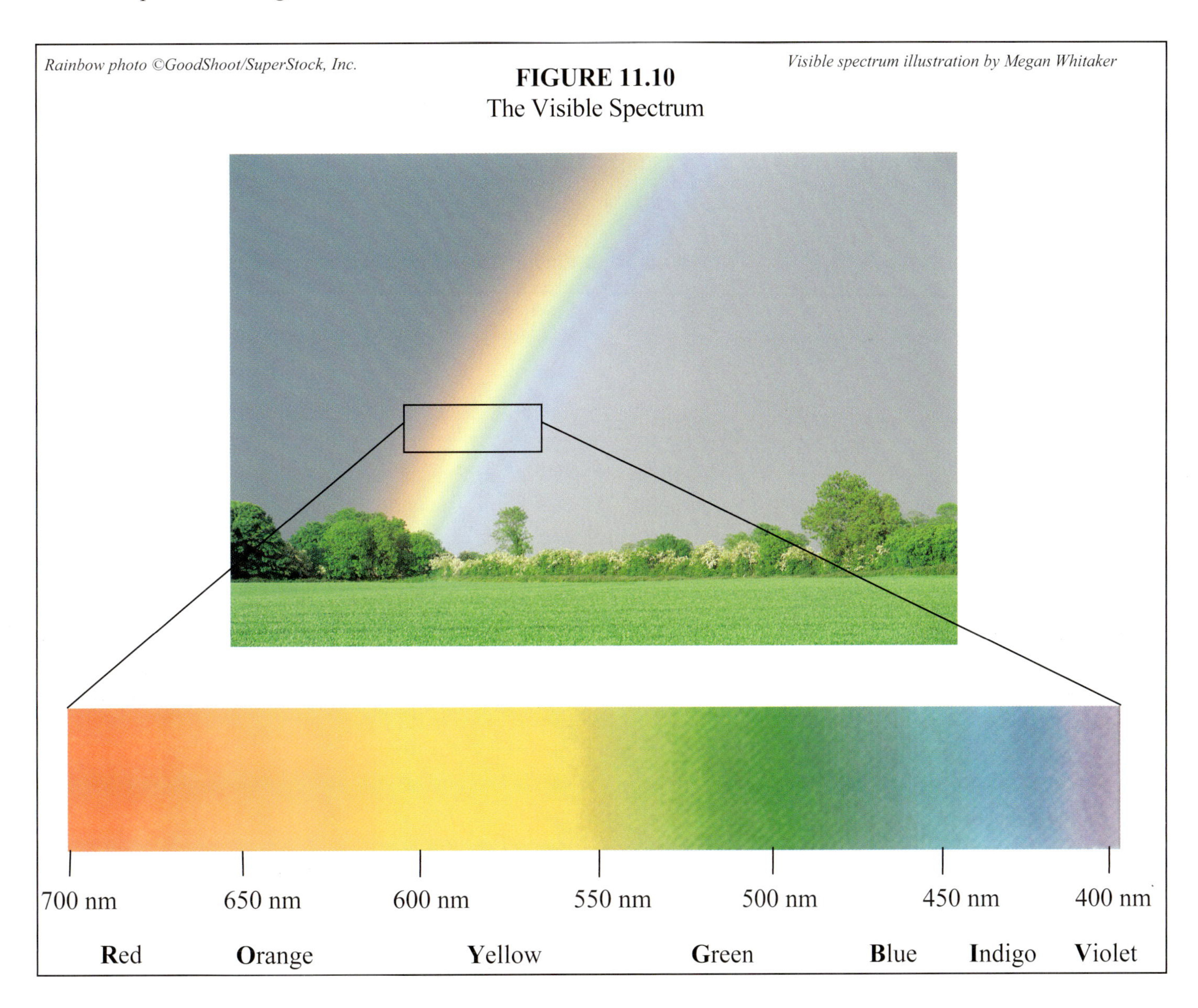

In the metric system, the prefix "nano," which means 10^{-9}, is abbreviated with an "n." Thus, red light, for example, has a wavelength around 700 nm, which is 700×10^{-9} m, or 7×10^{-7} meters. Obviously, then, the wavelengths of visible light are quite small. Now please remember that since we are treating light as a wave, it will obey Equation (11.1), which relates the frequency of the wave to its speed and wavelength. Since we know the speed of light (usually abbreviated as "c"), we could use Equation (11.1) to convert the wavelengths listed in either of the two figures we just studied into frequency. Alternatively, if we are given the frequency of an electromagnetic wave, we could calculate its wavelength. I will not go through an example of how to do that, because Example 11.1 already covers it. However, I will expect you to be able to calculate a light wave's frequency given its wavelength, and vice-versa.

So we see that there is strong evidence that light is a wave that oscillates electric and magnetic fields. This view of light became the standard view in physics until Albert Einstein appeared on the scene. I will discuss what happened then in the next section of this module.

ON YOUR OWN

11.7 A recently-developed laser used in the treatment of skin disorders produces light with a wavelength of 578 nm. According to Figure 11.10, what color is this light? What is its frequency?

11.8 As shown in Figure 11.9, the waves used to send radio signals are electromagnetic waves. I often listen to a radio station that transmits waves with a frequency of 1,240 kiloHertz. What is the wavelength of these waves?

Light as a Particle

Now remember, Young's experiments in the early 1800s gave clear evidence that light is a wave. Thus, that view of light became the dominant view of physics at that time. Many physical phenomena were explained assuming that light is a wave, so physicists at the turn of the century were relatively confident that light is, indeed, a wave. There were, however, a few "oddities" about how light worked that still puzzled physicists.

One of these little "oddities" is called the **photoelectric effect**. In this process, light is shined on a metal surface and, as a result, electrons fly off the metal. These electrons are typically called **photoelectrons**, because they come as a result of light shining on metal. Now, the fact that shining light on metal will liberate electrons was no problem to explain in terms of the idea that light is a wave. After all, waves have energy. Electrons in the metal can absorb that energy and, once they have enough energy to overcome the hold that the metal has on them, they break free of the metal. Typically, physicists use the term **work function** to refer to the energy with which a metal holds on to its electrons. If an electron could absorb enough energy so that it had more energy than the work function of the metal, the electron could break free of the metal.

All of that makes complete sense assuming that light is a wave. What didn't make sense at the time were the following *details* of what happens in the photoelectric effect:

1. **Electrons were ejected within a few billionths of a second after the light was turned on.**

2. **The intensity of the light had *no effect* on the energy of the electrons after they left the metal. The more intense the light, the greater *number* of electrons, but the energy of the electrons was independent of the light intensity.**

3. **Each metal had a unique "cutoff" frequency. If the light used was below this frequency, electrons would not be emitted.**

4. **The maximum kinetic energy of the emitted electrons was directly proportional to the *frequency* of the light used.**

None of the details discussed above were consistent with the idea that light is a wave. After all, if light is a wave, the energy that the light imparts should be proportional to its intensity, not its frequency. The larger the wave, the more energetic it is. After all, think of a sound wave. The sound from thunder is much more energetic than the sound from a stone hitting the ground because the thunder sound wave is much more intense (has a larger amplitude) than the sound wave from the stone hitting the ground. However, since each metal had a "cutoff" frequency under which light could not eject the electrons, and since the maximum kinetic energy of the electrons depended on the frequency of the light and not the intensity, the results of the photoelectric effect indicated that the frequency of light had something to do with its energy.

In addition, the fact that electrons were liberated from the metal so quickly made no sense in terms of thinking about light as a wave. After all, if the electrons needed to absorb energy from the wave, they would have to "soak up" that energy as they encountered the waves. That should take some time. The lower the intensity of the light, the longer time it should take. However, regardless of intensity, photoelectrons were emitted from the metal within just an instant after the light began to hit the metal.

Since these results were so contrary to the idea that light is a wave, you might think at first that physicists would stop thinking of light in that way. However, many experiments (such as Young's double-slit experiment) can only be understood if light is a wave. In the end, there was so much experimental evidence that light was a wave, physicists did not want to give up that notion.

Enter Albert Einstein. Einstein was willing to put aside the notion that light was a wave and simply said that he was willing to assume that a beam of light is something like a stream of particles, much like the water coming from a hose is made up of a stream of individual water molecules. He assumed that light came in little "electromagnetic bundles" called **photons** (foh' tahns), and that the energy of each photon was proportional to the frequency of the light, according to the following equation:

$$E = hf \qquad (11.4)$$

In this equation, "E" is the energy of the light; "f" is its frequency; and "h" is a fundamental constant in nature which is now known as **Planck's constant**, after Max Planck, a German physicist. The value of the constant is 6.63×10^{-34} J·sec.

This view explained all of the experimental data associated with the photoelectric effect. After all, if light was composed of individual photons, it was *collisions* between the photons and the electrons in the metal which transferred energy. When those collisions resulted in more energy going to the electron than the work function of the metal, the electron would break free of the metal. Thus, the electrons should be liberated from the metal almost as soon as the light was turned on, because energy would be transferred almost immediately by the collisions. (This explains observation #1.) Also, the intensity of the light would not mean more energy; it would simply mean more photons.

Thus, more electrons would be emitted, but their energy would not depend on the intensity of the light. (This explains observation #2.)

Since the energy of the photons depends on the frequency, there would be a frequency under which the photons would simply not be able to provide energy above the metal's work function. Thus, there would be a "cutoff" frequency under which the light could not liberate electrons from the metal. Since this would depend on the strength with which the metal holds on to its electrons, the "cutoff" frequency would be different for each type of metal. (This explains observation #3.) Finally, since the photon energy depends on the frequency, the maximum kinetic energy of the electrons should depend on the light frequency as well. (This explains observation #4.)

So, by assuming that light was actually composed of a stream of photons, Einstein could explain the photoelectric effect. It was this discovery that put Einstein into the scientific limelight. Now please realize that Einstein's explanations for the photoelectric effect were not just qualitative, as I discussed above. They could actually predict the data in a quantitative way. I want to give you an example of how Einstein's explanation gives actual numerical predictions for the photoelectric effect, but first, I would like to introduce a new energy unit which is very convenient to use when discussing the photoelectric effect.

When dealing with individual photons of light, the energy unit of a Joule is not a very useful unit. It is simply too big. The energy of a single photon of visible light, for example, is on the order of 10^{-19} Joules. Thus, it is more convenient to use a small unit for energy when dealing with light. That unit is called the **electron volt** (abbreviated as **eV**). You will learn the actual definition of the electron volt in a later module. For right now, I will just give you its relationship to the Joule:

$$1.000 \text{ eV} = 1.602 \times 10^{-19} \text{ J}$$

You need not memorize this relationship. If you need it on a test, I will give it to you. Just notice how small an electron volt is. It is only a tiny, tiny fraction of a Joule.

Now please realize that this is just another energy unit. There is no new physics here. If I have 3.204×10^{-19} J of energy, I can say that I have 2.000 eV of energy. Of course, with this new unit, I can give another value for Planck's constant. After all, Planck's constant has units of J·sec. Since eV is just another energy unit, we can convert Planck's constant from J·sec to eV·sec. When we do that, we find that Planck's constant is 4.14×10^{-15} eV·sec. Now once again, you need not memorize these values for Planck's constant. You will be given them if you need them on a test.

Now that you know this useful energy unit, let's see how Einstein's explanation of the photoelectric effect works.

EXAMPLE 11.4

Sodium metal has a work function of 2.28 eV. If blue light ($\lambda = 4.25 \times 10^{-7}$ m) is shined on sodium, what is the maximum kinetic energy of the electrons that are emitted from the metal?

Remember, the work function tells us how much energy it takes for an electron to escape the metal's grasp on it. If an electron has 2.28 eV, it can break free of the metal, but it will use all of its energy doing so. Thus, it will have no kinetic energy once it escapes. Therefore, the work function is the energy that the electron will lose as it escapes the metal. To determine how much energy the electrons get from the light, we must first figure out the light's frequency. According to Equation (11.1), the frequency can be calculated using the fact that light has a speed of 3.0 x 10^8 m/sec, which you are supposed to have memorized:

$$f = \frac{v}{\lambda} = \frac{3.0 \times 10^8 \ \frac{\cancel{m}}{sec}}{4.25 \times 10^{-7} \ \cancel{m}} = 7.1 \times 10^{14} \ \frac{1}{sec}$$

Now that we know the frequency, we can get the energy of each photon:

$$E = hf = (4.14 \times 10^{-15} \ eV \cdot \cancel{sec}) \cdot (7.1 \times 10^{14} \ \frac{1}{\cancel{sec}}) = 2.9 \ eV$$

Since the work function was given in eV, I decided to get the photon's energy in eV. I could have gotten the energy in Joules, but I would have had to convert at some point. By using Planck's constant in eV·sec, I can avoid the conversion.

Okay, so the photons shining on the metal have 2.9 eV of energy. If a collision occurs and a photon gives *all* of its energy to an electron, the electron will suddenly have 2.9 eV of extra kinetic energy. It can use 2.28 eV to break free of the hold that the metal has on it, leaving the following amount of energy as kinetic energy:

$$KE = 2.9 \ eV - 2.28 \ eV = 0.6 \ eV$$

Now, of course, not every collision will result in all of a photon's energy being transferred to an electron. Thus, electrons can have *less* kinetic energy than that, but the maximum possible electron energy is <u>0.6 eV</u>.

Notice how we analyze the photoelectric effect. The work function of the metal tells us how much energy an electron must expend just to get free of the metal. If the electron does not receive at least that much energy from a photon, it will not escape the metal. If it receives more than that amount of energy, it will escape the metal, but it will expend energy in doing so. Thus, the energy received by the electron minus the work function will be the energy that the electron has left over. We calculate the energy that the electron can get from a photon by using Equation (11.4).

Using this kind of reasoning, Einstein could predict the maximum kinetic energy of the electrons coming off a metal, and the prediction was always correct. He could also predict the minimum frequency for which electrons could be emitted from the metal. This provided strong evidence that Einstein's explanation of the photoelectric effect is correct.

Of course, Einstein's explanation leads to a major problem. If light is truly a stream of photons (as is assumed in the explanation of the photoelectric effect), why does it exhibit wave-like properties, such as the ability to produce interference patterns like those in Young's double-slit experiment? Even today, we do not have a really good answer to that question. Thus, we "fudge" the explanation by giving this phenomenon a name. We say that this illustrates the **particle-wave duality** of light. In some situations, we can best describe light as a wave. In other situations, we can best describe it as a particle. That's just what the experimental evidence shows.

Now believe it or not, the phenomenon of particle-wave duality is *not* a special property of light. Consider, for example, electrons. Electrons are clearly particles. However, if a beam of electrons is forced to pass through a screen with two slits it in, guess what? An interference pattern is produced! If you detect the electrons on the other side of the slits, you will find patches where there are a lot of electrons and patches where there are no electrons at all. Thus, electrons also have wave-like properties, despite the fact that we generally think of them as particles.

So, is light a particle or a wave? Are electrons particles or waves? What about a baseball? Is it a particle or a wave? Actually, there are no solid answers to these questions. The best we can do is say that light, electrons, and even baseballs exhibit both wave-like and particle-like characteristics. In general, it is convenient to think of light as a wave for most applications, and it is convenient to think of electrons and baseballs as particles for most applications. However, the reigning view in physics today is that *everything* in creation has dual characteristics of both a particle and a wave. Thus, although it might be convenient to think of something as a particle, you cannot forget that it does have wave-like characteristics. In the same way, although it is convenient to think of something else as a wave, you cannot forget that it has particle-like characteristics as well. If you take advanced physics, you will learn more about this interesting idea.

ON YOUR OWN

11.9 Cesium has a work function of 2.14 eV. If green light ($\lambda = 501$ nm) shines on cesium, what is the maximum kinetic energy that the electrons will have as they leave the metal?

11.10 A metal emits electrons when illuminated with light. However, if the light wavelength is greater than 195 nm, no electrons are emitted. What is the work function of the metal?

Biographies of Two Important Physicists

As a part of our discussion of light, I mentioned a few physicists. Two of them are particularly important in the development of physics as we understand it today. Thus, I think it is important for you to learn a little bit about them. I did not want to break up the narrative of the previous sections by giving you their biographies. However, now that we are done with the physics that I want to discuss in this module, I thought I would cover these two important men. Now don't worry about memorizing any of this information. I do not plan to test you on it. However, I do think that these two physicists are important enough to discuss.

I want to start with James Clerk Maxwell. As I said previously, Maxwell developed equations that unified the electric and magnetic force into the electromagnetic force. It is now understood that those equations tell us that the electromagnetic force is actually mediated by light and that light is an electromagnetic wave. As a result of this monumental accomplishment, science historians typically call Maxwell the founder of modern physics. A picture and brief biography of Maxwell are given below.

Original photograph in the possession of Sir Henry Roscoe, courtesy AIP Emilio Segrè Visual Archives

FIGURE 11.11
James Clerk Maxwell

James Clerk Maxwell was born in Edinburgh, Scotland in 1831. He was educated as a mathematician at the University of Edinburgh and the University of Cambridge. He spent many years as a university professor but devoted himself purely to research after his father's death in 1865. Although his most noted accomplishment was the unification of the electric and magnetic forces into a single, electromagnetic force, Maxwell made several other contributions to science. He was the first to develop an explanation for Saturn's rings, which was confirmed by the Pioneer spacecraft in the 1970s. He also made extensive contributions to our understanding of the behavior of gases. Maxwell was a devout Christian who frequently started classes and experiments with prayer. My favorite of his quotes is, "I believe…that Man's chief end is to glorify God and to enjoy him for ever." (Lewis Campbell and William Garnet, *The Life of James Clerk Maxwell* [London: MacMillan and Co. 1882], 158)

Of course, probably the most well-known scientist of the twentieth century is Albert Einstein. A picture of him and a brief biography are given below.

Image in the public domain

FIGURE 11.12
Albert Einstein

Albert Einstein was born in Ulm, Germany in 1879. He did not talk until the age of three, but by age twelve, he had taught himself Euclidean geometry! We have covered one of his lesser-known accomplishments: an explanation of the photoelectric effect. He is best known for his special and general theories of relativity, which form the basis of much of the physics that is done today. He won the Nobel Prize in physics in 1921. Although he was a pacifist, he decided to reject pacifism when Hitler came to power. In 1939, he worked with other physicists on a famous letter to President Franklin D. Roosevelt, discussing the possibility of making an atomic bomb and suggesting that Germany was already working on such a bomb. The letter bore only Einstein's signature, and although it caused Roosevelt to order the military to embark on making the atomic bomb, Einstein himself did none of the work that actually led to the construction of the bomb.

ANSWERS TO THE "ON YOUR OWN" PROBLEMS

11.1 This problem is very similar to the example. Given the wavelength, we can calculate the frequency with Equation (11.1). To do this, however, we need the speed. We get that from Equation (11.2).

$$v = (331.5+0.606 \cdot T)\ \frac{m}{sec} = (331.5+0.606 \cdot 15.0)\ \frac{m}{sec} = 340.6\ \frac{m}{sec}$$

Now that we have speed, we can use Equation (12.1)

$$f = \frac{v}{\lambda} = \frac{340.6\ \frac{\cancel{m}}{sec}}{1.0\ \cancel{m}} = 3.4 \times 10^{2}\ \frac{1}{sec}$$

The frequency, then, is <u>3.4×10^2 Hz</u>. Since this number is lower than 440 Hz, this sound has a <u>lower pitch than the "standard A."</u>

11.2 In this problem, we are asked to determine the temperature. The only equation we have with temperature in it is Equation (11.2). To use this equation, however, we would need to know the speed. Since we are given both frequency and wavelength, we can get that from Equation (11.1).

$$f = \frac{v}{\lambda}$$

$$501\ \frac{1}{sec} = \frac{v}{0.697\,m}$$

$$v = (501\ \frac{1}{sec}) \cdot (0.697\ m) = 349\ \frac{m}{sec}$$

Now that we have the speed, getting the temperature is a snap:

$$v = (331.5 + 0.606 \cdot T)\ \frac{m}{sec}$$

$$349\ \frac{\cancel{m}}{\cancel{sec}} = (331.5 + 0.606 \cdot T)\ \frac{\cancel{m}}{\cancel{sec}}$$

$$T = \frac{349 - 331.5}{0.606} = 3.0 \times 10^{1}$$

Notice how the units canceled here. The unit "m/sec" canceled because it was with each term on both sides of the equation. In the end, then, there are no units. However, the way this equation is designed, we know that temperature is always in degrees Celsius. Thus, the temperature is <u>3.0×10^1 °C</u>. Now remember how we have to do the significant figures here. Since the numerator has a difference, we must use the rules for addition and subtraction there. Since 349 goes out to the ones place, the answer must be reported to the ones place. Thus, the 17.5 must be rounded to 18. Then, when you divide, you must round to two significant figures, because 18 has only two significant figures.

11.3 To solve this problem, we will have to use Equation (11.3):

$$f_{observed} = \left(\frac{v_{sound} \pm v_{observer}}{v_{sound} \pm v_{source}} \right) \cdot f_{true}$$

The observer in this case is standing still, so his speed is zero. However, the source of the sound (the train's horn) is moving *away from* the observer. Thus, v_{source} = 35.0 m/sec. What about v_{sound}? We need to use the temperature to calculate that:

$$v = (331.5 + 0.606 \cdot T) \frac{m}{sec} = (331.5 + 0.606 \cdot 25.0) \frac{m}{sec} = 346.7 \frac{m}{sec}$$

Now all we have to do is determine whether to add or subtract v_{sound}. Well, the horn is moving away from the observer. Thus, he will hear a *lower* frequency, since the waves will be spread out as the train moves away from him. Since v_{source} is in the denominator of the fraction, we will have to *add* it to v_{sound} so that the denominator of the fraction gets bigger, making the value of the fraction smaller:

$$f_{observed} = \left(\frac{v_{sound} \pm v_{observer}}{v_{sound} \pm v_{source}} \right) \cdot f_{true}$$

$$f_{observed} = \left(\frac{346.7 \frac{\cancel{m}}{\cancel{sec}}}{346.7 \frac{\cancel{m}}{\cancel{sec}} + 35.0 \frac{\cancel{m}}{\cancel{sec}}} \right) \cdot (411 \text{ Hz})$$

$$f_{observed} = 373 \text{ Hz}$$

The man hears the train's horn with a frequency of 373 Hz.

11.4 We will use Equation (11.3) to solve this problem, but we will be solving for the speed of the observer, not the frequency. We still need to know the speed of sound, however. We can use Equation (11.2) for that:

$$v = (331.5 + 0.606 \cdot T) \frac{m}{sec} = (331.5 + 0.606 \cdot 25.0) \frac{m}{sec} = 346.7 \frac{m}{sec}$$

We also have to determine whether to add or subtract $v_{observer}$. Since the woman is moving towards the factory, the sound waves from the horn will be bunched up. This means she will hear the frequency as higher than the true frequency. Since $v_{observer}$ is in the numerator of the equation, this tells us that we must add $v_{observer}$. That will increase the value of the fraction, which will increase the frequency.

$$f_{observed} = \left(\frac{v_{sound} \pm v_{observer}}{v_{sound} \pm v_{source}} \right) \cdot f_{true}$$

$$511 \text{ Hz} = \left(\frac{346.7 \frac{m}{sec} + v_{observer}}{346.7 \frac{m}{sec}} \right) \cdot (494 \text{ Hz})$$

$$346.7 \frac{m}{sec} + v_{observer} = \frac{511 \text{ Hz}}{494 \text{ Hz}} \cdot 346.7 \frac{m}{sec}$$

$$v_{observer} = \frac{511 \cancel{\text{Hz}}}{494 \cancel{\text{Hz}}} \cdot 346.7 \frac{m}{sec} - 346.7 \frac{m}{sec} = 12 \frac{m}{sec}$$

The woman is traveling towards the factory at 12 m/sec.

11.5 As I explained in the example, we assume that the light reaches your eyes instantaneously; thus, the time delay is the time that it takes for the sound to travel the distance between you and where the lightning was formed. To determine that distance, we need to determine the speed:

$$v = (331.5 + 0.606 \cdot T) \frac{m}{sec} = (331.5 + 0.606 \cdot 20.0) \frac{m}{sec} = 343.6 \frac{m}{sec}$$

Now that we know the speed, we can determine the distance:

$$\text{distance} = \text{speed} \times \text{time} = (343.6 \frac{m}{\cancel{sec}}) \times (2.5 \cancel{sec}) = 860 \text{ m}$$

The lightning was formed 860 m away from your position.

11.6 Why do you see the smoke rising from the gun once it is fired? Because light reflects off it and hits your eyes. Since light travels so quickly, you see the smoke pretty much the instant the gun fires. However, the sound takes more time to travel. The time delay will be the result of that travel time. To figure out what the time delay is, then, we need the speed of sound at this temperature:

$$v = (331.5 + 0.606 \cdot T) \frac{m}{sec} = (331.5 + 0.606 \cdot 25) \frac{m}{sec} = 347 \frac{m}{sec}$$

Now that we have the speed, we can determine the time it takes for the sound to travel 402 meters:

$$\text{distance} = \text{speed} \times \text{time}$$

$$\text{time} = \frac{\text{distance}}{\text{speed}} = \frac{402 \text{ m}}{347 \frac{\text{m}}{\text{sec}}} = 1.16 \text{ sec}$$

You will hear the gun 1.16 seconds after you see the smoke.

11.7 In this problem, we are first asked the color of the light. All we have to do is look at Figure 11.10, which shows that wavelengths of just under 600 nm make up yellow light. You need not memorize the wavelengths of light and their corresponding color. I just wanted you to get used to the idea of identifying a wavelength as a color.

Next, we are asked to calculate frequency. This can be done with Equation (11.1), as long as we know the wave's speed. You are expected to memorize the fact that the speed of light is 3.0×10^8 m/sec. In order to get the units to work right, you will need to convert 578 nm into m. I assume you can do that by now:

$$f = \frac{v}{\lambda}$$

$$f = \frac{3.0 \times 10^8 \frac{\text{m}}{\text{sec}}}{5.78 \times 10^{-7} \text{ m}} = 5.2 \times 10^{14} \frac{1}{\text{sec}} = 5.2 \times 10^{14} \text{ Hz}$$

The frequency of the wave is 5.2×10^{14} Hz. Remember what this means: 5.2×10^{14} crests hit your eye every second when you look at yellow light!

11.8 In this problem, we are given frequency and asked to calculate wavelength. Once again, this uses Equation (11.1) with the fact that the speed of light is 3.0×10^8 m/sec. However, so that the units will work properly, we must convert 1240 kiloHertz into Hertz. I assume you can do that by now:

$$f = \frac{v}{\lambda}$$

$$\lambda = \frac{v}{f} = \frac{3.0 \times 10^8 \frac{\text{m}}{\text{sec}}}{1{,}240{,}000 \text{ Hz}} = \frac{3.0 \times 10^8 \frac{\text{m}}{\text{sec}}}{1{,}240{,}000 \frac{1}{\text{sec}}} = 240 \text{ m}$$

Notice that to show you how the units worked out, I had to replace the unit "Hz" with its definition, "1/sec." The wavelength of the radio waves that this station emits, then, is 240 m.

11.9 Remember, the work function tells us how much energy it takes for an electron to escape the metal's grasp on it. Therefore, the work function is the energy that the electron will lose as it escapes the metal. To determine how much energy the electrons get from the light, we must first figure out the light's frequency. We will use Equation (11.1) for that:

$$f = \frac{v}{\lambda} = \frac{3.0\times 10^{8}\ \frac{\cancel{m}}{sec}}{5.01\times 10^{-7}\ \cancel{m}} = 6.0\times 10^{14}\ \frac{1}{sec}$$

Now that we know the frequency, we can get the energy of each photon:

$$E = hf = (4.14\times 10^{-15}\ eV\cdot \cancel{sec})\cdot(6.0\times 10^{14}\ \frac{1}{\cancel{sec}}) = 2.5\ eV$$

Okay, so the photons shining on the metal have 2.5 eV of energy. If a collision occurs and a photon gives *all* of its energy to an electron, the electron will suddenly have 2.5 eV of extra kinetic energy. It can use 2.14 eV to break free of the hold that the metal has on it, leaving the following amount of energy as kinetic energy:

$$KE = 2.5\ eV - 2.14\ eV = 0.4\ eV$$

Now, of course, not every collision will result in all of a photon's energy being transferred to an electron. Thus, electrons can have *less* kinetic energy than that, but the maximum possible electron energy is <u>0.4 eV</u>.

11.10 To solve this problem, we have to think about what the work function means. It is the energy with which the metal holds its electrons. If the electrons get that energy *or more*, they will escape the metal. If they have less than that energy, they cannot escape the metal.

In the problem, electrons are emitted as long as the wavelength of light is less than or equal to 195 nm. However, when it gets greater than 195 nm, no more light is emitted. This must mean that light of 195 nm gives the electrons just enough energy to escape, but no extra. Thus, if we calculate the energy of light with a wavelength of 195 nm, we will have the work function of the metal. First, then, we need to calculate the light's frequency:

$$f = \frac{v}{\lambda} = \frac{3.0\times 10^{8}\ \frac{\cancel{m}}{sec}}{1.95\times 10^{-7}\ \cancel{m}} = 1.5\times 10^{15}\ \frac{1}{sec}$$

Now we can calculate the light's energy:

$$E = hf = (4.14\times 10^{-15}\ eV\cdot \cancel{sec})\cdot(1.5\times 10^{15}\ \frac{1}{\cancel{sec}}) = 6.2\ eV$$

This is the energy of the photons hitting the electrons in the metal. Since that is the *minimum* energy (maximum wavelength) for which electrons are emitted, that tells us the work function of the metal is <u>6.2 eV</u>.

REVIEW QUESTIONS

1. What is the difference between a transverse wave and a longitudinal wave? Which type of wave best represents sound that is transmitted through air? Which best represents light?

2. Two sounds of equal pitch are heard. The first is quiet, and the second is loud. Compare the frequencies of the waves. Compare the amplitudes of the waves. Compare their wavelengths.

3. Suppose a man is in a car that is stopped at a stoplight. Another car starts accelerating away from his car, blowing the horn. As the other car accelerates, what will happen to the pitch of the horn as heard by the man whose car is stopped at the stoplight?

4. A stationary siren emits a sound wave with a frequency of 511 Hz. If you hear the siren with a frequency of 555 Hz, are you moving towards the siren or away from it?

5. Name at least two differences between sound waves and light waves.

6. What is the particle-wave duality of light?

7. If light wave A has a longer wavelength than wave B, which wave has the higher frequency?

8. In the question above, which light wave has more energy?

9. Light travels faster in substance A than in substance B. Which substance has the highest density?

10. Light of a certain wavelength is shone on unknown metal A. Electrons are emitted from the metal, and the maximum kinetic energy of those electrons is measured. The same light is then shone on metal B, and electrons are also emitted. The maximum kinetic energy of these electrons is *lower* than the maximum kinetic energy of the electrons emitted from metal A. Which metal has the larger work function?

Cartoon by Speartoons

PRACTICE PROBLEMS

($1 \text{ nm} = 10^{-9}$ m, $h = 4.14 \times 10^{-15}$ eV·sec)

1. What is the speed of sound in air that has a temperature of 90.0 °C?

2. The frequency of a sound is 1621 Hz. If the temperature is 30.0 °C, what is the wavelength of the sound wave?

3. What is the temperature if a 545-Hz sound wave has a wavelength of 0.651 m?

4. You decide to measure the speed of sound with the help of a friend. The friend stands 450.0 m from you and bashes two cymbals together. You determine that the time between when you see the cymbals crash together and when you hear them is 1.29 seconds. What is the speed of sound, according to your measurement?

5. The thunderclap from a lightning flash is heard 4.2 seconds after the flash is observed. If the temperature is 20.0 °C, how far away was the lightning formed?

6. A flutist is playing a "middle C," which has a frequency of 278 Hz. The flutist is standing in the bed of a pickup truck, which is moving away from you with a speed of 25.6 m/sec. What frequency do you hear? (T = 25.0 °C)

7. A car's horn normally blows with a constant frequency of 551 Hz. You are standing in your yard and hear the horn with a frequency of 578 Hz. At what speed is the car moving? (T = 25.0 °C)

8. The wavelength of an indigo light wave is 442 nm. What is its frequency? What is its energy?

9. The work function of gold is 5.31 eV. What is the lowest frequency of light that will result in electrons being liberated from the metal? What is the wavelength of the light? If light of wavelength *lower* than that wavelength were shone on the metal, would electrons be liberated?

10. Light of wavelength 211 nm is shone on copper, which has a work function of 4.94 eV. What is the maximum kinetic energy of the electrons that are emitted from the metal?

MODULE #12: Geometric Optics

Introduction

In the previous module, you learned about the nature of light. In this module, you will learn about the propagation and behavior of light. A study of these phenomena is called **optics**. Now remember from the previous module that although light is often considered a wave, it can also be considered a particle. In this module, it will be most helpful to think about light as the latter. It will be easier to understand the concepts presented in this module by thinking of light as a beam of particles emanating from a source and traveling straight until forced to change path. This is easier for you to visualize, because that's what you're used to. If you turn on a flashlight, it emits a beam. That beam travels straight until it runs into a wall, mirror, or other obstruction. When talking about light in this manner, physicists say that a beam of light is a huge number of photons, all traveling in the same direction.

Now please understand that it is not *necessary* for you to think of light as a particle in order to understand what we will study in this module. It turns out that all of the phenomena that we will cover can also be understood in terms of the wave nature of light. However, picturing that in your mind will be difficult. Thus, to make it easy on you, I will usually discuss light in terms of its particle nature while we are in this module.

The Law of Reflection

We will start our discussion of optics with the Law of Reflection. Perform the following experiment to get an idea of what this law is all about.

EXPERIMENT 12.1
The Law of Reflection

Supplies:

- Safety goggles
- A flat mirror. The mirror can be very small, but it needs to be flat. You can always tell if a mirror is flat by looking at your reflection in it. If the image you see in the mirror is neither magnified nor reduced, the mirror is flat.
- A white sheet of paper
- A pen
- A protractor
- A ruler
- A flashlight
- Black construction paper or thin cardboard
- Tape
- A dark room

1. Take the construction paper and cut it into a circle that fits the face of the flashlight. Make it so that when the circle is taped to the face of the flashlight, little or no light will be able to escape.
2. At the edge of the circle, cut a small, thin slot. The circle of paper should look something like the drawing on the next page:

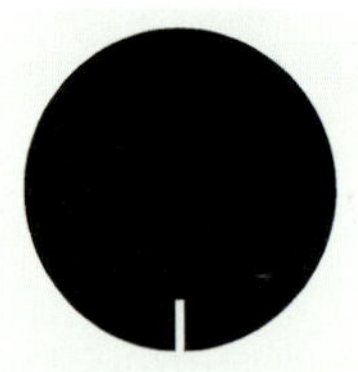

3. Now tape the circle to the face of the flashlight, so that light escapes only through the slot.
4. Lay the white piece of paper on a rectangular table or desktop so that its edge is even with the straight edge of the table.
5. Tape the paper down so that it does not move from this position.
6. Use the protractor the way you learned in geometry to make a line that is perpendicular to the edge of the table and is centered on the paper. In the end, your setup should look like this:

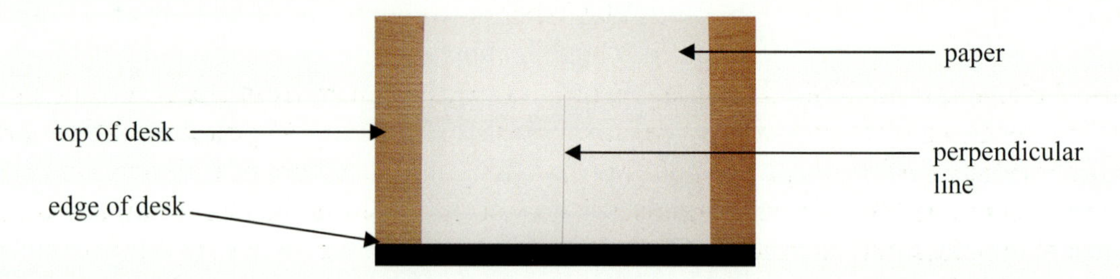

7. Push the mirror up against the edge of the table so that the line you drew is centered on the mirror and perpendicular to it. Tape the mirror to the edge of the table so it stays there.
8. Turn on the flashlight and turn out the lights.
9. Hold your flashlight so that the slot is on the bottom of the face, touching the paper. Play with the tilt of the flashlight until the light coming from the slot causes a beam on the paper which hits the mirror at the same point that the line touches the mirror. You should then see the beam reflect off the mirror back onto the paper, as show in the photograph below:

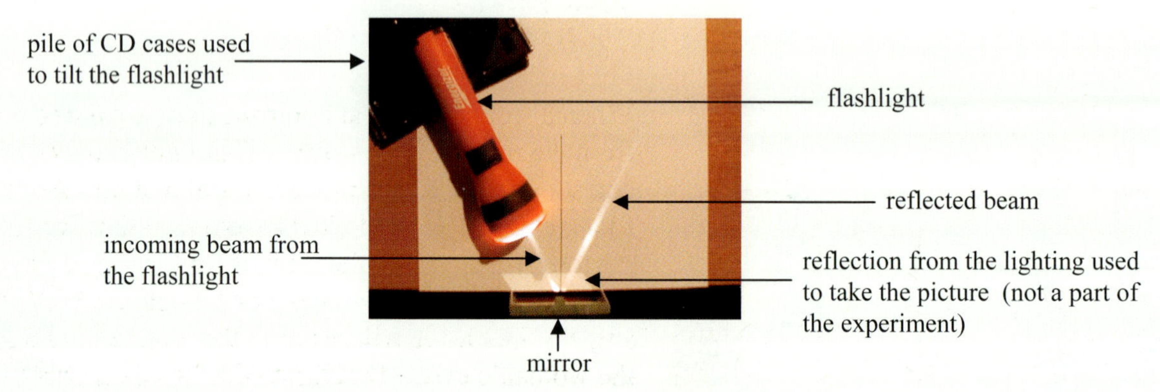

10. Use your pen to carefully trace the path of the beam as it travels from the flashlight and reflects off the mirror.
11. Turn on the lights.
12. Use your protractor to measure the angle of the line representing the path of the incoming beam relative to the perpendicular line that you originally drew.
13. Use your protractor to measure the angle of the line representing the reflected beam relative to the same perpendicular line.
14. Do the experiment twice more, changing the positioning of the flashlight so that the angle that the incoming beam makes with the perpendicular line is different each time. In each case, compare the angle made by the incoming beam to that of the reflected beam.
15. Clean up your mess.

What did you see in the experiment? Within experimental error, the angle that each incoming beam made with the perpendicular line should have equaled the angle that each reflected beam made with the perpendicular line. In general, that's what will happen when light reflects off a surface, and it's called the **Law of Reflection**.

The Law of Reflection - The angle of reflection equals the angle of incidence.

In the definition, the angle of reflection refers to the angle of the reflected light beam relative to a line perpendicular to the mirror, while the angle of incidence refers to the angle of the incoming light, relative to the same perpendicular line.

Flat Mirrors

Now believe it or not, this simple law is responsible for how mirrors work. To understand why, you first need to understand why we see things in the first place. As you look at the words on this page, light reflects off the page and up towards your eye. The pattern of reflected light is read by your eye and is converted to electrical impulses that are sent to your brain. Your brain then converts these electrical impulses into an image. That's how you see. That's also why you cannot see things without the aid of light. Your eyes cannot send anything to your brain unless light reflects off the thing that you are observing and then enters your eyes. Only then can a message be sent to your brain so that it can form an image.

With this in mind, consider a woman looking at herself in a mirror. Why does she see her foot, for example? Study Figure 12.1.

FIGURE 12.1
How a Reflection Is Made in a Flat Mirror

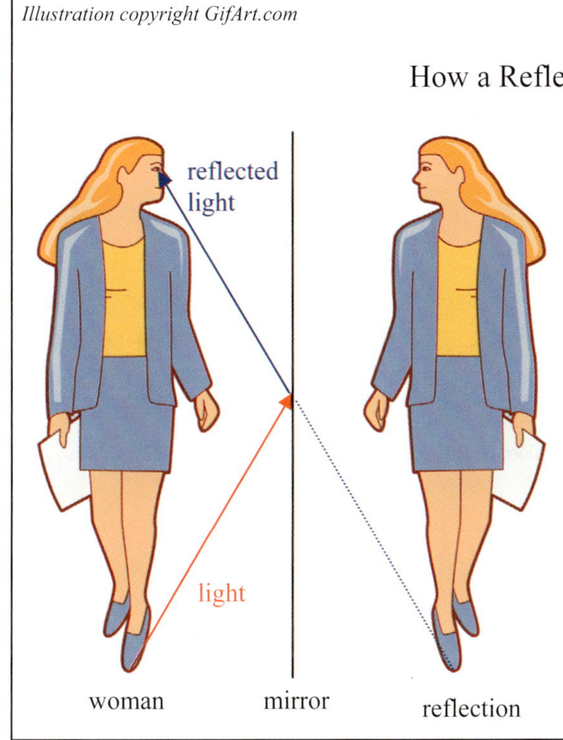

When a woman looks at a mirror, light is shining on her from the room's light source. That light bounces off her, and some of it hits the mirror. Light that bounces off her foot (for example) and hits the mirror will reflect off the mirror. According to the Law of Reflection, the angle of incidence will equal the angle of reflection. Well, a certain angle of incidence will result in the reflected light hitting the woman's eyes. That light will then cause her brain to form an image of her foot. If you extrapolate the reflected light ray backwards (illustrated by the dashed line in the figure), you will see that it appears to be coming from *behind* the mirror. Thus, that's where the image of her foot appears to be as well. Of course, this happens at every point on her body, so she sees a complete image of herself, but that image appears to be behind the mirror.

That's how images are formed in a mirror. The light that reflects off a mirror is detected by the eye, and the brain forms an image where the light appears to have originated. The image is, of course, fake. It is simply a result of the fact that reflected light appears to be coming from an extrapolated point behind the mirror. When an image is formed in this way, we call it a **virtual image**.

Virtual image - An image formed as the result of extrapolating light beams

It turns out that when the mirror is flat, the image that your brain constructs in the mirror is the same size as the object, and it is the same distance behind the mirror as the object is in front of the mirror. This is because of the law of reflection. Since the beam of light reflects off the mirror at the same angle that it hits the mirror, the reflected light beam is precisely opposite the original light beam. This forms an image that is precisely opposite the object in front of the mirror. Thus, the images that we see in a flat mirror are the same size as the objects themselves. They also appear to be as far behind the mirror as the original object is away from the mirror.

Spherical Mirrors

We know from experience that some mirrors magnify the objects put in front of them. Many women, for example, have a magnifying makeup mirror that allows them to see even tiny flaws in their makeup job. How do these mirrors work? Magnifying mirrors are not flat. Instead, they are curved so that the mirror bends away from the person looking at it (see the drawing below). When a mirror is curved in this way, it is called **concave**, and the nature of this curve changes the situation dramatically. Consider, for example, a beam of light hitting a concave mirror. It will be reflected as shown below.

FIGURE 12.2

A Light Beam Reflecting from a Concave Spherical Mirror

Illustration by Megan Whitaker

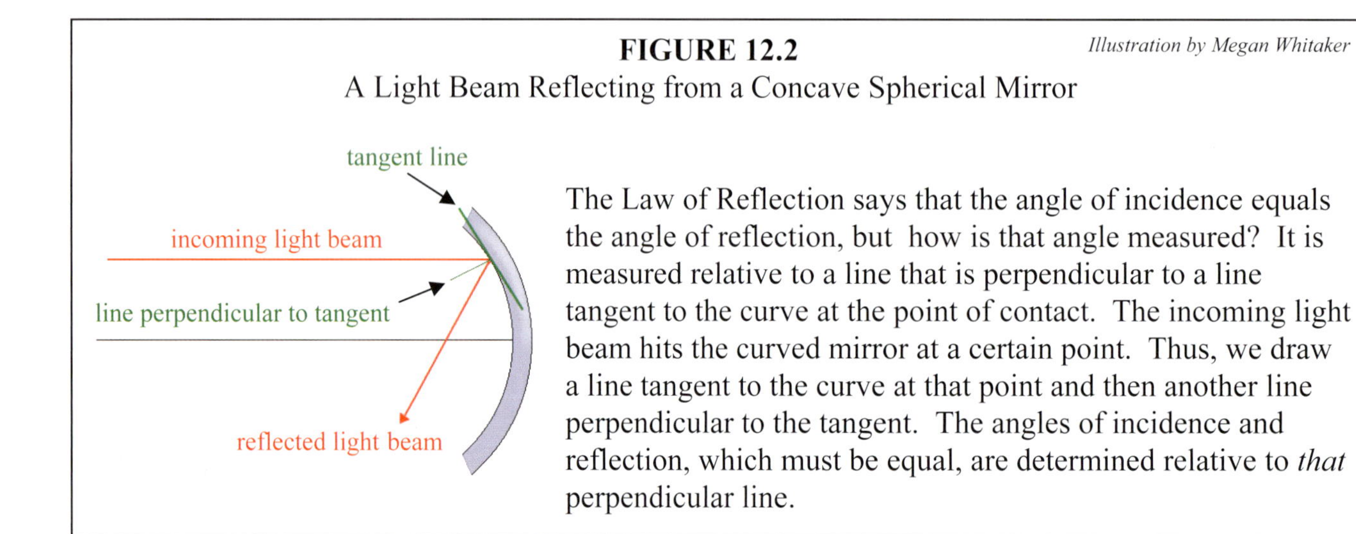

The Law of Reflection says that the angle of incidence equals the angle of reflection, but how is that angle measured? It is measured relative to a line that is perpendicular to a line tangent to the curve at the point of contact. The incoming light beam hits the curved mirror at a certain point. Thus, we draw a line tangent to the curve at that point and then another line perpendicular to the tangent. The angles of incidence and reflection, which must be equal, are determined relative to *that* perpendicular line.

Now, what happens if another beam of light is headed straight towards the mirror but below the center of the mirror? It will be reflected so that once again, the angle of incidence (defined by a line perpendicular to the tangent at the point of contact) is equal to the angle of reflection. Because the mirror is concave, however, the result will be as shown in Figure 12.3.

Illustration by Megan Whitaker

FIGURE 12.3
Two Light Beams Reflecting from a Concave Spherical Mirror

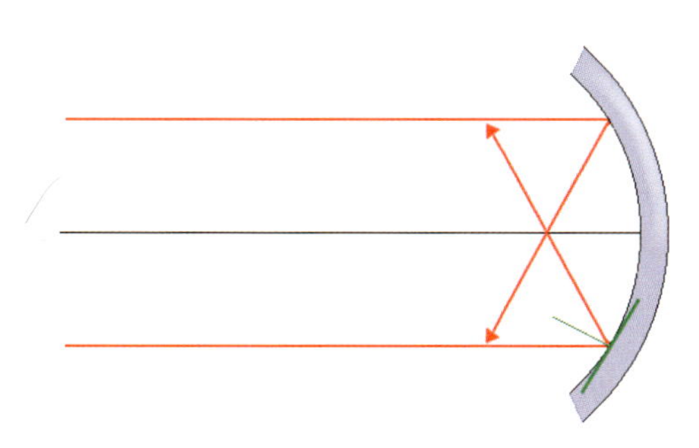

Now there are two light beams reflecting from the mirror. The top light beam is the same as the one drawn earlier. The bottom one is the new one. Note that this time, the line perpendicular to the tangent is pointed diagonally up and to the left. Thus, when the new incoming light beam reflects off the mirror, it must move diagonally upwards in order to keep the angle of incidence equal to the angle of reflection. As a result, the two light beams meet at a point that is level with the center of the mirror.

Do you see what happens? The light beams reflect off the curved surface so that they are directed to the same point. It turns out that *any* light beam traveling towards the mirror and parallel to the black line in the figure will be reflected so that it will pass through this point. We call that point the **focal point** of the mirror, because the mirror tends to focus light towards that point. Physicists call the black line in the figure the **optical axis** of the mirror. When I refer to light beams traveling parallel to the mirror's axis, I will call them "horizontal" light beams. If the mirror is tilted, they will not really be horizontal, but in all of the drawings in this text, the axis of the mirror will be horizontal, so light beams traveling parallel to it will also be horizontal.

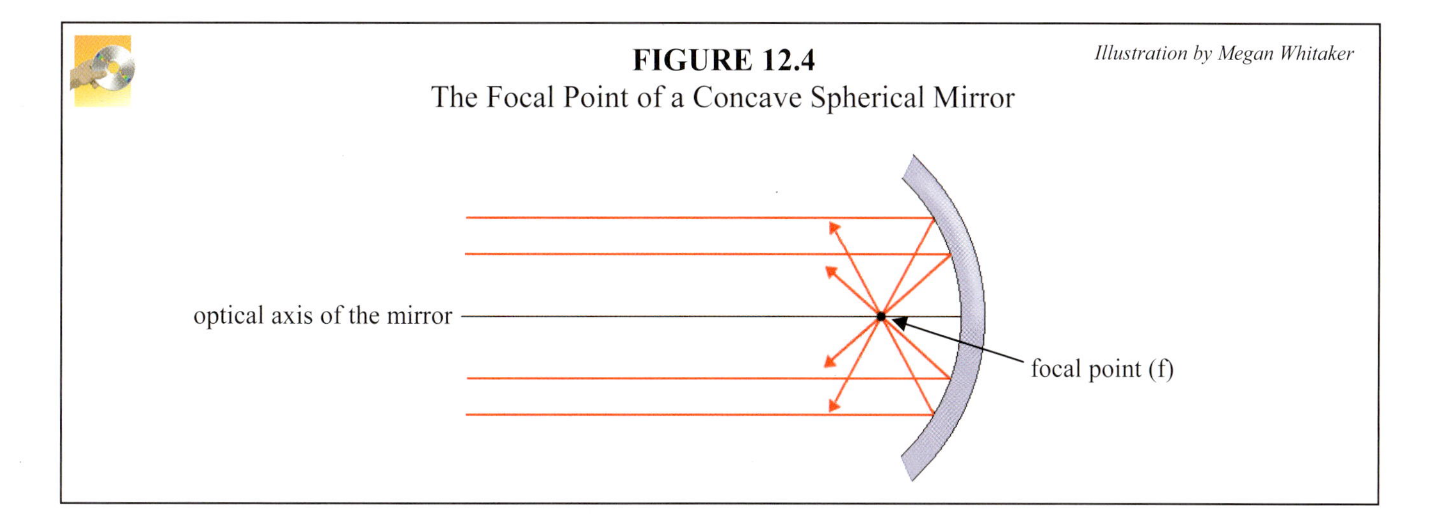

FIGURE 12.4
The Focal Point of a Concave Spherical Mirror

Illustration by Megan Whitaker

What determines where the focal point of a mirror is? Well, based on the drawings above, it should be clear that the focal point will depend on how curved the mirror is. The more curved the mirror, the closer the focal point will be to that mirror. In fact, a very simple equation relates the focal point to the radius of curvature of a spherical mirror:

$$f = \frac{R}{2} \qquad (12.1)$$

In this equation, "f" stands for the distance from the mirror to its focal point, and "R" represents the radius of curvature of the mirror.

Before I go into how you can use the focal point of such a mirror to predict the size and location of an image formed by the mirror, I need to tighten up my discussion a bit. Notice that throughout most of my discussion, I have called these mirrors "curved." It is important that you understand exactly *how* these mirrors are curved. For a curved mirror to focus all horizontal light beams to a single focal point, the mirror must be shaped as a **parabola**.

You should have covered parabolas in math class already. The general equation for a parabola that has its vertex at the origin of the Cartesian coordinate plane and is pointed to the left is as follows:

$$y^2 = -4dx \qquad (12.2)$$

Do you remember what "d" is called? It is called the *focus* of the parabola. What is the focus? Well, look at the graph of Equation (12.2).

FIGURE 12.5
Graph of a Parabola

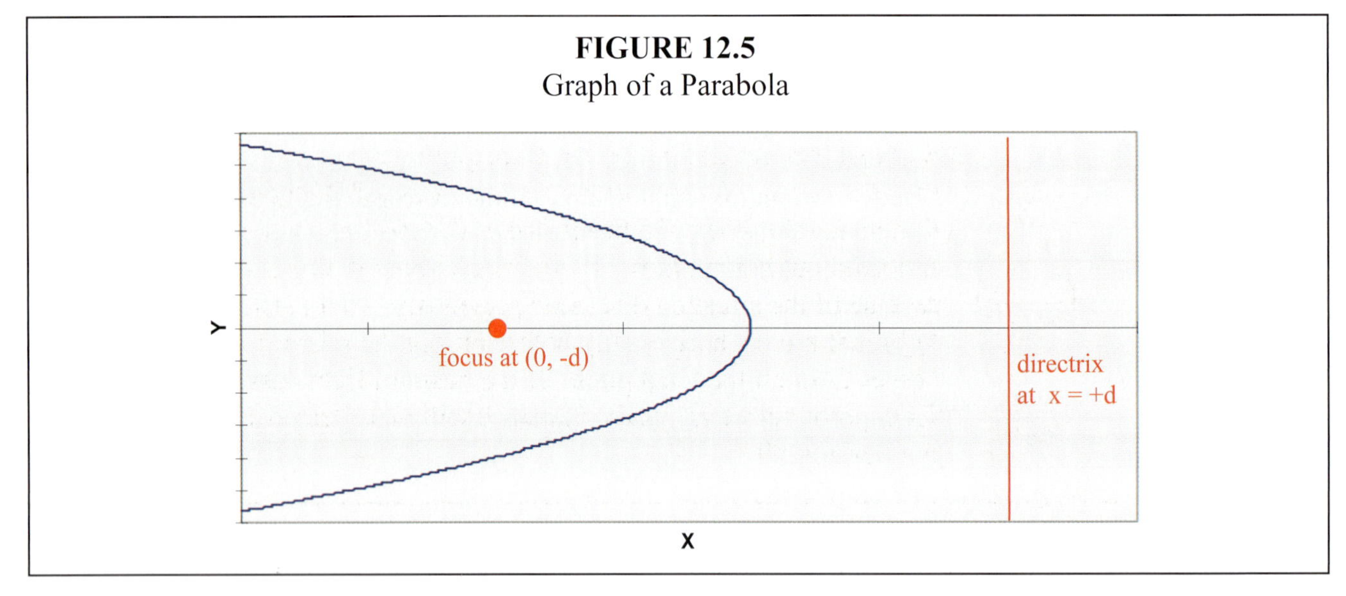

As you should have learned in mathematics, a parabola is defined as the set of all points that are equidistant from the focus and the directrix.

Now it turns out that for a curved mirror to focus all of the rays to a single point, it must be shaped like a parabola. If a mirror is parabolic, the focal point will be the same as the focus of the parabola. Notice, however, that in the figures, I have not labeled the mirrors I have been discussing as parabolic. Instead, I have labeled them as spherical. Why? Well, although parabolic mirrors direct horizontal light beams to their focus, they are hard to manufacture. Thus, we tend to use them only when the optics in a situation must be very precise. However, spherical mirrors are much easier to make and, as long as the arc of the sphere is not too large, it approximates the shape of a parabola. Thus, even though a spherical mirror does not focus light as perfectly as parabola, we nevertheless use spherical mirrors in our day-to-day life because they are so much easier to make than parabolic mirrors.

The fact that spherical mirrors do not focus horizontal light beams precisely at the focus leads to distortions in the image formed by the mirror. The closer light from the object strikes the outer edges of the mirror, the more noticeable the distortions are. This effect is called **spherical aberration**, and it is the result of the fact that a spherical mirror only approximates a parabolic mirror. If a parabolic mirror is used, there is no spherical aberration, and the image is sharper.

Ray Tracing in Concave Spherical Mirrors

Neglecting spherical aberrations, we know that a light beam traveling horizontally will be reflected through the focal point when it strikes a spherical mirror. However, what about light that is not traveling horizontally? If a beam of light travels diagonally, where will it be reflected? Well, as you might expect, it depends on exactly what angle the beam of light makes relative to the horizontal. However, there are two special ways that a beam of light traveling diagonally can approach a curved mirror. If light approaches a mirror from these two special paths, we can easily predict where it will be reflected.

Although I won't go through the geometry that proves it, the first special path should make sense, given what we have discussed above. It is given in the figure below.

Illustration by Megan Whitaker

FIGURE 12.6

A Light Beam Traveling through the Focal Point before Hitting a Concave Spherical Mirror

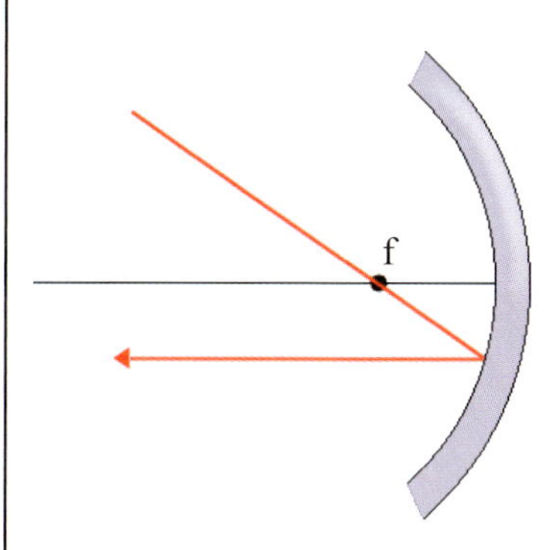

Suppose a beam of light passes through the mirror's focal point *before* it hits the mirror. When this happens, the Law of Reflection works out so that no matter where the beam strikes the mirror, it will always be reflected out horizontally. If you think about it, this is exactly the reverse of the situation discussed previously. In that situation, a beam of light that travels horizontally before hitting the mirror reflects so that it passes through the focal point. If the beam of light passes through the focal point before it hits the mirror, it will be reflected so that it travels out horizontally.

The second special path I want to discuss depends on the radius of curvature. Remember, a curved mirror looks like a portion of a circle. If I were to complete the entire circle and make a mark at its center, where would that fall in our drawings above? Well, the distance from the center of the circle to the mirror would be equal to the radius of curvature. Based on Equation (12.1), then, that point would be twice as far from the mirror as the focal point. What happens if a light beam travels through the radius of curvature before hitting a spherical mirror?

Illustration by Megan Whitaker

FIGURE 12.7

A Light Beam Traveling through the Radius of Curvature before Hitting a Concave Spherical Mirror

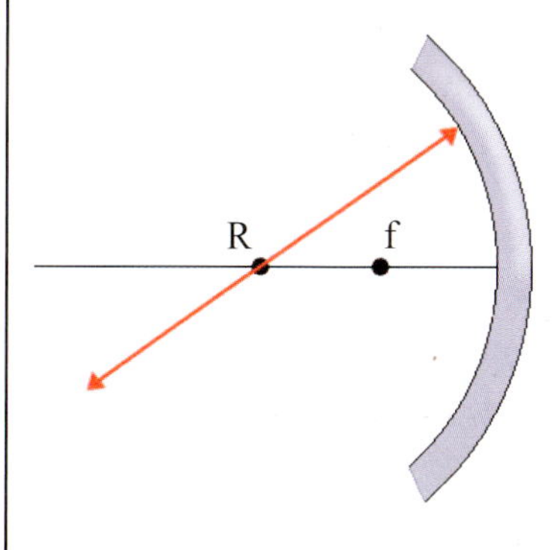

Suppose a beam of light passes through the mirror's radius of curvature before it hits the mirror. That means it would essentially be following the radius of the sphere. Geometry tells us that the radius of a sphere is always perpendicular to a line tangent to the sphere at the point of contact. Thus, the angle of incidence for such a beam would be zero degrees, because it runs along the line that is perpendicular to the tangent at the point of contact. The Law of Reflection tells us, then, that the angle of reflection must also equal zero degrees. What does this mean? It means that the beam of light will be reflected directly backwards.

Now if all of this seems a bit overwhelming, don't worry. You'll see how it all comes together in a moment. For right now, however, you just need to remember these important facts:

1. **When a beam of light travels horizontally and hits a spherical, concave mirror, it is reflected so that it passes through the mirror's focal point.**
2. **When a beam of light travels through the focal point of a spherical, concave mirror, it is reflected so that it travels horizontally.**
3. **When a beam of light travels through a spherical, concave mirror's radius of curvature, it will be reflected backwards along precisely the same path.**

What does all of this mean? Well, if we use these rules, we can trace the path of a few light beams as they travel from an object to a mirror. If we examine what happens to these light beams after they are reflected, we can determine what kind of image is seen in the mirror. To see how this is done, study the following example:

Illustrations by Megan Whitaker

EXAMPLE 12.1

An object is placed 20.0 cm from a concave, spherical mirror that has a 10.0-cm radius of curvature. How will the object appear in the mirror?

To answer this question, we first need to make a drawing of the situation. We know what a concave mirror is, and we also know that the focal point of a concave mirror is equal to half the radius of curvature. In addition, the object is twice as far from the lens as is "R." Thus, we can draw the following picture:

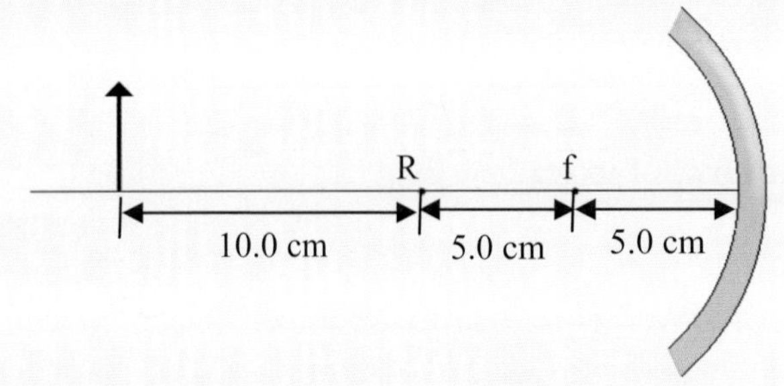

Notice that I have drawn an imaginary line through the center of the mirror and have placed the object (an upright arrow) on that line. Why do I represent the object as an upright arrow? You will see in a moment. Notice also how I have set up this picture. We know that the radius of curvature is 10.0 cm. Thus, "R" is at a distance of 10.0 cm from the mirror. Since Equation (12.1) tells us that the focal point is at a distance of R/2 from the mirror, it is 5.0 cm from the mirror. Finally, the object is 20.0 cm from the mirror, which is twice the distance that "R" is from the mirror.

What we will do now is draw three light beams that come off the object and hit the mirror. As long as the room in which this is taking place is well-lit, there will be light beams bouncing off our object in all directions. Thus, we can choose any paths that we want for our light beams. We will choose them very carefully, however. First, let's look at a light beam that bounces off the top of our object and travels horizontally towards the mirror. According to the discussion above, that light beam will reflect off the mirror so that it ends up traveling through the focal point. Including this light ray in our picture, then, gives us:

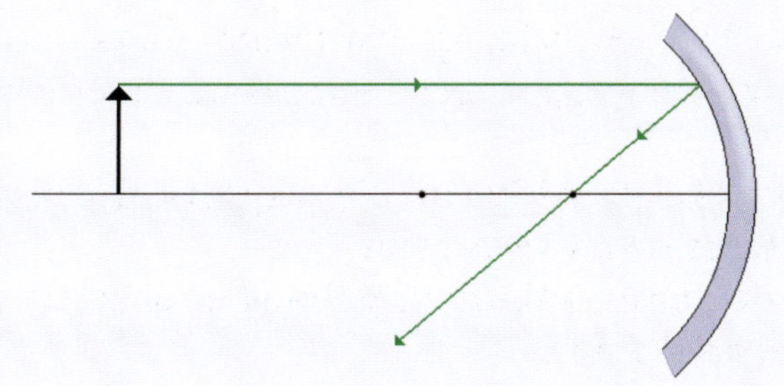

The next beam we will choose is similar to the first. It will originate from the same place (the top of the object), but this time it will travel towards the mirror so that it passes through the focal point. According to our rules, then, it will reflect off the mirror so that it begins traveling horizontally:

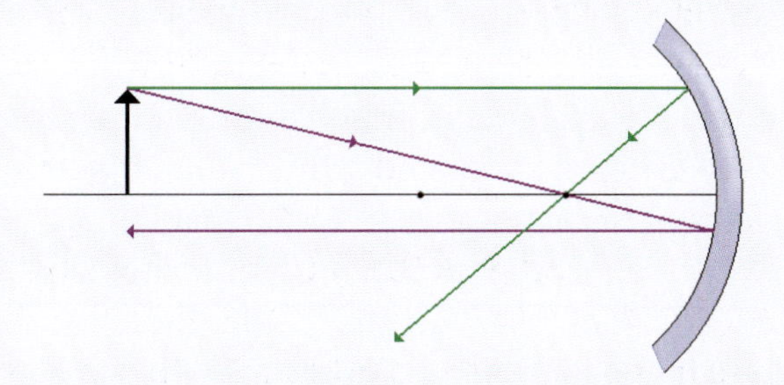

Finally, our last beam of light will originate from the top of the object and travel so that it travels along the radius of curvature. According to our rules, such a light beam will reflect directly backwards off the mirror:

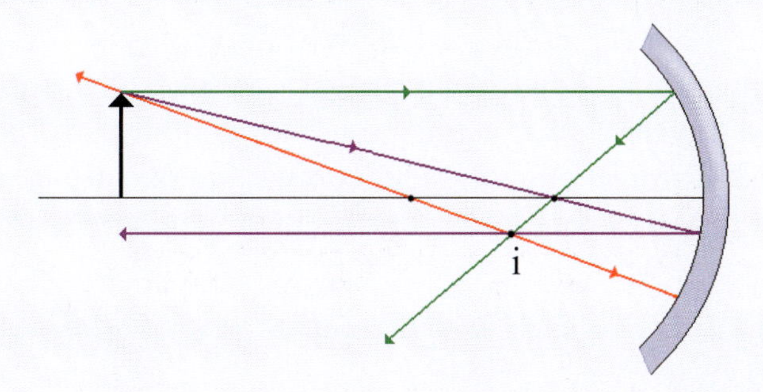

Look at what we have now. All three of the reflected light rays intersect at a point that I have labeled "i." What does this tell us? Well, remember, we see things because light bounces off them and our eyes detect the light. The brain then decodes the signals sent from our eyes and constructs a mental image of the object. Well, if you were looking at this object in this mirror, your eyes would see the light originating not from the object, but from the point labeled "i." As a result, you would think that the top of the object is actually at point "i."

Where will you think that the bottom of the object is located? Well, think about it. The bottom of the object rests on the imaginary line that runs through the center of the mirror. This line runs along the radius of curvature, so all light beams that come from there will be reflected directly backward, along the imaginary line. Thus, the light beams coming from the bottom of the object will all travel along that imaginary line. This means that the bottom of the object will appear to be on that line, directly under the top of the object. Thus, our final drawing looks like this:

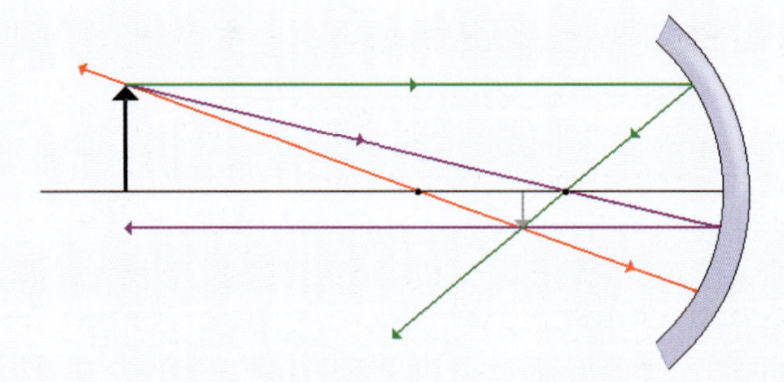

Now remember, if you look at this object in the mirror, you will not see the object. You will see the image, which is represented by the smaller, gray arrow. You see the image because your eyes receive the light from there.

What can we learn from this picture? Well, first we can see that the image which you would see in this situation is actually upside down compared to the actual object. In fact, that's why I used an arrow to represent the object. The arrow makes it easy to see that the image is inverted relative to the object. Also, you should be able to see that the image is significantly smaller than the original object. In the end, then, that's what we learn from this analysis. Under these conditions, when you view the object in the mirror, its image will be reduced and inverted.

This example may have seemed long and drawn-out, but let's review exactly what we did. Using an arrow to represent the object, we placed it at the specified position. Further, we made sure that the bottom of the object touched an imaginary horizontal line that was drawn through the center of the mirror. We also made a mark to represent the length of the radius of curvature and another to represent the focal point of the mirror. Finally, we chose three specific beams of light: one started from the top of the object and traveled towards the mirror horizontally; another started at the top of the object and moved towards the mirror through the focal point; and the last started at the top of the object and traveled along the radius of curvature. The point at which these three beams intersected told us where the top of the image was. From that point, the rest of the image was drawn so that the bottom touched the imaginary horizontal line which was drawn through the center of the mirror. This gave us a picture of what the image would look like.

This kind of exercise is called **ray tracing**. Light beams are often called light rays, and since this exercise traced where these light rays travel, we call it ray tracing. Ray tracing diagrams such as the one we produced in the example are very informative, and physicists who study optics use them quite frequently. Thus, you need to be able to draw such a diagram. It's not too hard, though. After all, you just need to know the three rules we developed right before the example. These three rules tell you which light rays to choose and where they go. Then, all you have to do is determine where your eyes would see the image.

The image that is formed by this mirror is fundamentally different from the image formed in a flat mirror. It is inverted, while images formed in a flat mirror are upright, but there is an even more fundamental difference than that. Remember, the reason you see yourself in a flat mirror is that the light rays from the mirror *appear* to be coming from behind the mirror, once you extrapolate them backwards. Look at the image formed in this mirror, however. The light rays do not have to be extrapolated backwards to form the image. The light rays actually *come from* the image. When this happens, we say that the mirror has formed a **real image**.

Real image - An image formed as the result of intersecting light beams

So, when you do ray tracing, if the light beams that you draw intersect to form the image, we say that the image is real. If you have to extrapolate the light rays backwards, we say that the image is virtual.

We'll do another ray tracing example in a moment, but let me remind you what got this whole discussion started. Remember that I originally was talking about magnifying makeup mirrors. In these mirrors, the image that you see is, in fact, larger than the original object. That's why we call it a magnifying mirror. In the example I just showed you, however, the image was smaller than the original object. Clearly, that's not magnification. I told you, however, that magnifying mirrors are, indeed, concave mirrors, just like the one in the example. How is this possible? Well, study the next example to find out.

Illustrations by Megan Whitaker

EXAMPLE 12.2

An object is placed 5.0 cm away from a concave, spherical mirror that has a 14-cm radius of curvature. Draw a ray-tracing diagram to show how the object will appear in the mirror.

In order to do a ray tracing diagram, we first draw the mirror, the focal point, "R," an imaginary horizontal line that passes through the center of the mirror, and an arrow to represent the object. The focal point is half the distance from the mirror as is "R," and the object is closer to the mirror than the focal point, but not by much. Thus, the drawing looks like this:

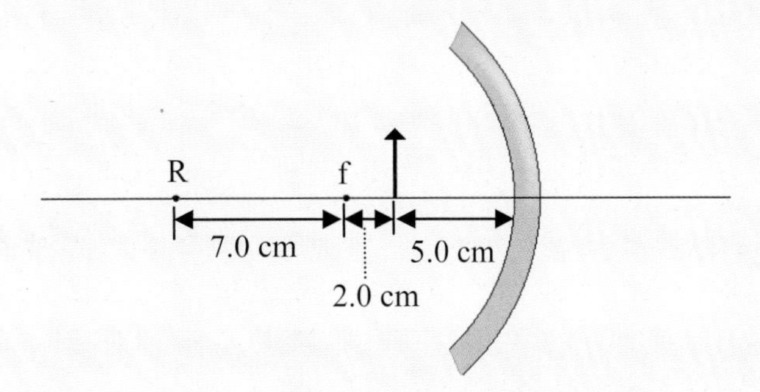

Now we choose three special rays that start at the top of the object and travel towards the mirror. The first travels towards the mirror horizontally and is reflected through the focus.

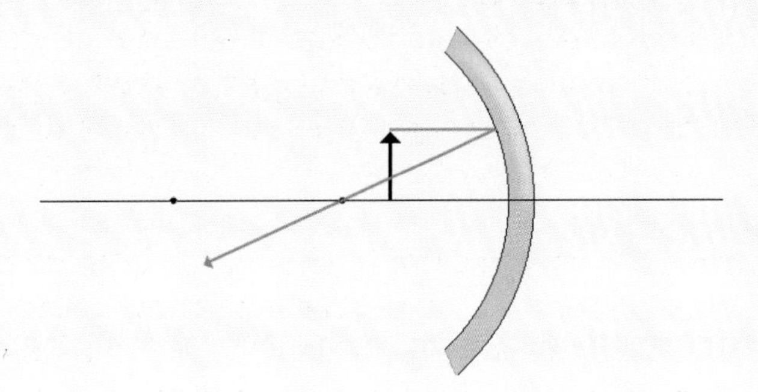

The next ray is supposed to leave the top of the object and travel towards the mirror through the focal point. Now wait a minute. The object is in front of the focal point. How do we draw a ray that leaves the object and travels towards the mirror through the focal point? Well, we can't. We can, however, draw a ray that leaves the object and travels towards the mirror *as if it came from* the focal point:

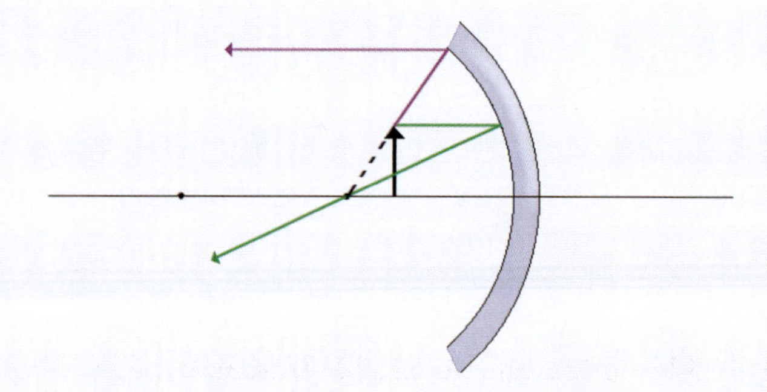

Do you see what I did here? The dashed portion of the ray shows that the ray travels as if it came from the focal point. In the same way, the last ray is supposed to travel from the object towards the mirror along the radius. We can draw that as long as we draw a ray that looks as if it came from the point we labeled "R."

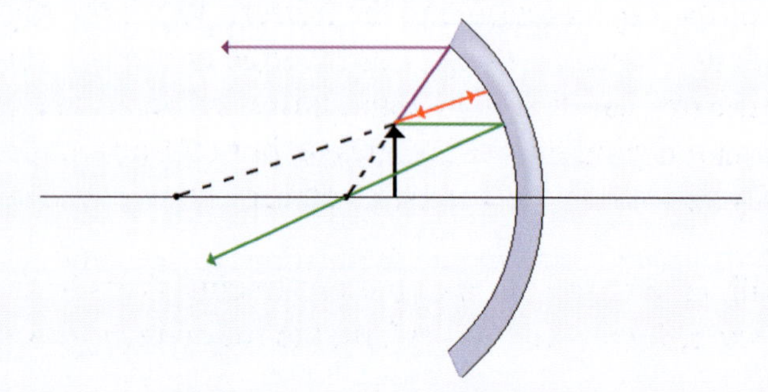

Hmmm... there seems to be a problem here, doesn't there? The reflected rays don't intersect anywhere, do they? They all travel off in their own directions. Does that mean there is no image? No, of course not! Remember how an image is formed in a flat mirror? The light rays are extrapolated backwards, assuming that they have been traveling in straight lines. Let's see what happens when we do that with these light rays:

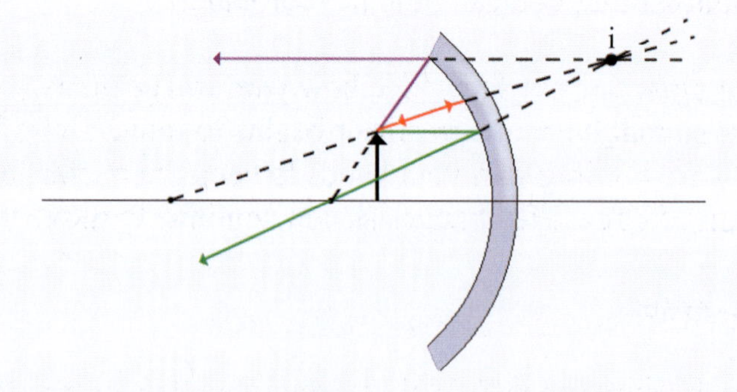

Do you see what happened? The extrapolated parts of the light rays intersected each other behind the mirror, at the point I labeled "i." This is where the image appears to be. We can finish drawing the image by continuing it down from that point until it hits the imaginary line:

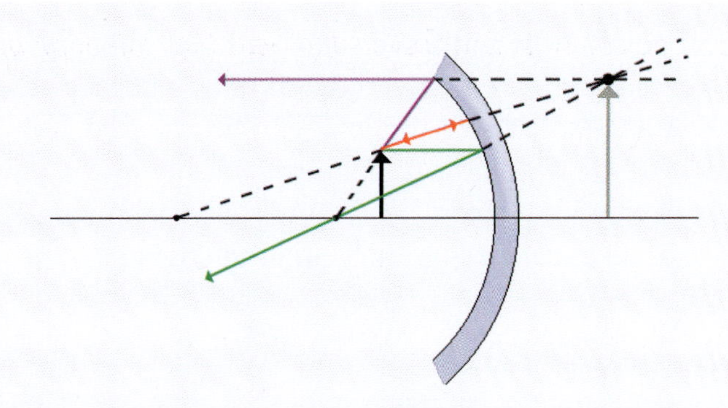

In this diagram, we see that the image is not upside down relative to the object. Also, we see that the image is bigger than object. This is how a magnifying mirror works.

So what's the difference between Example 12.1 and Example 12.2? Why, in one case, was the image inverted and small relative to the object, and in the other case it was upright and large? The difference was the *position* of the object relative to the mirror. When the object was farther away from the mirror than the radius of curvature, the image was inverted and small relative to the object. When the object was closer to the mirror than the focal point, however, the image was upright and large relative to the object. So, a magnifying mirror is concave, but in order for it to be magnifying and have an upright image, the object you want to view must be closer to the mirror than the focal point.

In a moment, you will get a chance to try your ray-tracing skills with a couple of "On Your Own" problems. First, however, I want you to perform the following quick experiment to see exactly what we have just studied.

EXPERIMENT 12.2

Real and Virtual Images in a Concave Mirror

Supply:

- A magnifying makeup mirror that you can hold in your hands

1. Hold the makeup mirror close to your face. See how your image is magnified?
2. Slowly extend your arm so that the makeup mirror begins to move away from your face. You may have to change how you are holding it in your hands to keep the mirror focused on your face.
3. Watch how the reflection of your face changes as you continue to move the mirror away from your face.
4. Return the mirror to its owner.

What did you see in the experiment? Well, as you moved the mirror away from your face, it should have gotten blurry. You should have eventually reached a point where your image was so blurry that you could not see it. Then suddenly, your face should have appeared upside down. As you continued to move the mirror away from you, your face should have remained upside down.

Can you guess what happened here? When you held the mirror close, your face was closer to the mirror than its focal point. Thus, the mirror was behaving as the mirror did in Example 12.2, and the image you saw was magnified and upright. When your face got so blurry that you couldn't see it, you were actually at the focal point of the mirror. When an object actually sits on the focal point, it cannot be seen in the mirror. As soon as you passed that point, your face was suddenly farther from the mirror than the focal point, and that caused the image to be inverted, as you will see when you solve one of the problems below. As you moved the mirror farther away, your face eventually passed "R," and the mirror began behaving like the one in Example 12.1, showing a reduced, inverted image. So you see that ray tracing does an excellent job of predicting the appearance of an image in a mirror!

ON YOUR OWN

12.1 An object is placed 4.5 cm away from a concave, spherical mirror with a 12.0-cm radius of curvature. Draw a ray-tracing diagram to illustrate how the image will appear in the mirror. Is it a real or virtual image? Is it upright or inverted compared to the object? Is it magnified or reduced?

12.2 An object is placed 7.5 cm away from a concave, spherical mirror with a 12.0-cm radius of curvature. Draw a ray-tracing diagram to illustrate how the image will appear in the mirror. Is it a real or virtual image? Is it upright or inverted compared to the object? Is it magnified or reduced?

Ray Tracing in Convex Spherical Mirrors

So far, I have concentrated on concave spherical mirrors, but there is another obvious way for a spherical mirror to be constructed. If the mirror is made so that the mirror bends *towards* the object, it is called a **convex** mirror. Now, of course, a convex mirror is just a concave mirror that is turned around, so the basic rules of ray tracing for spherical mirrors still apply. However, since the geometry of the situation is considerably different, it is worth going through an example and an "On Your Own" problem to make sure you understand how convex spherical mirrors work.

Illustrations by Megan Whitaker

EXAMPLE 12.3

An object is placed 14.0 cm away from a convex, spherical mirror whose radius of curvature is 10.0 cm. Draw a ray-tracing diagram to show how the object will appear in the mirror.

As is the case with all ray-tracing exercises, we must first draw the situation described in the problem:

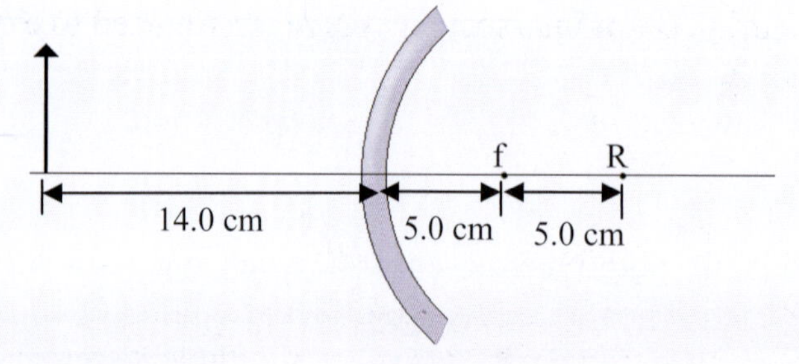

How do we draw our rays in this case? Well, a ray that travels horizontally before it hits the mirror will reflect *as if it came from the focal point*. This is much like the situation given in Example 12.2, where we had to draw a ray approaching the mirror as if it came from the focal point. In this case, we will simply draw a ray reflecting off the mirror as if it came from the focal point:

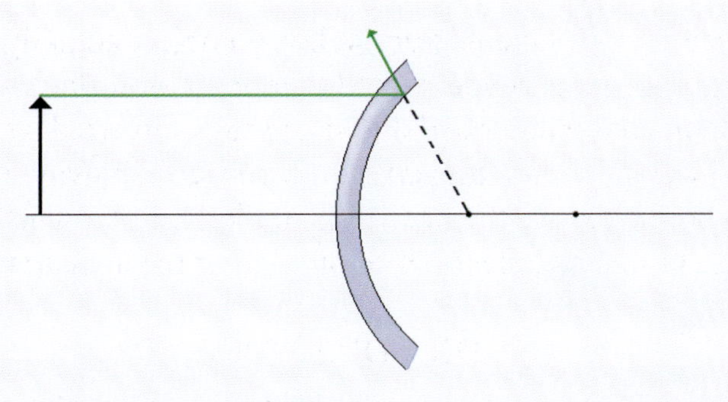

Note where I have the ray touching the mirror. These mirrors really should be very thin. I draw them thicker than they are to enhance the illustration. Thus, I must have the ray hit the inner curve of the mirror, as that's the curve from which all of the distances were determined.

The next ray that we must draw is supposed to come from the focal point and be reflected horizontally. Obviously, that cannot happen in this case, so we draw the ray *as if it is headed to the focal point*. Then, it will reflect horizontally:

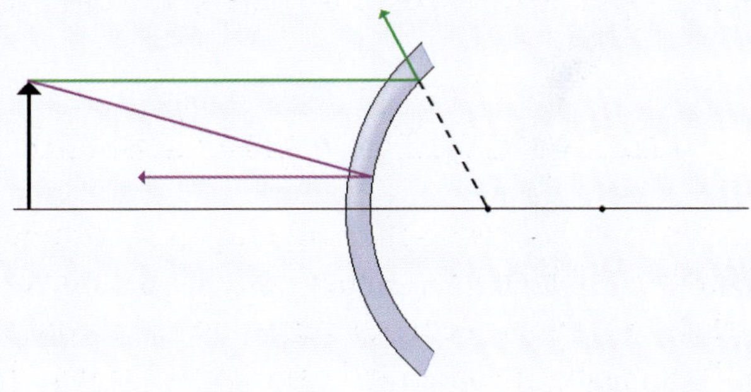

The final ray is supposed to come from "R" and be reflected straight back. However, that is also not possible, so once again we draw it *as if it is headed to "R,"* and then it is reflected straight back:

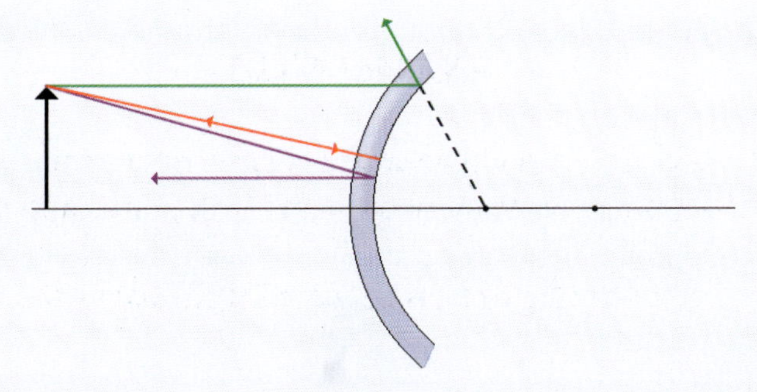

The reflected light rays clearly do not intersect anywhere, so we need to extrapolate them backwards to see if they intersect behind the mirror:

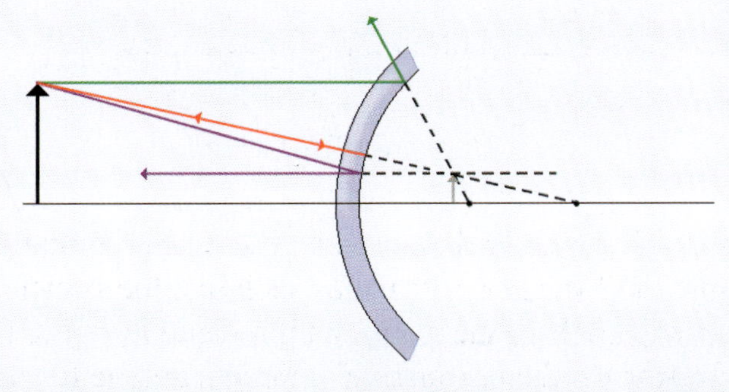

The image, then, is a virtual image; it is upright; and it is reduced as compared to the object.

In the end, then, convex mirrors are analyzed much like concave mirrors. However, since the object is on the other side of the mirror, we merely have to draw the horizontal ray's reflection as if it came from the focal point, draw another ray as if it is headed to the focal point, and draw the final ray as if it is headed to "R." Make sure you can work with convex mirrors by solving the following "On Your Own" problem.

ON YOUR OWN

12.3 An object is placed 2.0 cm from a convex, spherical mirror that has a 6.0-cm radius of curvature. Draw a ray-tracing diagram to determine how the image will appear in the mirror.

Snell's Law of Refraction

We have studied in great detail how light rays behave when they hit a reflective surface. There are, however, other types of obstacles that light can run into. For example, what happens when light hits a wall? Well, a wall is considered an **opaque** (oh payk') object, which means light cannot travel through it. Thus, the light stops. Actually, the light doesn't just stop. Some of it gets reflected according to the Law of Reflection. The rest gets absorbed by the wall. The main point, however, is that when light hits an opaque object, it travels no further along its original path.

Light can also run into a **transparent** (or semi-transparent) object. Transparent objects allow light to pass through them. A window is an example of a transparent object, because light can pass through it. What many people do not realize, however, is that whenever a ray of light hits a transparent object, a portion of the light ray travels through the object, but a portion of it is also reflected. Have you ever looked outside through a window and seen your own reflection? Well, the reason that you can look through the window into the outside world is because the window is transparent, and light from outside can pass through it to hit your eyes. However, the reason you see your reflection is that a portion of the light traveling from inside the room to the outside is reflected back into the room. That portion of the light strikes your eyes, forming your reflection.

Now when light travels through a transparent object, an interesting thing happens. The light actually bends a little. For example, if I were to shine a beam of light onto a thick piece of glass, here is what I would observe:

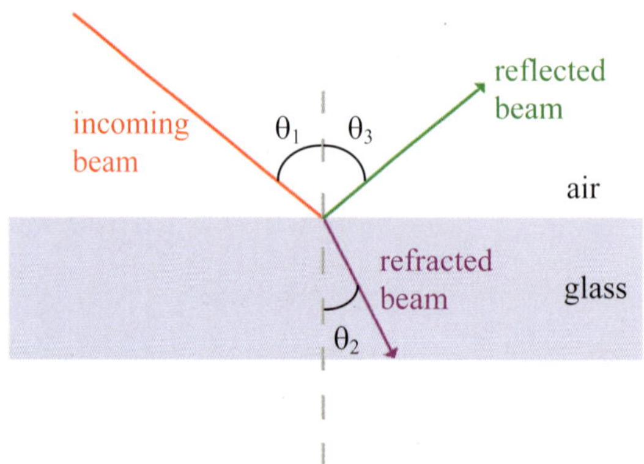

When the ray of light strike the glass, a portion of that light is reflected. The Law of Reflection tells us that the angle of incidence equals the angle of reflection; thus, $\theta_1 = \theta_3$. The portion of the light ray that travels through the glass, however, changes direction. It actually bends so that its angle relative to the perpendicular is smaller than it was before it hit the glass. This process is called **refraction** (rih frak' shun).

Refraction - The process by which a light ray bends when it encounters a new medium

Physicists, therefore, often say that when a ray of light strikes a transparent object, a portion of the light is *reflected*, and the rest of the light is *refracted.*

There are two questions a physicist must ask about refraction. First, why does it happen? Second, is there any way that I can predict θ_2? We'll tackle these two questions in order. In the previous module, I told you that light travels at different speeds depending on the medium in which it is traveling. In air, for example, light travels at 3.0×10^8 m/sec. In glass, however, its speed is significantly lower. This causes refraction, as illustrated in the figure below.

FIGURE 12.8
Refraction

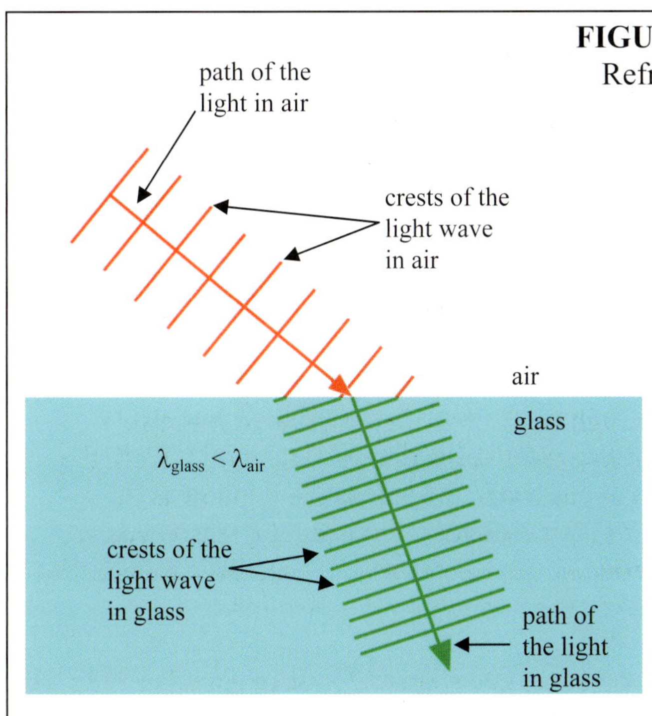

Consider the crest of a light wave. If it hits the boundary between air and glass at an angle, one portion of the crest will hit the boundary before the rest of the crest. As that portion of the crest enters the glass, it slows down. The problem is, the *frequency* of the light wave must be the same in both the air and the glass. Since the wave travels more slowly in the glass, the wavelength is shortened so that the frequency of the wave does not change. Thus, the portion of the light wave that is in the glass has a shorter wavelength than the portion of the wave that is still outside of glass. However, the crests of the light wave still need to line up. The only way the wavelength inside the glass can shorten while the crests inside and outside of the glass line up is if the light ray bends.

To answer the second question, we must turn to the Dutch mathematician Willebrod Snell, who discovered its answer in 1620. He did several experiments in which he measured θ_1 and θ_2 as illustrated in the picture on the previous page. In the end, he determined a simple relationship between them, which we now call **Snell's Law**:

$$n_1 \cdot \sin(\theta_1) = n_2 \cdot \sin(\theta_2) \tag{12.3}$$

Now, of course, this equation means nothing if we do not know what "n_1" and "n_2" represent.

As I said before, the fact that light bends when it begins to travel through a transparent medium is due to the fact that light travels at a different speed in that medium. In order to account for this, Snell came up with something called the **index of refraction**.

Index of refraction - The ratio of the speed of light in a vacuum to its speed in another medium

When physicists speak of a vacuum, they mean space that is completely empty. We can restate our definition mathematically:

$$n = \frac{c}{v} \tag{12.4}$$

where "c" is the speed of light in a vacuum, and "v" is the speed of light in the medium of interest. One thing to realize is that the speed of light in air is so close to the speed of light in a vacuum that we can consider them equal in this course. Thus, even though the index of refraction of air is just slightly larger than 1 (it is 1.0003), for the purposes of this course, we will just keep three significant figures. This is something you have to remember:

The index of refraction of air (to three significant figures) is 1.00.

In the previous module, I mentioned that in general, the denser the medium, the slower light travels in it. This is not an exact relationship, but it is a good rule of thumb. Well, since virtually every other medium is more dense than air, we can say that the "v" in equation (12.4) will be less than "c." As a result, a medium's index of refraction should be greater than one. The more dense the medium, the smaller the value of "v" (usually), which means the larger the value for "n." As you might expect, every substance has its own unique index of refraction. The index of refraction of ice, for example, is 1.31, while it is 2.42 for diamond. Thus, we can say that light travels more slowly through diamond than it does through ice.

So, now that we know what index of refraction is, we can finally understand Equation (12.3). Snell's Law says that if a light ray hits a transparent object, we can calculate the angle of refraction (θ_2) if we are given the angle of incidence (θ_1). To do this, we take the index of refraction of the medium it started out in (n_1) and multiply it by the sine of the angle of incidence (θ_1). That product must equal the index of refraction of the new medium (n_2) times the sine of the angle of refraction (θ_2). Study the following example to make sure you understand this.

EXAMPLE 12.4

A light beam shines on a pane of glass (n = 1.5). If the angle of incidence is 34.5°, what is the angle of refraction?

This problem can be diagrammed as follows:

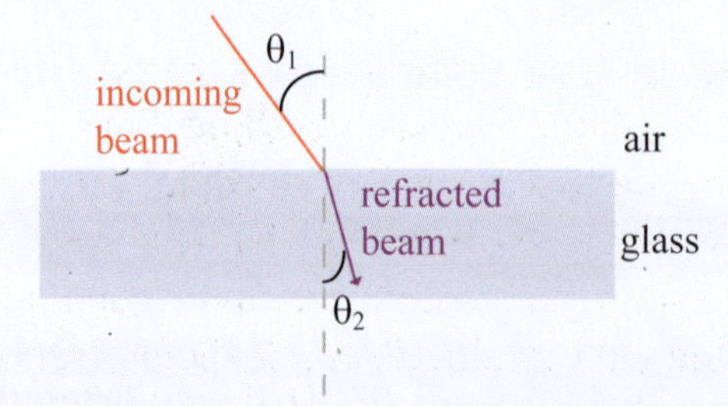

To determine the angle of refraction (θ_2), all we need to do is use Equation (12.3). We must realize, however, that n_1 is the index of refraction of the original medium, which is air. As a result, $n_1 = 1.00$.

$$n_1 \cdot \sin(\theta_1) = n_2 \cdot \sin(\theta_2)$$

$$(1.00) \cdot \sin(34.5°) = (1.5) \cdot \sin(\theta_2)$$

$$\sin(\theta_2) = \frac{(1.00) \cdot \sin(34.5°)}{1.5}$$

$$\theta_2 = 22°$$

When it enters the glass, the light ray bends, making an angle of 22° relative to the perpendicular.

Before you solve the "On Your Own" problems for this section, I want you to perform an experiment that will allow you to use Snell's Law to actually measure the index of refraction of a medium.

EXPERIMENT 12.3
Measuring the Index of Refraction of Glass

Note: A sample set of calculations is available in the solutions and tests guide. It is with the solutions to the practice problems.

Supplies:

- Safety goggles
- A sheet of plain white paper
- A pen
- A protractor
- A ruler
- A square or rectangular glass or clear plastic pan (Depending on how slanted the sides of your pan are, you might need a helper to hold the pan during the experiment. A piece of glass with two flat sides would work even better.)
- A flashlight with the same cover you used in Experiment 12.1

1. Take the plain white sheet of paper and draw a line down the length of the paper, right in the middle.
2. Use your protractor to draw another line perpendicular to the line you just drew. This line should be about 3 inches from one of the edges of the paper and it should span the entire width of the paper.
3. Use your protractor and ruler to draw a third line that starts at the edge of the paper nearest the line you just drew and travels through the intersection of the lines you drew in steps 1 and 2. This new line should make a 45 degree angle with each of the other lines.
4. Fix up your flashlight again so that it is just like what you used in Experiment 12.1.
5. Lay the glass pan on its side so the edge of the pan lies right along the line you drew in step 2.
6. Lay the flashlight on the table with its slot down. It should be pointed so that the beam of light that will come out of it will travel along the line you drew in step 3. In the end, your setup should look something like the photo on the next page:

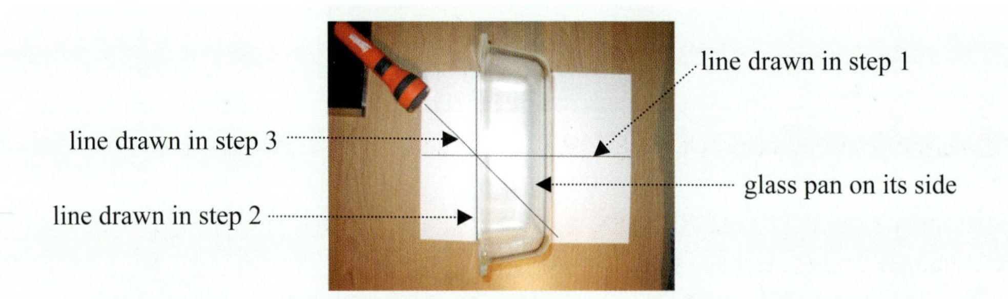

7. Turn on the flashlight and turn out the lights.
8. Position the flashlight so that the light beam shines on the paper and follows the line you drew in step 3 from the time it leaves the flashlight until it hits the pan.
9. Look down from directly above the pan.
10. Play with the tilt of the flashlight until you see three things:
    (1) a beam of light reflecting from where the beam from the flashlight hits the glass
    (2) a beam of light traveling in the glass with a refraction angle less than the incidence angle
    (3) a beam of light traveling down the line drawn in step 3 (This is light traveling under the pan, not in the glass. If your glass pan sits flush with the paper, you may not see this beam. That's fine.) Your experiment should look something like the picture below.

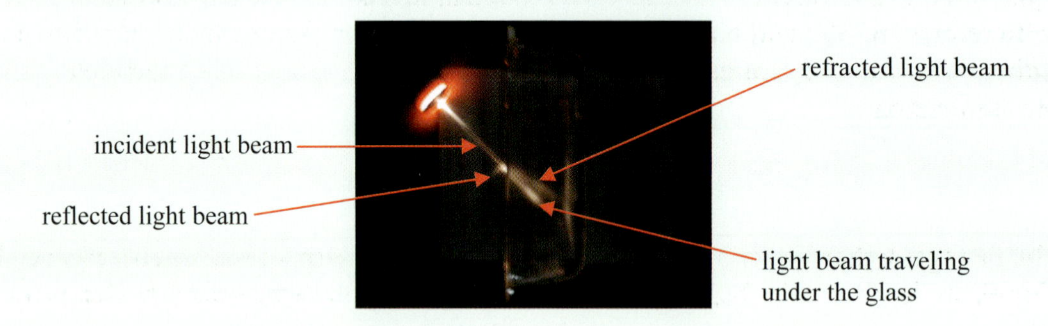

11. Use your pen to mark the point at which the refracted light beam leaves the other side of the glass.
12. Turn on the lights and turn off the flashlight.
13. Remove the pan and flashlight from the paper.
14. Use the ruler to draw a line from the point at which the incidence light beam struck the glass and the point at which the refracted light beam left the glass.
15. Use the protractor to measure the angle of the incident light beam (the line you drew in step 3) relative to the line you drew in step 1. This is θ_1 in Equation (12.3).
16. Use the protractor to measure the angle of the refracted light beam (the line you drew in step 14) relative to the line you drew in step 1. This is θ_2 in Equation (12.3).
17. You know the index of refraction of air, so you have n_1 in Equation (12.3). You have just measured θ_1 and θ_2 in Equation (12.3), so the only unknown is n_2, the index of refraction of glass. Solve for it.
18. Repeat this experiment twice more, using two different angles (other than 45 degrees) when you do step 3. That way, you have two more values for n_2, each for a different incidence angle.
19. Since the index of refraction of a substance does not depend on the incidence angle, the three values you calculated for n_2 should be equal. They probably will not be, because of experimental error. Average those three results, and you have an average measurement of the index of refraction for the glass that you used. It should probably be between 1.3 and 1.7, as that is the typical range for the index of refraction of glass.
20. Clean up your mess.

ON YOUR OWN

12.4 The index of refraction of water is 1.3. What is the speed of light in water?

12.5 A light ray travels through air and then encounters a diamond (n = 2.42). If the angle of incidence is 67.1°, what is the angle of refraction?

12.6 A light ray travels through glass (n = 1.5) and then encounters air. If the angle of incidence is 34.6°, what is the angle of refraction?

Before I leave this section, I want to make one quick point. Everyone knows that if you shine white light through a prism at the proper angle, the light will separate into the colors of the rainbow. In fact, water droplets suspended in the sky can act as tiny prisms, turning sunlight into a rainbow. This is called **dispersion**, and it is the result of the fact that the index of refraction of a medium is slightly different for different wavelengths of light. Thus, the *angle* of refraction is slightly different for each wavelength of light. As a result, a beam of white light is "spread out" into its various wavelengths when it is refracted. Although we know that this is true, we will not worry about it when dealing with refraction. We will make the simplifying assumption that the index of refraction of a given medium is the same regardless of the wavelength of light. Although not completely true, it is a reasonable assumption.

Converging Lenses

The fact that light rays tend to bend when they travel through transparent objects can be quite useful. For example, consider a light ray traveling through the following object, which is made of glass:

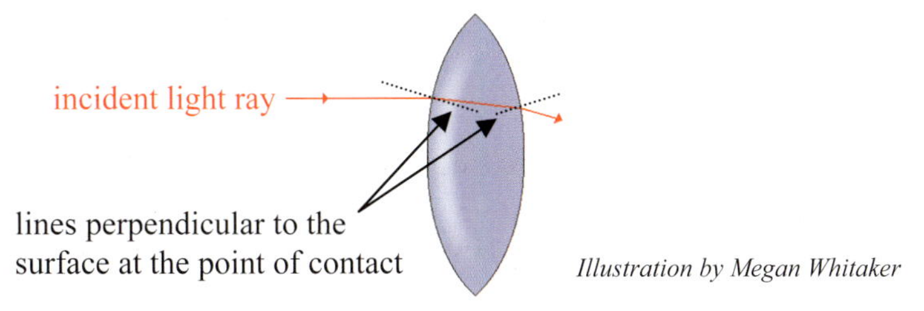

Illustration by Megan Whitaker

When the ray strikes the transparent object, it will be refracted. If you work out Snell's Law with n_1 equal to the index of refraction of air and n_2 equal to the index of refraction of glass, you will see that the angle of refraction is smaller than the angle of incidence. This means that the light will bend towards a line which is perpendicular to the surface at the point of contact.

The light ray will then travel through the glass, along the path indicated above. At some point, however, the light ray will reach the edge of the glass object and exit. At that point, the medium through which the light is traveling will change. Thus, the light will be refracted. If you work out Snell's Law in this case, things are a little different. In this case, n_1 will be the index of refraction of glass (because the light ray is currently in the glass), and n_2 will be the index of refraction of air (because air is the *new* medium). Working out Snell's Law will tell you that in this case, the angle of refraction will be greater than the angle of incidence. This means that the light ray will bend away from the perpendicular line when it exits the object.

If you were to work out Snell's Law for several light rays that travel horizontally towards the object, you would get the result pictured in Figure 12.9.

Illustration by Megan Whitaker

FIGURE 12.9

Horizontal Light Rays Hitting a Converging Lens

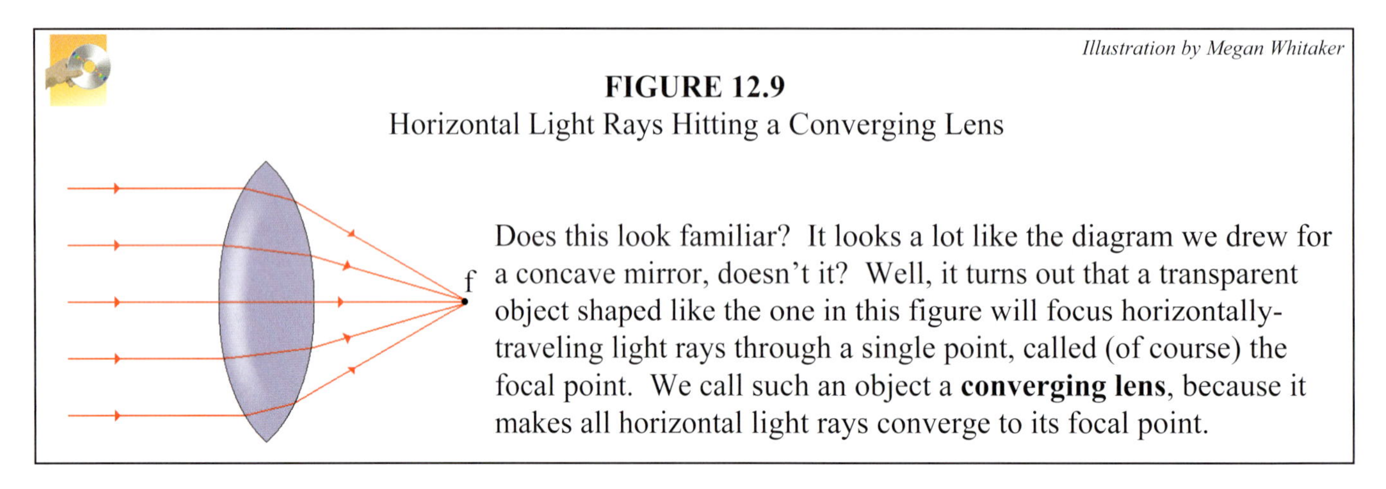

Does this look familiar? It looks a lot like the diagram we drew for a concave mirror, doesn't it? Well, it turns out that a transparent object shaped like the one in this figure will focus horizontally-traveling light rays through a single point, called (of course) the focal point. We call such an object a **converging lens**, because it makes all horizontal light rays converge to its focal point.

There are three things that you need to know about converging lenses. First, they are very thin. I drew the object above wide so that the illustration would be clear. Thus, in the illustrations, the lenses will be wide. Just remember that they really should be thin. Second, since light can travel through the lens, it has a focal point on each side, because light can enter the lens from either direction. The distance to the focal points of a lens is measured from the center of the lens. Finally, because lenses are spherical like most mirrors, they are subject to spherical aberrations as well.

Just as you learned with spherical mirrors, there are a few special rays whose paths through a converging lens can be easily predicted. Rather than going through all of the arguments that we can use to come up with these special rays and their paths, I will simply tell you about them:

1. **All light rays that enter the lens traveling horizontally will exit the lens traveling through the focal point.**
2. **All light rays that enter the lens traveling through the focal point will exit the lens horizontally.**
3. **All light rays that enter the lens traveling directly towards the center will experience no deflection as they exit the lens.**

As you might expect, these three rules can be used to draw ray-tracing diagrams that will allow us to predict the appearance of an object if it is viewed through a converging lens. This is very similar to what we did with a concave mirror, so the following example should not be hard to understand.

Illustrations by Megan Whitaker

EXAMPLE 12.5

An object is 10.0 cm from a converging lens that has 7.0-cm focal points. What is the appearance of the object if someone views it through the lens?

Just as we did when analyzing images from spherical mirrors, we need to make sure we draw a diagram that properly illustrates the situation. The object (which we will represent with an arrow) is farther from the lens than is the focal point, but not by much. Thus, the diagram should look like the drawing on the next page:

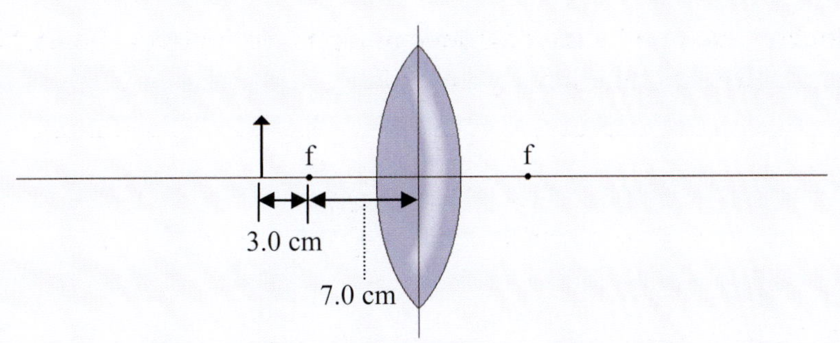

Notice that the distances are relative to the center of the lens and that there are focal points on each side of the lens. Now we use the rules that I listed previously in order to trace three rays that come off the top of the object. The first ray travels to the lens horizontally and leaves through the focal point.

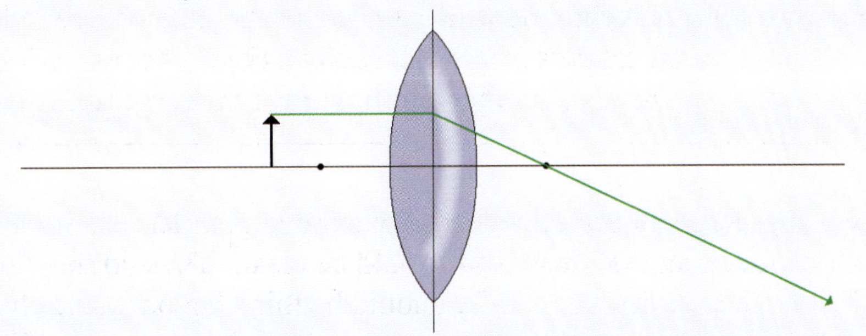

Note that we draw the ray so that it changes direction at the center of the lens. This is an attempt to show that despite the fact that the lens is drawn so that it is thick, in reality, the lens is very thin. The next ray travels to the lens through the focal point and exits the lens horizontally.

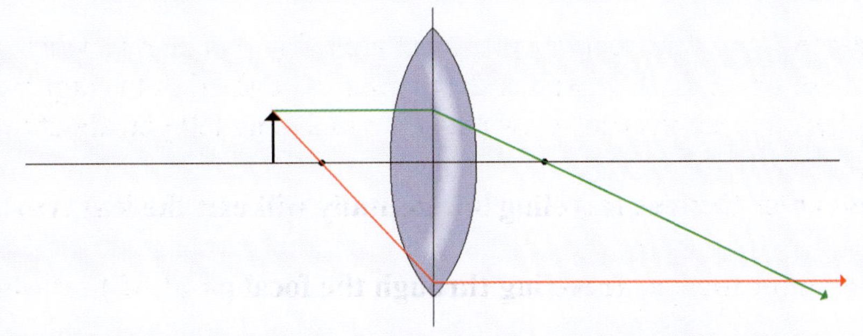

The final ray travels to the lens aimed at its center and exits without deflection.

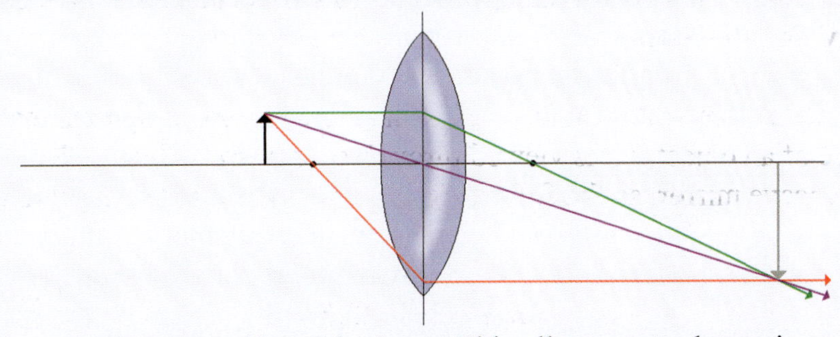

The three rays intersect on the other side of the lens. This allows us to determine what the image will look like. This diagram, then, tells us that if you look at the object through the lens, you will see a real, inverted, and magnified image compared to the object. Also, notice *where* the image appears. It will look to the observer like the object is, in fact, on the other side of the lens!

Notice that the only real difference between this example and the problems involving concave mirrors is the third light ray that we draw. All of the other concepts that we used to draw this diagram were, in fact, used to draw the ray-tracing diagrams for spherical mirrors. Thus, there is really nothing new to solving these problems. I will therefore not give you any more examples of drawing ray-tracing diagrams for converging lenses. Just solve the "On Your Own" problems to make sure that you really understand this.

ON YOUR OWN

12.7 A converging lens has focal points that are 5.0 cm from its center. If an observer looks through the lens at an object 10.0 cm from the lens, how will the image appear? Is it real or virtual? Is it upright or inverted? Is the image magnified, reduced, or essentially the same size as the object?

12.8 A converging lens has focal points that are 8.0 cm from its center. If an observer looks through the lens at an object 5.0 cm from the lens, how will the image appear? Is it real or virtual? Is it upright or inverted? Is the image magnified, reduced, or essentially the same size as the object?

Diverging Lenses

Another basic type of lens used in optics is the **diverging lens**. Like a converging lens, this lens uses the fact that light bends when it encounters glass. Also, like a converging lens, it is made of curves that are arcs of a sphere. However, unlike a converging lens, this lens bends horizontal light rays *away from each other*, as shown in Figure 12.10.

Illustration by Megan Whitaker

FIGURE 12.10
A Diverging Lens

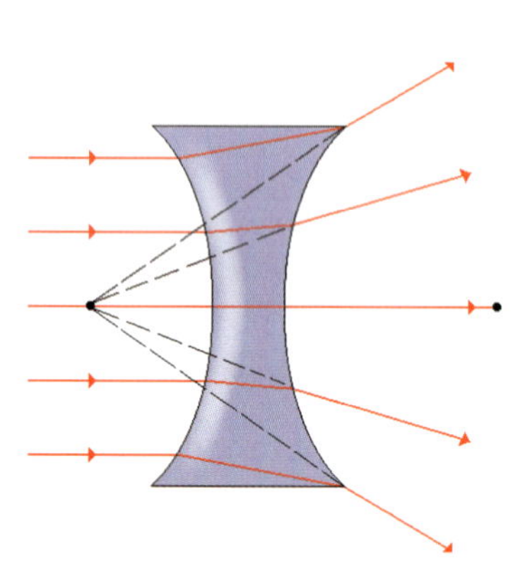

Because the spherical curves of a diverging lens are concave rather than convex, Snell's Law of refraction causes horizontal light rays to bend away from each other when they enter the lens. Just like a converging lens, a diverging lens also has focal points on both sides. Notice, however, the significance of the focal points in a diverging lens. Horizontal light rays are bent *as if they were coming from the focal point on the same side of the lens as that from which they entered.*

Not surprisingly, diverging lenses can be analyzed in a way that is similar to converging lenses, but the details of the analysis are a bit different. Study the following example to see what I mean.

Illustrations by Megan Whitaker

EXAMPLE 12.6

An object is 7.0 cm from a lens that has 5.0-cm focal points. What is the appearance of the object if someone viewed it through the lens?

As is always the case, we start our analysis by making a drawing of the situation. The object is outside the focal point of the lens, but not by much. The resulting diagram is given on the next page:

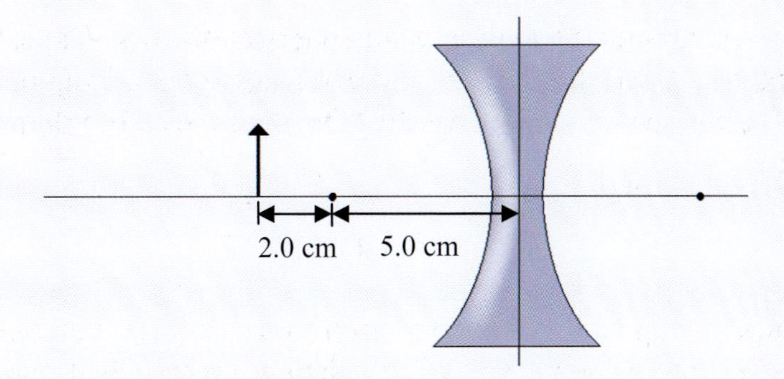

Note that like a converging lens, the measurements are taken from the center of the lens. This indicates that the lens is, in actuality, very thin. Now we draw our rays. The first ray travels horizontally to the lens. As shown in the figure above, it bends so that it appears to be coming from the focal point that is on the same side as the object.

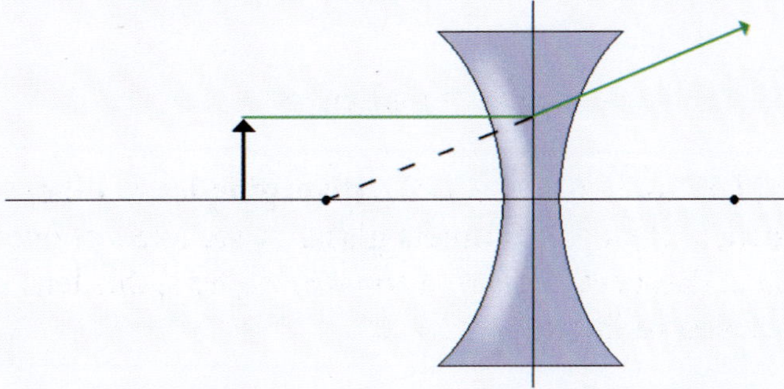

The next ray is the opposite of this one. It travels as if it is headed towards the focal point on the other side of the lens, and it comes out traveling horizontal.

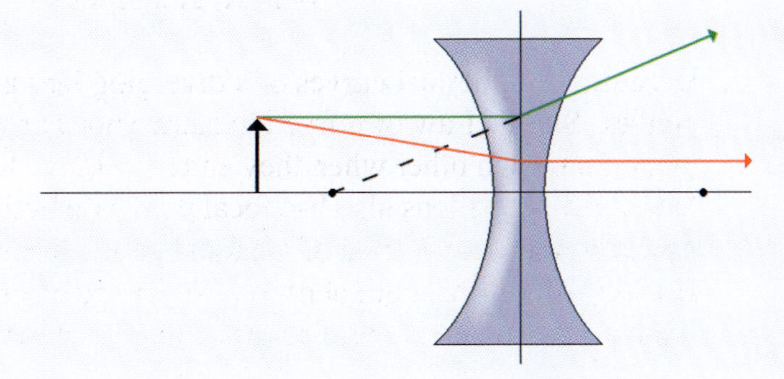

The final ray runs through the center of the lens with no deflection.

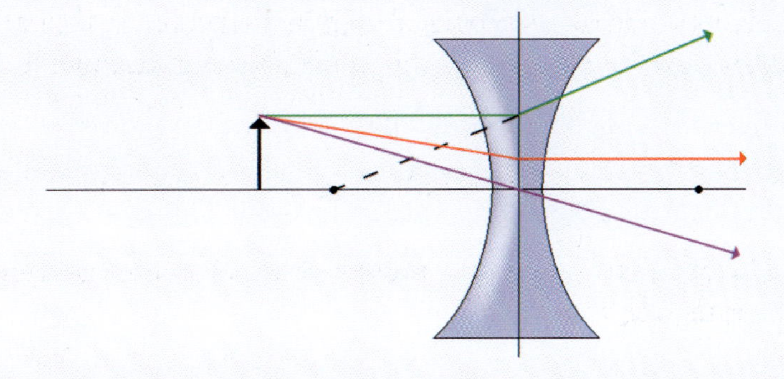

The light rays do not intersect, so we must extrapolate them. When we do that, we see that the intersect on the same side of the lens as the object.

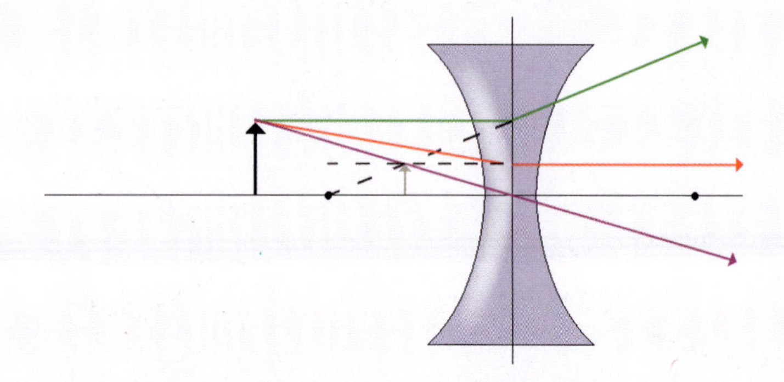

The image, then, is virtual, upright, and reduced. Don't be confused because one of the light rays did not need to be extrapolated in order to find the intersection point. The fact that the other two light rays had to be extrapolated means that the image is virtual.

Analyzing situations with diverging lenses, then, is not very different from analyzing situations involving converging lenses. Instead of the lens focusing horizontal light rays to the focal point on the other side of the lens, a diverging lens bends them as if they came from the focal point on the same side of the lens as the object. In addition, light rays that are traveling towards the focal point on the other side of the lens get bent so that they travel horizontally. Make sure you understand how to analyze these situations by solving the "On Your Own" problems that follow.

ON YOUR OWN

12.9 A diverging lens has focal points that are 7.0 cm from its center. If an observer looks through the lens at an object 16.0 cm from the lens, how will the image appear? Is it real or virtual? Is it upright or inverted? Is the image magnified, reduced, or essentially the same size as the object?

12.10 A diverging lens has focal points that are 8.0 cm from its center. If an observer looks through the lens at an object 5.0 cm from the lens, how will the image appear? Is it real or virtual? Is it upright or inverted? Is the image magnified, reduced, or essentially the same size as the object?

The Human Eye

Before ending this module, we must spend a few moments studying perhaps the single greatest thing in all of God's creation: the human eye. Now, there are many, many marvelous facets of the eye, but I just want to concentrate on one: the way it handles light. As I have said before, the reason that we see things is that light reflects off objects and enters our eyes. Our eyes then detect the light and send signals to the brain which forms an image in our mind. So, how do our eyes handle the light they receive? First, Figure 12.11 shows a simplified drawing of the human eye.

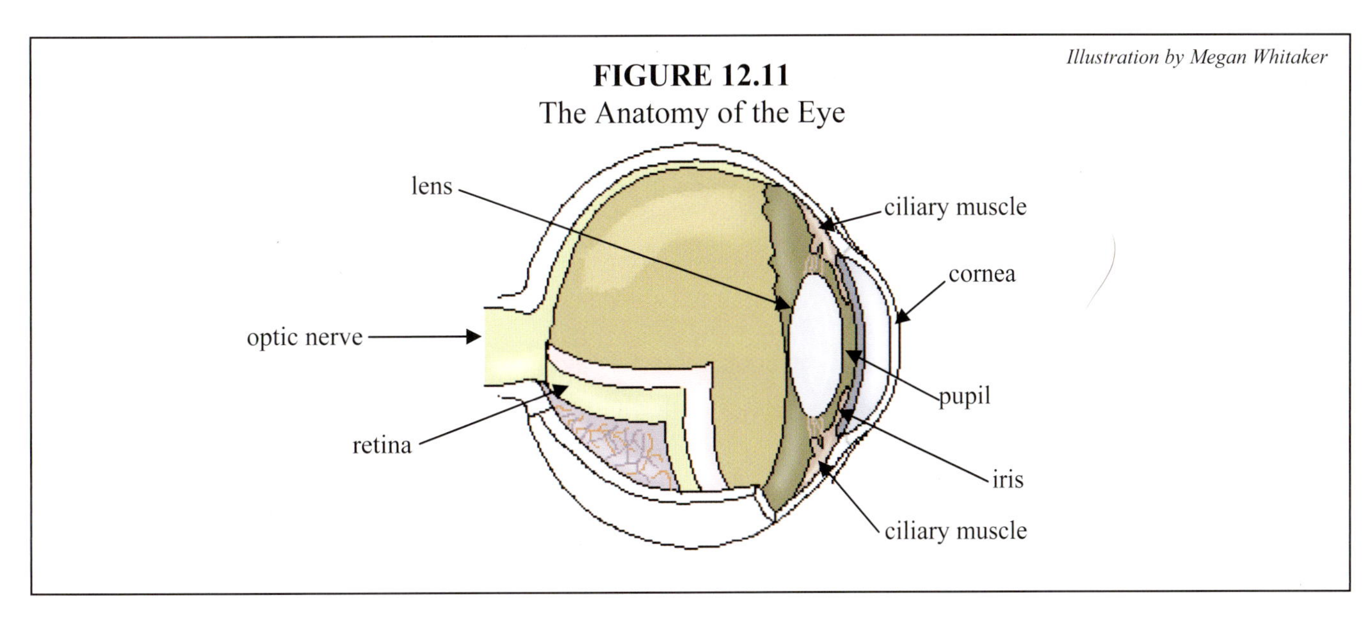

FIGURE 12.11
The Anatomy of the Eye

In simple terms, the eye contains several different optical elements that all work together. The eye is covered by a thin, transparent substance called the cornea. It protects the eye from abrasions and the like. It also bends light that enters it, because its index of refraction is greater than one. The iris is a cover that can open up wide or close down to just a small hole. This regulates how much light gets into the eye. The opening left by the iris is called the pupil. When you are in the presence of bright light, the iris closes down to allow only a small amount of light into the eye. This makes your pupil small. When there is little light, the iris opens wide, allowing a larger percentage of the light in. This makes your pupil large. Once light enters the pupil, it is focused by a converging lens. The light is focused on the retina, which is made up of light-sensitive cells. When these cells detect light hitting them, they send electrical messages down the optical nerve to the brain, which decodes the messages and forms them into mental images.

Now, here's the really neat thing about the way the eye focuses light. Have you noticed in the example and "On Your Own" problems that the position of the image formed by a converging lens depends on where the object is? If the object moves in relation to the lens, the image will form someplace else. Well, in order for the eye to work, the image must always be focused on one place: the retina. How does your eye accomplish this, given the fact that you can see objects regardless of their distance from your eye?

Believe it or not, the eye *can actually change the shape of the lens* in order to keep the image in the same place, regardless of where the object is! This is accomplished with the ciliary muscle. It squeezes or expands the lens, changing the lens's focal point. Thus, when an object moves in relation to the eye, the ciliary muscle changes the focal point of the lens to compensate, keeping the image focused on the retina! This is an amazing feat of physics. To give you some idea of just how amazing, think about modern-day cameras. Some are sophisticated enough to have autofocus. The camera can automatically adjust the position of the lens so that the image stays in focus on the film. Now remember, the eye changes the *shape* of the lens, while a camera changes the *position* of the lens. Nevertheless, the end result is the same: the image stays in focus regardless of where the object is.

Even the most sophisticated autofocus camera on earth, however, is still significantly slower in its autofocus capability than the eye, and the image's focus is still significantly less resolved! Thus, even the best that today's technology has to offer cannot come close to mimicking the marvelous design of the eye.

Even Charles Darwin admitted that the eye is such a sophisticated work of engineering that the very idea of it forming by evolution is preposterous. In his book, *The Origin of Species*, Darwin himself said:

> To suppose that the eye, with all its inimitable contrivances for adjusting the focus to different distances, for admitting different amounts of light, and for the correction of spherical and chromatic aberration, could have been formed by natural selection, seems, I freely confess, absurd in the highest degree. (Charles Darwin, *The Origin of Species* [London: Penguin Classics, 1985], 217)

Of course, even though Darwin admits that the whole idea of the eye forming as a result of evolution is "absurd," he still chose to believe that it did. Thus, Darwin had an enormous amount of faith in his theory! When you see incredible engineering like that of the eye, it seems much more reasonable to assume that such a marvel was designed by a very intelligent Designer!

Even the best of designs, however, can be ruined. Sometimes, due to flaws in genetics or due to overuse under the wrong types of circumstances, an eye can develop nearsightedness (myopia) or farsightedness (hyperopia). If a person has **myopia**, the combination of the lens and cornea focuses light too strongly. Thus, the focal length is too small, and the image is formed in front of the retina. This can be corrected with a diverging lens. After all, if the combination of the cornea and the lens focuses light too strongly, the way to fix that is to "unfocus" the light a bit. That's the job of a diverging lens.

It a person has **hyperopia**, the cornea and lens cannot bend light strongly enough and, as a result, the image is formed behind the retina. This can be corrected with a converging lens. After all, if light is not focused well enough, the way to fix it is to focus the light some more. That's the job of a converging lens.

As a person gets older, his ciliary muscles get less and less functional. It becomes hard for the lenses in his eyes to change focal length. A single corrective lens is not enough, in this case, to fix his vision. If the ciliary muscle cannot effectively change the focal length of the eye's lens, when an older person wants to see something that is up close, he needs a corrective lens with one focal length. However, to see things far away, he needs a corrective lens with another focal length. This is accomplished through bifocals or trifocals. A bifocal lens is made up of two different lenses, each with a different focal point. One lens has a focal point conducive to looking at objects up close, and the other for looking at objects far away. The older person then trains himself to look through one section of the lens when looking at close objects and the other section of the lens when looking at objects which are far away. Trifocals work in the same manner, but there are three different focal points: one for close objects, one for objects at a medium distance, and one for objects far away.

Hopefully, this brief discussion of the eye has gotten you to appreciate a little more the amazing design that is present in your body. Truly, the human eye (and the rest of the body) is an outstanding testament to the power and majesty of our Creator!

ANSWERS TO THE "ON YOUR OWN" PROBLEMS

12.1 In a ray-tracing diagram, we first have to construct the mirror and object relationship properly. Since the radius of curvature is 12.0 cm, the focal point is 6.0 cm, because Equation (12.1) tells us that the focal point is always half of the radius of curvature. Since the object is only 4.5 cm away from the mirror, we know that the object is closer to the mirror than the focal point. In fact, since 4.5 is 1.5 less than 6.0, and 1.5 is ¼ of 6.0, the object is ¼ closer to the mirror than the focal point. Our situation, then, looks like this:

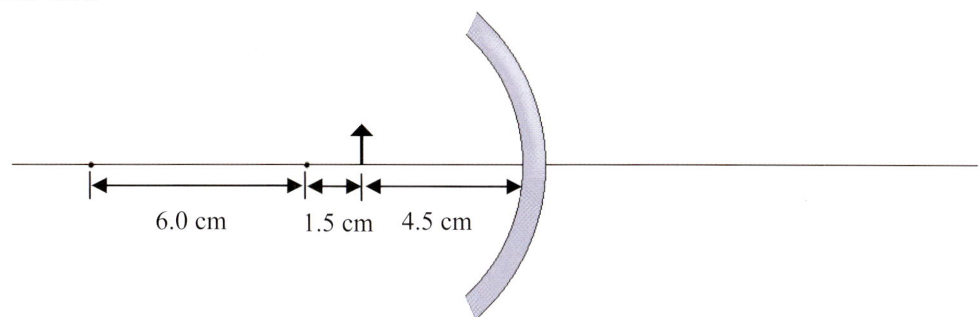

Now we can trace rays. Rather than tracing the rays one at a time, I will do them all at once. Remember what the rays are. They all start from the top of the object. The first travels to the mirror horizontally and then reflects back through the focal point. The next travels from the focal point to the mirror and reflects back horizontally. Since the object is inside the focal point, however, the only way to draw that ray is to draw it *as if it came* from the focal point. In the same way, the last ray travels *as if it came* from the radius of curvature and is reflected directly backwards. This is what it all looks like, then:

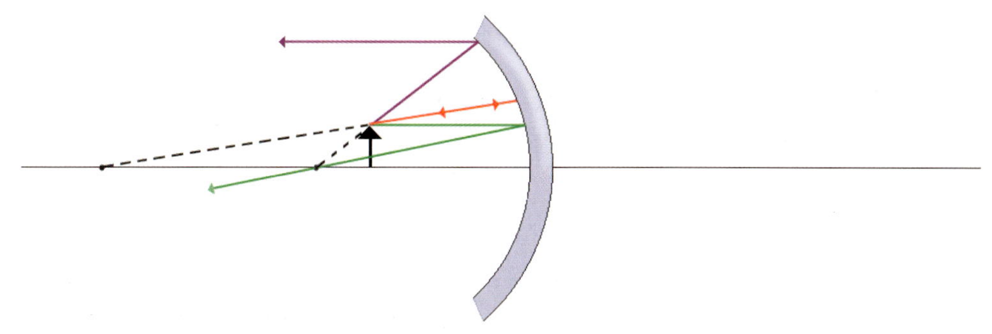

Since the reflected rays do not intersect, we must be dealing with a virtual image. Therefore, we will extrapolate the rays backwards until we can get them to intersect. That will be where the top of the image is. From there, we can draw the rest of the image, making sure that the bottom of the image touches the center line:

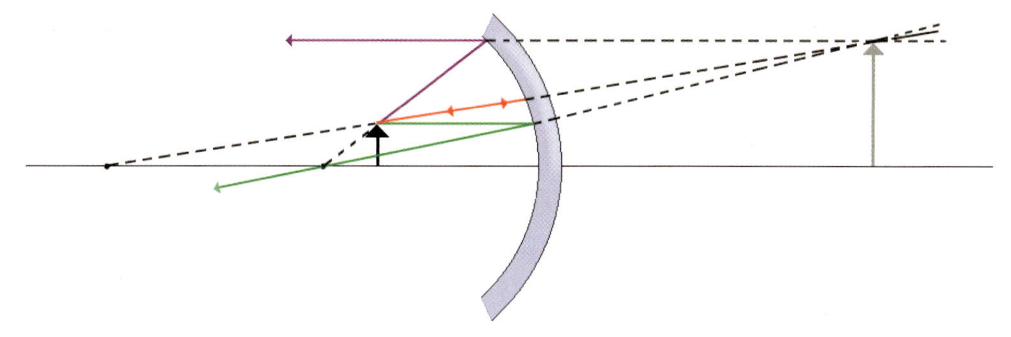

The image is formed by extrapolated light rays, so it's a virtual image. It is also pointing the same way as the original arrow, so it is upright. Finally, it is larger than the original arrow, so it is magnified.

12.2 Once again, we need to get the object/mirror relationship right. The focal point is 6.0 cm away, since it is half the radius of curvature. The object is ¼ of the way in between the focal point and "R." Thus, our picture looks like this:

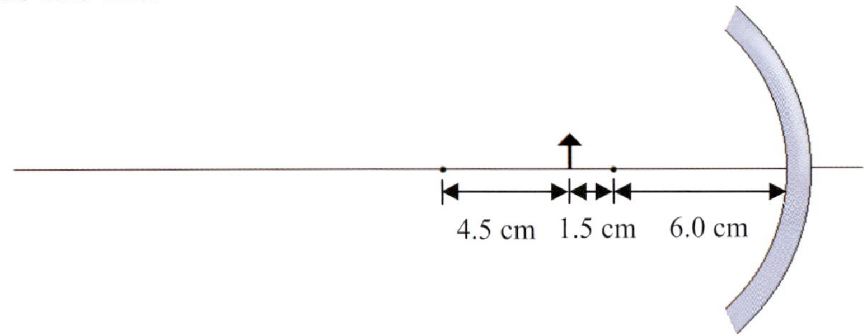

Drawing the three rays gives us:

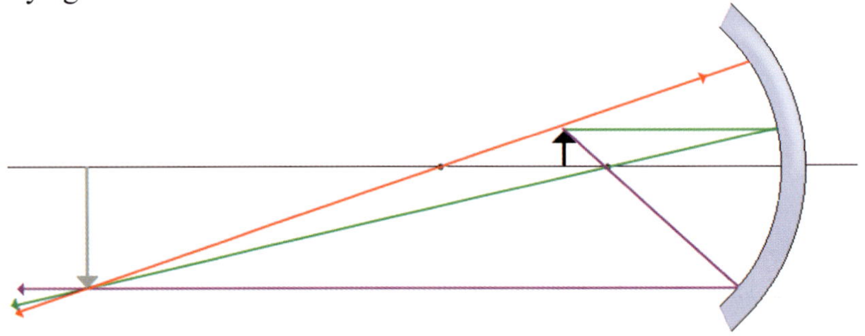

Since this image is formed by intersecting rays, it is a real image. It is upside down compared to the original arrow, so it is inverted. Finally, it is larger than the original arrow, so it is magnified.

Please note that as you make these ray-tracing diagrams, your three rays (or the extrapolated rays) may not intersect at exactly the same point. This could be because of small imperfections in your drawing, or you might actually be seeing the effect of spherical aberrations. In any event, don't worry if your rays do not intersect perfectly. Even with a relatively poor intersection, you will still be able to determine whether the image is real or virtual, upright or inverted, and magnified or reduced, and that's what's important.

12.3 Once again, we start with a picture of the situation. The focal point is at 3.0 cm, so the object is one-third closer to the mirror than is the focal point. Of course, since this is a convex mirror, the object is on the side opposite the focal point:

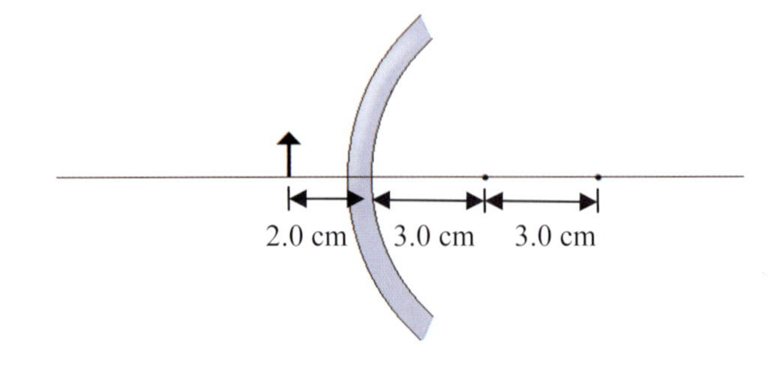

Drawing the three rays gives us:

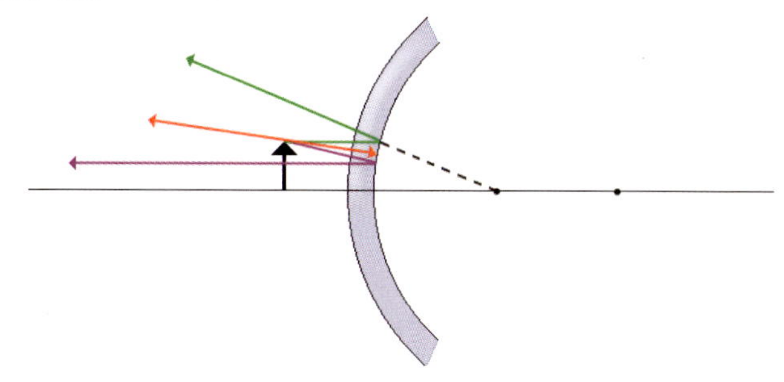

The rays do not intersect, so we will have to extrapolate them:

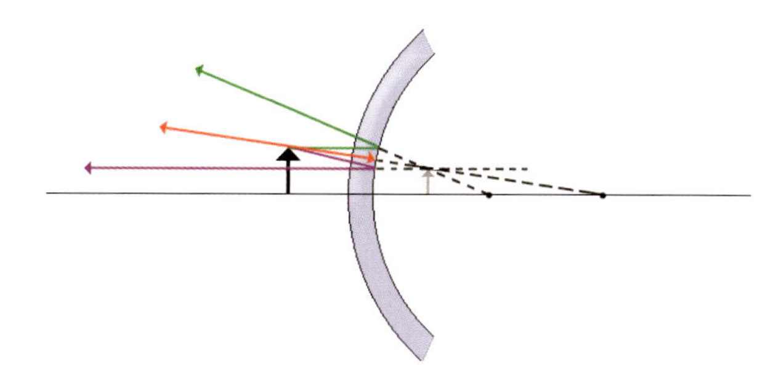

The image, then, is virtual, upright, and reduced.

12.4 Although there is no example illustrating how to solve this kind of problem, it's really pretty easy. After all, you are given the index of refraction for water and are asked to determine the speed of light in water. Well, Equation (12.4) relates these two things, so the solution is a snap:

$$n = \frac{c}{v}$$

$$1.3 = \frac{3.0 \times 10^8 \frac{\text{m}}{\text{sec}}}{v}$$

$$v = \frac{3.0 \times 10^8 \frac{\text{m}}{\text{sec}}}{1.3} = 2.3 \times 10^8 \frac{\text{m}}{\text{sec}}$$

The speed of light in water, then, is 2.3×10^8 m/sec.

12.5 This problem is just like the example. We are supposed to know the index of refraction of air, which means $n_1 = 1.00$. We are also given θ_1 (67.1°) and n_2 (2.42). From this information, we must determine θ_2.

$$n_1 \cdot \sin(\theta_1) = n_2 \cdot \sin(\theta_2)$$

$$(1.00) \cdot \sin(67.1°) = (2.42) \cdot \sin(\theta_2)$$

$$\sin(\theta_2) = \frac{(1.00) \cdot \sin(67.1°)}{2.42}$$

$$\theta_2 = 22.4°$$

The angle of refraction, then, is 22.4°.

12.6 This problem is slightly different from the example, but it still utilizes Equation (12.3). In this case, the ray starts out in glass. Thus, $n_1 = 1.5$. We are given the angle of incidence ($\theta_1 = 34.6°$). When the ray leaves the glass, it is in air, which means $n_2 = 1.00$. The only thing we don't know is θ_2, so we can solve for it:

$$n_1 \cdot \sin(\theta_1) = n_2 \cdot \sin(\theta_2)$$

$$(1.5) \cdot \sin(34.6°) = (1.00) \cdot \sin(\theta_2)$$

$$\sin(\theta_2) = \frac{(1.5) \cdot \sin(34.6°)}{1.00}$$

$$\underline{\theta_2 = 58°}$$

12.7 To solve this problem, we first have to draw the diagram consistent with the description. The object is twice as far from the lens as the focal point. This leads to the following diagram:

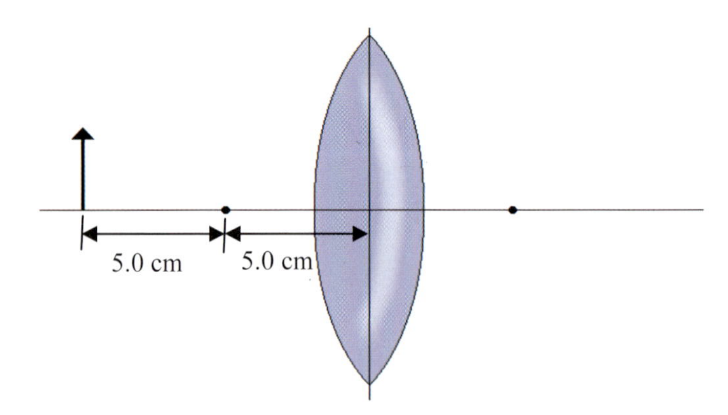

Now we can draw the three rays. Since these are the same as those in the example, I will just draw the final version of the diagram:

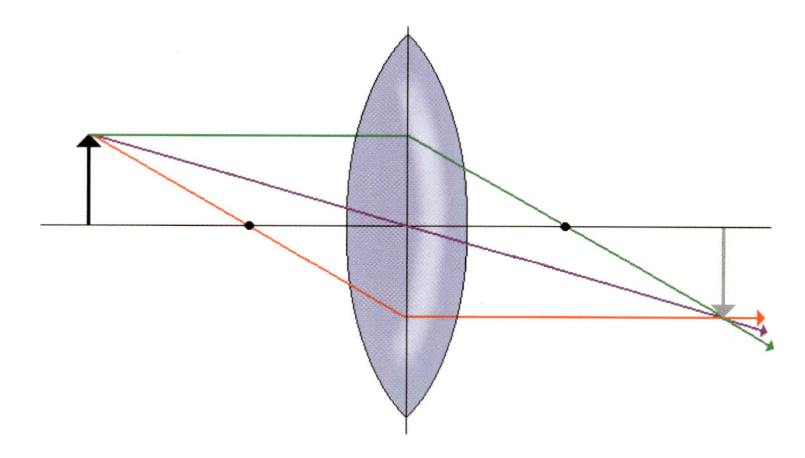

The image, then is real and inverted. It also appears to be about the same size as the object and appears on the other side of the lens.

12.8 This problem sets up a situation in which the object is closer to the lens than the focal point. Based on the measurements, the object is just slightly closer to the focal point than it is to the center of the lens:

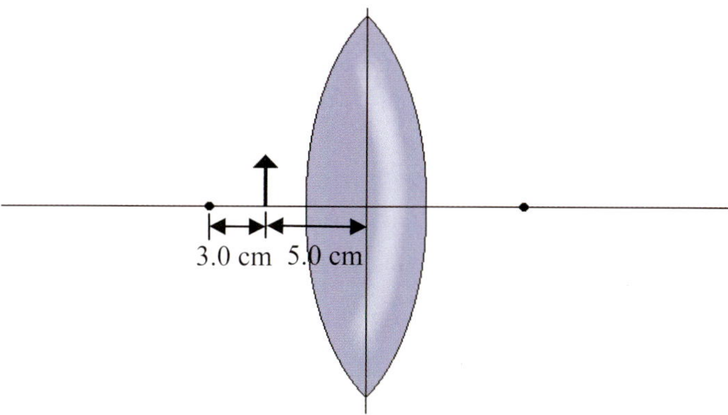

Now we draw the rays. The first is easy:

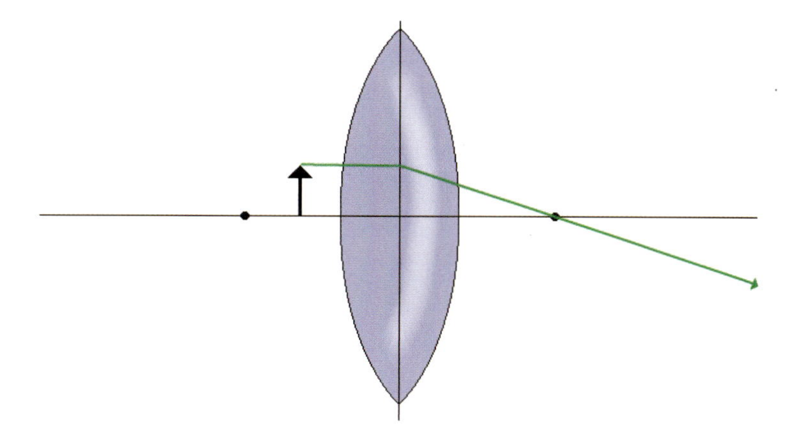

The next ray is harder to draw, though. After all, it is supposed to travel through the focal point to the lens. The object, however, is closer to the lens than the focal point. Thus, we must draw the ray *as if* it came from the focal point, just as we did in the case of the concave mirror.

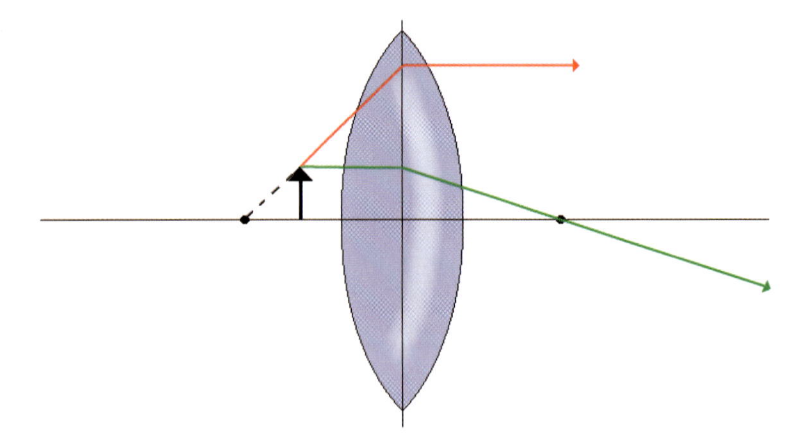

The last ray heads towards the center of the lens and is not deflected at all:

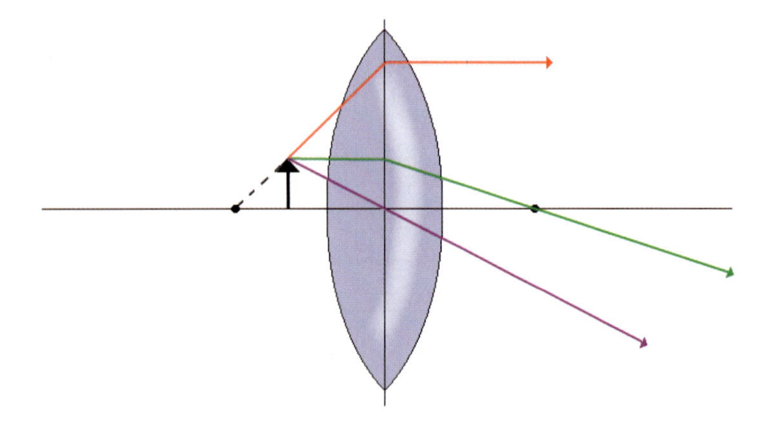

These rays do not intersect. Thus, just as we did with concave mirrors, we need to extrapolate the rays backwards to find the intersection point. That will determine the image.

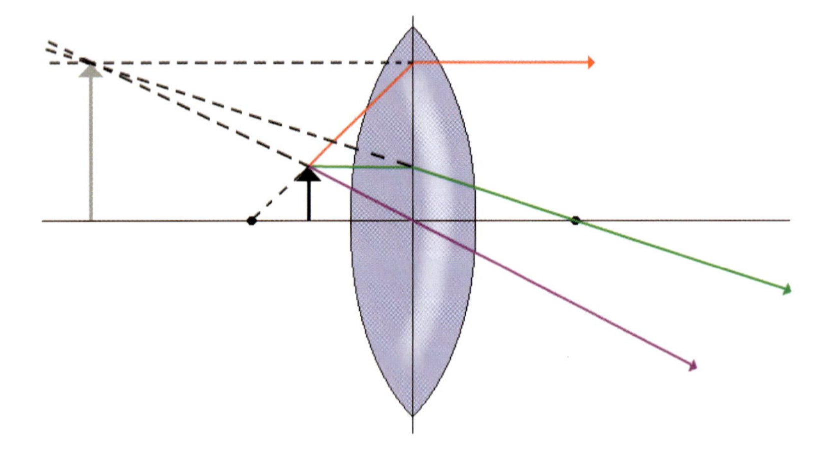

The image, then, is <u>virtual and upright</u>. The image also appears on the same side of the lens as the object and <u>is magnified</u>.

12.9 In this situation, the object is slightly more than twice the distance from the lens as compared to the focal point. Thus, the drawing looks like this:

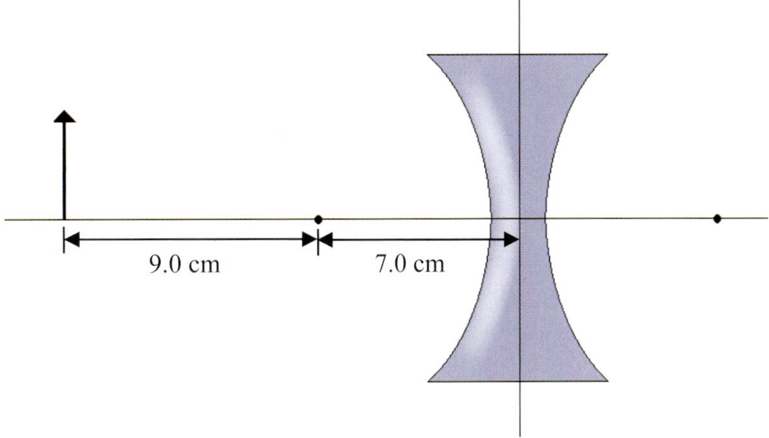

Now remember, for diverging lenses, horizontal rays pass through as if they were coming from the focal point on the same side of the lens; rays headed to the focal point on the other side of the lens come through traveling horizontally; and rays passing through the center are not deflected. The final drawing, then, looks like this:

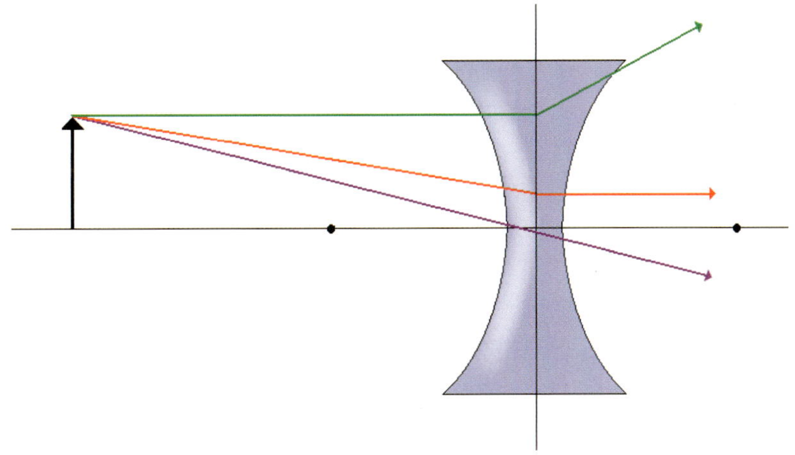

The light rays do not intersect, so they must be extrapolated backwards:

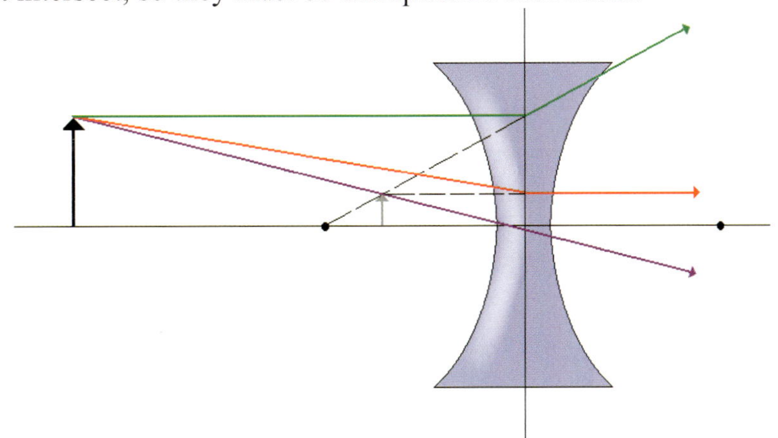

The image, then, is virtual, upright, and reduced.

12.10 In this case, the object is between the lens and the focal point. It is just slightly closer to the focal point than it is to the lens, however.

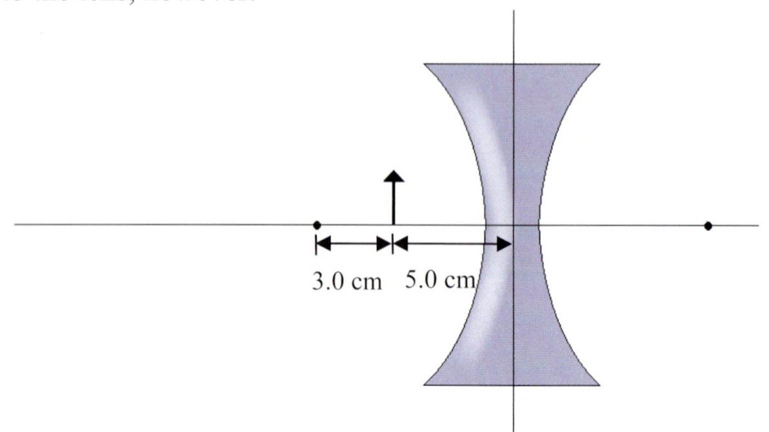

Now we draw our three rays:

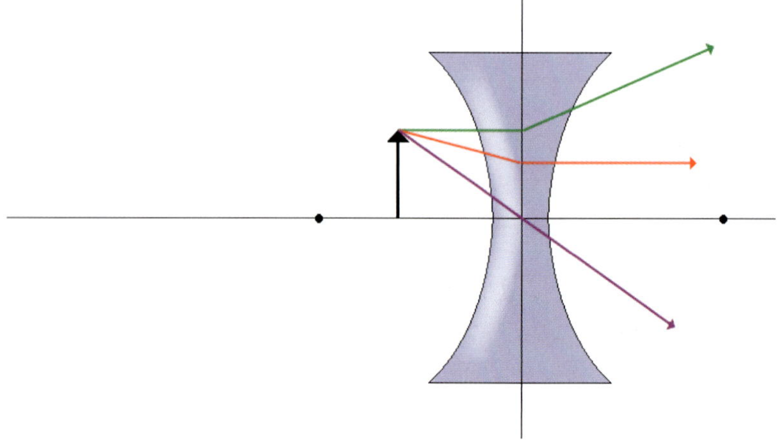

The rays do not intersect, so we must extrapolate them backwards:

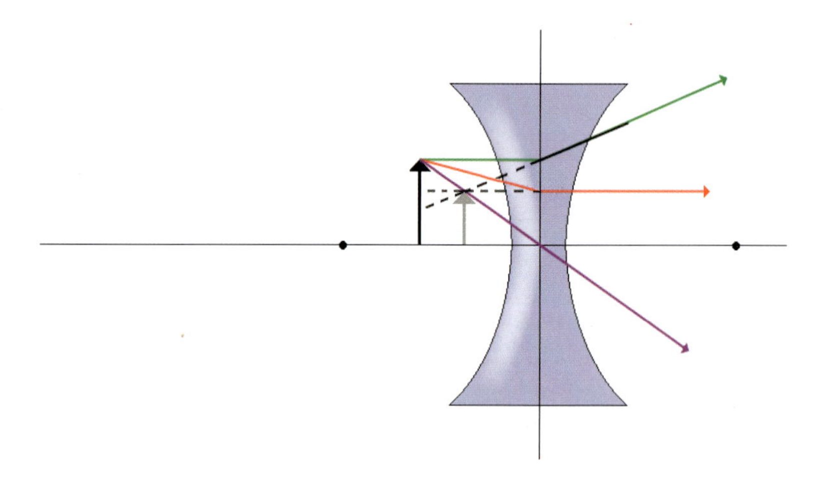

The image is <u>virtual, upright, and reduced</u>.

REVIEW QUESTIONS

1. Suppose you want to see your entire body in a flat mirror. Do you need a mirror that is as tall as you are? (HINT: Look closely at Figure 12.1.)

2. If you are doing a very precise optical experiment and need a curved mirror that will sharply focus all horizontal light rays to a single point, which kind of mirror should you use: a parabolic mirror or a spherical mirror?

a pencil in water

3. What is the difference between a virtual and real image?

4. Notice how the pencil looks disjointed above and below the water level in the picture on the right. What causes this?

5. Explain why you can see through a window but you can often see your reflection in the window as well.

6. In my church, there is a special window between the foyer and the nursery that allows parents to look into the nursery to check on their little children. However, from inside the nursery, the window looks like a mirror, and the children cannot see out of it. That way, the children will not see their parents when their parents are checking on them. Consider a light ray striking the window from *inside* the nursery. Will it be absorbed, reflected, refracted, or some combination of those three processes when it hits the window? Also, consider a light ray striking the window from *outside* of the nursery. Will it be absorbed, reflected, refracted, or some combination of those three processes when it hits the window?

7. The index of refraction of substance A is 75% that of substance B. In which substance will light travel with the greatest speed?

8. Given the statement in the question above, which substance most likely has the higher density?

9. When a photographer is taking a picture and does not use autofocus, she manually turns a cylinder around the lens. This focuses the image onto the film. What is the photographer doing to the lens when she turns the cylinder?

10. How does the human eye keep the image of the object it is viewing focused at the same place, even when the object moves?

PRACTICE PROBLEMS

1. An object is placed 3.5 cm away from a concave, spherical mirror whose radius of curvature is 5.0 cm. Draw a ray-tracing diagram to illustrate the image, and determine whether the image is real or virtual, upright or inverted, and magnified or reduced.

2. An object is placed 3.0 cm away from a spherical, concave mirror whose radius of curvature is 10.0 cm. Draw a ray-tracing diagram to illustrate the image, and determine whether the image is real or virtual, upright or inverted, and magnified or reduced.

3. An object is placed 5.0 cm away from a spherical, convex mirror whose radius of curvature is 6.0 cm. Draw a ray-tracing diagram to illustrate the image, and determine whether the image is real or virtual, upright or inverted, and magnified or reduced.

4. Light traveling in air is incident on a block of ice (n = 1.31) at an angle of 25°. What is the angle of refraction?

5. When light travels through a substance and encounters another substance with a lower index of refraction, there are some angles of incidence for which refraction is just not possible. For those angles, the light is totally reflected back into the first substance. This phenomenon is called "total internal reflection." Consider, for example, light traveling in glass (n = 1.5) and then encountering air. Angles of incidence ranging from θ_{tr} to 90° are totally reflected back into the glass. What is θ_{tr} for this situation? (HINT: Solve Snell's Law for θ_2 and consider what values of θ_1 make the equation undefined. Remember, the sine of an angle can never exceed 1.)

6. The index of refraction of ethyl alcohol is 1.4. What is the speed of light in ethyl alcohol?

7. Light is traveling in a special kind of glass known as "flint glass." If it encounters air at an angle of incidence equal to 20.1° and refracts at 35.7°, what is the index of refraction of the flint glass?

8. A converging lens has focal points that are 12.0 cm from its center. If an observer looks through the lens at an object 3.0 cm from the lens, how does the image appear? Is it real or virtual? Is it upright or inverted? Is the image magnified, reduced, or essentially the same size as the object?

9. A converging lens has focal points that are 8.0 cm from its center. If an observer looks through the lens at an object 12.0 cm from the lens, how does the image appear? Is it real or virtual? Is it upright or inverted? Is the image magnified, reduced, or essentially the same size as the object?

10. A diverging lens has focal points that are 8.0 cm from its center. If an observer looks through the lens at an object 5.0 cm from the lens, how does the image appear? Is it real or virtual? Is it upright or inverted? Is the image magnified, reduced, or essentially the same size as the object?

MODULE #13: Coulomb's Law and the Electric Field

Introduction

Have you ever walked across a carpet, touched something metal, and received an electrical shock? What about sliding out of a car and touching the metal handle? Have you ever gotten shocked that way? What causes these shocks? To some extent, everyone knows the answer to that question: it's static electricity. But what *is* static electricity, and why does walking across a carpet or sliding across the seat of a car produce it? Why don't you feel a shock until you touch something metal? In this module, we will try to learn about these phenomena through a study of **electrostatics** (ih lek' troh stat' iks):

Electrostatics - The study of electric charges at rest

To understand electrostatics, you need to understand a few things about electric charge. Most of the basics you probably already know, since you probably learned about electric charges in a previous course. There are two types: positive and negative. Most of the matter with which we are familiar has no net electric charge because it contains an equal amount of both positive and negative charges. Like charges (positive and positive or negative and negative) repel each other while unlike charges (positive and negative) attract one another. Finally, the fundamental unit of negative electric charge is the electron, and the fundamental unit of positive electric charge is the proton.

Like most subjects in physics, the study of electrostatics can get pretty deep. As a result, we will concentrate on just a few aspects of this fascinating subject. In this module you will learn how electric charges affect one another, how things become electrically charged, and the strength with which electric charges either repel or attract one another. In later modules, you will learn how these and other concepts can be applied to make useful things like electric circuits. I am getting ahead of myself, however. Let's get through the basics first.

The Basics of Electric Charge

The first thing you need to know is how things get electrically charged. As you should have learned in chemistry, matter is made up of atoms. Atoms, in turn, are made up of protons, neutrons, and electrons. The protons of an atom are positively charged and the electrons are negatively charged. The neutrons have no electric charge. The number of protons in an atom equals the number of electrons. Since there are an equal number of positive and negative charges, they cancel each other out, resulting in no net electric charge. When this happens, we say that the matter composed of these atoms is electrically neutral.

For a number of reasons that you should have learned in chemistry, sometimes an atom gains or loses one or more electrons. At that point, it is called an **ion** (eye' on). If the atom gains electrons, it has more negative charges than positive ones. As a result, the atom becomes a negatively-charged ion. If, on the other hand, an atom loses electrons, it has more positive charges than negative charges. This turns the atom into a positively-charged ion. Thus, matter becomes electrically charged when its atoms either gain or lose electrons. The best way to see how electric charge affects the behavior of matter is to perform the following experiment.

EXPERIMENT 13.1

Attraction and Repulsion

Supplies:

- Two balloons. Round balloons work best, but any kind will do.
- Thread
- Cellophane tape

1. Blow up the balloons and tie them off so that they each stay inflated.
2. Tie some thread to one of the balloons and attach the other end of the thread to the ceiling with some tape so that the balloon hangs from the thread. Make the length of the thread so that the balloon hangs at about the same height as your chest.
3. Take the balloon that is hanging by the thread and rub it in your hair a little. This will cause the balloon to pick up some electric charge. Now back away from the balloon and allow it to hang there.
4. Take the other balloon and rub it in your hair just a little.
5. Hold this balloon in both of your hands and slowly bring it close to the balloon that is hanging from the thread. What happens?
6. Play with the situation a bit, trying to see what kind of motion you can induce in the hanging balloon.
7. Vigorously rub the balloon that is in your hands in your hair. Spend significantly more time doing it this time as compared to what you did in step 4.
8. Once again, bring the balloon in your hands close to the balloon that is hanging on the thread. Note what happens, and note how the motion of the hanging balloon compares to its motion in step 5 and step 6.
9. Put away the balloon that is in your hands.
10. Take a piece of tape that is at least 15 cm long and tape it to the top of a table. Leave a little part of it unfastened, so that you can remove it in a moment. Be sure to ask your parents which table you should use for this, as what you will do in the next step can damage the finish on some tables.
11. Quickly rip the tape off the table and grasp it at both ends. Hold the tape near the balloon, with the sticky side facing the balloon. What happens this time?
12. Once again, play with the situation a bit to see what kind of motion you can induce in the hanging balloon.
13. Clean up your mess.

What did you see in the experiment? Well, in the first part of the experiment, you should have seen that the hanging balloon moved away from the balloon in your hands when you brought the two balloons close to one another. The more you rubbed the one balloon in your hair, the farther the hanging balloon tried to get away from the balloon in your hands. However, when you brought the piece of tape close to the hanging balloon, you should have seen the balloon move *towards* the tape.

Why did the hanging balloon behave the way that it did? When you rub a balloon in your hair, it picks up some stray electrons that are in your hair. This causes the balloon to pick up an overall negative charge. Since you rubbed both the hanging balloon and the balloon in your hand in your hair, both of them developed a negative charge. When you brought one close to the other, they began to repel each other, because charges that have the same sign repel each other. The repulsion was greater the closer the balloons got together. Also, when you rubbed the balloon in your hand more vigorously in your hair, it picked up more negative charge, which also increased the repulsion.

When you ripped the tape off the table, you were actually causing the tape to lose negative charges. This is because the tape leaves electrons behind in the sticky residue left on the table. Since the tape lost negative charges, it became positively charged. When you held that up to the balloon, the balloon was attracted to it, because negatively-charged matter attracts positively-charged matter.

The experiment should have given you a little experience with some concepts that you probably knew already. First of all, you probably already knew the repulsion/attraction rule regarding electric charges:

Like charges repel each other while opposite charges attract one another.

The multimedia CD has a "hair-raising" video that demonstrates this fact.

Also, it should be obvious that the more charge there is, the stronger the repulsion or attraction will be. Finally, the closer the charges are to one another, the stronger the repulsion or attraction will be.

In a moment, we'll put all of these facts together in a nifty mathematical law that will allow us to calculate the magnitude of the attractive or repulsive force that exists between electric charges. Before we do that, however, we need to get some terminology worked out. First of all, from an electricity point of view, we can separate all of matter into two basic groups: **conductors** and **insulators**. Conductors, as their name implies, can conduct charge. In other words, electric charges move freely from one place in a conductor to another. Insulators, on the other hand, do not conduct charge, so electric charges do not move freely in an insulator.

Conductor - A substance through which charge flows easily

Insulator - A substance through which charge cannot flow

Most metals (aluminum, silver, zinc, etc.) are conductors, while most non metals (plastics, ceramics, glass, etc.) are insulators. After the review questions at the end of this module, you will find my favorite joke about this topic.

As is often the case in nature, the boundary between these two classes of materials is not sharp. There are some materials through which charge can flow, but it takes a lot of work to accomplish the task. Also, there are a few substances known as semiconductors, which allow charge to flow only under certain conditions. These semiconductors are quite useful. Virtually anything that is computerized or uses radio waves is composed of semiconductors. In this course, however, we will not concentrate on these kinds of materials. We will focus on materials that are easily classified as either conductors or insulators.

Next, we need to explore a little bit about how things can become electrically charged. First, we must realize that charge must be conserved. As I said before, a substance becomes electrically charged when it gains or loses electrons. Well, these electrons must come from or go to somewhere; thus, if one thing becomes negatively charged, something else must become positively charged. For example, in Experiment 13.1, you charged a balloon by rubbing it in your hair. The balloon picked up electrons that were in your hair, causing it to become negatively charged. Did you (or someone watching you) notice what happened to your hair when you did this? Your hair began to stand on end a little. Since the balloon took electrons out of your hair, the atoms which make up your hair were suddenly short a few electrons. This left them with more protons than electrons, giving your hair a net positive charge. Since your hairs were positively charged, they repelled each other. In trying to get away from the other hairs with positive charge, some of your hairs stood up.

Do you see what happened? In order to get the balloon to become negatively charged, your hair had to become positively charged. That's because in order for something to gain electrons, something else must lose them. This leads us to the **Law of Charge Conservation**.

Law of Charge Conservation - The net amount of electric charge in the universe is constant.

In other words, since one substance must lose electrons for another to gain them, whenever a negative charge is formed, an equal positive charge must also form.

There are other ways of charging an object besides rubbing electrons on or off it. The two ways that we need to study are **conduction** and **induction**. Both of these charging methods are best demonstrated by experiment. If you do not perform this experiment, please read it through and follow it, because there are some concepts that I discuss only in the context of this experiment. You will be tested on these concepts, so you need to know them whether or not you actually perform the experiment.

EXPERIMENT 13.2

Making and Using an Electroscope

Supplies:

- Safety goggles
- A glass
- A plastic lid that fits over the glass. This lid can be larger than the mouth of the glass, but it cannot be smaller. The top of a margarine tub or something similar works quite well.
- A paper clip
- Two 5-cm x 1.5-cm strips of aluminum foil (the thinner the better)
- A balloon
- A pair of pliers
- Tape

1. Using your hands and the pliers, straighten out and then bend the paper clip so that it ends up looking something like this:

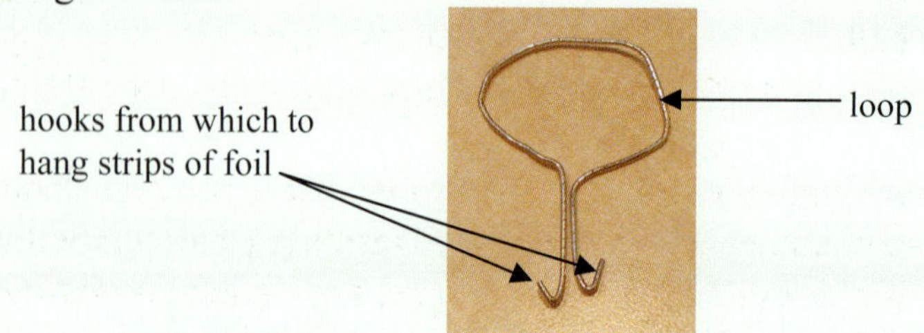

2. Cut a thin slot in the plastic lid. Slide the loop of the twisted paper clip into the slot, then twist it 90 degrees so that the slot holds the loop in place. The loop should stand perpendicular to the lid. You may need to use some tape to hold it in this position (see the photo on the next page).

3. Poke a hole near one end of each strip of foil and hang the foil strips on the tiny hooks that are at the bottom of the twisted paper clip.
4. Place the lid on top of the glass, so that the foil strips hang on the inside of the glass. You have just made an electroscope. It should look something like this:

5. Well, now that you've made the electroscope, what good is it? An electroscope detects the presence of electric charge. To see this, inflate the balloon and tie it off so that it stays inflated.
6. Rub the balloon in your hair to charge it.
7. Slowly bring the balloon close to the loop of the twisted paper clip without actually touching it. The foil strips should start to move. If they do not, your balloon is probably not charged well. Rub it more vigorously in your hair, or rub it in someone else's hair.
8. Note how the foil strips move as you bring the balloon closer to the loop. Don't actually touch the loop with the balloon!
9. Pull the balloon away from the loop and note how the foil strips move. Do this a couple of times so that you can describe the motion of the foil strips well.
10. Bring the balloon near the loop one more time. This time, however, allow the balloon to touch the paper clip. Note what happens to the foil strips.
11. Pull the balloon away. This time, the behavior of the foil strips should be noticeably different from what it was in step 9.
12. I want you to do one more thing. This might be a little tricky. Touch the loop with your finger. You should notice that the foil strips respond to your touch. Note what they do.
13. Take your finger away from the loop.
14. Bring the balloon close to the loop, but do not touch it with the balloon. When you see the foils move significantly, hold the balloon where it is and touch the paper clip with a finger from your other hand. Keep the finger resting on the paper clip. As soon as you touch the paper clip, the foils should move again.
15. Pull both the balloon and your finger away *at the exact same time*. The foils should move yet again, and behave similarly to what you saw in step 11. This doesn't always work the first time, because timing is essential. If it doesn't work the first time, try it a few more times. It should eventually work.
16. Clean up your mess.

What happened in the experiment? Well, let's start with what happened when you brought the balloon close to the loop but did not let it touch the loop (step 7). When you did that, the foils should have pulled away from each other. Why did this happen? The foil and the paper clip, like all forms of matter with which we are familiar, have both positive and negative charges in them. The number of positive charges and negative charges, however, are equal; thus, the foil and paper clip have no overall charge. The balloon, because it picked up some stray electrons when you rubbed it in your hair, had more negative charges than positive ones, so it had an overall negative charge. When the negatively-charged balloon came in close proximity to the electroscope, the negative charges in the paper clip and foil responded as shown in the figure below:

FIGURE 13.1

An Electroscope Responding to Charge

Illustration by Megan Whitaker

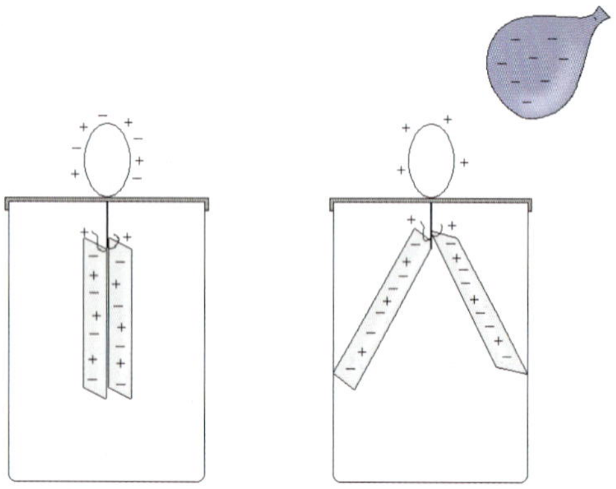

When brought in close proximity to the paper clip, the negative charge of the balloon repelled the negatively-charged electrons in the paper clip and the foil. Because the foil and the paper clip are conductors, the electrons were able to travel freely in them. Since they were repelled, they traveled away from the balloon, which caused the ends of the foil to be rich with electrons. The excess of negative charges in the foils caused each foil to become negatively charged. This made the foils repel each other, and that's why they pulled apart from each other in the experiment.

What happened when you moved the balloon away (step 9)? The foils should have relaxed back to their normal position. That's because once the negatively-charged balloon moved away, the electrons crammed together in the foil were able to travel back to their normal position, making everything neutral again. When that happened, the foils could hang down normally again. This is how an electroscope responds to electric charge that does not actually come into contact with it. The electrons in the electroscope move in response to the charge, which causes the foils inside the electroscope to move. This, then, is a way of detecting whether or not an object is electrically charged.

Okay, what happened in the next part of the experiment? In step 10, you let the balloon touch the loop, and then you pulled it away. When the balloon actually touched the loop, the foils probably relaxed a bit, but they still stayed pretty far apart from one another. This time, however, as you pulled the balloon away (step 11), the foils should not have relaxed much. Instead, they should have stayed far away from each other. Why did this happen?

When you touched the balloon to the paper clip, some of the balloon's extra electrons were able to flow into the paper clip and into the foils. Remember, those negative charges are repelled by the other negative charges on the balloon. Thus, they want to get away. However, they can't, because the balloon and the air around them are insulators. Thus, they cannot move anywhere. However, once the balloon touched the conductive paper clip, the negative charges could move, and they quickly moved into the paper clip and foils. This gave the system extra negative charges. When you pulled the balloon away, those extra electrons stayed. This caused a permanent negative charge to develop on the foils. Since the foils stayed negatively charged, they continued to be repelled from one another, so they continued to stay away from one another.

When you charge something by touching an electrically charged object to it and allowing the charges to flow between the electrically charged object and the object that you are charging, physicists say that you are **charging by conduction** (kun duk' shun).

Charging by conduction - Charging an object by allowing it to come into contact with an object that already has an electric charge

In other words, by allowing electric charges to be conducted between the object you are charging and an object that already has a charge, you are charging by conduction.

The foils should have stayed suspended away from each other until you touched them (step 12). At that point, they should have relaxed. Why? You are a conductor. Thus, the extra negative charges in the foil were able to travel into you and away from the foil and paper clip. You are so large compared to the foil and paper clip that you took essentially all of the extra negative charges away from the foil and paper clip. As a result, everything in the electroscope became neutral again, and the foil relaxed. When you get rid of the electric charge on an object, you are **discharging** the object.

Okay, we still have one aspect of the experiment to discuss. In the "tricky" part of the experiment (steps 13-15), you brought the balloon close to the loop without touching it. As you expected, the foils moved away from each other. Then, you touched the loop with a finger from the other hand. At that point, the foils relaxed. Then, when you pulled your finger and the balloon away at the same time, the foils should have pulled away from each other again and stayed that way. If they did not, that's okay. This part of the experiment is quite tricky. In any event, that's what *should* have happened. Why should it have happened? Study the figure below to find out.

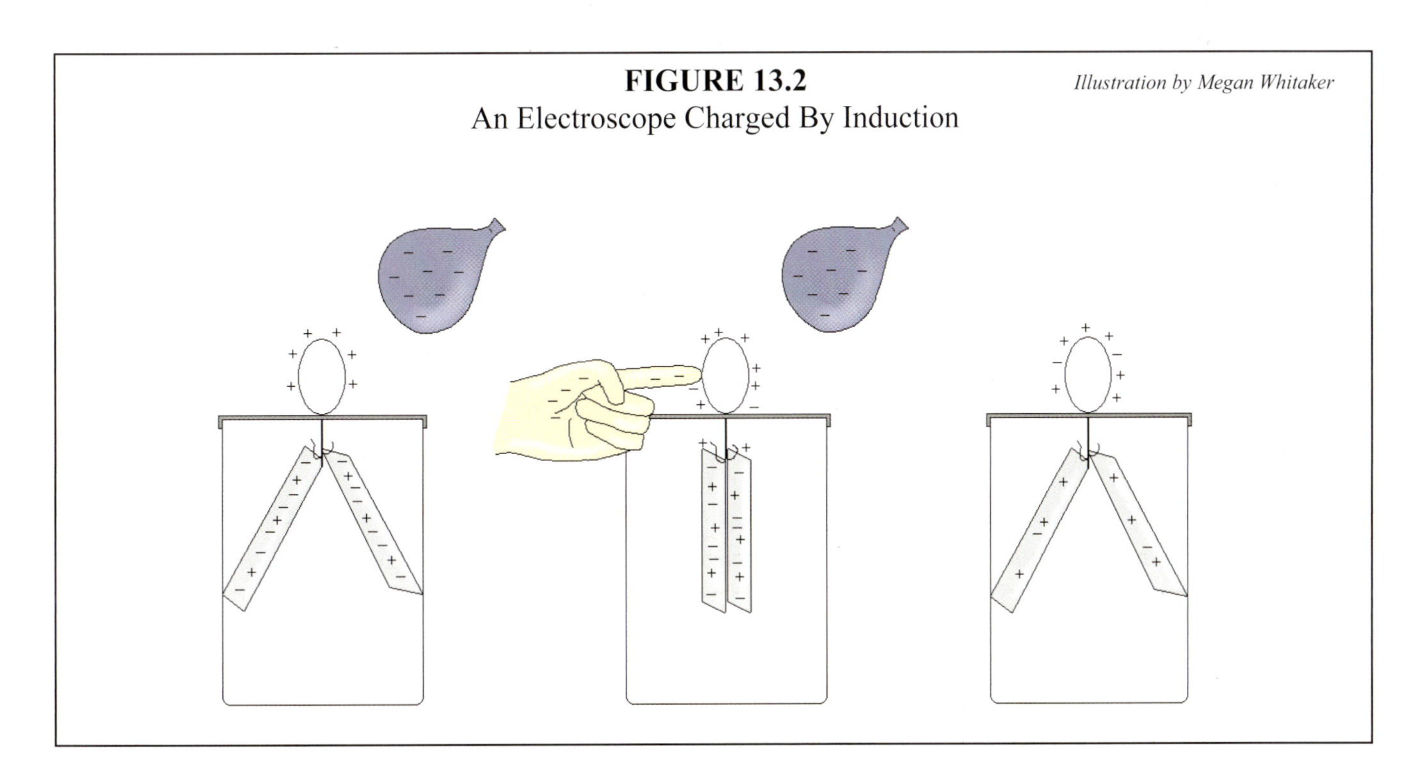

FIGURE 13.2
An Electroscope Charged By Induction

Illustration by Megan Whitaker

When the balloon was moved near the paper clip, the negative charges in the paper clip and foil moved away from the balloon, concentrating negative charge in the foils, causing the foils to repel each other. When you touched your finger to the paper clip, however, the negative charges traveled farther away from the balloon by traveling through your finger and into your body. Thus, you removed some of the negative charges from the foil and paper clip. This should have caused the foils to relax. When you moved your finger and the balloon away at the same instant, the paper clip and foil were left with fewer electrons than would have otherwise been there, because some of those electrons were in your body. This gave the foils and the paper clip an overall *positive* charge. Since the foils were positively charged, they repelled each other, and they moved away from each other again.

When you performed the experiment in this way, you induced the negative charges to leave the foil by giving them an escape route: your hand. This caused the foils to become positively charged when you took your hand and the balloon away. This, as you might imagine, is called **charging by induction** (in duk' shun).

Charging by induction - Charging an object without direct contact between the object and a charge

When you charged the electroscope in this way, the resulting charge was opposite that of the balloon. When you charged by conduction, you gave the foils some of the extra charge that was on the balloon. This caused the foils to become negatively charged. When you charged by induction, however, you forced negative charges out of the foils, causing them to become positively charged. Thus, charging by induction gave you the opposite charge as compared to charging by conduction.

ON YOUR OWN

13.1 When you walk on a carpeted floor, your body can sometimes become charged. This is evidenced by the fact that when you touch a metal object, you get shocked by the charge that passes from your body to the metal object. Does the carpet charge you by conduction or induction?

13.2 You want to use a positively-charged object to develop a negative charge on another object. Will you charge by conduction or induction?

Electrostatic Force and Coulomb's Law

In Experiment 13.1, you learned that the electrical force which exists between two charged objects depends on two things: the amount of charge and the distance between the objects. The closer the objects are, the greater the force; and the more charge there is, the greater the force. French engineer and physicist Charles Coulomb (koo' lohm) determined experimentally a mathematical formula that quantifies this relationship. Before I give you the formula, however, I must firm up our terminology. I have been throwing the term "electrical force" around quite a bit in this module, but it is not quite correct. There are many forces that arise from the phenomenon of electricity, so we must be more specific when talking about the force that we are studying here.

In both of the experiments you have done, you worked with what physicists call static charges. We call them static charges because, when the charge is in place, it does not move. For example, once you rubbed the balloon in your hair, the balloon developed a negative charge. That charge stayed on the balloon in Experiment 13.1. Even in Experiment 13.2, the charge did not move except when you transferred it to the electroscope. Once again, however, when the electroscope was charged, the charge did not move. Since the charges were static, the force that we observed in both experiments is called an **electrostatic force**.

Electrostatic force - The force that exists between two charges at rest

We will spend a lot of this module learning the details of the electrostatic force.

Coulomb's Law of electrostatic force between two charged objects is best expressed mathematically:

$$F = \frac{kq_1q_2}{r^2} \qquad (13.1)$$

In this equation, "F" is the magnitude of the electrostatic force, "q_1" is the charge of one of the two objects, "q_2" is the charge of the other object, and "r" is the distance between the centers of the objects. The symbol "k" represents a physical constant. We will get into the details of how to use this equation in a moment. First, however, notice how it reproduces the trends that you learned in Experiment 13.1. If the charge of either object increases, the force will increase. Since the distance between the objects is in the denominator of the fraction, a decreased distance will lead to an increased force. Thus, you can see that this equation qualitatively reproduces the observations you made in the experiment.

Now Equation (13.1) ought to look familiar to you. In Module #7, we discussed Newton's Law of Universal Gravitation:

$$F_g = \frac{Gm_1m_2}{r^2} \qquad (7.3)$$

Notice the similarity here. Both the gravitational force and the electrostatic force depend on the square of the distance between the objects, and they are both multiplied by a physical constant. Additionally, while the gravitational force depends on the mass of both objects in question, the electrostatic force depends on the charge of both objects in question. Why do these equations look so similar? We really don't know. It just turns out that way. I personally think that God enjoys symmetry. It turns out that the magnetic force is governed by an equation that is equally similar to each of these. An interesting "coincidence," don't you think?

Let's get into the details of Equation (13.1) now. First of all, we need to learn about measuring charge. To honor Charles Coulomb for his discovery of Equation (13.1), we call the standard unit of charge the Coulomb (abbreviated as "C"). Now it turns out that the Coulomb is a rather large unit. What I mean to say is that it is very rare to see a charge of even a few Coulombs. Most charges that physicists measure are milliCoulombs (0.001 Coulombs) or microCoulombs (0.000001 Coulombs). In fact, the electron's charge is very small, equaling -1.6×10^{-19} Coulombs. As you might expect, the proton's charge is $+1.6 \times 10^{-19}$ Coulombs. Thus, the proton and the electron have equal but opposite charges, as I mentioned previously.

It is rather interesting to note that the magnitude of the electron's charge and the proton's charge are precisely equal. To the limits which we can measure, there is absolutely no difference between the two. This is interesting because the proton is more than 1,800 times as massive as the electron. Thus, even though the proton is much more massive than the electron, each has exactly the same magnitude of charge. It turns out that we are quite "lucky" that this is the case. Electrostatic force calculations indicate that if these charges were different by as little as *one billionth of one percent*, the resulting electrostatic charge imbalance in the atoms that make up the human body would be so great that the human body would instantaneously explode! Think about that for a moment. Two incredibly different particles *just happen* to have exactly the same magnitude of electric charge. Furthermore, this *perfect* balance of charge *just happens* to be necessary for the stability of the complex molecular structures which make up living organisms. Another nice "coincidence," isn't it?

Now that we know the unit for charge, what is the "k" in Equation (13.1)? Often called the Coulomb constant, this physical constant has the following value:

$$k = 9.0 \times 10^{9} \ \frac{\text{Newton} \cdot \text{m}^2}{\text{C}^2}$$

You should be able to see why the Coulomb constant has the unit listed above. In the end, the electrostatic force must have a unit of Newton, since that's the SI unit of force. When you multiply k by both charges, the units of C^2 cancels. Also, when you divide by the distance squared, the unit m^2 cancels out, leaving the Newtons unit. Thus, the units for k result from the fact that Equation (13.1) must have a unit of Newton for the force.

Before I show you an example of how to use this equation, you need to know that even though Equation (13.1) does not have any vector notation in it, the force determined using this equation is, nevertheless, a vector. Force, after all, must be a vector, because force is what causes linear acceleration. Since acceleration must have a direction to it, force must also have a direction to it. Thus, the force in Equation (13.1) must be a vector. Why, then, are there no vectors in the equation? Just like the equation for the force due to gravity, Equation (13.1) only deals with magnitude. We have to use our brains to figure out the direction of the force. We already know that opposite charges attract one another while like charges repel one another. This tells us the direction of the force vector. You'll see how we do this in the example problem that follows.

EXAMPLE 13.1

A +3.0-mC charge is placed 2.2 m from a -2.0-mC charge. What is the electrostatic force exerted on each charge?

This is a straightforward application of Equation (13.1). There are, however, a couple of things that we need to learn about using this equation. First of all, even though the charges have signs, we do not put them into the equation. The signs are there simply to tell us whether or not the charges attract or repel one another. Since the signs are opposite, we know that the charges attract one another. So, we put the charges in without their signs. Second, we need to realize that the unit for "k" uses Coulombs. The charges that we have are in mC (milliCoulombs). We therefore must convert from mC to C. Since the prefix "milli" means one thousandth, the conversion is straightforward, and I will not do it for you.

$$F = \frac{kq_1q_2}{r^2}$$

$$F = \frac{(9.0 \times 10^9 \frac{\text{Newton} \cdot \cancel{m^2}}{\cancel{C^2}} \cdot (3.0 \times 10^{-3} \cancel{C}) \cdot (2.0 \times 10^{-3} \cancel{C})}{(2.2 \cancel{m})^2} = 1.1 \times 10^4 \text{ Newtons}$$

Now that we have the magnitude of the force, we need to determine its direction. That's where we have to (oh no!) think. Since the signs are opposite, the charges attract one another. Thus, the negative charge will be pulled to the positive charge and vice versa. The positive charge, then, experiences a 1.1 x 10^4 Newton force in the direction of the negative charge, and the negative charge experiences a 1.1 x 10^4 Newton force in the direction of the positive charge.

I want you to notice a few things about this example. First, notice that the two objects in the example each exert the same magnitude electrostatic force on each other. Just as with the gravitational force, it takes at least two objects to create the electrostatic force, and each object exerts an equal magnitude of force on the other. Second, notice the size of the force. It is pretty large. Let's spend a moment comparing the electrostatic force to the gravitational force. Consider an electron and a proton separated by 1.0 meter. If you put the mass of an electron and the mass of a proton into Equation (7.3) and calculate the gravitational force when they are 1.0 m apart, you will find that it is 1.0 x 10^{-67} Newtons. However, the electrostatic force between the electron and proton separated by that distance is 2.3 x 10^{-28} Newtons. In other words, the electrostatic force between an electron and proton is 2.3 x 10^{39} times greater than the gravitational force between them when they are separated by 1.0 meter. For reasons unknown to physicists, the electrostatic force, despite the fact that it behaves similar to the gravitational force, is significantly stronger than the gravitational force.

The last thing that you need to realize about the example is the nature of the answer we obtained. The fact is, this answer will probably only be correct for an instant. After all, since the charges are attracted to one another, as long as they are not being held in place, the charges will begin to move towards one another. That will change the value for "r," which will change the force. Since the charges will move towards one another, the value for "r" will decrease, making the value for the force increase. In other words, the force between these two objects will increase as they move towards each other. Since the answer to Example (13.1) is only correct for an instant, we usually call it the **instantaneous electrostatic force**. This reminds us that as soon as the objects move under the influence of that force, the magnitude of the force will change.

ON YOUR OWN

13.3 What is the instantaneous electrostatic force between two 1.4-mC charges placed 3.1 m apart?

13.4 Two charges (3.4 C and -1.2 C) exert an instantaneous electrostatic force of 1.1 x 10^{10} Newtons towards each other. At that instant, how far apart are the two charges?

Multiple Charges and the Electrostatic Force

Did you think that you were done with trigonometry? Well, you were wrong! In nature, you rarely find two isolated charges exerting a force on one another. Instead, there are usually multiple charges in close proximity to one another, and the total electrostatic force that results is the vector sum of each individual electrostatic force. Since this is so common in nature, we must at least get some experience analyzing situations in which multiple charged objects are present. That's where the trigonometry comes in. For example, suppose you had a situation involving three charged particles such as those pictured below:

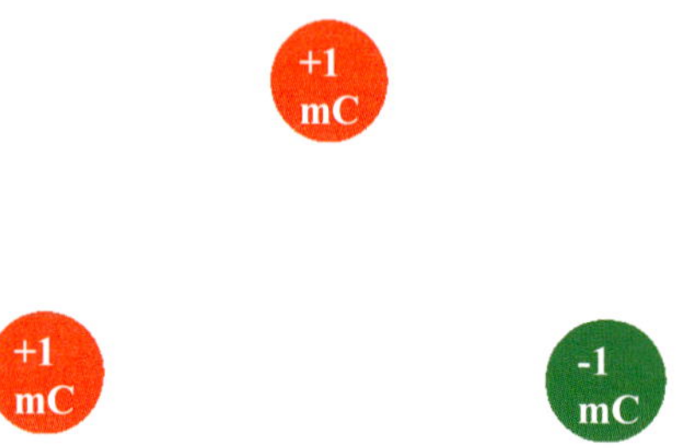

What force would the top charged particle experience? Well, the +1-mC charge at the bottom left would exert a repulsive force on it, while the -1-mC charge at the bottom right would exert an attractive force on it. The resulting vectors would look something like this:

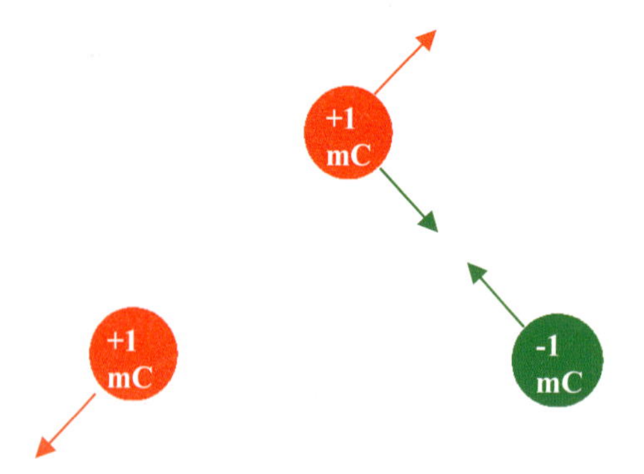

The red arrows represent the repulsive force that the two +1-mC charges exert on each other, while the green arrows represent the mutual electrostatic force between the +1-mC charge and the -1-mC charge. Realize, of course, that the two particles at the bottom of the diagram also exert an electrostatic force on each other. However, since in this discussion we are concentrating on the particle at the top of the diagram, that force does not matter to us.

Now, what is the force exerted on the particle at the top of the diagram? According to the diagram, there are two forces acting on that particle. The resulting force, then, will be the vector sum of those two forces. In other words, the total force exerted on that particle will look like this:

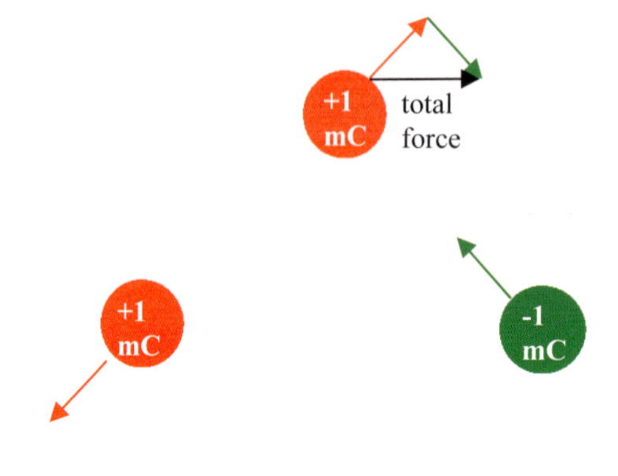

So, when we are faced with a situation that involves multiple particles, we must look at all of the forces acting on any one of those particles and compute the vector sum. That will give us the total force exerted on the particle of interest. Study the following examples to make sure you understand how to apply this idea.

EXAMPLE 13.2

Three charged particles (q_1 = +1.2 mC, q_2 = -3.4 mC, q_3 = -4.1 mC) are arranged according to the diagram below. What is the total instantaneous electrostatic force exerted on the -3.4-mC particle?

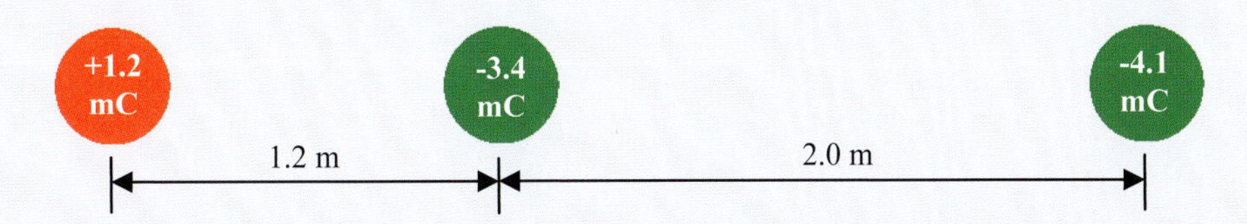

This problem is not so bad because all of the particles are in a straight line. As a result, all of the vectors will be pointing in only one dimension: the horizontal dimension. Thus, we do not need trigonometry to add them. In this problem, we are only interested in the -3.4-mC particle. Since that's the case, we do not need to worry about any forces except the ones acting on that particle. We know that the +1.2-mC particle will exert an attractive force on the -3.4-mC particle. The magnitude of that force will be:

$$F = \frac{kq_1q_2}{r^2}$$

$$F = \frac{(9.0 \times 10^9 \frac{\text{Newton} \cdot \cancel{\text{m}^2}}{\cancel{\text{C}^2}}) \cdot (1.2 \times 10^{-3} \cancel{\text{C}}) \cdot (3.4 \times 10^{-3} \cancel{\text{C}})}{(1.2 \cancel{\text{m}})^2} = 2.6 \times 10^4 \text{ Newtons}$$

Now we need to determine the direction of this force. Since the charges are opposite, the +1.2-mC charge attracts the -3.4-mC charge. As a result, the force vector acting on the -3.4-mC charge (the charge of interest) points towards the +1.2-mC charge.

The only other force acting on the -3.4-mC force is the repulsive force exerted by the -4.1-mC particle:

$$F = \frac{kq_1q_2}{r^2}$$

$$F = \frac{(9.0 \times 10^9 \frac{\text{Newton} \cdot \cancel{\text{m}^2}}{\cancel{\text{C}^2}}) \cdot (3.4 \times 10^{-3} \cancel{\text{C}}) \cdot (4.1 \times 10^{-3} \cancel{\text{C}})}{(2.0 \cancel{\text{m}})^2} = 3.1 \times 10^4 \text{ Newtons}$$

Now we need to determine direction. Since this is repulsive, the force vector acting on the -3.4-mC charge points away from the -4.1-mC charge. The two forces that we just calculated are the only forces acting on the -3.4-mC charge, so our force diagram looks like:

Now remember that there are a *lot* of other forces acting in this problem. However, the only thing that we are interested in is the -3.4-mC particle. Since the forces pictured here are the only ones acting on *that* particle, we are finished diagramming the problem.

Since both forces point in the same dimension, we are really only dealing with one-dimensional vectors. This means we don't need to use trigonometry. We can just use our convention that one-dimensional vectors pointed to the left are negative. Thus, the sum of these forces is:

$$\mathbf{F}_{\text{total}} = -2.6 \times 10^4 \text{ Newtons} + -3.1 \times 10^4 \text{ Newtons} = -5.7 \times 10^4 \text{ Newtons}$$

The total instantaneous electrostatic force acting on the -3.4-mC particle, then, is <u>5.7×10^4 Newtons, pointing towards the +1.2-mC particle</u>.

Three charges are arranged at the vertices of a 45-45-90 right triangle (see diagram below). What is the instantaneous electrostatic force acting on the +6.0-C charge?

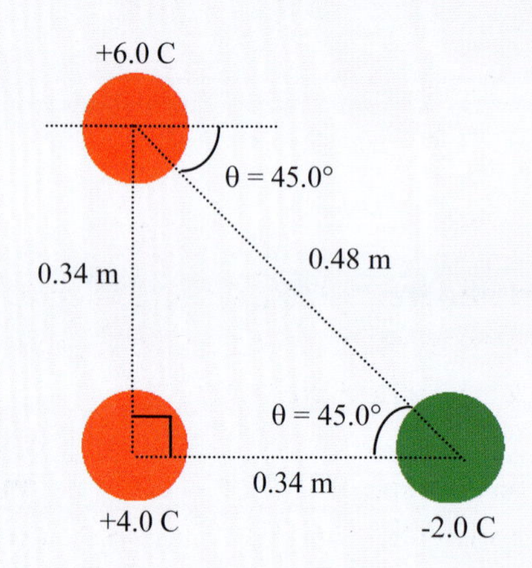

Since we are only interested in the +6.0-C charge, we need concern ourselves only with forces that affect that charge. Well, the +4.0-C charge will repel the +6.0-C charge with a force of:

$$F = \frac{kq_1q_2}{r^2}$$

$$F = \frac{(9.0 \times 10^9 \frac{\text{Newton} \cdot \cancel{\text{m}^2}}{\cancel{\text{C}^2}}) \cdot (4.0\cancel{\text{C}}) \cdot (6.0\cancel{\text{C}})}{(0.34\cancel{\text{m}})^2} = 1.9 \times 10^{12} \text{ Newtons}$$

Since this force is repulsive, it will push the +6.0-C charge directly away from the +4.0-C charge, resulting in a force vector that points straight up.

The other force acting on the +6.0-C charge is from the -2.0-C charge. We can calculate its magnitude:

$$F = \frac{kq_1q_2}{r^2}$$

$$F = \frac{(9.0 \times 10^9 \frac{\text{Newton} \cdot \cancel{m^2}}{\cancel{C^2}}) \cdot (2.0 \cancel{C}) \cdot (6.0 \cancel{C})}{(0.48 \cancel{m})^2} = 4.7 \times 10^{11} \text{ Newtons}$$

Since this force is attractive, the force vector points from the 6.0-C charge to the -2.0-C charge. Our force diagram, then, looks like this:

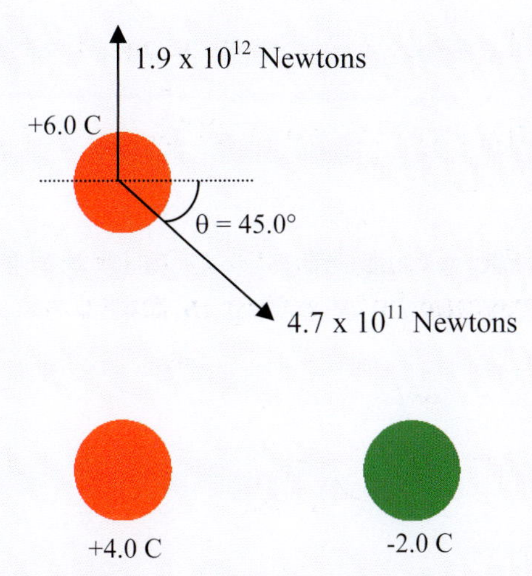

To get the total instantaneous electrostatic force, then, we just need to add these two vectors together. Unlike the first example problem, however, these vectors do not point in only one dimension. Thus, we will need to use trigonometry (ugh!) to add them together.

When we add two-dimensional vectors, we must know their angles. Now remember, the angle of a vector is always determined counterclockwise from the positive x-axis. This means that the $1.9 \text{ x } 10^{12}$-Newton vector has an angle of 90.0°. The $4.7 \text{ x } 10^{11}$-Newton vector, however, has an angle of 315.0°. Now we can add these vectors. If you forget how to do this, go back and review Module 3, particularly Examples 3.5 and 3.6. In order to make the notation easier, I will call the $1.9 \text{ x } 10^{12}$-Newton vector "vector A," and the other vector will be "vector B." The sum of these two vectors will be called "vector C." In order to add vectors A and B, we must first split them up into their components.

$$A_x = (1.9 \times 10^{12} \text{ Newtons}) \cdot \cos(90.0°) = 0$$

$$A_y = (1.9 \times 10^{12} \text{ Newtons}) \cdot \sin(90.0°) = 1.9 \times 10^{12} \text{ Newtons}$$

$$B_x = (4.7 \times 10^{11} \text{ Newtons}) \cdot \cos(315.0°) = 3.3 \times 10^{11} \text{ Newtons}$$

$$B_y = (4.7 \times 10^{11} \text{ Newtons}) \cdot \sin(315.0°) = -3.3 \times 10^{11} \text{ Newtons}$$

Now we can add the components just like they are one-dimensional vectors:

$$C_x = A_x + B_x = 0 + 3.3 \text{ x } 10^{11} \text{ Newtons} = 3.3 \text{ x } 10^{11} \text{ Newtons}$$

$$C_y = A_y + B_y = 1.9 \text{ x } 10^{12} \text{ Newtons} + -3.3 \text{ x } 10^{11} \text{ Newtons} = 1.6 \text{ x } 10^{12} \text{ Newtons}$$

All that's left to do now is convert these x- and y-components into vector magnitude and direction:

$$\text{magnitude} = \sqrt{C_x^{\ 2} + C_y^{\ 2}} = \sqrt{(3.3 \times 10^{11} \text{ Newtons})^2 + (1.6 \times 10^{12} \text{ Newtons})^2} = 1.6 \times 10^{12} \text{ Newtons}$$

$$\theta = \tan^{-1}\left(\frac{C_y}{C_x}\right) = \tan^{-1}\left(\frac{1.6 \times 10^{12} \cancel{\text{Newtons}}}{3.3 \times 10^{11} \cancel{\text{Newtons}}}\right) = 78°$$

Since both the x and y components of the vector are positive, we know that the vector is in quadrant I of the Cartesian coordinate plane. This means that we need not do anything to our answer in order to properly define the vector angle. The instantaneous electrostatic force on the 6.0 C charge, then, is <u>1.6 x 10^{12} Newtons at an angle of 78°</u>.

ON YOUR OWN

13.5 What is the instantaneous electrostatic force on the -3.1-mC charge in the diagram below?

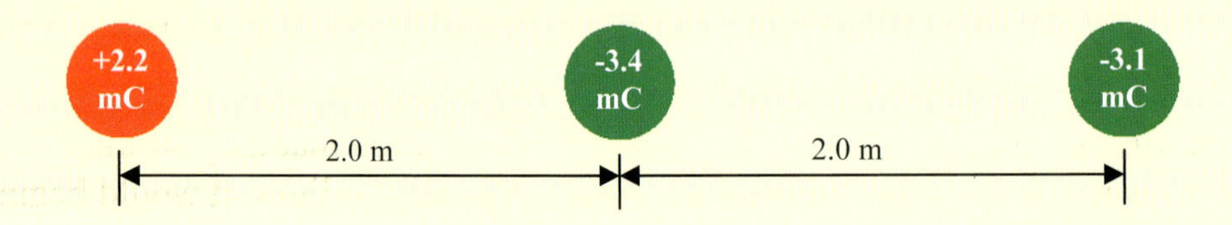

13.6 Three positive charges are placed at the vertices of an equilateral triangle as pictured below. What is the instantaneous electrostatic force on the +2.0-C charge?

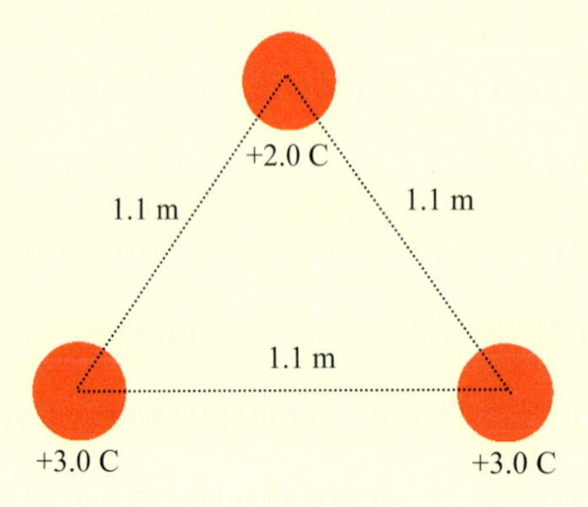

The Electric Field

Now that you have some experience using Coulomb's Law of electrostatic force, it is time to look at the electrostatic force in another way. Suppose you had a positive charge sitting alone in space. Suppose further that this charge is fastened to something sturdy so that it will not move. What would happen if you brought another positive charge (let's call it a "test charge") near the stationary charge and then let the test charge go? Well, the positive charges would repel each other. Because the first charge cannot move, however, it will simply stay put. The test charge will move away quickly, traveling away from the stationary charge in a straight line.

If you think about this for a moment, it is almost like the stationary charge has "appendages" which push the other charge away from it. We could even picture these appendages, as shown in the figure below.

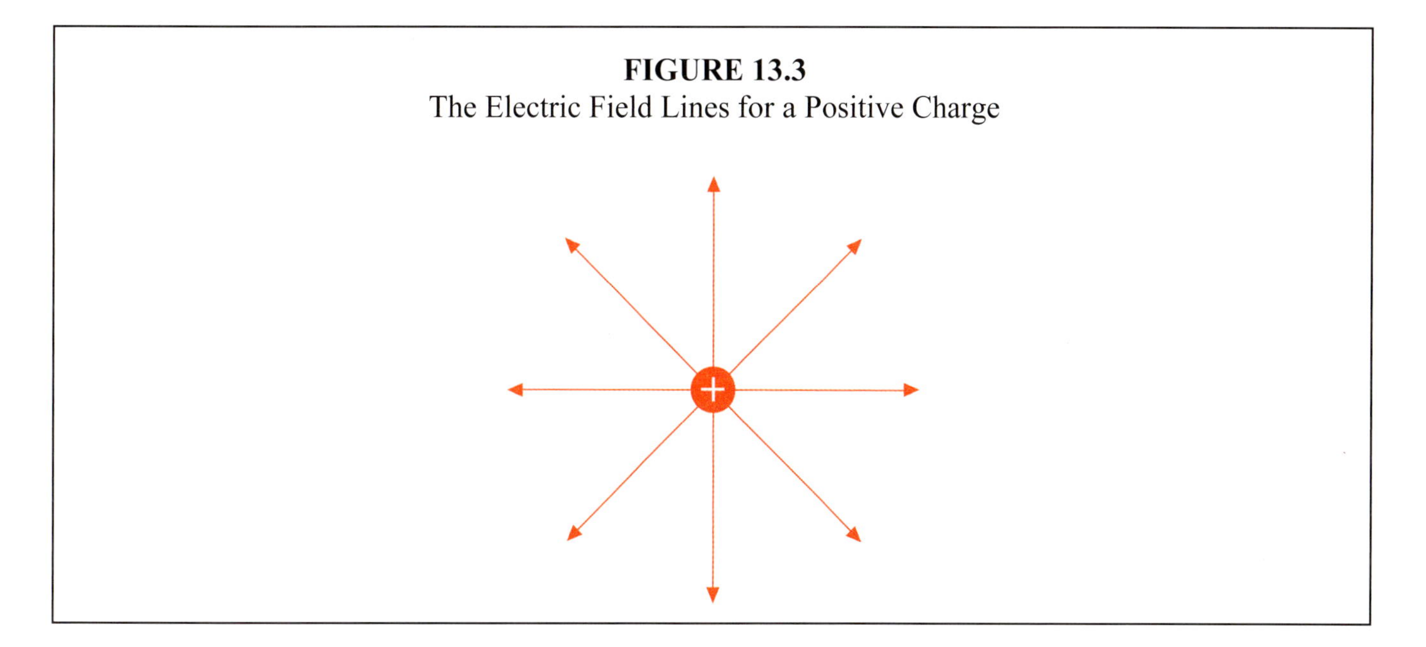

FIGURE 13.3
The Electric Field Lines for a Positive Charge

The arrows represent the appendages, and they point in the direction that they would shove a positive particle. Thus, if I were to put a positive particle on any one of these lines, it would begin to travel in the direction pointed out by the arrow. What's the point of this drawing? We'll see in a moment.

Before we see what all of this means, let's suppose we replace the stationary positive charge with a stationary negative charge. What would happen if I placed a positive test charge near that stationary negative charge? Well, since the unlike charges exert an attractive force on one another, the charges will want to move together. We've made it so that the stationary charge cannot move, however. As a result, the positive particle will move towards the negative charge. In this case, it's like the stationary negative charge has appendages that pull the positive charge towards it. We could draw those appendages as shown in Figure 13.4.

FIGURE 13.4
The Electric Field Lines for a Negative Charge

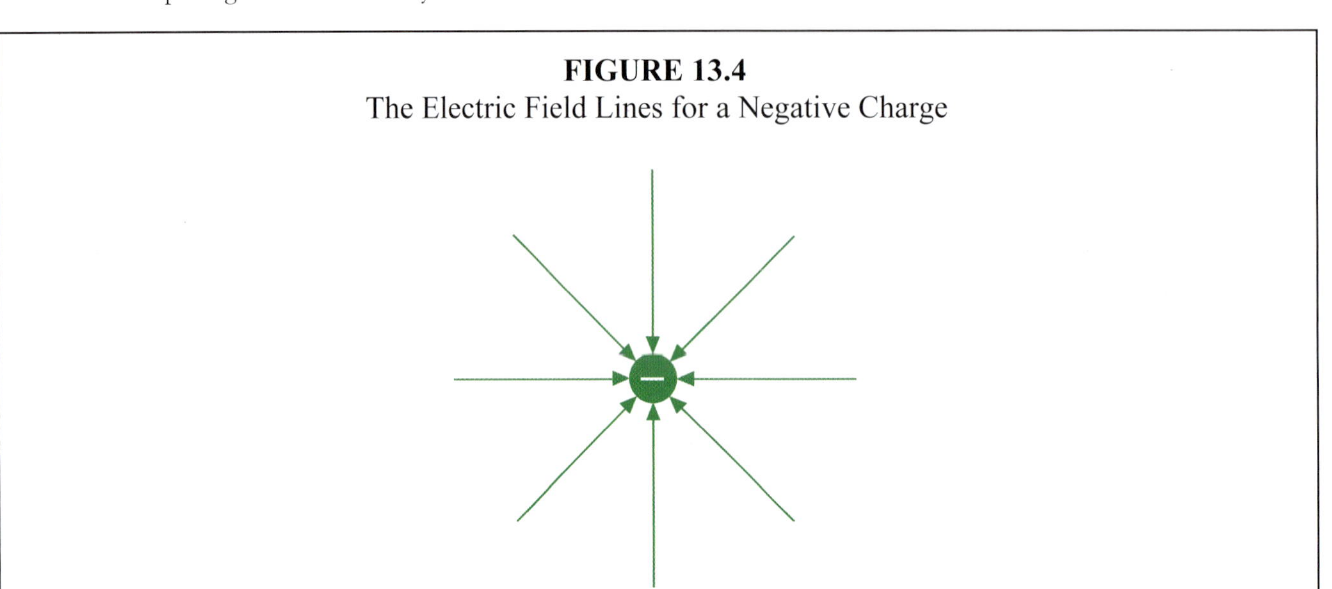

What am I trying to do here? Well, physicists have developed an idea called the **electric field**. This field exerts force on any other charged particles that come into it. The way we picture this field is by drawing arrows that either point straight out from the charge (if the charge is positive) or point straight into the charge (if it is negative). Thus, what I drew in the last two figures was, in fact, the electric field associated with the stationary charge that I was discussing. We call the arrows in the diagram the **electric field lines**.

Now remember, electric fields are a means for us to visualize the electrostatic force. When drawn properly, they can provide a lot of information about situations involving charges. What do electric field lines tell us? Well, first, they tell us which way a charge will travel when placed in the field. A positive charge will follow the arrows, while a negative charge will travel opposite of the direction pointed out by the arrows.

Not only do the electric field lines tell you what direction a charged particle will travel when in an electric field, they will also give you an idea of the force that the charge will experience. When you look at an electric field, the density of lines tells you the relative strength of the field. The denser the lines at any given point, the stronger the electric field. Thus, when a particle is placed in an electric field, the force it experiences is proportional to the density of the electric field lines.

Since these electric field lines are an easy way to visualize the electrostatic force, it's important for you to know how to draw them. Basically, there are just three rules:

1. **The relative number of electric field lines drawn is proportional to the charge of the particle creating the field.**

2. **The arrows that form the electric field lines point out of positive charges and into negative charges.**

3. **Electric field lines can never cross.**

If you follow these rules, you can draw electric field lines that are very informative to other physicists. For example, look at Figures 13.3 and 13.4. By looking at the lines, you can tell that in Figure 13.3, a positive particle will move away from the stationary charge and a negative particle (because it goes opposite the direction pointed by the arrows) will travel towards the stationary charge.

Big deal, huh? After all, without those electric field lines drawn, you could have already told me that a positive particle will move away from a positive stationary charge and a negative particle will move towards a positive stationary charge. So what good are these electric field lines? Well, situations involving multiple charges can get really complicated. Electric field lines help you sort out the situation. For example, look what happens when two stationary positive charges are placed close together.

FIGURE 13.5
Electric Field Lines for Two Stationary Positive Charges

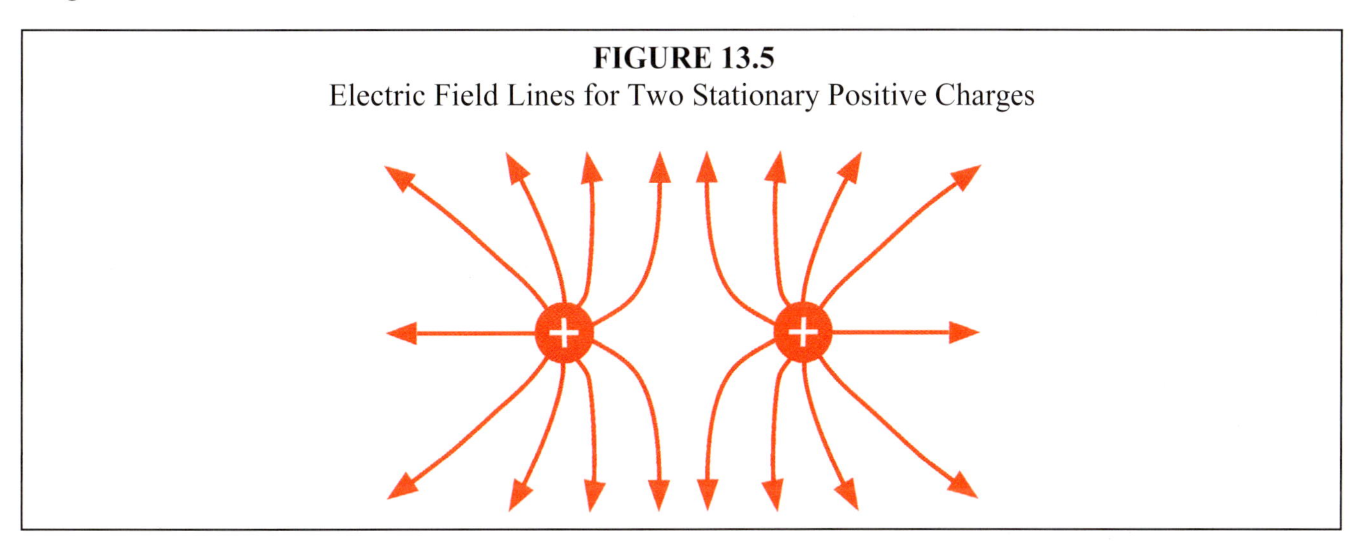

Before I tell you how I drew this picture, let's see what the picture means. First, it tells us that a positive charge placed anywhere near these two stationary positive charges will travel along a path shown by the nearest field line, in the direction pointed out by the arrow. A negative charge placed in the same vicinity will follow the same path, but in the opposite direction. Also, since the density of field lines is essentially the same near either charge, we know that a charged particle will experience the same acceleration whether it is placed near the left stationary charge or the right one. Now remember, the electric field drawn here does not affect the two charges that are in the diagram. Those charges are fixed and do not move. Instead, the electric field lines tell us how a *third* charge would be affected if it were placed in the electric field and allowed to move freely. This is an important distinction.

Now, how did I draw the picture? Well, I decided that the stationary charges should be identical. As a result, the same number of lines come out of each charge. Also, since these are positive stationary charges, the lines must come out of the charges. If there were a negative stationary charge present, the lines would go into that charge. Since there is none, however, the lines simply come out of the positive stationary charges. The reason that the lines curve is that they are not allowed to cross. However, they must point away from the positive charges. Thus, they curve away from each so that they do not cross but still point away from the positive charges. This leads to the electric field shown in the diagram.

You will be required to both draw and interpret electric field lines. Study the examples and perform the "On Your Own" problems that follow in order to make sure that you can do this.

EXAMPLE 13.3

A stationary charge of -0.30 C sits in close proximity to a stationary charge of +0.60 C. Draw a diagram of the resulting electric field.

Since the drawing is for illustration only, we really don't need to know how far the charges are from each other. If we wanted to calculate the magnitude of the electrostatic force between them, we would need to know how far apart they are. Remember, however, that the electric field we will draw does not affect these two charges. The field we draw is the *result* of these two charges being close to one another and fixed so that they do not move. The field lines will tell us what a *third* charge will do if placed close to the charges.

To draw the diagram, I first have to realize that I have two different charges. Thus, one of my charges (the positive one) will have field lines coming out of it, and the other will have field lines going into it. Furthermore, the positive charge (which I will draw red) will have twice as many field lines coming out of it as the negative charge will have going into it. In general, we like to draw the field lines coming out of a positive charge so that they end up going into a negative charge. We couldn't do that in the previous diagram, because there were no negative charges. In this case, however, we have negative charges, so I will make sure that half of the electric field lines leave the positive charge and end up in the negative charge. I can only do that with half of the lines, however, because the negative charge is only half as strong as the positive. In the end, then, the drawing looks like this:

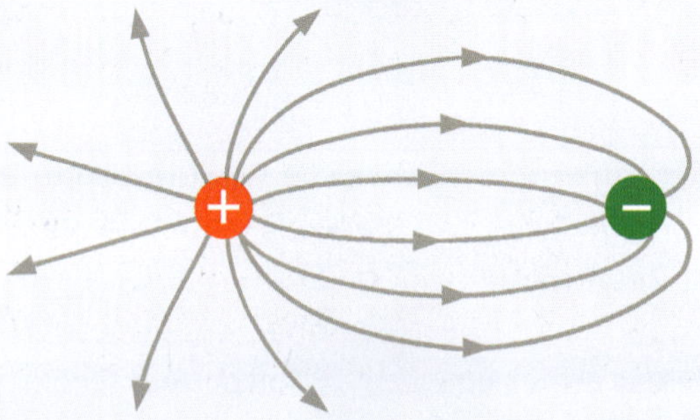

Notice how this is drawn. I started with 12 lines coming out of the positive charge. Why 12? There's really no reason. There is no set number of lines that I should have started with. It's generally a good idea to keep your drawing from getting too cluttered, though. Twelve lines should be enough to tell you what's going on. The negative charge bends these lines towards it, because electric field lines travel into negative charges. Therefore, I bent the lines closest to the negative charge so that they end up going into the negative charge. Since the negative charge was only half as strong as the positive charge, however, it could only take the first six lines. The six lines farthest from the negative charge, then, were still bent, but not enough to go into the negative charge. Thus, they continue on without entering the negative charge.

Using the diagram above, where would you put a charged particle so that it moves with the greatest acceleration?

If another charge were placed on the diagram, the electrostatic force (which determines the acceleration) it feels will be proportional to the density of the field lines. Looking at the diagram, the field lines are densest near the positive charge, to the right. Thus, placing the particle near the positive charge (to the right) would give it the greatest acceleration.

ON YOUR OWN

13.7 Two stationary charges of equal magnitude but opposite sign are placed close to one another. Draw a diagram of the resulting electric field. Where would a third charge experience the greatest acceleration?

Calculating the Strength of the Electric Field

Although qualitative pictures of the electric field are useful, it is always good to have a way to calculate the electric field quantitatively. Well, it turns out that there is a pretty simple equation that allows you to do just that:

$$\mathbf{E} = \frac{\mathbf{F}}{q_o} \tag{13.2}$$

In this equation, "**E**" is the electric field at a given point in space; "**F**" is the electrostatic force experienced by a positive test charge that is placed in the field; and "q_o" is the magnitude of the positive test charge. Please note that the electric field is a vector, and it is pointed in the same direction as the electrostatic force. Now remember, if there are two or more stationary charges making up the electric field (such as the situation in Example 13.3), the electrostatic force at any point is the vector sum of the electrostatic force from each individual charge. As a result, the electric field is also the vector sum of the electric field from each individual charge. Also, please note that unlike Equation (13.1), in this equation, *you do include the sign of the charge.* You will see how that works in a moment.

What is the unit associated with the electric field? Well, the electrostatic force is given in Newtons, as long as we are using SI units. In addition, the SI unit for charge is the Coulomb. Thus, the SI unit for the electric field is Newtons/C. What does the unit mean? It means that the value for the electric field tells you how many Newtons of force the field will generate per Coulomb of charge that is placed into the field.

In a moment, I will run through an example of how to calculate the electric field, but first I want to manipulate Equation (13.2) a bit. To calculate the electric field generated by a stationary charge (let's call it "Q"), we would first calculate the electrostatic force between that charge and a positive test charge (q_o). We would then divide that force by the magnitude of the positive test charge. Let's do that now:

$$E = \frac{F}{q_o} = \frac{kQ\cancel{q_o}/r^2}{\cancel{q_o}} = \frac{kQ}{r^2} \tag{13.3}$$

Think about what I did here. I cannot calculate the direction of the electric field, because I must use Equation (13.1) to calculate the electrostatic force, and it has no vectors in it. Thus, this equation gives us only the strength of the electric field. However, since we know that electric field lines point away from positive charges and towards negative charges, we can determine the direction of the electric field by knowing the sign of "Q." Also, notice what happened. The "**F**" in Equation (13.2) is between the stationary charge and the test charge. Since, in the end, we divide the force by the test charge, the test charge drops out, and the only charge left is the stationary charge. Thus, the electric field generated by a stationary charge depends only on the magnitude of the charge and the distance from that charge.

In the end, then, for a single stationary charge, the electric field is pretty easy to calculate:

$$E = \frac{kQ}{r^2} \tag{13.4}$$

Just as is the case for Equation (13.1), this equation deals only with magnitude, and you do not use the sign of the charge in the equation. In addition, the direction of the electric field can be determined by the sign of the charge involved. If Q is positive, the electric field points directly away from Q. Remember, electric field lines always point away from a positive charge. Similarly, if Q is negative, the electric field points directly towards Q, because electric field lines always point towards a negative charge. Finally, if there is more than one charge forming the electric field, the total electric field is the vector sum of all of the individual electric fields generated by the individual electric charges.

All of this probably seems a bit confusing at this point. However, a couple of example problems ought to clear everything up.

EXAMPLE 13.4

Consider the two stationary charges given in the previous example (-0.30 and +0.60 C). If these two charges are 1.60 m apart, what are the magnitude and direction of the electric field at a point halfway in between the two?

A point that is halfway between the two charges would be 0.80 m away from each charge. Thus, we need to calculate the electric field at the point given by the black dot below:

E = ?

0.80 m 0.80 m

Since there are two stationary charges creating the electric field, we will have to add their individual electric fields in order to determine the total electric field. Now, we have two equations for calculating the electric field: Equation (13.2) and Equation (13.4). Which do we use? Well, the magnitude of a test charge is not given in the problem, so Equation (13.2) would be hard to use. Thus, we will have to use Equation (13.4). Remember, each stationary charge produces its own field, so we will have to calculate each charge's field individually. Let's start with the positive charge. We will call its electric field "E_1."

$$E_1 = \frac{kQ}{r^2} = \frac{(9.0\times10^9 \frac{\text{Newton}\cdot\cancel{\text{m}^2}}{\text{C}^{\cancel{2}}})\cdot(0.60\ \cancel{\text{C}})}{(0.80\ \cancel{\text{m}})^2} = 8.4\times10^9\ \frac{\text{Newtons}}{\text{C}}$$

Now remember, electric field lines always point directly away from positive charges. Thus, this electric field is pointed directly to your right. Now let's calculate the electric field from the negative charge:

$$E_2 = \frac{kQ}{r^2} = \frac{(9.0\times10^9\ \frac{\text{Newton}\cdot\cancel{\text{m}^2}}{\cancel{\text{C}^2}})\cdot(0.30\ \cancel{\text{C}})}{(0.80\ \cancel{\text{m}})^2} = 4.2\times10^9\ \frac{\text{Newtons}}{\text{C}}$$

Since electric field lines always point towards negative charges, this electric field points directly to your right as well. Thus, the two electric fields look like this:

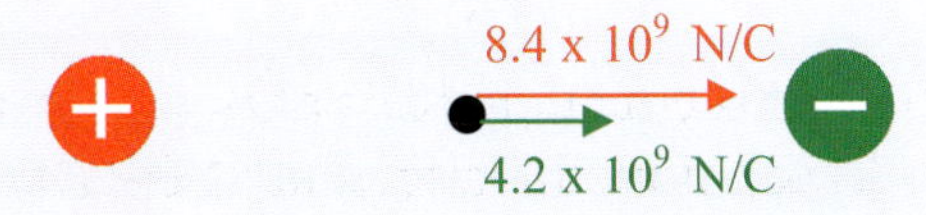

Since the two vectors are parallel, we don't need to use trigonometry to add them. We can just add them directly:

$$\mathbf{E}_{\text{total}} = \mathbf{E}_1 + \mathbf{E}_2 = 8.4\times10^9\ \frac{\text{Newtons}}{\text{C}} + 4.2\times10^9\ \frac{\text{Newtons}}{\text{C}} = 1.26\times10^{10}\ \frac{\text{Newtons}}{\text{C}}$$

Note that since we are adding here, we determine the significant figures by decimal place. Each of the individual electric fields are reported to the hundred millions place (0.1×10^9), so the answer must be reported to the hundred millions place (0.01×10^{10}). The electric field, then, is 1.26 x 10^{10} Newtons/C, pointed directly to the right. Note that this answer agrees with the qualitative diagram of the electric field that we drew in the previous example. The diagram has the electric field in between the two charges pointed to the right, as does this result.

Suppose a -0.0060-mC charge is placed directly between the two stationary charges. What is the instantaneous electrostatic force it would experience?

We could solve this problem using Equation (13.1), but that won't be necessary. We already know the electric field at the point directly in between the two charges. Thus, we can quickly calculate the force using Equation (13.2):

$$\mathbf{E} = \frac{\mathbf{F}}{q_o}$$

$$\mathbf{F} = q_0\mathbf{E} = (-6.0\times10^{-6}\ \cancel{\text{C}})\cdot1.26\times10^{10}\ \frac{\text{Newtons}}{\cancel{\text{C}}} = -7.6\times10^4\ \text{Newtons}$$

Make sure you understand what I did here. Remember, as I mentioned when I gave you Equation (13.2), in this equation you *do include* the sign of the charge. That's why the negative is in there. Also, notice that I had to convert mC into C in order for the units to work out.

What does the negative sign mean? It just means that the force is pointed opposite the electric field. Thus, since the electric field is pointed to the right, the charge experiences a 7.6 x 10^4-Newton force that is pointed to the left.

ON YOUR OWN

13.8 Two +0.40-mC charges are placed 2.00 m apart. Calculate the magnitude and direction of the electric field at the point shown in the drawing below:

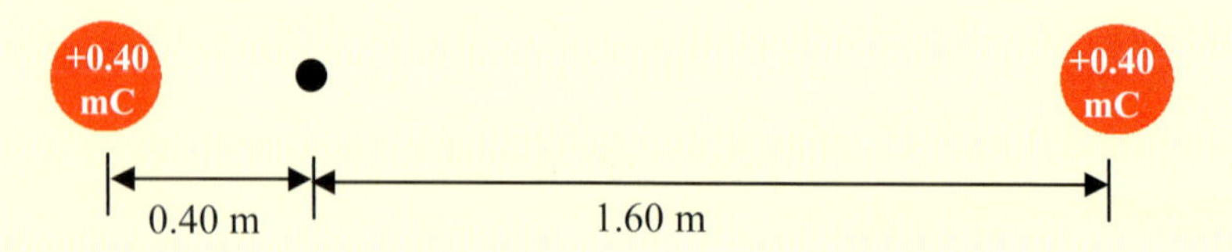

What is the force a 0.56-mC charge would experience at that point?

13.9 What is the electric field at the midpoint between the charges?

Applying Coulomb's Law to the Bohr Model of the Atom

In chemistry, you should have learned a little about the Bohr model of the atom. Although not entirely correct, this model does give us a reasonably good idea of what an atom looks like. In general, The Bohr model says that an atom consists of a small nucleus containing positively-charged protons as well as neutrons, which have no charge. Orbiting around the nucleus in fixed, circular orbits are negatively-charged electrons. A picture of the Bohr model for the helium atom (which contains 2 protons, 2 neutrons, and 2 electrons) is shown in Figure 13.6.

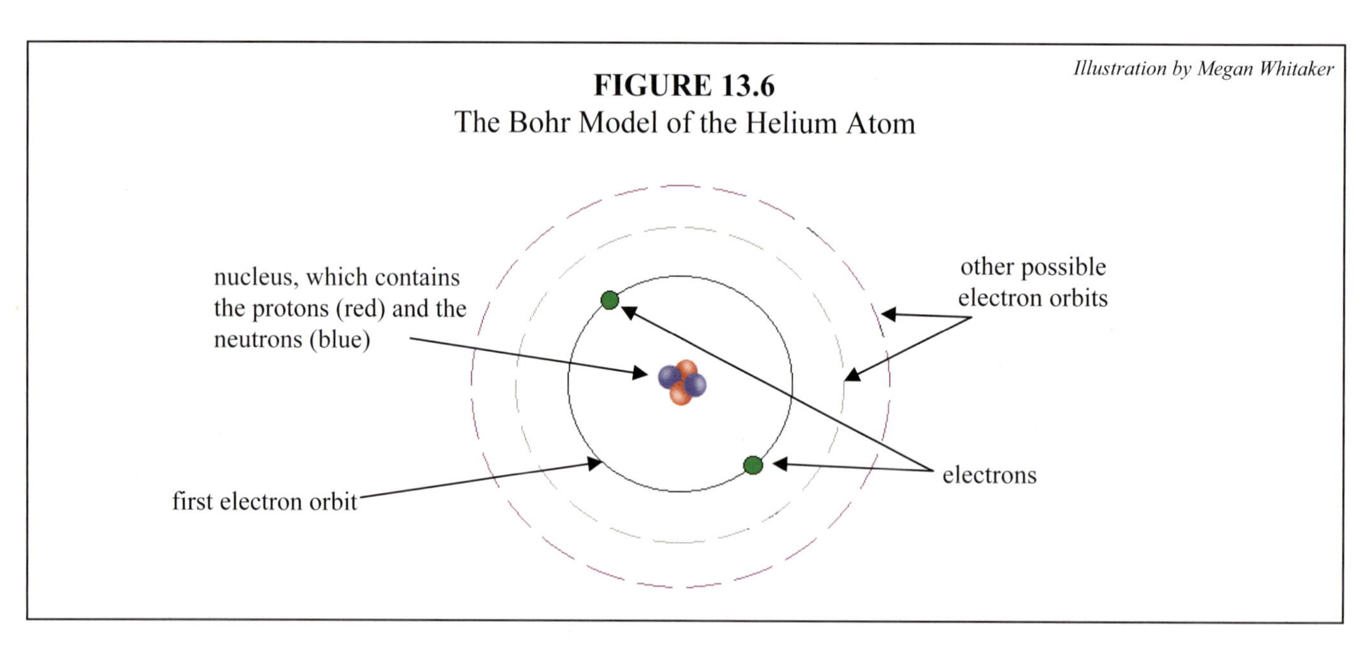

Illustration by Megan Whitaker

FIGURE 13.6
The Bohr Model of the Helium Atom

The electrons stay in orbit around the nucleus because the positively-charged protons within the nucleus attract them. As was already pointed out, the delicate balance that exists between the charges of the proton and the electron help keep the atom stable. Notice that although I have drawn both electrons in the atom in a single orbit, there are many other possible orbits that they can occupy. If the electrons are given enough energy, they can move into one of the other possible orbits. Alternatively, if they are in one of those orbits, they can move to an orbit closer to the nucleus by releasing some energy, usually in the form of light.

Now although the intricacies of the Bohr model are well worth studying, I want to look at something very specific. You were told in chemistry that the electrons orbit around the nucleus. Do you know how fast they orbit? Well, with the physics you have learned in this module, along with some review, you can calculate the electron's speed. Let's look at the simplest atom: hydrogen.

EXAMPLE 13.5

In a hydrogen atom that is in its lowest energy state, the electron orbits a single proton with an orbital radius of 5.29 x 10^{-11} m. At what speed does the electron travel? ($q_{electron}$ = -1.6 x 10^{-19} C, $m_{electron}$ = 9.1 x 10^{-31} kg, q_{proton} = +1.6 x 10^{-19} C)

As I said, to solve this problem, we must do a little review. The hydrogen atom discussed here looks like this:

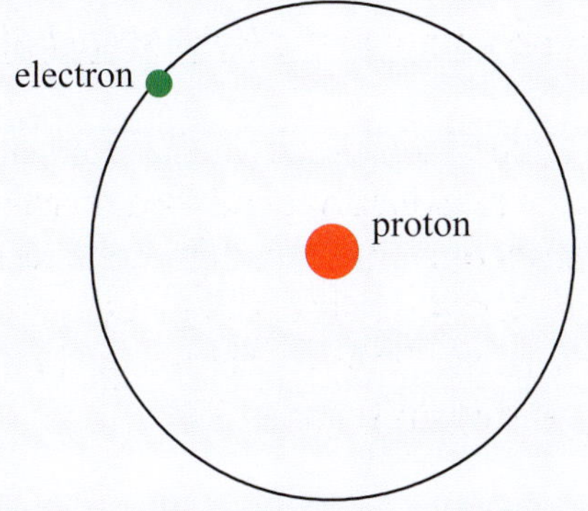

Now, remember why things travel in a circle. They experience centripetal force. Way back in Module #7, we were given an equation that allows us to calculate the magnitude of this force:

$$F_c = \frac{mv^2}{r} \qquad (7.1)$$

where "m" is the mass of the orbiting object, "v" is the speed, and "r" is the radius of the orbit. Back in Module #7, when we wanted to learn things about an orbiting object, we had to figure out what was supplying that centripetal force. In the case of a child twirling a ball on a string, the tension in the string supplied the centripetal force. In the case of a car turning on a curve, friction supplied the centripetal force. Well, what supplies the centripetal force here? The electrostatic force! The only thing keeping the electron traveling in the orbit is the electrostatic attraction between it and the nucleus. Thus, the electrostatic force is supplying the centripetal force. This means that the electrostatic force and the centripetal force are the same. We can therefore take the expression for centripetal force found in Equation (7.1) and set it equal to the expression for the electrostatic force found in Equation (13.1). The result is:

$$\frac{mv^2}{r} = \frac{kq_1q_2}{r^2}$$

Notice that we know everything here except "v." The problem gave us the mass of the electron (m), the orbital radius (r), and the charges of both the proton and the electron (q_1 and q_2). We can therefore plug those numbers in and rearrange the equation to solve for v:

$$\frac{mv^2}{\cancel{r}} = \frac{kq_1q_2}{r^{\cancel{2}}}$$

$$v^2 = \frac{kq_1q_2}{rm} = \frac{(9.0\times10^9\ \frac{\text{Newton}\cdot\text{m}^{\cancel{2}}}{\cancel{\text{C}^2}})\cdot(1.6\times10^{-19}\ \cancel{\text{C}})\cdot(1.6\times10^{-19}\ \cancel{\text{C}})}{(5.29\times10^{-11}\ \cancel{\text{m}})\cdot(9.1\times10^{-31}\ \text{kg})}$$

$$v^2 = 4.8\times10^{12}\ \frac{\text{Newton}\cdot\text{m}}{\text{kg}} = 4.8\times10^{12}\ \frac{\frac{\cancel{\text{kg}}\cdot\text{m}}{\text{sec}^2}\cdot\text{m}}{\cancel{\text{kg}}} = 4.8\times10^{12}\ \frac{\text{m}^2}{\text{sec}^2}$$

$$v = 2.2\times10^6\ \frac{\text{m}}{\text{sec}}$$

The electron, in this case, whirls around the proton at a whopping 2.2×10^6 m/sec!

So, electrons orbit the nucleus of an atom really quickly. It turns out that this kind of calculation can only be done with atoms or ions that have only one electron. Why? Well, although the Bohr model of the atom is a reasonable "first step" in picturing the details of atomic structure, it is fundamentally flawed. As a result, the model really doesn't work for atoms or ions that have more than one electron in them. The currently accepted model of the atom (the quantum mechanical model) works for atoms and ions with more than one electron. However, the mathematics involved in the quantum mechanical model require skills that are well beyond the scope of high school. In the end, then, the calculations that I will expect you to be able to perform will all deal with atoms or ions that have only one electron.

ON YOUR OWN

($q_{electron} = -1.6 \times 10^{-19}$ C, $m_{electron} = 9.1 \times 10^{-31}$ kg, $q_{proton} = 1.6 \times 10^{-19}$ C)

13.10 A He^+ ion has two protons but only one electron. If the orbital radius of that one electron is 1.3×10^{-11} m, at what speed does it orbit the nucleus?

This module represents your first introduction into the world of electricity and magnetism. As with all subjects we discuss in this course, we have only begun to scratch the surface here. You can see that electricity is a complicated subject and can be confusing to a student who has not already been exposed to it. Hopefully, in the final three modules, you will get more and more comfortable wrestling with these concepts.

ANSWERS TO THE "ON YOUR OWN" PROBLEMS

13.1. When you walk across a carpet, you can pick up electrons from the carpet, just like the balloons in Experiments 13.1 and 13.2 picked up electrons from your hair. When you touch a metal object, these electrons jump from your hand to the object, which causes you to feel a shock. Since you are in contact with the carpet and charge is going from the carpet to you, you are being charged by conduction.

13.2. You want to get the other object to reach a charge opposite of the object that is already charged. You must accomplish this by induction, because charging by conduction will just give you the same charge.

13.3 This is a simple application of Equation (13.1). We must remember, however, to convert the charges from mC to C.

$$F = \frac{kq_1q_2}{r^2}$$

$$F = \frac{(9.0 \times 10^9 \frac{\text{Newton} \cdot \cancel{\text{m}^2}}{\cancel{\text{C}^2}} \cdot (1.4 \times 10^{-3} \cancel{\text{C}}) \cdot (1.4 \times 10^{-3} \cancel{\text{C}})}{(3.1 \cancel{\text{m}})^2} = 1.8 \times 10^3 \text{ Newtons}$$

Since both charges are positive, they exert a repulsive force on one another. Thus, the force is 1.8×10^3 Newtons away from each other.

13.4 In this problem, we know the charges and the force and we need to determine the distance between the charges. Equation (13.1) relates these variables:

$$F = \frac{kq_1q_2}{r^2}$$

$$1.1 \times 10^{10} \text{ Newtons} = \frac{(9.0 \times 10^9 \frac{\text{Newton} \cdot \text{m}^2}{\text{C}^2} \cdot (3.4 \text{ C}) \cdot (1.2 \text{ C})}{r^2}$$

$$r^2 = \frac{(9.0 \times 10^9 \frac{\cancel{\text{Newton}} \cdot \text{m}^2}{\cancel{\text{C}^2}} \cdot (3.4 \cancel{\text{C}}) \cdot (1.2 \cancel{\text{C}})}{1.1 \times 10^{10} \cancel{\text{Newtons}}}$$

$$r = 1.8 \text{ m}$$

The distance between the objects, then, is 1.8 m.

13.5 In this problem, we are only interested in the -3.1-mC charge. As a result, we only consider the forces which act on that particular charge. The -3.4-mC charge exerts a repulsive force that has a magnitude of:

$$F = \frac{kq_1q_2}{r^2}$$

$$F = \frac{(9.0\times10^9 \frac{\text{Newtons}\cdot\cancel{\text{m}^2}}{\cancel{\text{C}^2}})\cdot(3.1\times10^{-3}\ \cancel{\text{C}})\cdot(3.4\times10^{-3}\ \cancel{\text{C}})}{(2.0\ \cancel{\text{m}})^2} = 2.4\times10^4 \text{ Newtons}$$

The other force acting on the -3.1 mC charge is the attractive force exerted by the +2.2-mC charge.

$$F = \frac{kq_1q_2}{r^2}$$

$$F = \frac{(9.0\times10^9 \frac{\text{Newtons}\cdot\cancel{\text{m}^2}}{\cancel{\text{C}^2}})\cdot(3.1\times10^{-3}\ \cancel{\text{C}})\cdot(2.2\times10^{-3}\ \cancel{\text{C}})}{(4.0\ \cancel{\text{m}})^2} = 3.8\times10^3 \text{ Newtons}$$

The fact that the first force is repulsive and the second is attractive will give us the directions of the force vectors, making our force diagram look like this:

Since the force vectors both point in the same dimension, we can treat them as one-dimensional vectors. This means we can take care of direction with positives and negatives and then simply add the magnitudes together. Using the convention that vectors pointing to the left are negative, the total force is:

$$\mathbf{F}_{\text{total}} = 2.4 \times 10^4 \text{ Newtons} + -3.8 \times 10^3 \text{ Newtons} = 2.0 \times 10^4 \text{ Newtons}$$

A positive force means that the final vector points to the right. Thus, the final instantaneous electrostatic force is <u>2.0×10^4 Newtons to the right</u>.

13.6 In an equilateral triangle, the interior angles are all 60.0 degrees, and the sides are all the same length. This means that the particles are all 1.1 m away from each other. If we are only interested in the +2.0-C charge, we only need concern ourselves with forces which act on that charge. Each of the +3.0-C charges exerts a repulsive force on the +2.0-C charge. Since both charges are the same size (3.0 C) and the same distance from the +2.0-C charge (1.1 m), they exert the same magnitude of force:

$$F = \frac{kq_1q_2}{r^2}$$

$$F = \frac{(9.0\times10^9\ \frac{\text{Newtons}\cdot\cancel{m^2}}{\cancel{C^2}})\cdot(3.0\ \cancel{C})\cdot(2.0\ \cancel{C})}{(1.1\ \cancel{m})^2} = 4.5\times10^{10}\ \text{Newtons}$$

Since each +3.0-C charge exerts a repulsive force, the vectors point directly away from those charges:

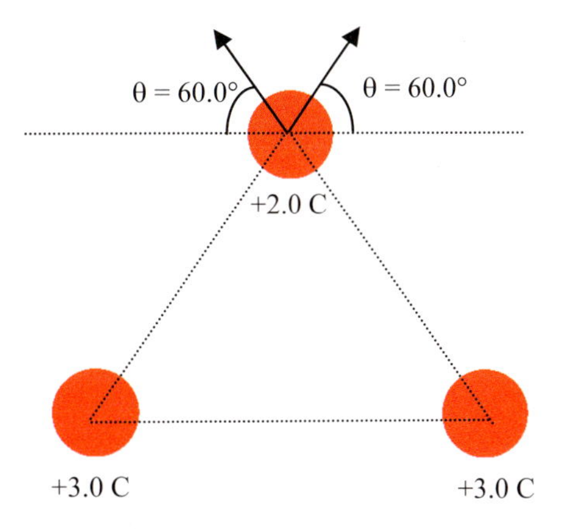

Since these vectors do not point in the same dimension, we will have to add them with trigonometry. First, however, let's define the angles properly. Vector angles are always defined counterclockwise from the positive x-axis. This means that the first angle (60.0°) is okay. The angle for the vector on the left, however, is 120.0°. Now we can add these vectors:

$$A_x = (4.5\times10^{10}\ \text{Newtons})\cdot\cos(60.0°) = 2.3\times10^{10}\ \text{Newtons}$$

$$A_y = (4.5\times10^{10}\ \text{Newtons})\cdot\sin(60.0°) = 3.9\times10^{10}\ \text{Newtons}$$

$$B_x = (4.5\times10^{10}\ \text{Newtons})\cdot\cos(120.0°) = -2.3\times10^{10}\ \text{Newtons}$$

$$B_y = (4.5\times10^{10}\ \text{Newtons})\cdot\sin(120.0°) = 3.9\times10^{10}\ \text{Newtons}$$

$$C_x = A_x + B_x = 2.3 \text{ x } 10^{10}\ \text{Newtons} + -2.3 \text{ x } 10^{12}\ \text{Newtons} = 0\ \text{Newtons}$$

$$C_y = A_y + B_y = 3.9 \text{ x } 10^{10}\ \text{Newtons} + 3.9 \text{ x } 10^{10}\ \text{Newtons} = 7.8 \text{ x } 10^{10}\ \text{Newtons}$$

Do you see what has happened here? Since the forces have equal but opposite x-components, they cancel each other out. The y-components, however, point in the same direction, so they add to one another. In the end, then, the total instantaneous electrostatic force is one-dimensional. It is 7.8 x 10^{10} Newtons pointing straight up.

13.7 In this problem, the stationary charges are equal in magnitude. This means that the positive charge will have 12 lines coming out of it and all 12 of those lines will end up going into the negative charge. I will place the positive charge on the left, although it really doesn't matter. The diagram, then, looks like this:

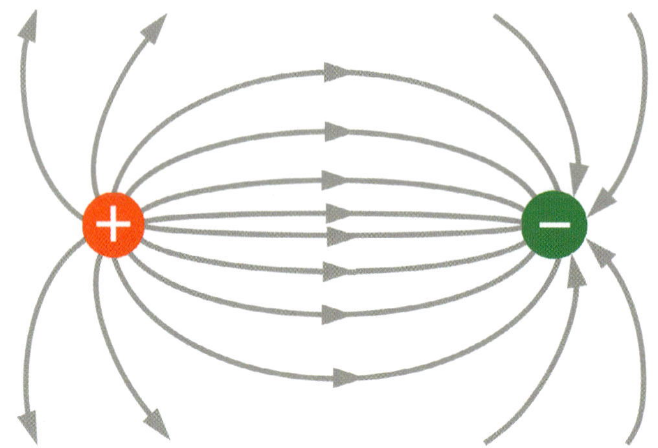

Note that to save vertical space, I did not draw the outer four lines in their entirety. I just drew the part of each line that leaves the positive charge and the part that enters the negative charge. Your drawing, of course, can have unbroken lines, if you want. A charge will experience the greatest electrostatic force (and thus the greatest acceleration) wherever the line density is greatest. Clearly, the line density is greatest nearest to and on a straight line between the charges.

13.8 In order to calculate the electric field at this point, we need to calculate the electric field from each charge and add those two fields together. Here is the electric field for the charge on the left:

$$E_1 = \frac{kQ}{r^2} = \frac{(9.0\times10^{9}\ \frac{\text{Newton}\cdot\text{m}^2}{\text{C}^2})\cdot(4.0\times10^{-4}\ \text{C})}{(0.40\ \text{m})^2} = 2.3\times10^{7}\ \frac{\text{Newtons}}{\text{C}}$$

This electric field points to the right, as electric field lines point directly away from positive charges. The electric field for the charge on the right is:

$$E_2 = \frac{kQ}{r^2} = \frac{(9.0\times10^{9}\ \frac{\text{Newton}\cdot\text{m}^2}{\text{C}^2})\cdot(4.0\times10^{-4}\ \text{C})}{(1.60\ \text{m})^2} = 1.4\times10^{6}\ \frac{\text{Newtons}}{\text{C}}$$

This electric field will point to the left, since electric field lines point directly away from positive charges. As result, we have the following diagram:

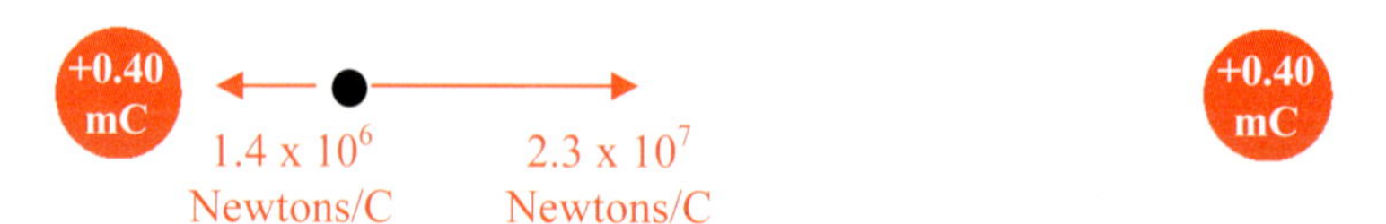

Since the vectors are in the same dimension, we can just add them, using the convention that vectors pointing to the right are positive and those pointing to the left are negative:

$$\mathbf{E}_{\text{total}} = \mathbf{E}_1 + \mathbf{E}_2 = 2.3\times10^{7}\ \frac{\text{Newtons}}{\text{C}} - 1.4\times10^{6}\ \frac{\text{Newtons}}{\text{C}} = 2.2\times10^{7}\ \frac{\text{Newtons}}{\text{C}}$$

Since the sum is positive, it means that the electric field vector points to the right. Thus, the electric field is 2.2×10^{7} Newtons/C pointing to the right.

Now that we have the electric field at that point, we can calculate the electrostatic force for any charge that is put there. The question asks for the force experienced by a +0.56-mC charge:

$$\mathbf{E} = \frac{\mathbf{F}}{q_o}$$

$$\mathbf{F} = q_0\mathbf{E} = (5.6\times10^{-4}\ \cancel{\text{C}})\cdot 2.2\times10^{7}\ \frac{\text{Newtons}}{\cancel{\text{C}}} = 12{,}000\ \text{Newtons}$$

Since the force is positive, it also points to the right. Thus, the force is 12,000 Newtons pointed to the right.

13.9 You can go through the mathematics, or you can just recognize that the electric field from each charge will be equal in magnitude at the midpoint between the two charges. However, the electric field from one charge will be opposite that of the other charge, because electric field lines always point away from positive charges. Thus, the two electric fields will add to zero.

13.10 Don't worry that the ion in this problem has two protons. The protons are so close together that they can be considered a single source that has double the charge of a proton. Thus, the positive charge from the nucleus is 3.2×10^{-19} C (twice the charge of a single proton). The important thing is that there is only one electron. Thus, we can solve this problem just as we did the previous one:

$$\frac{mv^2}{\cancel{r}} = \frac{kq_1q_2}{r^{\cancel{2}}}$$

$$v^2 = \frac{kq_1q_2}{rm} = \frac{(9.0\times10^{9}\ \frac{\text{Newton}\cdot\text{m}^{\cancel{2}}}{\cancel{\text{C}^2}})\cdot(3.2\times10^{-19}\ \cancel{\text{C}})\cdot(1.6\times10^{-19}\ \cancel{\text{C}})}{(1.3\times10^{-11}\ \cancel{\text{m}})\cdot(9.1\times10^{-31}\ \text{kg})}$$

$$v^2 = 3.9\times10^{13}\ \frac{\text{m}^2}{\text{sec}^2}$$

$$v = 6.2\times10^{6}\ \frac{\text{m}}{\text{sec}}$$

This electron, then, circles the nucleus at 6.2×10^{6} m/sec.

REVIEW QUESTIONS

1. The three particles that make up an atom are the proton, the neutron, and the electron. State the electric charge for each. Don't worry about the magnitudes. Just give me the signs.

2. Matter is made up of atoms and therefore has both positive and negative charges in it. Nevertheless, most matter has no electric charge. Why?

3. Compare and contrast charging by induction and charging by conduction.

4. You have a negatively-charged object and you want to give another object a negative charge. Will you charge by induction or conduction?

5. What is the difference between insulators and conductors? In general, in which class do metals belong? In which class do non metals belong?

6. In an electric field drawing, what is the significance of the directions in which the arrows point?

7. In an electric field drawing, what is the significance of the density of the electric field lines?

8. Two positive charges are placed near each other. If the charges are free to move, what will happen? How will the force that they exert on one another change as time goes on?

9. A physicist is doing some experiments on charges. She measures the electrostatic force that exists between two charges. In her notebook, she writes down that charge 1 exerts a 1,001-Newton force on charge 2. What is the magnitude of the force that charge 2 exerts on charge 1?

10. Two charges are held in place close to one another, and the electrostatic force that they exert on one another is measured. If the distance between the charges is then doubled, what happens to the magnitude of the force?

My Favorite Electricity Joke

A man lives in a foreign country, and his job is to operate the train that connects one town to another. He is not very good at his job, and he is also greedy. Since his income does not meet his expenses, he decides to steal from his passengers' fares. At first, he steals only a little. However, as he gets more and more greedy, he steals more and more. Eventually, he is caught. The company is furious. Once he has been tried and found guilty, the company asks for the death penalty. The court refuses, choosing to banish him from the country instead.

The man moves on to another country, certain that he can dedicate himself to a new life. However, the only thing that he really understands is operating trains, so before long, he is a train operator in this new country. Unfortunately, the old habits come back, and after a while, he starts stealing from the passengers' fares again. Once again, he is eventually caught and taken to trial. Once he is found guilty, the judge says that he sees no hope for reform, since this is the second time the man has been caught doing this. Thus, the judge sentences the man to death.

On the day of the execution, he is placed on the electric chair, and the chair is turned on. Much to the surprise of everyone there, the man is not even hurt. He just sits there, as if nothing is happening. The instrument panel says that the electric chair is working, but the man is completely unaffected! The chair is turned off and on several more times, but the man doesn't even flinch! Finally, one of the guards asks the man why the electric chair isn't even hurting him, and the man says, **"Well, I've always been a pretty poor conductor**." Think about it…

PRACTICE PROBLEMS

$$k = 9.0 \times 10^9 \frac{\text{Newtons} \cdot \text{m}^2}{\text{C}^2}$$

1. Two charged particles (q_1 = 6.7 C, q_2 = -3.1 C) are 1.2 m apart. What is the instantaneous electrostatic force between them?

2. Two charges (q_1 = 0.32 mC, q_2 = 0.55 mC) exert a mutual repulsive force of 2.5 x 10^5 Newtons. How far apart are they?

3. Two particles of equal but unknown charge are held 45 cm apart. If their electrostatic repulsion is 2.2 x 10^5 Newtons, what is the magnitude of their charge?

4. Three charges (none of them are forced to stay stationary) are arranged as follows:

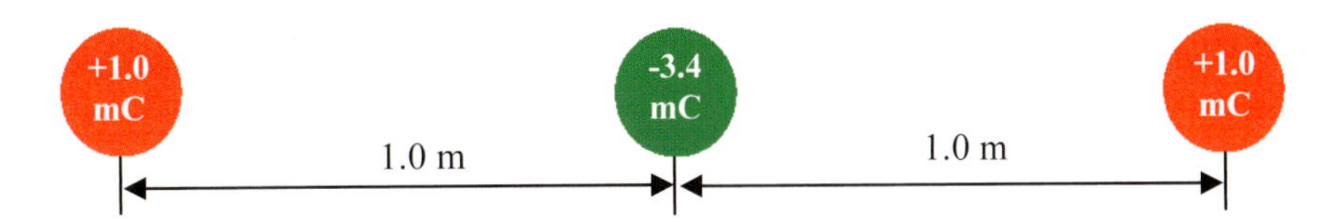

Explain why the instantaneous electrostatic force exerted on the -3.4-mC charge is zero. Even though the force exerted on the -3.4-mC charge is zero, explain why this system does not stay in this arrangement.

5. Three charges are arranged as follows:

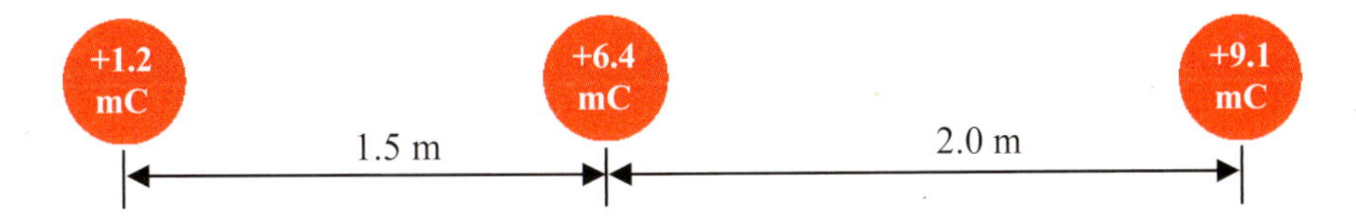

a. What is the instantaneous electrostatic force exerted on the +6.4-mC charge?
b. If you were to make the 1.2-mC and 9.1-mC charges stationary and completely remove the 6.4-mC charge, what would be the electric field at the point where the 6.4-mC charge was?

6. Three charges are arranged as follows:

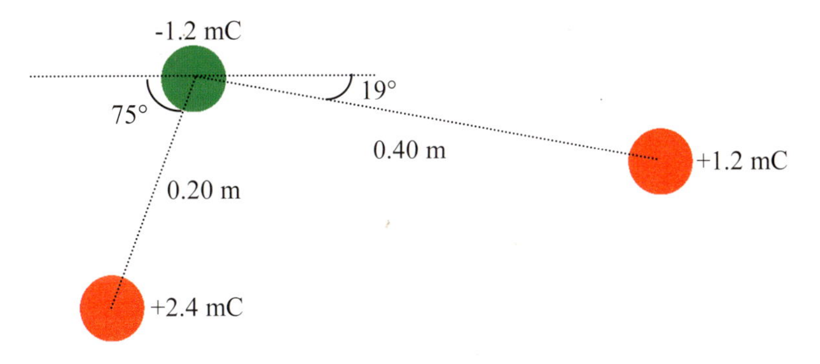

What is the instantaneous electrostatic force on the -1.2-mC object?

7. Draw the electric field generated by these two stationary charges:

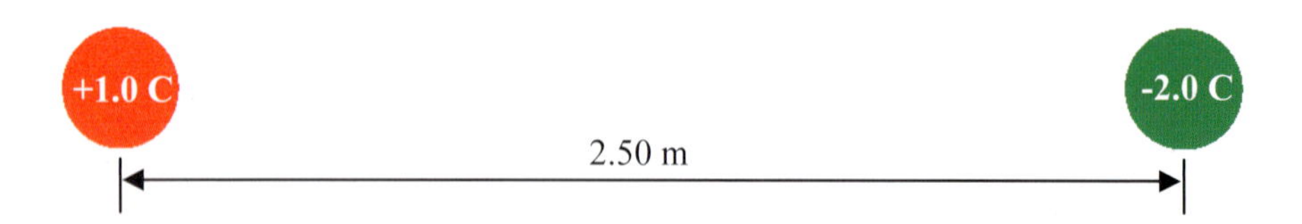

If a negatively-charged particle were placed in this field, in what direction would it travel, and where would it experience the greatest acceleration?

8. Calculate the electric field at the midpoint between the charges drawn in problem #7. What electrostatic force would a -0.66-mC charge experience if it were placed at that point?

9. Draw the electric field generated by the three stationary charges shown below. They sit at the vertices of an equilateral triangle and they are all of equal positive charge.

What is the magnitude of the electric field at the very center of the triangle?

10. The second orbit for the only electron in a He^+ ion has a radius of 5.29×10^{-11} m. What is the speed of an electron in that orbit? Remember, a helium ion has two protons, each with a charge of $+1.6 \times 10^{-19}$ C, and one electron with a charge of -1.6×10^{-19} C. The mass of the electron is 9.1×10^{-31} kg.

MODULE #14: Electric Potential

Introduction

Now that you've learned about electric fields and the electrostatic force, it is time for you to learn a bit more about how electric charges move in response to the electrostatic force. This is a very important subject in the study of electricity, because the kinetic energy of charged particles in motion can be used to do work. In fact, that's how electricity does things like lighting a light bulb, warming up a heating coil, and running an electric motor. The kinetic energy of electrons as they move through wires is converted into whatever form of energy is desired. In the case of a light bulb, for example, the resistance of the filament converts kinetic energy into heat. When the filament gets hot enough, it glows with a bright white light. Thus, the kinetic energy of the moving electrons is transformed first into heat energy and then into light energy.

Of course, in order to really understand how to use the motion of charged particles to power useful things like light bulbs and electric motors, we first need to learn a few things about how charges move. We will start with a discussion of potential energy and charged particles. After all, motion occurs when potential energy is converted into kinetic energy. So, to study charges in motion, we first need to see where the kinetic energy of that motion comes from.

Electric Potential

Consider a positively-charged particle held in place so that it cannot move. If I were to place a negative particle near this fixed charge, the negative particle would immediately be attracted to it. As a result, the negative particle would begin to accelerate towards the positive one. As the negative particle got nearer to the positive particle, the force of attraction would increase, causing the negative particle's acceleration to increase. Thus, the longer the negative particle is allowed to move, the faster it will travel towards the positive particle.

What does this tell us? Well, if the negatively-charged particle starts at rest and suddenly begins moving, we can conclude that it must have gained kinetic energy. The First Law of Thermodynamics tells us that energy cannot just be created out of the blue. Instead, energy can only be transformed from one form to another. So, there must be some form of potential energy that is converted to kinetic energy as the negative particle approaches the fixed positive charge. This potential energy comes from something we call the **electric potential**, which can be described with the following simple equation:

$$V=\frac{kQ}{r} \tag{14.1}$$

In this equation, "V" stands for the electric potential, "k" is the Coulomb constant that we used in the previous module, "Q" represents the magnitude of the fixed charge, and "r" is the distance between the fixed charge and the moving one.

In a moment, we will begin to learn the details of how to use Equation (14.1). Before we do, however, there are a few things I want to point out. First, notice that the electric potential depends only on the charge that creates the electric field. Remember from the previous module that a single particle produces an electric field, and the lines of the field tell us what will happen to *a different* particle if it moves into that field. In the same way, the electric potential depends only on a single charge. In this module, we will always use a fixed charge, so we can say that the electric potential depends only on the fixed charge.

Second, notice the units here. The Coulomb constant has the SI unit of $\frac{N \cdot m^2}{C^2}$, charge has the SI unit of C, and the SI unit for distance is meters. When you plug these units into Equation (14.1), you get a unit of $\frac{N \cdot m}{C}$. Notice that this is *not* an energy unit. That's because electric potential is *not quite* potential energy. There is one thing that we must do to the electric potential in order to turn it into potential energy. We'll get to that in a moment. For right now just keep this in mind:

Electric potential is *not* potential energy.

Thankfully, physicists have named the unit for electric potential. We now call it the "volt," in honor of Count Alessandro Volta, an Italian physicist who invented the first primitive battery.

Finally, Equation (14.1) looks very similar to Equation (13.1), the equation for the electrostatic force. There are a few differences. The distance is squared in Equation (13.1) but not in Equation (14.1). Also, there are two charges in Equation (13.1) and only one in Equation (14.1). The biggest difference, however, is not readily apparent. When using Equation (13.1), I told you not to put the sign of the charge into the equation. Instead, we just put the magnitude of the charge in the equation, and we used our minds to figure out the direction of the force. Precisely the *opposite* is true of Equation (14.1). This is a potential source of confusion, so I really need to emphasize it:

When using Equation (14.1), put the sign of the charge directly into the equation.

So, you must keep Equations (13.1) and (14.1) separate in your mind. Even though they look similar, they are used to calculate different things.

Now we are ready to use Equation (14.1). Study the following example and complete the "On Your Own" problem. Although this might seem pretty elementary right now, it gets more complicated, so we need to make sure you can do this before going any further.

EXAMPLE 14.1

Consider a +3.2-mC positive charge that cannot move. What is the electric potential 45 cm from the charge?

First of all, realize that the charge we put in Equation (14.1) must be in Coulombs. Thus, we need to convert 3.2 mC into 3.2×10^{-3} C. Likewise, the distance needs to be converted from 45 cm into 0.45 m. Now that we have SI units for everything, we can use Equation (14.1):

$$V = \frac{kQ}{r}$$

$$V = \frac{\left(9.0 \times 10^9 \frac{\text{N} \cdot \text{m}^{\cancel{2}}}{\cancel{\text{C}^2}}\right) \cdot (+3.2 \times 10^{-3} \cancel{\text{C}})}{0.45 \cancel{\text{m}}} = 6.4 \times 10^7 \frac{\text{N} \cdot \text{m}}{\text{C}} = 6.4 \times 10^7 \text{ volts}$$

Notice how the units work out. A Coulomb and a meter both cancel from the constant, leaving $\frac{\text{N} \cdot \text{m}}{\text{C}}$ which is the definition of a volt. The electric potential, then, is 6.4×10^7 volts.

At this point, of course, this number should mean absolutely nothing to you, because we really haven't seen how electric potential relates to potential energy. Since we put the positive charge in Equation (14.1), we know that this is a positive electric potential. Nevertheless, you really have no idea what a positive electric potential is. Well, never fear. That's what we'll learn in the next section.

ON YOUR OWN

14.1 A -2.1-mC charge is anchored so that it cannot move. A +4.1-C charge is placed 61 cm from it. What is the electric potential at that point?

Electric Potential, Potential Energy, and Potential Difference

Now if the electric potential does not tell us potential energy, what does? Notice that in Example 14.1 and "On Your Own" problem 14.1, we only considered one charge: the fixed charge. However, we learned in the previous module that an electrostatic *force* is generated between two charges. Thus, in order to see the effects of the electrostatic force, another charge must enter into the picture. The acceleration that this second charge experiences due to its electric attraction or repulsion to the fixed charge depends on the magnitudes of *both* charges under consideration. If the acceleration depends on both charges, so will the speed and hence the kinetic energy. Well, the kinetic energy comes from the potential energy, so the potential energy must depend on both charges.

This reasoning tells us how to convert electric potential to potential energy. If we want to calculate the potential energy, then, we must take the other charge into account. Here's how we do it:

$$PE = qV \qquad (14.2)$$

In this equation, "PE" is the potential energy, "V" represents the electric potential from the fixed charge, and "q" is the magnitude of the *other* charge. This way, the potential energy takes the magnitudes of both charges into account.

The multimedia CD has a video demonstrating potential energy in an electrical system.

When we put units into this equation, it becomes obvious why we need to multiply by the magnitude of the other charge in order to calculate potential energy:

$$PE = qV$$

$$PE = (\cancel{C}) \cdot (\frac{N \cdot m}{\cancel{C}}) = N \cdot m = J$$

Since the Coulomb unit from "q" cancels with the Coulomb on the bottom of the fraction, the result is a N·m, which is the same as a Joule. Since potential energy must have a unit of Joules, you can now see that this equation calculates potential energy. In the end, then, that's how we calculate the potential energy due to the electrostatic force. We calculate the potential, and then multiply the potential by the charge that will actually be moving. Once again, we must put the sign of the charge in Equation (14.2). You will see why this is important in the next section.

So now we see how electric potential relates to potential energy. The larger the electric potential, the larger the potential energy. That's not the only way to affect the magnitude of the potential energy, however. The potential energy is also increased when the magnitude of the mobile charge increases. Make sure you understand this relationship by solving the following "On Your Own" problem.

ON YOUR OWN

14.2 A +7.9-mC charge is placed 1.1 m away from a -2.3-C stationary charge. What is the potential energy in this situation?

Potential Difference and the Change in Potential Energy

Now that we have the basics of potential energy in situations involving charge, we can finally analyze the motion of a charged particle as it travels through an electric field. Consider, for example, the "On Your Own" problem that you just performed. The potential energy that you calculated will only be valid for an instant, right? After all, since the +7.9-mC charge is free to move, it will begin moving towards the fixed -2.3-C charge. Once the +7.9-mC charge moves, it will get closer to the fixed charge. That will change the value of "r," which will, in turn, change the value of the electric potential, which will change the potential energy.

How will the potential energy change? Well, that's the reason we have to make sure to keep the signs of the charges in Equations (14.1) and (14.2). In this case, we have a negative fixed charge. When we put that charge into Equation (14.1), the electric potential will be negative. When you use Equation (14.2) to calculate the potential energy, the potential energy will also be negative, because the negative potential will be multiplied by a positive freely-moving charge. In the end, then, the potential energy will be negative.

As the freely-moving charge approaches the stationary charge, the magnitude of the electric potential (and thus the potential energy) increases. Thus, as the freely-moving positive charge approaches the stationary negative charge, the potential energy becomes more negative. In other words, the *change* in the electric potential is negative. That means the change in potential energy is negative, which means the potential energy decreased. This is what we would expect. Since the freely moving charge starts traveling faster and faster, we know that it is converting potential energy into kinetic energy. Thus, we know that the potential energy decreases, while the kinetic energy increases.

This is why we must keep the charges in Equations (14.1) and (14.2). Suppose we changed the situation a little. Suppose that the stationary charge was negative and the freely-moving charge was also negative. Now suppose that I give the freely-moving charge an initial velocity towards the stationary charge. What will happen to the potential energy? Well, let's first think about what we know will happen. Since the two negative charges repel each other, the freely-moving charge will begin to slow down, eventually coming to a halt and reversing its course. In other words, the freely-moving charge's kinetic energy will decrease. This means that the potential energy will increase.

Now, is that what our equations tell us? Yes, as long as we put the charges in the equation. You see, since the fixed charge is negative, the electric potential is also negative. However, when the potential energy is calculated, that negative electric potential is multiplied by a negative charge. In the end, then, the potential energy is positive. So, as the freely-moving charge gets closer to the stationary charge, the potential energy will increase, which means the kinetic energy must decrease. In the end, then, we keep the signs of the charges in Equations (14.1) and (14.2) in order to determine the *change in* the potential energy.

This brings me to another very important concept in the study of charges moving in electric fields: the **potential difference**. When we study the electric potential in a situation, the value of the electric potential at a single point is not all that interesting. However, if we look at the *difference* in the electric potential from one point to another, we can learn a great deal. After all, as the electric potential changes, so does the potential energy of the situation. Thus, the *change in* potential energy is determined by the *change in* the electric potential. In other words, the change in potential energy depends on the potential difference. This can be summed up mathematically as:

$$\Delta PE = q\Delta V \tag{14.3}$$

Now I know that this just looks like Equation (14.2) with two delta signs put in. However, this is the more useful of the two equations, because it allows us to determine the change in potential energy in a situation, which will allow us to determine how the kinetic energy changes in a situation. That, of course, will allow us to determine the speed of the particle as it moves.

Your head is probably swimming a bit at this point, so I want to try to sum up all of this with an example problem.

EXAMPLE 14.2

A -5.6-mC charge that is free to move is placed 68 cm from a fixed +4.8-mC charge. If the freely-moving charge moves 15 cm towards the stationary charge, what is the change in potential energy? Did the potential energy increase or decrease?

To calculate the change in the potential energy, I will need to determine the change in the electric potential. Thus, let's start by using Equation (14.1) to calculate the electric potential at each point. The stationary charge is +4.8 mC, so once we convert it to Coulombs, we use it in the equation. Also, the initial distance between the charges is 68 cm, which must be converted to meters.

$$V = \frac{kQ}{r}$$

$$V_{initial} = \frac{\left(9.0 \times 10^{9} \frac{N \cdot m^{\not{2}}}{C^{\not{2}}}\right) \cdot (4.8 \times 10^{-3} \not{C})}{0.68 \not{m}} = 6.4 \times 10^{7} \frac{N \cdot m}{C} = 6.4 \times 10^{7} \text{ volts}$$

Now remember, this is just the initial electric potential. We can use the same method to determine the final electric potential. If the freely-moving charge moves 15 cm closer to the stationary charge, the distance between the charges will be only 53 cm. That's the final value for r:

$$V = \frac{kQ}{r}$$

$$V_{final} = \frac{\left(9.0 \times 10^{9} \frac{N \cdot m^{\not{2}}}{C^{\not{2}}}\right) \cdot (4.8 \times 10^{-3} \not{C})}{0.53 \not{m}} = 8.2 \times 10^{7} \frac{N \cdot m}{C} = 8.2 \times 10^{7} \text{ volts}$$

Now that we have initial and final electric potential, we can subtract the two to get ΔV:

$$\Delta V = V_{final} - V_{initial} = 8.2 \times 10^{7} \text{ volts} - 6.4 \times 10^{7} \text{ volts} = 1.8 \times 10^{7} \text{ volts}$$

That's the potential difference between the moving charge's initial and final positions. However, that's not the answer to the question. The question asks for the change in *potential energy*. Using Equation (14.3), we can now calculate that. Of course, to use Equation (14.3), we must use the charge of the freely-moving particle (q). This was given as -5.6 mC, which we will have to convert to Coulombs in order to be able to use it in the equation:

$$\Delta PE = q\Delta V = (-5.6 \times 10^{-3} \not{C}) \cdot (1.8 \times 10^{7} \frac{N \cdot m}{\not{C}}) = -1.0 \times 10^{5} \text{ N} \cdot \text{m} = -1.0 \times 10^{5} \text{ J}$$

Now what does this tell us? The change in potential energy is negative. That tells us that the potential energy decreased by 1.0×10^{5} J.

Do you see what we accomplished by leaving the charges in the equations? When we subtract the initial electric potential from the final electric potential, we got a positive answer, which told us that the electric potential *increased.* However, when we took that potential difference and multiplied it by a negative charge, we got a negative answer, which told us that the potential energy *decreased.*

I really want to make sure you understand what we did here, so let's go over it again. Using Equation (14.1), we can calculate the electric potential at any given point. Although that is nice, it is not very useful. However, if we then calculate the electric potential at another point and take the difference, we have a value for the potential difference between those two points. If we then multiply that potential difference by the charge of a particle that moves between those two points, we know how much the potential energy changes while the particle moved, and as we will see in the next section, that is a *very* useful thing to know.

The potential difference is such a useful quantity that it is how we rate batteries. If you look at a standard "D-cell" flashlight battery, for example, you will read that it is a 1.5-volt battery. What does that mean? It means that the difference in electric potential between the negative side of the battery and the positive side of the battery is 1.5 volts. Thus, if a charged particle moves from the negative side of the battery to the positive side of the battery, it will have moved through a potential difference of 1.5 volts. Alternatively, if a charged particle moves from the positive side of the battery to the negative side, it experiences a change in electric potential of -1.5 volts. Thus, you can determine the change in the potential energy by taking the charge of the particle and multiplying it by the potential difference. Please note that it does not matter what path the charged particle takes to do this. Regardless of *how* the particle moves, if it starts at the negative side of the battery and ends at the positive side, the electric potential changes by 1.5 volts. Alternatively, if it starts at the positive side of the battery and moves to the negative side, the electric potential changes by -1.5 volts, regardless of *how* the particle made the trip.

The potential difference also allows us to learn the definition of the **electron volt (eV)**. Remember that in Module 11, I introduced the electron volt as a new energy unit that is convenient to use when dealing with things on the atomic scale. At that time, I told you how the electron volt related to the Joule, but I did not tell you the definition of an electron volt. I can do that now:

Electron volt - The decrease in potential energy when an electron moves through a potential difference of exactly 1 volt

Now think about what this definition says. An electron ($q = -1.6 \times 10^{-19}$ C) moves through a potential difference of exactly 1 volt. Well, if $\Delta V = 1$ volt, the change in potential energy is easy to calculate:

$$\Delta PE = q\Delta V = (-1.6 \times 10^{-19} \cancel{C}) \cdot (1 \frac{N \cdot m}{\cancel{C}}) = -1.6 \times 10^{-19} \ N \cdot m = -1.6 \times 10^{-19} \ J$$

Thus, the potential energy in that situation decreases by 1.6×10^{-19} J. Think about what that means. It means that 1 eV is 1.6×10^{-19} J. If you look back in Module 11, you will see that this is the relationship I gave you between electron volts and Joules. Now you understand where it comes from!

ON YOUR OWN

14.3 A -1.1-mC charge that is free to move is placed 45 cm from a stationary -1.8-mC charge. If the freely-moving charge moves an additional 25 cm away from the stationary charge, what is the change in potential energy? Did it increase or decrease?

14.4 A -5.6-mC charge that has been given a large initial velocity moves from the positive side to the negative side of a 9.0-volt battery. What is the change in potential energy? Did it increase or decrease?

Conservation of Energy in an Electric Potential

Now that we know how to analyze the change in potential energy in a situation involving electric charge, we can begin to analyze the energy balance in electrical systems. We already know from Module #8 that the total amount of energy (kinetic + potential) in a system cannot change. Thus, we know that if potential energy increases, that increase must have come from kinetic energy. As a result, the object we are studying must have slowed down. Alternatively, if the potential energy decreases, the kinetic energy increases, speeding up the object. We can therefore take all of the techniques that we learned in Module #8 and apply them to situations involving electric charges. Let's see how this works by studying a few example problems.

EXAMPLE 14.3

A freely-moving -15.1-mC charged particle (m = 2.34 kg) is placed 13 cm from a fixed -2.0-C charge. If the particle starts at rest, at what speed will it be moving when it is 95 cm away from the fixed charge?

To solve this problem, we must look at the energetics of the situation. When the particle is at rest, its kinetic energy is zero. Its potential energy, however, is not zero. Since the total energy must stay the same, if the potential energy were to change, the kinetic energy would have to change so as to keep the total energy constant. Thus, let's calculate how the potential energy changes as a result of the change in position, and that should tell us something about the kinetic energy, which will, in turn, tell us something about the speed.

To calculate the change in potential energy, we will start by determining the potential difference between the initial and final positions. To do this, we first must calculate the potential at the initial position (0.13 m from the fixed charge):

$$V = \frac{kQ}{r}$$

$$V_{initial} = \frac{\left(9.0 \times 10^{9} \frac{\text{N} \cdot \text{m}^{\cancel{2}}}{\text{C}^{\cancel{2}}}\right) \cdot (-2.0 \cancel{\text{C}})}{0.13 \cancel{\text{m}}} = -1.4 \times 10^{11} \frac{\text{N} \cdot \text{m}}{\text{C}} = -1.4 \times 10^{11} \text{ volts}$$

Next, we calculate the potential at the final position (0.95 m from the fixed charge):

$$V = \frac{kQ}{r}$$

$$V_{final} = \frac{\left(9.0 \times 10^{9} \frac{\text{N} \cdot \text{m}^{\cancel{2}}}{\text{C}^{\cancel{2}}}\right) \cdot (-2.0 \cancel{\text{C}})}{0.95 \cancel{\text{m}}} = -1.9 \times 10^{10} \frac{\text{N} \cdot \text{m}}{\text{C}} = -1.9 \times 10^{10} \text{ volts}$$

Now that we have the potential at each point, we can calculate the potential difference:

$$\Delta V = V_{final} - V_{initial} = (-1.9 \times 10^{10} \text{ volts}) - (-1.4 \times 10^{11} \text{ volts}) = 1.2 \times 10^{11} \text{ volts}$$

This gives us the information we need to calculate the change in the potential energy:

$$\Delta PE = q\Delta V = (-0.0151 \text{ €})\cdot(1.2\times10^{11} \ \frac{N\cdot m}{\text{€}}) = -1.8\times10^{9} \ N\cdot m = -1.8\times10^{9} \ J$$

What does this tell us? It tells us that the particle's potential energy *decreased* by 1.8×10^9 J. Now remember, the total energy (KE + PE) must remain the same. Thus, if the potential energy *decreased* by 1.8×10^9 J, the kinetic energy must have *increased* by 1.8×10^9 J. Otherwise, the sum of kinetic plus potential energies would not stay the same. Well, if the kinetic energy started out at zero (the particle was not moving) and increased by 1.8×10^9 J, the final kinetic energy must be 1.8×10^9 J. Now that we know the final kinetic energy, we can calculate the speed, because Equation (8.3) relates the speed of an object to its mass and kinetic energy:

$$KE = \frac{1}{2}mv^2$$

$$1.8\times10^{9} J = \frac{1}{2}\cdot(2.34 \ kg)\cdot v^2$$

$$v = 3.9\times10^{4} \ \frac{m}{sec}$$

When the particle is 95 cm away from the stationary charge, then, its speed is $\underline{3.9 \times 10^4 \text{ m/sec}}$. If you're not quite sure how the units worked out in the equation above, just remember that a Joule is the same as a $\frac{kg\cdot m^2}{sec^2}$. That's why we used kg for the mass unit.

Now this example might have seemed a bit long, but the only really new thing in it is the calculation of the change in potential energy with Equations (14.1) and (14.3). Notice that everything else is a repeat of Module #8. We realized that the total energy of the situation could not change, so any change in potential energy would have to change the kinetic energy. Once we found out that the potential energy decreased, we realized that this would have to cause a corresponding increase in the kinetic energy, which then allowed us to determine the speed. Make sure you understand this by studying the following example problems.

EXAMPLE 14.4

A -2.3-mC charged particle (m = 1.44 kg) is placed 1.50 m from a +1.8-mC stationary charge. If the particle starts from rest, how fast will it be moving after it has traveled 75 cm?

Once again, to solve a problem like this one, we simply look at the energetics involved. First, we need to calculate the electric potential at the initial position:

$$V = \frac{kQ}{r}$$

$$V_{initial} = \frac{\left(9.0\times10^{9}\ \frac{N\cdot m^{\cancel{2}}}{C^{\cancel{2}}}\right)\cdot(1.8\times10^{-3}\ \cancel{C})}{1.50\ \cancel{m}} = 1.1\times10^{7}\ \frac{N\cdot m}{C} = 1.1\times10^{7}\ \text{volts}$$

Next, we calculate the electric potential at the final position. This is a bit tricky, because the problem doesn't tell us the final position. We have to figure that out. The problem says that after starting 1.50 m from the stationary charge, the freely-moving charge travels 75 cm (0.75 m). Well, since the freely-moving charge is negative and starts out at rest, it will travel *towards* the stationary positive charge. Thus, the freely-moving charge travels 0.75 m closer to the stationary charge. If it started 1.50 m from the stationary charge and moves 0.75 m closer, it will end up 1.50 m - 0.75 m = 0.75 m from the stationary charge. That's "r" in Equation (14.1):

$$V = \frac{kQ}{r}$$

$$V_{final} = \frac{\left(9.0\times10^{9}\ \frac{N\cdot m^{\cancel{2}}}{C^{\cancel{2}}}\right)\cdot(1.8\times10^{-3}\ \cancel{C})}{0.75\ \cancel{m}} = 2.2\times10^{7}\ \frac{N\cdot m}{C} = 2.2\times10^{7}\ \text{volts}$$

Now that we have the potential at each point, we can calculate the potential difference:

$$\Delta V = V_{final} - V_{initial} = (2.2 \text{ x } 10^{7} \text{ volts}) - (1.1 \text{ x } 10^{7} \text{ volts}) = 1.1 \text{ x } 10^{7} \text{ volts}$$

This gives us the information we need to calculate the change in the potential energy:

$$\Delta PE = q\Delta V = (-2.3\times10^{-3}\ \cancel{C})\cdot(1.1\times10^{7}\ \frac{N\cdot m}{\cancel{C}}) = -2.5\times10^{4}\ N\cdot m = -2.5\times10^{4}\ J$$

What does this tell us? It tells us that the particle's potential energy *decreased* by 2.5 x 10^4 J. Now remember, the total energy (KE + PE) must remain the same. Thus, if the potential energy *decreased* by 2.5 x 10^4 J, the kinetic energy must have *increased* by 2.5 x 10^4 J. Well, if the kinetic energy started out at zero (the particle was not moving) and increased by 2.5 x 10^4 J, the final kinetic energy must be 2.5 x 10^4 J. Now that we know the final kinetic energy, we can calculate the speed using Equation (8.3):

$$KE = \frac{1}{2} mv^2$$

$$2.5 \times 10^4 J = \frac{1}{2} \cdot (1.44 \text{ kg}) \cdot v^2$$

$$v = 190 \frac{m}{sec}$$

After the particle has traveled 75 cm, then, its speed is 190 m/sec.

A -0.21-mC particle (m = 3.1 kg) is given an initial velocity of 15 m/sec away from a +0.45-mC stationary charge. If the particle's initial position is 33 cm from the stationary charge, how far away will the particle be able to travel before being pulled back towards the stationary charge?

Before we solve this problem, let's make sure we understand why the particle will be moving away from the stationary charge but will eventually be pulled back towards it. Since the particle's initial velocity is directed away from the stationary charge, it will begin moving in that direction. However, since the two charges are opposite, they attract one another. Thus, the electrostatic force will pull the particle back towards the stationary charge. The particle will not turn around immediately, however, because it has kinetic energy, allowing it to travel away from the stationary charge. However, as the particle travels farther and farther away, the potential energy will get larger and larger. This takes kinetic energy away from the particle. Eventually, all of the kinetic energy will be taken away, and the particle will stop. At that point, then, the attractive electrostatic force can begin to pull it in.

So, to solve this problem, we need to find out where the kinetic energy of the particle equals zero. That's the point at which the particle stops and turns around. Well, if we determine the initial kinetic energy, we can determine how much must be lost to bring the kinetic energy to zero:

$$KE = \frac{1}{2} mv^2$$

$$KE_{initial} = \frac{1}{2} \cdot (3.1 \text{ kg}) \cdot (15 \frac{m}{sec})^2 = 350 \text{ J}$$

If the particle is going to stop, its final kinetic energy must be zero. Thus, the kinetic energy must *decrease* by 350 J. If the kinetic energy *decreases* by 350 J, the potential energy must *increase* by 350 J to keep the total energy constant. Thus, we know that the change in potential energy must be +350 J. This allows us to calculate the change in electric potential needed:

$$\Delta PE = q\Delta V$$

$$350 \text{ J} = (-2.1 \times 10^{-4} \text{ C}) \cdot (\Delta V)$$

$$\Delta V = \frac{350\text{ J}}{-2.1\times10^{-4}\text{ C}} = \frac{350\text{ N}\cdot\text{m}}{-2.1\times10^{-4}\text{ C}} = -1.7\times10^{6}\ \frac{\text{N}\cdot\text{m}}{\text{C}} = -1.7\times10^{6}\text{ volts}$$

That's the change in electric potential that the particle must experience. Well, we have all of the information that we need to calculate the initial potential:

$$V = \frac{kQ}{r}$$

$$V_{\text{initial}} = \frac{\left(9.0\times10^{9}\ \frac{\text{N}\cdot\text{m}^{\cancel{2}}}{\cancel{\text{C}^{2}}}\right)\cdot(4.5\times10^{-4}\ \cancel{\text{C}})}{0.33\ \cancel{\text{m}}} = 1.2\times10^{7}\ \frac{\text{N}\cdot\text{m}}{\text{C}} = 1.2\times10^{7}\text{ volts}$$

If the initial potential is 1.2 x 10^7 volts, and the potential must change by -1.7 x 10^6 volts, then the final potential must be:

$$\Delta V = V_{\text{final}} - V_{\text{initial}}$$

$$-1.7\times10^{6}\text{ volts} = V_{\text{final}} - 1.2\times10^{7}\text{ volts}$$

$$V_{\text{final}} = 1.0\times10^{7}\text{ volts}$$

Now that we know the final electric potential, we can determine the final distance from the stationary charge:

$$V = \frac{kQ}{r}$$

$$1.0\times10^{7}\text{ volts} = \frac{\left(9.0\times10^{9}\ \frac{\text{N}\cdot\text{m}^{2}}{\text{C}^{2}}\right)\cdot(4.5\times10^{-4}\text{C})}{r}$$

$$r = \frac{\left(9.0\times10^{9}\ \frac{\cancel{\text{N}}\cdot\text{m}^{\cancel{2}}}{\cancel{\text{C}^{2}}}\right)\cdot(4.5\times10^{-4}\ \cancel{\text{C}})}{1.0\times10^{7}\ \frac{\cancel{\text{N}\cdot\text{m}}}{\cancel{\text{C}}}} = 0.41\text{ m}$$

So, we finally see that the particle can only travel an additional 8 cm, until it is 0.41 m away from the charge. At that point, the particle stops and begins to head towards the stationary charge.

Even though these problems are long, they should not be difficult. After all, they are almost entirely review from Module #8! Just remember to think in terms of energy. A change in potential energy will cause a corresponding change in kinetic energy, or vice versa. Try your hand at this with the following "On Your Own" problems.

ON YOUR OWN

14.5 A +6.5-C charged particle (m = 42.3 kg) is placed 1.4 m from a +3.4-C stationary charge. If it starts from rest, how fast will the particle be traveling when it is 2.5 m away from the stationary charge?

14.6 A -7.1-mC charged particle (m = 1.1 kg) is placed 65 cm from a +5.2-mC stationary charge. If the particle starts at rest, how fast will it be moving after it has traveled 15 cm?

14.7 A +8.8-mC charged particle (m = 3.5 kg) is shot with an initial velocity of 255 m/sec towards a +4.5-mC stationary charge. If the particle starts out 1.2 meters from the stationary charge, how close will it come to the charge before turning around and moving away?

Capacitors

Now although the study of electric potential caused by a single stationary charge is important, there are many other ways to create an electric potential. For example, suppose you brought many charges together in one place and forced them all to stay in the same general location. Each of the charges would create its own electric field, and the total electric field would be the enormous vector sum of all of the individual electric fields. That total electric field would result in an electric potential. Now, of course, this could get very complicated. However, there are certain situations like this in which the net result is, believe it or not, quite simple. One of those situations occurs when you have a **capacitor** (kuh pass' uh tur).

Capacitor - A device that stores charge

Because capacitors store charge they have many uses. We will see one of the uses in the final section of this module.

There are many different types of capacitors, but we will concentrate on the most common, the **parallel-plate capacitor**. As its name implies, this capacitor consists of two conductive plates that are parallel to one another, but separated by a small distance. The plates are full of opposite charges, as illustrated in the figure below.

FIGURE 14.1
A Parallel Plate Capacitor

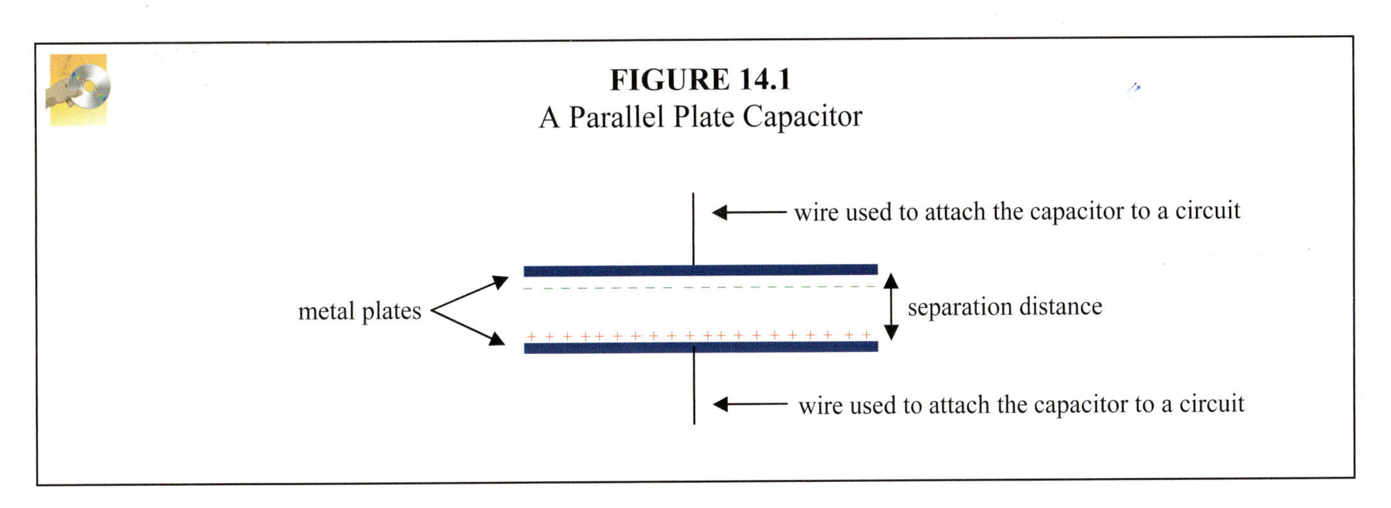

Since the opposite charges attract one another, they want to get as close as possible. However, these charges are not free to move towards one another, as the substance between them is an insulator. As a result, they stay on the conductive plates, trying to get as close to the other plate as possible. The smaller the separation distance, the closer the opposite charges can get to one another. In a capacitor, the total amount of charge on each plate must be the same. Thus, if a total of 10 C of charge is stored on one plate, the other plate must have a total charge of -10 C. Get some experience with parallel plate capacitors by performing the following experiment.

EXPERIMENT 14.1

Making a Parallel-Plate Capacitor and Storing Charge

Supplies:

- Safety goggles
- Two Styrofoam® plates (Paper plates will not work nearly as well.)
- Aluminum foil
- Tape
- Scissors
- A metal paper clip
- Balloon

1. Cut out two circles of aluminum foil that are as large as the bottom of a Styrofoam® plate.
2. Tape one of the circles in the center of the top of one plate.
3. Tape the other circle to the center of the *other side* (the bottom) of the plate. Now you have two circles of foil that are parallel to one another, separated by the thickness of the plate.
4. Cut a strip of aluminum foil that is one-half inch wide and 6 inches long.
5. Tape that strip of foil to the foil circle on the bottom of the plate. Be sure that the foil strip is in good contact with the foil circle, as charges must be able to travel from the foil circle and into the strip. The strip needs to stick out beyond the plate.
6. Place this plate on top of the other plate. The foil strip that is taped to the bottom of the first plate needs to stick out from in between the two plates.
7. Tape the paper clip to the foil circle on the top side of the plate. The metal of the paper clip needs to be in good contact with the foil circle, and it should be taped near the outer edge of the foil circle, so that it extends beyond the foil circle. In the end, your setup should look something like this:

8. Blow up the balloon and tie it off so that it stays inflated.
9. Rub the balloon vigorously in your hair so as to charge the balloon.
10. Place your empty hand (the one not holding the balloon) on the foil strip that extends from the bottom of the plate.

11. Bring the balloon in contact with the foil circle that is on top of the plate. Move the balloon around on the foil, so that any charge that is on the balloon ends up touching the foil.
12. Repeat steps 9-11 nine more times.
13. Put the balloon on the ground away from the plate.
14. Darken the room as much as possible, but make sure you can still see the plate.
15. Place one hand on the foil strip extending from the bottom of the plate, and slowly bring a finger from the other hand to the paper clip. The instant before you touch the paper clip you should feel something. You might even see and hear it as well. Record your experience in your lab notebook.
16. Repeat steps 9-11 ten times.
17. Place the balloon on the floor and walk away from the plate. Be sure to tell everyone in the house to stay away from the plate as well.
18. In about an hour, come back to the plate and repeat steps 14-15. Once again, record your experience in your laboratory notebook.
19. Clean up your mess.

What happened in the experiment? Well, the foil circles taped to opposite sides of the plate made a capacitor. After all, you had two parallel conductors separated by a distance. Charges could not travel from one conductor to the other, because there was an insulator (the Styrofoam®) between them. This actually brings up an important point. Although parallel-plate capacitors are usually drawn with nothing but air in between the plates, that is not the only way to make a parallel-plate capacitor. As long as any insulator is placed in between the plates, you still have a parallel-plate capacitor, since charges cannot move from one plate to another. As a point of terminology, an insulator is often called a **dielectric** (dy' uh lek' trik). Thus, if something other than air is between the plates of a parallel-plate capacitor, we often say that the plates are separated by a dielectric.

Okay, what did you do with the parallel-plate capacitor that you made? Well, you first charged a balloon by rubbing it in your hair. This allowed the balloon to pick up electrons, making it negative. When you then brought the balloon into contact with the foil on top of the capacitor, the negative charges flowed into the foil on top of the capacitor. As a result, that foil became negatively charged. That negative charge could not travel to the foil on the other side of the plate, because the plate was an insulator. However, the negative charge on the top foil did repel the electrons in the bottom foil. Since that bottom foil was connected to a foil strip that was touching your hand, the electrons could travel out of the bottom foil and into your hands. Thus, the bottom foil developed a positive charge in response to the top foil's negative charge.

As you continued to charge the balloon and then bring the balloon into contact with the top foil, the top foil became more and more negatively charged. The increased negative charge on the top foil repelled more electrons in the bottom foil, and since those electrons were free to leave that foil through the strip and into your hand, the bottom foil became more and more positively charged. For every unit of negative charge on the top foil, then, a unit of positive charge developed on the bottom foil.

When you were finally done charging the foil, you put the balloon away and placed one hand on the foil strip. With the other hand, you touched the paper clip. At that point, you should have felt a jolt of electricity. If the room was dark enough, you probably even saw a spark and heard a snap. Why did that happen? Think about it. The capacitor *stored* the charge that you put on it. Thus, one foil was positive, and the other foil was negative. When you touched both the foil strip and the paper clip, you connected the positive foil to the negative foil with your body. Since you are a conductor (which is why electricity can be dangerous), the charges flowed through your body. That shock you felt was the result of the motion of the charges from the paper clip to your hand.

What happened in the last part of the experiment, where you left the capacitor alone for an hour? It still gave you a shock, didn't it? That's because the capacitor *stored* the charge. Thus, even though you left it for an hour, the charge was still there when you went back. Please note that this would *not* happen if the charge were left on the balloon. True, you were able to charge the balloon. However, had you left the balloon alone for an hour, the vast majority of the charge would have been gone when you returned. That's because random collisions between molecules in the air and the balloon would have allowed those molecules to pick up electrons from the balloon. As a result, the balloon would have lost charge. However, because the charges on the negative side of the capacitor were attracted to the charges on the positive side of the capacitor, the charges tended to stay on the capacitor even when molecules from the air collided with the capacitor.

Now since parallel-plate capacitors contain a large number of positive and negative charges, you might think it nearly impossible to calculate the electric potential produced by such a device. Actually, the electric fields from all of the individual charges add up in such a way as to make a rather simple overall electric field, as shown in Figure 14.2.

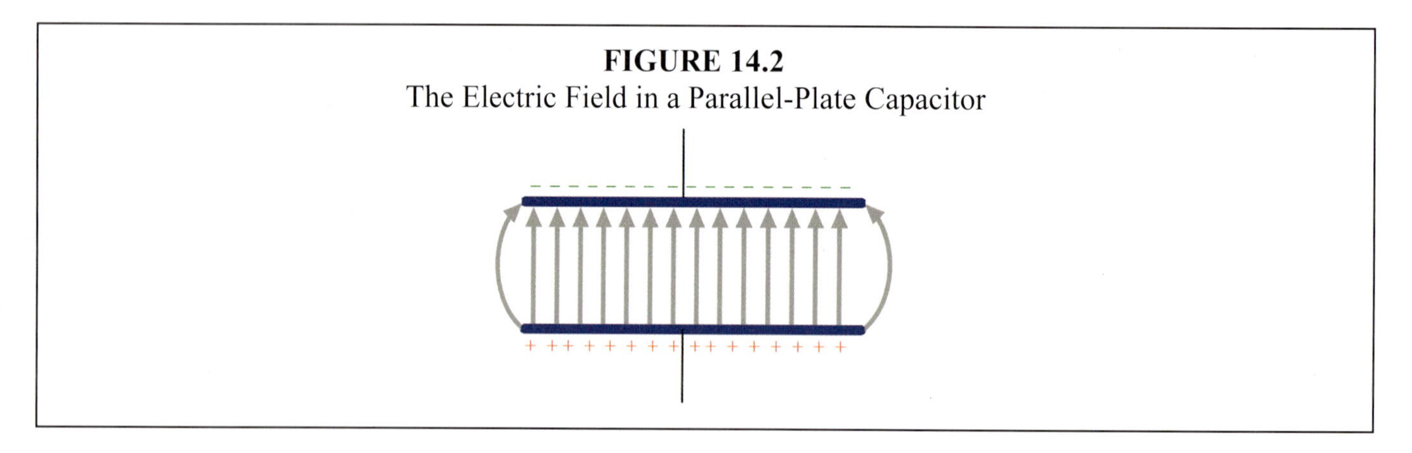

FIGURE 14.2
The Electric Field in a Parallel-Plate Capacitor

As you might expect, a relatively simple electric field results in an easy-to-calculate electric potential. The equation to calculate this electric potential is:

$$V = \frac{Q}{C_q} \tag{14.4}$$

Where "V" is the electric potential, "Q" is the total amount of positive charge stored in the capacitor, and "C_q" is called the **capacitance** (kuh pass' uh tens) of the capacitor.

Now, of course, to understand this equation, we must also understand what capacitance is. There is a detailed formula with which you can describe the capacitance of a capacitor, but I don't think it's all that useful for the purposes of this discussion. The capacitance simply tells us how much charge a capacitor can hold at a given voltage. It depends on things like the size of the plates in the capacitor and the separation distance. If, for example, the plates are large, there is a lot of room for charge to be stored. As a result, the capacitance is high. If the separation distance is large, then the charges don't attract one another much, so the capacitance is low. All of this is important to physicists, but I don't want you to worry about it too much. Just understand that the capacitance is a number that is proportional to how much charge a capacitor holds at a given voltage. The larger the capacitance, the more charge it can store.

What are the units for capacitance? Well, we know that the SI unit for electric potential is the volt, and the SI unit for charge is the Coulomb. Thus, in order to get the units to work out right in this equation, capacitance has a unit of $\frac{\text{C}}{\text{volt}}$. Physicists call this unit the **farad** (abbreviated with an "F" and pronounced fair' ud) in honor of Michael Faraday, a pioneer in the study of electricity and magnetism. As with the Coulomb, this is a rather large unit. Most capacitors that we use in electronics have capacitances on the order of microfarads (one millionth of a farad).

I don't want you to get all worried about the idea of capacitance. It is a number that describes a capacitor. Thus, when you use a capacitor, its capacitance is easy to find out. As a result, I will always just give you the capacitance of any capacitor that we deal with, or I will give you enough information to calculate it from Equation (14.4). The main thing I want you to understand is how to use this equation.

EXAMPLE 14.5

A 2.6 x 10^{-6}-farad capacitor is charged up so that there is 2.5 mC of total charge on its positive plate. What is the electric potential of this capacitor?

This is a straightforward application of Equation (14.4), as long as we convert 2.5 mC into Coulombs:

$$V = \frac{Q}{C_q}$$

$$V = \frac{2.5 \times 10^{-3}\ \text{C}}{2.6 \times 10^{-6}\ \text{F}} = \frac{2.5 \times 10^{-3}\ \cancel{\text{C}}}{2.6 \times 10^{-6}\ \frac{\cancel{\text{C}}}{\text{volt}}} = 9.6 \times 10^{2}\ \text{volts}$$

The electric potential of the capacitor, then, is 9.6 x 10^2 volts.

ON YOUR OWN

14.8 If a 4.6 x 10^{-6}-F capacitor must be used to generate a 125-volt electric potential, how much charge must it store ?

An Application of Capacitors

All of the previous discussion must have seemed rather useless up to this point. However, now that you know how to use Equation (14.4), you can start to see why I wanted you to study this in the first place. A capacitor is one way of creating an electric potential that is localized to a certain region. As you can see from Figure 14.2, the electric field of a parallel-plate capacitor is mostly confined to the region between the plates. The electric field does extend beyond the plates, but compared to the strength of the field between the plates, the strength of the field outside the plates is rather small. Thus, for practical purposes, we can concentrate on the electric potential between the plates.

Before we see how this potential is used, let's make sure you understand exactly *what* the potential of a capacitor means. If I were to tell you that a certain capacitor has an electric potential of 5 volts, what would that mean? It would mean that the potential difference between the positive and negative plate is 5 volts. In other words, a freely-moving charge would experience a change in electric potential of 5 volts as it traveled from one plate to the other. For example, if a freely-moving negative charge (not one that is in the capacitor, one that is free to move between the plates) were to start at the negative plate and move to the positive plate, it would experience a change in electric potential of 5 volts.

Okay, so the particle experiences a change in potential of 5 volts. Is this a 5-volt *increase* in potential or a 5 volt *decrease* in potential? In physics we adopt the convention that moving from the negative plate to the positive plate is an *increase* in potential, whereas moving from the positive to the negative plate is a *decrease* in electric potential. This is something that you need to remember:

When a particle travels from the negative plate in a capacitor to the positive plate, it experiences an increase in electric potential. If it moves the opposite way, it experiences a decrease in electric potential.

When we go through the example and "On Your Own" problems, you will see why this is important to remember.

What good does all of this do us? Well, it turns out that one thing we use capacitors for is to accelerate charged particles. Consider, for example, a freely-moving positive charge that is placed right next to the positive plate of a capacitor. What will happen to the charge? Because it is repelled by the positive plate and attracted to the negative plate, it will begin to accelerate towards the negative plate. What determines how fast the charge will be moving by the time it reaches the negative plate? The electric potential of the capacitor. The greater the change in electric potential, the greater the change in potential energy. Well, the greater the change in potential energy, the greater the change in kinetic energy, which will result in a greater change in the speed. Thus, by calculating the potential of a capacitor, we can calculate the speed at which a particle moves from one plate to another. Let me show you how this is done.

EXAMPLE 14.6

If a freely-moving electron ($m = 9.1 \times 10^{-31}$ kg, $q = -1.6 \times 10^{-19}$ C) starts at rest and travels from the negative plate of a 3.1×10^{-6}-F capacitor to the positive plate, how fast will it be moving? Assume that the capacitor has 0.55 mC of stored charge.

In order to solve this problem, we need to look at the energetics of the situation. Thus, we need to figure out how the potential energy change as the electron travels from the negative plate to the positive plate. That change in potential energy will cause a corresponding change in kinetic energy, which will determine the electron's speed. To determine the change in potential energy, we must first determine the change in electric potential. For a capacitor, that is given by Equation (14.4):

$$V = \frac{Q}{C_q}$$

$$V = \frac{5.5\times10^{-4}\,C}{3.1\times10^{-6}\,F} = \frac{5.5\times10^{-4}\,\cancel{C}}{3.1\times10^{-6}\,\frac{\cancel{C}}{\text{volt}}} = 1.8\times10^{2}\ \text{volts}$$

By the convention you memorized above, the fact that the electron moves from the negative plate to the positive one means this is an *increase* in potential, so the change in potential is $+1.8 \times 10^2$ volts.

This change in potential, then, will lead to a change in potential energy, given by Equation (14.2):

$$\Delta PE = q\Delta V = (-1.6 \times 10^{-19}\ C) \cdot (1.8 \times 10^2\ \text{volts}) = -2.9 \times 10^{-17}\ J$$

Since the result of the equation is negative, we know that the electron's potential energy *decreased.* Well, if the potential energy of the electron *decreased* by 2.9×10^{-17} J, its kinetic energy must have *increased* by the same amount. Since the kinetic energy started at zero and increased by 2.9×10^{-17} J, we know that the final kinetic energy of the electron is 2.9×10^{-17} J. We can now use Equation (8.3) to determine the final speed:

$$KE = \frac{1}{2}mv^2$$

$$2.9\times10^{-17}\ J = \frac{1}{2}\cdot(9.1\times10^{-31}\ kg)\cdot v^2$$

$$v = 8.0\times10^{6}\ \frac{m}{sec}$$

The electron, therefore, accelerates to the incredible speed of 8.0×10^6 m/sec.

Do you see how we did this problem? Once again, the major concepts all come from Module #8. The only difference between this problem and the ones in Module #8 is the way in which potential energy is calculated. Try the "On Your Own" problems that follow in order to make sure you understand how to do this.

ON YOUR OWN

14.9 A proton ($m = 1.67 \times 10^{-27}$ kg, $q = +1.6 \times 10^{-19}$ C) is placed at the edge of the positive plate of a 3.4×10^{-6}-F capacitor. If the capacitor holds 12 mC of charge on its positive plate, how fast will the proton be moving when it reaches the negative plate of the capacitor?

14.10 Capacitors are used by physicists to accelerate ions so that their behavior can be studied when they collide with other ions. A physicist wants to get an ammonium ion ($m = 2.8 \times 10^{-26}$ kg, $q = +1.6 \times 10^{-19}$ C) traveling with a speed of 1.2×10^{6} m/sec in order to collide it with another ion. If the physicist has a 2.3×10^{-6}-F capacitor, what charge must it hold for the experiment to work?

How a Television Makes Its Picture

Now that you understand a little bit about capacitors, you know enough to understand how a television makes its picture. The screen of a television set is the front of a device known as a **cathode** (kath' ohd) **ray tube (CRT)**. In short, a CRT is a glass container in which most of the air has been removed. A schematic of such a CRT is shown in the figure below:

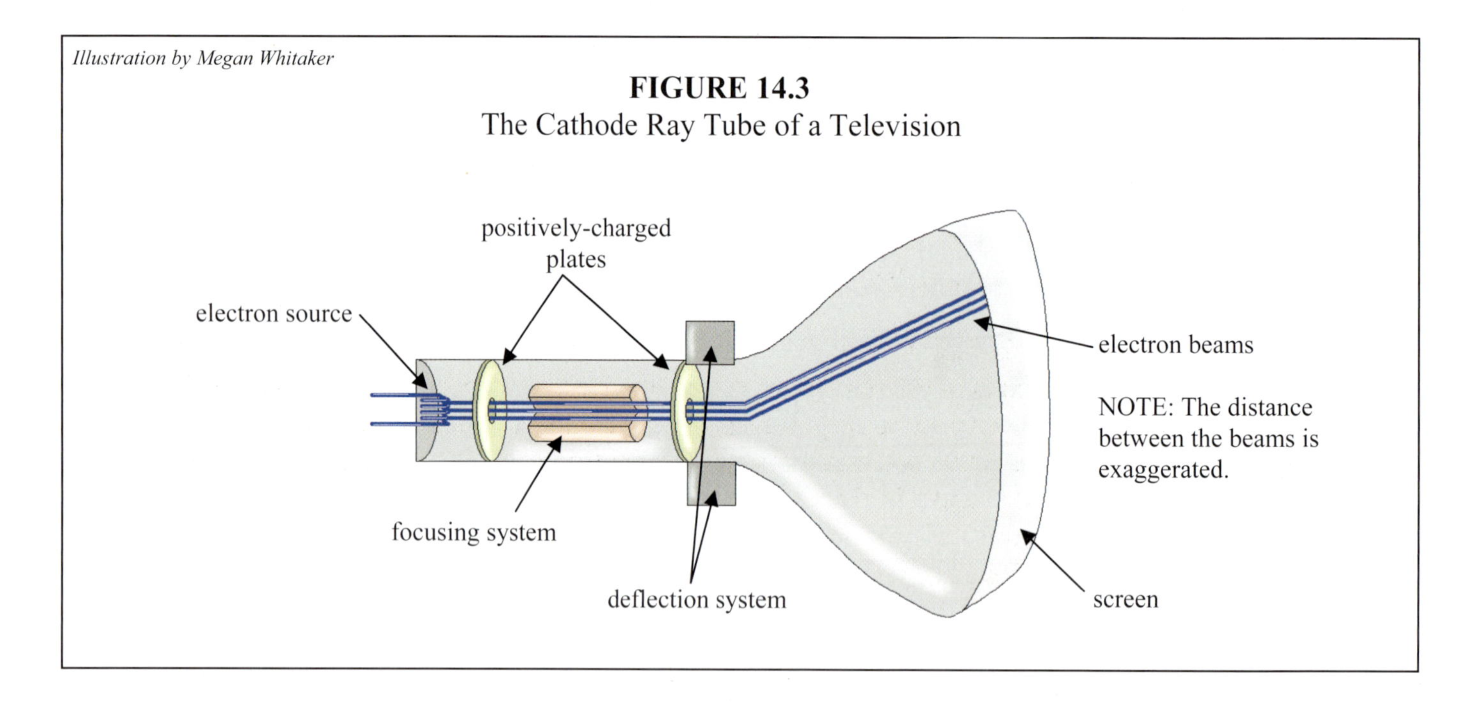

FIGURE 14.3
The Cathode Ray Tube of a Television

In a television set, there is an electron source that produces electrons which are free to move in the CRT. This electron source is negatively charged. Right in front of the electron source is a positively-charged plate. Thus, you have two oppositely charged plates separated by a distance. This is much like a parallel-plate capacitor. The electrons accelerate from the source to the first plate because of the electric potential. They then travel through a system that focuses them into three tight beams, and they are further accelerated by a second positively-charged plate. The second positively-charged plate has more charge than the first, so that the electrons travel through a positive electric potential as they travel from the first positively-charged plate to the second. As a result, their potential energy decreases, and their kinetic energy increases.

Once they are fully accelerated using these two positively-charged plates, they pass through a deflection system. The deflection system bends the beams of electrons so that each beam hits a specific point on the screen. The screen is made up of little dots known as pixels. These pixels come in groups of three, and each member of the group is made of a slightly different substance. One substance emits a red light when an electron beam hits it, another emits a green light, and the third emits a blue light. The three beams hit the three pixels, and the resulting three colors are mixed, producing a unique color. The screen is covered with these groups of three pixels, and by aiming the electron beams at different parts of the screen, those parts will light up with the color that is formed when the red, green, and blue colors emitted by the pixels mix.

Believe it or not, your television can sweep the electron beams back and forth across the screen so quickly that to you, it looks like the whole screen is lit up at once. In fact, the screen is being lit up one group of pixels at a time, but your eyes and brain together cannot process information quickly enough to distinguish this. Thus, as far as you are concerned, the whole screen is lit up all of the time. The television, however, is sweeping the beam across the screen, hitting just the right regions at just the right time to make a coherent picture. If you want to convince yourself that this is happening, sit in front of the television with a pencil in your hand. Look away from the television and wave the pencil in front of your face, watching it with your eyes. You just see a blur, right? Now, look at the television and do the same thing. What happens now? You see little "snapshots" of the pencil as it moves in front of your eyes. This is because the movement of the pencil is picking up the sweeping of the electron beams across the television screen.

How does the television know where to deflect the electron beams so that the desired picture is formed? That information is carried on the signal that the television receives from the antenna, satellite, or cable to which it is connected. The signal tells the deflection system exactly where to deflect the beams at each instant so that the desired picture is produced.

The part of the television that relates to this module is the system that accelerates the electrons. You see, in order for the pixels in the screen to light up, the electrons need to hit the pixels with a lot of energy. Thus, the electrons must be accelerated. This is achieved through what is essentially a three-plate capacitor. The electric potential between the electron source (the negative "plate") and the first positively-charged plate results in the electrons being accelerated enough to be focused into three tight beams. The electric potential between the first positively-charged plate and the second positively-charged plate (which has significantly more positive charge than the first) accelerates the electrons to their final speed.

The total potential difference between the electron source and the second positively-charged plate is generally on the order of 20,000 volts in a color television set. You can work out the calculation yourself, but that means that the electrons are traveling at about 83 *million* meters/sec, which is nearly 28% of the speed of light! It turns out that this calculation is slightly in error due to the effects of special relativity, which go beyond the scope of this course. Nevertheless, it gives you some idea of how incredibly fast the electrons in a television set travel.

Although CRTs are incredibly useful, they are slowly becoming obsolete for televisions and computer monitors. That's because a new technology known as "liquid crystal display" (LCD) is becoming more popular. The details of an LCD are beyond the scope of this course, but the main advantage it provides is a more compact display device. Because beams of electrons are swept across a CRT screen, the CRT must be deep enough so that the electrons travel a reasonable distance after leaving the deflection system. That way, even a modest deflection angle can result in a large displacement for the electron beams. An LCD does not use electron beams, so there is no physical reason for the LCD to be as deep. As a result, an LCD television or monitor is significantly thinner than a television or monitor that uses a CRT. As time goes on, more and more televisions and computer monitors will probably be LCDs.

ANSWERS TO THE "ON YOUR OWN" PROBLEMS

14.1 Notice that there are two charges in the problem. However, to calculate the potential, you need only one charge. So which charge do we use? The important charge in terms of the electric potential is the fixed one, because that's the one we say generates the electric field. As a result, we will use the -2.1 mC charge. The other charge was put in the problem as extra information that you do not need.

Now that we know which charge to use, we must remember three things: include the negative sign in the equation, convert milliCoulombs to Coulombs, and convert centimeters to meters. Once all of that is done, Equation (14.1) looks like this:

$$V = \frac{kQ}{r}$$

$$V = \frac{\left(9.0 \times 10^{9} \frac{N \cdot m^{\not{2}}}{C^{\not{2}}}\right) \cdot (-2.1 \times 10^{-3} \not{C})}{0.61 \not{m}} = -3.1 \times 10^{7} \frac{N \cdot m}{C} = -3.1 \times 10^{7} \text{ volts}$$

The electric potential, then, is <u>-3.1×10^{7} volts</u> at that point.

14.2 To calculate the potential energy, we start by calculating the electric potential. This is done using Equation (14.1). Since the -2.3-C charge is the stationary one, that's what we use in the equation:

$$V = \frac{kQ}{r}$$

$$V = \frac{\left(9.0 \times 10^{9} \frac{N \cdot m^{\not{2}}}{C^{\not{2}}}\right) \cdot (-2.3 \not{C})}{1.1 \not{m}} = -1.9 \times 10^{10} \frac{N \cdot m}{C} = -1.9 \times 10^{10} \text{ volts}$$

Now that we have the electric potential, we can use Equation (14.2) to determine the potential energy. Since the +7.9-mC charge is the freely-moving one, we use that charge in this equation. Keep in mind, though, that we must convert it to Coulombs before we use it.

$$PE = qV$$

$$PE = (+7.9 \times 10^{-3} \not{C}) \cdot (-1.9 \times 10^{10} \frac{N \cdot m}{\not{C}}) = -1.5 \times 10^{8} \text{ N} \cdot \text{m} = -1.5 \times 10^{8} \text{ J}$$

The potential energy, then, is -<u>1.5×10^{8} J</u>.

14.3 To solve this problem, we simply need to calculate the initial and final electric potentials. Then, we can subtract them to determine the change in electric potential. After that, we can use Equation

(14.3) to determine the change in potential energy. We will have to convert milliCoulombs to Coulombs in order to do this, as well as converting centimeters to meters. The initial distance between the charges is 0.45 m, and the final distance is 0.70 m, since the freely-moving charge moved 25 cm away. The initial electric potential is:

$$V = \frac{kQ}{r}$$

$$V_{initial} = \frac{\left(9.0\times10^{9}\ \frac{N\cdot m^{\not 2}}{C^{\not 2}}\right)\cdot(-1.8\times10^{-3}\ \not C)}{0.45\ \not m} = -3.6\times10^{7}\ \frac{N\cdot m}{C} = -3.6\times10^{7}\ \text{volts}$$

The final electric potential is:

$$V = \frac{kQ}{r}$$

$$V_{final} = \frac{\left(9.0\times10^{9}\ \frac{N\cdot m^{\not 2}}{C^{\not 2}}\right)\cdot(-1.8\times10^{-3}\ \not C)}{0.70\ \not m} = -2.3\times10^{7}\ \frac{N\cdot m}{C} = -2.3\times10^{7}\ \text{volts}$$

Now we can determine the potential difference:

$$\Delta V = V_{final} - V_{initial} = -2.3 \times 10^{7}\ \text{volts} - (-3.6 \times 10^{7}\ \text{volts}) = 1.3 \times 10^{7}\ \text{volts}$$

Now that we have the potential difference, we can use Equation (14.3) to determine the change in potential energy:

$$\Delta PE = q\Delta V = (-1.1\times10^{-3}\ \not C)\cdot(1.3\times10^{7}\ \frac{N\cdot m}{\not C}) = -1.4\times10^{4}\ N\cdot m = -1.4\times10^{4}\ J$$

Since the change in potential energy is negative, the potential energy decreased by 1.4×10^4 J.

14.4 This problem is simpler than the previous problem, as long as you remember what the voltage of a battery means. Since it is a 9.0-volt battery, a particle that travels from the positive side to the negative side experiences a potential difference of -9.0 volts. Thus, $\Delta V = -9.0$ volts. With that information, calculating the change in potential energy is a snap:

$$\Delta PE = q\Delta V = (-5.6\times10^{-3}\ C)\cdot(-9.0\,\text{volts}) = (-5.6\times10^{-3}\ \not C)\cdot(-9.0\ \frac{N\cdot m}{\not C}) = 0.050\ N\cdot m = 0.050\ J$$

Since the change in potential energy is positive, the potential energy increased by 0.050 J.

14.5 To solve a problem like this one, we simply look at the energetics involved. First, we need to calculate the initial electric potential:

$$V = \frac{kQ}{r}$$

$$V_{initial} = \frac{\left(9.0\times10^{9}\ \frac{N\cdot m^{\not 2}}{C^{\not 2}}\right)\cdot(3.4\ \not C)}{1.4\ \not m} = 2.2\times10^{10}\ \frac{N\cdot m}{C} = 2.2\times10^{10}\ \text{volts}$$

Next, we calculate the final electric potential:

$$V = \frac{kQ}{r}$$

$$V_{final} = \frac{\left(9.0\times10^{9}\ \frac{N\cdot m^{\not 2}}{C^{\not 2}}\right)\cdot(3.4\ \not C)}{2.5\ \not m} = 1.2\times10^{10}\ \frac{N\cdot m}{C} = 1.2\times10^{10}\ \text{volts}$$

The potential difference is the final potential minus the initial potential:

$$\Delta V = V_{final} - V_{initial} = 1.2 \text{ x } 10^{10} \text{ volts} - 2.2 \text{ x } 10^{10} \text{ volts} = -1.0 \text{ x } 10^{10} \text{ volts}$$

This allows us to use Equation (14.3) to determine the change in potential energy:

$$\Delta PE = q\Delta V = (6.5\ \not C)\cdot(-1.0\times10^{10}\ \frac{N\cdot m}{\not C}) = -6.5\times10^{10}\ N\cdot m = -6.5\times10^{10}\ J$$

This tells us that the potential energy *decreased* by 6.5 x 10^{10} J. Since we know that the total energy must stay the same, we know that the kinetic energy *increased* by 6.5 x 10^{10} J. Well, the initial kinetic energy was zero (the particle started from rest), so its final kinetic energy is 6.5 x 10^{10} J. We can use that to calculate the final speed of the particle:

$$KE = \frac{1}{2}mv^2$$

$$6.5\times10^{10}\ J = \frac{1}{2}\cdot(42.3\ kg)\cdot v^2$$

$$v = 5.5\times10^{4}\ \frac{m}{sec}$$

When the particle has traveled to 2.5 m away from the stationary charge, its speed is <u>5.5 x 10^4 m/sec</u>.

14.6 First, we need to calculate the initial electric potential:

$$V = \frac{kQ}{r}$$

$$V_{initial} = \frac{\left(9.0 \times 10^{9} \frac{N \cdot m^{\cancel{2}}}{C^{\cancel{2}}}\right) \cdot (5.2 \times 10^{-3} \cancel{C})}{0.65 \cancel{m}} = 7.2 \times 10^{7} \frac{N \cdot m}{C} = 7.2 \times 10^{7} \text{ volts}$$

Since the particle starts at rest, it will travel towards the stationary charge, because opposite charges attract. Thus, it will move 15 cm closer to the stationary charge, making its final position 0.50 m from the stationary charge. This makes the final electric potential:

$$V = \frac{kQ}{r}$$

$$V_{final} = \frac{\left(9.0 \times 10^{9} \frac{N \cdot m^{\cancel{2}}}{C^{\cancel{2}}}\right) \cdot (5.2 \times 10^{-3} \cancel{C})}{0.50 \cancel{m}} = 9.4 \times 10^{7} \frac{N \cdot m}{C} = 9.4 \times 10^{7} \text{ volts}$$

The potential difference is the final potential minus the initial potential:

$$\Delta V = V_{final} - V_{initial} = 9.4 \text{ x } 10^{7} \text{ volts} - 7.2 \text{ x } 10^{7} \text{ volts} = 2.2 \text{ x } 10^{7} \text{ volts}$$

This allows us to use Equation (14.3) to determine the change in potential energy:

$$\Delta PE = q\Delta V = (-7.1 \times 10^{-3} \cancel{C}) \cdot (2.2 \times 10^{7} \frac{N \cdot m}{\cancel{C}}) = -1.6 \times 10^{5} \text{ N} \cdot \text{m} = -1.6 \times 10^{5} \text{ J}$$

This tells us that the potential energy *decreased* by 1.6 x 10^5 J. Since we know that the total energy must stay the same, we know that the kinetic energy *increased* by 1.6 x 10^5 J. Well, the initial kinetic energy was zero (the particle started from rest), so its final kinetic energy is 1.6 x 10^5 J. We can use that to calculate the final speed of the particle:

$$KE = \frac{1}{2} mv^{2}$$

$$1.6 \times 10^{5} \text{ J} = \frac{1}{2} \cdot (1.1 \text{ kg}) \cdot v^{2}$$

$$v = 540 \frac{m}{sec}$$

When the particle has traveled 15 cm, its speed is 540 m/sec.

14.7 To solve this problem, we need to find out where the kinetic energy of the particle equals zero. That's the point at which the particle stops and changes direction. Let's start by determining how much kinetic energy the particle has initially:

$$KE_{initial} = \frac{1}{2}mv^2 = \frac{1}{2} \cdot (3.5\text{ kg}) \cdot (255\,\frac{\text{m}}{\text{sec}})^2 = 1.1 \times 10^5\text{ J}$$

In order to get to the point where the kinetic energy is zero, the potential energy will have to *increase* by 1.1 x 10^5 J. At that point, all kinetic energy will have been converted to potential energy. We can use that increase in potential energy to determine the potential difference through which the particle must travel:

$$\Delta PE = q\Delta V$$

$$\Delta V = \frac{\Delta PE}{q} = \frac{1.1 \times 10^5\text{ J}}{8.8 \times 10^{-3}\text{ C}} = \frac{1.1 \times 10^5\text{ N}\cdot\text{m}}{8.8 \times 10^{-3}\text{ C}} = 1.3 \times 10^7\,\frac{\text{N}\cdot\text{m}}{\text{C}} = 1.3 \times 10^7\text{ volts}$$

Now that we know ΔV, we just have to calculate the initial potential and then use it and ΔV to calculate the final potential:

$$V = \frac{kQ}{r}$$

$$V_{initial} = \frac{\left(9.0 \times 10^9\,\frac{\text{N}\cdot\text{m}^{\not 2}}{\not{\text{C}}^{\not 2}}\right) \cdot (4.5 \times 10^{-3}\,\not{\text{C}})}{1.2\,\not{\text{m}}} = 3.4 \times 10^7\,\frac{\text{N}\cdot\text{m}}{\text{C}} = 3.4 \times 10^7\text{ volts}$$

Now we can see what the final electric potential must be:

$$\Delta V = V_{final} - V_{initial}$$

$$1.3 \times 10^7\text{ volts} = V_{final} - 3.4 \times 10^7\text{ volts}$$

$$V_{final} = 4.7 \times 10^7\text{ volts}$$

We can use the final electric potential to determine the final distance between the moving particle and the stationary one:

$$V = \frac{kQ}{r}$$

$$4.7 \times 10^7\text{ volts} = \frac{\left(9.0 \times 10^9\,\frac{\text{N}\cdot\text{m}^2}{\text{C}^2}\right) \cdot (4.5 \times 10^{-3}\text{ C})}{r}$$

$$r = \frac{\left(9.0\times10^{9}\ \frac{\cancel{N}\cdot m^{\cancel{2}}}{\cancel{C^2}}\right)\cdot(4.5\times10^{-3}\ \cancel{C})}{4.7\times10^{7}\ \frac{\cancel{N\cdot m}}{\cancel{C}}} = 0.86\text{ m}$$

So, we finally see that the particle can only travel until it is 0.86 m away from the charge. At that point, the particle stops and heads away from the stationary charge.

14.8 This is another straightforward application of Equation (14.4):

$$V = \frac{Q}{C_q}$$

$$125\text{ volts} = \frac{Q}{4.6\times10^{-6}\text{ F}}$$

$$Q = (125\ \cancel{\text{volts}})\cdot(4.6\times10^{-6}\ \frac{C}{\cancel{\text{volts}}}) = 5.8\times10^{-4}\text{ C}$$

This means that on its positive plate, the capacitor must store 5.8×10^{-4} C total charge.

14.9 As the proton moves from one plate of the capacitor to the other, it experiences a change in electric potential. We can use Equation (14.4) to determine the amount of the change:

$$V = \frac{Q}{C_q} = \frac{1.2\times10^{-2}\text{ C}}{3.4\times10^{-6}\text{ F}} = \frac{1.2\times10^{-2}\ \cancel{C}}{3.4\times10^{-6}\ \frac{\cancel{C}}{\text{volts}}} = 3.5\times10^{3}\text{ volts}$$

The fact that the proton moves from the positive plate to the negative one means that this is a decrease in potential, so the change in potential is -3.5×10^3 volts.

This change in potential, then, will lead to a change in potential energy, given by Equation (14.3):

$$\Delta PE = q\Delta V = (1.6 \times 10^{-19}\text{ C})\cdot(-3.5 \times 10^{3}\text{ volts}) = -5.6 \times 10^{-16}\text{ J}$$

Since the result of the equation is negative, this tells us that the proton's potential energy decreased. Well, if the potential energy of the proton decreased by 5.6×10^{-16} J, then its kinetic energy must have increased by the same amount. Since the kinetic energy started at zero and increased by 5.6×10^{-16} J, we know that the final kinetic energy of the proton is 5.6×10^{-16} J. We can now use Equation (8.3) to determine the final speed:

$$KE = \frac{1}{2}mv^2$$

$$5.6 \times 10^{-16}\ J = \frac{1}{2} \cdot (1.67 \times 10^{-27}\ kg) \cdot v^2$$

$$\underline{v = 8.2 \times 10^5\ \frac{m}{sec}}$$

14.10 This problem is just a little different from the example and the previous problem. In this case, we know the final speed. What we need to figure out is the potential. That way, we can determine the charge on the capacitor. Since we know that the ion ends up with a speed of 1.2 x 10^6 m/sec, we can determine its final kinetic energy:

$$KE = \frac{1}{2}mv^2 = \frac{1}{2} \cdot (2.8 \times 10^{-26}\ kg) \cdot (1.2 \times 10^6\ \frac{m}{sec})^2 = 2.0 \times 10^{-14}\ J$$

Where did that kinetic energy come from? Well, it came from a decrease in the potential energy. Thus, we know that ΔPE = -2.0 x 10^{-14} J. Using Equation (14.3), then, we can determine the change in electric potential:

$$\Delta PE = q\Delta V$$

$$-2.0 \times 10^{-14} J = (1.6 \times 10^{-19} C) \cdot \Delta V$$

$$\Delta V = \frac{-2.0 \times 10^{-14}\ J}{1.6 \times 10^{-19}\ C} = -1.3 \times 10^5\ volts$$

The negative sign just tells us that the ion travels from the positive plate to the negative plate. Thus, the potential of the capacitor is just 1.3 x 10^5 volts. Now we can determine the charge that must be stored:

$$V = \frac{Q}{C_q}$$

$$1.3 \times 10^5\ volts = \frac{Q}{2.3 \times 10^{-6}\ F}$$

$$Q = (1.3 \times 10^5\ \cancel{volts}) \cdot (2.3 \times 10^{-6}\ \frac{C}{\cancel{volts}}) = 0.30\ C$$

In order to achieve the proper speed, then, the physicist must store 0.30 C of charge on the positive plate of the capacitor. Had you kept the negative sign throughout the problem, you would have gotten -0.30 C of charge, which is simply the charge stored on the negative plate.

REVIEW QUESTIONS

1. Are electric potential and potential energy the same? Why or why not?

2. A positive particle is near a stationary negative charge. Is the electric potential positive or negative? What about the potential energy of the positive particle?

3. A negative particle is near a stationary positive charge. Is the electric potential positive or negative? What about the potential energy of the negative particle?

4. A D-cell battery is rated at 1.5 volts. If a proton travels from the positive side of the battery to the negative side, what potential difference does it experience?

5. A negative particle is placed a certain distance away from a positive stationary charge. If the particle suddenly moves towards the stationary charge, does it experience an increase or decrease in potential energy?

6. A positive particle is placed a certain distance away from a positive stationary charge. If the particle moves towards the stationary charge due to the action of an outside force, does it experience an increase or decrease in potential energy?

7. A negative charge is placed near a stationary charge. If the electric potential is positive, is the stationary charge positive or negative?

8. A negative particle moves from the negative plate of a capacitor to the positive plate. Is its change in electric potential positive or negative? Does its potential energy increase or decrease?

9. Draw the path of the electron beam as it passes through the capacitor drawn below. Assume the charge stored in the capacitor is pretty large.

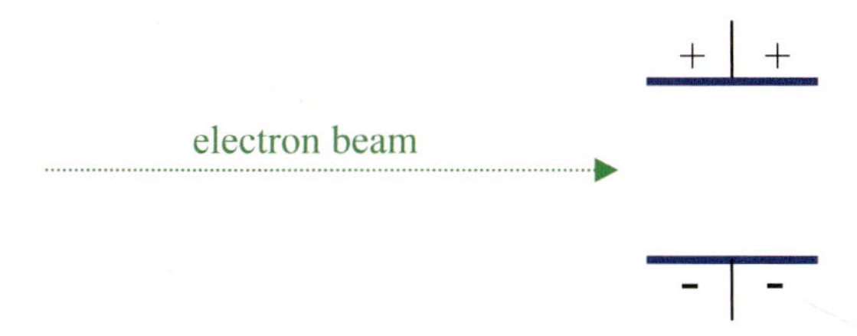

10. Why are there three beams of electrons in a color television set? How many beams would you expect in a black-and-white television set?

PRACTICE PROBLEMS

$$k = 9.0 \times 10^9 \frac{\text{Newtons} \cdot \text{m}^2}{\text{C}^2}$$

1. A +2.3-mC charge is placed 35 cm away from a stationary charge. If it experiences an electric potential of -4.6 x 10^7 volts, what is the value of the fixed charge?

2. In a 3.4 x 10^3-volt potential, a particle has a potential energy of -12 J. What is the particle's charge?

3. A -5.1-mC charge that is free to move is placed 2.20 m from a stationary +5.8-mC charge. If the freely-moving charge moves 2.00 m towards the stationary charge, what is the change in potential energy? Did the potential energy increase or decrease?

4. A -6.5-mC charge (m = 0.0015 kg) travels from the negative side of a 1.5-volt battery to the positive side. If it started from rest, what is its speed when it reaches the positive side of the battery?

5. A +1.5-C charged particle (m = 42.3 kg) is placed 1.4 m from a +4.4-C stationary charge. If it starts from rest, how fast will the particle be traveling when it is 2.0 m away from the stationary charge?

6. A -9.2-mC charged particle (m = 3.5 kg) is placed 75 cm from a +2.5-mC stationary charge. If the particle starts at rest, how fast will it be moving after it has traveled 25 cm?

7. A +3.8-mC charged particle (m = 5.0 kg) is shot with an initial velocity of 245 m/sec towards a +1.5-mC stationary charge. If the particle starts out 1.2 meters from the stationary charge, how close will it come to the charge before stopping and moving away?

8. What is the electric potential of a 3.2 x 10^{-6}-F capacitor when it stores 0.50 C of charge?

9. A proton (m = 1.7 x 10^{-27} kg, q = +1.6 x 10^{-19} C) is placed at the edge of the positive plate of a 1.4 x 10^{-6}-F capacitor. If the capacitor holds 2.2 mC of charge on its positive plate, how fast will the proton be moving when it reaches the negative plate of the capacitor?

10. An electron (m = 9.1 x 10^{-31} kg, q = -1.6 x 10^{-19} C) moves from the negative plate to the positive plate of a 5.1 x 10^{-6}-F capacitor. If it starts from rest and reaches a speed of 3.2 x 10^5 m/sec, how much charge is stored on the capacitor?

MODULE #15: Electric Circuits

Introduction

In the previous module, we studied charges in motion. Those charges moved freely in space; the only thing that restricted their movements was the electrostatic force caused by other charges in the vicinity. Although that kind of situation is important to study, there is another application of charges in motion that we need to study. In this module, we will study how charges move when they are restricted to the confines of a conductive material.

Remember, when a substance is a conductor, charges move freely within it. Thus, even though the charges cannot leave the metal, they can move around inside the metal. This is a useful situation, because since the charges are confined to the metal, we can exert control over where those charges move. By shaping the metal and controlling the electric field, we can force charges to move where we want them to move. When done properly, this allows us to utilize the kinetic energy of those charges while they are in motion. That's what we will study in this module.

Batteries, Circuits, and Conventional Current

In order to get charges moving anywhere, we first must have an electric field. In the previous module, we utilized stationary charges to create electric fields. A much more popular way to create an electric field, however, is with the use of a battery. Now you should have learned about batteries in chemistry, so I don't want to spend a lot of time on how a battery works. Instead, I want to concentrate on what a battery does. In brief, a battery is a container that holds two chemical reactants which are separated from each other. One of the chemicals wants to give up its electrons, and the other wants to take those electrons. When the two sides of the battery are connected by a conductor, electrons will flow from the chemical that wants to get rid of them to the chemical that wants to receive them. The battery "goes dead" when all of the molecules of the first reactant have given up all of the electrons that they can. When electric charge travels through a conductor from one side of a battery to another, we call the system an **electric circuit**.

As we discussed previously, a battery is rated by its voltage. Most cylindrical batteries are rated at 1.5 volts, while the small, rectangular batteries with electric posts at the top are typically 9.0 volts. As you learned in the previous module, this voltage refers to the potential difference between the positive and negative sides of the battery. Thus, if a charge travels from one side of the battery to the other, it experiences a change in electric potential which results in a change in potential energy. This change in potential energy causes a corresponding change in kinetic energy, which can be calculated by taking the electric potential and multiplying it by the charge of an electron. Since the charge of an electron never changes, the electric potential is directly proportional to the increase in the kinetic energy of the electrons as they travel from one side of the battery to the other. Thus, the larger the voltage of the battery, the more kinetic energy the electrons have as they travel.

Now remember why we want the electrons to travel from one end of the battery to the other. We want to convert their kinetic energy into work. Thus, the more voltage the battery has, the more work each electron can do. If you think about it, however, the voltage of a battery is not the only factor that determines how much work gets done in an electric circuit. Not only does the kinetic energy of each electron influence the work done in an electric circuit, but so does the *number* of electrons that travel through the circuit. After all, when ten electrons travel through an electric circuit, ten times more work will get done than when only one electron travels through the circuit. Therefore, in order to determine how much work can get done in an electric circuit, we need to determine how many electrons flow through it.

Physicists study the number of electrons that flow through a circuit with the concept of **electric current**.

Electric current - The amount of charge that travels past a fixed point in an electric circuit each second

This definition can be expressed mathematically as follows:

$$I = \frac{\Delta Q}{\Delta t} \tag{15.1}$$

where "I" stands for current, "ΔQ" represents the total charge traveling past a given point in the circuit, and "Δt" is the time it takes for the charge to travel. Looking at the equation, it is pretty easy to see that the unit for electric current is $\frac{\text{Coulombs}}{\text{sec}}$. Physicists call this unit the **Ampere** (abbreviated "amp" or "A") in honor of Andre Ampere, a physicist who studied electricity and magnetism.

Now let's learn about how this current flows in an electric circuit. Suppose I take a metal wire and connect it to each side of a battery. The result might look something like what is drawn below.

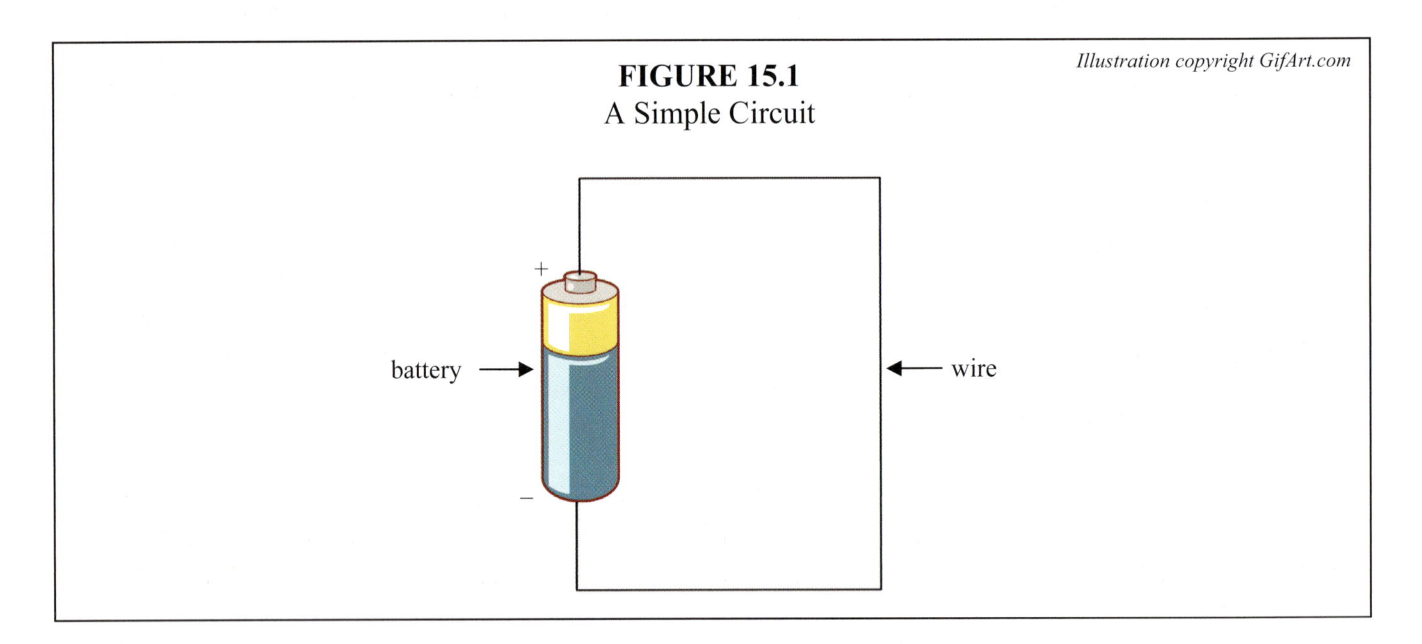

FIGURE 15.1
A Simple Circuit

This is a very basic electric circuit. Electrons will flow through the wire from the negative side of the battery to the positive side. In order to reduce the time it takes to draw circuits, physicists have developed symbols to represent the major components of an electric circuit. A battery is symbolized with two parallel lines, one longer than the other. The positive side of the battery is usually represented by the longer line. In the end, then, the simple circuit above can be drawn as shown in Figure 15.2.

FIGURE 15.2
An Abbreviated Drawing of the Circuit in Figure 15.1

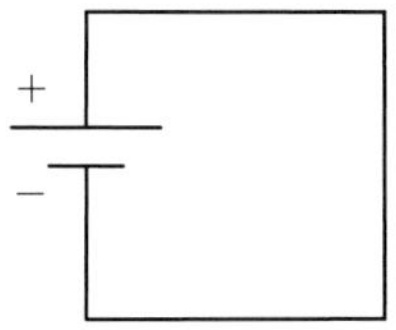

Since the longer line of the battery is supposed to represent the positive side, there is really no need to label the sides of the battery as positive and negative as I have in the drawing above. Thus, in all future circuit drawings, I will not. I just put them in this drawing to emphasize that the longer line represents the positive side of the battery, while the smaller line represents the negative side.

The first thing you need to be able to do when you look at a circuit like this is determine how the current flows. Unfortunately, Benjamin Franklin caused no end of confusion over this very point. Although he is best remembered for his political endeavors as one of America's founding fathers, Franklin was one of the most respected scientists in his time. His theories and experiments regarding electricity were known and admired the world over. Many of his ideas laid the groundwork for our modern theories of electricity.

You see, batteries were invented long before anyone really understood what electricity was. As a result, scientists tried to study batteries and the electricity they produced to better understand what electricity was. Franklin theorized that batteries had a positive and a negative side and that electricity was made of particles which flowed from one side of a battery to the other. These ideas were readily accepted and highly regarded by scientists around the world.

Now, although Franklin's concept of positive and negative sides of a battery helped revolutionize the way scientists studied electricity, it has also caused a bit of confusion. You see, Franklin thought that the positive side of a battery had too many of these mysterious particles and the negative side had too few. Thus, he said that electricity must flow from the positive side of the battery to the negative side. Since his ideas were highly regarded, this idea was accepted the world over. Scientists from everywhere began drawing electric circuits assuming that the electricity flowed from the positive side of the battery to the negative side.

As is usually the case with technological inventions, people began finding uses for electricity long before science figured out what electricity really was. Thus, engineers began designing electric circuits, and they, too, drew the circuits assuming that electricity flowed from the positive side of the battery to the negative side. As time went on, however, science slowly showed the error of this assumption. The electron was discovered, and it was determined that when you hook up a battery in a circuit made of metal wires, the resulting current is made up of electrons. Thus, in our circuit drawing above, the current actually flows from the negative side of the battery to the positive side.

This conclusion, however, contradicted circuit drawings that had been made over the years. Engineers had always drawn current in these kinds of circuits as flowing from the positive side of the battery to the negative side. They didn't want to stop doing it just because science had shown that the reverse was true. As a result, people just kept on drawing electric current in such circuits as starting from the positive side of the battery and flowing to the negative side, even though they knew it was wrong. This came to be known as **conventional current**, and it is still the way circuits are drawn today.

Conventional current - Current that flows from the positive side of the battery to the negative side

Now I must point out that while conventional current is an inaccurate representation of what is really happening in circuits such as those drawn above, there are many instances in which electric current really does result from the flow of positively-charged particles. In these instances, then, electric current does flow from a positive source to a negative source. However, in circuits such as the ones we will study in this course, the actual electric current is a flow of electrons. Nevertheless, we will still use conventional current when we draw them, as that makes our work consistent with everyone else's work.

As a quick side note, I want to point out that from an electrical standpoint, whether the current is made up of negative charges or positive charges really doesn't matter. The end result is the same. Whether I have a current of negative charges moving to the left or the same current of positive charges moving to the right, all of the measurable aspects of the circuit remain the same. As a result, there is no physical reason why we shouldn't use conventional current in all of our diagrams.

The point to this long, drawn-out discussion is that in our original circuit diagram, the current can be pictured as follows.

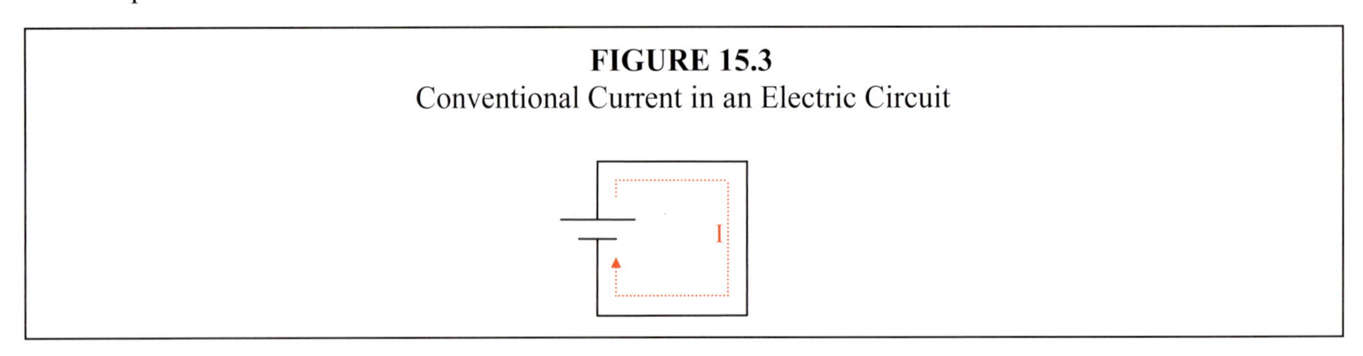

FIGURE 15.3
Conventional Current in an Electric Circuit

In this diagram, the dashed line shows the path of the current in the wire, and "I" labels the dashed line as representing current. Of course, even though the dashed line is not drawn right on top of the lines representing the wire, it is still understood that the current is actually flowing in the wire, not where the dashed line is drawn.

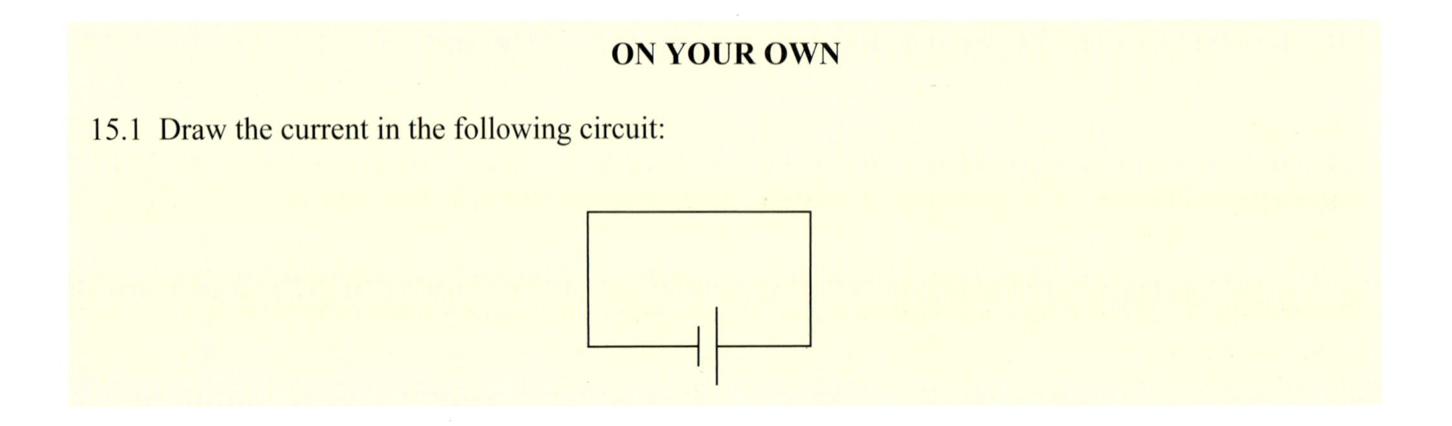

ON YOUR OWN

15.1 Draw the current in the following circuit:

Resistance

Since you now know how current flows in an electric circuit, the next logical step is to determine *how much* current flows in an electric circuit. To understand the factors that influence this, we will start with a simple experiment.

EXPERIMENT 15.1
Current and Resistance

Supplies:

- Safety goggles
- A 1.5-volt battery (Any AA-, C-, or D-cell battery will work. Do not use any battery other than one of those, though, because a higher voltage can make the experiment dangerous.)
- Aluminum foil
- Scissors

1. Cut a small strip of aluminum foil that is about 1.3 times the length of the battery and only about one cm wide.
2. Lay the foil across the battery and, using your thumb and forefinger, pinch the foil so that it makes contact with both ends of the battery, as shown below:

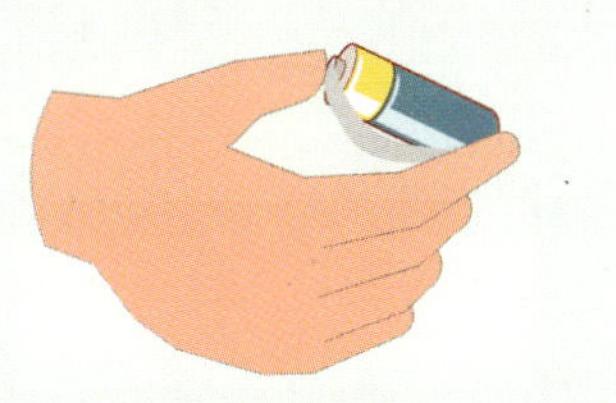

Illustration copyright GifArt.com

3. Hold the foil there for a few moments. Note what you feel. Do not hold the foil for too long, as it can get painful!
4. Put everything away.

What did you feel? You should have felt the foil getting hot. The heat you felt comes from the electrons moving in the foil. Since aluminum is a conductor, as soon as you touched it to both ends of the battery, electrons could flow through the aluminum from the negative side of the battery to the positive side. The aluminum, however, resisted the flow of the electrons to some extent, reducing the total current. The resistance of the aluminum is something much like friction, and thus kinetic energy was converted into heat, which you felt in your fingers.

Each conductor resists current flow differently, so we say that each conductor has its own **resistance**.

Resistance - The ability of a material to impede the flow of charge when it is subject to a potential difference

It turns out that the *type* of conductor is not the only thing that determines resistance. The resistance of a wide conductor, for example, is lower than the resistance of a thin conductor made out of the same material. Also, the longer the conductor, the larger the resistance.

The relationship between the current in a conductor and that conductor's resistance is summarized in **Ohm's Law**.

$$V = IR \tag{15.2}$$

This equation, first discovered in 1827 by German physicist George Simon Ohm, tells us that the current in a conductor (I) times that conductor's resistance (R) is equal to the potential difference (V) to which the conductor is exposed. Believe it or not, I still remember this equation with a mnemonic: Twinkle, twinkle little star, voltage equals "I" times "R."

Now we already know that potential difference is measured in volts and current in amps, but what is resistance measured in? Well, the SI unit for resistance is called the "Ohm" (for obvious reasons) and is abbreviated with the Greek letter omega (Ω). It is rather hard to explain what an Ohm is, so instead, let's look at how an Ohm relates to volts and amps, the units we already know. Equation (15.2) tells us, for example, that an A·Ω is a volt, and an Ohm is the same as a $\frac{V}{A}$.

Electric Heaters

With the knowledge of current, electric circuits, and resistance under your belt, you can finally begin to understand how the simplest electrical devices work. For example, consider an electric heater. This could be a space heater used to warm up a room, a coil on an electric stove, or even the wires on the inside of a toaster. When such a device is turned on, the heater begins to glow, emitting a large amount of heat. This works because the material used to make the heater has a certain amount of resistance. The result of this resistance is that kinetic energy is converted into heat and light energy. The light energy causes the heater to glow, and the heat energy warms the room, cooks the food, or browns the bread.

Using Equation (15.2), we can analyze the amount of voltage, resistance, and current in an electric heater. Before we do that, however, I want you to see the circuit diagram that a physicist would draw in order to represent an electric heater.

FIGURE 15.4
Circuit Diagram for an Electric Heater

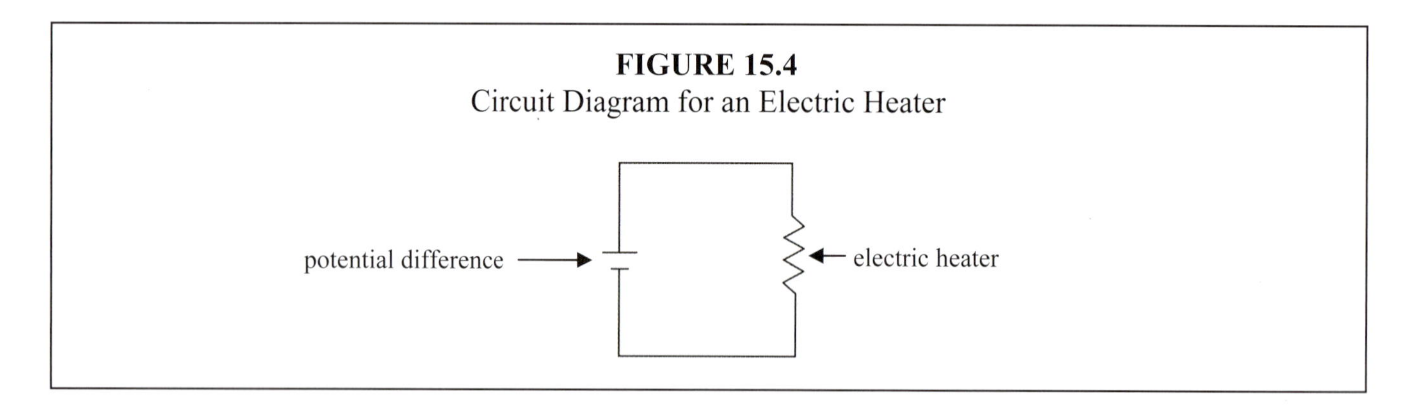

In this diagram, there are two symbols. First, there is the standard symbol for a battery. Even though most space heaters plug into wall sockets instead of batteries, we will go ahead and use the battery symbol to denote any source of potential difference. We will see in the next module that there are differences between the potential difference supplied by a battery and that supplied by a wall socket, but I'm getting ahead of myself. The second symbol in the diagram (the jagged line), is the standard symbol for a resistor. Thus, this simple circuit diagram tells us that we have a voltage source attached to a resistor. That's the basic construction of an electric heater.

Now before we start to analyze circuit diagrams, there is one thing that you need to understand. In circuit diagrams, the lines that connect the battery to the resistor (or resistors) are considered to be conductors which are free of resistance. Of course, this is not true. After all, you saw in Experiment 15.1 that a simple aluminum strip resisted the flow of charge. However, in general, the resistance of the conductors that connect the battery to the resistor (or resistors) in a circuit is so small compared to the resistor (or resistors), that their resistance can be ignored. Thus, although no conductor is free of resistance, we will treat the conductors in an electric circuit as if they have no resistance.

Okay, now that we have a few basics related to electric circuits, let's analyze a reasonably simple circuit.

EXAMPLE 15.1

A typical space heater has a resistance of 11 Ohms. When plugged into a wall socket, this heater is exposed to the equivalent of a 120-volt potential difference. Draw a circuit for the heater, determine the current, and draw the direction of the current flow.

The circuit diagram is easy, since I already drew one. We need to do a couple of things to that drawing, however. First, we have to label everything. This is standard practice, and I will expect you to do it on all circuit diagrams that you make. Also, we need to draw in the current flow. We learned before that we always draw conventional current. Thus, the current flows from the positive side of the battery (the long parallel line) to the negative side (the short parallel line).

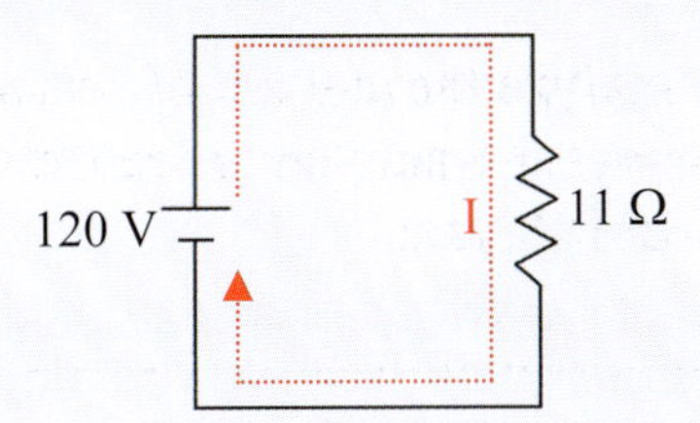

Now we need to figure out *how much* current is in the circuit. Using Equation (15.2), that's a pretty easy task, since we already know the voltage (120 V) and the resistance (11 Ω).

$$V = IR$$

$$120\text{ V} = I\cdot(11\ \Omega)$$

$$I = \frac{120\text{ V}}{11\ \Omega} = 11\text{ A}$$

Notice that since an Ohm is the same as a $\frac{V}{A}$, then a $\frac{V}{\Omega}$ must be an amp. Thus, 11 amps run through a space heater.

Now since you don't have a lot of experience with electricity so far, you probably don't have an appreciation for what this number means. Is 11 amps of current a lot of current, or only a little current? It turns out that 11 amps is a *lot* of current. I want to illustrate this to you in two different ways.

First, if 11 amps were to be sent through your body, it would almost certainly be lethal. Indeed, the range of current reported in electric chairs used to execute people is generally 7 to 12 amps. This is why electric space heaters have (or at least should have) lots of protective devices that shut off power whenever anything unusual happens, such as when the heater tips over.

The other way to illustrate how much current is in 11 amps is by using the definition of an amp to calculate how many electrons pass by a point in the circuit every second. Remember, an amp is defined as a Coulomb per second. Well, we know that each electron has a charge of -1.6×10^{-19} C. Thus, we know how many Coulombs per second pass through the heater and we also know how many Coulombs are in each electron. Therefore, we can use a simple conversion process to tell us how many electrons per second pass by a point in the circuit:

$$\frac{11\,\cancel{\text{C}}}{\text{sec}} \times \frac{1\ \text{electron}}{1.6 \times 10^{-19}\ \cancel{\text{C}}} = 6.9 \times 10^{19}\ \frac{\text{electrons}}{\text{sec}}$$

This tells us that a whopping 6.9×10^{19} electrons pass by a point in the circuit each second!

Now please understand that this means 6.9×10^{19} electrons leave the negative side of the potential difference each second, and 6.9×10^{19} electrons reach the positive side of the potential difference each second. There cannot be a "pileup" of electrons in the circuit. Thus, for every electron that leaves the negative side of the potential difference, there must be an electron reaching the positive side. This ensures that charge is conserved in the circuit, which must be the case.

Of course, Ohm's Law is applicable to many other things besides an electric heater. I just used that in the example because it's a practical application. Nevertheless, there are some materials that conduct electricity but do not follow Ohm's Law. These materials, called **non-Ohmic** substances, are a bit difficult to explain at this point. You do need to be aware that they exist, however. Nevertheless, most conductors obey Ohm's Law, so it is important that you know how to use it.

ON YOUR OWN

15.2 An electrician measures 0.023 amps of current in the following circuit. Draw the current flow and determine the resistance of the resistor:

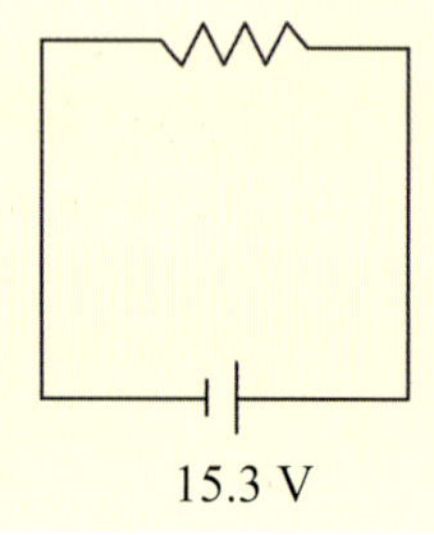

Electric Power

Even though we can now calculate the amount of current in a space heater, that still doesn't tell us everything we need to know. Sure, we know that 11 amps is a lot of current, but how warm does that make the space heater? To answer that question, we need to determine how much heat energy the space heater produces. Now it turns out that it is easier to calculate the *power* that the space heater uses, and then if we want to calculate the energy output, we can use the definition of power to do that.

Remember that we discussed power in Module #8. We learned that power, measured in Watts, tells us how much energy is used per second. For example, the typical teenage male uses an average of 148 Watts of power in the activities of everyday life. This means that every second, the average teenage male burns 148 Joules of energy. A typical light bulb uses 100 Watts, meaning that 100 Joules of energy are used up every second. This can give us an idea of how hot the space heater gets. The higher the power, the more energy output per second. Thus, the higher the power, the hotter the heater.

We already have an equation for power, Equation (8.5), but it doesn't do us much good in this case because we don't know how to calculate the quantities in that equation when electric circuits are involved. Fortunately, however, there is a rather simple equation that relates power in an electric circuit to the things that we already know:

$$P = IV \qquad (15.3)$$

In this equation, "P" is the power, "I" represents the current, and "V" stands for the electric potential in the circuit.

Since we already know that power is measured in Watts, this equation also tells us something about the units we are working with. According to the equation, you get power when you multiply current (measured in amps) with electric potential (measured in volts). This tells us that a Watt is the same as an A·V.

Now it turns out that there is another equation for power as well. Ohm's Law tells us that voltage is equal to current times resistance. If we take that expression for voltage and plug it into Equation (15.3) we get:

$$P = I \cdot V = I \cdot (I \cdot R) \qquad (15.3a)$$

$$P = I^2 \cdot R \qquad (15.4)$$

Just as the previous equation told us that a Watt is the same as an A·V, this equation tells us that a Watt is also the same as an $A^2 \cdot \Omega$.

It is nice to have two equations for power because if you don't know the voltage of the circuit, you can use Equation (15.4) instead of Equation (15.3). Both equations are right, of course. What you know will simply determine which you use. Study the following example problems to see how this works.

EXAMPLE 15.2

In the previous example, we learned that a space heater uses a 120-volt wall socket to run and draws 11 A of current. How much power does it use?

In this case, we have the voltage and the current, so we will use Equation (15.3) in order to determine the power:

$$P = IV = (11\ A)\cdot(120\ V) = 1.3 \times 10^3\ A\cdot V = 1.3 \times 10^3\ \text{Watts}$$

The space heater, then, draws 1.3×10^3 Watts of power. Remember, the average light bulb is 100 Watts, so this space heater draws the same power as 13 light bulbs!

A motor draws 1.19 A of current and has a resistance of 101 Ω. Assuming that the motor is 100% efficient, what is its power output?

In this case, we need to use Equation (15.4) because we have current and resistance, not current and voltage.

$$P = I^2R = (1.19\ A)^2\cdot(101\ \Omega) = 143\ \text{Watts}$$

Now remember, this power is really the power that the motor *draws* from the power source. To get how much power the motor outputs, we have to make an assumption about how efficient the motor is in converting the energy it gets from the power source to kinetic energy of motion to move whatever it is supposed to be moving. Of course, the problem says to assume that the motor is 100% efficient, which is, in fact, impossible. However, making that assumption, every Watt of power drawn from the source gets converted to a Watt of power to move whatever is being moved. Thus, the motor outputs 143 Watts.

If you remember back to Module #8, we did some problems in which we knew the power of an engine and used that power to determine how far the engine could move something in a certain length of time. If that engine happens to be electric, this is how we can calculate that power. This type of analysis, then, can be very useful when applied in the real world. Of course, a more realistic number for efficiency would have to be used, but you get the idea.

ON YOUR OWN

15.3 A heater is rated to draw 13 amps and has a resistance of 9.3 Ω. Assuming 100% efficiency, how many Watts of power does it produce?

15.4 An electric motor runs on a 1.5-volt battery and draws 0.32 A of current. How much power does it use?

Switches and Circuits

With the tools you have learned so far, you can look at a simple circuit and determine the amount of electric current in a circuit and determine the direction in which that current flows. With that knowledge under your belt, I want you to do a quick experiment.

EXPERIMENT 15.2

Building a Simple Circuit to Turn on a Light Bulb

Supplies:

- Safety goggles
- A tabletop to which you can tape things or a flat piece of cardboard (The tabletop can't be metal.)
- Insulated wire, 20-14 gauge (It is best to use wire that has several small conductors twisted together. This is typically called "braided wire.")
- Scissors
- Tape
- 1.5-volt battery. (D-cells are easiest to work with, but any flashlight battery will do. Make sure the voltage is 1.5 volts. A higher voltage could make the experiment dangerous.)
- Flashlight

1. Tape the battery to the tabletop (or cardboard) with several pieces of tape so that it is secure.
2. Cut two long strips of wire.
3. Use your scissors to strip about ½ inch of insulation from both ends of each wire. The best way to do this is to put the wire in your scissors and squeeze the scissors gently. You should feel an increase in resistance as the scissors begin to touch the wire. Squeeze the scissors until you feel that resistance and then back off. Continue squeezing and backing off as you slowly turn the wire round and round (see drawing on the right). Be careful. You can cut yourself if you are not paying proper attention! You will eventually have a cut that goes through the insulation all the way around the wire. At that point, you can simply pull the insulation off. It will take some practice to get this right, but you *can* do it. Make sure that there is at least a quarter inch of bare conductor sticking out of both ends of each wire.

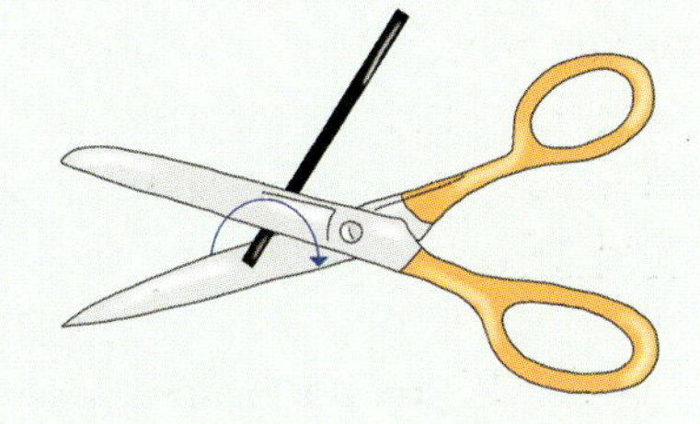

4. Tape the exposed conductor from the end of one wire to the positive side of the battery. Make sure there is good contact between the bare conductor and the metal that makes up the positive side.
5. Tape the exposed conductor from the end of the other wire to the negative side of the battery. Make sure there is good contact between the bare conductor and the metal that makes up the negative side.
6. Screw the end off the flashlight.
7. Examine the back side of the assembly that holds the light bulb. It should look something like the photograph below:

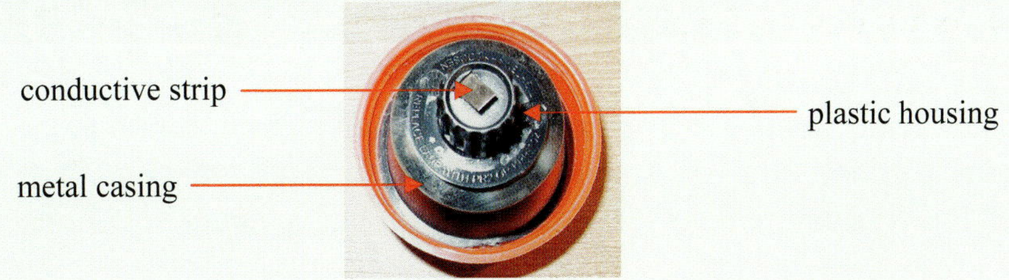

The conductive strip pointed out in the photograph is electrically connected to one end of the filament in the light bulb. The metal casing is connected to the other end of the filament. The plastic housing separates the two.

8. Cut two more lengths of wire and once again use the scissors to strip about ½ inch of insulation from both ends of each wire.
9. Tape the end of one of the wires to the conductive strip on the light bulb assembly. Make sure there is good contact between the bare conductor in the wire and the conductive strip.
10. Tape the end of the other wire to the metal casing on the light bulb assembly. Make sure there is good contact between the bare conductor in the wire and the metal casing. Also, make sure that the conductor in this wire doesn't get anywhere near the conductive strip or the conductor in the other wire.
11. Now connect the bare conductors from one of the wires attached to the battery to one of the wires attached to the light bulb assembly. The best way to do this is to twist the conductors together so that they intermingle with one another and then tape them to the tabletop (or cardboard). If you cannot twist the conductors together, just tape them down to the tabletop so that they are in good contact with one another.
12. Make sure that the bare conductors of the other two wires do not come into contact with one another. Your setup should look something like this:

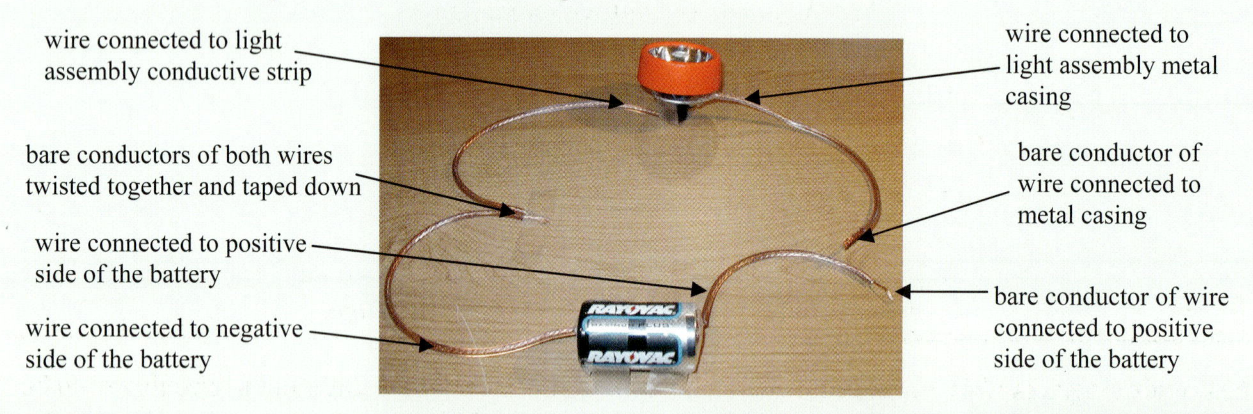

At this point, you do not have a complete circuit. One portion of your circuit is connected to one side of the battery, but the other portion is not connected to the other side of the battery. Thus, as it stands right now, there is no potential difference across which electrons can travel. In order to get current to flow, you must have a full conductive path from one side of the battery to the other.

13. To complete the circuit, grab the remaining wire coming from the battery by the insulation. Touch the bare conductor of the wire to the bare conductor of the remaining wire from the light bulb assembly. You should see the light bulb turn on. If you do not see the light bulb turn on, you either have a dead battery or a bad connection. Check to make sure all of the conductors are in good contact with what they are supposed to be touching.
14. Pull the wire back so that the conductors are no longer touching. Watch the light turn off.
15. Make the conductors of the wire touch again, and watch the light turn on.
16. You have just made a simple circuit with a switch.
17. Clean up your mess, but save everything. You are going to use this basic setup again.

The circuit you made in this experiment was pretty simple, but don't worry. You will make a more complex circuit later on. First, let's make sure we understand what happened in the experiment. Before you touched the conductors together in step 13, you had a conductor connecting one end of the battery to the filament of the light bulb. However, the other end of the filament was not connected to the other side of the battery. Thus, current could not flow in the circuit. Why not? In order to get charges to flow, there must be a potential difference. That way, there is potential energy that can be converted into kinetic energy. Well, until you connected the circuit so that both sides of the battery were connected by a conductor, there was no potential difference. As a result, current would not flow. Once the two sides of the battery were connected, there was a potential difference, and as a result, potential energy could be converted to kinetic energy, making current flow.

As I told you in the experiment, what you essentially made was a switch. When you touched the conductors from the remaining two wires, the circuit was complete, and current could flow, lighting the bulb. When you moved the wire so that the conductors no longer touched, there was once again no potential difference, so current could no longer flow, and the light bulb turned off. This is the principle behind a switch.

In circuit diagrams, we usually represent a switch as shown in Figure 15.5.

FIGURE 15.5
A Simple Circuit with a Switch

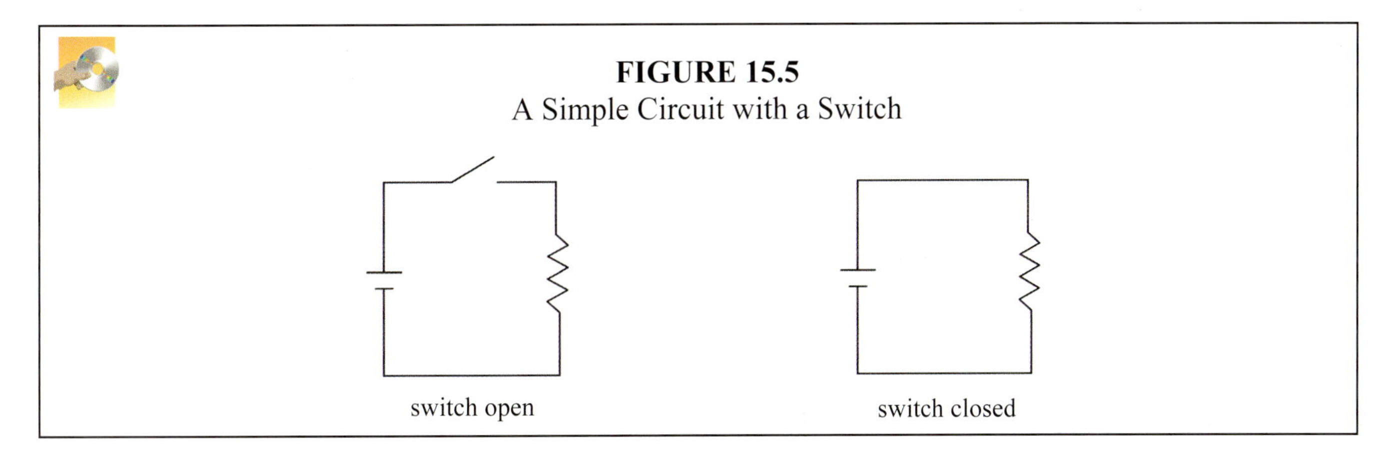

The circuit diagrams above could represent the circuit you made. The battery in the circuit would be the battery in the experiment, and the resistor in the circuit would be the light bulb. The diagonal line on top of the diagram on the left would be the switch. When the diagonal line is up (as is drawn in the figure on the left), the switch is "open" When the diagonal line is no longer there, the switch is "closed." Physicists call circuits such as the one drawn on the left side of the figure **open circuits**.

Open circuit - A circuit that does not have a complete connection between the two sides of the battery. As a result, current does not flow.

When you finished the setup (as pictured in step 12) you had an open circuit. As a result, current did not flow. However, when you touched the two conductors together in step 13, you closed (completed) the circuit. As a result, current could flow and the light bulb turned on.

You should now see that when you flip a light switch, you are changing the circuit that powers the light. If the light is off, the circuit is open. By flipping the switch, you are actually connecting the two sides of the power source and current can therefore flow. That turns on the light. If you flip the switch again, you are breaking that connection, and the current flow stops, turning the light off. That's how a switch works.

Series and Parallel Circuits

The idea that a break in an electric circuit can stop the electron flow and thus stop the electrical device or devices in the circuit from working is, most of the time, quite useful. However, sometimes it's a real problem. Consider, for example the following situation:

FIGURE 15.6
Light Bulbs in Series

Illustration by Megan Whitaker

In this circuit, *neither* of the light bulbs will light. Why? Well, a light bulb lights up because electrons flow through the filament of the bulb. The resistance of the filament produces heat, and the filament gets so hot that it glows with a bright light. Now, the reason a light bulb burns out is that eventually, the filament breaks due to wear and tear. Notice how the filament (red wire) in the bulb on the right is broken. That's why the bulb is labeled as burnt out. Because the filament is broken, the electrical connection from one side of the light bulb to the other is broken. Well, since the light bulb is a part of the circuit, this breaks the electrical connection of the entire circuit. Thus, a burnt-out light bulb is the same as an open switch! So, even though the second light bulb is a working light bulb, it will not light because current cannot flow through the circuit.

When lights (or any electrical devices) are hooked up in this way, we say that they are in **series**. All devices on such a circuit, called a **series circuit**, will cease to function when any one of the devices ceases to function. This, of course, is an undesirable situation. Luckily, such a situation can be avoided. Consider the following situation:

FIGURE 15.7
Light Bulbs in Parallel

Illustration by Megan Whitaker

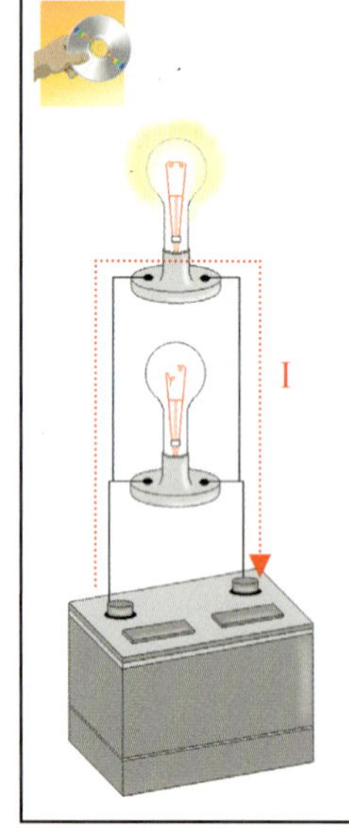

In this case, the working light bulb lights, even though it is in the same circuit as the burnt-out light bulb. Why? Look at the situation. Since the burnt-out light bulb is like an open switch, current cannot flow through it. But, as you look at the drawing, you see that current can still flow from one side of the battery to the other. It does not have to go through the burnt-out light bulb. Instead, it can take the longer route through the working light bulb. Since the working light bulb still has a conductor that connects one side of the bulb to the other, there is still a complete circuit through that bulb. Thus, current can flow from one side of the battery to the other, as shown by the dashed red line. This allows the working light bulb to light, even though the other light bulb is burnt out.

When devices are hooked as shown in Figure 15.7, we say that they are wired in **parallel**. A **parallel circuit**, then, allows the devices to work independently of each other. If one stops working, the others can still work, because current can still flow through them. Next time you hang lights on your Christmas tree, look at the box. Somewhere on the box, it might say "each lamp burns independently." You see, when Christmas tree lights were first made, the lights were hooked up in series. The problem with that, of course, was when one light bulb burnt out, they all stopped working. I'm sure if you ask your parents, you will hear about how they had to try and get a string of lights working again by changing out light bulbs one by one until they found the burnt-out one. Once they replaced the one burnt-out bulb, all of the bulbs would suddenly light. Christmas tree lights that burn independently, however, are hooked up in parallel. When one burns out, the others still light. That's the difference between parallel and series circuits.

Series and parallel circuits differ in more ways than how they react to a burnt-out light bulb. In order to learn about another important difference, perform the following experiment.

EXPERIMENT 15.3

Series and Parallel Resistors

Supplies:

- Safety goggles
- The complete experimental setup from Experiment 15.2
- Insulated wire like that used in Experiment 15.2
- Scissors
- Tape
- 5 spoons (They cannot be silver. They must be stainless steel flatware. Most people use this kind of flatware as their "everyday" spoons.)

1. Set up the experiment just like you did in Experiment 15.2.
2. Strip 1.5 inches of insulation from the end of a wire. You may have to do this in two or three steps, as it is easier to strip smaller sections of insulation.
3. If you are using braided wire, twist the conductor so that the individual strands are all twisted together.
4. Cut the twisted conductor off the end so that you have 1.5 inches of bare conductor. If you need to twist it some more to keep the individual wires from falling off, go ahead and do so. If you lose a few wires, it's no big deal.
5. Repeat steps 2-4 three more times so that you have four twisted pieces of bare conductor, each 1.5 inches long.
6. Repeat steps 2-4 twice more, this time making the length of the bare conductor 2.5 inches.
7. Cut a 4-inch piece of wire, and strip about ½ inch of insulation from each side.
8. Take two of the 1.5-inch strips of bare conductor and tape one to one end of a spoon and the other to the other end of a spoon. The result should look like the photograph below:

9. Take two different spoons and lay them parallel to one another, facing the same direction.
10. Use the two 2.5-inch strips of bare conductor to connect the spoons. Use one conductor on one end of the spoons and the other conductor on the other end. Be sure to tape the conductors to the

spoons, but leave the conductor in between bare. The result should look like the photograph below:

11. Finally, with the two remaining spoons, use the 4-inch piece of wire to connect the bowl of one spoon to the handle of another.
12. Tape one of the remaining 1.5-inch strips of conductor to one end of this two-spoon assembly and the other strip of conductor to the other end. The result should look something like the photograph below:

Each one of the spoon assemblies you made is a resistor. The spoons are made of a mixture of iron and carbon, which has significantly more resistance than copper. Now you want to see how these resistor assemblies affect the current in the circuit.

13. To get started, turn out the lights. Make sure there is enough ambient light so that you can still see what you are doing.
14. Ignore the three resistor assemblies and simply close the switch on the circuit like you did in Experiment 15.2. Don't be afraid to touch the bare conductor as you do this. You are not using enough voltage or current to make it dangerous. Press the two conductors together hard so that they make good contact. Look at the light and get a good idea in your mind about how bright it is.
15. Take the spoon assembly you made in step 8 and flip it over so that the bare conductors are touching the tabletop.
16. Touch the bare end of the wire coming from the bulb to one bare conductor on the spoon, and touch the bare end of the wire coming from the battery to the bare conductor on the other end of the spoon. Once again, press the conductors together hard so that there is good contact between them. In this situation, the spoon is now a part of the circuit.
17. Observe the brightness of the light and compare it to the brightness in step 14. You can always go back and repeat step 14 to remind yourself of how bright the light was.
18. Now try the resistor assembly you made in step 10. Flip it over so that the bare conductors connecting the spoons are touching the tabletop.
19. Touch the bare end of the wire coming from the light bulb to one of the bare conductors connecting the spoons, and touch the bare end of the wire coming from the battery to the other conductor that is connecting the spoons. Once again, press the conductors together hard so that there is good contact between them. In this situation, the two spoons are in the circuit, but they are hooked up in parallel.
20. Observe the brightness of the light bulb and compare it to what you observed in step 17. You can always go back and repeat steps 15-17 so that you can compare the two.
21. Finally, use the resistor assembly you made in step 12. Once again, flip it over so that the bare conductors are touching the tabletop.
22. Touch the bare end of the wire coming from the bulb to one bare conductor on the two-spoon setup, and touch the bare end of the wire coming from the battery to the bare conductor on the

other end of the two-spoon setup spoon. Once again, press the conductors together hard so that there is good contact between them. In this situation, the two spoons are a part of the circuit, but they are in series.

23. Observe the brightness of the light. It is possible that the light will not come on at all. That's fine.
24. Clean up your mess. You will not be using this setup again. Make sure in particular that all of the copper and tape are off the spoons. Wash the spoons thoroughly before you put them away.

Before I discuss what you saw in the experiment, I want to say one word of caution. In this experiment, you were touching bare conductors that were a part of a circuit which had current flowing in it. As I told you in the experiment, in this case you could do that, because the voltage and current (and therefore the power) were low. However, you should *never* do that unless someone who knows exactly what he is doing tells you! Touching a conductor in a live circuit can be lethal if you don't know what you are doing!

Okay, now we can talk about the experiment. What did you notice about the brightness of the bulb when you put the single spoon in the circuit as compared to when no spoons were in the circuit? You should have noticed that the bulb was dimmer when the spoon was in the circuit. That's because the spoon has a significantly higher resistance than the wires in your circuit. As a result, it reduced the current of the circuit. Remember, V = IR, and "V" never changed throughout the experiment, because you used the same battery the entire time. Thus, if "R" increased, "I" had to decrease. This is evidenced by the fact that the bulb got dimmer when the spoon was made a part of the circuit.

Now compare that to what happened when you used the two spoons in series (steps 21-23). When you used the two spoons in series, the bulb was probably incredibly dim. In fact, you might not have been able to see it light at all. Why? Once again, the spoons are resistors. When resistors are in series in a circuit, their individual resistances add together. Thus, the resistance of the two spoons in series was twice that of the single spoon. This reduced the current even more severely. In fact, it might have reduced the current so much that the bulb would not light.

Finally, compare that result to what happened when you used the two spoons hooked up in parallel (steps 18-20). If you did the experiment correctly, you should have noticed that the bulb burned *more brightly* with the two spoons hooked in parallel compared to the *one spoon* (steps 15-17). Why in the world did that happen? It turns out that when resistors are hooked in parallel, their resistances do not add directly. Instead, the resistance of two resistors in parallel is actually *lower* than the resistance of just one of the resistors. You will see why this is the case mathematically in a moment. Physically, you can think of it this way. A single resistor with current flowing through it is like a single pipe with water flowing through it. For a given pressure, only a certain amount of water can get through. However, if we put *two* pipes in the system in parallel, more water could get through. In the same way, two resistors in parallel allow for more current flow than a single resistor. Since the current is higher, the resistance must be lower.

The experiment, then, demonstrated that when you hook up two resistors in series, the resistance is larger than a single resistor. However, if you hook up two resistors in parallel, the resistance is lower than a single resistor. Of course, the resistance is even less if you get rid of the resistors entirely!

The Mathematics of Series and Parallel Circuits

Let's quantify these observations with some equations, starting with resistors in series.

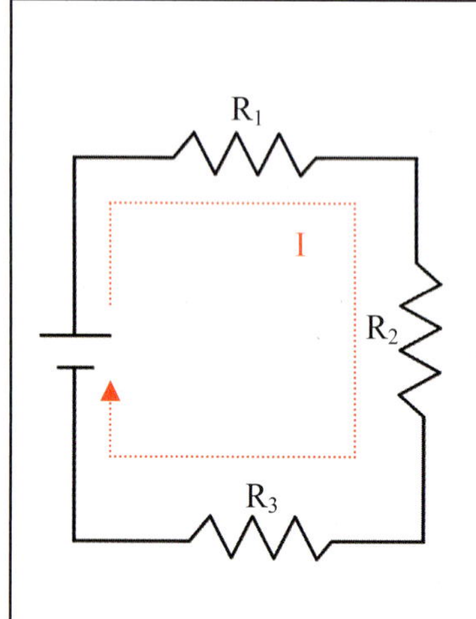

FIGURE 15.8
Resistors in Series

In this circuit, the resistors are hooked up in series. We can tell this because in order for the current to travel from one end of the battery to the other, it must pass through *all three* resistors. Since all current passes through all resistors, we know that if one resistor breaks, current will stop flowing. We also know from the experiment that the total resistance of the circuit is greater than the resistance of any one resistor. In fact, the resistances just add together. Thus, the total resistance of the circuit is the sum of the individual resistors.

Since the total resistance of the circuit is simply the sum of the three individual resistors, I would have the same circuit if I just put one resistor in there, as long as its resistance was equal to the sum of the three resistors. This is the concept of **effective resistance**. When many resistors are in a circuit, we can say that those many resistors could be replaced by a single resistor, the **effective resistor**. So, I can take resistors in series and replace them all with an effective resistor whose resistance is given by:

$$R_{effective} = R_1 + R_2 + R_3 + \ldots \qquad \text{(for resistors in series) (15.5)}$$

The three dots simply tell us that no matter how many resistors are hooked up in series, the effective resistance is equal to the sum of all the individual resistances.

We could take the circuit in Figure 15.8, then, and replace it with a circuit that looks like this:

$R = R_1 + R_2 + R_3$

As long as the resistance of this single resistor is equal to the sum of the three resistances in the previous circuit, the current would be the same in each circuit.

Resistors hooked up in parallel can also be replaced by effective resistors, but the whole situation gets a lot more complicated. That should be expected, since the experiment gave such an interesting result. When resistors are hooked up in parallel, the effective resistance is actually lower than the resistance of the individual resistors. How does this work? Examine Figure 15.9.

FIGURE 15.9
Resistors in Parallel

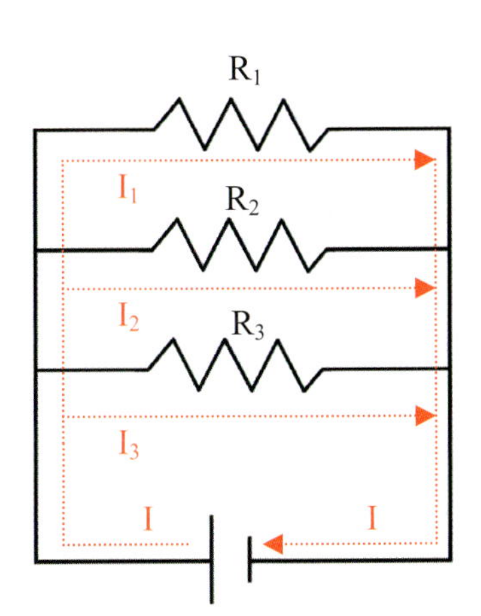

When resistors are hooked up in parallel, the current has a "choice" as to which resistor it flows through. If the total current in the circuit is "I," then some portion of that current (I_1) will flow through R_1. Another portion of the current (I_2) will flow through R_2. Finally, the rest of the current (I_3) will flow through R_3. First, since the current leaving the positive side of the battery must equal the current entering the negative side, we know that $I_1 + I_2 + I_3 = I$. We also know that I_1, I_2, and I_3 all experience the same potential difference, since they all originally left the positive side of the battery and they all end up entering the negative side of the battery.

Now remember, the potential difference experienced by each portion of current that travels through the resistors is the same. Let's call it "V." Well, Ohm's Law tells us that V = IR. Thus, we can determine the portion of the current that goes through each resistor:

$$I_1 = \frac{V}{R_1} \qquad I_2 = \frac{V}{R_2} \qquad I_3 = \frac{V}{R_3} \tag{15.6}$$

Well, since we know that $I_1 + I_2 + I_3 = I$, we can say that:

$$I = \frac{V}{R_1} + \frac{V}{R_2} + \frac{V}{R_3} \tag{15.7}$$

Which is the same as:

$$I = V \cdot \left(\frac{1}{R_1} + \frac{1}{R_2} + \frac{1}{R_3}\right) \tag{15.8}$$

Now, suppose I make this substitution into Equation (15.8):

$$\frac{1}{R_{effective}} = \frac{1}{R_1} + \frac{1}{R_2} + \frac{1}{R_3} + \ldots \quad \text{(for resistors in parallel)} \tag{15.9}$$

What would I get? I would get:

$$I = \frac{V}{R_{effective}} \tag{15.10}$$

This equation is just a rearrangement of Ohm's Law, using "$R_{effective}$" as the total resistance of the circuit. Thus, resistors in parallel can also be replaced by a single, effective resistor, as long as the value of the resistor is given by Equation (15.9)! Now, of course, the equation is more complex than the one for resistors in series, but the idea is the same. If all of this seems a bit overwhelming, don't worry about it. A couple of examples should clear the whole thing up.

EXAMPLE 15.3

Consider the circuit diagrammed below. Redraw the circuit with an effective resistor that can replace the individual ones. Determine the resistance of the effective resistor.

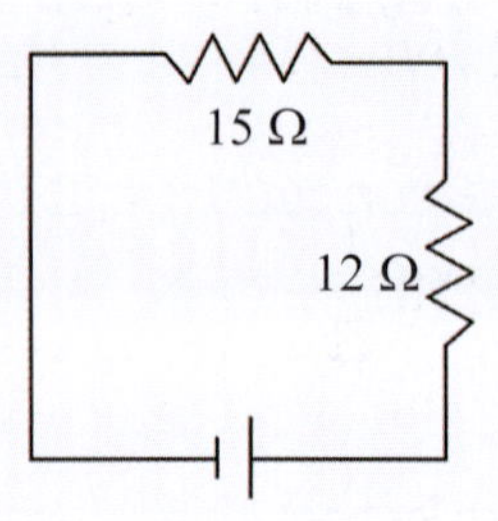

The resistors in this diagram are hooked up in series. We can see this because the current has no choice but to travel through each resistor in order to get to the other end of the battery. This makes the circuit a series circuit. In order to determine the effective resistance, then, we simply use Equation (15.5):

$$R_{effective} = R_1 + R_2$$

$$R_{effective} = 12\ \Omega + 15\ \Omega = 27\ \Omega$$

These two resistors in series, then, work the same as a single resistor with a value of 27 Ω.

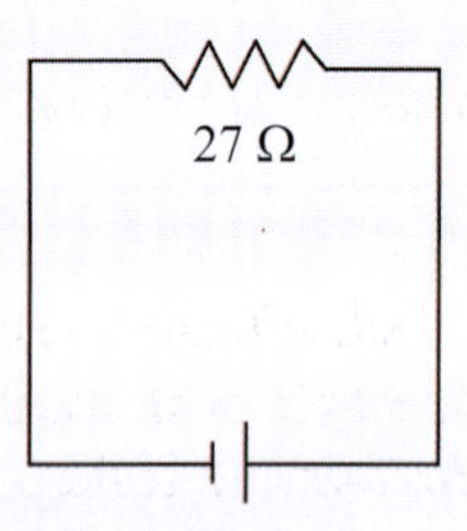

Replace the two resistors in the circuit diagrammed below with an effective resistor. What is its resistance?

15 Ω

12 Ω

In this problem, the resistors are hooked up in parallel. We can tell this because as the current leaves the positive end of the battery, it has a choice of paths to take to the negative end. It could either go through the 12 Ω resistor and then to the negative side of the battery, or it could go through the 15 Ω resistor and then to the negative side of the battery. Rather than being forced to travel through both resistors, the current has a choice of which resistor to travel through. This means that the resistors are hooked up in parallel. Since they are hooked up in parallel, we need to use Equation (15.9) to determine the effective resistance.

$$\frac{1}{R_{effective}} = \frac{1}{R_1} + \frac{1}{R_2}$$

$$\frac{1}{R_{effective}} = \frac{1}{12\Omega} + \frac{1}{15\Omega} = 0.083\frac{1}{\Omega} + 0.067\frac{1}{\Omega}$$

$$\frac{1}{R_{effective}} = 0.150\frac{1}{\Omega}$$

$$R_{effective} = 6.67\ \Omega$$

These two resistors in parallel, then, could be replaced by a single resistor with a resistance of 6.67 Ω.

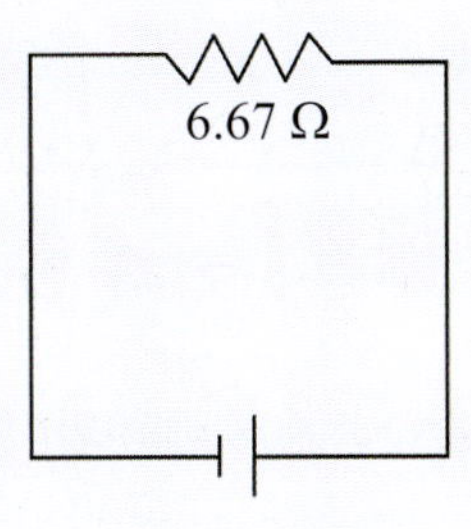

Notice that this resistance is, in fact, less than that of the two individual resistors. This is what you observed in Experiment 15.3.

In order to reduce many resistors to their effective resistance, then, we first have to determine whether or not the resistors are hooked up in series or in parallel. If they are in series, we can use Equation (15.5) to determine the effective resistance. If they are in parallel, we use Equation (15.9). We determine whether we have a series or parallel circuit by seeing whether or not the current must pass through all of the resistors in question. If it must pass through all of the resistors, the resistors are in series. If, instead, the current has a choice as to which resistor it travels through, the resistors are in parallel. Now even though these might seem like pointless exercises right now, there is an amazingly useful thing that we can learn once we master the technique. So, for right now, just do the "On Your Own" problems that follow so that you know you have the technique down. In the next three sections, you can begin to see why this stuff is useful.

ON YOUR OWN

15.5 Replace the resistors in the circuit below with an equivalent resistor. What is the resistance of that equivalent resistor?

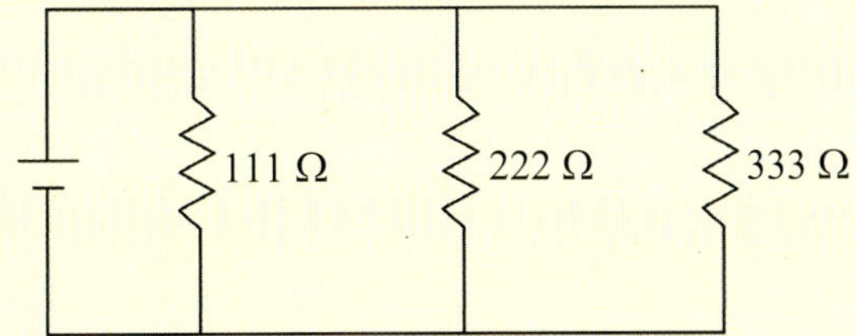

15.6 Replace the resistors in the circuit below with an equivalent resistor. What is the resistance of that equivalent resistor?

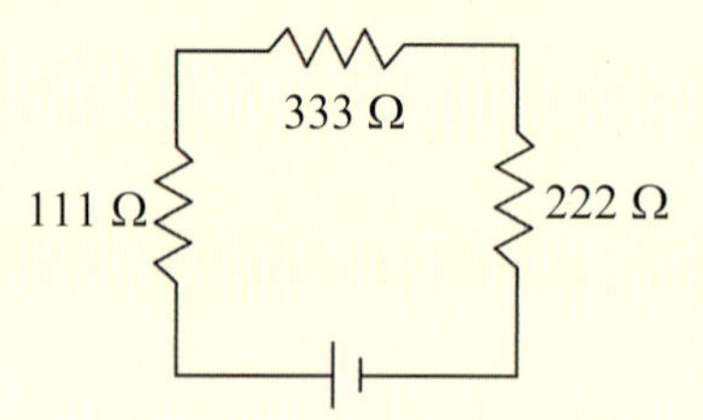

Fuses and Circuit Breakers

Remember, nearly any device that runs off electricity converts kinetic energy into some form of energy (heat, light, etc.) that it can use to perform its function. Thus, most electrical devices can be viewed as a resistors. So the resistors that we represent with our resistance symbol can, in fact, represent toasters, heaters, light bulbs, etc.

Thus, when an electrician looks at the wiring of a house, for example, she sees each appliance as a resistor. When wiring the house, she must be very careful to make sure that the right number of resistors exist on each individual circuit of the house. If too many resistors exist in a parallel circuit, or if too few resistors exist in a series circuit, the circuit may draw too much current. The more current a circuit draws, the hotter it gets. If a circuit gets too hot, a fire can result. In fact, a large number of house fires are caused by a circuit that has too much current running through it. As a result, electricians try to limit the current available to just what the circuit needs.

How is this accomplished? There are two different ways: **fuses** and **circuit breakers**. A fuse is made up of a thin conductor that is very sensitive to heat. The conductor is typically encased in glass and put in a circuit. As more current runs through the conductor, it gets hotter and hotter. If too much current runs through the fuse, the conductor will break. This is sometimes called "blowing a fuse." Once this happens, the fuse acts like an open switch, because current can no longer pass through it. Thus, when a fuse blows, the entire circuit ceases to function, because the open-switch effect of the broken fuse keeps current from running through the circuit. Study Figure 15.10 to see how this works.

FIGURE 15.10
A Fuse in a Circuit

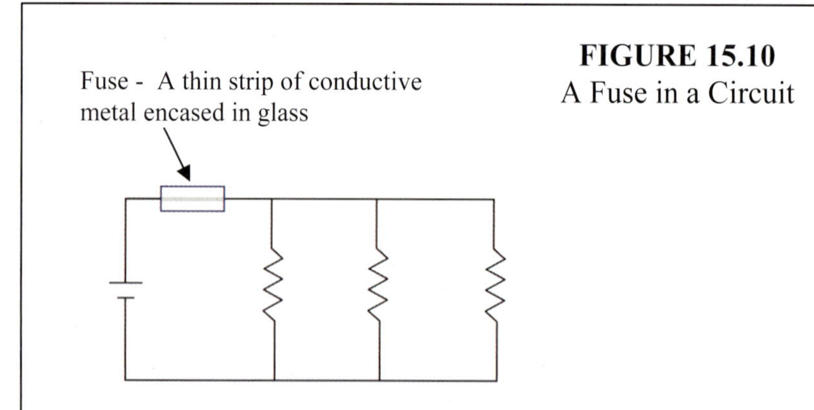

This circuit works fine. The metal in the fuse conducts electricity, so current can run from one side of the battery, through the parallel circuit, to the other side of the battery.

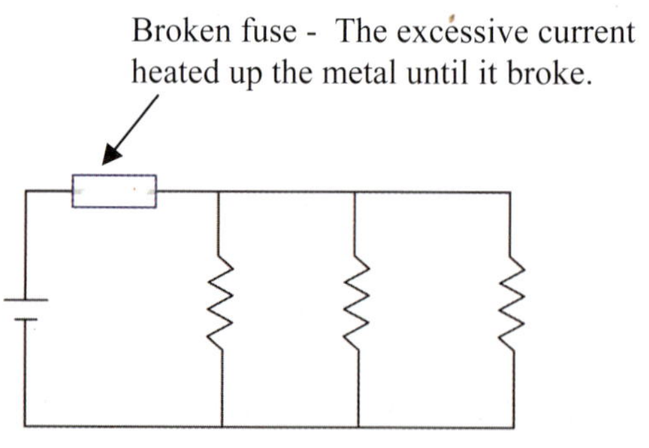

This circuit does not operate. Since the metal in the fuse is broken, the fuse no longer conducts electricity. Thus, the circuit is now open, and no current can flow.

Circuit breakers work on the same principle, but their construction is different. A circuit breaker is a special kind of switch that is made of a material that bends as it heats up. When the switch is closed, electricity flows. If too much electricity flows, however, the material that makes up the switch begins to bend. The hotter the material gets, the more it bends. When the metal bends to a predetermined point, it flips the switch open. At that point, current can no longer flow because the switch is open. When this happens, we say that the circuit breaker has been "tripped."

Whether you use a fuse or a circuit breaker, the effect is the same. If too much current runs through the circuit, a break in the circuit is made. This stops any more current from flowing through the circuit. In this way, the circuit is protected from drawing too much current and therefore getting too hot. The main difference between fuses and circuit breakers is that once a fuse breaks, it is no good anymore. You must throw the fuse away and replace it with a new fuse if you want the circuit to work again. A circuit breaker, however, is reusable. If a circuit breaker trips, you can simply reset it again. Of course, if you do not fix the problem that caused too much current to flow in the first place, it will simply trip again.

Now you might be asking yourself why circuits cause household fires if we have fuses and circuit breakers. After all, if these devices stop current from flowing before the circuit gets too hot, how is it possible for circuits to cause fires? Well, often it is because people by-pass these safety precautions. For example, if you have an older home, your electricity is still probably controlled by a fuse box. In order to keep from having to replace fuses when they break, some people simply stick pennies in the fuse box where the fuse should go. The penny, since it is made out of copper, will conduct electricity and allow the circuits in the house to run. Since copper does not break until it reaches really high temperatures, however, it does not protect against too much current being drawn. Thus, if you overload a circuit, the penny will not break. Instead, the circuit itself will get too hot and cause a fire. Replacing fuses with pennies (or other conductors) is very dangerous. It should never be done!

Another thing that people do to bypass these safety precautions is to replace a fuse or circuit breaker with one that is rated for a much higher current. For example, many household circuits run on fuses or circuit breakers rated for 10 amps. If more than 10 amps run through the fuse, it breaks. If more than 10 amps runs through the circuit breaker, it trips. You can, however, go down to the hardware store and buy fuses and circuit breakers rated for much higher currents (60 amps, for example). Often, because a fuse keeps breaking or a circuit breaker keeps tripping, a person will replace the fuse or circuit breaker with one that has a higher rating. As a result, the fuse will no longer break (or the circuit breaker will no longer trip), but the circuit then has more current than it can stand. Once again, the circuit can heat up and cause a fire.

The lesson to be learned is that fuses and circuit breakers exist for a reason. If your fuse keeps breaking or your circuit breaker keeps tripping, this should tell you that you are overloading a circuit. You fix this problem *not by bypassing the safety of the fuse or circuit breaker*, but by finding out what is overloading the circuit and fixing that problem. Electronic appliances, for example, can break down and start drawing way too much current. Thus, a tripped circuit breaker might be telling you that one of your appliances needs repair. Also, if you have way too many things plugged into an outlet, you might be drawing more current than the wires in the circuit can handle. You fix this problem by unplugging some of the things that you have plugged in.

Current and Power in Series and Parallel Circuits

Now that you know about fuses and circuit breakers, it is time to learn how an electrician determines how much current a circuit should be given. You learned in a previous section that if you know the voltage and resistance of a circuit, you can calculate the current that should run through it and the power that it will consume. You also just recently learned how you can take several resistors and reduce them to an effective resistor for the entire circuit. We can now put all of these things together so as to analyze a circuit and determine how much current it should be allowed to have and how much power it will draw.

EXAMPLE 15.4

An electrician is making a circuit that is diagrammed below. She needs to decide how big a fuse to put on the circuit so that it can draw enough current to function but not so much current that it will overheat. She has fuses rated for 2 amps, 4 amps, and 6 amps. Which fuse should she use? How much power will the circuit draw?

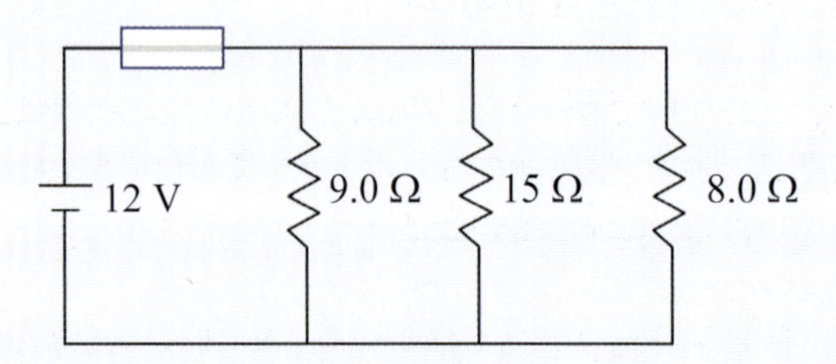

In order to calculate the current in a circuit, I need to use Equation (15.2). To use that equation, however, I need to know the resistance. Which resistance do I use? Well, if I want to calculate the current that travels through the *entire* circuit, then I need the resistance of the *entire* circuit. In other words, I need the *effective* resistance of all three resistors. Thus, I need to reduce these three resistors to their effective resistance. Then I have the resistance of the entire circuit, and that's what I use in Equation (15.2).

First, then, we reduce these three resistors to an effective resistor. You can see that the resistors are in parallel, because the current has its choice of resistors. Since they are in parallel, their equivalent resistance is given by Equation (15.9).

$$\frac{1}{R_{\text{effective}}} = \frac{1}{R_1} + \frac{1}{R_2} + \frac{1}{R_3}$$

$$\frac{1}{R_{\text{effective}}} = \frac{1}{9.0\ \Omega} + \frac{1}{15\ \Omega} + \frac{1}{8.0\ \Omega}$$

$$\frac{1}{R_{\text{effective}}} = 0.30\ \frac{1}{\Omega}$$

$$R_{\text{effective}} = 3.3\ \Omega$$

This circuit, then, has an overall resistance of 3.3 Ω. That's what we use to calculate the current.

$$V = IR$$

$$12\text{ V} = I \cdot (3.3\ \Omega)$$

$$I = \frac{12\text{ V}}{3.3\ \Omega} = 3.6\frac{\text{V}}{\Omega} = 3.6\text{ A}$$

So the circuit should draw 3.6 amps of current. If the electrician used the 2-amp fuse, the circuit would never work because the fuse would keep blowing. The electrician could use the 6-amp fuse, because it would always allow 3.6 A to pass through it. However, if something went wrong and the circuit started drawing too much current, the 6-amp fuse would not blow for a while. This might damage the appliances represented by the resistors or, worse, overheat the circuit and cause a fire. Thus, using the 4-amp fuse would allow the circuit to run while at the same time offering the best protection against too much current being drawn.

Now that we have the current drawn by the circuit, we can easily calculate the power. We actually could use either Equation (15.3) or (15.4) to make the calculation, since we have current, voltage, and resistance. I will use Equation (15.3):

$$P = IV = (3.6\text{ A})\cdot(12\text{ V}) = 43\text{ A}\cdot\text{V} = 43\text{ Watts}$$

When running properly, then, this circuit draws 43 Watts of power.

I hope that you now see why you had to learn how to reduce a circuit to its equivalent resistance. Since most circuits in the real world have several appliances (which to an electrician means several resistors), an electrician must be able to reduce the circuit to its equivalent resistance to calculate the current that will flow through it. This will tell the electrician what kind of fuse to put on the circuit and how much power the circuit will draw.

ON YOUR OWN

15.7 An electrician has fuses that are rated at 5 amps, 10 amps, 15 amps, and 20 amps. Which fuse should be placed on the circuit below?

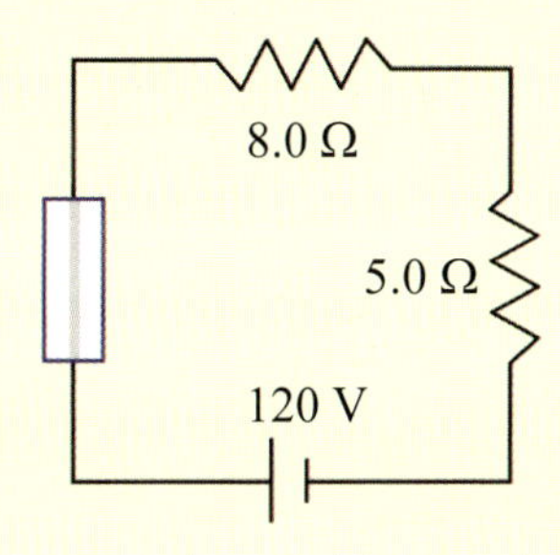

15.8 How much power will the following circuit draw?

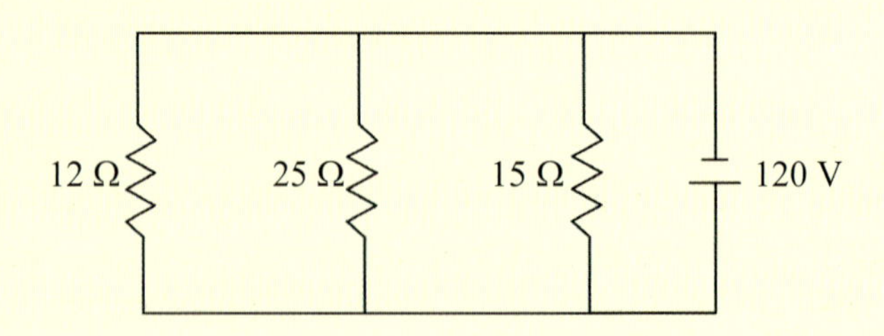

Analyzing More Complicated Circuits

Now it turns out that things are always more complicated than this in the real world. In real electric circuits, there are many other components. Capacitors are used to store charge and make the voltage smooth in most electronic circuits. Also, there are components such as transistors and integrated circuit chips that are used in modern-day circuitry. In other words, things get a lot more difficult. Even if we limit ourselves to looking just at resistors, things still get more complex. In the real world, circuits are not just series circuits or just parallel circuits; they are a mixture of the two. To give you a small taste of just how complicated this can get, in this final section we will predict the current and power used by circuits that only have resistors in them, but the resistors are arranged in a more complex way.

EXAMPLE 15.5

Determine the power drawn by the following circuit.

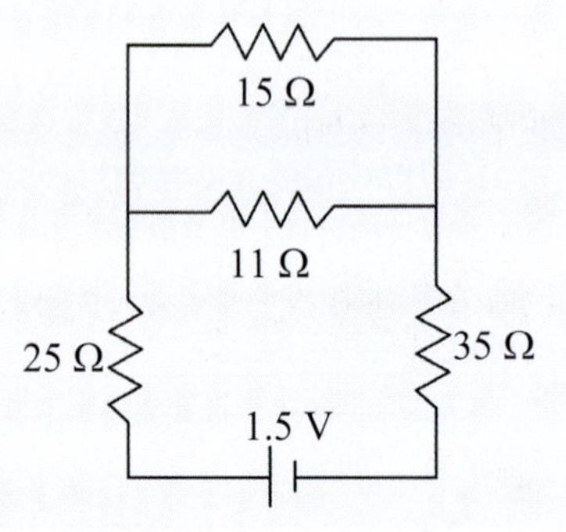

In order to determine the power consumption of a circuit, we must determine the current draw for the entire circuit. To do this, we need to reduce the circuit to a single, effective resistance and then use Equation (15.2). The problem here is, how do we reduce the circuit? If you look at it, there is a mixture of parallel and series resistors here. How can we get this all down to a single, effective resistance?

We start by figuring out which resistors are in series and which are in parallel. We do this by following the path of the conventional current. Starting at the positive end of the battery and moving around the circuit to the negative end, we see that all of the current must go through the 25-Ω resistor. Thus, this is a series resistor. However, after the current moves through the 25-Ω resistor, it can either branch up and go through the 15-Ω resistor or branch right and go through the 11-Ω resistor. Thus, the

15-Ω and 11-Ω resistors are hooked up in parallel. Regardless of which path the current takes, however, it all comes back together and goes through the 35-Ω resistor. Thus, the 35-Ω resistor is a series resistor, because all of the current travels through it.

This analysis, then, tells us that there are two parallel resistors and two series resistors. To reduce this circuit, we need to convert it to all series resistors. The way we do this is to get rid of the parallel resistors by reducing them to their effective resistance. Thus, for this circuit, we need to reduce the 15-Ω and 11-Ω resistors to their effective resistance. To do that, we use Equation (15.9). I will assume that by now you are comfortable using this equation. Therefore, I will just tell you that the effective resistance of these two parallel resistors is 6.33 Ω. We can therefore simplify the circuit by replacing these two parallel resistors by a single 6.33-Ω resistor:

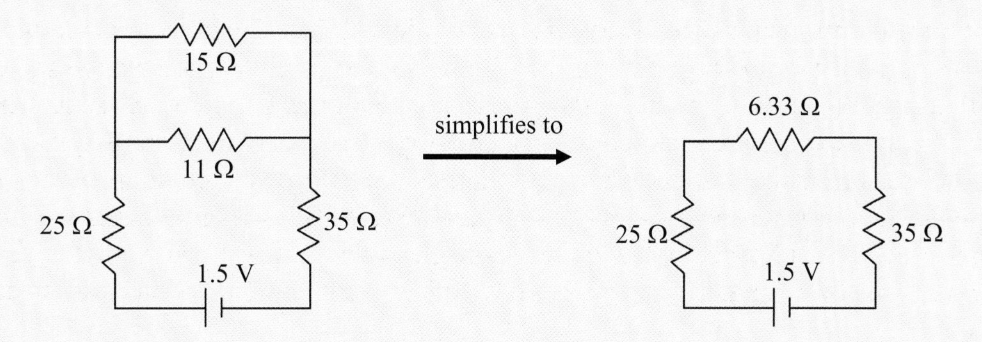

Now this is an easy circuit to deal with. All three of *these* resistors are in series, so we can just use Equation (15.5) to determine the effective resistance of the circuit. Once again, I'll assume you know how to use that equation, so the effective resistance is 66 Ω. We can now use that to determine the current drawn by the circuit:

$$V = IR$$

$$1.5\ \text{V} = I \cdot (66\ \Omega)$$

$$I = \frac{1.5\ \text{V}}{66\ \Omega} = 0.023\ \frac{\text{V}}{\Omega} = 0.023\ \text{A}$$

Now we can finally determine the power.

$$P = IV = (0.023\ \text{A})\cdot(1.5\ \text{V}) = 0.035\ \text{A}\cdot\text{V} = 0.035\ \text{Watts}$$

The circuit draws <u>0.035 Watts</u> of power.

Although this was a long problem, there was nothing new here. We simply used the techniques that we have already learned in a new way. In these more complicated circuits, we need to reduce all parallel resistors to their effective resistors so that, in the end, we are left with just a bunch of resistors in series. After that, it is relatively easy to solve the problem. Try your hand at this with these last two "On Your Own" problems.

ON YOUR OWN

15.9 If an electrician has 1-amp, 3-amp, and 5-amp fuses, which should be used for this circuit?

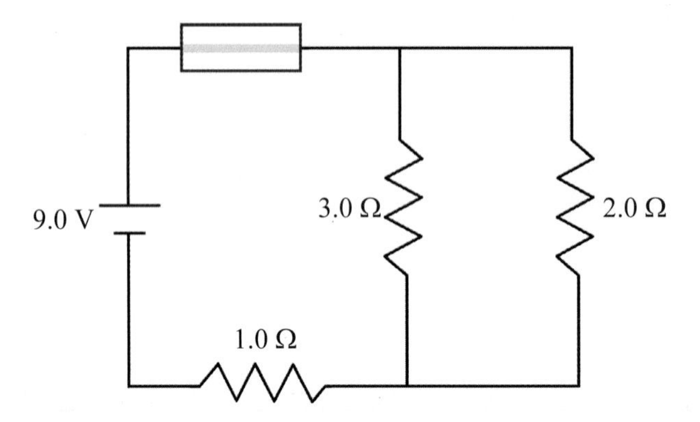

15.10 How much power is used by the following circuit?

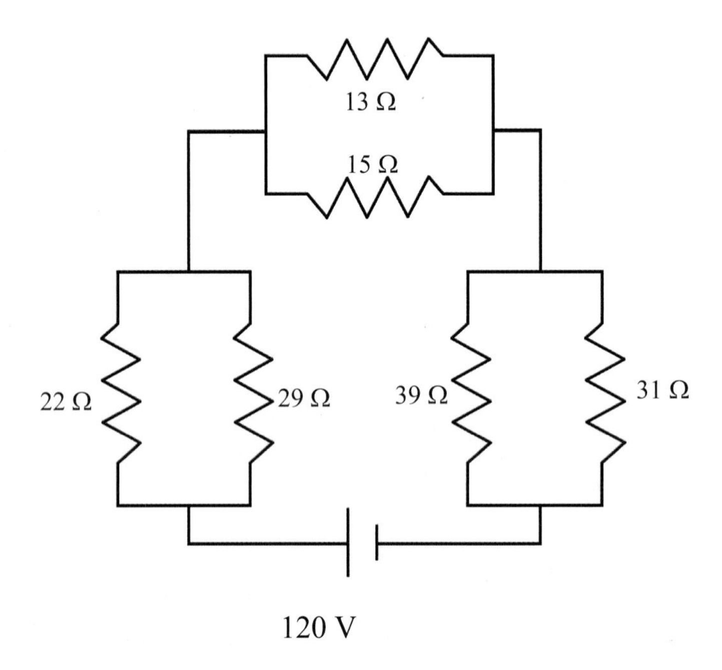

If you enjoyed what you studied in this module, then a career as an electronics expert or an electrical engineer might be right for you. Be warned, however, that what you studied here just barely scratches the surface of this interesting and complicated field. Because of its complexity, a career in electronics can be both financially and intellectually rewarding. If you liked what you saw in this module, you might want to investigate it further.

ANSWERS TO THE "ON YOUR OWN" PROBLEMS

15.1 As was mentioned in the text, even though it is wrong for these kinds of circuits, we still say that current flows from the positive side of the battery to the negative side. Since the long line in the battery symbol represents the positive side and the short line represents the negative side, the current flow is as follows:

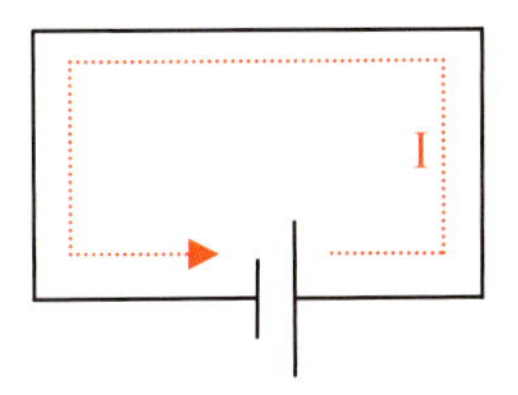

15.2 Drawing the current is easy. We just make it travel from the positive side of the battery to the negative side:

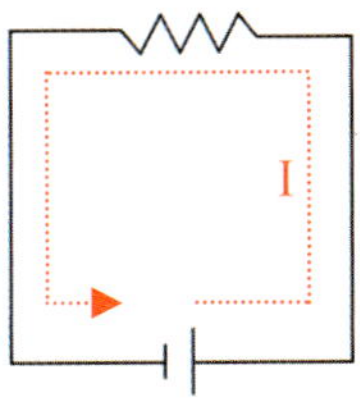

To figure out the resistance, we use Ohm's Law:

$$V = IR$$

$$15.3\text{ V} = (0.023\text{ A})\cdot(\text{R})$$

$$R = \frac{15.3\text{ V}}{0.023\text{ A}} = 670\frac{\text{V}}{\text{A}} = 670\ \Omega$$

Whatever this resistor is, then, it has <u>670 Ω</u> of resistance.

15.3 To solve this problem, we need to use Equation (15.4) because we have current (13 A) and resistance (9.3 Ω), not current and voltage.

$$P = I^2R = (13\text{ A})^2\cdot(9.3\ \Omega) = 1.6 \times 10^3\text{ Watts}$$

This is the power that the heater *draws* from the power source. However, if we assume that the heater is 100% efficient, it is also the heat produced. Thus, the heating power is <u>1.6×10^3 Watts</u>.

15.4 We can use Equation (15.3) for this problem, since we are given the potential (1.5 V) and the current (0.32 A):

$$P = IV = (0.32\text{ A})\cdot(1.5\text{ V}) = 0.48\text{ A}\cdot\text{V} = 0.48\text{ Watts}$$

The power is <u>0.48 Watts</u>.

15.5 This is a parallel circuit. We know it is parallel because the current can travel through any one resistor but not all three. Since the current has a "choice," the resistors must be arranged in parallel. Parallel resistors can be replaced by an equivalent resistor as long as we use Equation (15.9) to calculate that resistance.

$$\frac{1}{R_{\text{effective}}} = \frac{1}{R_1} + \frac{1}{R_2} + \frac{1}{R_3}$$

$$\frac{1}{R_{\text{effective}}} = \frac{1}{111\ \Omega} + \frac{1}{222\ \Omega} + \frac{1}{333\ \Omega} = 0.00901\ \frac{1}{\Omega} + 0.00450\ \frac{1}{\Omega} + 0.00300\ \frac{1}{\Omega}$$

$$\frac{1}{R_{\text{effective}}} = 0.01651\ \frac{1}{\Omega}$$

$$R_{\text{effective}} = 60.57\ \Omega$$

The effective resistance, then, is 60.57 Ω. This is less than the resistance of each individual resistor, which is the way parallel circuits work. Note the significant figures here. Since the resistances have three significant figures, their inverses must also have three. However, after that, we are adding. Thus, we look at decimal place. Since the numbers all have their last significant figure in the hundred thousandths place, the answer must also have its last significant figure in the hundred thousandths place. When we then invert that to get $R_{\text{effective}}$, we must once again count significant figures. At that point, there are four significant figures, so the answer must have four as well.

15.6 This is a series circuit, as evidenced by the fact that the current must travel through all of the resistors to get to the other end of the battery. The effective resistance can be calculated using Equation (15.5):

$$R = R_1 + R_2 + R_3$$

$$R = 111\ \Omega + 222\ \Omega + 333\ \Omega = 666\ \Omega$$

These three resistors, then, can be replaced by a single resistor of 666 Ω.

15.7 In order to determine the kind of fuse to use, we must determine the current that the circuit is supposed to draw. To do that, we must reduce this circuit to its effective resistance. We can do this using Equation (15.5), because the resistors are wired in series.

$$R_{\text{effective}} = R_1 + R_2 = 8.0\ \Omega + 5.0\ \Omega = 13.0\ \Omega$$

This is the effective resistance of the circuit. We can use this in Equation (15.2) to determine the current that the circuit should draw:

$$V = IR$$

$$120 \text{ V} = \text{I} \cdot (13.0 \, \Omega)$$

$$\text{I} = \frac{120 \text{ V}}{13.0 \, \Omega} = 9.2 \frac{\text{V}}{\Omega} = 9.2 \text{ A}$$

This circuit, then, draws 9.2 amps of current. In order for the circuit to work but still provide the maximum protection against too much current draw, the fuse used should be the smallest rating greater than the necessary current. Thus, a 10-amp fuse should be used.

15.8 To determine the power, we need either "I" and "V" so that we can used Equation (15.3), or we need to know "I" and "R" so that we can use Equation (15.4). Thus, we need to figure out "I." Since we need to know the power that the entire circuit is drawing, we need to know "I" for the entire circuit. The only way we can do that is to reduce these three resistors to a single, effective resistance. Then we can use Equation (15.2) to calculate "I" for the entire circuit.

We will need to use Equation (15.9) to reduce these resistors, since they are hooked up in parallel:

$$\frac{1}{\text{R}_{\text{effective}}} = \frac{1}{\text{R}_1} + \frac{1}{\text{R}_2} + \frac{1}{\text{R}_3}$$

$$\frac{1}{\text{R}_{\text{effective}}} = \frac{1}{12 \, \Omega} + \frac{1}{25 \, \Omega} + \frac{1}{15 \, \Omega}$$

$$\frac{1}{\text{R}_{\text{effective}}} = 0.190 \frac{1}{\Omega}$$

$$\text{R}_{\text{effective}} = 5.26 \, \Omega$$

With this effective resistance, we can now use Equation (15.2) to determine the current in the entire circuit:

$$\text{V} = \text{IR}$$

$$120 \text{ V} = \text{I} \cdot (5.26 \, \Omega)$$

$$\text{I} = \frac{120 \text{ V}}{5.26 \, \Omega} = 23 \frac{\text{V}}{\Omega} = 23 \text{ A}$$

Now we can finally calculate the power drawn by the circuit. At this point, we can use either Equation (15.3) or (15.4), because we have I, V, and R for the entire circuit. I'll use Equation (15.3):

$$\text{P} = \text{IV} = (23 \text{ A}) \cdot (120 \text{ V}) = 2.8 \times 10^3 \text{ V} \cdot \text{A} = 2.8 \times 10^3 \text{ Watts}$$

The circuit, then, uses 2.8×10^3 Watts of power.

15.9 When working these complex circuits, we first need to determine where the parallel resistors are so that we can reduce them to their effective resistance. As the current moves from the positive end of the battery, it first encounters the fuse. Then it has a choice. It can go to your right through the 2.0-Ω resistor or go down through the 3.0-Ω resistor. These two resistors, then, are in parallel. Regardless of the choice that the current makes here, however, all of the current must go through the 1.0-Ω resistor, so that is a series resistor. To solve this circuit, then, we first must reduce the two parallel resistors to their equivalent resistance. I assume you know how to use Equation (15.9) by now, so I'll just tell you that the equivalent resistance of these two resistors is 1.2 Ω. The circuit, therefore, reduces to:

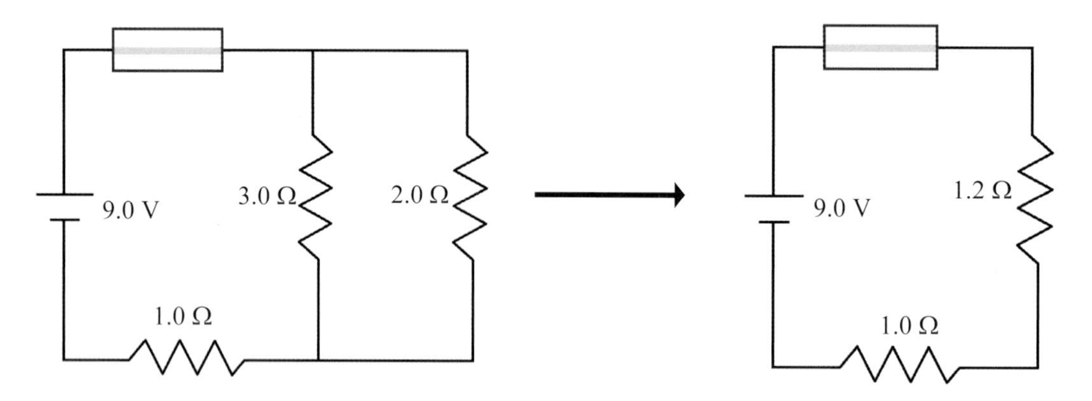

Now we have a simple series circuit. We can use Equation (15.5) to learn that the effective resistance of this circuit is 2.2 Ω. We can now use that resistance to determine the current that this circuit will draw:

$$V = IR$$

$$9.0\text{ V} = I \cdot (2.2\ \Omega)$$

$$I = \frac{9.0\text{ V}}{2.2\ \Omega} = 4.1\frac{\text{V}}{\Omega} = 4.1\text{ A}$$

Since the circuit draws 4.1 A, we need to use the 5-amp fuse.

15.10 This is a complicated circuit, but we will do the same thing that we did with the previous circuit. As the current leaves the positive side of the battery, it immediately has a choice between traveling through the 22-Ω resistor or the 29-Ω resistor. Thus, these are two parallel resistors. After that, the current comes back together, but it is immediately hit with another choice between the 13-Ω or the 15-Ω resistor. That means that those resistors are parallel. After that, the current has yet another choice between the 39-Ω resistor and the 31-Ω resistor. We therefore have three sets of parallel resistors. We need to simplify this circuit by reducing each set of parallel resistors to its effective resistance.

Using Equation (15.9), we find that the equivalent resistance of the first two parallel resistors is 13 Ω; the equivalent resistance of the second set of parallel resistors is 6.94 Ω, and the equivalent resistance of the last two parallel resistors is 17 Ω. Thus, we can reduce the entire circuit to three series resistors, as shown below:

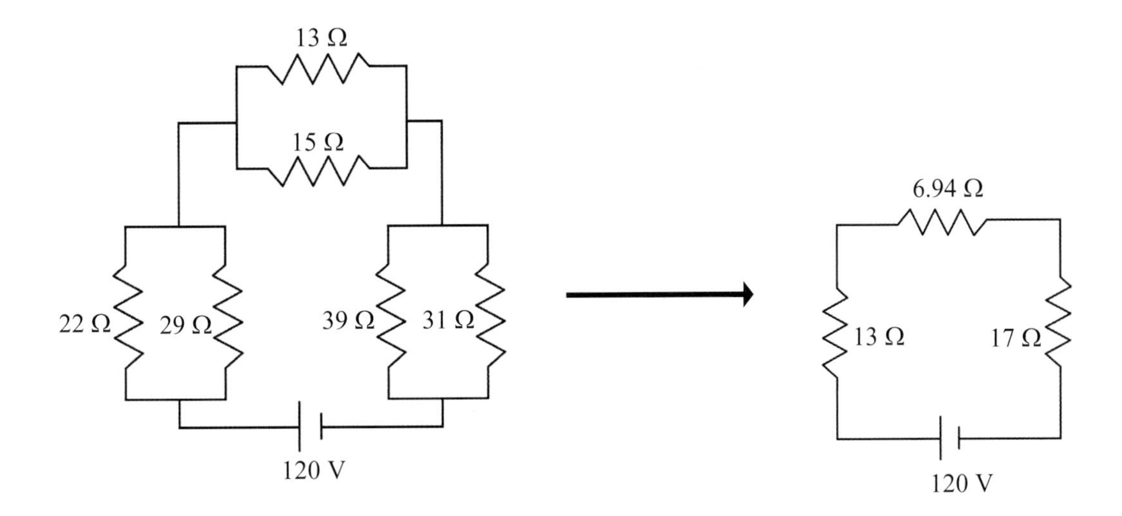

Now the circuit is really simple. It is a series circuit with an effective resistance of 37 Ω. Now we can determine the total current:

$$V = IR$$

$$120\text{ V} = I\cdot(37\ \Omega)$$

$$I = \frac{120\text{ V}}{37\ \Omega} = 3.2\,\frac{\text{V}}{\Omega} = 3.2\text{ A}$$

Now that we have the current, we can determine the power.

$$P = IV = (3.2\text{ A})\cdot(120\text{ V}) = 380\text{ V}\cdot\text{A} = 380\text{ Watts}$$

This circuit draws <u>380 Watts</u> of power.

REVIEW QUESTIONS

1. What is conventional current and why do we use it?

2. What is electric current? What units do we use to measure it?

3. Explain the concept of resistance in your own words.

4. How does a switch turn a light bulb on and off?

5. What is the difference between a series circuit and a parallel circuit?

6. You observe a string of lights and notice that some are lit while others are burnt out. Is this string of lights a series circuit or a parallel circuit?

7. Two identical bulbs are placed in series in a circuit. Later on, the circuit is rearranged. It uses the same battery, but now, the bulbs are placed in parallel. Compare the brightness of the bulbs when they are in series to their brightness when they are in parallel.

8. Why do we use fuses? Explain how they work.

9. What is the main difference between a fuse and a circuit breaker?

10. What does "effective resistance" mean?

PRACTICE PROBLEMS

1. What is the current in the following circuit? Draw its path.

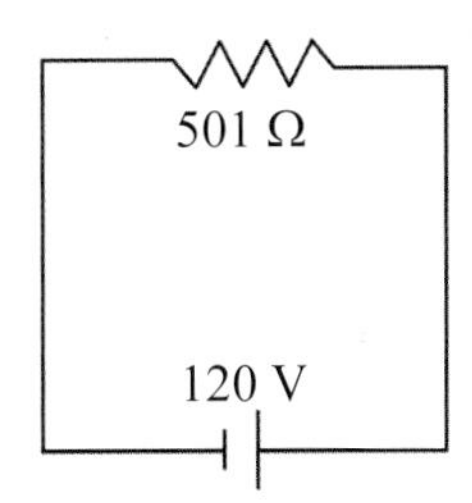

2. A circuit runs on a 9.0-volt battery. If a physicist measures the current to be 0.50 amps, what is the resistance of the circuit?

3. A motor draws 315 Watts of power. If it runs on a 120-volt power source, what current does it draw?

4. A 15-Ω resistor draws 1.2 amps of current. How much power does it use?

5. Determine the effective resistance of the following circuit:

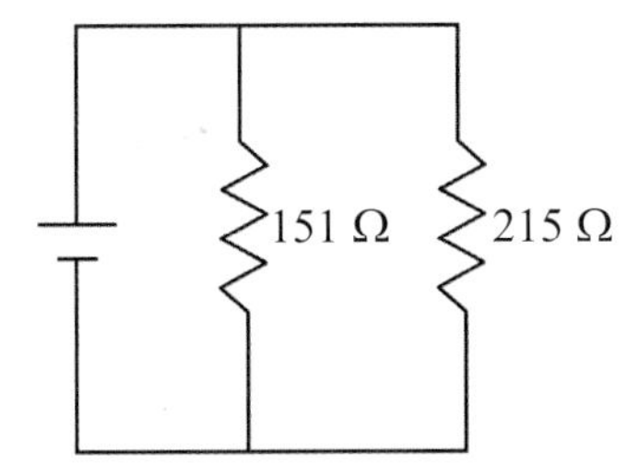

6. Determine the effective resistance of the following circuit:

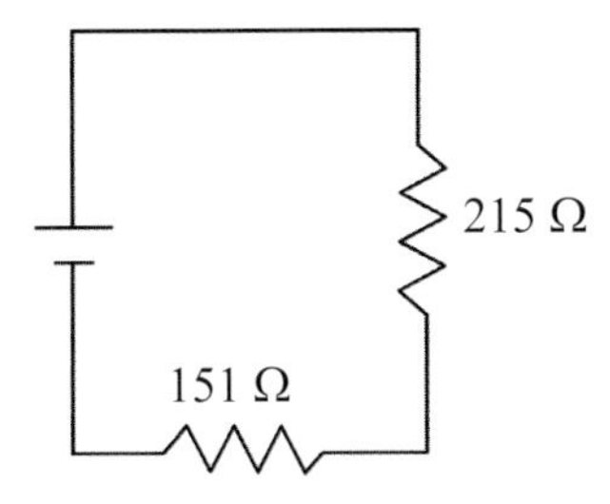

7. Determine the power drawn by the following circuit:

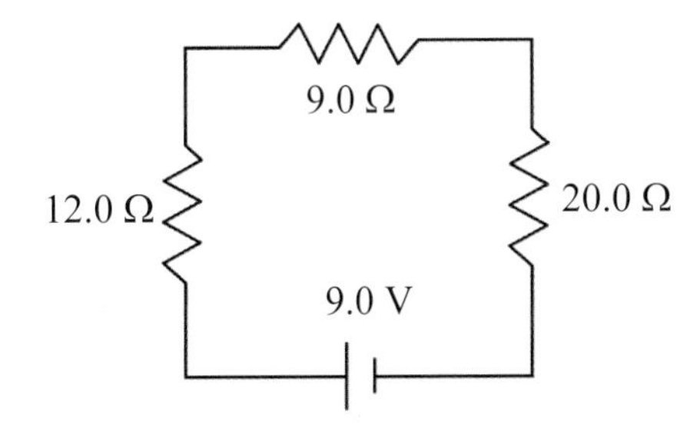

8. Determine the power drawn by the following circuit:

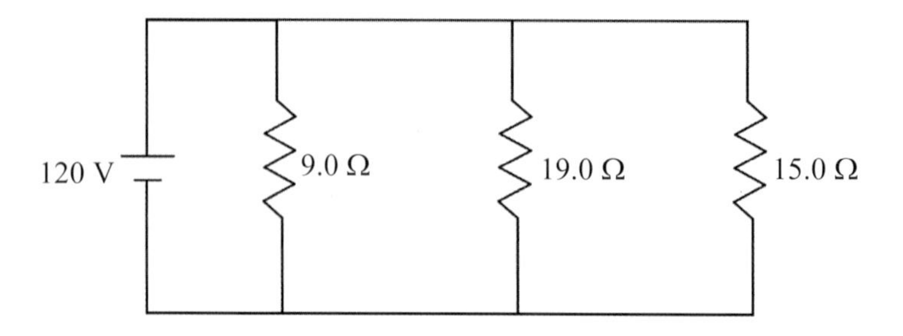

9. If an electrician has 1-, 5-, and 10-amp fuses, which should be used to protect this circuit?

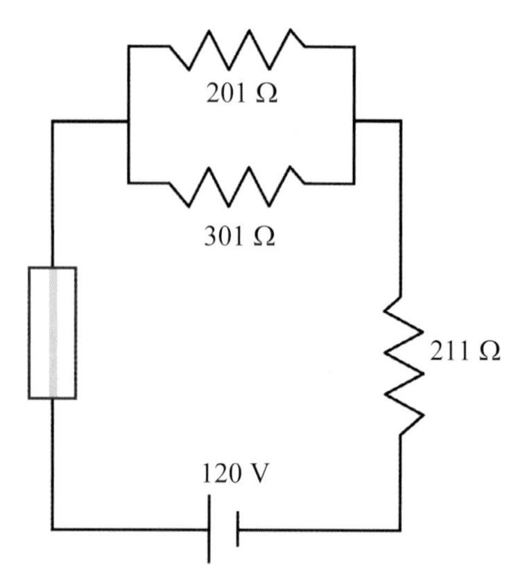

10. What is the power drawn by this circuit?

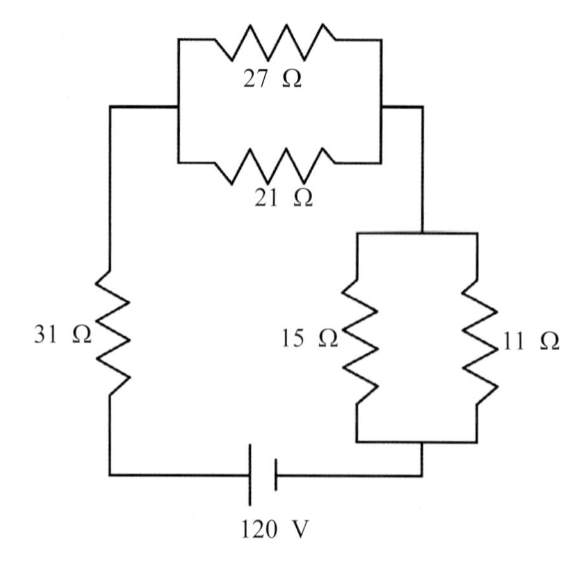

MODULE #16: Magnetism

Introduction

We've spent three modules learning about electricity, but there is one thing that we've left out. A study of electricity goes hand-in-hand with a study of magnetism. That's what I want to concentrate on in this module. Now we all have some experience with magnets. Every family I know has lots of magnets stuck to the refrigerator. I'm sure you've played with these magnets at one time or another and have gotten some sense of how magnets behave. In this module, however, we will learn about magnets in a little more detail. We will see *how* magnets become magnets, what magnetic fields look like, and the use of magnets for the production of electricity.

As you begin to go through this module, you will notice that it is different from the others that you have read. There is very little math in this one. There are also very few "On Your Own" problems. Finally, you will see no practice problems at the end of this module, only some review exercises. Why? Well, the mathematics that you need to really understand magnetism goes well beyond what I want to do here. Thus, there is no way that I can hold you accountable for the math attached to this subject. Nevertheless, the subject of magnetism is both incredibly fascinating and terribly useful, so I feel compelled to give you at least an introduction to it.

Permanent Magnets

It's best to begin a study of magnetism with something that's familiar to you: the permanent magnet. As I said in the introduction, I'm sure you have some magnets around your house. If you've played with them at all, you'll recognize that there are two sides to a magnet. In physics, we call them the **north pole** and **south pole** of the magnet. In many ways, the poles of a magnet are much like charges. For example, if you point the north pole of a magnet toward the south pole of another magnet, the two poles will attract each other. If, however, you point the north pole of a magnet to the north pole of another one, they will repel each other. Thus, as I'm sure you're already aware, opposite poles attract one another and like poles repel one another, just as opposite charges attract one another and like charges repel one another.

Like magnetic poles repel one another, while opposite magnetic poles attract one another.

One big difference between poles and charges is that you can never find just one pole of a magnet hanging around. In Modules 13 and 14, we talked a lot about single positive charges or single negative charges existing by themselves in space. As far as we know, however, this cannot happen with magnets. Whenever you have a magnet, you have a north pole and a south pole. Thus, magnets are called **dipoles**, because magnetic poles always come in pairs: one north and one south. Now I say that this is the case "as far as we know," because there are scientists out there looking for an isolated magnetic pole. Scientists call this the search for a magnetic **monopole**. If such a thing were ever found, it would radically alter our understanding of magnetism, so that's why some scientists look for it. I am doubtful, however, that they will ever find one.

So, as far as we know, magnets always come with two poles. The image we will use to represent a magnet, then, is a long bar, split down the middle. One half will be called the "south pole" of the magnet and the other will be the "north pole." Although magnets come in all shapes and sizes, the drawing in the figure below will be our representation of a magnet. In physics terms, we usually call this a "bar magnet."

FIGURE 16.1
A Bar Magnet

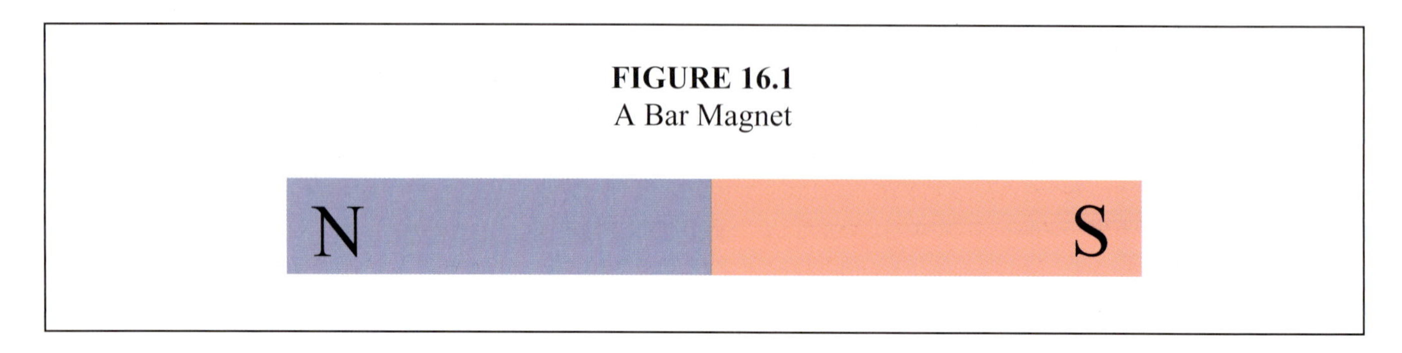

Now you might think to yourself, "What if I take a magnet and split it right down the middle? Wouldn't I have the north pole of the magnet in one hand and the south pole in the other? Couldn't I then take those poles and separate them far away from each other, thereby making two magnetic monopoles? Well, that's a good idea, but it just doesn't work. Look at Figure 16.2 to see why.

FIGURE 16.2
Splitting a Bar Magnet

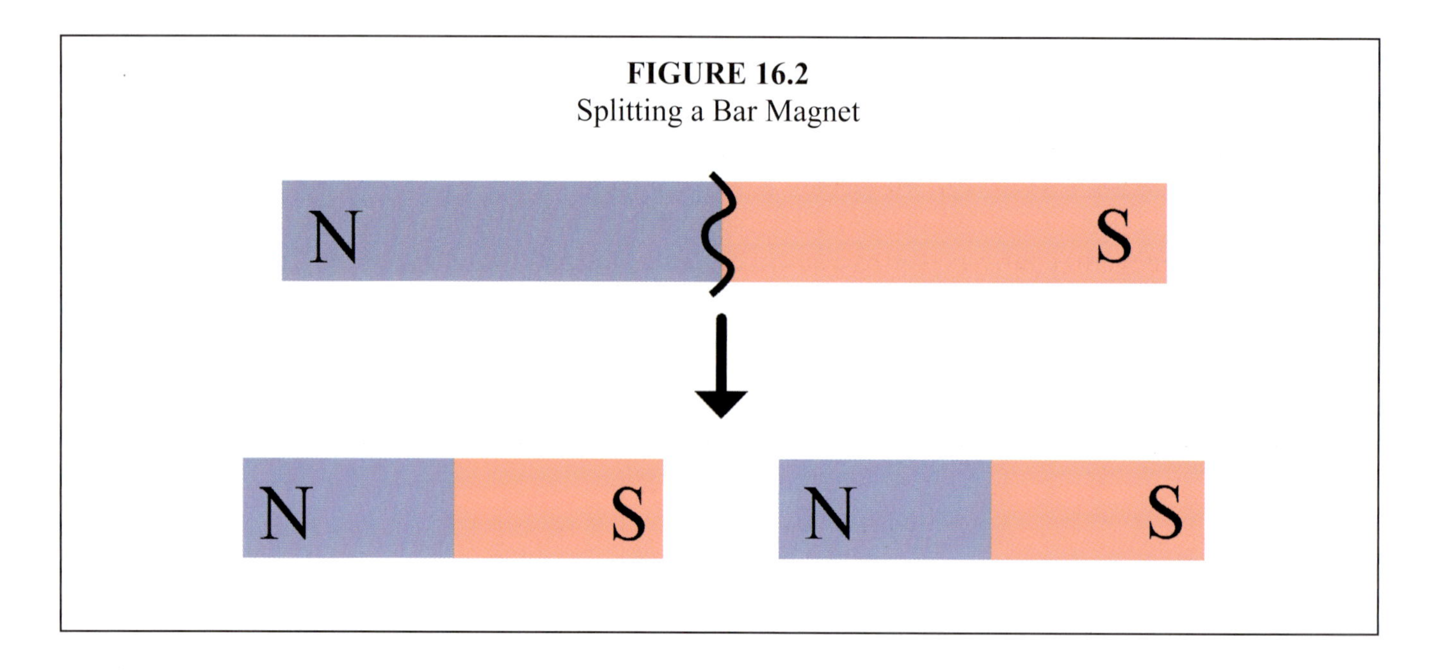

As soon as you split a bar magnet, half of the north pole will turn into a south pole and half of the south pole will turn into a north pole. The result, then, is two smaller magnets, each of which has a north and a south pole. Now there actually turns out to be a reason for this kind of behavior. Hopefully, you will understand this reason before we finish the module, but there are a few other things we need to learn first. Nevertheless, at this point we can at least say that magnetic poles always come in groups of two.

A magnet will always exist as a dipole, meaning that any magnet will always have a north and a south pole.

Magnetic Fields

There are, of course, other similarities between magnets and charges. For example, the strength of the electrostatic force between two charged particles is inversely proportional to the square of the distance between them [see Equation (13.1)]. The same can be said of the magnetic force. Suppose I put the north pole of a magnet near the south pole of another magnet and measured the attractive force between the two poles. If I then doubled the distance between the poles, the strength of the attractive force would decrease by a factor of four. If, on the other hand, I decreased the distance between the poles by a factor of three, the attractive force between them would increase by a factor of nine. Thus, in terms of varying the distance between the objects in question, the electrostatic force and the magnetic force behave the same.

Another similarity between magnetic and electrostatic forces is the fact that they both can be described using fields. If you remember back to Module #13, you learned rules for drawing lines that represented the force field of a charge or group of charges. Well, the same thing can be done with magnets. Drawing magnetic field lines gives you a good picture for the force a magnetic field produces. As with electrical field lines, we have basic rules that govern their drawing:

1. **Magnetic field lines come out of north poles and go into south poles.**

2. **The direction of the magnetic field lines tells you the direction that the north pole of a magnet will point if it is placed in the field and is free to move, and the density of lines tells you the relative strength of the magnetic field.**

3. **Magnetic field lines can never cross.**

Using these rules, we can draw simple magnetic fields. For example, this is the magnetic field produced by a single bar magnet.

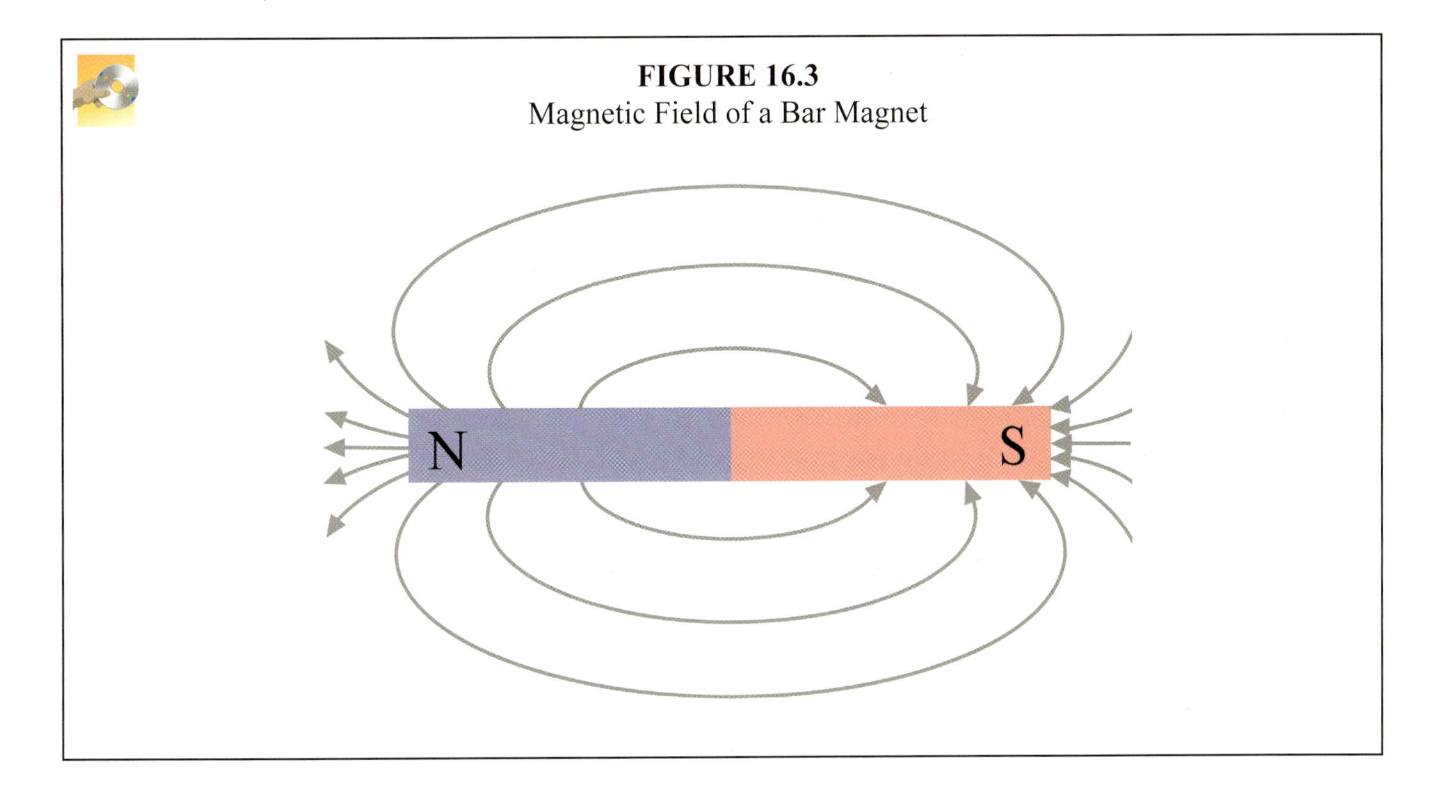

FIGURE 16.3
Magnetic Field of a Bar Magnet

Since every magnet must have both a north and a south pole, you can see that every line which comes out of a north pole will end up in a south pole. You might have, at one time, done the experiment where you sprinkle iron filings on a paper that has a magnet underneath it. The iron filings fall right into these lines, mapping out the magnetic field. Of course, if there is another south pole closer to the north pole, then the lines will go there instead. Look at the example and do the "On Your Own" problem to make sure you can draw magnetic fields.

EXAMPLE 16.1

Draw the magnetic field lines for the situation below.

Magnetic field lines go out from north poles and in to south poles. Since we do not worry about pole strength, we will assume that both north poles have the same number of lines going out of them. The thing to remember, however, is that these field lines will leave the north pole and go to the *nearest* south pole. Thus, the lines coming out the north pole on the right side of the left magnet will go directly to the south pole of the other magnet. Lines will come out of the other side of the north pole as well, however. Since the south pole of the same magnet is closest to the north pole on that side of the magnet, the lines will leave the north pole and enter the south pole of the same magnet. The final result, then, will look like this:

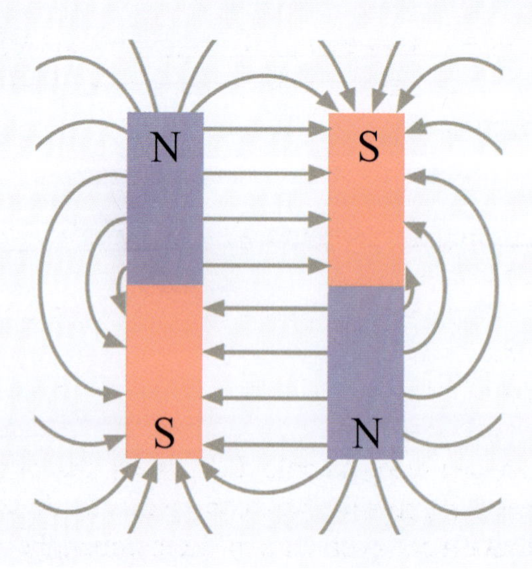

ON YOUR OWN

16.1 Draw the magnetic field of the "horseshoe magnet," pictured below.

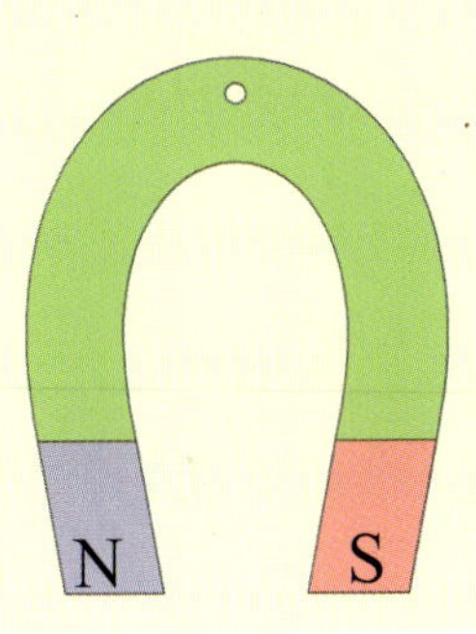

How Magnets Become Magnetic

Now although you might have known much of this already, I expect you don't know what makes a magnet magnetic. For that matter, you probably don't know why some materials react to magnets and some do not. For example, if you put iron near a magnet, the iron will be attracted to it. A piece of wood, however, is not attracted to a magnet, no matter how close you bring the two together. What makes the iron react to the magnet and the wood not?

The first hint of an answer to this question came from Hans Christian Oersted, a Danish physicist. In 1819, he discovered the fact that electricity produces a magnetic field. His experiment was really quite simple. If you have a compass, you can do the experiment yourself.

EXPERIMENT 16.1
Oersted's Experiment

Supplies:

- A compass
- Insulated wire like you used in the experiments from the previous module
- A 1.5-volt battery (Any size cell will do, just make sure it is a 1.5-volt battery. A battery of higher voltage could make the experiment dangerous.)
- Tape

1. Cut a long piece of wire and, using the technique you used in Experiments 15.2 and 15.3, strip ½ inch of insulation from both ends.
2. Lay the wire on a table or desktop so that it passes over the compass. Arrange the wire so that the needle and the wire are parallel.
3. Tape the wire to the tabletop on both sides of the compass. This will ensure that the wire stays over the compass.
4. Take both ends of the wire and get ready to hold the bare conductor against the ends of the battery.
5. As you watch the compass needle, press the bare conductors on the ends of the wire against the ends of the battery so as to make a complete circuit. Remember, the only reason you can do this is because the voltage and current in the circuit are low. *In general, you should not touch the conductor in a circuit with current flowing.*

6. What happened to the needle?
7. Release the wire so that current stops flowing. What happens to the needle? Repeat steps 5-7.
8. Clean up your mess.

What happened in the experiment? If you made a complete circuit with the battery, the compass needle should have swung so that it was perpendicular to the wire. When you released the wire from the battery ends, the needle should have swung back. When you made the electrical connection again, the needle should have swung back around again.

This rather simple experiment demonstrated to Oersted a fundamental principle of physics: moving charged particles make magnetic fields. Now you might wonder how this applies to a permanent magnet like the ones you have experience with. It is easy to see that there are moving charged particles in an electric circuit. After all, that's what current is. Where are the moving charged particles in a refrigerator magnet, however? Well, think about it. The refrigerator magnet is made up of atoms, right? In chemistry, you should have learned that an atom is made up of a positively-charged nucleus and negatively-charged electrons that orbit around the nucleus. The electrons, since they are orbiting the nucleus, are certainly in motion. They are also charged particles. The electrons in an atom, then, are the charged particles in motion that cause a magnet to be magnetic.

Now right away you should question this. In fact, *all matter with which you and I are familiar is made up of atoms*. Since the electrons in an atom are the charged particles in motion that cause magnetism, why aren't *all materials* magnets? First, you must remember that an atom can have many, many electrons in it, depending on what element it is. Each individual electron in the atom, because of its motion, is creating its own magnetic field. Depending on how many electrons are in an atom, and depending on how those electrons are arranged, these magnetic fields can cancel each other out. Thus, even though each electron is generating a magnetic field, another electron in the atom might generate an opposite field. The net result of such a situation, then, would be an atom with no magnetic field.

It turns out that many atoms are like this. Their electron configuration is such that the magnetic fields of the individual electrons cancel each other out. As a result, the material composed of these atoms is not magnetic. A substance made up of such atoms is called **diamagnetic** (dye' uh mag neh' tik).

<u>Diamagnetic substance</u> - A substance that is made up of atoms with no net magnetic field

Even though diamagnetic substances have no net magnetic field, they do *react* to magnetic fields. When a diamagnetic substance is exposed to a magnetic field, the motion of its electrons changes. As a result, the substance produces a weak magnetic field that is opposite the direction of the field to which it is exposed. It turns out that all substances actually have this property, but it is only apparent when the atoms of the substance have no net magnetic field. When the atoms in a substance have a magnetic field, other effects (which I will discuss below) overpower this one.

There are some atoms that have an electron configuration which results in one or more electrons whose magnetic fields have nothing to cancel them. As a result, the atom is magnetic, and the material composed of these atoms is magnetic, to one degree or another.

What do I mean when I say, " to one degree or another?" Well, when magnetic atoms are grouped together in a substance, one of three things can happen. First, the atoms can be arranged so that their magnetic fields cancel each other out. Just as electrons that are arranged properly can cancel out the magnetic fields of other electrons and cause an atom to be nonmagnetic, atoms that are arranged properly can cancel out the magnetic fields of other atoms and cause the entire substance to be nonmagnetic.

Now even though substances such as these have no net magnetic field, they can *become* magnetic, with the help of an external magnetic field. When such a substance is placed in a magnetic field, the individual atoms, because they are magnetic, will interact with the field. This will cause some of them to move, and the magnetic fields of those atoms will cease canceling the magnetic fields of other atoms. Thus, while the external magnetic field exists, the substance is weakly magnetic. As soon as the external field is gone, however, the atoms will reorient themselves so that their magnetic fields once again cancel, and the substance will no longer be magnetic. We call these substances **paramagnetic** (pehr' uh mag neh' tik).

<u>Paramagnetic substance</u> - A substance whose magnetic atoms are arranged so that their individual magnetic fields cancel out. Under the influence of an external magnetic field, however, these substances produce a magnetic field that is in the same direction of the external magnetic field.

The last type of substance I want to discuss is a **ferromagnetic** (fehr' oh mag neh' tik) substance. This kind of substance is made up of magnetic atoms which are arranged so that their individual magnetic fields add to each other rather than cancel each other out. When this happens, we say that the atoms' magnetic fields are aligned. As a result, substances like these are strongly magnetic.

<u>Ferromagnetic substance</u> - A substance whose magnetic atoms are aligned so that their magnetic fields add to one another

Iron, cobalt, nickel, gadolinium, and dysprosium are ferromagnetic.

Now once again, you should be asking yourself a question. "If iron is ferromagnetic, why isn't everything made of iron a magnet?" Bar magnets are usually made of iron, but so are many nails. The bar magnets are definitely magnets, but an iron nail is not. Why? Study Figure 16.4 to find out.

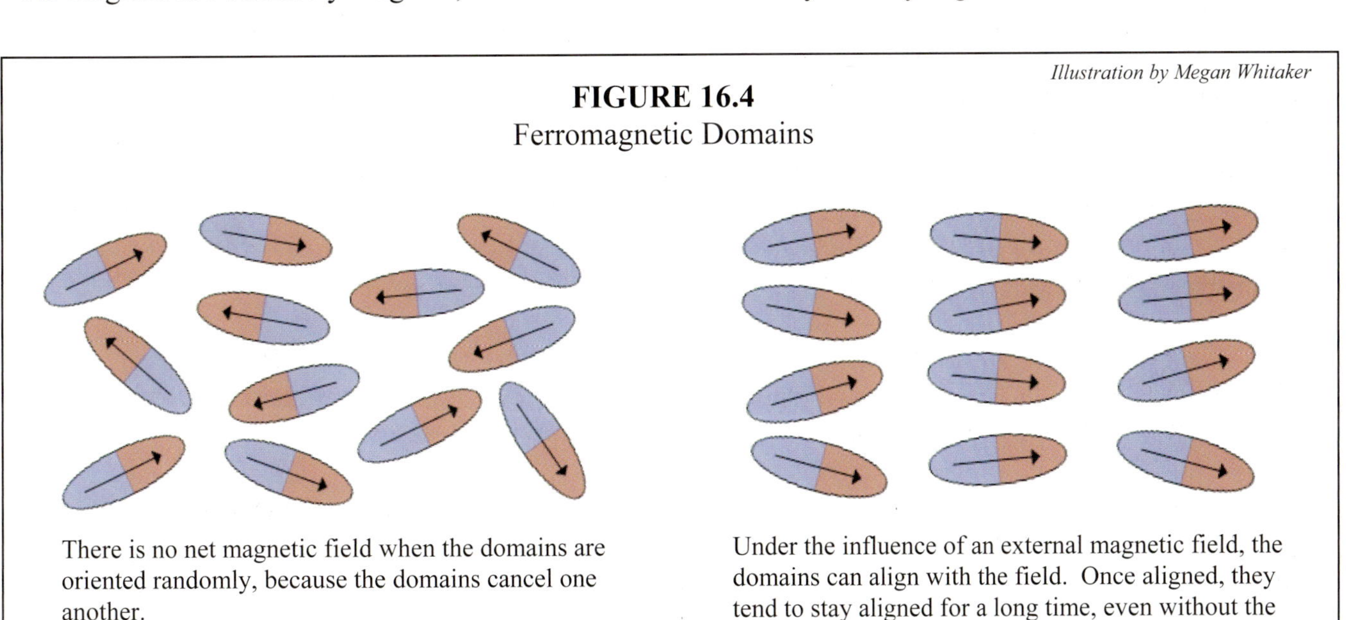

Illustration by Megan Whitaker

FIGURE 16.4
Ferromagnetic Domains

There is no net magnetic field when the domains are oriented randomly, because the domains cancel one another.

Under the influence of an external magnetic field, the domains can align with the field. Once aligned, they tend to stay aligned for a long time, even without the presence of the external magnetic field.

When atoms align their magnetic fields in a material, they do so in small clusters called **domains**. Each domain acts as a small magnet. Now, if all of these domains are aligned, as illustrated on the right side of the figure, then the substance emits a strong magnetic field. This is the case in a permanent magnet. There are many ferromagnetic materials, however, whose domains do not align. Instead, the domains within the substance are arranged so that the magnetic fields cancel, as illustrated on the left side of the figure. Thus, the substance is not a magnet. However, much like a paramagnetic substance, when such a ferromagnetic substance is subjected to an external magnetic field, the domains will align, producing a strongly magnetic substance. If the substance is rigid, these domains will tend to stay aligned even when the external magnetic field is removed. This, in the end, causes a substance that was not a magnet to become one. If the substance is soft and pliable, however, the domains will probably move back to their original configuration, and the substance will no longer be magnetic.

This whole discussion may have been a bit confusing, so I want to sum it all up. Each of the electrons that orbits the nucleus of any atom produces a magnetic field. In some atoms, the electrons are arranged so that these magnetic fields cancel. Substances made up of these atoms are called **diamagnetic** substances. If, however, the electrons in an atom do not cancel out each other's magnetic fields, that atom is magnetic. A substance made of such atoms is either paramagnetic or ferromagnetic. If the atoms are completely misaligned so that their magnetic fields cancel out, then the substance is **paramagnetic**. If, instead, the atoms align themselves in little clusters that each behave as a tiny magnet, then the substance is **ferromagnetic**. If these clusters, called domains, happen to align, then the ferromagnetic substance becomes a magnet.

Paramagnetic substances can have their atoms weakly aligned in the presence of an external magnetic field. Thus, these substances respond weakly to a magnet. The domains of ferromagnetic substances align strongly in the presence of an external magnetic field, and therefore ferromagnetic substances respond strongly to magnets. Finally, if the ferromagnetic substance is rigid enough and the external field is strong enough, a ferromagnetic substance's domains might stay aligned even after the external field is removed. At that point, the ferromagnetic substance is a magnet. This is, in fact, how most commercial magnets are made. Naturally-formed magnets are hard to find, but ferromagnetic material like iron is easy to find. Thus, to make a magnet, a rigid ferromagnetic substance is placed in a strong magnetic field. When removed, the magnetic domains stay aligned, and the substance is a magnet.

To give you a little more experience with these classifications of materials, perform the following experiment.

EXPERIMENT 16.2

Diamagnetic, Paramagnetic, and Ferromagnetic Compounds

Supplies:

- A 1.5-volt battery (Any size cell will do, just make sure it is a 1.5-volt battery. A battery of higher voltage could make the experiment dangerous.)
- Two iron nails (One should be large.)
- Insulated wire like that used in the previous experiment
- A metal paper clip
- A wooden matchstick or toothpick

1. Lay the paper clip, matchstick, and the smaller of the two nails on a table or desk.
2. Take the larger nail and touch it to the paper clip, then pull it away. Did the paper clip stick to the nail as if the nail was a magnet?
3. Take a piece of insulated wire that is about three times the length of the nail and strip ½ inch of insulation from each end.
4. Wrap the wire around the larger nail several times, leaving a few centimeters of wire on each end. Your nail and wire should look something like this:

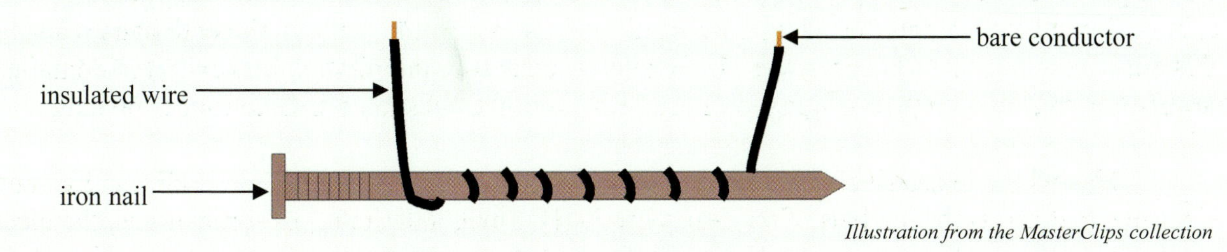

Illustration from the MasterClips collection

5. Use your thumb and forefinger to touch the bare conductor at the ends of the wire to the ends of the battery. That way, electricity will flow through the wire.
6. Wait a few moments and then, with the electricity still flowing, touch the nail to the paper clip. What happens?
7. With the electricity still flowing, touch the nail to the matchstick. What happens?
8. Release the wire from the battery and unwrap the nail.
9. Touch the nail to the paper clip now. What happens?
10. Stretch out the paper clip so that it is relatively straight.
11. Wrap the wire around the straightened paper clip like you did the nail.
12. Once again, touch the bare conductor on the ends of the wire to the battery to start the electricity flowing.
13. Touch the paper clip to the other nail (the one you did not wrap with wire). What happens?
14. Clean up your mess.

What happened in the experiment? When you touched the nail to the paper clip in step 2, the paper clip was not attracted to the nail. Despite the fact that the nail was made of iron, it was not a magnet. However, when you wrapped the wire around the nail and got an electric current running through the wire, the paper clip should have been attracted to the nail. Thus, when you touched the nail to the paper clip in step 6, the paper clip should have stuck to the nail.

What's the explanation here? Well, the electricity flowing through the wire produced a magnetic field, as you learned in Experiment 16.1. The magnetic domains in the iron nail aligned, making the nail a magnet. Thus, the paper clip was attracted to it. The matchstick was not attracted to it, however, so the matchstick is diamagnetic.

What happened in step 9 when you touched the nail that was wrapped in wire to the paper clip? Sometimes this doesn't work, but most likely, the paper clip was still attracted to the nail. Why? Remember, the nail is made of iron, which is ferromagnetic. While in the presence of a magnetic field, its domains aligned. That alignment was not lost when the magnetic field was turned off, so the nail remained a magnet. Thus, the paper clip should have been attracted to it.

What happened when you wrapped the paper clip in wire and started electric current running through the wire? Was the smaller nail attracted to the paper clip (step 13)? No. The paper clip was attracted to the magnet in step 6; however, because the atoms in the paper clip were really misaligned to begin with, the magnetic field could only align them a little. This was enough to make the paper clip attracted to the nail when the nail was a magnet. However, the external magnetic field caused by the electricity just couldn't align the atoms enough to make the paper clip itself a magnet. This tells us that the paper clip is paramagnetic.

The Earth's Magnetic Field

You probably already know that the earth itself is a huge magnet. We know this because a compass, which is just a small magnet that is balanced and free to rotate, always points to the north. If I were to draw the magnetic field of the earth, it would look like Figure 16.5.

Illustration from the MasterClips collection

FIGURE 16.5
The Earth's Magnetic Field

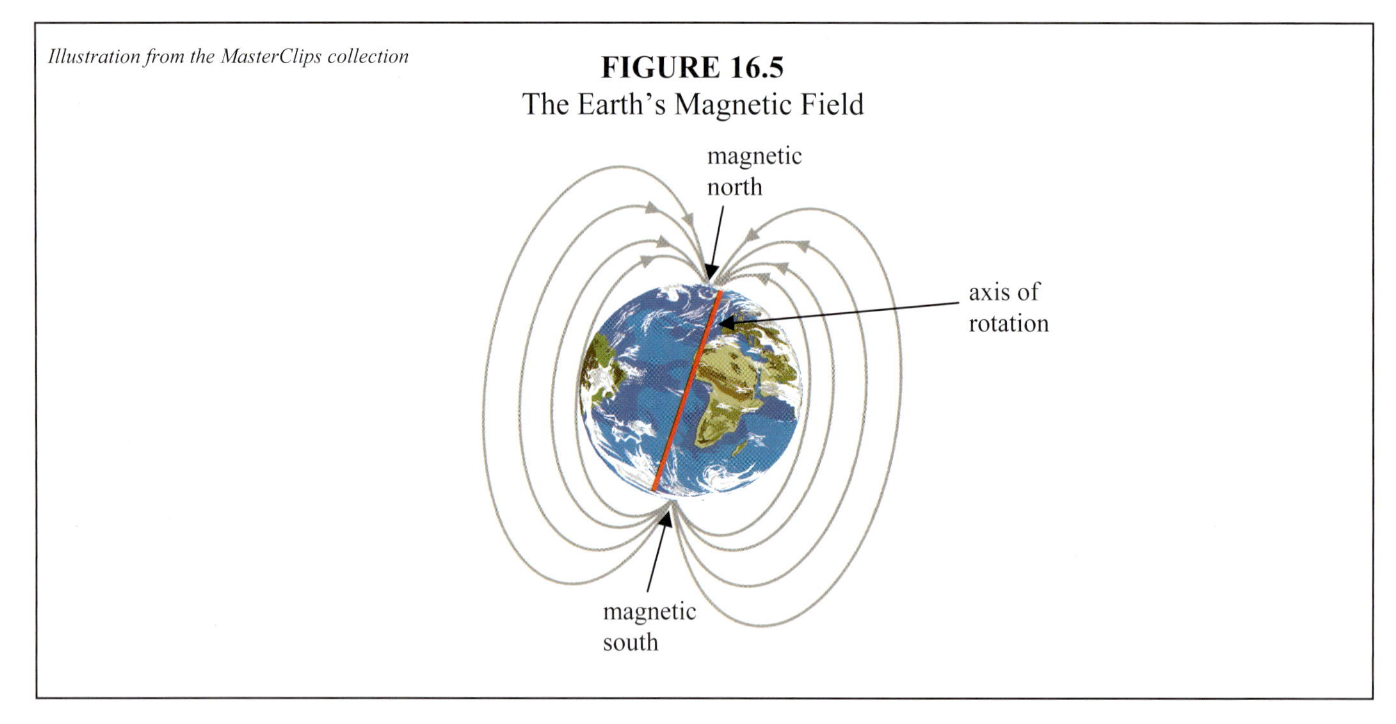

There are two things that you need to be aware of when it comes to the earth's magnetic field and compasses. First, the earth's magnetic field is not aligned with its axis of rotation. As a result, magnetic north is not the same as geographic north. That's why, in the figure, magnetic north is a little to the left of the northern tip of the earth's rotational axis, which is the northern tip of the earth. As a result, the compass needle doesn't truly point north. It points a little to one side of north. Second, magnetic north is actually the south pole of the earth's magnetic field. If you think about this, it makes perfect sense. The north pole of the compass needle always points north. But the north pole of the compass needle is attracted to the *south* pole of the earth's magnetic field. So the north pole of the compass needle will point to the south pole of the earth's magnetic field. Thus, *the south pole of the earth's magnetic field is actually magnetic north*!

Even though you probably knew that the earth has a magnetic field, you probably don't know why. The earth is not made of some magnetic substance. Its magnetic field is not caused by permanent magnets scattered throughout its surface. In fact, we're not entirely sure what causes this magnetic field. We have a pretty good idea, though.

Based on many different experimental studies, scientists are convinced that the earth has a solid core (center) that is surrounded by a molten outer core which is mostly nickel and iron. That by itself would not generate the magnetic field that we observe for the earth. It is assumed, therefore, that this molten core has many, many charged particles in it, and the entire core is swirling. The swirling motion of the charged particles in the outer core produces a current, which generates the magnetic field that we observe.

There are, of course, a few mysteries to this idea. First of all, what got this swirling motion started in the first place? We are not sure. Also, we know that over the past 200 years, the magnetic field of the earth has been decreasing. Why? We're not really sure about that, either. There are two main theories that attempt to answer these questions.

One theory, called the "dynamo theory," says that the earth's rotation and heat transfer are supposed to circulate the outer core. The positive and negative charges in the outer core are assumed to circulate unevenly, which produces a net electric current. That current produces the magnetic field. If the dynamo theory is true, the earth's magnetic field fluctuates because of the uneven nature of the positive and negative charges' circulation. In fact, it fluctuates so much that it can even reverse, so that the south and north poles switch places. If that ever happened, the compass needle would point south, not north. According to this theory, then, the decay that we have been measuring over the past 200 years is just a part of the natural fluctuation of the magnetic field.

There is some evidence for the dynamo theory. For example, geologists have found evidence that the earth's magnetic field has reversed in the past. However, no working model that accurately reproduces the behavior of the earth's magnetic field and the magnetic fields (or lack of them) for the other planets in our solar system has been produced, despite fifty years of research into the problem. For example, when the spacecraft Voyager flew by Mercury and measured that it had a magnetic field, this surprised those who believed in the dynamo theory. Mercury's rotation was assumed to be too small to produce a magnetic field via a dynamo. Even though dynamo theorists have tried to produce a dynamo model of Mercury's magnetic field, the actual field is much larger than anything that the dynamo theory can produce.

The other theory, called the "rapid decay theory," assumes that the earth was created only a few thousand years ago, and that the process by which the earth was created started the outer core in motion. After that, the motion of the outer core has been slowly decreasing due to the electrical resistance of the outer core. Thus, according to the rapid decay theory, the decrease that we have been measuring over the past 200 years is just a portion of the long decrease that has been occurring since the time of creation.

There is strong evidence to support the rapid decay theory. For example, making some assumptions about how the earth was created and that the creation was only a few thousand years ago, the rapid decay theory accurately predicts both the strength of the earth's magnetic field today as well as the observed decay over the past 200 years. In addition, the rapid decay theory accurately predicts the presence or absence of magnetic fields of the other planets in the solar system, and for those that have a magnetic field, it accurately predicts their strengths. In fact, when these calculations were done, the magnetic fields of both Uranus and Neptune had not been measured. Nevertheless, the originator of the rapid decay theory (D. Russell Humphreys) made predictions of what they should be, based on his theory. When the spacecraft Voyager later made measurements of these magnetic fields, the predictions of the rapid decay theory were shown to be correct, while the predictions of various versions of the dynamo theory were not.

The rapid decay theory also allows for reversals of the magnetic field, but they must result from cataclysmic tectonic activity. Humphreys postulates that such activity could have happened during the time of Noah's Flood. Such cataclysmic activity would result in rapid reversals of the magnetic field under the assumptions of the rapid decay theory. However, under the assumptions of the dynamo theory, the reversals should be very slow. Interestingly enough, starting in 1989, two geophysicists studying lava flows found evidence for rapid reversals of the earth's magnetic field. They continued their work, and by 1995, they had compiled an impressive set of evidence that the earth's magnetic field has reversed rapidly in the past. The rapidity of the reversals is not compatible with the dynamo theory, but it is compatible with the rapid decay theory.

A detailed discussion of both the dynamo theory and the rapid decay theory goes beyond the scope of this course. However, if you are interested, the course website mentioned in the "Student Notes" section of this book has links to detailed discussions of both models. Up to this point, we do not know which model is correct. However, I find the evidence for the rapid decay theory much more compelling than that for the dynamo theory.

Even though we don't completely understand the earth's magnetic field, we know that it is absolutely essential for life to exist on this planet. There are certain high-energy charged particles that come from the sun. These particles, called cosmic rays, are detrimental to most biological life forms. Because the moving charged particles create a magnetic field, they are repelled by the earth's magnetic field. As a result, they veer away from and do not hit the earth. This protects all of life on earth from these devastating rays. Clearly, then, the earth has a magnetic field because it *needs* one, and the Designer put it there. One of these days, we will hopefully understand better *how* He did it.

The Magnetic Field of a Current-Carrying Wire

The two experiments that you have done in this module clearly show that when current flows through a conductor, a magnetic field is generated. The shape of this magnetic field is quite interesting. Suppose we put compasses around a wire and then allowed a current to pass through the wire. Here's what we would see:

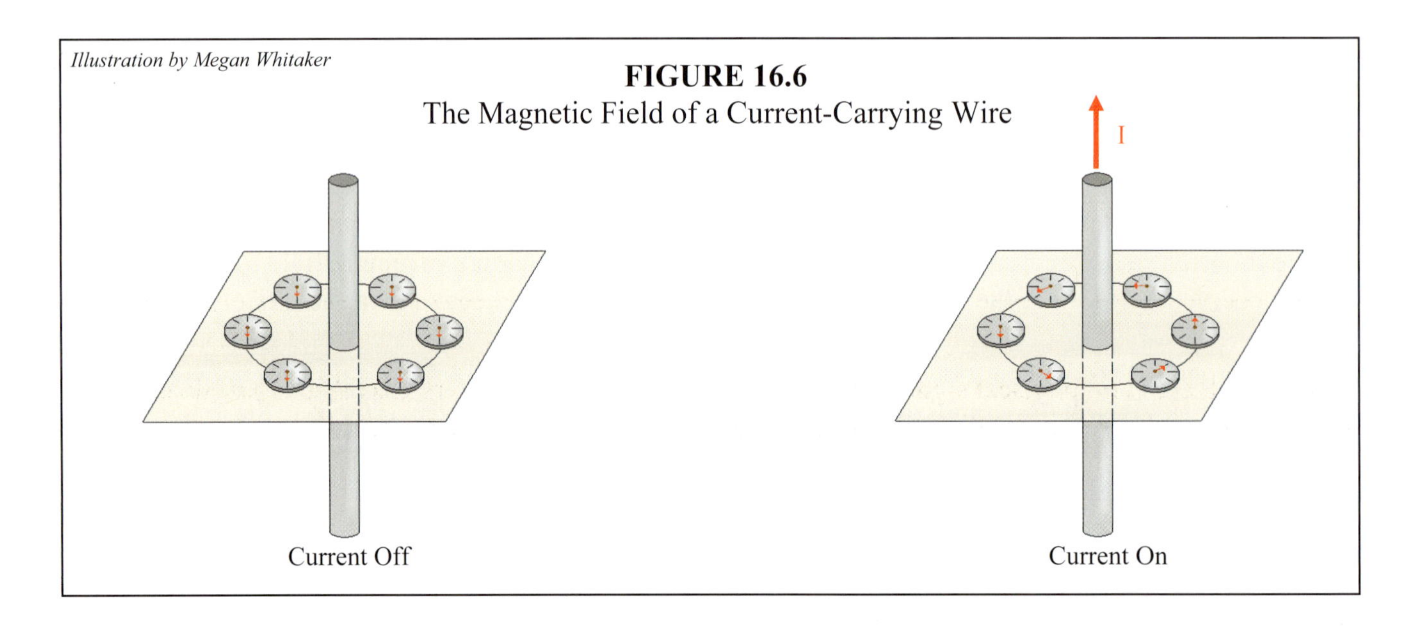

FIGURE 16.6
The Magnetic Field of a Current-Carrying Wire

When the wire is not carrying electric current, the needles on the compasses all point north because of the earth's magnetic field. However, when current passes through the wire, the magnetic field generated by the current is much stronger in the vicinity of the compasses than the earth's magnetic field. As a result, the compass needles spin in response to the magnetic field, much like you observed in Experiment 16.1. Notice how they point. The compass needles point in a circle that is going counterclockwise. Thus, the magnetic field of a current-carrying wire is circular, with the wire at the center of the circle.

It turns out that predicting the direction of the magnetic field from a current-carrying wire is quite easy. All you have to do is learn the **right-hand rule**.

Right-hand rule - The direction of the magnetic field lines for a current-carrying wire can be determined by pointing the thumb of your right hand in the direction of the current. Your fingers will curl in the direction of the magnetic field lines.

Notice that if you apply the right-hand rule to the figure above, you will see how I determined the direction of the magnetic field lines.

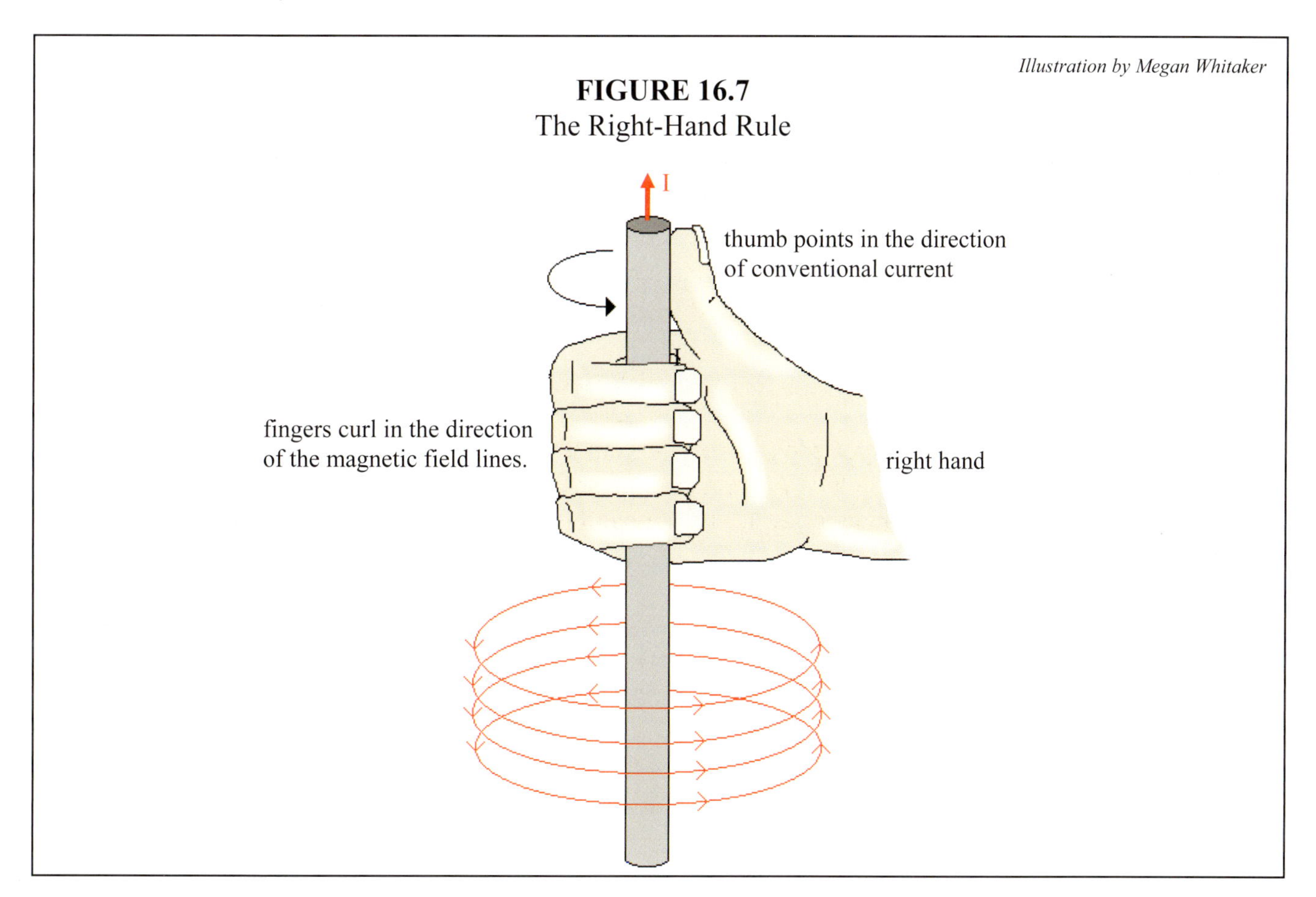

FIGURE 16.7
The Right-Hand Rule

Of course, there is no physical significance to the fact that the right-hand rule works. It just does. It's an easy way to determine the direction of a current-carrying wire's magnetic field lines.

Now I need to point out that some physics courses teach this concept using a "left-hand rule." I actually like the left-hand rule a bit better, as it is more illustrative of what is actually happening. However, the vast majority of physics courses teach the right-hand rule, so that's how I am teaching it here. In case you are interested, the left-hand rule is very similar, but you use your left hand (surprise!) instead of your right hand, and you point your thumb in the direction of the *electron flow*. Remember, for wires made of metal conductors, conventional current is wrong. Electricity traveling through such wires is actually the result of electrons flowing, and they flow opposite of conventional current. Thus, if you want to follow the electrons, you would point in the direction opposite of conventional current. To still get the proper direction of the magnetic field, then, you would have to use your left hand. Obviously, either rule works, but most likely, you will use the right-hand rule in college courses, so you might as well start now. Study the example and solve the "On Your Own" problem to make sure you know how to apply this rule.

Illustrations by Megan Whitaker

EXAMPLE 16.2

Draw the magnetic field lines for the following current-carrying wire.

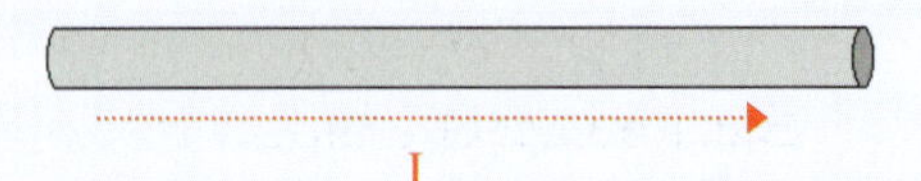

We know that for all current-carrying wires, the magnetic field lines are circular, with the wire at the center. Now all we have to do is determine the direction. To do this, we use the right-hand rule. If we point our right-hand thumb in the direction of the current, our fingers point as shown below:

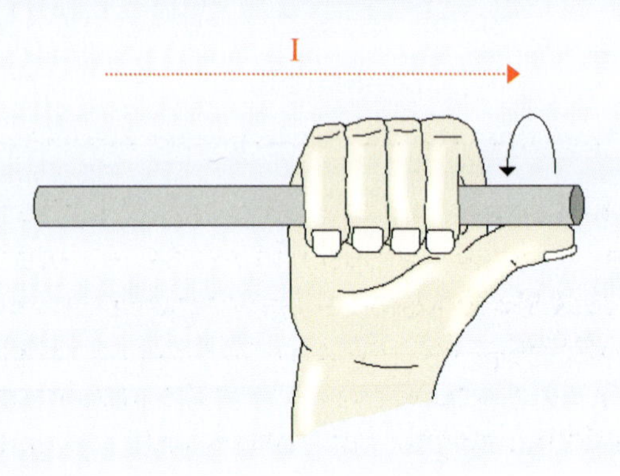

The field lines, then, look like this:

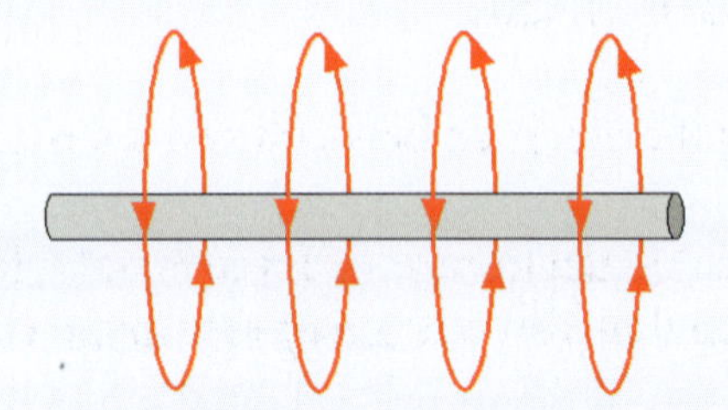

ON YOUR OWN

16.2 Draw the magnetic field lines for the following current-carrying wire.

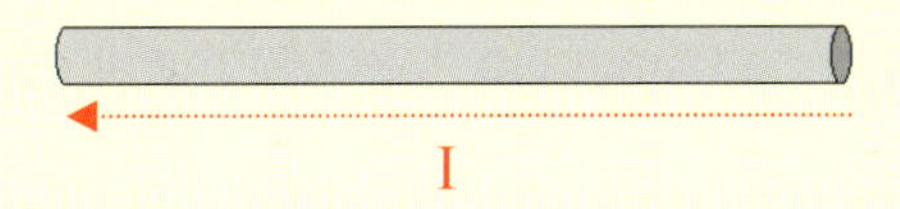

Faraday's Law of Electromagnetic Induction

So far, we have learned that charged particles in motion produce magnetic fields. In the previous section, we saw in particular what this magnetic field looks like when the charges travel through a wire. Well, if charges traveling through a wire can produce a magnetic field, doesn't it make sense that a magnetic field can produce charges moving through a wire? In fact, it can. This was first demonstrated by the great English scientist Michael Faraday in 1831. Consider, for example, the illustration below.

FIGURE 16.8
Electromagnetic Induction

Illustration by Megan Whitaker

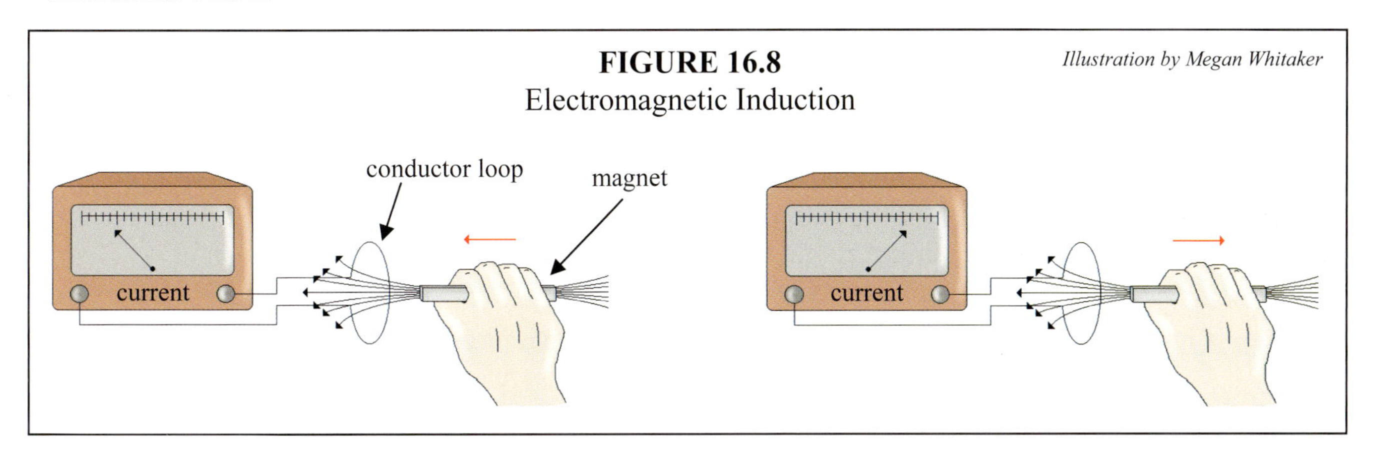

In the figure, an ammeter (a device that measures current) is hooked to each end of a conductor that has been formed into a loop. A magnet is pushed towards the loop and away from the loop. As the magnet is being pushed towards the loop, the ammeter measures a current in one direction. When the magnet is pulled away from the loop, the ammeter measures a current in the other direction. Now it turns out that it really doesn't matter whether you move the magnet or the loop of conductor. You will get the same effect if you hold the magnet still and move the loop towards or away from it.

Faraday developed a law that describes this effect. However, in order to understand his law, you have to learn some terminology. First, you need to know that potential difference is often referred to as **electromotive force** (abbreviated as **emf**).

Electromotive force - The potential difference between two sides of a battery when no current flows

I don't like to use this term because the electromotive force is not a force at all. It is a potential difference. Nevertheless, it is a term that you will encounter in the course of studying electricity and magnetism.

Second, you need to learn the concept of **magnetic flux**. As you might imagine, the strength of the magnetic field used in an experiment like the one depicted in Figure 16.8 affects the amount of current that will be generated. In addition, the area enclosed by the loop of conductor also plays a role. The larger the area, the greater the current produced. Finally, the angle between the magnetic field and the area enclosed by the conductor plays a role. In the experiment illustrated in the figure, the maximum current will be measured when the magnet is perpendicular to the surface enclosed by the loop. To take these effects into account, we define the magnetic flux with the following equation:

$$\phi_m = BA \cdot \cos(\theta) \tag{16.1}$$

where "ϕ_m" stands for the magnetic flux, "B" is the strength of the magnetic field, "A" is the area enclosed by the loop of conductor, and "θ" is the angle between the direction of the magnetic field and a vector normal (perpendicular) to the area enclosed by the loop.

I am not going to require you to use this equation. I just want you to understand what it means. When the strength of the magnetic field increases, the equation says that the magnetic flux increases. When the area enclosed by the loop of conductor increases, the magnetic flux increases. Finally, if you consider a vector perpendicular to the area enclosed by the loop, the magnetic flux is at its maximum when that vector and the magnetic field are parallel (so that $\theta = 0°$). Since the cosine of 90° is zero, it also tells us that if the magnetic field is perpendicular to that same vector, the magnetic flux is zero.

So, what does this have to do with the situation depicted in Figure 16.8? Well, Faraday developed a law that we now call **Faraday's Law of Electromagnetic Induction**.

Faraday's Law of Electromagnetic Induction - The electromotive force in a loop of wire is proportional to the rate of change of magnetic flux through the loop.

You can restate Faraday's Law of Electromagnetic Induction mathematically as:

$$\text{emf} = \frac{\Delta\phi_m}{\Delta t} \tag{16.2}$$

Thus, in order to create a potential difference from one end of a loop of wire to another, you must change the magnetic flux through that loop of wire. The faster the magnetic flux changes, the greater the potential difference and, as a result, the greater the flow of current.

If you think about Faraday's Law, you can see why the experiment shown in Figure 16.8 works the way it does. In that experiment, A is constant (the loop never changes size), and the average value of θ is constant, because the loop and magnet stay at the same angle relative to each other. However, as the magnet is pushed towards the loop, the strength of the magnetic field experienced by the loop increases, because the magnet gets closer to the loop. As a result, the magnetic flux changes, and an emf is produced, which causes current to flow.

Now think about what happens as the magnet is pulled away from the loop. Once again, A and the average value of θ don't change, but the magnetic field experienced by the loop decreases as the magnet is pulled away from the loop. Thus, the magnetic flux changes, and an emf is produced, which causes current to flow.

Think, for a moment, about the *sign* of the change in flux in the experiment. As the magnet is pushed towards the loop, the magnetic flux *increases*. Thus, the *change in flux* is positive. However, as the magnet is pulled away, the flux *decreases*. Thus, the *change in flux* is *negative*. This explains why the current travels in one direction when the magnet is pushed towards the loop and in the other direction when the magnet is pulled away from the loop. Since the sign of the change in flux is opposite in these two situations, the sign of the emf is opposite in these two situations. As a result, the direction of the current flow is opposite as well.

In the next section, we are going to see the usefulness of Faraday's Law of Electromagnetic Induction. However, before we do that, please read a brief biography about the genius who brought us this law, and then solve the "On Your Own" problems that follow.

Image from the National Portrait Gallery, London, courtesy AIP Emilio Segre Visual Archives

FIGURE 16.9
Michael Faraday

Michael Faraday was born in 1791 in Newington, Surrey, England. As the son of a blacksmith, he had little formal education. However, while working as an apprentice to a bookbinder, he began reading scientific works and experimenting with electricity. Although best known for his work in electricity and magnetism (such as his Law of Electromagnetic Induction), he was also a brilliant chemist and is credited with the discovery of benzene. Throughout his scientific career, Faraday was a devout Christian. He belonged to a group called the Sandemanians, who believed in a strict literal interpretation of the Scriptures. My favorite quote from him is, "Since peace is alone in the gift of God; and as it is He who gives it, why should we be afraid? His unspeakable gift in His beloved Son is the ground of no doubtful hope." (Christian History Institute, "Michael Faraday: At Play in the Fields of the Lord," *Glimpses* 72 [1995]: 1)

ON YOUR OWN

16.3 Suppose the experiment depicted in Figure 16.8 was changed so that the magnet was not moved towards or away from the conductor loop. Instead, suppose the magnet was simply rotated about its center. Under those conditions, would the ammeter read any current? Why or why not?

16.4 Consider the experiment depicted in Figure 16.8 once again. Suppose the experimenter first moves the magnet slowly and measures the current produced. Then, suppose he moves the magnet quickly and measures the current produced. Will the current reading be different between these two cases? If the readings are different, which will be larger?

Using Faraday's Law of Electromagnetic Induction

A modern-day electrical power plant makes use of Faraday's Law to generate electrical power. In most power plants, magnets are spun inside large coils of wires. As the magnets spin, the relative angle between the magnetic field and the vector normal to the surface enclosed by the coil [θ in Equation (16.1)] changes. Because the angle changes, the magnetic flux changes, and that produces an emf.

Except in the case of solar power stations (they work on a completely different principle), *all* modern-day power stations work on this principle. The only difference between one type of power station and another is how the magnets are made to spin in the coil of wire. Hydroelectric power plants, for example, hook the magnets to something that resembles a waterwheel. The current from the river (or the waterfall) by the power plant turns the waterwheel, which turns the magnets. The spinning magnets produce a changing magnetic flux in the wire coils, producing electricity.

The other types of power plants (coal, gas, wood, and nuclear) all use the same method for turning the magnets in the coil of wire. In these plants, water is heated until it boils. A turbine (much like a fan) is placed above the boiling water. As steam rises from the boiling water, it hits the blades of the turbine, causing the turbine to turn. The turbine is hooked to the magnets in the coils of wire so that when the turbine turns, the magnets do, too. Once the steam rises above the turbine, it is funneled into a pipe system where it is cooled back into its liquid phase. The water is then pumped back into the heating system, and the whole process repeats.

Well, if nuclear, coal, wood, and gas power plants all use the same method for creating electricity, what's the difference between them? The only difference between these plants is *the way in which they heat the water*! A coal-burning power plant boils the water by burning coal; a wood-burning power plant burns wood; a gas power plant burns natural gas; and a nuclear power plant uses a controlled nuclear reaction. Once the water has been heated, however, all of these power plants look essentially the same. In each case, steam turns turbines, which turn a series of magnets in coils of wires, which then produce current.

The reason we have so many different types of power plants is because each of these methods for heating water has its drawbacks. Coal is cheap, but it is full of pollutants that get dumped into the air. Natural gas burns cleanly and efficiently, but it is relatively expensive. Wood can burn reasonably clean and, since you can always grow more trees, there is an endless supply of it. The problem is that it is very inefficient. In order to keep a constant supply of wood, a wood-burning power plant needs many acres of dedicated forest. This is just not practical for large-scale applications. Finally, nuclear power is limitless, cheap, and incredibly efficient. It is also very dangerous, however, both because the reaction can go out of control and cause some serious damage, and because its byproducts are highly radioactive. In the end, there is no really ideal power source. Even hydroelectric power has been shown to damage the aquatic ecosystem in the river upon which the power plant is built. Thus, each type of power plant has its drawbacks.

I want to examine the energy conversions that go on in an electrical power plant. After all, when current is produced in a conductor, electrons have gained kinetic energy. The First Law of Thermodynamics tells us that this energy cannot just appear out of nowhere. So where did it come from?

In a hydroelectric plant, the kinetic energy of the water as it flows down the river or waterfall is used to turn the waterwheel. Thus, the kinetic energy of the water is converted into the kinetic energy of the waterwheel. The kinetic energy of the waterwheel is then converted into the kinetic energy of the turning magnets. Finally, the kinetic energy of the turning magnets is converted into the kinetic energy of the electrons in the coil of wires, making electrical current.

In the other types of power plants, two more steps of energy conversion are necessary. First of all, the potential energy of the heating source must be converted into the kinetic energy of heat. For example, the potential energy contained in the chemical bonds which make up coal is converted to the kinetic energy of the heat that is produced when the coal burns. That kinetic energy is then used to increase the motion of the water molecules until they break away from their liquid phase and become gas. After these two conversions, the process is essentially the same as that used in a hydroelectric plant. The kinetic energy of the steam is turned into the kinetic energy of the turbines, which is turned into the kinetic energy of the magnets, which is finally turned into the kinetic energy of the electrons in the current.

So you see, even though we say that electrical power plants "produce" energy, they really don't. Energy is simply converted from one form to another. In the end, then, an electrical power plant is just a big conversion factory. It uses Faraday's Law of Electromagnetic Induction to convert an available form of energy into electrical energy.

Alternating Current

When we discussed Faraday's Law, you learned that the current in a loop of wire will flow in one direction when the magnetic flux is increasing and then flow in the opposite direction when the magnetic flux is decreasing. Well, consider magnets turning inside coils of wires. As the magnet turns, the relative angle between the magnetic field and a vector normal to the area enclosed by the wires will change. This changes the flux. While the angle is changing from 0° to 180°, for example, the magnetic flux will decrease, because the cosine in Equation (16.1) goes from 1 to -1 over that range. As the angle varies from 180° to 360°, however, the flux will increase, because the cosine varies from -1 to 1 over that range. Thus, sometimes the change in flux is negative, and sometimes it is positive. This produces current that sometimes flows in one direction and sometimes flows in the other direction. We call this kind of current **alternating current** (abbreviated **AC**).

Alternating current - Electrical current that changes direction back and forth in a circuit

If you were able to see the current that comes out of a wall socket, you would see it move one way in a circuit and then turn around and move the other way. In the United States, the current from power plants changes direction 60 times each second. In Europe, the current from power plants changes direction 50 times each second.

Please realize that the behavior of this kind of current is different from that produced by batteries. In a battery, the negative side is always negative and the positive side is always positive. Thus, current always flows one way. If we are talking about conventional current, it always flows from the positive end of the battery to the negative end. This kind of current is called **direct current** (abbreviated **DC**).

<u>Direct current</u> - Current that always flow in the same direction around a circuit

Now remember how an electrician looks at a circuit. The appliances that you plug into wall sockets are, to an electrician, just glorified resistors. In the end, then, a circuit plugged into a wall socket at home is a bunch of resistors attached to a power source. Well, the current in this circuit is constantly changing direction back and forth, but the resistors that represent your appliances never change. Since Ohm's Law tells us that voltage equals current times resistance, if the resistance of the circuit is not changing, then the voltage of the circuit must change right along with the current. In fact, if you graphed the voltage of a circuit plugged into your wall socket as well as the current, you would see graphs like those presented in Figure 16.10.

FIGURE 16.10
Voltage and Current in a Household Circuit

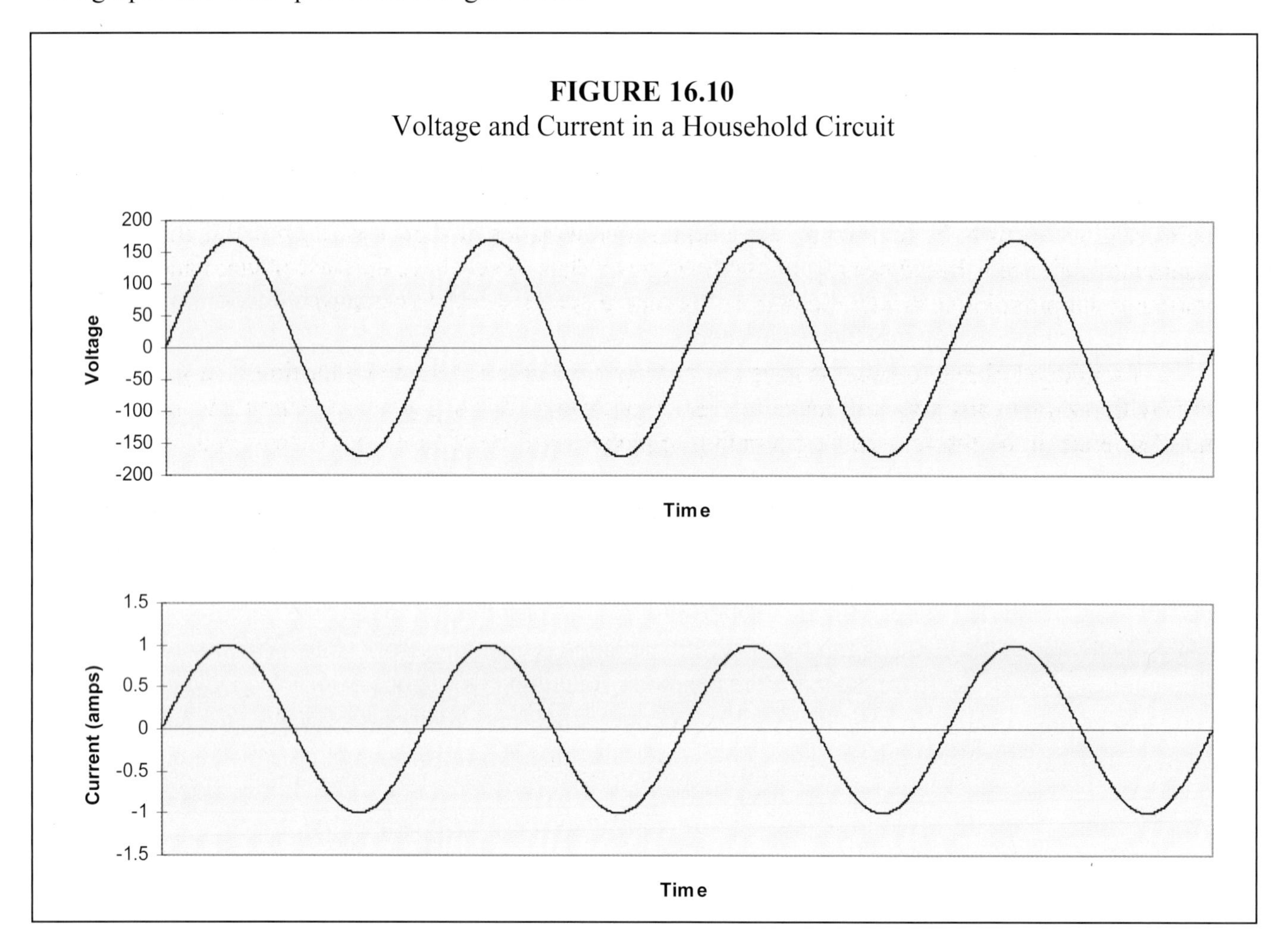

Notice two things about these graphs. The voltage and current repeat themselves over and over. As I said before, in the United States the voltage and current repeat themselves 60 times each second. This is called the **cycle** of the circuit. Also, notice that the maximum voltage is 170 volts. That is, in fact, the maximum voltage that comes out of your household wall sockets. If you are familiar with household electricity at all, you have probably heard that a wall socket provides a potential difference of 120 volts. That's true, but you have to realize what that means.

Since the voltage of an alternating current circuit is always changing, you cannot say at any given instant what the voltage is. As a result, the voltage reported for an alternating current circuit is some sort of average voltage. In the case of wall sockets, we report the **effective voltage**, which is often called the "rms voltage." The effective voltage is calculated as the square root of the average of the square of the voltage. In other words, to calculate the effective voltage, the voltage is squared, and the average of that squared voltage is calculated over a long period of time. Then, the square root of that average is taken. If you do this for the potential difference in a wall socket in the United States, you will find that the effective voltage is 120 volts, while its maximum voltage is 170 volts.

Some Final Thoughts

This discussion of magnetism wraps up your first course in physics. Many students find physics to be the toughest of all subjects. Therefore, if you struggled through it, don't worry so much. The main thing is that you did get through it. Not everything worth doing in life is easy. This course might be a very good illustration of this fact.

You should have noticed that, more than any other science, physics is heavily math-oriented. This is because, as I said in the introduction, physics is the most fundamental of all sciences. Since God's creation seems to be built on a foundation of mathematics, it is necessarily the case that the most fundamental science is based heavily on mathematics. Thus, if you have enjoyed physics and are thinking about pursuing it further, be warned that there is a *lot* of math involved.

If you don't plan to pursue physics any further, you have not wasted your time. Any career that works with nature in any way will make reference to physics. Thus, if you are planning to study any kind of science in the future, a basic grounding in physics will be of enormous help to you.

Even if you aren't planning on studying the sciences further, physics is still a good thing to know. Since everything in your daily life is governed by physics, a basic knowledge of physics will help you better understand the wonderful creation God has given you!

ANSWERS TO THE "ON YOUR OWN" PROBLEMS

16.1 Magnetic field lines go out of the north pole and into the south pole. Since these poles are so close, the field lines don't have to travel very far.

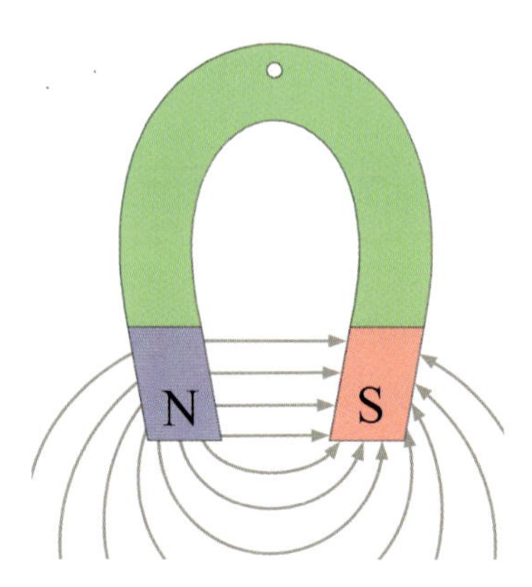

16.2 The magnetic field lines form circles around the current-carrying wire. The only thing we need to do is figure out the direction in which they point. To do this, we use the right-hand rule, pointing our right-hand thumb in the direction of the conventional current.

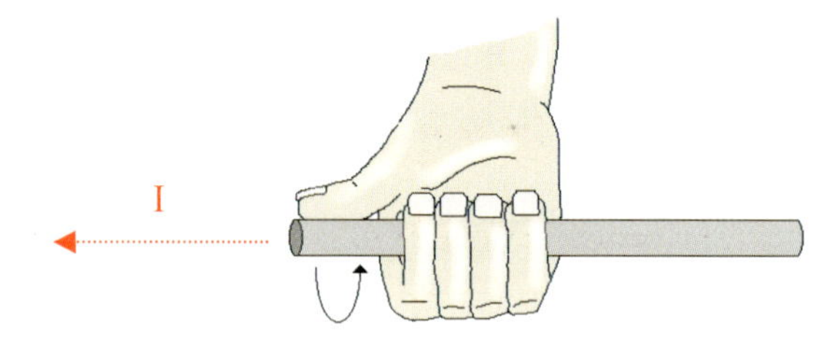

The field lines point in the direction that our fingers curl.

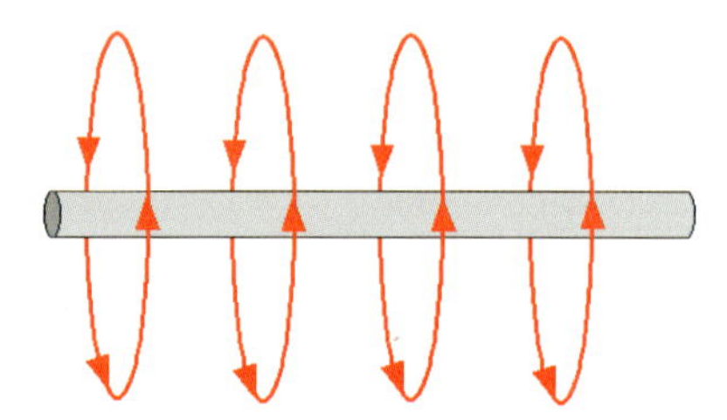

16.3 <u>Yes, the ammeter will read current</u>. The strength of the magnetic field in the area enclosed by the loop will change as the orientation of the magnet changes, and so will the angle between the magnetic field and the vector perpendicular to the area enclosed by the loop. Thus, the magnetic flux will definitely be changing, which means a current will be produced.

16.4 The emf (and as a result the current) depends on the *change in magnetic flux*. The faster you move the magnet, the greater the change in magnetic field, and therefore the greater the change in the magnetic flux. As a result, <u>the current readings will be different. The larger one will occur when the magnet moves more quickly</u>.

REVIEW QUESTIONS

1. The north pole of one magnet is put near the north pole of another. Will these poles attract or repel one another?

2. What is the significance of the direction in which magnetic field lines point?

3. Draw the magnetic field for the situation given below. Draw the field lines only for the area enclosed in the dashed box.

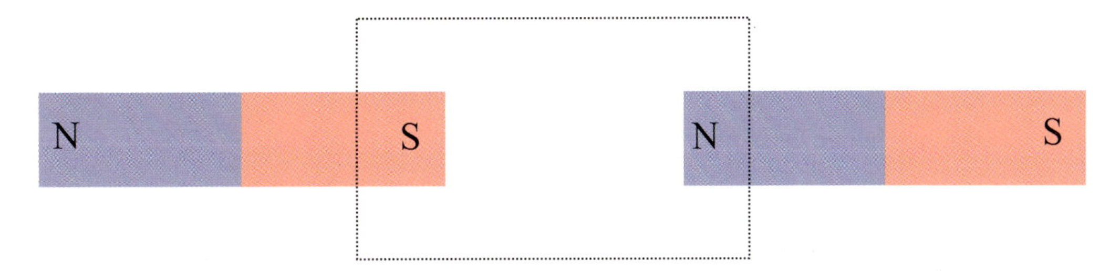

4. What is the difference between a monopole and a dipole? Is there such a thing as an electric monopole? Is there such a thing as a magnetic monopole?

5. What causes magnetism?

6. What are the three classifications of substances from a magnetic point of view?

7. Contrast the three classifications of substances. What must a substance have to belong to each classification?

8. Why are some atoms magnetic and others not?

9. Are all ferromagnetic substances permanent magnets? Why or why not?

10. A substance is not attracted to a very powerful magnet. How would you classify the substance?

11. A substance is attracted to a rather weak magnet. How would you classify the substance?

12. A substance is weakly attracted to a powerful magnet. How would you classify the substance?

13. Draw the magnetic field lines for the following current-carrying wire:

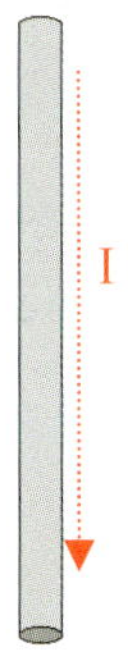

14. State Faraday's Law of Electromagnetic Induction.

15. In which of the two situations below will the ammeter read the largest current? Assume that the angle between the surface enclosed by the loop and the magnet are the same in each case.

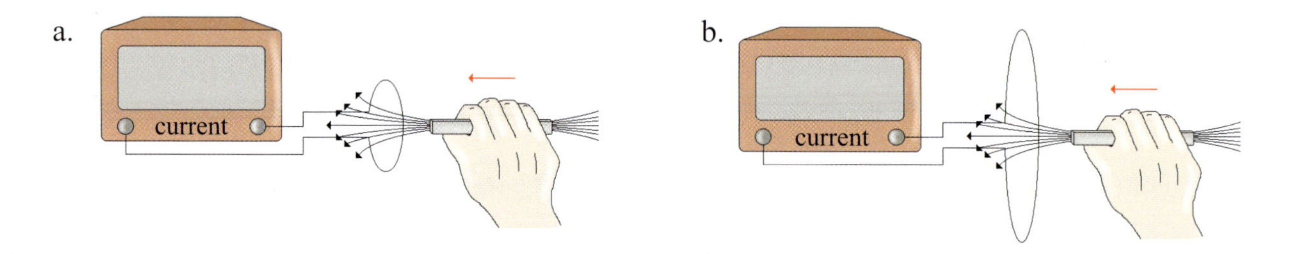

16. Describe the energy conversion steps in a hydroelectric power plant.

17. Describe the energy conversion steps in a wood-burning power plant.

18. What is the difference between a direct current circuit and an alternating current circuit?

19. Would a battery power a direct current circuit or an alternating current circuit?

20. In Europe, the standard electrical outlet produces alternating current and is referred to as a 220-volt electrical outlet. Does this mean that the maximum potential from the socket is 220 volts? Why or why not?

GLOSSARY

The numbers in parentheses indicate the page number where the term was first discussed

Acceleration - The time rate of change of an object's velocity (23)

Accuracy - An indication of how close a measurement is to the true value (3)

Air resistance - The drag that air produces on objects traveling through it (53)

Alternating current - Electrical current that changes direction back and forth in a circuit (541)

Ampere - The SI unit of electric current, which is equal to a Coulomb/sec (488)

Amplitude - The maximum distance away from equilibrium that an object in periodic motion travels (326)

Angle of a vector - The counterclockwise angle between a two-dimensional vector and the positive x-axis. The angle tells you the direction of the vector. (73)

Angle of incidence - When light hits a surface, the angle of incidence is the angle that the light ray makes with a line that is perpendicular to the surface. (387)

Angle of reflection - When light hits a surface, the angle of reflection is the angle that the reflected light ray makes with a line that is perpendicular to the surface. (387)

Angular momentum - The momentum associated with rotational motion (305)

Angular velocity - The rate at which the position angle of an object changes in rotational motion (305)

Average acceleration - The acceleration of an object over an extended period of time (27)

Average velocity - The velocity of an object over an extended period of time (13)

Axis of rotation - The point (or line) in space that is the center of rotational motion (188)

Black hole - An object so dense that not even light can escape its gravitational pull (241)

Blood plasma- The fluid portion of the blood, which is mostly water (230)

Capacitance - A measure of how much charge a capacitor can store at a given potential difference (472)

Capacitor - A device that stores charge (469)

Centrifugal force - A fictional force that results from a poor physical analysis of situations involving motion along a curve (228)

Centrifuge- A machine that spins objects at a high rate of speed for the purpose of separating components in a mixture (229)

Centripetal acceleration - The acceleration caused by centripetal force (219)

Centripetal force - The force necessary to make an object move in a circle. It is directed towards the center of the circle. (219)

Charging by conduction - Charging an object by allowing it to come into contact with an object that already has an electric charge (429)

Charging by induction - Charging an object without direct contact between the object and a charge (430)

Chemical energy - Energy associated with the chemical bonds of a molecule (267)

Circuit breaker - A switch that opens when too much current passes through it, limiting the current that can flow in a circuit (508)

Circumference - The distance around a circle (236)

Cochlea - Snail-shaped structure in the ear that contains the organ of hearing (354

Coefficient of friction - A number (usually less than 1) that is used to calculate the force due to friction. It depends on the object in question and the surface upon which the object rests or moves (160)

Coefficient of kinetic friction - The coefficient of friction used when an object is already in motion (160)

Coefficient of static friction - The coefficient of friction used when an object is at rest (160)

Coherent light waves - Light waves in which the spatial relationship of the crests and troughs of one wave compared to those of the other does not change (368)

Compressions - The sections of a longitudinal wave that are bunched up, which correspond to the crests of the wave (353)

Conductor - A substance through which charge flows easily (425)

Conservation of energy equation - The total mechanical energy in a system is equal to the kinetic energy plus the kinetic energy and cannot change (259)

Constructive interference - The phenomenon that occurs when two or more waves overlap so that their crests and troughs add together to make larger crests and troughs (369)

Conventional current - Current that flows from the positive side of the battery to the negative side (490)

Converging lens - A transparent object that focuses light rays traveling parallel to its optical axis to a single point, called the focal point (406)

Crest of a wave - The highest point in the oscillation of a wave (352)

Derivation - A process by which one starts with a known fact, such as an equation, and uses logic and reasoning to come up with a different fact, such as another equation (38)

Destructive interference - The phenomenon that occurs when two or more waves overlap so that their crests and troughs cancel one another out (369)

Diamagnetic substance - A substance that is made up of atoms with no net magnetic field (528)

Dielectric - A substance that does not conduct electricity (471)

Diffraction - The spreading of waves around an obstacle (368)

Dipole - Consisting of two poles, such as a magnet (523)

Direct current - Current that always flow in the same direction around a circuit (542)

Discharging - Ridding an object of its charge (429)

Displacement - The change in an object's position (6)

Diverging lens - A transparent object that bends light rays traveling parallel to its optical so that they appear to be coming from a single point, called the focal point (408)

Doppler effect - The change in pitch caused by relative motion between the source of a sound and the observer of the sound (359)

Dynamic equilibrium - When an object moves with a constant velocity, it is said to be in dynamic equilibrium. (177)

Dyne (unit) - A unit of force, which is a $[\text{g}\cdot\text{cm}]/\text{sec}^2$ (146, 147)

Electric current - The amount of charge that travels past a fixed point in an electric circuit each second (488)

Electric field- A representation of the electrostatic force exerted by a stationary charge on a positive test charge (440)

Electrical energy - Energy associated with the motion (or potential motion) of charged particles (267)

Electromotive force (emf) - The potential difference between two sides of a battery when no current flows (537)

Electron volt - The decrease in potential energy when an electron moves through a potential difference of exactly 1 volt (463)

Electrostatic force - The force that exists between two charges at rest (431)

Electrostatics - The study of electric charges at rest (423)

Energy - The ability to do work (251)

Equilibrium position - The position of an object when there are no net forces acting on it (321)

External auditory meatus - The "canal" that leads to the inner parts of the ear (354)

Factor-label method - Method used in this course to convert between units (2)

Faraday's Law of Electromagnetic Induction - The electromotive force in a loop of wire is proportional to the rate of change of magnetic flux through the loop. (538)

Ferromagnetic substance - A substance whose magnetic atoms are aligned so that their magnetic fields add to one another (529)

Focal point - The point at which a concave spherical mirror or a converging lens will direct light rays that travel parallel to the optical axis. Diverging lenses and convex spherical mirrors direct these light rays so that they appear to have come from the focal point. (389)

Force - A push or a pull exerted on an object in an effort to change that object's velocity (146)

Free fall - The motion of an object when it is falling solely under the influence of gravity (51)

Frequency - The number of times per second an object in uniform circular motion travels around the circle (235)

Friction - A force that opposes motion, resulting from the contact of two surfaces (143)

Fuse - A thin conductor that breaks when a certain amount current passes through it, limiting the current that can flow in a circuit (508)

Galaxy - A collection of star systems (241)

Gravity - The attractive force that exists between all objects which have mass (230)

Heat - Energy that is transferred from one object to another as a result of a difference in temperature (267)

Hooke's Law - The force required to stretch a spring is proportional to the distance by which it is stretched. (320)

Hyperopia - A condition in which the eye does not focus light strongly enough. As a result, objects far from the eye can stay in focus, but objects near the eye do not (412)

Impulse - The change in momentum experienced by an object, given by $\mathbf{F} \cdot \Delta t$ (290)

Impulsive force - A force applied over a short time interval (290)

Index of refraction - The ratio of the speed of light in a vacuum to its speed in another medium (402)

Infrasonic waves - Sound waves with a frequency below 20. People cannot hear these waves (364)

Instantaneous acceleration - The acceleration of an object at one moment in time (27)

Instantaneous velocity - The velocity of an object at one moment in time (13)

Insulator - A substance through which charge cannot flow (425)

Joule (unit) - The SI unit of energy, which is a Newton·m (253)

Kinetic energy - Energy in motion (254)

Kinetic friction - Friction that opposes motion once the motion has already started (156)

Law of Angular Momentum Conservation - If the sum of the torques on a system is equal to zero, the angular momentum never changes. (307)

Law of Charge Conservation - The net amount of electric charge in the universe is constant. (426)

Law of Momentum Conservation - When the sum of the forces working on a system is zero, the total momentum in the system cannot change. (294)

Law of Universal Gravitation - Every particle in the universe attracts every other particle in the universe with a force proportional to the product of the particles' masses and inversely proportional to the square of the distance between the particles. (231)

Lever arm - The length of an imaginary line drawn from the axis of rotation to the point at which the force is being applied (191)

Longitudinal wave - A wave whose propagation is parallel to its oscillation (353)

Magnitude of a vector - The length of a vector, which represents the amount (72)

Mass - A measure of how much matter is in an object (148)

Mechanical energy - Energy associated with the movement (or potential movement) of objects (267)

Medium - A term used to refer to a substance in which something travels (351)

Milky Way - The galaxy in which earth is found (241)

Momentum - The product of an objects mass and its momentum (287)

Myopia - A condition in which the eye focuses light too strongly. As a result, objects near the eye can stay in focus, but objects far from the eye do not (412)

Newton (unit) - The SI unit of force, which is a $[\text{kg}\cdot\text{m}]/\text{sec}^2$ (146)

Newton's First Law (The Law of Inertia) - An object in motion (or at rest) will tend to stay in motion (or at rest) until it is acted upon by an outside force. (143)

Newton's Second Law - When an object is acted on by one or more outside forces, the vector sum of those forces is equal to the mass of the object times the resulting acceleration vector. (146)

Newton's Third Law - For every action, there is an equal and opposite reaction. (166)

Non-Ohmic substance- A substance that does not obey Ohm's Law (494)

Normal force - A force that results from the contact of two bodies and is perpendicular to the surface of contact (154)

Ohm- The SI unit for resistance, which is the same as a V/A (492)

Ohm's Law - The potential difference in a circuit is the product of the current in the circuit and the resistance of the circuit (492)

Opaque object - An object through which light cannot pass (400)

Open circuit - A circuit that does not have a complete connection between the two sides of the battery. As a result, current does not flow. (499)

Optical axis - A line that hits the center of a mirror or lens and is perpendicular to the surface of the mirror or lens at the point of contact (389)

Optics - The study of the propagation and behavior of light (385)

Parabolic motion - Motion that occurs when an object moves in two dimensions but has zero acceleration in one of those dimensions and a constant, non-zero acceleration in the other (111)

Parallel circuit - A circuit in which current has a "choice" as to which devices it flows through (500)

Paramagnetic substance - A substance whose magnetic atoms are arranged so that their individual magnetic fields cancel out. Under the influence of an external magnetic field, however, these substances produce a magnetic field that is in the same direction of the external magnetic field. (529)

Particle-wave duality - The idea that anything in nature has both wave-like characteristics and particle-like characteristics (376)

Period - The time it takes for an object in uniform circular motion to travel through one full circle (235)

Periodic motion - Motion that repeats itself regularly (317)

Photoelectric effect - The term that refers to the fact that when light is shone on certain metals, electrons can be liberated from the metal (372)

Photoelectrons- Electrons liberated from a metal via the photoelectric effect (372)

Pitch - An indication of how high or low a sound is, which is primarily determined by the frequency of the sound wave (355)

Position - A vector quantity that describes where an object is relative to a reference point (6)

Potential energy - Energy that is stored, ready to do work (254)

Pound (unit) - The English unit for weight, which is a $[\text{slug}\cdot\text{ft}]/\text{sec}^2$ (149)

Power - The amount of energy expended per second (275)

Precision - An indication of the scale on the measuring device that was used (3)

Radians (unit) - A unit for measuring angle. An angle of π radians is equal to an angle of 180° (339)

Radius of curvature - Assuming that a curve is an arc from a circle, the radius of curvature is the radius of that circle. (226)

Rarefactions - The sections of a longitudinal wave that are spread out, which correspond to the troughs of the wave (353)

Reaction time - The time elapsed between recognizing an event and reacting to it (54)

Real image - An image formed as the result of intersecting light beams (394)

Recoil velocity - The velocity that an object develops in response to launching another object, which is a result of the Law of Momentum Conservation (301)

Refraction - The process by which a light ray bends when it encounters a new medium (401)

Resistance - The ability of a material to impede the flow of charge when it is subject to a potential difference (491)

Restoring force - A force, directed towards the system's equilibrium position, which is applied as a result of the system's displacement from equilibrium (321)

Right-hand rule - The direction of the magnetic field lines for a current-carrying wire can be determined by pointing the thumb of your right hand in the direction of the current. Your fingers will curl in the direction of the magnetic field lines. (535)

Rotational equilibrium - The state in which the sum of the torques acting on an object is zero (194)

Rotational motion - Motion around a central axis such that an object could repeatedly pass the same point in space relative to that axis (188)

Scalar quantity - A physical measurement that does not contain directional information (6)

Series circuit - A circuit in which current must flow through each of the electrical devices (500)

SI units - System that contains the standard metric units used in science (2)

Significant figure - A digit in a measurement that is either non-zero, a zero that is between two significant figures, or a zero at the end of the number and to the right of the decimal (3)

Simple harmonic motion - Periodic motion whose period is independent of its amplitude (326)

Slope of a graph - A measure of how steeply a curve is rising or falling. It is calculated by taking the change in the y-coordinate and dividing by the change in the x- coordinate. (17)

Slug (unit) - The English unit for mass (149)

Sonic waves - Sound waves with a frequency between 20 and 20,000 Hz. These are sound waves that people can hear (364)

Spherical aberrations - Distortions in the image formed by a mirror or lens that is spherical in shape rather than parabolic (390)

Static equilibrium - When an object is at rest, it is said to be in static equilibrium. (177)

Static friction - Friction that opposes the initiation of motion (156)

System Internationale (SI) - System that contains the standard metric units used in science (2)

Tension - The force from a tight string, rope, or chain. This force is directed away from the object to which the string, rope, or chain is anchored. (178)

Terminal velocity - The velocity a falling object has when, due to air resistance, its acceleration is reduced to zero. This is the maximum velocity a falling object subject to air resistance can achieve.(60)

The First Law of Thermodynamics- Energy cannot be created or destroyed. It can only change form. (258)

The Law of Inertia (Newton's First Law) - An object in motion (or at rest) will tend to stay in motion (or at rest) until it is acted upon by an outside force. (143)

The Law of Reflection - The angle of reflection equals the angle of incidence. (387)

Torque - The tendency of a force to cause rotational acceleration. The magnitude of the torque is equal to the length of the lever arm times the component of the force that is applied perpendicular to it. (191)

Translational equilibrium - An object is said to be in translational equilibrium when the sum of the forces acting on it is equal to zero. (177)

Translational motion - Motion from one point to another which does not involve repeatedly passing the same point in space (188)

Transparent object - An object through which light can pass (400)

Transverse wave - A wave whose propagation is perpendicular to its oscillation (352)

Trough of a wave - The lowest point in the oscillation of a wave (352)

Two-dimensional motion - Motion that occurs in a plane (71)

Two-dimensional vector - An arrow whose length represents the magnitude of a vector quantity and whose angle represents the direction of the vector quantity (72)

Tympanic membrane - The ear drum (354)

Ultrasonic waves - Sound waves with a frequency greater than 20,000 Hz. People cannot hear these sound waves(364)

Uniform circular motion - Motion in a fixed circle at a constant speed (217)

Vector quantity - A physical measurement that contains directional information (6)

Velocity - The time rate of change of an object's position (6)

Virtual image - An image formed as the result of extrapolating light beams (388)

Volume - An indication of how loud a sound is, which is primarily determined by the amplitude of the sound wave (355)

Watt (unit) - The SI unit of power, which is a Joule/sec (275)

Wave - A disturbance that propagates in a medium (351)

Wave interference - The phenomenon that occurs when two or more waves overlap. Their crests and troughs mix to make a different wave (368)

Wavelength - The distance between crests (or troughs) of a wave (352)

Weight - A measure of the strength with which gravity pulls on an object (148)

Wile (proposed unit) - Proposed name for the SI unit of momentum, which is a kg·m/sec (275)

Work - The product of the displacement of an object and the component of the applied force that is parallel to the displacement (251)

Work function - The energy that an electron in a metal must have in order to break free of the metal (372)

APPENDIX A
FORMULAE & LAWS

FORMULAE

Acceleration (**a**)	$\mathbf{a} = \dfrac{\Delta \mathbf{v}}{\Delta t}$
Angular Momentum	$L = mvr$
Centripetal Acceleration ($\mathbf{a_c}$) Direction of the acceleration is towards the center of the circular motion.	$a_c = \dfrac{v^2}{r}$
Centripetal Force ($\mathbf{F_c}$)	$F_c = \dfrac{mv^2}{r}$
Doppler Effect	$f_{observed} = \left(\dfrac{v_{sound} \pm v_{observer}}{v_{sound} \pm v_{source}} \right) \cdot f_{true}$
Effective Resistance of Resistors in Parallel	$\dfrac{1}{R_{effective}} = \dfrac{1}{R_1} + \dfrac{1}{R_2} + \dfrac{1}{R_3} + \ldots$
Effective Resistance of Resistors in Series	$R_{effective} = R_1 + R_2 + R_3 + \ldots$
Electric Current	$I = \dfrac{\Delta Q}{\Delta t}$
Electric Field (**E**)	$\mathbf{E} = \dfrac{\mathbf{F}}{q_o}$ and $E = \dfrac{kQ}{r^2}$
Electric Potential (V)	$V = \dfrac{kQ}{r}$
Electric Potential of a Capacitor	$V = \dfrac{Q}{C_q}$
Electric Potential Energy (PE)	$PE = qV$ and $\Delta PE = q\Delta V$
Electric Power (P)	$P = IV$ and $P = I^2 \cdot R$
Electrostatic Force (**F**)	$F = \dfrac{kq_1q_2}{r^2}$
Energy of a Photon	$E = hf$
Faraday's Law of Electromagnetic Induction	$emf = \dfrac{\Delta \phi_m}{\Delta t}$
Focal Length of a Spherical Mirror	$f = \dfrac{R}{2}$
Frequency (f) and Period (T) in Periodic Motion	$f = \dfrac{1}{T}$

Frequency (f) of a Wave	$f = \frac{v}{\lambda}$
Frictional force (f)	$f = \mu F_n$
Gravitational Force ($\mathbf{F_g}$) Equal force is exerted on each mass. Direction of the force on one mass is towards the other mass.	$F_g = \frac{Gm_1m_2}{r^2}$
Hooke's Law	$\mathbf{F} = -k\Delta\mathbf{x}$
Impulse ($\Delta\mathbf{p}$ sometimes abbreviated **I**) The force (**F**) is the *average* force over the time interval.	$\Delta\mathbf{p} = \mathbf{F}\Delta t$
Kinetic Energy (KE)	$KE = \frac{1}{2}mv^2$
Law of Momentum Conservation Assumes that the sum of the forces on the system is 0.	$\Delta\mathbf{p} = 0$
Magnetic Flux (ϕ_m)	$\phi_m = BA \cdot \cos(\theta)$
Magnetic Force ($\mathbf{F_m}$)	$F_m = \frac{Kp_1p_2}{r^2}$
Momentum (**p**)	$\mathbf{p} = m\mathbf{v}$
Newton's Second Law	$\Sigma\mathbf{F} = m\mathbf{a}$ and $\mathbf{F} = \frac{\Delta\mathbf{p}}{\Delta t}$
Objects on an Incline The value for θ is defined as the angle between the incline and the ground.	Component of weight perpendicular to the incline: $w \cdot \cos(\theta)$ Component of weight parallel to the incline: $w \cdot \sin(\theta)$
Ohm's Law	$V = IR$
One-Dimensional Motion Equations Assume constant acceleration and linear motion.	$\mathbf{v} = \mathbf{v}_o + \mathbf{a}t$ $\mathbf{v}^2 = \mathbf{v}_o^2 + 2\mathbf{a} \cdot \Delta\mathbf{x}$ $\Delta\mathbf{x} = \mathbf{v}_o t + \frac{1}{2}\mathbf{a}t^2$
Period (T) of a Mass / Spring System	$T = 2\pi\sqrt{\frac{m}{k}}$
Period (T) of a Simple Pendulum Assumes you are near the surface of the earth and that the angle of displacement is small.	$T = 2\pi\sqrt{\frac{\ell}{g}}$
Potential Energy (PE) From Gravity Assumes you are near the surface of the earth.	$PE = mgh$
Potential Energy (PE) of a Stretched Spring	$PE = \frac{1}{2}k(\Delta x)^2$
Power (P)	$P = \frac{\Delta W}{\Delta t}$

Range Equation Assumes two-dimensional projectile motion with no air resistance. The initial and final heights of the projectile must be equal. The value for θ can range from 0° to 90°.	$\text{range} = \frac{v_o^2 \cdot \sin(2\theta)}{g}$
Slope	$\text{slope} = \frac{\text{change in y - coordinate}}{\text{change in x - coordinate}} = \frac{\text{rise}}{\text{run}}$
Snell's Law of Refraction	$n_1 \cdot \sin(\theta_1) = n_2 \cdot \sin(\theta_2)$
Speed	$\text{speed} = \frac{\Delta d}{\Delta t}$
Speed of Sound in Air The value for temperature (T) must be in degrees Celsius.	$v = (331.5 + 0.606 \cdot T)\ \frac{m}{sec}$
Tangent (tan)	$\tan(\theta) = \frac{\text{opposite side of the triangle}}{\text{adjacent side of the triangle}}$
Torque (τ)	$\tau = F_{\perp} \cdot r$
Total Mechanical Energy (TE)	$TE = PE + KE$
Total Mechanical Energy (TE) in a Mass / Spring System	$KE_{\text{mass/spring}} + PE_{\text{mass/spring}} = \frac{1}{2} kA^2$
Vector Analysis Equations We define θ counterclockwise from the positive x-axis. **The value for θ may not be defined properly. You must follow Figure 3.1 to get the properly-defined value for θ.	$\text{magnitude of A} = \sqrt{A_x^2 + A_y^2}$ $\theta = \tan^{-1}(\frac{A_y}{A_x})^{**}$ $A_y = A \cdot \sin(\theta)$ $A_x = A \cdot \cos(\theta)$
Velocity (**v**)	$\mathbf{v} = \frac{\Delta \mathbf{x}}{\Delta t}$
Weight (**w**) Assumes you are near the surface of the earth.	$\mathbf{w} = m\mathbf{g}$
Work (W)	$W = F_{\parallel} \cdot x$

LAWS DISCUSSED IN THE COURSE

Newton's Laws of Motion

Newton's First Law (The Law of Inertia) - An object in motion (or at rest) will tend to stay in motion (or at rest) until it is acted upon by an outside force.

Newton's Second Law - When an object is acted on by one or more outside forces, the vector sum of those forces is equal to the mass of the object times the resulting acceleration vector.

Newton's Third Law - For every action, there is an equal and opposite reaction.

Newton's Law of Universal Gravitation

Every particle in the universe attracts every other particle in the universe with a force proportional to the product of the particles' masses and inversely proportional to the square of the distance between the particles.

First Law of Thermodynamics

Energy cannot be created or destroyed. It can only change form.

Laws of Momentum Conservation

When the sum of the forces working on a system is zero, the total momentum in the system cannot change.

If the sum of the torques on a system is equal to zero, the angular momentum never changes.

Hooke's Law

The force required to stretch a spring is proportional to the distance by which it is streched.

Law of Reflection

The angle of reflection equals the angle of incidence.

Snell's Law of Refraction

When light passes from one medium into a second medium, the sine of the angle of refraction is equal to the index of refraction of the initial medium times the sine of the incidence angle divided by the index of refraction of the final medium.

Law of Charge Conservation

The net amount of electric charge in the universe is constant.

Coulomb's Law

The electrostatic force between two charges is proportional to the product of their charges and inversely proportional to the square of the distance between them.

Ohm's Law

The potential difference in a circuit is equal to the product of the current times the resistance.

Faraday's Law of Electromagnetic Induction

The electromotive force in a loop of wire is proportional to the rate of change of magnetic flux through the loop.

APPENDIX B
EXTRA PRACTICE PROBLEMS

EXTRA PRACTICE PROBLEMS FOR MODULE #1

1. A person standing on the edge of a 30 foot cliff throws a ball straight up into the air. If the ball travels 50 feet up and then falls to the bottom of the cliff, what would be the total distance traveled? What would be the displacement of the ball?

2. It takes an infant 3.5 minutes to crawl the 22 meters from the playpen to the refrigerator and 11 meters back towards the playpen. What is the infant's average speed and average velocity?

3. A runner's average velocity over a race of 600 meters is 0.61 m/sec. Since the race was run on an oval track, the runner's displacement was only 62 meters. How much time did it take the runner to run the race?

4. While driving, a man slows his car from 72 mph to 65 mph in 1.8 seconds. What is his acceleration in miles/sec^2?

5. What would be the final velocity of a car that is initially moving at 62 m/sec and accelerates at -5.0 m/sec^2 for 9.2 sec?

Questions 6-7 use the following position-versus-time graph that describes a car's motion.

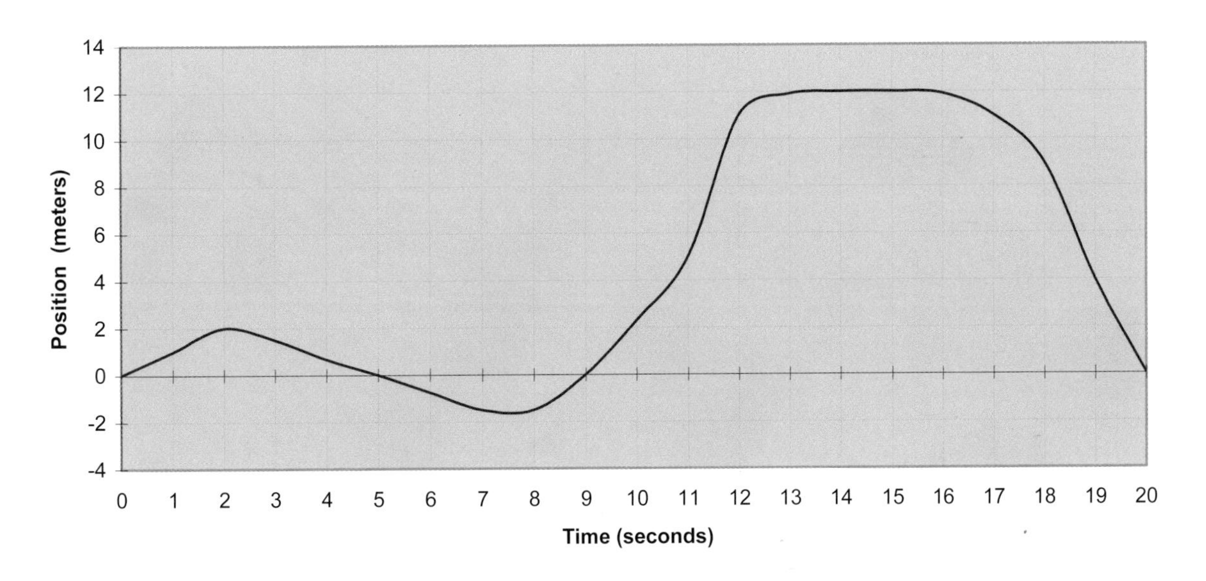

6. How many times does the car switch directions and during what time intervals is the car moving the slowest?

7. What is the instantaneous velocity at 5 seconds?

Questions 8-10 use the following velocity-versus-time graph that describes a car's motion.

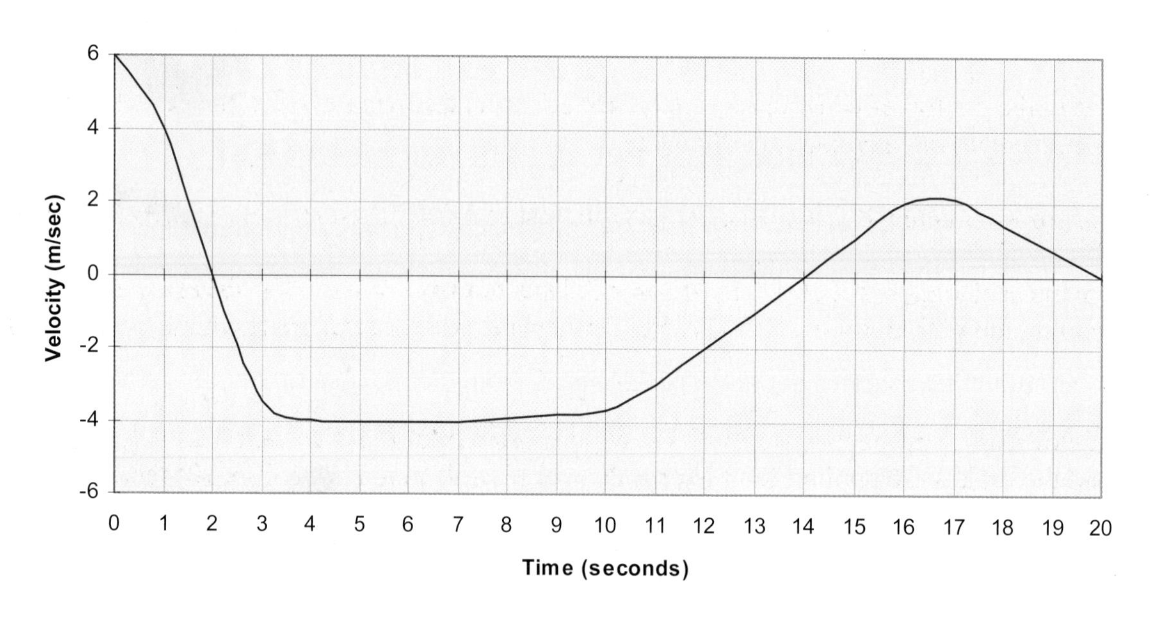

8. Over what time intervals is the car speeding up?

9. During what time intervals is the acceleration equal to zero?

10. What is the acceleration at 13.5 seconds?

EXTRA PRACTICE PROBLEMS FOR MODULE #2

1. A physics student drops a rock from a 55 m cliff. How long does it take to hit the ground?

2. How far will a car travel while it accelerates for 3.2 seconds at a rate of 2.5 m/sec^2? Prior to the acceleration, the car was traveling at 9.8 m/sec.

3. How long does it take a bullet to accelerate from rest to 3,050 ft/sec at a rate of 2.96×10^5 ft/sec^2?

4. An airplane accelerates at -3.1 m/sec^2 down a 3.8 km runway. What is the fastest speed with which the pilot could land with in order to assure that the airplane will stop before reaching the end of the runway?

5. A train slows as it approaches a curve. What is the acceleration of the train if its velocity was 62 mph and the brakes are applied for 12 seconds over a displacement of 8.00×10^2 feet?

6. A cheetah accelerates in a straight line from a velocity of 10.0 m/sec to 15.2 m/sec in 2.1 seconds. How much distance was covered during this acceleration?

7. Upon discovering a hole in the ground, a girl drops a rock and times how long it takes for the rock to hit the bottom. Neglecting air resistance and the time for sound to travel, how deep is the hole if it takes 5.3 seconds for the rock to hit the bottom?

8. Neglecting air resistance, what would be the final velocity of a sandbag that is dropped from a hot air balloon at 1140 ft?

9. What would be the velocity if the sandbag in #8 was dropped from a balloon that was descending at a rate of 250 ft/min?

10. A skyscraper, the X-Seed 4000, has been proposed to be built in Tokyo, Japan. It will have an estimated height of 4.0×10^3 m. If a coin was thrown upward from the edge of this tower with a velocity of 2.0×10^2 m/sec, how long will it take for the coin to reach the ground? Please neglect air resistance.

EXTRA PRACTICE PROBLEMS FOR MODULE #3

Use the following vectors to answer questions 1 – 3.

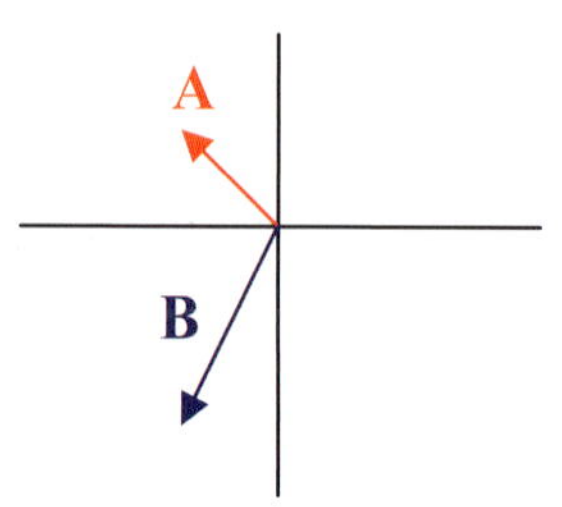

1. Sketch **A** + **B**.

2. Sketch **A** – **B**.

3. Sketch **B** – **A**.

4. What are the x- and y-components of a vector that has a 42.9 m magnitude and a direction of 252°?

5. An airplane's velocity has an x-component of -155 mph and a y-component of -45 mph. What are the magnitude and direction of the plane's velocity?

6. Determine the sum of these two vectors: **A** = 28.9 m at 165°, **B** = 34.6 m at 305°.

7. Determine the sum of these two vectors: **A** = 3.00×10^2 mph at 15.0°, **B** = 2.00×10^2 mph at 16.0°.

8. A bird flies 258 m due east ($\theta = 0.000°$) and then flies 189 m in a NW direction ($\theta = 135.000°$). What is the bird's final displacement?

9. A swimmer tries to cross the English channel. She swims with a velocity of 3.2 km/hr at 290°. If the current is moving against her at 0.89 km/hr at 32°, what will be her actual velocity vector?

10. A boat is traveling southwest ($\theta = 205°$) at 15 knots against a 4.0-knot current that is flowing north ($\theta = 90.0°$). What is the boat's actual velocity? (Note: The unit "knots" means "nautical miles per hour." Thus, it is a velocity unit.)

EXTRA PRACTICE PROBLEMS FOR MODULE #4

1. A map detailing the location of a buried treasure says to walk 525 m due west ($\theta = 180.0^{o}$) then turn to a heading of 50.0° and walk for 375 m. If you follow the instructions, what is your final displacement?

2. An airplane travels at 305 mph for 45.0 minutes at 80.0°. Then, the airplane travels at 262 mph for 2.50 hours at a 45.0°. What is the plane's resultant displacement vector?

3. From a 60.0 m cliff, a physics student throws a rock horizontally ($\theta = 0.000^{o}$) at 18 m/sec. How far does the rock travel horizontally in the first 3.5 seconds?

4. For the situation in problem #3, how far does the rock travel vertically in the first 3.5 seconds?

5. A baseball player makes contact with the ball at 2.5 feet above the ground, and the ball hits the fence 2.5 feet above the ground. If the velocity of the ball was 160 ft/sec when it left the bat, and the launch angle was 15°, how far did the ball travel?

6. At what speed must a cannonball be launched when aimed at a 2.0×10^{1} degree angle for it to travel 1.00 mile (5,280 feet)? Assume that the ground is level.

7. At what launch angle must a water balloon launcher be aimed at if you are going to hit your friend's fort, which is 250 m away? Assume that the balloon will leave the launcher with a speed of 59 m/sec and that the fort is at the same height as the launch point of the water balloon.

8. A javelin thrower sends her javelin with an initial velocity of 23.5 m/sec at an angle of 25.0°. If the distance of her throw was measured to be 45.7 m, at what height did she release the javelin?

9. How far will a bomb travel horizontally if it is dropped 9,144 m from a bomber that is traveling horizontally at a speed of 235 m/sec?

10. When an archer aims at an angle of 10.0° and gives the arrow an initial velocity of 240 m/sec, how long is the arrow in the air and how far does it travel horizontally? Assume the arrow is 5.0 feet above the level ground when he shoots it and that the arrow hits the ground at the end of its flight.

EXTRA PRACTICE PROBLEMS FOR MODULE #5

1. What is the coefficient of kinetic friction between a 810 kg crate and the pavement if the frictional force is 2.7×10^3 Newtons?

2. An astronaut takes a bathroom scale with him in order to measure the acceleration due to gravity on a planet. While on earth, he measures his weight in his space suit to be 220 pounds. If the astronaut measures his weight in his space suit to be 556 pounds on the planet, what is the acceleration due to gravity for the planet?

3. A 15-kg crate is being pulled across a wooden floor by a 88 N force. What is the acceleration experienced by the crate? ($\mu_k = 0.30$, $\mu_s = 0.48$)

4. What is the mass of a rock if it took 320 pounds of force to start it moving? ($\mu_k = 0.22$, $\mu_s = 0.40$)

5. A 660-Newton girl is on an elevator. What force is the elevator exerting on her to get her to accelerate upwards at 3.0 m/sec^2?

6. What force must a child exert on a 580-gram toy to keep it moving at a constant velocity? ($\mu_k = 0.20$, $\mu_s = 0.37$)

7. What is the acceleration of a 0.32 kg hockey puck traveling across the ice? Assume that the only forces acting on the puck are friction, gravity, and the normal force. ($\mu_k = 0.03$, $\mu_s = 0.10$)

8. What force would cause a 1.7×10^3-pound crate to accelerate at 1.6 ft/sec^2? Neglect friction.

9. Repeat problem #8, this time taking friction into account. ($\mu_k = 0.30$, $\mu_s = 0.48$)

10. A tractor is pulling a rock that is 6.10 tons (1 ton = 2000 pounds, an exact number). With what force does the tractor need to pull in order to keep the rock moving at a constant velocity? ($\mu_k = 0.30$, $\mu_s = 0.48$)

EXTRA PRACTICE PROBLEMS FOR MODULE #6

1. What are the tensions T_1 and T_2 in the strings holding up the 5.0-kg sign in the diagram below?

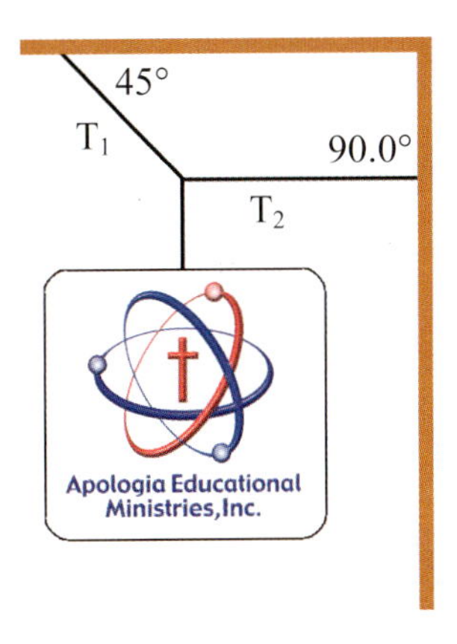

2. A stoplight hangs from two wires. If the tension on each wire is 10.3 lbs, what is the mass of the stoplight?

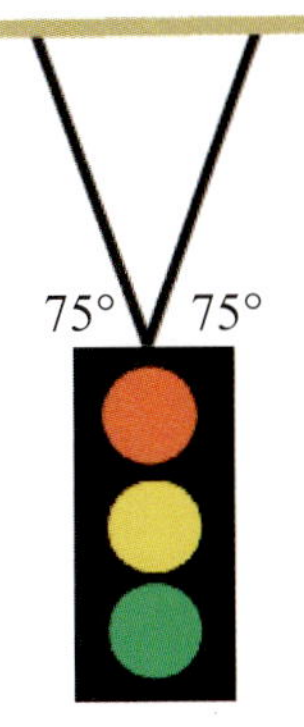

3. A flat tire has stranded a motorist. The lugnuts were put on with a torque of 250 N·m. The driver can place the force at 52.0 cm from the axis of rotation. What force must the driver provide to remove the lugnut?

4. A 660-Newton girl walks onto an elevator with bathroom scales. What will the scale read if the elevator causes her to accelerate up at 3.0 m/sec^2?

5. What mass should be placed at 0.50 m from the fulcrum in order to achieve static rotational equilibrium?

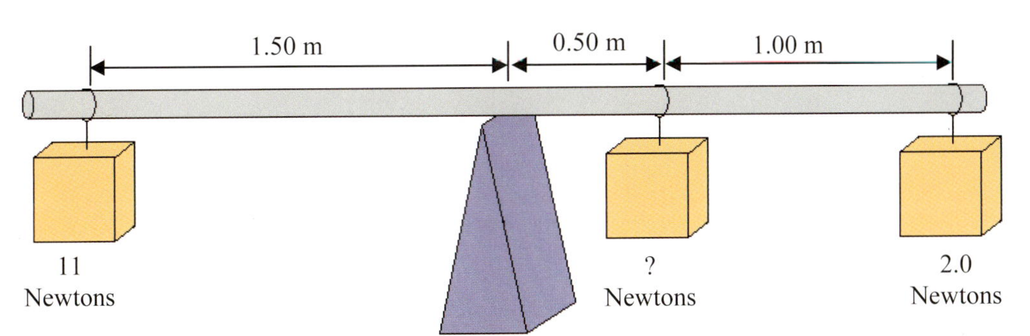

6. Where must the 10.0 g mass be placed on the rod shown below in order to balance all of the masses?

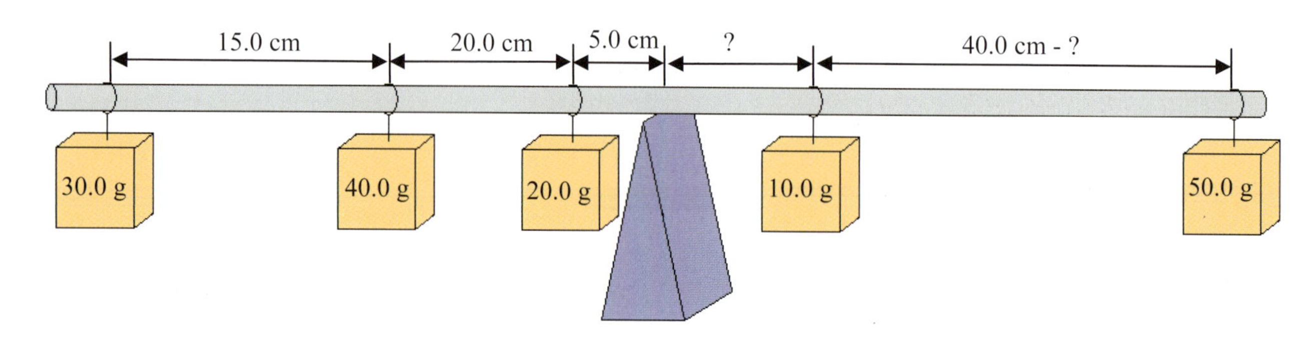

7. A piano is being pushed up a frictionless ramp that makes a 25° angle with the floor. If the piano weighs 510 pounds, and the force pushing the piano up the ramp is 240 pounds, what is the acceleration of the piano?

8. What is the acceleration of a 290-Newton child sliding down a playground slide that makes a 55° angle with the ground ($\mu_k = 0.15$)?

9. A semi is pulling two trailers with a force of 6,600 pounds. If the weight of the rear trailer is 2.00 tons, and the weight of the front trailer is 3.00 tons (1 ton = 2000 lb, exact numbers), what is the tension on the chain between the trailers, and how fast are the trailers accelerating? Ignore friction.

Illustration by Megan Whitaker

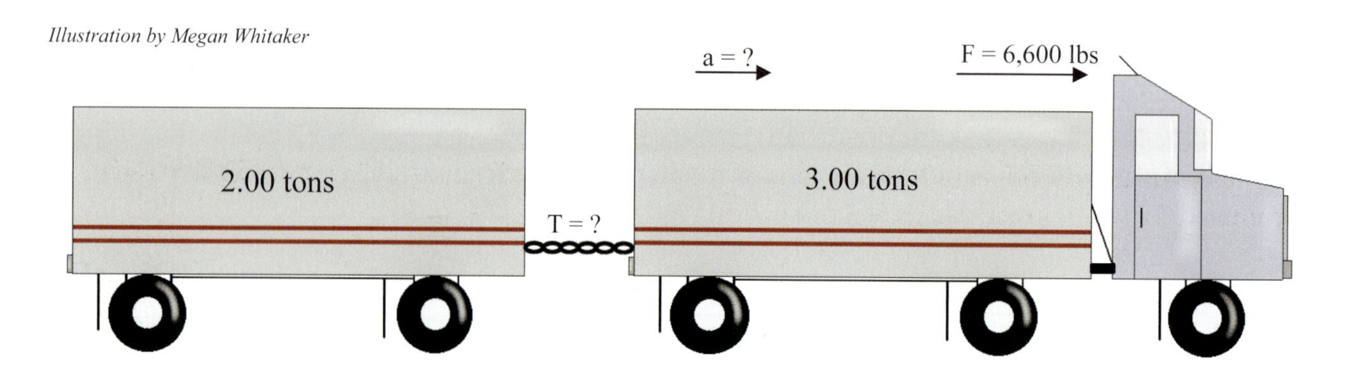

10. A man pulls two boxes across a floor with a force of 1,750 Newtons. Taking friction into account, what is the acceleration of the boxes and the tension in the rope between them? ($\mu_k = 0.35$)

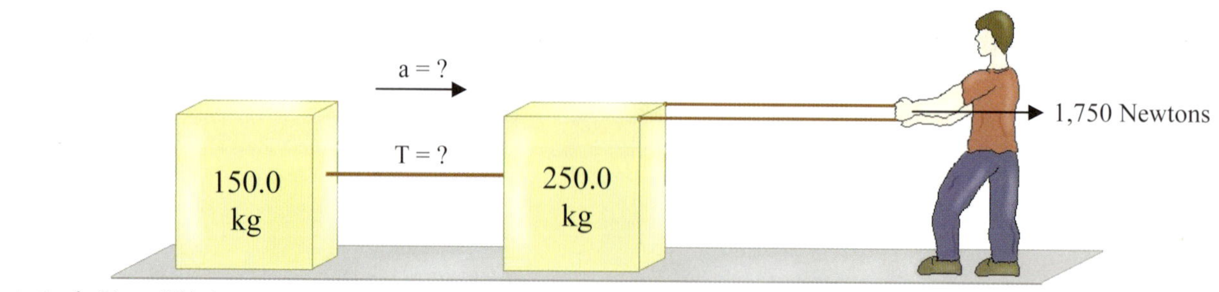

Illustration by Megan Whitaker

EXTRA PRACTICE PROBLEMS FOR MODULE #7

$$(G = 6.67 \times 10^{-11} \ \frac{\text{Newton} \cdot \text{m}^2}{\text{kg}^2})$$

1. What is the velocity of the outer rim of a 12 cm diameter CD if the player spins the CD with a centripetal acceleration of 9.0 m/sec^2?

2. A race car takes a curve at 175 mph without slipping. Assuming that the car took the curve as fast at is possibly could, what is the radius of curvature? ($\mu = 0.35$)

3. Fishing line is rated by how much tension can be placed on it before it breaks. For example, a 10.0-lb test line means that the fishing line can withstand 10.0 pounds of tension before breaking. How fast can a child twirl a 2.00 lb toy airplane above his head before the 10.0-lb test line breaks? The plane is flying in a 5.0 foot diameter circle.

4. For the following situations, describe what will happen to the centripetal force required to keep the object moving in a circle:

   a. The mass doubles and the other variables are held constant.
   b. The velocity is reduced by four times, and the other variables are held constant.
   c. The radius increases by 10 times, the mass decreases by three times, and the velocity is held constant.

5. A race car driver completes a lap every 26.890 seconds. What is the driver's frequency, and how long will it take to complete 18 laps?

6. The Hubble Space Telescope (weight = 24500 lb) orbits the earth (mass = 5.98×10^{24} kg) at a speed of 17,000 miles per hour. What is the radius of its orbit?

7. Triton, the largest moon of Neptune, has an orbital period of 5.88 days. Calculate the mass of Neptune if Triton's orbital radius is 3.548×10^8 m.

8. What is the distance between two objects ($m_1 = 62.0$ kg, $m_2 = 51.8$ kg) if there is a gravitational force of 2.37×10^{-9} N between them?

9. Assuming Mercury's orbit around the sun is a perfect circle, how long does it take Mercury to make one orbit around the sun? The mass of the sun is 2.0×10^{30} kg, and the radius of Mercury's orbit is 5.791×10^7 km.

10. What is the centripetal force experienced by a 11,110-kg satellite as it orbits at an altitude of 612 km above the earth? The mass of the earth is 5.98×10^{24} kg, and its radius is 6.38×10^6 m.

EXTRA PRACTICE PROBLEMS FOR MODULE #8

1. A child pulls on a string that is attached to a car. If the child does 80.2 J of work while pulling the car 25.0 m, with what force is the child pulling? Assume the force is parallel to the displacement.

2. A train is sitting still at the top of 990-m mountain. It then coasts down the mountain and back up another mountain that is 520 m high. What is the train's speed at the top of the second hill?

3. A truck is traveling at a constant speed of 28 m/sec. It then coasts down a 65 m hill to a valley that is 15 m above the reference point used to measure the hill. What is its speed in the valley?

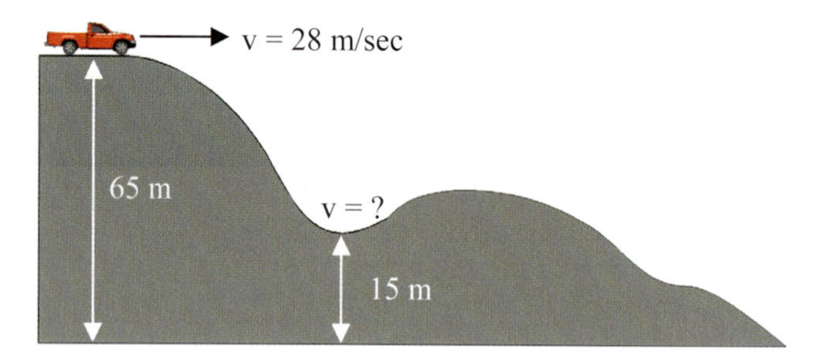

4. A car at rest on top of a 21.7-m hill coasts down the hill and then back up to the crest of another hill. If the car was traveling at 16.8 m/sec at the top of the second hill, how high is the hill?

5. A ball hanging on a rope is given a shove. If the ball reaches a height of 3.1 m, what was the ball's initial speed?

6. How much work will gravity do on a 48-pound stone if it falls from George Washington's nose (470 feet above the ground) on Mount Rushmore?

7. How far will a 50.0-kg ice skater move across the ice if she starts to glide with a speed of 3.80 m/sec? Assume that $\mu_k = 0.10$.

8. As a box slides down a ramp, friction does 23.0 J of work. At the bottom of the ramp, the box has 3.8 J of kinetic energy and is traveling at 2.8 m/sec. How high is the ramp?

9. How long does it take a machine that has 228 W of power to lift a 2.0×10^3 kg crate 250 m at a constant velocity?

10. Suppose the frictional energy of a 1.5-kg box sliding down a 2.0 m high ramp could be harnessed to burn a 60.0-Watt light bulb. How long could this light bulb burn if the box reaches a speed of 2.0 m/s at the bottom of the ramp?

EXTRA PRACTICE PROBLEMS FOR MODULE #9

1. When Tiger Woods hits a 0.04593 kg golf ball, the golf ball is usually traveling around 281 kilometers per hour. What average force does he exert on a golf ball if the club head and golf ball are in contact for 0.030 seconds?

2. It takes a telephone pole 0.010 seconds to stop a bullet. What is the mass of a bullet that is traveling at 340 m/s if the telephone pole uses a force of 88 N to stop it?

3. Two toy cars (m_1=0.150 kg, m_2=0.120 kg) are held together by a compressed spring. What is the velocity of the second car if, when the spring is released, the first car obtains a speed of 5.00 m/s? Assume the spring adds no mass to the two cars.

4. What is the recoil velocity of a 2.25-pound BB gun that fires a 0.177 caliber copper BB at 320 ft/sec? Here is some additional information that you will need for this problem: The caliber is the diameter of the sphere BB in inches, the volume of a sphere is calculated with $V = (4/3)\cdot\pi\cdot r^3$ and the density of copper is 0.323 pounds/in^3.

5. A child is standing still on her ice skates waiting for her father to come and pick her up into the air. The father, who has a mass of 71.0 kg, comes skating in at 2.00 m/s and picks her up. What is the mass of the daughter if they continue on at 1.20 m/s after he picks her up?

6. A 910-kg boulder is rolling at a rate of 5.0 m/s toward a 530 kg boulder. At what velocity is the lighter boulder rolling if both boulders come to a complete stop on colliding?

7. Two racecars going in the same direction collide and stick together. Both are 691 kg, but one is initially traveling at 89 m/sec while the other is initially traveling at 78 m/sec. What will the speed of the two cars be after the collision?

8. What is the speed of an 3.1-kg object that is traveling in a 4.0 m-radius with an angular momentum of 240 kg$\cdot$m^2/sec?

9. A 25-kg child is sitting on the edge of a 2.5-m radius frictionless merry-go-round and traveling at a speed of 2.8 m/sec. If the child moves toward the center of the merry-go-round and stops at a distance of 0.60 m from the center, what is the child's new speed?

10. A boy is twirling a 1.8-kg toy airplane on a string. The radius of the circle is 68 cm, and the speed of the plane is 4.1 m/s. One of the wings (m_{wing}=0.5 kg) suddenly falls off, but the boy keeps the airplane spinning with the same radius. What is the new velocity of the airplane?

EXTRA PRACTICE PROBLEMS FOR MODULE #10

1. An 8.0-cm long spring has a spring constant of 250 N/m. How much force is required to stretch the spring to a length of 11.0 cm?

2. A toy manufacturer wants to keep the acceleration of a projectile down to reasonable 15 m/sec^2. What is the mass limit the toy manufacturer could use with a spring that has a spring constant of 49 N/m and is compressed 3.5 cm?

3. What is the frequency of a 0.85-kg block bouncing on the end of a 3.8 N/m spring?

4. What is the mass of an object on a spring that vibrates back and forth with a period of 0.95 sec? The spring constant is 18.2 N/m.

5. What is the spring constant of a spring that has a 1.25-kg mass bouncing on it with a frequency 25 Hz?

6. A spring, k = 25.6 N/m, is compressed 32.8 cm. How much work was done?

7. An object is on a spring (k = 6.90 N/m). The maximum speed of the object is 0.58 m/sec, and the maximum displacement is 93.0 cm. What is the mass of the object?

8. A spring, k = 1.51 N/m, has a 2.46 kg object hanging on it. If the object is pulled 11.8 cm from its equilibrium position, what will be the object's speed at 10.0 cm away from equilibrium position?

9. What is the frequency of a 22.50 meter pendulum?

10. The first astronaut to land on Mars sets up a 1.00 meter pendulum. If the period of this pendulum is 3.28 s, what is the acceleration due to gravity for Mars?

EXTRA PRACTICE PROBLEMS FOR MODULE #11

($1 \text{ nm} = 10^{-9}$ m, $h = 4.14 \times 10^{-15}$ eV·sec)

1. Yellow light can have wavelengths ranging anywhere in between 600 and 550 nm. What is the range of frequencies that yellow light can have?

2. At what temperature will sound travel 400.0 m/s?

3. What is the frequency of a sound that has a wavelength of 0.890 m and is traveling through air with a temperature of 20.0 °C?

4. During a thunderstorm in which the temperature was 25.0 °C, lightning strikes the ground 5.0 km away from a girl. How long does it take for the sound to reach her?

5. The thunderclap from a lightning flash is heard 2.1 seconds after the flash is observed. If the temperature is 15.5 °C, how far away was the lightning formed?

6. A horn emits a "standard A," which has a frequency of 440.0 Hz. A car headed towards you at 19.1 m/sec honks the horn. What frequency do you hear? (T = 20.0 °C)

7. A car's horn normally blows with a constant frequency of 512 Hz. You are standing in your yard and hear the horn with a frequency of 495 Hz. At what *velocity* is the car moving? (T = 25.0 °C)

8. The wavelength of an indigo light wave is 442 nm. What is the energy of its photons?

9. Light of wavelength 191 nm is shone on copper, which has a work function of 4.94 eV. What is the maximum kinetic energy of the electrons that are emitted from the metal?

10. What is the lowest frequency of light that will result in electrons being liberated from copper (work function = 4.94 eV)? What is the wavelength of the light? If light of wavelength *larger* than that wavelength were shined on the metal, would electrons be liberated?

EXTRA PRACTICE PROBLEMS FOR MODULE #12

1. Draw a ray tracing diagram to illustrate what an image will look like if it is 2.0 cm away from a concave, spherical mirror whose radius of curvature is 5.0 cm.

2. Draw a ray tracing diagram to illustrate what an image will look like if it is 15.0 cm away from a concave, spherical mirror whose radius of curvature is 7.0 cm. Is the image real or virtual, reduced or magnified, and upright or reversed?

3. Draw a ray tracing diagram to illustrate what an image will look like if it is 11.0 cm away from a convex, spherical mirror whose radius of curvature is 7.0 cm.

4. A light beam traveling in 20.0 °C water (n=1.33) is incident on a sapphire crystal at an angle of 55.0°. If the angle of refraction is 38.0°, what is the index of refraction for sapphire?

5. Based on the index of refraction you calculated in the problem above, what is the speed of light in sapphire?

6. What is the angle of refraction if light traveling in air is incident on glass at an angle of 51.0°? The index of refraction for this glass is 1.42.

7. An object is placed 5.0 cm from a diverging lens that has focal points 2.5 cm from its center. Draw a ray-tracing diagram to illustrate the resulting image. Will the image be real or virtual, upright or inverted, and magnified or reduced?

8. An object is placed 4.0 cm from a diverging lens that has focal points 6.0 cm from its center. Will the image be real or virtual, upright or inverted, and magnified or reduced?

9. You are looking at an object through a converging lens whose focal points are 6.0 cm from its center. If the object is 8.0 cm away from the center of the lens, is its image real or virtual, reduced or magnified, and upright or reversed?

10. Draw a ray-tracing diagram for an object that is 5.0 cm from a converging lens whose focal points are 16.0 cm from its center.

EXTRA PRACTICE PROBLEMS FOR MODULE #13

$$k = 9.0 \times 10^9 \ \frac{\text{Newtons} \cdot \text{m}^2}{\text{C}^2}$$

1. Two charged particles are 2.0 cm apart, and there is a 7.8×10^6 N repulsive force between them. What are the charges of the two particles if one particle has twice of the charge as the other?

2. What is the attractive force between two particles (q_1 = 2.0 mC, q_2 = -21.1 mC) if they are 2.5 meters away from each other?

3. How many centimeters are there between two charged particles (q_1 = -90.1 mC, q_2 = -120.3 mC) that are exerting 3.33×10^6 N on each other?

4. Three charges are arranged as follows:

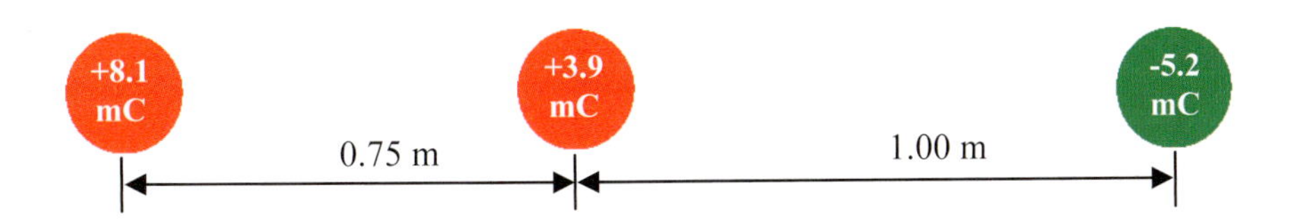

What is the instantaneous force exerted on the 3.9 mC charge?

5. Three charges are arranged as follows:

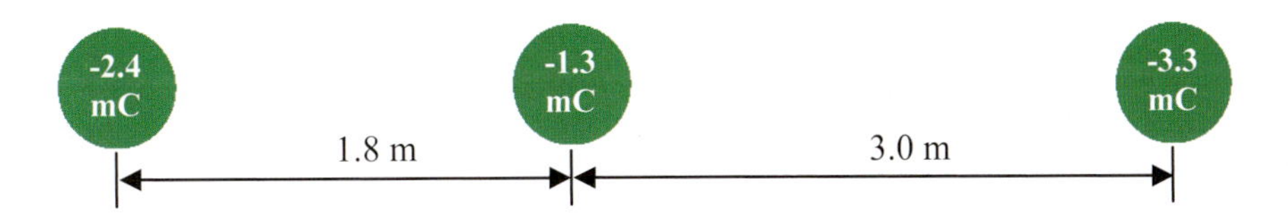

What is the instantaneous force exerted on the -1.3 mC charge?

6. Three charges are arranged as follows:

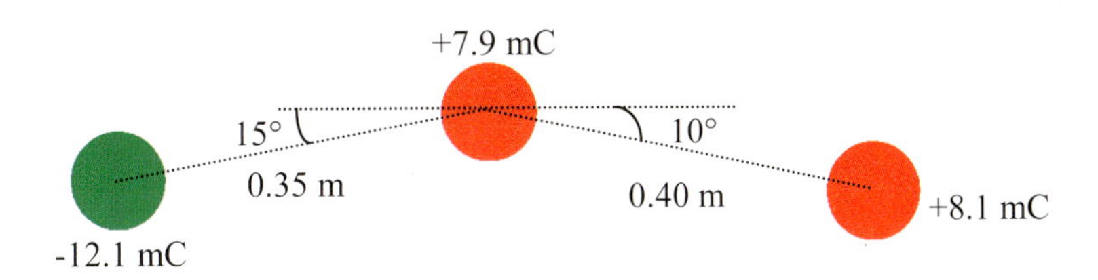

What is the instantaneous force on the 7.9 mC charge?

7. Draw the electric field generated by these two stationary charges:

8. Suppose the charges in problem #7 are 0.500 m apart. What is the magnitude of the electric field at the midpoint between them? What electrostatic force would a +0.50 mC charge experience if placed there?

9. Two equal but oppositely-charged particles are placed 1.00 m away from each other. If the magnitude of the electric field at the midpoint between them is 5.4×10^6 N/C, determine the charge on each particle.

10. What is the mass of an electron orbiting a He^+ nucleus at a radius of 2.38×10^{-10} m at a speed of 1.46×10^6 m/sec? ($q_{electron} = -1.6 \times 10^{-19}$ C and $q_{proton} = 1.6 \times 10^{-19}$ C)

EXTRA PRACTICE PROBLEMS FOR MODULE #14

$$k = 9.0 \times 10^9 \ \frac{\text{Newtons} \cdot \text{m}^2}{\text{C}^2}$$

1. Determine the electric potential 2.5 m from a stationary 4.0 C charge. What is the potential energy of a -7.0 C charge placed at that point?

2. What is the change in potential energy experienced by a 0.0018 mC charge that is originally placed 13 cm away from a fixed 0.0074 mC charge and is allowed to move to 85 cm away?

3. A 4.1-mC charge is moving towards a stationary -9.5-mC charge. What is the change in potential energy experienced by this positive charge if it starts 1.8 m away from the fixed charge and moves to only 0.52 m away from the fixed charge?

4. A -33.3-mC charged particle (m = 550 g) is placed 2.4 m from a stationary -9.8-mC charge. If the particle starts at rest, how fast will it be moving after it has traveled 1.5 m?

5. A 5.6-C charged particle (m = 2.0 kg) is placed 42.0 cm from a stationary -3.7-C charge. If the particle starts at rest, how fast will it be moving after it has traveled 24.0 cm?

6. A -10.5-mC charged particle (m = 890 g) is shot directly at a stationary -8.1-mC charge with a velocity of 175 m/s. If the particle starts out 2.3 m from the stationary charge, how close will it come to the charge before stopping and moving the other direction?

7. A 7.0-C charged particle (m = 8.0 kg) is shot directly at a stationary 5.0-C charge with a velocity of 61000 m/s. If the particle starts out 4.0 m from the stationary charge, how close will it come to the charge before stopping and moving the other direction?

8. What is the capacitance of a capacitor that stores 9.09×10^8 C of charge at 9.00 V?

9. An electron ($m = 9.1 \times 10^{-31}$ kg, $q = -1.6 \times 10^{-19}$ C) is placed at the edge of the negative plate of a 9.0×10^{-6}-F capacitor. If the capacitor holds 8.5 mC of charge, how fast will the electron be moving when it reaches the positive plate of the capacitor?

10. A proton ($m = 1.7 \times 10^{-27}$ kg, $q = 1.6 \times 10^{-19}$ C) moves from the positive plate to the negative plate of a 4.2×10^{-3}-F capacitor. If it starts from rest and reaches a speed of 1.5×10^8 m/sec, how much charge is stored in the capacitor?

EXTRA PRACTICE PROBLEMS FOR MODULE #15

1. What is the current in the following circuit?

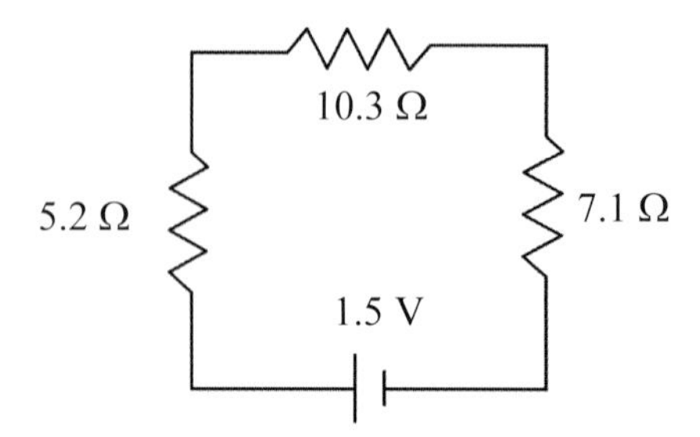

2. What is the resistance of a wire that can carry 32 A under 220 V?

3. What is the voltage behind a 15.0 A current flowing to a heater that draws 3.3×10^3 Watts of power?

4. What is the current of a circuit that has 34.0 Ω of resistance and uses 4.05 Watts?

5. What is the current in the following circuit ?

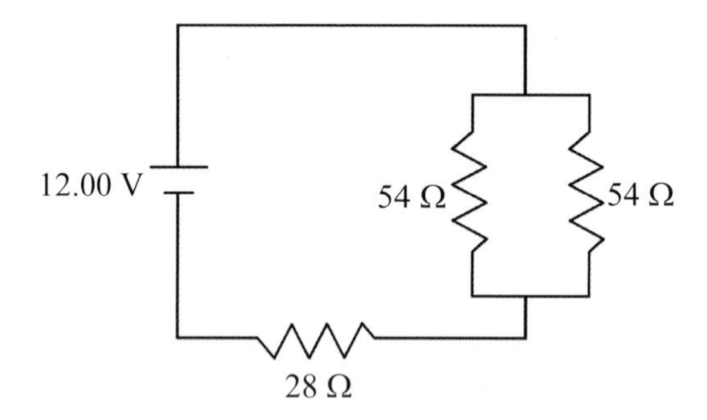

6. What is the current in the following circuit?

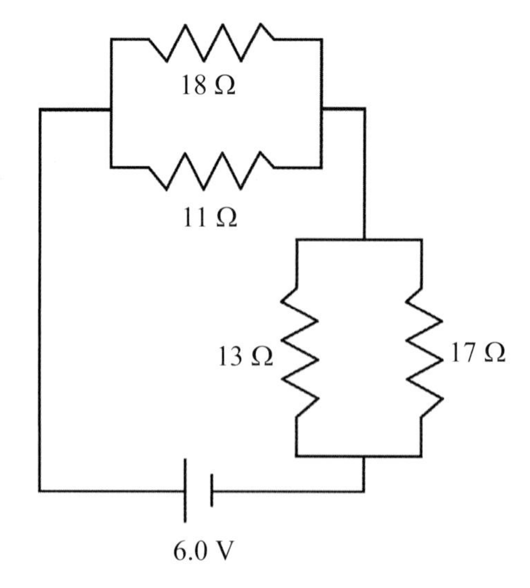

7. What is the power drawn by the following circuit?

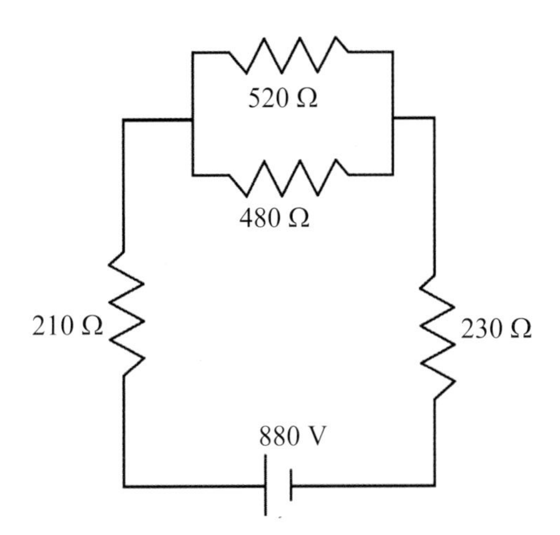

8. What is the power drawn by the following circuit?

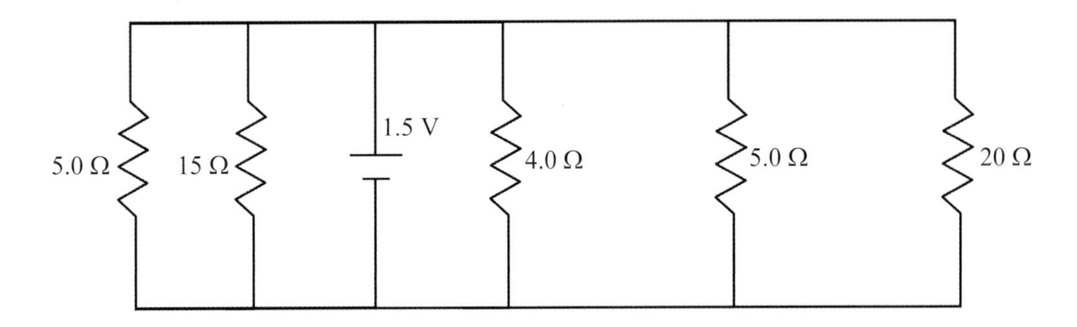

9. If an electrician has 1, 5, and 10 Amp fuses, which should be used to protect this circuit?

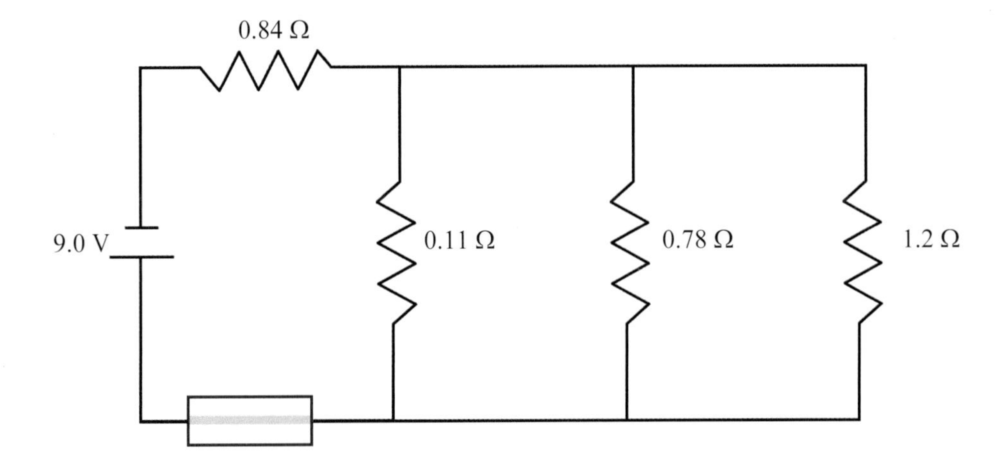

10. What is the power drawn by this circuit?

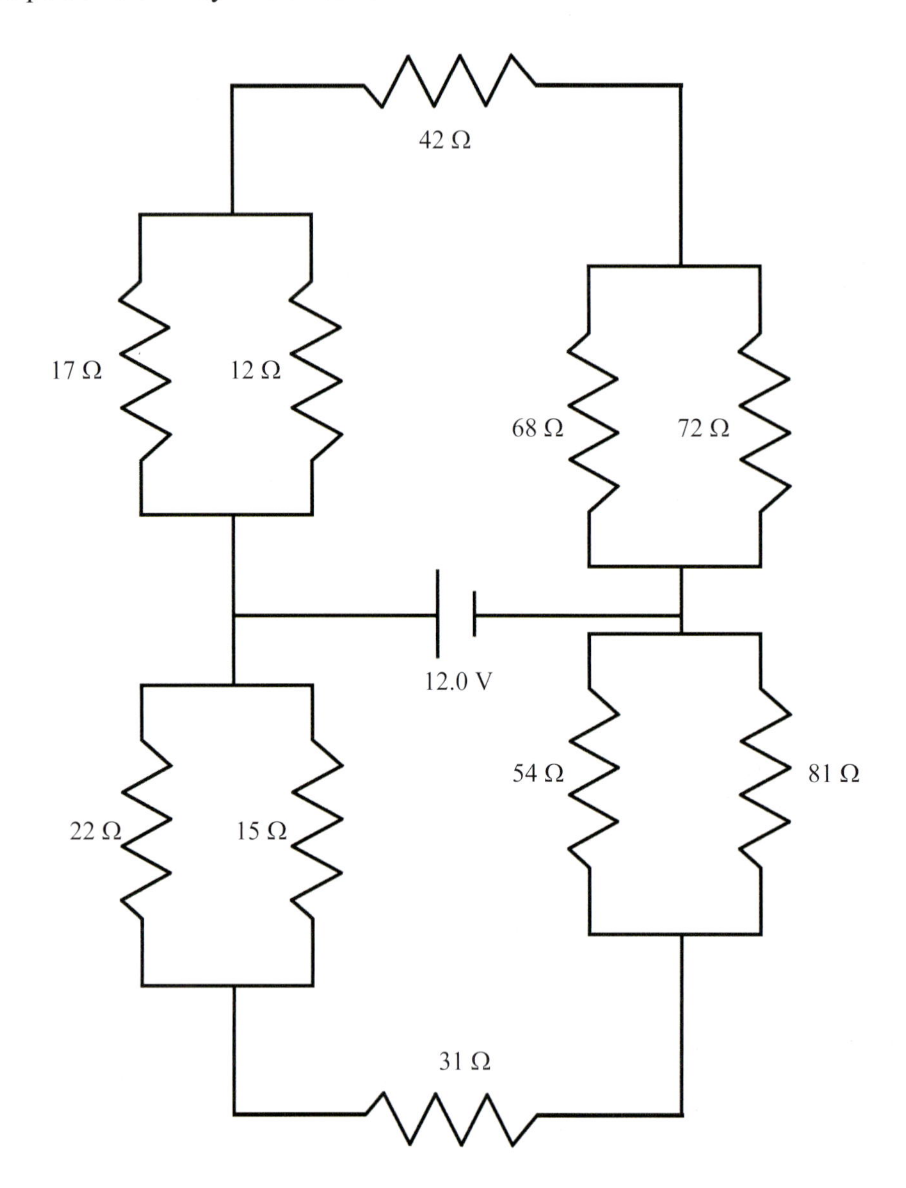

NOTE: There are no practice Problems for Module #16, so there are no extra practice problems for that module.

APPENDIX C
A COMPLETE LIST OF LAB SUPPLIES

Module #1

- Safety goggles
- A wooden board, about 1 meter long (Any long, flat surface that you can prop up on one end will do. It needs to be as smooth as possible.)
- A pencil (Anything that you can use to mark the board will do.)
- A stopwatch (A watch with a second hand will do.)
- A pile of books between 18 and 27 centimeters thick
- A ball that will easily roll down the board
- Masking tape or electrical tape
- An uncarpeted floor

Module #2

- Safety goggles
- A large (at least 21 cm by 27 cm), heavy book
- A small (about 10 cm by 10 cm) piece of paper
- A ruler, preferably metric
- Another person to help you
- Four small (about 10 cm by 10 cm) pieces of paper, all the same size

Module #3

- Safety goggles
- Modeling clay (Play-Doh® or Silly Putty® will work as well.)
- A pencil (It should be at least 6 inches long and have a sharp point on the end.)
- A wooden ruler (It needs to have a flat end at zero inches.)
- A plain 8.5-in. x 11-in. sheet of paper
- A pen
- A protractor
- Ruler
- Map given in book and on course website

Module #4

- Safety goggles
- A rubber band
- A stopwatch
- A person to help you
- A toy car or a ping-pong ball
- A meterstick
- A flat table in a room with plenty of space on at least one side of the table

Module #5

- Safety goggles
- Someone to help you
- A small beanbag, or any object that does not bounce when dropped
- A sidewalk, driveway, or long, flat yard outside
- Three rocks
- A plastic tub (like the kind that holds margarine) with lid
- A board that is at least a meter long and wider than the plastic tub listed above
- A rubber band (Thin rubber bands work better than thick ones.)
- Scissors
- Aluminum foil
- Liquid soap (Dish detergent or body wash will work as well.)
- A washcloth
- A ruler
- Sand or kitty litter

Module #6

- Safety goggles
- A building with an elevator (In other words, it's time for a field trip!)
- A bathroom scale (A digital scale may not work. It needs to be a scale that responds quickly.)
- A large nut and bolt. The nut should be at least 1/4 inch in diameter and the bolt should fit the nut.
- A board or something that has a hole big enough for the bolt to fit through
- Either one wrench that has a long handle or three wrenches that are of different lengths but nevertheless all fit the nut
- Another wrench or a pair of pliers that can hold onto the head of the bolt
- A board that is at least half a meter long
- A block of wood or metal that can fit on the surface of the board. If you don't have a block of metal or wood, try a small cardboard box.
- A meter stick

Module #7

- Safety goggles
- A pen that can be disassembled
- String
- A ruler
- Four washers
- One washer that is a bit smaller than the four washers used above
- A marker
- A stopwatch

Module #8

- Safety goggles
- A piece of string at least 25 cm long
- A ruler
- A desk or table
- Some tape
- Some books
- A mass (a ball or a nut, for example) to hang on the string. It should be heavy enough to keep the string taught when it hangs at the end.
- The inner cardboard tube from a roll of paper towels
- Scissors
- A golf ball or marble
- Someone to help you
- A desk, table, or counter that you can tape things to. Make sure the finish won't come off with tape, or that your parents don't care if it does. There should be plenty of space on at least one side.

Module #9

- Safety goggles
- Two helpers (wearing old clothes or large aprons)
- Old clothes or an apron for yourself
- Two eggs
- A reasonably large board or some other hard, flat surface such as a big cookie sheet
- A fitted bed sheet
- Four ping pong balls
- Some thread
- Some tape
- A ruler or meterstick
- Several books

Module #10

- A spring with loops on each end (These are available at any hardware store. You need one that is 3-5 cm long. You should be able to stretch it to a length of 10-15 cm with your hands. There is a test at the beginning of the experiment to let you know whether or not your spring is acceptable.)
- 2 paper clips
- A Ziploc® plastic bag
- Some sand, kitty litter, or fine gravel
- A mass scale (You can get these at any supermarket, as dieters use them to weigh out their food. It should probably have a range of 0 - 1,000 grams, but a smaller range will work.)
- A few heavy books
- Two rulers (One of these needn't be a ruler. It just needs to be a long, flat object from which you can hang the spring.)
- A stopwatch

Module #11

- Safety goggles
- 16 inches of copper pipe, ½-inch diameter preferred (This is available at any hardware store. A good hardware store will even cut it for you. You need two 6-inch pieces and one 4-inch piece.)
- Freezer
- Hacksaw or pipe cutter (You won't need these if the store cuts the pipe for you.)
- File (Depending on the way the pipes are cut, you may not need this.)
- Hot tap water
- Large bowl
- Warm gloves
- A car with a horn and a parent to drive the car
- The street you live on or a country road
- A bicycle

Module #12

- Safety goggles
- A flat mirror. The mirror can be very small, but it needs to be flat. You can always tell if a mirror is flat by looking at your reflection in it. If the image you see in the mirror is neither magnified nor reduced, the mirror is flat.
- A white sheet of paper
- A pen
- A protractor
- A flashlight
- A ruler
- Black construction paper or thin cardboard
- Tape
- A dark room
- A magnifying makeup mirror that you can hold in your hands
- A square or rectangular glass or clear plastic pan (Depending on how slanted the sides of your pan are, you might need a helper to hold the pan during the experiment. A piece of glass with two flat sides would work even better.)

Module #13

- Safety goggles
- Two balloons. Round balloons work best, but any kind will do.
- Thread
- Cellophane tape
- A glass
- A plastic lid that fits over the glass. This lid can be larger than the mouth of the glass, but it cannot be smaller. The top of a margarine tub or something similar works quite well.
- A paper clip
- Two 5-cm x 1.5-cm strips of aluminum foil (the thinner the better)
- A pair of pliers

Module #14

- Safety goggles
- Two Styrofoam® plates (Paper plates will not work nearly as well.)
- Aluminum foil
- Tape
- Scissors
- A metal paper clip
- Balloon

Module #15

- Safety Goggles
- A 1.5 -volt battery (Any AA, C, or D-cell battery will work. Do not use any battery other than one of those, though, because a higher voltage can make the experiment dangerous.)
- Aluminum foil
- Scissors
- A tabletop to which you can tape things or a flat piece of cardboard (The tabletop can't be metal.)
- Insulated wire, 20-14 gauge (It is best to use wire that has several small conductors twisted together. This is typically called "braided wire")
- Tape
- Flashlight
- 5 spoons (They cannot be silver. They must be stainless steel flatware. Most people use this kind of flatware as their "every day" spoons.)

Module #16

- A compass
- Insulated wire like you used in the experiments from the previous module
- A 1.5- volt battery (Any size cell will do, just make sure it is a 1.5-volt battery. A battery of higher voltage could make the experiment dangerous.)
- Tape
- Two iron nails (One should be large.)
- Insulated wire like that used in the previous experiment
- A metal paper clip
- A wooden match stick or toothpick

INDEX

-A-

-B-

-C-

-D-

-E-

-F-

-G-

-H-

-I-

-J-

-K-

-N-

-O-

-P-

-Q-

-R-

-S-

-T-

-U-

-V-

-W-

-Y, Z-

Metric Prefixes

Power	Prefix	Abbreviation	Power	Prefix	Abbreviation
10^{-18}	atto	a	10^{1}	deka	da
10^{-15}	femto	f	10^{2}	hecto	h
10^{-12}	pico	p	10^{3}	kilo	k
10^{-9}	nano	n	10^{6}	mega	M
10^{-6}	micro	μ	10^{9}	giga	G
10^{-3}	milli	m	10^{12}	tera	T
10^{-2}	centi	c	10^{15}	peta	P
10^{-1}	deci	d	10^{18}	exa	E

Significant Figures

A digit within a number is considered to be a significant figure if:

I. It is non-zero OR
II. It is a zero that is between two significant figures OR
III. It is a zero at the end of the number *and* to the right of the decimal point

When using measurements in mathematical equations, you must follow these rules:

Adding and Subtracting with Significant Figures: When adding and subtracting measurements, round your answer so that it has the same precision as the ***least precise*** measurement in the equation.

Multiplying and Dividing with Significant Figures: When multiplying and dividing measurements, round the answer so that it has the ***same number of significant figures as the measurement with the fewest significant figures***.